HANDBOOK OF COMPOSITES

VOLUME 2

HANDBOOK OF COMPOSITES

VOLUME 2

Series Editors

A. KELLY

Vice-Chancellor's Office
University of Surrey
Guildford
Surrey GU2 5XH
England

the late YU.N. RABOTNOV

Academy of Sciences of the USSR
Department of Mechanics and Mathematics
Moscow State University
117234 Moscow
U.S.S.R.

NORTH-HOLLAND – AMSTERDAM • NEW YORK • OXFORD • TOKYO

STRUCTURES AND DESIGN

Volume Editors

C.T. HERAKOVICH

Civil Engineering Department
University of Virginia
Charlottesville, VA 22903
U.S.A.

Y.M. TARNOPOL'SKII

Institute of Polymer Mechanics
Academy of Sciences of the Latvian SSR
226006 Riga
U.S.S.R.

1989

NORTH-HOLLAND – AMSTERDAM • NEW YORK • OXFORD • TOKYO

ISBN: 0 444 70264 4

Publishers:
ELSEVIER SCIENCE PUBLISHERS B.V.
P.O. BOX 103
1000 AC AMSTERDAM
THE NETHERLANDS

Sole distributors for the U.S.A. and Canada:
ELSEVIER SCIENCE PUBLISHING COMPANY, INC.
655 AVENUE OF THE AMERICAS
NEW YORK, NY 10010
U.S.A.

Library of Congress Cataloging in Publication Data
Structures and design/volume editors, C.T. Herakovich, Y.M. Tarnopol'skii.
p. cm. -- (Handbook of composites; v. 2)
Includes index.
ISBN 0-444-70264-4
1. Fibrous composites. 2. Metallic composites. I. Herakovich, Carl T. II. Tarnopol'skii, Y.M. III. Series.
TA481.5.S77 1988
620.1'18--dc19 88-10138
CIP

PRINTED IN THE NETHERLANDS

Preface to the Series

During 1978 Dr. W.H. WIMMERS, at that time Managing Director of the North-Holland Publishing Company, made arrangements with the Publishers Atomizdat and his own Company, the North-Holland Publishing Company, for the publication simultaneously within the Soviet Union and in the West of Handbooks concerning technical subjects of importance. The idea was that Handbooks consisting of a number of volumes should be edited jointly by an Editor from the Soviet Union and an Editor from the West. These Editors would agree a format for each particular volume and would encourage Authors of individual chapters to write these, the Authors to be chosen from the Soviet Union and from the Western World.

The first Editors for the series of Handbooks of Composite Materials were chosen to be Yuri Nikolaevich Rabotnov and myself. Yuri Rabotnov died in May 1985. He was a full member of the Academy of Sciences of the U.S.S.R. from 1958, and Head of the Laboratory for the Strength of Materials, Institute of Mechanics, Academy of Sciences from 1946. He was also Professor at Moscow University and well known throughout the world for his outstanding research work on creep and plasticity and on local stability.

It was hoped, by our publishing initiative, that scientists and technologists in the West and within the Soviet Union would learn something more of one another than they presently glean from the currently published literature.

It would, of course, have been nice to include within this idealistic framework Authors from the so-called Third World to which, of course, the People's Republic of China belongs but a journey of a thousand miles starts with a single step and the first arrangements were possible with the publishing house Atomizdat and with North-Holland.

These volumes are offered by us as a first step along the road towards the idea of publication throughout the world of tracts of importance. There have been a number of difficulties to overcome in reconciling the different cultural systems and the different ways of operating of the scientific communities of the Soviet Union, those in Western Europe and in the United States. At times the interaction has been slow and uncertain. However, we are grateful to all the Authors for their patience and their ready acceptance of Editorial comment.

I am very glad to see this volume appear in print and hope there will be many others like it.

A. KELLY

Preface to Volume 2

The intent of this volume is to provide a survey of analysis techniques, design considerations, and design methods for fibrous composite structures. The volume is divided into two parts. Part 1 includes six chapters which are general in nature having broad application to all composite structures. Part 2 includes six chapters which are more specific in nature having to do with particular structural elements, loading configurations, or fabrication methods.

The six chapters of Part 1 include coverage of optimum design, unsymmetric laminates, nonlinear analysis, edge effects, damage, and reliability. Applications to plates and shells abound throughout the chapters. The six chapters of Part 2 are concerned with woven fabric structures, beams and plates, pressure vessels, flywheels, and stability critical compression members and panels. All chapters reflect the interest and expertise of the individual authors consistent with the broad aim of the volume.

The chapter on optimum design provides a brief introduction to the general theory followed by applications to beams, plates, and shells. Static and dynamic loadings are considered. The chapter on unsymmetric laminates presents the theory necessary for understanding the deformation characteristics of unsymmetric laminates as a function of temperature. Geometrically nonlinear effects are included in the theory. The chapter on nonlinear analysis includes discussion of models for elastic and inelastic nonlinear material response followed by extensive applications to laminated shells. The chapter also presents variational formulations based upon the principle of virtual displacements.

The chapter on free edge effects presents a discussion of the underlying physics behind the development of interlaminar stresses followed by typical examples of interlaminar stresses in cross-ply, angle-ply, and quasi-isotropic laminates. The chapter concludes with a discussion of the influence of interlaminar stresses on laminate failure. The chapter on service induced damage is concerned with the development of damage associated with longterm, cyclic loading. Examples of damage development, the mechanics of damage, and life prediction are discussed. The chapter on reliability discusses the basic theory of structural reliability, statistical fracture mechanics, tensile and compressive loading of unidirectional composites, and reliability in the presence of cracks.

The first chapter in Part 2 provides an overview of the application of woven fabric composites to aerospace structures. Fabrication methods, the rational for fabrics, typical mechanical properties, design considerations, and applications are induced. The chapter on composite beam structures includes a brief presentation of lamination theory and several energy methods. This is followed by strength considerations for laminated beams under axial, transverse, and torsional loadings. The chapter also includes a rather extensive discussion of time-dependent effects. Composite pressure vessels is the subject of the next chapter. This chapter is concerned primarily with designing for strength considerations. The approach includes optimal design, structural analysis, and consideration of the fabrication process. In the chapter on composite flywheels various types of structures are analyzed and studied for increased energy storage. Special attention is given to discs and chord flywheels.

The final two chapters are concerned with stability of composite structures. The first of these two chapters considers compression members such as channels, zees, and hat-stiffened panels. The second chapter considers both flat and curved composite panels with stiffeners that are representative of aerospace structural configurations. Theoretical and experimental data for initial buckling, postbuckling, and failure characteristics are presented.

It is the hope of the editors that this volume will not only provide the student, interested reader, designer, or researcher with fundamental principles of design and analysis techniques for composite structures, but also an appreciation for the far-reaching potential of composites in engineering structures. It is further hoped that the volume will convey that while design and analysis with laminated, anisotropic materials is not without pitfalls (such as edge effects) and added complexity, the analytical tools and basic understanding are at hand to properly design using composites for more efficient engineering structures.

The editors are indebted to all of the authors for contributing their time, talents, and expertise to this volume and to Professor Marek-Jerzy Pindera of the Engineering Science and Mechanics Department at Virginia Polytechnic Institute and State University for meticulous editing of the Soviet chapters for publication in English. Thanks also go to Doctor Vladimir Kulakov and Mrs. Elga Ozolina of Institute of Polymer Mechanics (Riga) for their contribution to the appearance of the *Handbook of Composites*.

Finally, the editors fervently hope that the publication of this volume represents a small step forward in the development of closer and more open relationships between East and West.

CARL T. HERAKOVICH
Charlottesville, Virginia, U.S.A.

YURI TARNOPOL'SKII
Riga, Latvia, U.S.S.R.

Contents

List of Contributors

N.A. Alfutov, *Moscow Higher Technical School, 5, 2nd Bauman Street, 107005 Moscow, U.S.S.R.* (Chapter 3)

R.R. Arnold, *Anamet Laboratories, Inc., Hayward, CA 94545-3811, U.S.A.* (Chapters 11 and 12)

J.A. Bailie, *Structures Department, Missile Systems Division, Lockheed Missiles and Space Company, Sunnyvale, CA 94086, U.S.A.* (Chapter 7)

V.V. Bolotin, *U.S.S.R. Academy of Sciences, Institute of Mechanical Engineering, Gribovedova 4, 101830 Moscow, U.S.S.R.* (Chapter 6)

F.Ya. Bulavs, *Department of Structural Mechanics, Riga Polytechnic Institute, 226355 Riga, U.S.S.R.* (Chapter 8)

V.A. Bunakov, *Moscow Institute of Aviation Technology, 27 Petrovka Street, 103767 Moscow, U.S.S.R.* (Chapter 9)

C.T. Herakovich, *Civil Engineering Department, University of Virginia, Charlottesville, VA 22903, U.S.A.* (Chapter 4)

M.W. Hyer, *Department of Engineering Science and Mechanics, Virginia Polytechnic Institute and State University, Blacksburg, VA 24061, U.S.A.* (Chapter 2)

K.T. Kedward, *Alcoa Defense Systems, Inc., San Diego, CA 92127, U.S.A.* (Chapters 11 and 12)

I.F. Obraztsov, *Moscow Aviation Institute, 4 Volokolamskoe Shosse, 125871 Moscow, U.S.S.R.* (Chapter 1)

B.G. Popov, *Moscow Higher Technical School, 5, 2nd Bauman Street, 107005 Moscow, U.S.S.R.* (Chapter 3)

G.G. Portnov, *Institute of Polymer Mechanics, Academy of Sciences of the Latvian SSR, 23 Aizkraukles Street, 226006 Riga, U.S.S.R.* (Chapter 10)

V.D. Protasov, *Moscow Institute of Aviation Technology, 27 Petrovka Street, 103767 Moscow, U.S.S.R.* (Chapter 9)

K.L. Reifsnider, *Department of Engineering Science and Mechanics, Virginia Polytechnic Institute and State University, Blacksburg, VA 24061, U.S.A.* (Chapter 5)

A.M. Skudra, *Department of Structural Mechanics, Riga Polytechnic Institute, 226355 Riga, U.S.S.R.* (Chapter 8)

E.E. Spier, *Rohr Industries, Inc., Chula Vista, CA 92012, U.S.A.* (Chapter 11)

V.V. Vasil'ev, *Moscow Institute of Aviation Technology, 27 Petrovka Street, 103767 Moscow, U.S.S.R.* (Chapter 1)

Part 1

CHAPTER 1

Optimal Design of Composite Structures

I.F. Obraztsov

Moscow Aviation Institute
4 Volokolamskoe Shosse
125871 Moscow
U.S.S.R.

V.V. Vasil'ev

Moscow Institute of Aviation Technology
27 Petrovka Street
103767 Moscow
U.S.S.R.

Contents

HANDBOOK OF COMPOSITES, VOL. 2 – Structure and Design
Edited by C.T. HERAKOVICH and Y.M. TARNOPOL'SKII

List of Symbols

A_{ij} – stretching stiffnesses
a – length of a rectangular plate
B_{ij} – bending-stretching coupling stiffnesses
B – longitudinal stiffness of a bar
b – width of a rectangular plate
$C_{mn}^{(i)}$ – stiffnesses of the ith layer
C – shearing stiffness of a bar
$c = a/b$ – aspect ratio of a plate
D_{ij} – bending stiffnesses
$E_{\parallel}, E_{\perp}$ – Young's moduli in the directions parallel and perpendicular to the fibers
E_1, E_2 – Young's moduli in the directions x_1, x_2
E – Energy
$e_{\parallel}, e_{\perp}, e_{\parallel\perp}$ – normal strains in the directions parallel and perpendicular to the fibers and corresponding to in-plane shear strain
e_1, e_2, e_{12} – strains in the x_1, x_2 coordinate system
e – relative energy characteristic
F_{ij} – material strength tensor
$G_{\parallel\perp}$ – in-plane shear modulus relative to $\parallel$ and $\perp$ directions
G_{12} – in-plane shear modulus relative to the directions x_1, x_2
h – overall thickness
h_i – thickness of the ith layer
$\bar{h}_i = h_i/h$ – relative thickness of the ith layer
k – number of layers
k_i, k_{ij} – plate curvatures
l – length
M_i, M_{ij} – moments and stress couples
M, M_g – mass
m – mass per unit length
N_i, N_{ij} – forces and stress resultants
p_i – body forces
p – normal pressure
$\bar{q}$ – meridional in-surface body forces
t – time
t_i – coordinates of plate and shell layers
U – strain energy
u_i – displacements
v – beam deflection
w – plate deflection

$\gamma_i = t_i/h$ – relative coordinates of plate and shell layers
$\mu_{\parallel\perp}$, $\mu_{\perp\parallel}$ – Poisson's ratios associated with uniaxial loading along and across the fibers
μ_{12}, μ_{21} – Poisson's ratios associated with uniaxial loading in the directions x_1, x_2
ρ – density
$\sigma_{\parallel}$, $\sigma_{\perp}$ – normal stresses with respect to the ply-symmetry axes
σ_1, σ_2 – normal stresses with respect to Cartesian coordinates
$\tau_{\parallel\perp}$, τ_{12} – shear stresses with respect to the ply-symmetry axes and Cartesian coordinates
φ – angle between the fibers and x_1 axis
ω – angular velocity
ω_{ij} – natural frequencies of vibration

Subscripts

c – critical, composite
m – metal, mass
u – ultimate

1. Introduction

Modern composites with polymeric or metallic matrices and glass, graphite, organic or boron fibers currently are widely used in different fields of engineering. Intensive application of composites in load-carrying structural elements is attributed to a number of special features distinguishing these materials from conventional ones.

The main feature which has focused the designers' attention on the application of composites is their extremely high specific strength (4–5 times the strength of steel, aluminum and titanium alloys) and, in some cases, high specific rigidity. It should be emphasized that the high values of ultimate stress and modulus of elasticity of a unidirectional layer are realized only in one direction—along the fibers. In tension or compression across the fibers, as well as in shear, ultimate strength of a fibrous composite is relatively low for it is largely controlled by the matrix properties. In connection with this, the efficiency of composites in structural members is conditioned mainly by the correspondence of the material's structural parameters (fiber orientation, thickness of layers, etc.) to the field of external loads. It should also be noted that composite materials do not exist separately from the resulting structures and their specific fabrication methods, the latter providing for structural elements with directional properties. Modern methods of the manufacture of composite members on the one hand provide opportunities that are hardly realizable in the technology of conventional materials (e.g. a desirable thickness variation is readily obtained) and, on the other hand, impose restrictions on the element shape and material structure (e.g. equilibrium conditions for the tape on the mandrel surface must be satisfied during winding) which should be taken into account. Thus, the design of a composite structural element, besides the conventional determination of its dimensions and design configuration, involves estimation of the material structural parameters corresponding to the service conditions and consideration of fabrication methods.

The present study is concerned primarily with the problem of optimum design of composite beams, plates and shells.

Let us consider the formulation of the problem of optimum design of composite components. As optimality criteria, the general condition of minimum mass, or the conditions of maximum carrying capacity or rigidity under constraints of mass will be employed. As was mentioned above, the design variables in the case under consideration include fiber orientations and thicknesses of the layers. It is significant that the number of layers, as a rule, is also unknown and the optimal combination of material parameters is not unique for a given stress state. As an example to illustrate this important assertion, let us consider a plane problem, for a plate, consisting for a number of unidirectional layers with fiber orientations φ_i and thicknesses h_i ($i = 1, 2, 3 \dots k$). The plate is loaded along its edges by stress resultants N_1, N_2, N_{12} that are parallel to the plane x_1, x_2 (Fig. 1). Allowing for the fiber properties only and neglecting the load carrying capacity and rigidity of the matrix (i.e. assuming the so-called netting model of the composite) one can obtain the following equilibrium equations:

$$N_1 = h \sum_{i=1}^{k} \sigma_{\|}^{(i)} \bar{h}_i \cos^2 \varphi_i\,, \quad N_2 = h \sum_{i=1}^{k} \sigma_{\|}^{(i)} \bar{h}_i \sin^2 \varphi_i\,,$$
$$N_{12} = h \sum_{i=1}^{k} \sigma_{\|}^{(i)} \bar{h}_i \sin \varphi_i \cos \varphi_i\,, \tag{1.1}$$

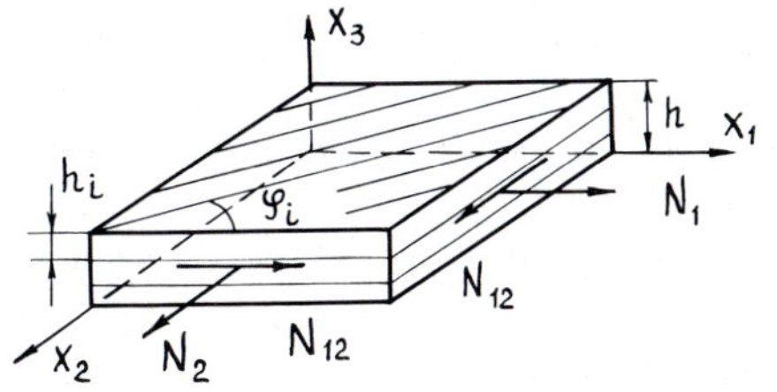

FIG. 1. A multilayer composite in a plane state of stress.

where

$$\bar{h}_i = h_i/h\ , \quad h = \sum_{i=1}^{k} h_i\ .$$

The first two equations (1.1) are added to obtain the total thickness

$$h = (N_1 + N_2)\Big/\sum_{i=1}^{k} \sigma_{\parallel}^{(i)} h_i\ . \tag{1.2}$$

The problem under consideration is to find $2k$ parameters φ_i and h_i providing the minimum value of the total thickness (all the layers are assumed to be made of one and the same material) by taking into account the following constraints:

$$|\varphi_i| \leq \pi/2\ , \quad \bar{h}_i \geq 0\ , \quad \sum_{i=1}^{k} \bar{h}_i = 1\ , \quad |\sigma_{\parallel}^{(i)}| \leq \sigma_u \tag{1.3}$$

where σ_u is the ultimate strength of the material along the fibers in tension or compression. By introducing new variables v_i, which convert the inequality strength constraints on $\sigma_{\parallel}^{(i)}$ into equalities according to the relation $\sigma_{\parallel}^{(i)}/\sigma_u = \sin v_i$ along with the Lagrangian multipliers λ and η_i, the following augmented function can be constructed:

$$F = h + \lambda\Big(\sum_{i=1}^{k} \bar{h}_i - 1\Big) + \sum_{i=1}^{k} \eta_i[\sigma_{\parallel}^{(i)} - \sigma_u \sin v_i]\ .$$

For F to be stationary with respect to $\bar{h}_i$, $\sigma_{\parallel}^{(i)}$, η_i and v_i it is required that

$$\sigma_{\parallel}^{(i)} = \sigma_u\ , \quad h = (N_1 + N_2)/\sigma_u\ , \tag{1.4}$$

$$\begin{gathered} \sum_{i=1}^{K} \bar{h}_i(N_2 \cos^2 \varphi_i - N_1 \sin^2 \varphi_i) = 0\ , \\ \sum_{i=1}^{k} \bar{h}_i[(N_1 + N_2) \sin \varphi_i \cos \varphi_i - N_{12}] = 0\ . \end{gathered} \tag{1.5}$$

In the above derivation equations (1.1) and (1.2) were taken into account. The first two constraints in (1.3) can be rewritten in terms of stress resultants as

$$0 \leq \frac{N_1}{N_1 + N_2} \leq 1\ , \quad \left|\frac{N_{12}}{N_1 + N_2}\right| \leq \tfrac{1}{2}\ . \tag{1.6}$$

Thus, if the inequalities (1.6) are satisfied, it follows from the first requirement in (1.4) that the optimal system of fibers is a system of uniform strength. Note that $2k$ structural parameters φ_i and h_i are connected by two conditions (1.5) only, so that their optimal values are not unique. The second requirement in (1.4) shows that the total thickness of the plate is the same for all the solutions of (1.5). This simple example illustrates the main difficulty encountered in the application of computer programs based on various numerical programming methods (e.g. gradient, penalty function method, etc.) for optimization of composite structures. As is known, these methods imply the application of an iterative technique which is used to refine material parameters starting from some initial values given a priori. Obviously, a variable result completely depending on these initial values would be obtained for the example considered earlier. With this in mind, particular emphasis is placed on analytical rather than numerical methods in this study, for the former provide more general analysis of the existence and properties of optimal structures.

Let us consider some specific constraints that should be taken into account in designing composite structural elements. First of all, note that, in addition to conventional geometrical restrictions, the total thickness of the laminate must be divisible by the thickness of an individual ply in the case of uniform materials. Furthermore, constraints on strains are rather significant for composites, since they are widely used to increase the rigidity of engineering structures. However, the most important and specific constraints are those imposed by the strength criteria which will be briefly discussed below. In the present analysis, the most popular model of a composite describing the material as a system of elementary layers with angles φ_i and thicknesses h_i $(i = 1, 2, 3 \ldots k)$ is employed. Each individual layer is considered to behave macroscopically as a homogeneous, orthotropic, linearly elastic material whose mechanical properties (elastic constants $E_\parallel$, $E_\perp$, $G_{\parallel\perp}$, $\mu_{\parallel\perp}$ and ultimate stresses $\sigma^{\mathrm{u}}_\parallel$, $\sigma^{\mathrm{u}}_\perp$, $\tau^{\mathrm{u}}_{\parallel\perp}$) are determined from experimental data on unidirectional samples. Furthermore, the layers are assumed to be perfectly bonded and the stresses in each layer can readily be obtained by using the conventional analysis methods. To estimate the static strength of the layer, one of the developed phenomenological anisotropic failure criteria described by TSAI and HAHN (1975) and ROWLANDS (1975) is employed. Henceforth, the following universal quadratic polynomial will be applied:

$$F_{11}\sigma_\parallel^2 + F_{12}\sigma_\parallel\sigma_\perp + F_{22}\sigma_\perp^2 + F_{33}\tau_{\parallel\perp}^2 \leq 1 \,. \tag{1.7}$$

In particular, for the von Mises–Hill criterion $F_{11} = -F_{12} = 1/(\sigma^{\mathrm{u}}_\parallel)^2$, $F_{22} = 1/(\sigma^{\mathrm{u}}_\perp)^2$, $F_{33} = 1/(\tau^{\mathrm{u}}_{\parallel\perp})^2$. If the normal stresses $\sigma_\parallel$, $\sigma_\perp$ are tensile (compressive), then the corresponding tensile (compressive) ultimate stresses are substituted in inequality (1.7). As is known, for a unidirectional layer $\sigma^{\mathrm{u}}_\parallel \gg \sigma^{\mathrm{u}}_\perp$ and $\sigma^{\mathrm{u}}_\parallel \gg \tau^{\mathrm{u}}_{\parallel\perp}$, so that the limit surface has the form shown in Fig. 2(a). With this in

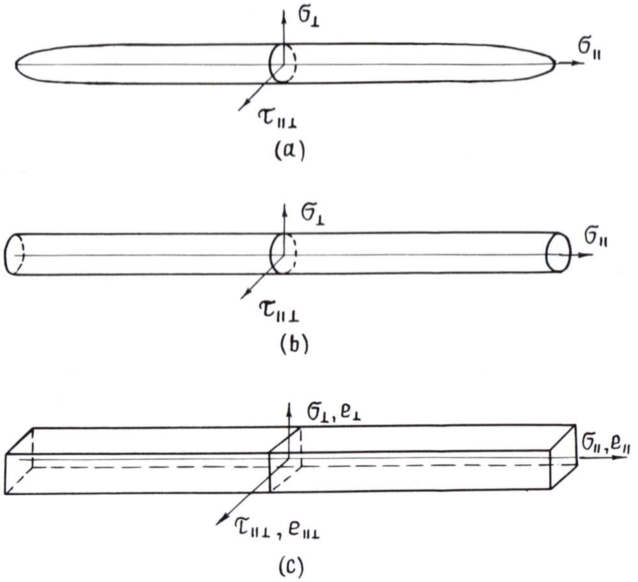

FIG. 2. Failure surface of a unidirectional layer defined by inequalities (1.7) – a; (1.8), (1.9) – b; (1.10), (1.11) – c.

mind, the general failure criterion (1.7) can be simplified and reduced to the form

$$(\sigma_\perp/\sigma_\perp^u)^2 + (\tau_{\|\perp}/\tau_{\|\perp}^u)^2 \leqslant 1\,, \tag{1.8}$$

$$(\sigma_\|/\sigma_\|^u)^2 \leqslant 1\,. \tag{1.9}$$

Inequality (1.8) defines a cylindrical limit surface (Fig. 2(b)) which corresponds to resin crazing or loss of integrity of a unidirectional layer, resulting in cracks parallel to the fibers. For some types of materials (e.g. for $\pm\varphi$ angle-ply laminates) resin crazing leads to total fracture of the laminate. However, this phenomenon does not result in the failure of a properly designed laminate with fiber-controlled ultimate strength or a laminate consisting of more than two layers with different angles of fiber orientation. Despite the fact that resin crazing is frequently considered to have detrimental fatigue and environmental effects on epoxy composites, it practically does not influence their static strength which is determined by inequality (1.9) predicting fracture of the fibers. The limit surface can be even more simplified and reduced to the surface of a rectangular parallelepiped shown in Fig. 2(c). In this case, the failure criterion has the form

$$|\sigma_{\|}| \leq \sigma_{\|}^{u}, \quad |\sigma_{\perp}| \leq \sigma_{\perp}^{u}, \quad |\tau_{\|\perp}| \leq \tau_{\|\perp}^{u}, \tag{1.10}$$

which can also be expressed in terms of strains

$$|e_{\|}| \leq e_{\|}^{u}, \quad |e_{\perp}| \leq e_{\perp}^{u}, \quad |e_{\|\perp}| \leq e_{\|\perp}^{u}. \tag{1.11}$$

Criteria (1.10) and (1.11), being more simple, have some additional advantages in comparison with the general criterion (1.7) because they identify the stress responsible for failure of the layer. The first inequalities in (1.10) and (1.11) correspond to failure of the fibers; the second and third determine the loss of integrity due to normal or shearing stresses, respectively.

Thus, optimum design of composite structures can be implemented under constraints on the loss of integrity or the load carrying capacity of the fibers allowing for material degradation which may occur without precipitating ultimate failure. To implement the latter design model of the composite, one can use the netting model which neglects the load carrying capacity and rigidity of the matrix and represents the material as a system of fibers. This model essentially simplifies the problem of design and will be frequently employed below.

The following sections of the present study will be devoted to some problems of the optimum design of composite beams, plates and shells considered within the framework of the assumptions described above.

2. Thin-walled composite bars of closed cross-section

2.1. Governing equations

One of the most promising fields of application of high-performance fiber composites in load-carrying structures is in the fabrication of truss elements. The uniaxial stress state of these elements, being almost perfectly suited to the material's directional properties, provides an ideal application of composites in truss components. An in-depth review of the state-of-the art in structural analysis and design methods of composite truss and beam elements has been given by KNOELL and ROBINSON (1975).

Thin-walled bars of closed cross-section whose cross-sectional dimensions are small in comparison with the length can be treated on the basis of the Bernoulli–Euler hypothesis, assuming that plane sections remain plane after loading. Taking into account the relatively low shear modulus of unidirectional composites, the sections are not assumed to remain normal to the neutral axis. According to these hypotheses the following constitutive equations can be obtained for the composite bar of an arbitrary cross-section:

$$
\begin{aligned}
N_1 &= S_1 u_1' + S_2 \theta_2' + S_3 \theta_3' ,\\
N_2 &= K_2(u_2' + \theta_3) + K_{23}(u_3' + \theta_2) + C_2 \theta_1' ,\\
N_3 &= K_3(u_3' + \theta_2) + K_{32}(u_2' + \theta_3) + C_3 \theta_1' ,\\
M_1 &= D_1 \theta_1' + C_2(u_2' + \theta_3) + C_3(u_3' + \theta_2) ,\\
M_2 &= D_2 \theta_2' + D_{23} \theta_3' + S_2 u_1' ,\\
M_3 &= D_3 \theta_3' + D_{32} \theta_2' + S_3 u_1' .
\end{aligned}
\tag{2.1}
$$

In the above, N_i and M_i $(i = 1, 2, 3)$ are the forces and moments shown in Fig. 3; u_i and θ_i are the corresponding displacements of neutral point 0 and angles

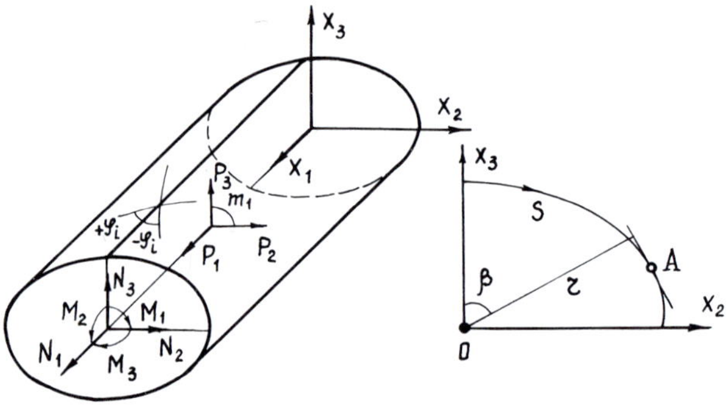

FIG. 3. A thin-walled composite bar.

of rotation of the cross-section; the prime denotes differentiation with respect to x_1;

$$
\begin{aligned}
&S_1 = \oint Bh \,\mathrm{ds} , \quad S_2 = \oint Bhx_3 \,\mathrm{ds} , \quad S_2 = \oint Bhx_2 \,\mathrm{ds} ,\\
&K_2 = \oint Ch\dot{x}_2^2 \,\mathrm{ds} , \quad K_3 = \oint Ch\dot{x}_3^2 \,\mathrm{ds} ,\\
&K_{23} = K_{32} = \oint Ch\dot{x}_2\dot{x}_3 \,\mathrm{ds} ,\\
&D_1 = \oint Chr^2 \,\mathrm{ds} , \quad D_2 = \oint Bhx_3^2 \,\mathrm{ds} , \quad D_3 = \oint Bhx_2^2 \,\mathrm{ds} ,\\
&D_{23} = D_{32} = \oint Bhx_2x_3 \,\mathrm{ds} ,\\
&C_2 = \oint Ch\dot{x}_2 r \,\mathrm{ds} , \quad C_3 = \oint Ch\dot{x}_3 r \,\mathrm{ds} ,
\end{aligned}
\tag{2.2}
$$

specify generalized elastic stiffnesses of the bar; $x_2(s)$, $x_3(s)$ represent parametric equations of the midcontour of the cross-section; $\dot{x}_2 = \mathrm{d}x_2/\mathrm{d}s = \cos\beta$, $\dot{x}_3 = \mathrm{d}x_3/\mathrm{d}s = -\sin\beta$ (see Fig. 3); h is the thickness of the wall.

Internal forces and moments must satisfy the equilibrium equations (Fig. 3)

$$
\begin{gathered}
N_i' + p_i = 0 \quad (i = 1, 2, 3)\,, \\
M_1' + m_1 = 0\,, \quad M_2' - N_3 = 0\,, \quad M_3' - N_2 = 0\,.
\end{gathered}
\tag{2.3}
$$

Upon integrating equations (2.3), it is readily verified that

$$
\begin{gathered}
N_i = N_i^0 - \int_0^{x_1} p_i\,\mathrm{d}x_1 (i = 1, 2, 3)\,, \quad M_1 = M_1^0 - \int_0^{x_1} m_1\,\mathrm{d}x_1\,, \\
M_2 = M_2^0 + N_3^0 dx_1 - \int_0^{x_1} \mathrm{d}x_1 \int_0^{x_1} p_3\,\mathrm{d}x_1\,, \\
M_3 = M_3^0 + N_2^0 x_1 - \int_0^{x_1} \mathrm{d}x_1 \int_0^{x_1} p_2\,\mathrm{d}x_1\,,
\end{gathered}
\tag{2.4}
$$

where N_i^0 and M_i^0 are constants of integration corresponding to the forces and moments acting in the initial cross-section $x_1 = 0$.

Constitutive equations (2.1) in conjunction with the solution (2.4) give the displacement for the neutral point 0 (u_i) and the angles of rotation (θ_i). According to the Bernoulli–Euler hypothesis, the following relations defining the displacements, for an arbitrary point, can be obtained:

$$
U_1 = u_1 + \theta_2 x_3 + \theta_3 x_2\,, \quad U_2 = u_2 + \theta_1 x_3\,, \quad U_3 = u_3 - \theta_1 x_2\,. \tag{2.5}
$$

The displacement along the contour of the cross-section has the form

$$
U_s = U_2 \dot{x}_2 + U_3 \dot{x}_3\,. \tag{2.6}
$$

Consequently, according to equation (2.5) and (2.6), the longitudinal and shearing strains are related to the displacement and rotation components u_i, θ_i by

$$
\begin{aligned}
e_1 &= \partial U_1/\partial x_1 = u_1' + \theta_2' x_3 + \theta_3' x_2\,, \\
e_{1s} &= \frac{\partial U_s}{\partial x_1} + \frac{\partial U_1}{\partial s} \\
&= \dot{x}_2(u_2' + \theta_3) + \dot{x}_3(u_3' + \theta_2) + (\dot{x}_2 x_3 - \dot{x}_3 x_2)\theta_1'\,.
\end{aligned}
\tag{2.7}
$$

Generalized elastic stiffnesses (2.2) depend on fiber orientations, thicknesses of the layers and circumferential rigidity of the cross-sectional contour. If the rigidity is equal to zero, i.e. if the contour is absolutely flexible, then

$$e_s = -\mu_{s1} e_1 \tag{2.8}$$

and

$$B = C_{11} - \frac{C_{12}^2}{C_{22}}, \quad C = C_{33}, \quad \mu_{s1} = \frac{C_{12}}{C_{22}}, \tag{2.9}$$

where

$$C_{mn} = \sum_{i=1}^{k} \bar{h}_i C_{mn}^{(i)}, \tag{2.10}$$

$$\begin{aligned} C_{11}^{(i)} &= \bar{E}_{\|}^{(i)} \cos^4 \varphi_i + 2\bar{E}_{\|}^{(i)} \mu_{\|\perp}^{(i)} \sin^2 \varphi_i \cos^2 \varphi_i \\ &\quad + \bar{E}_{\perp}^{(i)} \sin^4 \varphi_i + G_{\|\perp}^{(i)} \sin^2 2\varphi_i , \\ C_{12}^{(i)} &= C_{21}^{(i)} = [\bar{E}_{\|}^{(i)} + \bar{E}_{\perp}^{(i)}] \sin^2 \varphi_i \cos^2 \varphi_i \\ &\quad + \bar{E}_{\|}^{(i)} \mu_{\|\perp}^{(i)} (\sin^4 \varphi_i + \cos^4 \varphi_i) - G_{\|\perp}^{(i)} \sin^2 2\varphi_i , \\ C_{22}^{(i)} &= \bar{E}_{\|}^{(i)} \sin^4 \varphi_i + 2\bar{E}_{\|}^{(i)} \mu_{\|\perp}^{(i)} \sin^2 \varphi_i \cos^2 \varphi_i \\ &\quad + \bar{E}_{\perp}^{(i)} \cos^4 \varphi_i + G_{\|\perp}^{(i)} \sin^2 2\varphi_i , \\ C_{33}^{(i)} &= [\bar{E}_{\|}^{(i)} + \bar{E}_{\perp}^{(i)} - 2\bar{E}_{\|}^{(i)} \mu_{\|\perp}^{(i)}] \sin^2 \varphi_i \cos^2 \varphi_i + G_{\|\perp}^{(i)} \cos^2 2\varphi_i , \\ \bar{E}_{\|,\perp}^{(i)} &= E_{\|,\perp}^{(i)} / [1 - \mu_{\|\perp}^{(i)} \mu_{\perp\|}^{(i)}], \quad E_{\|}^{(i)} \mu_{\|\perp}^{(i)} = E_{\perp}^{(i)} \mu_{\perp\|}^{(i)} . \end{aligned} \tag{2.11}$$

In the above $E_{\|}^{(i)}$, $E_{\perp}^{(i)}$ are Young's moduli in the directions parallel and perpendicular to the fibers; $G_{\|\perp}^{(i)}$ is the in-plane shear modulus of the ith layer; $\mu_{\|\perp}^{(i)}$ and $\mu_{\perp\|}^{(i)}$ are Poisson's ratios associated with uniaxial loading across and along the fibers; $\bar{h}_i = h_i/h$. It is assumed that the wall of the bar contains symmetrical pairs of layers consisting of the same material and being of the same thickness with material axes of symmetry alternately oriented at angles $+\varphi_i$ and $-\varphi_i$ to the bar generator. Thus h_i denotes the total thickness of two layers with angles $\pm\varphi_i$ (Fig. 3).

If the contour of the cross-section is absolutely rigid (e.g. if the bar is stiffened by a system of circumferential ribs or internal core), the circumferential strain $e_s = 0$ and

$$B = C_{11}, \quad C = C_{33} \tag{2.12}$$

The stresses and the corresponding strains in the ith layer are related by the generalized Hooke's law

$$\sigma_{\parallel}^{(i)} = \bar{E}_{\parallel}^{(i)}[e_{\parallel}^{(i)} + \mu_{\parallel\perp}^{(i)} e_{\perp}^{(i)}] ,$$

$$\sigma_{\perp}^{(i)} = \bar{E}_{\perp}^{(i)}[e_{\perp}^{(i)} + \mu_{\perp\parallel}^{(i)} e_{\parallel}^{(i)}] , \tag{2.13}$$

$$\tau_{\parallel\perp}^{(i)} = G_{\parallel\perp}^{(i)} e_{\parallel\perp}^{(i)} ,$$

where the strains of the layer are connected with the average strains of the bar by the following geometrical relations

$$e_{\parallel}^{(i)} = e_1 \cos^2 \varphi_i + e_s \sin^2 \varphi_i + e_{1s} \sin \varphi_i \cos \varphi_i ,$$

$$e_{\perp}^{(i)} = e_1 \sin^2 \varphi_i + e_s \cos^2 \varphi_i - e_{1s} \sin \varphi_i \cos \varphi_i , \tag{2.14}$$

$$e_{\parallel\perp}^{(i)} = (e_s - e_1) \sin 2\varphi_i + e_{1s} \cos 2\varphi_i .$$

Thus, to calculate the stress–strain state of a bar with a given form, loading conditions, and material properties, one should find internal forces and moments from (2.4), obtain the generalized rigidities with the aid of relationships (2.9) or (2.12), determine displacements and angles of rotation of a cross-section from (2.1) and then estimate the stresses in the layers according to (2.7), (2.8), (2.13) and (2.14). The ultimate combination of stresses for an arbitrary layer can be predicted by one of the failure criteria discussed in Section 1.

As has been mentioned already, the simplified netting model of the composite is frequently used in the design of composite members. In this case the elastic stiffnesses of the layer should be calculated by putting $E_{\perp}^{(i)} = G_{\parallel\perp}^{(i)} = \mu_{\parallel\perp}^{(i)} = 0$, so that the equations in (2.11) take the following form:

$$C_{11}^{(i)} = E_{\parallel}^{(i)} \cos^4 \varphi_i , \quad C_{22}^{(i)} = E_{\parallel}^{(i)} \sin^4 \varphi_i ,$$

$$C_{12}^{(i)} = C_{33}^{(i)} = E_{\parallel}^{(i)} \sin^2 \varphi_i \cos^2 \varphi_i . \tag{2.15}$$

Constitutive relations (2.13) reduce to

$$\sigma_{\parallel}^{(i)} = E_{\parallel}^{(i)} e_{\parallel}^{(i)} , \tag{2.16}$$

where $e_{\parallel}^{(i)}$ is specified by the first of equations (2.14).

2.2. *Bars of circular cross-section*

Expressions (2.2) for generalized elastic stiffnesses can be simplified considerably by taking advantage of the symmetry of the bar cross-section. In particular, if there is one axis of symmetry, say x_2 (Fig. 4(a)), then $S_2 = K_{23} = K_{32} =$

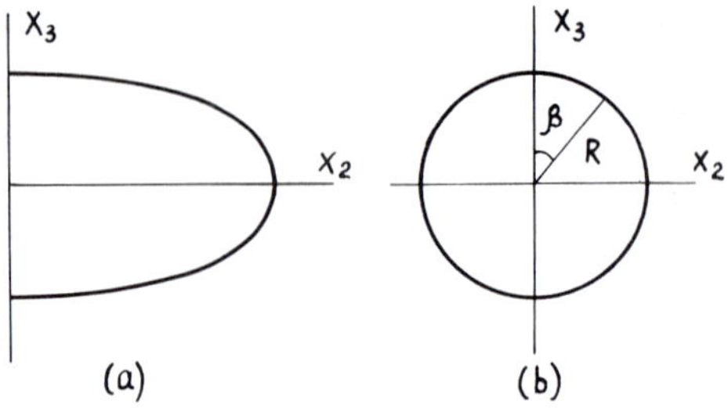

FIG. 4. Cross-sections with one (a) and two (b) axes of symmetry.

$D_{23} = D_{32} = C_2 = 0$. In the case of double symmetry about the axes x_2, x_3 (Fig. 4(b)), in addition to the above one should set $S_3 = C_3 = 0$. For the latter case, the cross-section equations (2.1) reduce to

$$u_1' = \frac{N_1}{BF}, \quad u_2' + \theta_3 = \frac{N_2}{CL_2}, \quad u_3' + \theta_2 = \frac{N_3}{CL_3},$$
$$\theta_1' = \frac{M_1}{CI}, \quad \theta_2' = \frac{M_2}{BI_2}, \quad \theta_3' = \frac{M_3}{BI_3}, \tag{2.17}$$

where F is the cross-sectional area and I, I_2, I_3 are moments of inertia of the cross-section with respect to axes the x_1, x_2, x_3, respectively. For the bar with a circular cross-section

$$x_2 = R\sin\beta, \quad x_3 = R\cos\beta, \quad s = R\beta,$$
$$\dot{x}_2 = \cos\beta, \quad \dot{x}_3 = -\sin\beta, \quad r = R$$

and

$$F = 2\pi Rh, \quad I = 2\pi R^3 h, \quad I_2 = I_3 = \pi R^3 h,$$
$$L_2 = L_3 = \pi Rh.$$

Relationships (2.17) are in good agreement with the experimental results.

Let us consider some simply stress states such as tension, compression, torsion and bending that are typical for truss and frame elements.

(1) For the bar axially loaded by tensile (N_1) or compressive ($-N_1$) force, (2.7), (2.8), (2.17) give

$$e_1 = \pm\frac{N_1}{2\pi RBh}, \quad e_s = \mp\frac{N_1 C_{12}}{2\pi RBC_{22}h}.$$

The longitudinal stiffness calculated for $\pm\varphi$ angle-ply laminates with the aid of equations (2.9) to (2.11) is shown in Fig. 5. It is obvious that the 0° orientation

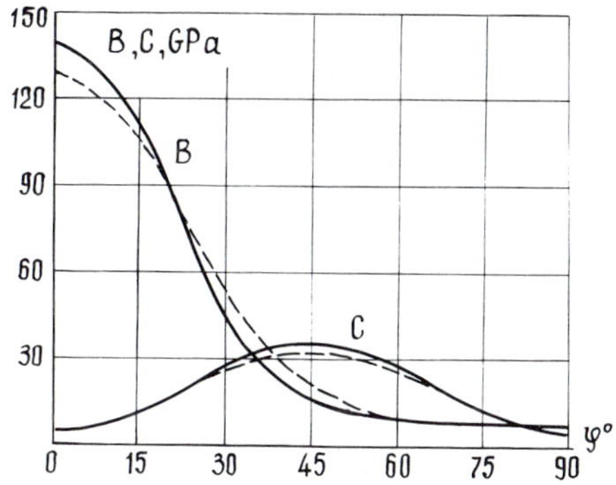

FIG. 5. Longitudinal (B) and shearing (C) stiffness for graphite–epoxy thin-walled bars. The curves shown are for $\pm\varphi$ orientation (———) and $\pm\varphi/90°$ lamination (– – – –).

provides the maximum value of B. However, this is true only for tension. In compression, the circumferential strain (e_s) is positive and when it reaches its ultimate value $e_\perp^u$, which is relatively low for unidirectional composites, the tube splits into a number of longitudinal strips and fails due to buckling of the strips. As was shown by TARNOPOL'SKII (1979), this mode of failure can be prevented by the incorporation of circumferential overwraps of relatively small thickness. The influence of the 90° layer of relative thickness $\bar{h}_{90} = 0.1$ in the $\pm\varphi$ layers on the rigidity of the bar is shown in Fig. 5 by dashed lines. It can be seen that the reduction of the longitudinal rigidity coefficient due to an additional circumferential layer is less than 10%.

When the truss element loading is compressive, instability failure must also be considered. For a simply supported composite bar the critical Euler load is given by

$$N_c = \frac{\pi^2 BI}{l^2[1 + (\pi^2 BI)/(CLl^2)]} .$$

For preliminary design-analysis estimates, especially for long, slender elements, the shear contribution can be ignored by putting $C \to \infty$. Then the critical load is controlled by the longitudinal rigidity coefficient B whose maximum value corresponds to $\varphi = 0$. Thus, in the case under consideration, the composite material of the optimal structure consists of longitudinal layers in conjunction with circumferential overwraps. Sometimes, in order to satisfy special attachment and technology requirements, the longitudinal plies are replaced by helical ones (e.g. in struts with metal fittings). In such cases, the circumferential layer is quite necessary for providing high strength capacity and it does not greatly influence the rigidity coefficients (as shown in Fig. 5). It should be noted that high strength joints, which transfer loads into the local joint structure by normal pressure between the composite layers and metal fittings, are, as a rule, characterized by comparatively high compliance. For the sake of rigidity, longitudinally reinforced bars with bonded adhesive joints are more effective, but their strength capacity is relatively low.

Thus, the structural parameters of composite truss members are known qualitatively a priori. The material consists of longitudinal or helical and circumferential layers, the angle of helical fibers and the thickness of circumferential layers being as small as possible. The only unknown variables are the thickness of longitudinal or helical layers and the moment of inertia of a cross-section required to meet strength, rigidity and stability constraints, i.e. the same variables that characterize the conventional truss components. Consequently, numerous and well established analytical and numerical methods for the design of metal trusses can be readily applied to composite structures.

(2) For the bar loaded by forces statically equivalent to a twisting moment M_1 it follows from equations (2.7), (2.17) that

$$e_{1s} = R\theta_1' = \frac{M_1}{2\pi R^2 Ch} .$$

It can be seen from Fig. 5, which shows the dependence of the shear rigidity coefficient C on φ, that the optimal fiber orientation is 45°. In order to prevent delamination and local buckling of the helical layers compressed along the fibers, a circumferential ply should be added. The effect of this additional ply on the shear rigidity coefficient is shown in Fig. 5.

(3) For the bar loaded by forces inducing in its cross-section the bending moment M_2 and transverse shear force N_3, one can obtain from (2.7), (2.8) and (2.17) (see, Fig. 4b)

$$\begin{aligned} e_1 &= \theta_2' R \cos\beta = \frac{M_2 \cos\beta}{\pi R^2 Bh} , \\ e_s &= -\frac{C_{12}}{C_{22}} e_1 , \\ e_{1s} &= -(u_3' + \theta_2) \sin\beta = -\frac{N_3 \sin\beta}{\pi RCh} . \end{aligned} \tag{2.18}$$

For a qualitative analysis let us use the netting model of the composite described by (2.15) and (2.16) and consider a bar composed of helical layers of relative overall thickness $\bar{h}_\varphi$ oriented at $\pm\varphi$ and longitudinal layers of relative thickness $\bar{h}_0$. Then, (2.18) takes the form:

$$e_1 = \frac{M_2 \cos\beta}{\pi R^2 E_\| h_0} , \quad e_{1s} = -\frac{4N_3 \sin\beta}{\pi R E_\| h_\varphi \sin^2 2\varphi} .$$

It can be seen from these relations that the optimal structure must contain helical plies oriented at ±45° resisting the transverse shear force and 0° plies resisting the bending moment. The thicknesses of ±45° and 0° layers are readily obtained provided that the force and moment are known.

3. Composite beams

High-modulus composites are increasingly being used to increase the rigidity of the beams loaded by transverse distributed and concentrated forces. The stress–strain state of a beam is governed by the following set of equations based on the classical Bernoulli–Euler hypothesis (Fig. 6):

$$Dv^{\mathrm{IV}} = q\,,$$
$$M = -Dv''\,,\quad Q = -Dv'''\,,\quad \sigma_1 = \frac{E_1 M}{D}(x_2 - e)\,,\quad \tau_{12} = -\frac{QS}{bD}\,, \tag{3.1}$$

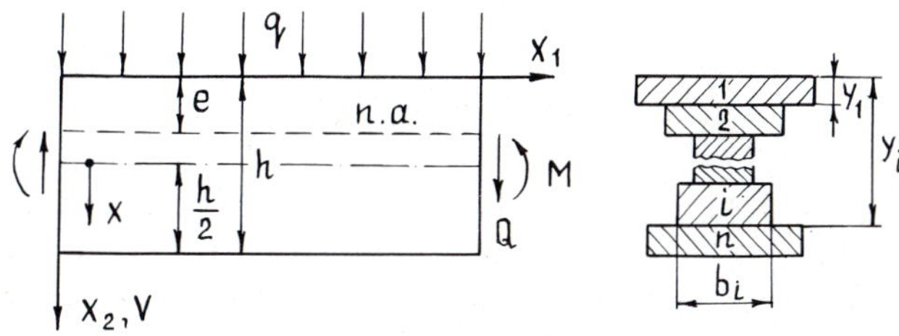

FIG. 6. A laminated beam.

where M and Q denote bending moment and transverse shear stress resultant; v designates deflection; σ_1, τ_{12} are flexure and shear stresses;

$$D = I_2 - eI_1\,,\quad e = I_1/I_0\,, \tag{3.2}$$

are bending stiffness and the distance between the top edge of the beam and neutral axis, respectively;

$$I_n = \int_0^h bE_1 x_2^n \,\mathrm{d}x_2 \quad (n = 0, 1, 2)\,,\quad S = \int_0^{x_2} bE_1(x_2 - e)\,\mathrm{d}x_2\,. \tag{3.3}$$

It is assumed that the width (b) and modulus of elasticity (E_1) of the beam are arbitrary functions of x_2. For a laminated beam consisting of k layers, b_i and $E_1^{(i)}$ are constant through the thickness of the ith layer. Integration in a piecewise manner from layer to layer yields:

$$I_n = \frac{1}{n+1}\sum_{i=1}^{k} E_1^{(i)} b_i (y_i^{n+1} - y_{i-1}^{n+1}) \quad (n = 0, 1, 2)\,,$$

where $y_0 = 0$, $y_k = h$, h being the height of the beam; y_i is the distance between the top surface of the beam and the bottom surface of the ith layer (Fig. 6).

As was mentioned, composite materials are mainly used in beams to increase the bending stiffness. Let us prove that the maximum value of bending stiffness is obtained for beams symmetrically laminated with respect to the midplane $x_2 = h/2$. Introducing the coordinate $x = x_2 - (h/2)$ (see Fig. 6), one can rewrite equations (3.2) and (3.3) in the following form:

$$D = \bar{I}_2 - (\bar{I}_1^2/I_0)\,, \tag{3.4}$$

where

$$\bar{I}_1 = \int_{-h/2}^{k/2} bE_1 x \,\mathrm{d}x\,, \quad \bar{I}_2 = \int_{-h/2}^{h/2} bE_1 x^2 \,\mathrm{d}x\,.$$

Denote $bE_1 = f(x)$ and decompose the function $f(x)$ as follows: $f(x) = f_s(x) + f_a(x)$ where $f_s(-x) = f_s(x)$ is an even function and $f_a(-x) = -f_a(x)$ is an odd function. Then

$$\bar{I}_1 = \int_{-h/2}^{h/2} f_a x \,\mathrm{d}x\,.$$

It follows from equation (3.4) that D reaches its maximum value if $\bar{I}_1 = 0$, i.e. if $f_a(x) \equiv 0$ and consequently $f(x) \equiv f_s(x)$. Thus, the symmetrical distribution of layers with respect to the midplane provides the maximum value of the bending stiffness D.

As an example, let us consider a metal beam of rectangular cross-section with identical top and bottom composite cover plates of relative thickness αh (Fig. 7). The rigidity of this beam in tension and bending and its mass per unit

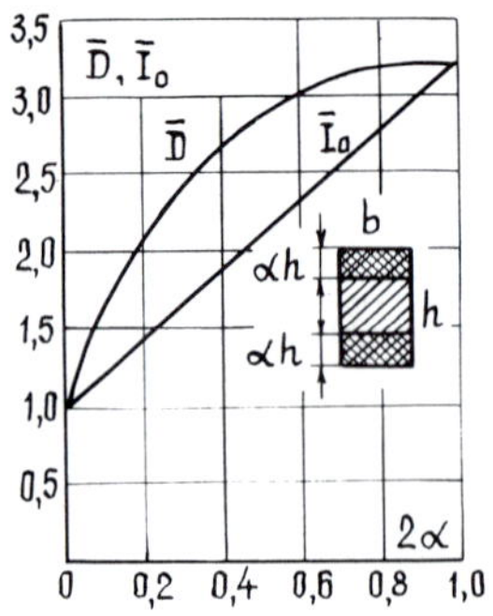

FIG. 7. Variation of relative tensile ($\bar{I}_0$) and bending ($\bar{D}$) stiffnesses for an aluminum beam with relative thickness of boron–aluminum cover plates.

length are specified by the following formulas:

$$I_0 = E_m bh[1 + 2\alpha(\beta - 1)] ,$$

$$D = \tfrac{1}{12} E_m bh^3[1 + 2\alpha(\beta - 1)(4\alpha^2 - 6\alpha + 3)] ,$$

$$g = \rho_m bh[1 - 2\alpha(1 - \gamma)] ,$$

where $\beta = E_c/E_m$, $\gamma = \rho_c/\rho_m$; ρ is specific gravity; indices 'c' and 'm' indicate the respective characteristics of composite and metal, respectively. In the case $\alpha = 0$, i.e. for a metal beam, $I_m = E_m bh$, $D_m = E_m bh^3/12$, $M_{gm} = \rho_m bh$. Fig. 7 represents graphical plots of $\bar{I}_0 = I_0/I_m$, $\bar{D} = D/D_m$ ($\bar{M}_g = M_g/M_{gm} = 1$) as functions of 2α and illustrates the efficiency of stiffening of an aluminum beam with boron fibers ($\beta = 3.22$; $\gamma = 1$).

For a metal beam of symmetric I-shaped section with composite cover plates (Fig. 8) we find

$$D = D_m + b\alpha\delta(E_c - E_m)\left[\frac{h^2}{4} - \alpha\delta\left(h - \frac{2\alpha\delta}{3}\right)\right] ,$$
$$M_g = M_{gm} - b\alpha(\rho_m - \rho_c) , \tag{3.5}$$

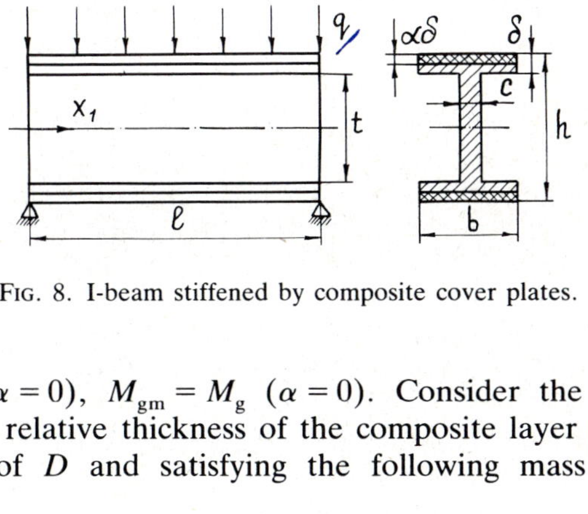

FIG. 8. I-beam stiffened by composite cover plates.

where $D_m = D$ ($\alpha = 0$), $M_{gm} = M_g$ ($\alpha = 0$). Consider the following design problem: find the relative thickness of the composite layer (α) providing the maximum value of D and satisfying the following mass and constitutive constraints:

$$M_g \leq M_{gm} , \quad e_m = \frac{\bar{M}}{D}\left(\frac{h}{2} - \alpha\delta\right) \leq e_m^u , \tag{3.6}$$

$$e_c = \frac{\bar{M}h}{2D} \leq e_c^u , \quad \tau = \frac{\bar{Q}E_c}{2D}\alpha\delta(h - \alpha\delta) \leq \tau_u . \tag{3.7}$$

It is assumed that the initial metal beam ($\alpha = 0$) whose bending stiffness is to be increased satisfies all the constitutive constraints. Since for modern composites and metal alloys $\rho_m \leqslant \rho_c$, the first constraint in (3.6) is satisfied automatically. The second constraint in (3.6), where e_m is the strain attained at the compressive and tensile outer fibers of the metal part of the beam and e_m^u designates the ultimate strain of the metal, is also satisfied since e_m $(\alpha = 0) \leqslant e_m^u$. Constraints (3.7) are the strength conditions with respect to the maximum normal strain (or stress) in the composite layers and the maximum shear stress at the composite–metal bond surface including the corresponding ultimate strain e_c^u in tension or compression (the minimum value is substituted in the formula), ultimate shear stress at the bond surface τ_u and maximum (through the length of the beam) values of the bending moment ($\bar{M}$) and the transverse shear stress resultant ($\bar{Q}$).

Thus, we should find α that maximizes the bending stiffness (3.5) and satisfies inequalities (3.7).

Approximate but rather good results can be obtained by assuming $\alpha\delta \ll h$ in (3.5) and (3.7). Under this assumption inequalities (3.7) specify the admissible change of parameter α as follows:

$$\alpha_{\min} \leqslant \alpha \leqslant \alpha_{\max} , \tag{3.8}$$

where

$$\alpha_{\min} = \frac{\bar{M} - M^*}{bh\delta e_c^u (E_c - E_m)} , \quad \alpha_{\max} = \frac{\tau_u D_m}{\delta h E_c (\bar{Q} - Q^*)} \tag{3.9}$$

and

$$M^* = \frac{2}{h} D_m e_c^u , \quad Q^* = \tau_u bh\left(1 - \frac{E_m}{E_c}\right) .$$

According to the relationship between $\bar{M}$, $\bar{Q}$ and M^*, Q^* obtained from (3.9) the following particular cases are possible:

(1) If $\bar{M} < M^*$, $\bar{Q} < Q^*$, behaviour constraints are insignificant and $0 \leqslant \alpha \leqslant 1$.

(2) If $\bar{M} > M^*$, $\bar{Q} < Q^*$, the first strength condition (3.7) is essential and $\alpha_{\min} \leqslant \alpha \leqslant 1$.

(3) If $\bar{M} < M^*$, $\bar{Q} > Q^*$, the second strength condition (3.7) is essential and $0 \leqslant \alpha \leqslant \alpha_{\max}$.

(4) If $\bar{M} > M^*$, $\bar{Q} > Q^*$, both constraints (3.7) should be taken into account and α must satisfy inequalities (3.8).

To obtain the maximum value of the bending stiffness, the maximum admissible value of α should be taken. The results may be verified and improved by using the exact equations (3.5) and (3.7).

As an example, consider a simply supported beam of $l = 1$ m span length and loaded by a uniformly distributed load $q = 8 \times 10^4$ N/m (Fig. 8). Maximum values of the bending moment and shear stress resultant of this beam are as follows: $\bar{M} = ql^2/8$, $\bar{Q} = ql/2$. The cross-section constant over the span and designed according to the yield limit of aluminum alloy ($E_m = 70.6$ GPa; $\rho_m = 2700$ kg/m^3; $e_m^u = 0.36\%$) has the following dimensions: $h = 0.1$ m; $b = 0.5h$; $\delta = 0.1h$; $c = 0.05h$. For a beam stiffened by boron fibers in an aluminum matrix ($E_c/E_m = 3.22$; $\rho_c/\rho_m = 1$; $e_c^u = 0.5\%$; $\tau_u = 74$ MPa) case (1) is applicable, i.e. the constitutive constraints are not significant. Assuming for example $\alpha = 0.9$, we obtain $D/D_m = 2.95$; $M_g/M_{gm} = 1$. For a beam stiffened by high-modulus graphite fibers in an epoxy matrix ($E_c/E_m = 2.7$; $\rho_c/\rho_m = 0.59$; $e_c^u = 0.29\%$; $\tau_u = 3.8$ MPa) case (4) is applicable and $0.062 \leqslant \alpha \leqslant 0.111$. Assuming for example $\alpha = 0.1$, we find $D/D_m = 1.19$; $M_g/M_{gm} = 0.98$.

For I beams with thin webs and rigid flanges the effects of shear deformation can be significant in determining beam stresses and displacements. The governing equations allowing for these effects and based on the assumption of uniform distribution of shear strain through the height of the web are as follows:

$$M' = Q\,, \quad Q' = -q\,,$$
$$M = D\theta'\,, \quad Q = K(\theta + v')\,, \tag{3.10}$$

where θ is the angle of rotation of a cross-section, and $K = cG_m(h - 2\delta)$ specifies the shear stiffness of the I beam (see Fig. 8). The solution to (3.10) is given by:

$$Q = -\int_0^{x_1} q\,dx_1 + Q_0\,, \quad M = -\int\!\!\int_0^{x_1} q\,dx_1^2 + Q_0 x_1 + M_0\,,$$

$$\theta = -\frac{1}{D}\left(-\int\!\!\int\!\!\int_0^{x_1} q\,dx_1^3 + Q_0\frac{x_1^2}{2} + M_0 x_1\right) + \theta_0\,,$$

$$v = \frac{1}{D}\left(\int\!\!\int\!\!\int\!\!\int_0^{x_1} q\,dx_1^4 - Q_0\frac{x_1^3}{6} - M_0\frac{x_1^2}{2}\right)$$
$$+ \frac{1}{K}\left(-\int\!\!\int_0^{x_1} q\,dx_1^2 + Q_0 x_1\right) - \theta_0 x_1 + v_0\,.$$

The subscript 0 is used to indicate the values of the corresponding functions at the end $x_1 = 0$. For a free end, $M_0 = Q_0 = 0$, for a simply supported end, $M_0 = v_0 = 0$, and for a clamped end $\theta_0 = v_0 = 0$.

Shear deformation of the web decreases the apparent bending rigidity of a beam. For example, the effective bending rigidity of the beam shown in Fig. 8 is given by

$$D^e = D\Big/\left(1 + \frac{48D}{Kl^2}\right).$$

This result is obtained by matching the maximum deflection obtained from (3.10) with that from (3.1).

For the aluminum I beam considered above $D^e_m = 0.6D_m$. For an aluminum beam stiffened by boron fibers ($\alpha = 0.9$) $D^e = 0.34D = 1.67D^e_m$ and for a beam reinforced by high-modulus graphite fibers $D^e = 0.56D = 1.11D^e_m$. Thus, the reduction of the effective bending rigidity due to the effect of shear deformation is 1.77 times larger in the case of boron fibers and 1.07 times larger in the case of graphite fibers.

4. Composite panels

4.1. Plane stress state

4.1.1. General relations

Properly designed plane panels, composed of a system of composite layers with different fiber orientations, are characterized by high specific strength and rigidity and are increasingly being introduced into a variety of structures in aeronautical engineering applications where high stiffness and strength to weight ratios are of paramount importance.

Let us consider a plate consisting of k unidirectional layers with fiber orientation φ_i and thicknesses h_i ($i = 1, 2, 3 \ldots k$) loaded by stress resultants N_1, N_2, N_{12} as it is shown in Fig. 1. The equilibrium equations can be written as follows:

$$N_1 = h \sum_{i=1}^{k} \bar{h}_i[\sigma_{\parallel}^{(i)} \cos^2 \varphi_i + \sigma_{\perp}^{(i)} \sin^2 \varphi_i - \tau_{\parallel\perp}^{(i)} \sin 2\varphi_i] ,$$

$$N_2 = h \sum_{i=1}^{k} \bar{h}_i[\sigma_{\parallel}^{(i)} \sin^2 \varphi_i + \sigma_{\perp}^{(i)} \cos^2 \varphi_i + \tau_{\parallel\perp}^{(i)} \sin 2\varphi_i] , \tag{4.1}$$

$$N_{12} = h \sum_{i=1}^{k} h_i\{[\sigma_{\parallel}^{(i)} - \sigma_{\perp}^{(i)}] \sin \varphi_i \cos \varphi_i + \tau_{\parallel\perp}^{(i)} \cos 2\varphi_i\} .$$

Assuming that the layers are perfectly bonded together, the strains in the directions of the axes of elastic symmetry of the jth layer are found by simple transformation of the strains associated with the chosen pair of axes x_1 and x_2

$$e_{\parallel}^{(j)} = e_1 \cos^2 \varphi_j + e_2 \sin^2 \varphi_j + e_{12} \sin \varphi_j \cos \varphi_j ,$$

$$e_{\perp}^{(j)} = e_1 \sin^2 \varphi_j + e_2 \cos^2 \varphi_j - e_{12} \sin \varphi_j \cos \varphi_j , \tag{4.2}$$

$$e_{\parallel\perp}^{(j)} = (e_2 - e_1) \sin 2\varphi_j + e_{12} \cos 2\varphi_j .$$

For an orthotropic layer, stresses are expressed in terms of strains via the following constitutive equations:

$$\sigma_{\|}^{(j)} = \bar{E}_{\|}^{(j)}[e_{\|}^{(j)} + \mu_{\|\perp}^{(j)} e_{\perp}^{(j)}]\,, \quad \sigma_{\perp}^{(j)} = \bar{E}_{\perp}^{(j)}[e_{\perp}^{(j)} + \mu_{\perp\|}^{(j)} e_{\|}^{(j)}]\,,$$
$$\tau_{\|\perp}^{(j)} = G_{\|\perp}^{(j)} e_{\|\perp}^{(j)}\,, \quad \bar{E}_{\|,\perp}^{(j)} = E_{\|,\perp}^{(j)}/[1 - \mu_{\|\perp}^{(j)}\mu_{\perp\|}^{(j)}]\,. \tag{4.3}$$

For given stress resultants N_1, N_2, N_{12} and material parameters, (4.1) to (4.3) allow us to obtain averaged strains e_1, e_2, e_{12} and stresses $\sigma_{\|}^{(j)}, \sigma_{\perp}^{(j)}, \tau_{\|\perp}^{(j)}$ in each layer. Failure of a unidirectional layer can be predicted using one of the phenomenological strength criteria described in Section 1, e.g.

$$F_j = F_{11}^{(j)}[\sigma_{\|}^{(j)}]^2 + F_{12}^{(j)}\sigma_{\|}^{(j)}\sigma_{\perp}^{(j)}$$
$$+ F_{22}^{(j)}[\sigma_{\perp}^{(j)}]^2 + F_{33}^{(j)}[\tau_{\|\perp}^{(j)}]^2 \leqslant 1\,. \tag{4.4}$$

4.1.2. Unidirectional layer of minimum mass

Before discussing the general case of a laminated panel, let us consider the problem of the optimum design of a single unidirectional layer and establish the optimum angle of fiber orientation. In this case $k = 1$ and the stresses $\sigma_{\|}$, $\sigma_{\perp}$, $\tau_{\|\perp}$ (index i denoting the number of the layer is omitted) are statically determinate. Assuming that the stress resultants increase in proportion to some parameter p, i.e. that $N_1 = pN_1^0$, $N_2 = pN_2^0$, $N_{12} = pN_{12}^0$ $(0 \leqslant p \leqslant \bar{p})$ and inverting equations (4.1) (note that $k = 1$ and $\bar{h}_1 = 1$), one can find

$$\sigma_{\|} = \frac{p}{h}(N_1^0 \cos^2\varphi + N_2^0 \sin^2\varphi - N_{12}^0 \sin 2\varphi)\,,$$
$$\sigma_{\perp} = \frac{p}{h}(N_1^0 \sin^2\varphi + N_2^0 \cos^2\varphi + N_{12}^0 \sin 2\varphi)\,, \tag{4.5}$$
$$\tau_{\|\perp} = \frac{p}{h}[(N_2^0 - N_1^0)\sin\varphi\cos\varphi + N_{12}^0 \cos 2\varphi]\,.$$

Upon substitution of equations (4.5) into the failure criterion (4.4) one obtains the following results:

$$(h/p)^2 \geqslant (h/\bar{p})^2 = 0.25[(F_{11} + F_{12} + F_{22})(N_1^0 + N_2^0)^2$$
$$+ F_{33}(N_1^0 - N_2^0)^2 + 4F_{33}(N_{12}^0)^2] + [0.5(N_1^0 - N_2^0)\cos 2\varphi$$
$$+ N_{12}^0 \sin 2\varphi]\{(F_{11} + F_{22} - F_{12} - F_{33})[0.5(N_1^0 - N_2^0)\cos 2\varphi$$
$$+ N_{12}^0 \sin 2\varphi] + (F_{11} - F_{22})(N_1^0 + N_2^0)\}\,. \tag{4.6}$$

The conditions of maximum strength (or minimum thickness) can be written in the form

$$\frac{\mathrm{d}}{\mathrm{d}\varphi}(h/\bar{p}) = 0\,, \quad \frac{\mathrm{d}^2}{\mathrm{d}\varphi^2}(h/\bar{p}) > 0\,. \tag{4.7}$$

Substituting (4.6) into conditions (4.7), performing differentiation and expressing in the obtained relations N_1^0, N_2^0, N_{12}^0 in terms of $\sigma_\parallel$, $\sigma_\perp$, $\tau_{\parallel\perp}$ via (4.5), one obtains

$$\begin{aligned} &\tau_{\parallel\perp}[(F_{11}+F_{22}-F_{12}-F_{33})(\sigma_\parallel-\sigma_\perp)+(F_{11}-F_{22})(\sigma_\parallel+\sigma_\perp)] = 0\,,\\ &4(F_{11}+F_{22}-F_{12}-F_{33})\tau_{\parallel\perp}^2-(\sigma_\parallel-\sigma_\perp)[(F_{11}+F_{22}-F_{12}\\ &\quad -F_{33})(\sigma_\parallel-\sigma_\perp)+(F_{11}-F_{22})(\sigma_\parallel+\sigma_\perp)] > 0\,. \end{aligned} \tag{4.8}$$

This result for Hill's criterion was obtained by BRANDMAIR (1970). For fiber reinforced epoxy composites with $\sigma_\parallel \gg \sigma_\perp$, $\sigma_\parallel^{\mathrm{u}} \gg \sigma_\perp^{\mathrm{u}}$, $\sigma_\parallel^{\mathrm{u}} \gg \tau_{\parallel\perp}^{\mathrm{u}}$ conditions (4.8) have the following approximate form:

$$\tau_{\parallel\perp}F_{33}\sigma_\parallel = 0\,, \quad 4\tau_{\parallel\perp}^2(F_{22}-F_{33}) + \sigma_\parallel^2 F_{33} > 0\,.$$

It can be seen that these conditions are satisfied if $\tau_{\parallel\perp} = 0$. Thus, the fibers in the layer of minimum mass must be oriented along the directions of maximum principal stresses. In this case the third equation (4.5) gives

$$\tan 2\varphi = \frac{2N_{12}^0}{N_1^0 - N_2^0}\,, \tag{4.9}$$

while the first and second equations in (4.5) enable us to find the stresses in the optimal layer

$$\sigma_{\parallel,\perp} = \frac{p}{2h}(I_1 \pm I_2)\,, \tag{4.10}$$

where

$$I_1 = N_1^0 + N_2^0\,, \quad I_2 = [(N_1^0 - N_2^0)^2 + 4(N_{12}^0)^2]^{1/2}\,.$$

Eliminating φ from (4.6) with the aid of (4.9) permits the minimum value of the thickness to be expressed as

$$\begin{aligned} (h_{\mathrm{m}}/\bar{p})^2 = 0.25[&(F_{11}+F_{22}+F_{12})(I_1^2+I_2^2)\\ &+2(F_{11}-F_{22})I_1I_2]\,. \end{aligned} \tag{4.11}$$

By introducing stress resultants in the directions of the axes of elastic symmetry of the layer as $\sigma_\parallel h = N_\parallel = pN_\parallel^0$, $\sigma_\perp h = N_\perp = pN_\perp^0$, one can obtain from (4.10)

$$I_{1,2} = N_\parallel^0 \pm N_\perp^0\,. \tag{4.12}$$

Let us perform one formal transformation. As can be seen from (4.11) and (4.12), the minimum thickness h_m depends on the stress state of the layer. Assuming that $I_1 \neq 0$, let us formally minimize $(h_m/\bar{p})$ with respect to (I_2/I_1), i.e.

$$\frac{d}{d(I_2/I_1)}\left(\frac{h_m}{\bar{p}I_1^2}\right) = 0 .$$

The result is as follows:

$$\frac{I_2}{I_1} = \frac{F_{22} - F_{11}}{F_{11} + F_{22} - F_{12}} .$$

Taking into account (4.12) and (4.11), we can write the following relationship specifying the optimal stress distribution

$$\frac{N_\perp^0}{N_\parallel^0} = \frac{\sigma_\perp}{\sigma_\parallel} = \frac{2F_{11} - F_{12}}{2F_{22} - F_{12}} \tag{4.13}$$

and the corresponding expression for the absolute minimum value of the thickness

$$\left(\frac{h_{am}}{\bar{p}}\right)^2 = \frac{(N_1^0 + N_2^0)^2(4F_{11}F_{22} - F_{12}^2)}{4(F_{11} + F_{22} - F_{12})} . \tag{4.14}$$

It should be noted that in this case equation (4.13) specifies the relationship between the given stress resultants N_1, N_2, N_{12}. Thus (4.14) is valid for special loading only. However, in the case of a laminated composite, minimization with respect to stresses in the layers is more reasonable, for these stresses cannot be directly expressed in terms of N_1, N_2, N_{12}.

4.1.3. Optimality criteria for laminated composites

Let us consider a composite plate consisting of k layers. By adding the first two equations in (4.1) one can obtain the following expression for the thickness:

$$h = (N_1 + N_2) \Big/ \sum_{i=1}^{k} \bar{h}_i[\sigma_\parallel^{(i)} + \sigma_\perp^{(i)}] . \tag{4.15}$$

The stresses in the layers $\sigma_\parallel^{(i)}$, $\sigma_\perp^{(i)}$ must satisfy the strain compatibility equations which follow from (4.2) and (4.3) and allow us to express stresses in the jth layer in terms of stresses in any, the first, say, layer as

$$L_{ij} = a_1[\sigma_{\|}^{(1)} - \sigma_{\|}^{(j)}] + a_2[\sigma_{\perp}^{(1)} - \sigma_{\perp}^{(j)}] = 0\,,$$

$$L_{2j} = b_1[\sigma_{\|}^{(1)} \cos 2\varphi_1 - \sigma_{\|}^{(j)} \cos 2\varphi_j] - b_2[\sigma_{\perp}^{(1)} \cos 2\varphi_1$$

$$-\sigma_{\perp}^{(j)} \cos 2\varphi_j] - c[\tau_{\|\perp}^{(1)} \sin 2\varphi_1 - \tau_{\|\perp}^{(j)} \sin 2\varphi_j] = 0\,, \qquad (4.16)$$

$$L_{3j} = b_1[\sigma_{\|}^{(1)} \sin 2\varphi_1 - \sigma_{\|}^{(j)} \sin 2\varphi_j] - b_2[\sigma_{\perp}^{(1)} \sin 2\varphi_1$$

$$-\sigma_{\perp}^{(j)} \sin 2\varphi_j] - c[\tau_{\|\perp}^{(1)} \cos 2\varphi_1 - \tau_{\|\perp}^{(j)} \cos 2\varphi_j] = 0\,,$$

where

$$a_1 = (1 - \mu_{\perp\|})/E_{\|}\,, \quad a_2 = (1 - \mu_{\|\perp})/E_{\perp}\,,$$

$$b_1 = (1 + \mu_{\perp\|})/E_{\|}\,, \quad b_2 = (1 + \mu_{\|\perp})/E_{\perp}\,, \quad c = 1/G_{\|\perp}\,.$$

As was mentioned in Section 1, the optimal design in the space of material variables k, φ_i, h_i is not unique. Let us find, then, the optimal stress distribution providing the minimum value of thickness (4.15) and satisfying both strain compatibility equations (4.16) and failure criteria (4.4). The strength constraints are adjoined by using the Lagrangian multipliers $\lambda_1^{(j)}$ and the slack variables v_j convert the inequality constraints on stresses in the layers into equality constraints of the form $2F_j = 1 + \sin v_j$. The strain compatibility equations are adjoined by the Lagrangian multipliers $\lambda_2^{(j)}$, $\lambda_3^{(j)}$ and $\lambda_4^{(j)}$. Necessary conditions for the optimal solution can now be derived in a standard way by introducing the augmented function

$$\Phi = h + \sum_{j=1}^{k} \lambda_1^{(j)}(1 + \sin v_j - 2F_j) + \sum_{j=2}^{k} [\lambda_2^{(j)} L_{1j} + \lambda_3^{(j)} L_{2j} + \lambda_4^{(j)} L_{3j}]\,.$$

Stationarity of Φ with respect to stresses, slack variables and Lagrangian multipliers requires that

$$\bar{h}_1 f + 2\lambda_1^{(1)}[2F_{11}\sigma_{\|}^{(1)} + F_{12}\sigma_{\perp}^{(1)}] - a_1 \sum_{i+2}^{k} \lambda_2^{(i)} - b_1 \sum_{i=2}^{k} \xi_1^{(i)} = 0\,, \qquad (4.17)$$

$$\bar{h}_1 f + 2\lambda_1^{(1)}[2F_{22}\sigma_{\perp}^{(1)} + F_{12}\sigma_{\|}^{(1)}] - a_2 \sum_{i=2}^{k} \lambda_2^{(i)} + b_2 \sum_{i=2}^{k} \xi_1^{(i)} = 0\,, \qquad (4.18)$$

$$4\lambda_1^{(1)} F_{33} \tau_{\|\perp}^{(1)} + c \sum_{i=2}^{k} [\lambda_3^{(i)} \sin 2\varphi_1 - \lambda_4^{(i)} \cos 2\varphi_2] = 0\,, \qquad (4.19)$$

$$\bar{h}_j f + 2\lambda_1^{(j)}[2F_{11}\sigma_{\|}^{(j)} + F_{12}\sigma_{\perp}^{(j)}] + \alpha_1 \lambda_2^{(j)} + b_1 \xi_j^{(j)} = 0\,, \qquad (4.20)$$

$$\bar{h}_j f + 2\lambda_1^{(j)}[2F_{22}\sigma_{\perp}^{(j)} + F_{12}\sigma_{\|}^{(j)}] + a_2 \lambda_2^{(j)} - b_2 \xi_j^{(j)} = 0\,, \qquad (4.21)$$

$$4\lambda_1^{(j)} F_{33} \tau_{\|\perp}^{(j)} + c[\lambda_3^{(j)} \sin 2\varphi_j - \lambda_4^{(j)} \cos 2\varphi_j] = 0\,, \qquad (4.22)$$

$$\lambda_1^{(j)}[1 - (2F_j - 1)^2]^{1/2} = 0\,, \qquad (4.23)$$

$$L_{1j} = 0\,, \quad L_{2j} = 0\,, \quad L_{3j} = 0\,, \qquad (4.24)$$

where

$$f = \frac{N_1 + N_2}{\left\{ \sum_{i=1}^{k} \bar{h}_i [\sigma_{\|}^{(i)} + \sigma_{\perp}^{(i)}] \right\}^2}, \quad \xi_{\rm m}^{(j)} = \lambda_3^{(j)} \cos 2\varphi_{\rm m} + \lambda_4^{(j)} \sin 2\varphi_{\rm m} .$$

The obtained set (4.17) to (4.24) consists of $(7k-3)$ equations involving $(7k-3)$ unknowns, i.e. $3k$ stresses $\sigma_{\|}^{(j)}$, $\sigma_{\perp}^{(j)}$, $\tau_{\|\perp}^{(j)}$ k multipliers $\lambda_1^{(j)}$, and $3(k-1)$ multipliers $\lambda_2^{(j)}$, $\lambda_3^{(j)}$, $\lambda_4^{(j)}$. It should be emphasized that the solution of these equations defines the optimal stress distribution in the material of minimum thickness rather than the structural parameters $\bar{h}_i$ and φ_i which may be found from equilibrium equations (4.1).

To compare the results with those obtained in Section 4.1.2 let us consider again the case of a unidirectional layer. By putting $k = j = 1$, the set of equations (4.17) to (4.24) reduces to (4.17) to (4.19) and (4.23). The strain compatibility equations (4.16) being satisfied automatically for a single layer, the corresponding Lagrangian multipliers λ_2, λ_3, λ_4 may be set to zero. Then, (4.19) shows that $\tau_{\|\perp} = 0$, (4.23) takes the form $F = 1$, (4.17) and (4.18) allow us to obtain condition (4.13) and (4.15) reduces to (4.14). Thus, the set of variational equations specifies the stress distribution corresponding to the absolute minimum value of the thickness.

In view of the difficulties of obtaining a direct solution of the system of equations (4.17) to (4.24) in the general case of a laminated composite, a semi-inverse method will be applied. It was established earlier that the fibers in a layer of minimum thickness should be oriented in the directions of maximum principal stresses. With this in mind, let us put $\tau_{\|\perp}^{(j)} = 0$ in the set of variational equations. It can be seen from (4.19) and (4.22) that the following two cases are possible:

$$\lambda_3^{(j)} = \lambda_4^{(j)} = 0 , \tag{4.25}$$

$$\lambda_4^{(j)} / \lambda_3^{(j)} = \tan 2\varphi_j = \tan 2\varphi_1 . \tag{4.26}$$

The next two sections are devoted to the discussion of these cases.

4.1.4. *Panels of uniform strength*

Let $\tau_{\|\perp}^{(j)} = 0$ and let the conditions (4.25) be fulfilled. Then it follows from the strain compatibility equations (4.16) that for all the layers

$$\sigma_{\|}^{(j)} = \sigma , \quad \sigma_{\perp}^{(j)} = n\sigma , \tag{4.27}$$

where

$$n = E_{\perp}(1 + \mu_{\perp\|}) / E_{\|}(1 + \mu_{\|\perp}) . \tag{4.28}$$

Thus, the case under consideration corresponds to the laminate of uniform stress distribution between the layers. Since (4.16) are satisfied, multipliers $\lambda_2^{(j)}$ may also be set to zero. Taking into account (4.27), one can obtain the following variational equations:

$$\frac{(N_1 + N_2)\bar{h}_j}{\sigma^2(1+n)^2} + 2\lambda_1^{(j)}\sigma(2F_{11} + nF_{12}) = 0\,, \tag{4.29}$$

$$\frac{(N_1 + N_2)\bar{h}_j}{\sigma^2(1+n)^2} + 2\lambda_1^{(j)}\sigma(2nF_{22} + F_{12}) = 0\,,$$
$$\lambda_1^{(j)}[1 - (2F_j - 1)^2]^{1/2} = 0\,. \tag{4.30}$$

It can be seen from (4.15) that in the case under consideration $N_1 + N_2 \neq 0$, i.e. there exists a number j such that $\lambda_1^{(j)} \neq 0$. For the jth layer it follows from (4.30) that $F_j = 1$ and the strength criterion (4.4) allows us to obtain the stress

$$\sigma = (F_{11} + nF_{12} + n^2F_{22})^{-1/2}\,. \tag{4.31}$$

(4.15), (4.27) and (4.31) permit us now to find the thickness of the laminate

$$h = \frac{N_1 + N_2}{1 - n}(F_{11} + nF_{12} + n^2F_{22})^{1/2}\,. \tag{4.32}$$

If $\lambda_1^{(j)} \neq 0$ for $j = 1, 2, 3 \ldots k$, equations (4.30) give $F_j = 1$, i.e. all the layers fail simultaneously. In this case, subtraction of the first equation in (4.29) from the second one yields:

$$n = n_0 = \frac{2F_{11} - F_{12}}{2F_{22} - F_{12}}\,, \tag{4.33}$$

where n is specified by equation (4.28). Condition (4.33), being similar to (4.13), provides the absolute minimum of the thickness (4.32). Indeed, assuming that the thickness satisfies the minimum condition in the form $\partial h/\partial n = 0$, we obtain $n = n_0$. In general, (4.33) cannot be fulfilled for modern composites. However, if the matrix load carrying capacity and rigidity is ignored ($\sigma_\perp^{\mathrm{u}} = 0$, $E_\perp = 0$), it follows from (4.28) and (4.33) that $n = n_0 = 0$. Thus, the absolute minimum of the thickness is achieved in uniformly stressed systems with fibers oriented in the directions of principal stresses.

Substituting the stresses in (4.27) into equilibrium equations (4.1) and taking into account that $\tau_{\parallel\perp}^{(j)} = 0$, one can obtain the following conditions defining the optimal structural parameters:

$$\sum_{i=1}^{k} \bar{h}_i \cos^2\varphi_i = \frac{N_1 - nN_2}{(1-n)(N_1 + N_2)}\,,$$
$$\sum_{i=1}^{k} \bar{h}_i \sin\varphi_i \cos\varphi_i = \frac{(1+n)N_{12}}{(1-n)(N_1 + N_2)}\,. \tag{4.34}$$

Stress resultants must satisfy the following constraints which can be derived from equations (4.34):

$$\frac{n}{1+n} \leqslant \frac{N_1}{N_1+N_2} \leqslant \frac{1}{1+n}, \quad \left|\frac{N_{12}}{N_1+N_2}\right| \leqslant \frac{1-n}{2(1+n)} \tag{4.35}$$

For the netting model of the composite, equations (4.34) and inequations (4.35) reduce to expressions (1.5) and (1.6), which were discussed in Section 1, with the corresponding conclusions being true for the case under consideration.

Application of the obtained results to filament wound membrane shells of resolution whose loading is in accord with the inequalities in (4.35) will be considered in Section 5.

4.1.5. *Two-layer composites reinforced in the directions of principal stresses*

Consider a material satisfying conditions (4.26). Obviously, the above conditions can be fulfilled only for two layers with orientations of $\varphi_1 = \varphi$ and $\varphi_2 = \varphi + (\pi/2)$. In this case, the strain compatibility equations (4.16) (note that $\tau_{\parallel\perp} = 0$) allow us to express the stresses $\sigma_{\parallel}^{(2)}$, $\sigma_{\perp}^{(2)}$ in the second layer in terms of the stresses $\sigma_{\parallel}^{(1)}$, $\sigma_{\perp}^{(1)}$ in the first layer, i.e.

$$\begin{aligned} \sigma_{\parallel}^{(2)} &= \frac{1}{1-\mu_{\parallel\perp}\mu_{\perp\parallel}}\left((\mu_{\parallel\perp}-\mu_{\perp\parallel})\sigma_{\parallel}^{(1)} + \frac{E_{\parallel}}{E_{\perp}}(1-\mu_{\parallel\perp}^2)\sigma_{\perp}^{(1)}\right), \\ \sigma_{\perp}^{(2)} &= \frac{1}{1-\mu_{\parallel\perp}\mu_{\perp\parallel}}\left((\mu_{\perp\parallel}-\mu_{\parallel\perp})\sigma_{\perp}^{(1)} + \frac{E_{\perp}}{E_{\parallel}}(1-\mu_{\perp\parallel}^2)\sigma_{\parallel}^{(1)}\right). \end{aligned} \tag{4.36}$$

With the aid of (4.36), the failure criteria (4.4) for both layers can be written in terms of the stresses $\sigma_{\parallel}^{(1)}$, $\sigma_{\perp}^{(1)}$. The corresponding limit interaction curve for graphite fibers in an epoxy matrix is shown in Fig. 9. Eqs. (4.23) are satisfied by letting $\lambda_1^{(1)} = 0$, $F_2 = 1$ (curve 2) or by $\lambda_1^{(2)} = 0$, $F_1 = 0$ (curve 1). The case $\lambda_1^{(1)} = \lambda_1^{(2)} = 0$, $F_{1,2} < 1$ is obviously impossible (the system of variational equations has no solution in this case) and the case $\lambda_1^{(1)} \neq 0$, $\lambda_1^{(2)} \neq 0$, $F_{1,2} = 1$ corresponds to the simultaneous failure of the layers (point a in Fig. 9). The

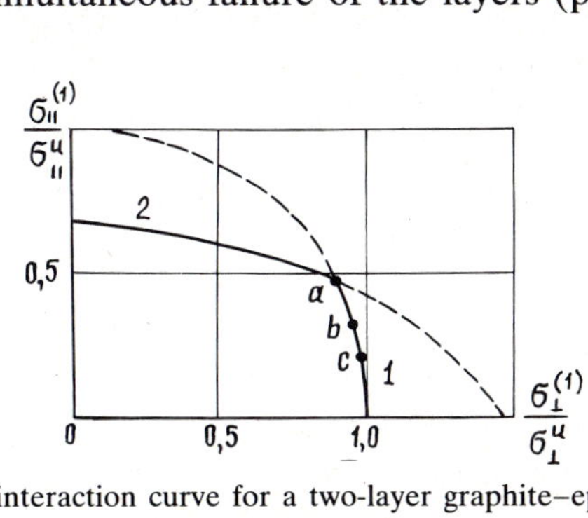

FIG. 9. Limit interaction curve for a two-layer graphite–epoxy composite.

optimal stress distribution can be obtained from (4.17), (4.18), (4.20) and (4.21) by putting $\xi^{i}_{\mathrm{m}}=0$, $\lambda_2^{(j)}=0$, $k=2$; the corresponding angle φ_1 and thicknesses h_1, h_2 can be found from (4.1) as

$$\tan 2\varphi_1 = \frac{2N_{12}}{N_1 - N_2} = \alpha \,,$$
$$h_{1,2} = \frac{(N_1 + N_2)[\sigma_{\parallel}^{(2,1)} - \sigma_{\perp}^{(2,1)}] + (N_1 - N_2)[\sigma_{\parallel}^{(2,1)} + \sigma_{\perp}^{(2,1)}]\sqrt{1+\alpha^2}}{2[\sigma_{\parallel}^{(1)}\sigma_{\parallel}^{(2)} - \sigma_{\perp}^{(1)}\sigma_{\perp}^{(2)}]} \,. \tag{4.37}$$

The results of calculations for a two-layer material loaded by $N_1 = N$, $N_2 = 2N$ are shown in Fig. 9. The point (a) corresponds to the case $N_{12} = 0$ ($\varphi_1 = 0$), point (b) – to the case $N_{12} = 0.5N$ ($\varphi_1 = -13.5°$), while point (c) to $N_{12} = N$ ($\varphi_1 = -22°$). The total thickness of the laminate is the same in all three cases.

Let us consider one important particular class of composite materials for which the constitutive equations are analogous to the generalized Hooke's law for an isotropic material. Such a class of materials may be useful if one has to substitute metal for composite to obtain some additional weight savings over a designed metal structure.

The constitutive equations connecting stress resultants and averaged strains have the form:

$$N_1 = h(C_{11}e_1 + C_{12}e_2 + C_{13}e_{12}) \,,$$
$$N_2 = h(C_{21}e_1 + C_{22}e_2 + C_{23}e_{12}) \,, \tag{4.38}$$
$$N_{12} = h(C_{31}e_1 + C_{32}e_2 + C_{33}e_{12}e_{12}) \,.$$

In the above

$$C_{mn} = \sum_{i=1}^{k} \bar{h}_i C_{mn}^{(i)} \,, \quad \bar{h}_i = h_i/h$$

the elastic stiffness $C_{11}^{(i)}$, $C_{12}^{(i)} = C_{21}^{(i)}$, $C_{22}^{(i)}$, $C_{33}^{(i)}$ are specified by (2.11) and the rest are as follows:

$$C_{13}^{(i)} = C_{31}^{(i)} = \sin\varphi_i \cos\varphi_i \{\bar{E}_{\parallel}^{(i)}[1-\mu_{\parallel\perp}^{(i)}]\cos^2\varphi_i - \bar{E}_{\perp}^{(i)}[1-\mu_{\perp\parallel}^{(i)}]\sin^2\varphi_i - 2G_{\parallel\perp}^{(i)}\cos 2\varphi_i\} \,,$$
$$C_{23}^{(i)} = C_{32}^{(i)} = \sin\varphi_i \cos\varphi_i \{\bar{E}_{\parallel}^{(i)}[1-\mu_{\parallel\perp}^{(i)}]\sin^2\varphi_i - \bar{E}_{\perp}^{(i)}[1-\mu_{\perp\parallel}^{(i)}]\cos^2\varphi_i + 2G_{\parallel\perp}^{(i)}\cos 2\varphi_i\} \,. \tag{4.39}$$

In the case of fibers lying along the directions of principal stresses we have

$\tau_{\parallel\perp}^{(j)} = 0$ and, consequently, $e_{\parallel\perp}^{(j)} = 0$. Thus according to the third equation of (4.2)

$$\tan 2\varphi_j = \frac{e_{12}}{e_1 - e_2} . \tag{4.40}$$

Now, consider a composite consisting of two layers with the orientations φ_1 and $\varphi_2 = \varphi_1 + (\pi/2)$ satisfying (4.40). Let us assume that the thicknesses of the layers satisfy the following relations

$$\frac{\bar{h}_2}{\bar{h}_1} = \frac{\bar{E}_{\parallel}^{(1)} - \bar{E}_{\perp}^{(1)}}{\bar{E}_{\parallel}^{(2)} - \bar{E}_{\perp}^{(2)}} , \quad \bar{h}_1 + \bar{h}_2 = 1 . \tag{4.41}$$

As may be verified with the aid of expressions in (4.39) for the material under consideration $C_{32} = -C_{31}$. Finally, using (4.40) to eliminate e_{12} from the first two equations in (4.38) and $(e_1 - e_2)$ from the third equation, after some manipulations the desired form of constitutive equations can be written as follows:

$$N_1 = \bar{E}h(e_1 + \mu e_2) , \quad N_2 = \bar{E}h(e_2 + \mu e_1) , \quad N_{12} = Ghe_{12} , \tag{4.42}$$

where

$$\bar{E} = \frac{E}{1 - \mu^2} , \quad G = \frac{E}{2(1 + \mu)} , \tag{4.43}$$

$$\bar{E} = \frac{\bar{E}_{\parallel}^{(1)} E_{\parallel}^{(2)} - \bar{E}_{\perp}^{(1)} \bar{E}_{\perp}^{(2)}}{\bar{E}_{\parallel}^{(1)} + \bar{E}_{\parallel}^{(2)} - \bar{E}_{\perp}^{(1)} - \bar{E}_{\perp}^{(2)}} ,$$
$$\mu = \frac{E_{\parallel}^{(1)} E_{\parallel}^{(2)} [\mu_{\parallel\perp}^{(1)} + \mu_{\parallel\perp}^{(2)}] - E_{\perp}^{(1)} E_{\perp}^{(2)} [\mu_{\perp\parallel}^{(1)} + \mu_{\perp\parallel}^{(2)}]}{E_{\parallel}^{(1)} E_{\parallel}^{(2)} - E_{\perp}^{(1)} E_{\perp}^{(2)}} . \tag{4.44}$$

In order to develop expressions (4.44), the equations in (4.41) have been used.

If the layers possess the same elastic properties ($E_{\parallel}^{(1)} = E_{\parallel}^{(2)} = E_{\parallel}$, etc.), equations (4.41), (4.43) and (4.44) reduce to the form

$$\frac{\bar{h}_1}{\bar{h}_2} = 1 , \quad \mu = \frac{E_{\parallel} \mu_{\parallel\perp} + e_{\perp} \mu_{\perp\parallel}}{E_{\parallel} + E_{\perp}} ,$$
$$E = \frac{1}{2(\bar{E}_{\parallel} + \bar{E}_{\perp})} [\bar{E}_{\parallel}^2 (1 - \mu_{\parallel\perp}^2) + \bar{E}_{\perp}^2 (1 - \mu_{\perp\parallel}^2) + 2\bar{E}_{\parallel} \bar{E}_{\perp} (1 - \mu_{\parallel\perp} \mu_{\perp\parallel})] . \tag{4.45}$$

Expressions (4.44) and (4.45) indicate the effective elastic constants of the two-layer composite and may be used to evaluate the efficiency of the composite panel in comparison with a metal one. Since the constitutive equations (4.42) for the composite panel under consideration have the form of Hooke's law for an isotropic material and the equilibrium and geometrical equations do not depend on material properties, one is able to determine the angle φ_1 by using formulas (4.37) or (4.40) and stress resultants or strains corresponding to the inital isotropic panel. The thickness of the panel can be obtained then by taking into account the required strength or rigidity constraints.

Consider the first case. The limit interaction curve indicating ultimate combinations of stresses in the two-layer composite in terms of the stresses $\sigma_{\parallel}^{(1)}$, $\sigma_{\perp}^{(1)}$ in the first layer has the form shown in Fig. 9. By putting $h_1 = h_2$ according to the first equation in (4.45), one can obtain from the second equation in (4.37), after elimination of $\sigma_{\parallel}^{(2)}$, $\sigma_{\perp}^{(2)}$ with the aid of equations (4.36), the following relation:

$$\frac{\sigma_{\parallel}^{(1)}}{\sigma_{\perp}^{(1)}} = \frac{(N_1 + N_2)A_1 - (1+\alpha^2)^{1/2}(N_1 - N_2)A_2}{-(N_1 + N_2)A_3 + (1+\alpha^2)^{1/2}(N_1 - N_2)A_4}, \tag{4.46}$$

where

$$A_{1,2} = \mp\tfrac{1}{2}\left[1 + \frac{\mu_{\perp\parallel} - \mu_{\parallel\perp} \mp \dfrac{E_{\parallel}}{E_{\perp}}(1-\mu_{\parallel\perp}^2)}{1-\mu_{\parallel\perp}\mu_{\perp\parallel}}\right],$$

$$A_{3,4} = \tfrac{1}{2}\left[1 + \frac{\dfrac{E_{\perp}}{E_{\parallel}}(1-\mu_{\perp\parallel}^2) \mp (\mu_{\parallel\perp} - \mu_{\perp\parallel})}{1-\mu_{\parallel\perp}\mu_{\perp\parallel}}\right].$$

(4.46) corrresponds graphically to a straight line in the $\sigma_{\parallel}^{(1)}$, $\sigma_{\perp}^{(1)}$ plane and the actual stresses can be determined as coordinates of the point of interaction of this straight line and the limit curve shown in Fig. 9. The stresses in the second layer and the thickness of the laminate $h = 2h_1 = 2h_2$ may be obtained from (4.36) and the second equation in (4.37), respectively.

As an example of application of the described approach, consider an infinite plate containing a circular hole shown in Fig. 10(a). The stress concentration factor is calculated according to the following formula:

$$k_\sigma = \max_{x_1, x_2} F(x_1, x_2)/F(x_1 \to \infty, x_2 \to \infty),$$

where F is specified by the strength criterion (4.4). Fiber orientations obtained from the first equation in (4.37) and the well-known solutions for stress resultants in isotropic plates are shown in Figs. 9(b), (c), (d). In the case of

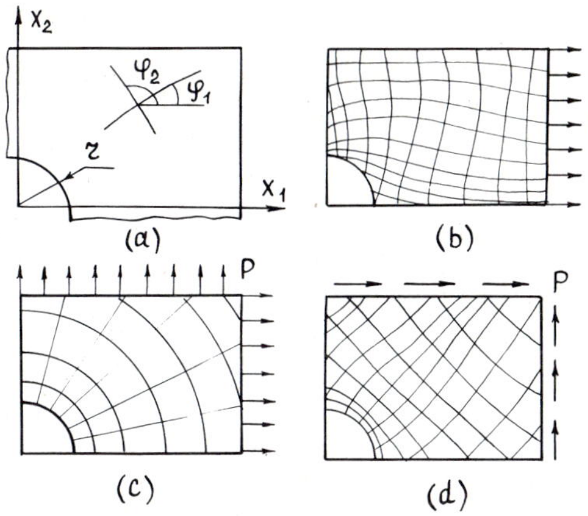

FIG. 10. Fiber orientations for a plate with (a) a circular hole; (b) subjected to uniaxial; (c) biaxial tension; and (d) pure shear.

uniaxial tension (Fig. 9(b)) we have $k_\sigma = 6.73$. Note that for a 0, 90° cross-ply laminate with each layer having the same thickness the corresponding solution for an orthotropic plate gives $k_\sigma = 29.6$. For an equal biaxial loading and fiber orientations shown in Fig. 9(c) $k_\sigma = 2.13$ and for a cross-ply laminate $k_\sigma = 10.48$. In pure shear, (Fig. 9(d)) we have in the case of the optimal laminate $k_\sigma = 22.57$; for a cross-ply laminate $k_\sigma = 101.3$ and for an angle-ply ($\varphi_1 = 45°$, $\varphi_2 = 135°$) laminate $k_\sigma = 29.86$. Thus, the stress concentration occurring in the optimal laminates is essentially lower than that occurring in conventional laminates.

Now consider the problem of substituting a metal panel for a composite one having the same rigidity, e.g. an aluminum panel ($E_m = 72$ GPa; $G_m = 27$ GPa; $\mu_m = 0.3$; $\rho_m = 2700\ \mathrm{kg/m^3}$) for a graphite–epoxy one whose unidirectional layer is characterized by the specific gravity $\rho = 1400\ \mathrm{kg/m^3}$ and material constants: $E_\parallel = 140$ GPa; $E_\perp = 12$ GPa; $G_{\parallel\perp} = 3$ GPa; $\mu_{\perp\parallel} = 0.28$, failure strength in tension $\sigma_\parallel^u = 1.1$ GPa; $\sigma_\perp^u = 0.054$ GPa; in compression $\sigma_\parallel^u = 0.9$ GPa; $\sigma_\perp^u = 0.108$ GPa, and in shear $\tau_{\parallel\perp}^u = 0.03$ GPa. The effective elastic moduli (4.45) of the optimal composite plate ($E = 76.6$ GPa; $G = 36.57$ GPa) are higher than those of a metal plate and the Poisson's ratio is equal to 0.04. Let us assume that the composite panel has the same rigidity in tension or compression as the aluminum one. Then the thickness of the composite panel is found to be $h = 0.94h_m$, its in-plane shearing rigidity is 1.27 times the rigidity of the aluminum panel and the mass of the composite panel is theoretically half the mass of the metal one.

As an example to illustrate this approach, consider a delta-wing structure shown in Fig. 11 and similar to that investigated by HANG (1978). Preliminary design of this wing made of aluminum alloy and loaded by uniformly distributed pressure has been performed by the finite element method, longitudinal and transversal flanges and webs of wing ribs being modelled as beam and plate

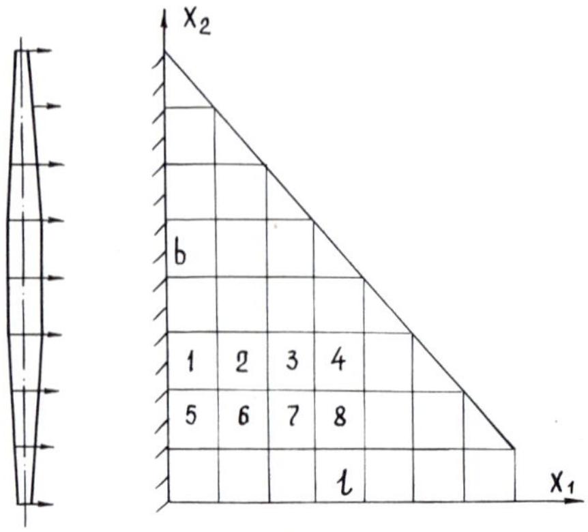

FIG. 11. A delta-wing structure ($l = 1.8$ m; $b = 2.4$ m).

elements. The skin panels stiffened by closely spaced stringers, whose effects may be smeared over the panels, were modelled as orthotropic plates. Actual thicknesses of the metal skin (h_m) and their reduced values allowing for the stringers (h_s) as well as safety factors (n) found from the finite element analysis are listed in Table 1. For brevity's sake, only eight of the panels indicated in Fig. 11 are presented in the table.

Now, let us substitute aluminum skin panels for graphite–epoxy composite panels whose effective elastic constants are specified by (4.45) and were calculated above. Fiber orientations (φ_1) obtained by substitution of the stress resultants corresponding to metal panels into the first formula in (4.37), and thicknesses (h) providing the same rigidity with respect to tension or compression in the x_1 direction are shown in Table 1. It should be noted that the stress resultants in composite panels are not quite the same as in metal panels since the structure is redundant and the effective elastic constants of composite panels differ slightly from the corresponding characteristics of aluminum alloy. Thus the obtained values of φ_1 are not exact, however, the safety factors (n) calculated by the finite element method on the basis of Hill's failure criterion and shown in Table 1 indicate the results to be quite satisfactory. Note that the thicknesses listed in Table 1 exceed the theoretical values since they are taken to be divisible by the thickness of an elementary ply. The mass of the metal skin is 2.2 times that of the composite skin.

The two-layer panels considered above have an essential disadvantage because the obtained fiber orientations are valid only for a single distribution of stress resultants, i.e. for one case of loading. If the loading conditions are changed, the directions of the fibers do not coincide with the directions of principal stresses and the crazing of the matrix leads to the failure of the laminate. It can be readily proven that three or more systems of fibers with different orientations are capable of withstanding an arbitrary combination of external stress resultants acting in a plate. Keeping this in mind, let us substitute the optimal two-layer panel with the most popular four-layer lami-

TABLE 1. Parameters of metal and composite panels.

Panel no.	Metal panels		Composite panels reinforced in the directions of principal stresses				Composite panels with fiber orientations of 0.90°, ±45°				Stiffened composite panels	
	h_m ($\times 10^{-3}$ m)	h_s ($\times 10^{-3}$ m)	n	φ_1^0	h ($\times 10^{-3}$ m)	n	h_0 ($\times 10^{-3}$ m)	h_{90} ($\times 10^{-3}$ m)	h_{45} ($\times 10^{-3}$ m)	n	h_{sk} ($\times 10^{-3}$ m)	h_{st} ($\times 10^{-3}$ m)
1	0.40	5.8	1.4	2	5.28	1.20	1.92	1.92	2.40	2.1	4.32	1.92
2	0.35	5.3	3.8	8	4.80	3.30	1.80	1.80	2.16	5.6	3.96	1.80
3	0.30	4.2	10.9	12	3.84	9.40	1.44	1.44	1.68	13	3.12	1.44
4	0.25	3.7	28.5	28	3.36	24.30	1.32	1.32	1.44	36	2.76	1.32
5	0.40	5.8	1.4	1	5.28	1.12	1.92	1.92	2.40	2.0	4.32	1.92
6	0.35	5.3	3.1	8	1.80	2.60	1.80	1.80	2.16	4.2	3.96	1.80
7	0.30	4.2	12.6	16	3.84	11.0	1.44	1.44	1.68	16	3.12	1.44
8	0.25	3.7	40.0	30	3.36	35.0	1.32	1.32	4.08	15	2.76	1.32

nate which is characterized by the fiber orientations $\varphi_1 = 0$, $\varphi_2 = 90°$, $\varphi_3 = \pm 45°$ (with respect to the spars of the wing shown in Fig. 11) and the corresponding thicknesses of the layers $h_1' = h_0$, $h_2' = h_{90}$, $h_3' = h_{45}$. In order to determine the thicknesses let us employ the following conditions

$$E_1 h' = Eh\,, \quad E_2 h' = Eh\,, \quad G_{12} h' = Gh \tag{4.47}$$

which provide the panel of total thickness $h' = h_0 + h_{90} + h_{45}$ with the same rigidity in the x_1, x_2 plane as the two-layer optimal panel with effective moduli E, G. Taking into account that

$$E_1 = C_{11} - \frac{C_{12}^2}{C_{22}}\,, \quad E_2 = C_{22} - \frac{C_{12}^2}{C_{11}}\,, \quad G_{12} = C_{33}\,, \quad \sum_{i=1}^{3} \bar{h}_i' = 1\,,$$

where

$$C_{mn} = \sum_{i=1}^{3} \bar{h}_i' C_{mn}^{(i)}\,, \quad \bar{h}_i' = h_i'/h'$$

and $C_{mn}^{(i)}$ are specified by equations (2.11), we obtain the three equations in (4.47) for h_0, h_{90}, h_{45}. The results of calculations, as well as the safety factors (n) found by the finite element analysis, are shown in Table 1. Note that h_0 and h_{90} are taken to be divisible by the thickness of an elementary ply (δ) and h_{45} is divisible by 2δ. In this case the mass of the metal skin is found to be 1.83 times that of the composite one.

A similar approach may be applied to the design of optimal structures exposed several types of loading. The maximum (in all cases) value of thickness h calculated for a given panel, each type of loading being considered independently, is substituted then in equations (4.47).

4.2. Buckling and load carrying capacity of composite panels

4.2.1. Basic equations

The laminated plate considered here is assumed to consist of an arbitrary number (k) of thin layers (Fig. 12). Each individual layer being homogeneous

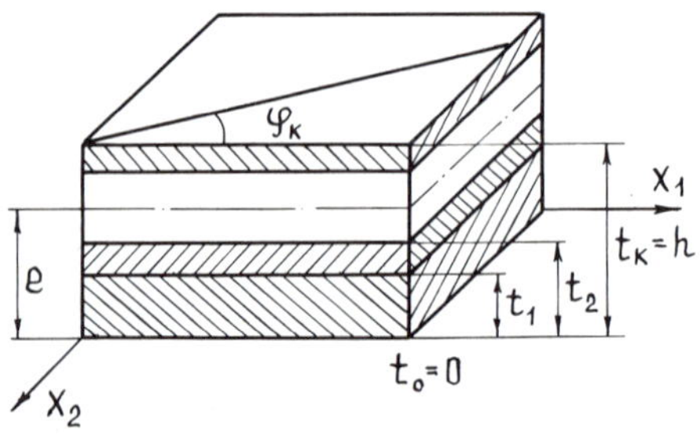

FIG. 12. A laminated plate.

through the thickness and orthotropic. The constitutive equations expressing stress resultants and moments in terms of in-plane strains and curvature changes of the reference surface can be written as

$$
\begin{aligned}
N_1 &= A_{11}e_1 + A_{12}e_2 + B_{11}k_1 + B_{12}k_2 ,\\
N_2 &= A_{21}e_1 + A_{22}e_2 + B_{21}k_1 + B_{22}k_2 ,\\
N_{12} &= A_{33}e_{12} + B_{33}k_{12} ,\\
M_1 &= B_{11}e_1 + B_{12}e_2 + D_{11}k_1 + D_{12}k_2 ,\\
M_2 &= B_{21}e_1 + B_{22}e_2 + D_{21}k_1 + D_{22}k_2 ,\\
M_{12} &= B_{33}e_{12} + D_{33}k_{12} .
\end{aligned}
\tag{4.48}
$$

In the above, membrane or stretching (A_{mn}), bending–stretching (B_{mn}) and bending (D_{mn}) stiffnesses have the form (Fig. 12)

$$
\begin{aligned}
A_{mn} &= A_{nm} = \sum_{i=1}^{k} C_{mn}^{(i)}(t_i - t_{i-1}) ,\\
B_{mn} &= B_{nm} = \tfrac{1}{2} \sum_{i=1}^{k} C_{mn}^{(i)}(t_i^2 - t_{i-1}^2) - eA_{mn} ,\\
D_{mn} &= D_{nm} = \tfrac{1}{3} \sum_{i=1}^{k} C_{mn}^{(i)}(t_i^3 - t_{i-1}^3) - 2eB_{mn} - e^2A_{mn} ,
\end{aligned}
\tag{4.49}
$$

where e denotes the coordinate of the reference plane with respect to which moments M_1, M_2, M_{12} are calculated. Coefficients $C_{mn}^{(i)}$ are specified by the equation in (2.11). It is also assumed that the ith layer of thickness $h_i = t_i - t_{i-1}$ contains pairs of plies consisting of the same material and thickness and with material axes of symmetry oriented at the angles $\pm\varphi_i$ to the x_1 direction, i.e. each pair of $+\varphi_i$ and $-\varphi_i$ plies is considered as one balanced layer.

The von Karman-type large-deflection equations for strains and changes of curvature can be expressed in terms of displacements (u_1, u_2) of the reference plane in the x_1, x_2 directions and the lateral deflection (w) as

$$
\begin{aligned}
e_1 &= \frac{\partial u_1}{\partial x_1} + \frac{1}{2}\left(\frac{\partial w}{\partial x_1}\right)^2 , \quad e_2 = \frac{\partial u_2}{\partial x_2} + \frac{1}{2}\left(\frac{\partial w}{\partial x_2}\right)^2 ,\\
e_{12} &= \frac{\partial u_1}{\partial x_2} + \frac{\partial u_2}{\partial x_1} + \frac{\partial w}{\tau x_1}\frac{\partial w}{\partial x_2} ,\\
k_1 &= -\frac{\partial^2 w}{\partial x_1^2} , \quad k_2 = -\frac{\partial^2 w}{\partial x_2^2} , \quad k_{12} = -2\frac{\partial^2 w}{\partial x_1 \partial x_2} .
\end{aligned}
\tag{4.50}
$$

The stress resultants and the moments satisfy the von Karman-type equilibrium equations

$$\frac{\partial N_1}{\partial x_1} + \frac{\partial N_{12}}{\partial x_2} = 0\,, \quad \frac{\partial N_{12}}{\partial x_1} + \frac{\partial N_2}{\partial x_2} = 0\,,$$

$$\frac{\partial^2 M_1}{\partial x_1^2} + 2\frac{\partial^2 M_{12}}{\partial x_1 \partial x_2} + \frac{\partial^2 M_2}{\partial x_2^2} - N_1 k_1 - N_{12}k_{12} - N_2 k_2 + p = 0\,, \tag{4.51}$$

where p is the lateral pressure. Finally, let us write the expression for the elastic strain energy in a rectangular plate as follows:

$$U = \tfrac{1}{2}\int_0^a\int_0^b (N_1 e_1 + N_2 e_2 + N_{12}e_{12} + M_1 k_1 + M_2 k_2 + M_{12}k_{12})\,\mathrm{d}x_1\,\mathrm{d}x_2\,, \tag{4.52}$$

where a and b are the dimensions of the plate in the x_1, x_2 directions.

4.2.2. Homogeneous plates

Single-layer plates or plates consisting of a number of identical layers can be treated as homogeneous plates, assuming that the material properties are constant through the thickness. In this case, by putting $e = h/2$, i.e. using the midplane as the reference plane, the stiffnesses (4.49) can be essentially simplified and reduced to

$$A_{11} = \bar{E}_1 h\,, \quad A_{12} = A_{21} = \mu_{12}\bar{E}_1 h\,, \quad A_{22} = \bar{E}_2 h\,, \quad A_{33} = G_{12}h\,,$$

$$\bar{E}_{1,2} = E_{1,2}/(1-\mu_{12}\mu_{21})\,, \tag{4.53}$$

$$D_{11} = \frac{\bar{E}_1 h^3}{12}\,, \quad D_{12} = D_{21} = \mu_{12}\frac{\bar{E}_1 h^3}{12}\,, \quad D_{22} = \frac{\bar{E}_2 h^3}{12}\,, \quad D_{33} = \frac{G_{12}h^3}{12}\,,$$

$$B_{mn} = 0$$

To illustrate the design of composite plates consider an angle-ply $\pm\varphi$ laminate, which may approximately be treated as homogeneous since the bending–stretching coupling effect is very weak in a multi-ply laminate with plies having alternating orientations of $+\varphi$ and $-\varphi$. Let an angle-ply rectangular plate be subjected to uniaxial compression as is shown in Fig. 13. Prior to buckling, the initial (pre-buckling) stress resultant N is uniformly distributed across the width of the plate (Fig. 14a). Pre-buckling stresses in the unidirectional plies are defined by the equations in (4.1) to (4.3) and in the case under

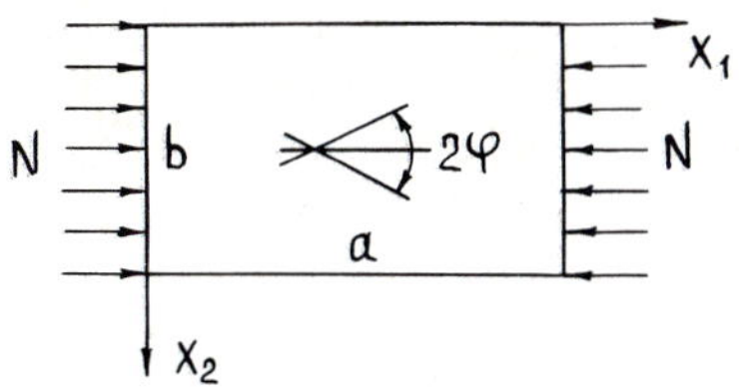

FIG. 13. An angle-ply $\pm\varphi$ plate under uniaxial compression.

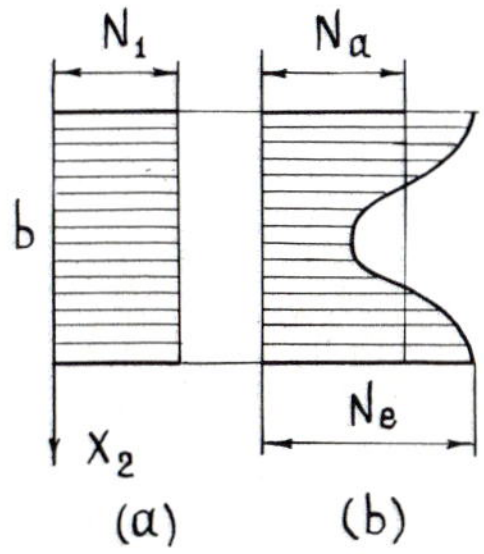

FIG. 14. Distribution of longitudinal stress resultant across the width of the plate (a) before and (b) after buckling.

consideration take the following form:

$$\sigma_{\parallel} = -\frac{N}{h} c_1 , \quad \sigma_{\perp} = -\frac{N}{h} c_2 , \quad \tau_{\parallel\perp} = \frac{N}{h} c_{12} , \tag{4.54}$$

where

$$c_1 = \frac{\bar{E}_{\parallel}}{E_1} [\cos^2 \varphi + \mu_{\parallel\perp} \sin^2 \varphi - \mu_{21}(\sin^2 \varphi + \mu_{\parallel\perp} \cos^2 \varphi)] ,$$

$$c_2 = \frac{\bar{E}_{\perp}}{E_1} [\sin^2 \varphi + \mu_{\parallel\perp} \cos^2 \varphi - \mu_{12}(\cos^2 \varphi + \mu_{\parallel\perp} \sin^2 \varphi)] ,$$

$$c_3 = \frac{G_{\parallel\perp}}{E_1} (1 - \mu_{21}) \sin 2\varphi .$$

Fig. 15 presents a graphical plot of the ultimate load $\bar{N}_s = N_s/N_s^0$, where $N_s^0 = N_s\ (\varphi = 0)$ with N_s calculated with the aid of the equations in (4.54) and Hill's failure criterion as a function of φ for a square graphite–epoxy plate with $a = b = 0.2$ m and $h = 2.6 \times 10^{-3}$ m.

To analyze the buckling and post-buckling behaviour of a plate simply supported along its edges, assume the following trigonometric expression for the normal deflection:

$$w = w_{mn} \sin \lambda_m x_1 \sin \lambda_n x_2 , \tag{4.55}$$

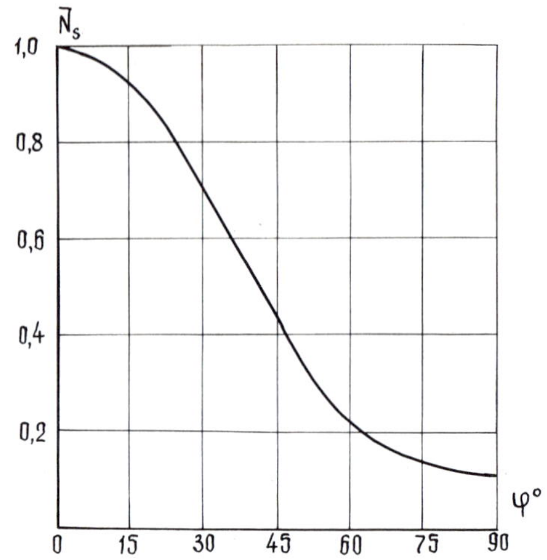

FIG. 15. Ultimate pre-buckling stress resultant for uniaxially compressed composite plates of $\pm\varphi$ orientation.

where $\lambda_m = \pi m/a$, $\lambda_n = \pi n/b$; w_{mn} and m, n being the unknown amplitude and the numbers of half waves in the x_1 and x_2 direction, respectively. Using a displacement formulation one can substitute the stress resultants (4.48) and then the strains and changes of curvature (4.50) into the in-plane equilibrium equations (4.51) and obtain two equations in terms of the displacements u_1, u_2 and the deflection w. Integration of these equations with account for expression (4.55) yields:

$$\begin{aligned} u_1 &= -\left(\frac{N}{E_1 h} + \frac{\lambda_m^2}{8} w_{mn}^2\right)x_1 - \frac{w_{mn}^2}{16\lambda_m}(\lambda_m^2 - \mu_{12}\lambda_n^2)\sin 2\lambda_m x_1 \\ &\quad + \frac{\lambda_m}{16} w_{mn}^2 \sin 2\lambda_m x_1 \cos 2\lambda_n x_2\,, \\ u_2 &= \left(\mu_{21}\frac{N}{E_1 h} - \frac{\lambda_n^2}{8} w_{mn}^2\right)x_2 + \frac{w_{mn}^2}{16\lambda_n}(\lambda_n^2 - \mu_{21}\lambda_m^2)\sin 2\lambda_n x_2 \\ &\quad + \frac{\lambda_n}{16} w_{mn}^2 \cos 2\lambda_m x_1 \sin 2\lambda_n x_2\,. \end{aligned} \tag{4.56}$$

Displacements (4.56) satisfy the following boundary conditions corresponding to uniaxial compression of a simply supported plate (Fig. 13):

$$\begin{aligned} &\int_0^b N_1(x_1 = 0; a)\,dx_2 = -Nb\,, \\ &\int_0^a N_2(x_2 = 0; b)\,dx_1 = 0\,. \end{aligned} \tag{4.57}$$

Deflection (4.55) and displacements (4.56) do not naturally satisfy the third equilibrium equation (4.51). So let us construct an approximate solution employing the variational principle of the minimum total potential energy

$$\Pi = U - Nb \int_0^a \frac{\partial u_1}{\partial x_1}\, dx_1 ,$$

where U is specified by expression (4.52). Stationarity of Π with respect to w_{mn} requires that

$$w_{mn} = \frac{16(N - N_{mn})}{\lambda_n^2\left(E_1 \dfrac{\lambda_m^2}{\lambda_n^2} + E_2 \dfrac{\lambda_n^2}{\lambda_m^2}\right)} , \tag{4.58}$$

where

$$N_{mn} = \frac{\pi^2}{a^2 m^2}\,(D_1 m^4 + D_2 m^2 n^2 + D_3 n^4) \tag{4.59}$$

and

$$D_1 = D_{11} , \quad D_2 = 2c^2(D_{12} + 2D_{33}) , \quad D_3 = c^4 D_{22} , \quad c = a/b .$$

Eq. (4.59) yields the eigenvalues of the problem under consideration and the lowest one gives the buckling load N_c. Expression (4.59) for N_{mn} permits exact minimization with respect to m and n resulting in $n = 1$ and

$$\begin{aligned} &m = 1, \text{ if } 0 < c < \sqrt[4]{\bar{D}} , \\ &m = 2, \text{ if } \sqrt[4]{\bar{D}} < c < \sqrt[4]{36\bar{D}} , \\ &m = 3, \text{ if } \sqrt[4]{36\bar{D}} < c < \sqrt[4]{144\bar{D}} \text{ etc} \end{aligned} \tag{4.60}$$

where $\bar{D} = D_{11}/D_{22}$. Inequalities (4.60) are not convenient for the problem of optimum design. More useful result can be obtained by the geometric programming method

$$m^2 = \frac{c^2}{\sqrt{\bar{D}}} \left(\frac{\sqrt{\bar{D}}}{c^2}\right)^r \left(\frac{1+r}{1-r}\right)^{(1-r)/2} , \tag{4.61}$$

where

$$r = \begin{cases} r_1 = \dfrac{\bar{D} - c^4}{\bar{D} + c^4} & \text{if } r_1 > 0 \\ 0 & \text{if } r_1 \leqslant 0 . \end{cases}$$

Since the value of m at buckling specified by (4.61) is not in general an integer, (4.59) gives a lower bound of critical load in this case. For instance, let us consider a graphite–epoxy square parallel-ply ($\varphi = 0$) plate with $a = b = 0.1$ m, $h = 2 \times 10^{-4}$ m. According to formulas (4.61) and (4.59) we have $m = 0.99$, $bN_c = 11.5$ N, while the exact result is $m = 1$, $bN_c = 11.7$ N.

Fig. 16 displays the variation of the dimensionless critical load $\bar{N}_c = N_c/N_s^0$ with φ for the plate whose strength dependence on φ in the pre-buckled stress state is shown in Fig. 15. Comparison of Figs. 15 and 16 suggests that one might find some optimal fiber orientation in the interval $0 \leqslant \varphi \leqslant 45°$ providing the minimum value of the thickness and satisfying both the strength ($N \leqslant N_s$) and stability ($N \leqslant N_c$) conditions. This optimal orientation can be determined through the use of any of various available techniques for numerical solution of non-linear constrained minimization problems. As an example, the results obtained for homogeneous angle-ply $\pm\varphi$ plates by penalty techniques are shown in Table 2.

Now, let us consider some peculiarities of optimum design of composite panels. According to the conventional formulation of this problem described above, the longitudinal edges of the plate $x_2 = 0$ and $x_2 = b$ (see Fig. 13) are assumed to be unrestricted with respect to the displacement u_2. However, in

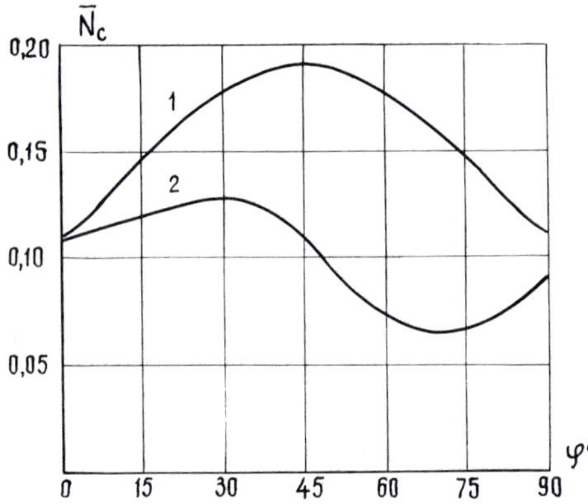

FIG. 16. Buckling stress resultant for simply supported uniaxially loaded composite plates of $\pm\varphi$ orientation with movable (1) and immovable (2) longitudinal edges.

TABLE 2. Optimal parameters of rectangular ($a/b = 2$) graphite–epoxy plates subjected to uniaxial compression ($N = 10^6$ N/m).

b (m)	Homogeneous plates		Sandwich plates					
	h ($\times 10^{-3}$ m)	φ_1^0	h ($\times 10^{-3}$ m)	t_1 ($\times 10^{-3}$ m)	φ_1^0	t_2 ($\times 10^{-3}$ m)	φ_2^0	Δ (%)
0.1	4.34	13.18	3.42	0.82	45.21	2.6	1.28	21.2
0.2	6.60	16.53	5.95	1.78	45.23	3.57	1.28	19
0.4	10.00	20.00	9.22	3.71	45.22	5.51	1.28	7.82

many practical instances this displacement is restricted by the adjacent panels. This implies that the second boundary condition (4.57) should be replaced by $u_2\ (x_2=0)=u_2\ (x_2=b)=0$. In this case, the transverse stress resultant $N_2=-\mu_{12}N$ is present in the plate due to the Poisson's effect, i.e. the plate undergoes biaxial compression. This effect is usually not considered to contribute to the critical load of metal panels, however, it may be significant for composite panels, as is shown in Fig. 16 (curve 2). It can be seen from this figure that the additional compression in the x_2 direction essentially reduces the critical load and changes the optimal fiber orientation for plates with simply supported immovable longitudinal edges.

Although buckling is an important measure of the load carrying capacity of a structure, it is well known that a panel under uniaxial compression can carry considerable load beyond buckling. In the case of $N>N_c$ the deflection is specified by (4.55) and (4.58), the numbers of half waves (m, n) being unknown and dependent on the load level. The numbers of half waves, corresponding to the given external load, may be estimated through energy considerations discussed by AZIKOV and VASIL'EV (1982). Taking into account (4.58), the total potential energy can be written as follows (for simplicity, $n=1$)

$$\Pi=\frac{N^2ab}{2E_1h}+\frac{(N-N_{mn})^2ab}{h[E_1+E_2(c/m)^4]}\ .$$

Let us suppose that the two stress states corresponding to the same potential energy and different values of load and half wave number are possible, i.e. $\Pi(N^{(1)}, m^{(1)})=\Pi(N^{(2)}, m^{(2)})$. Intuitively, one would assume m corresponding to the lower value of N, i.e. $m=m^{(2)}$, if $N^{(1)}>N^{(2)}$, and $m=m^{(1)}$, if $N^{(2)}>N^{(1)}$. Thus, m being determined, stress resultants can be obtained with the help of (4.48), (4.50), (4.53), (4.55) and (4.56) and take the form

$$\begin{aligned} N_1&=-N-\frac{E_1}{8}\lambda_m^2w_{mn}^2\cos 2\lambda_nx_2\ ,\\ N_2&=\frac{E_2}{8}\lambda_n^2w_{mn}^2\cos 2\lambda_mx_1\ ,\\ N_{12}&=0\ , \end{aligned} \tag{4.62}$$

where w_{mn} is specified by (4.58).

The solution given by (4.58) and (4.62) agrees reasonably well with the experimental results obtained by BANKS and HARVEY (1979) for uniaxially compressed plates. It may be regarded as the first approximation to the exact solution which can be found by the perturbation method. In this method, displacements are decomposed into basic non-linear parts constructed in the way described above and additional parts, which in the case of simply supported edges can be represented in the form of double Fourier series. Minimizing the total potential energy with respect to the coefficients of these series and neglecting the non-linear terms of the series (the additional displacements are assumed to be small in comparison with the basic ones) yields an infinite set of linear algebraic equations for the unknown coefficients. The agreement between the solution obtained by the perturbation method and the approximate one with an accuracy of better than 2% and 10% for w and N_1, respectively, confirms that the approach described above may be considered to be valid for the problem of design of composite panels.

According to the first equation in (4.62), the distribution of the stress resultant N_1 across the width of the plate is not uniform (Fig. 14(b)). The longitudinal load carried by the plate may be calculated as $P = fbN_e$. Here, (fb) is the reduced width; N_e denotes the maximum value of N_1 at the edge; i.e. $N_e = N_1\ (x_2 = 0; b)$; f is the reduction factor specified by the relationship

$$f = N_a / N_e\,, \tag{4.63}$$

where

$$N_a = \frac{1}{b}\int_0^b N_1\,dx_2\,,$$

is the averaged stress resultant (see Fig. 14(b)). With the aid of (4.62), (4.58) and (4.63) one can obtain

$$f = \frac{1}{3+\alpha}\left(1 + \alpha + 2\,\frac{N_c}{N_e}\right), \tag{4.64}$$

where

$$\alpha = (E_2 n^4 c^4)/(E_1 m^4)$$

and N_c is the critical value (see Fig. 16). It can be seen from (4.64) that the reduction factor decreases as N_e increases, the maximum value of N_e being limited by the ultimate strength (see Fig. 15). Fig. 17 illustrates the dependence

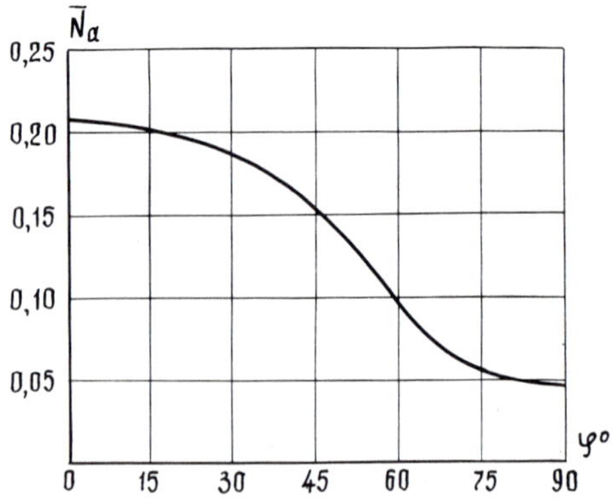

FIG. 17. Ultimate averaged post-buckling stress resultant for composite plates of $\pm\varphi$ orientation.

of the ultimate averaged stress resultant $\bar{N}_a = N_a/N_e^0$, where $N_a = P_u/b$, $N_e^0 = N_e$ $(\varphi = 0)$, on the fiber orientation of the angle-ply laminate considered earlier. It follows from this figure that the optimal orientation corresponding to the ultimate load carrying capacity of the plate beyond buckling coincides with the orientation providing the maximum strength in the pre-buckled state (see Fig. 15) and not with that providing the maximum value of the critical load (see Fig. 16). Naturally, this conclusion is valid only for angle-ply laminates, but it should be taken into account while designing composite plates that will be subjected to compressive loads.

4.2.3. *Laminated plates*

As was proved in Section 3, the maximum value of the bending stiffness corresponds to the beam symmetrically laminated with respect to its midplane. This conclusion is true for laminated plates as well. As numerous calculations show, midplane symmetrically laminated plates possess higher bending stiffness, critical load and carrying capacity in comparison to unsymmetrically laminated plates. A plate is said to be symmetrically laminated if for each layer located above the midplane there is an identical layer located at the same distance below the midplane. For such plates, assuming $e = h/2$ in (4.49), one can obtain

$$A_{mn} = A_{nm} = \sum_{i=1}^{k} C_{mn}^{(i)}(t_i - t_{i-1}) ,$$

$$B_{mn} = B_{nm} = 0 , \tag{4.65}$$

$$D_{mn} = D_{nm} = \tfrac{1}{3} \sum_{i=1}^{k} C_{mn}^{(i)}(t_i^3 - t_{i-1}^3) - \frac{h^2}{4} A_{mn} .$$

Thus the bending–stretching coupling stiffnesses are identically equal to zero for all symmetrically laminated plates and their behaviour is similar to that of homogeneous plates. Equations derived in Section 4.2.2 are valid for the plates under consideration if one determines the bending stiffnesses (D_{mn}) from (4.65) and the elastic constants according to the following formulas:

$$E_1 h = A_{11} - \frac{A_{12}^2}{A_{22}} , \quad E_2 h = A_{22} - \frac{A_{12}^2}{A_{11}} , \quad G_{12} h = A_{33} ,$$

$$\mu_{12} = \frac{A_{12}}{A_{11}} , \quad \mu_{21} = \frac{A_{12}}{A_{22}} .$$

Eqs. (4.54) specify the stresses in the ith layer if one sets $\varphi = \varphi_i$, $\bar{E}_\parallel = \bar{E}_\parallel^{(i)}$, $\bar{E}_\perp = \bar{E}_\perp^{(i)}$, $G_{\parallel\perp} = G_{\parallel\perp}^{(i)}$, $\mu_{\parallel\perp} = \mu_{\parallel\perp}^{(i)}$. Eqs. (4.59) and (4.60) or (4.61) allow us to obtain the critical load and expressions (4.62) and (4.64) specify the stress resultants and the reduction factor after buckling.

As an example, consider a sandwich graphite–epoxy plate with t_1, φ_1; t_2, φ_2; $t_3 = t_2 + t_1$, $\varphi_3 = \varphi_1$ (see Fig. 12 for the notation). The results of minimization of the overall thickness (h) with respect to these parameters by the penalty function method under strength and stability constraints are shown in Table 2. It can be seen from the table that the central layer has the fiber orientation of $\varphi_2 \approx 0$ and the fibers of the outer layers are oriented at $\varphi_1 \approx 45°$. This result seems to be quite natural since these values correspond to the maximum values of the ultimate stress resultant (Fig. 15) and the critical load (Fig. 16), respectively. The quantity Δ characterizes the reduction in mass in comparison with homogeneous plates whose parameters are also listed in Table 2. A further increase in the number of layers does not lead to an essential reduction in mass.

For uniaxially compressed simply supported plates consisting of layers arbitrarily located through the thickness one can obtain the following expression for the eigenvalues:

$$N = \frac{1}{\lambda_m^2}\left(a_{33} - \frac{a_{11}a_{23}^2 + a_{22}a_{13}^2 - 2a_{12}a_{13}a_{23}}{a_{11}a_{22} - a_{12}^2}\right), \tag{4.66}$$

where

$$\begin{aligned}
&a_{11} = A_{11}\lambda_m^2 + A_{33}\lambda_n^2, \quad a_{12} = \lambda_m\lambda_n(A_{12} + A_{33}),\\
&a_{22} = A_{22}\lambda_n^2 + A_{33}\lambda_m^2,\\
&a_{13} = \lambda_m[B_{11}\lambda_m^2 + (B_{12} + 2B_{33})\lambda_n^2],\\
&a_{23} = \lambda_n[B_{22}\lambda_n^2 + (B_{12} + 2B_{33})\lambda_m^2],\\
&a_{33} = D_{11}\lambda_m^4 + 2(D_{12} + 2D_{33})\lambda_m^2\lambda_n^2 + D_{22}\lambda_n^4.
\end{aligned} \tag{4.67}$$

The critical load can be found by minimizing N with respect to $\lambda_m = \pi m/a$, $\lambda_n = \pi n/b$. For design purposes, the following approximate formulas can be derived by the geometric programming method:

$$N_c = \frac{\pi^2}{b^2}\left[\sqrt{D_1D_3}\left(\frac{1}{c^2}\sqrt{\frac{D_1}{D_3}}\right)^r\left(\frac{1+r}{2}\right)^{-\frac{1+r}{2}}\left(\frac{1-r}{2}\right)^{-\frac{1-r}{2}} + D_2\right], \tag{4.68}$$

$$n = 1, \quad m = c^2\sqrt{\frac{D_3}{D_1}}\left(\frac{1}{c^2}\sqrt{\frac{D_1}{D_3}}\right)^r\left(\frac{1+r}{1-r}\right)^{\frac{1-r}{2}}, \quad c = \frac{a}{b} \tag{4.69}$$

where

$$r = \begin{cases} r_1 = \dfrac{D_1 - c^4 D_3}{D_1 + c^4 D_3} & \text{if } r_1 > 0 \\ 0 & \text{if } r_1 \leq 0 , \end{cases}$$

$$D_1 = D_{11} - \frac{b_1}{a_1} ,$$

$$D_2 = 2c^2(D_{12} + 2D_{33}) - \frac{c^2}{a_1^2}(a_1 b_2 - a_2 b_1) - \frac{f}{c^4 a_1}(a_1 d_2 - a_2 d_1 c^2) ,$$

$$D_3 = D_{22}c^4 - \frac{d_1}{a_1} - \frac{f}{a_1}(a_1 b_5 - d_1 a_3) ,$$

$$a_1 = A_{11}A_{33} , \quad a_2 = A_{11}A_{22} - A_{12}(A_{12} + 2A_{33}) , \quad a_3 = A_{22}A_{33} ,$$

$$b_1 = B_{11}^2 A_{33} ,$$

$$b_2 = B_{11}^2 A_{22} + B_{12}^2 A_{11} + A_{11}(B_{12} + 2B_{33})^2 - 2A_{12}B_{11}(B_{12} + 2B_{33}) ,$$

$$b_3 = 2(B_{12} + 2B_{33})[B_{11}A_{22} + B_{22}A_{11} - A_{12}(B_{12} + 2B_{33})] - 2B_{11}(B_{22}A_{12} + B_{12}A_{33}) ,$$

$$b_4 = A_{22}(B_{12} + 2B_{33})^2 - 2A_{12}B_{22}(B_{12} + 2B_{33}) + A_{12}B_{22}^2 + A_{22}B_{12}^2 ,$$

$$b_5 = B_{22}^2 A_{33} ,$$

$$d_1 = \frac{c^4}{a_1^2}[a_1(a_1 b_3 - a_3 b_1) - a_2(a_1 b_2 - a_2 b_1)] ,$$

$$d_2 = \frac{c^6}{a_1^2}[a_1^2 b_4 - a_3(a_1 b_2 - a_2 b_1)] ,$$

$$f = \begin{cases} \dfrac{D_{11}}{a_1 D_{22} + a_2 (D_{11} D_{22})^{1/2} + a_3 D_{11}} & \text{if } c \geq 1 \\ \dfrac{c^4}{a_1 + a_2 c^2 + a_3 c^4} & \text{if } c < 1 . \end{cases}$$

As an example, consider a plate composed of six graphite–epoxy composite layers. Let us assume the reference-plane location (e) to be given by

$$e = \sum_{i=1}^{6} C_{11}^{(i)}(t_i^2 - t_{i-1}^2) \Big/ \sum_{i=1}^{6} C_{11}^{(i)}(t_i - t_{i-1})$$

In this case the longitudinal bending–stretching coupling stiffness is equal to zero and the remaining stiffnesses are listed in Table 3. According to (4.68) and (4.69) we have $N_c = 10.5$ kPa m, $m = 2.2$. The result of direct minimization of equation (4.66) is $N_c = 10.9$ kPa m, $m = 2$. Note that neglecting all bending–stretching coupling stiffnesses ($B_{mn} = 0$) is (4.66) leads to a 27% reduction in the critical load.

TABLE 3. Stiffnesses of a six-layer graphite–epoxy composite plate.

mn	A_{mn} ($\times 10^{-7}$ Pa m)	B_{mn} ($\times 10^{-4}$ Pa m^2)	D_{mn} (Pa m^3)
11	9.58	0	10.6
12	2.98	−5.4	5.74
22	3.78	−6.46	7.12
33	3.38	−5.9	6.45

Finally, let us consider again the delta-wing shown in Fig. 11 and discussed in Section 4.1.5. In that section the metal skin of the wing was substituted by smooth graphite–epoxy panels having approximately the same rigidity as the metal ones. Comparison of the averaged stress resultants in the compressed panels, found by means of the finite elements method, with the corresponding critical values shows that the stability constraints are not satisfied. Consequently, let us substitute smooth composite panels with stiffened panels whose skin contains all the piles of the smooth panels with the fibers oriented at ±45° and 90° while the plies with longitudinal (0°) orientation are replaced by stringers of the cross-section shown in Fig. 18. The space between the stringers (l) and the

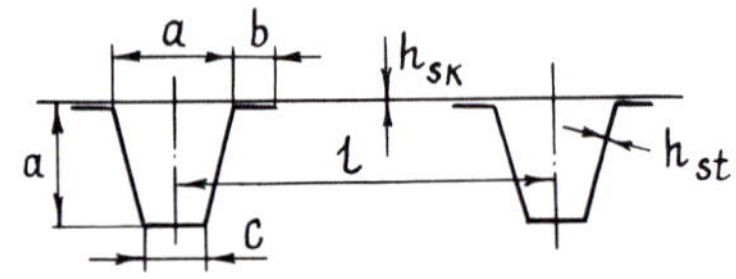

FIG. 18. Cross-section of a stiffened panel.

dimensions of the cross-section (a, b, c, h_{st}) must satisfy local skin and stringer instability and general panel instability constraints, which can be written with the aid of (4.68). Note that in order to predict general instability, a stiffened plate may be treated as a laminated plate consisting of layers with special mechanical properties corresponding to the skin of the plate, flanges and webs of the stringers. The results of calculations are as follows: $l = 0.125$ m, $a = 0.03$ m, $b = 0.5a$, $c = 0.83a$; the thicknesses of the skins (h_{sk}) and the stringers (h_{st}) are shown in Table 1.

5. Membrane shells

5.1. Governing equations

Axisymmetrically loaded membrane shells of revolution made of fibrous composites by means of the winding technique are increasingly being introduced into different fields of engineering. Consider a shell consisting of unidirectional layers, identical in thickness and material properties and having alternating orientations of $+\varphi$ and $-\varphi$ (Fig. 19). The condition of equilibrium yields the following meridional, circumferential and shearing averaged stress resultants:

$$
\begin{aligned}
N_1 &= -R\,\frac{[1+(z')^2]^{1/2}}{rz'a}\,,\\
N_2 &= -\frac{a}{z'}\,[1+(z')^2]^{1/2}\left\{pr - \frac{Rz''}{z'a^2[1+(z')^2]}\right\},\\
N_{12} &= 0\,,
\end{aligned}
\tag{5.1}
$$

where

$$
R = \frac{Q}{2\pi} + a^2\int_{r_0}^{r}(p+z'q)r\,\mathrm{d}r\,,
$$

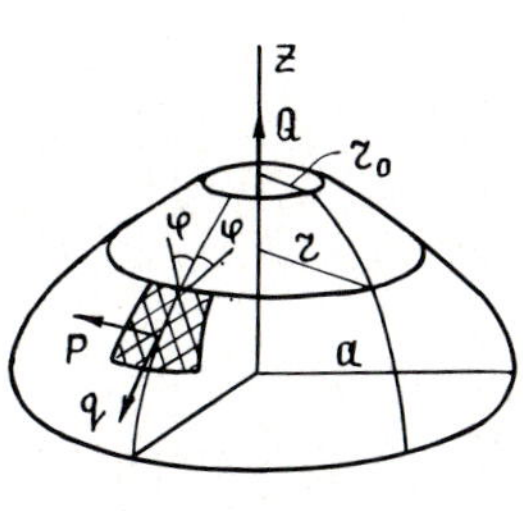

FIG. 19. A composite shell of revolution.

Q is the axial force uniformly distributed along the contour of the polar opening of a radius r_0; p and q denote internal pressure and meridional force per unit area, respectively; z and r are non-dimensional coordinates normalized to the radius of the equator (a); and the prime indicates differentiation with respect to r.

Since the stress resultants in (5.1) satisfy the inequalities given in (4.35) (only this type of loading is considered) there exists a uniformly stressed system of fibers providing, as was shown in Section 4.1.3, the minimum value of the shell mass. According to (4.27), (4.28) and (4.31) the stresses acting in this system are as follows:

$$\sigma_{\parallel} = \sigma = (F_{11} + nF_{12} + n^2F_{22})^{-1/2} , \quad \sigma_{\perp} = n\sigma , \quad \tau_{\parallel\perp} = 0 , \tag{5.2}$$

where

$$n = E_{\perp}(1 + \mu_{\perp\parallel})/E_{\parallel}(1 + \mu_{\parallel\perp}) . \tag{5.3}$$

Taking into account (5.3), one can readily obtain from equilibrium conditions (4.1) for the case under consideration ($k = 2$; $\bar{h}_1 = \bar{h}_2$; $\varphi_1 = -\varphi_2 = \varphi$)

$$N_1 = \sigma h[n + (1 - n)\psi] , \quad N_2 = \sigma h[1 - (1 - n)\psi] , \quad N = 0 , \tag{5.4}$$

where $\psi = \cos^2 \varphi$. Combining (5.1) and (5.4) we obtain

$$\sigma h = aR[1 + (z')^2]^{1/2}/z'r[n + (1 - n)\psi] , \tag{5.5}$$

$$\frac{rz''}{z'[1 + (z')^2]} = \frac{pr^2a^2}{R} - f , \tag{5.6}$$

where

$$f = \frac{1 - (1 - n)\psi}{n + (1 - n)\psi} .$$

Let us differentiate (5.5) with respect to r. Taking into account that $\sigma =$ constant and eliminating z'' with the aid of (5.6), yield

$$\frac{\mathrm{d}}{\mathrm{d}r} \{rh[n + (1 - n)\psi]\} - h[1 - (1 - n)\psi] = -\frac{rqa}{\sigma} [1 + (z')^2]^{1/2} . \tag{5.7}$$

Thus, the design problem of a uniformly stressed shell reduces to the solution of the system of equations (5.6) and (5.7) involving three unknown functions, namely $z(r)$, $\psi(r)$ and $h(r)$. In order to solve the problem one of these functions should be given. Consider the following two cases:

(1) Filament wound shells whose thickness is dependent on r according to relationship

$$h = h_1 \cos \varphi_1/r \cos \varphi , \tag{5.8}$$

where $h_1 = h(1)$ and $\varphi_1 = \varphi(1)$ are the thickness and the winding angle at the equator. In this case, (5.6) and (5.7) allow us to obtain the shape of the shell $z(r)$ and the winding angle $\varphi(r)$.

(2) Shells made of tapes with an arbitrary width or thickness variation which are laid along the meridians. In this case

$$\psi = 1 \tag{5.9}$$

and (5.6) and (5.7) allow us to determine $z(r)$ and $h(r)$.

5.2. Uniform internal pressure and axial force

Let $p = \text{constant}$, $q = 0$ and the thickness satisfy (5.8). Then the solution of (5.7) can be written as follows:

$$r^2\psi^n[1-(1-n)\psi]^{1-n} = \psi_1^n[1-(1-n)\psi_1]^{1-n}\,, \tag{5.10}$$

where $\psi_1 = \cos^2\varphi_1$.

Consider the most popular approach based on the netting theory of membrane composite shells. With the assumption that $n = 0$, solution (5.10) reduces to

$$r\sin\varphi = \sin\varphi_1 \tag{5.11}$$

and specifies a geodetic line on a surface of revolution. Taking into account (5.11), (5.6) can be written as follows:

$$\frac{z''}{z'[1+(z')^2]} = \frac{2r}{r^2-\beta^2} - \frac{r_0^2}{r(r^2-\beta^2)}\,, \tag{5.12}$$

where

$$r_0 = \sin\varphi_1\,, \quad \beta^2 = r_0^2 - \frac{Q}{\pi p a^2}\,.$$

The shape satisfying equation (5.12) is called an isotensoid and, subject to the boundary conditions $z(1) = 0$, $z'(1) \to -\infty$, can be expressed in terms of elliptic integrals in the following way:

$$z = \frac{k_2-\beta^2}{(1-k_2)^{1/2}}F(k,\theta) + (1-k_2)^{1/2}E(k,\theta)\,, \tag{5.13}$$

where

$$k_{1,2} = \frac{1-2\beta^2}{2}\left\{-1 \pm \left[\frac{1}{1-r_0^2}\left(1+\frac{3-4\beta^2}{1-2\beta^2}r_0^2\right)\right]^{1/2}\right\},$$

$$k = \left(\frac{1-k_1}{1-k_2}\right)^{1/2}, \quad \theta = \arcsin\left(\frac{1-r^2}{1-k_1}\right)^{1/2}.$$

Isotensoids (5.13) corresponding to various relationships between the axial force (Q) and internal pressure (p) are shown in Fig. 20. In particular, curves 1 and 2 illustrate the shapes of shells loaded by an axial force; curves 3 and 4 correspond to internal pressure vessels with closed (capped) and open center hole, respectively. The thickness of the shell at the equator can be calculated as

$$h_1 = \frac{\pi p a^2(1-r_0^2)+Q}{2\pi a\sigma_\parallel^u\cos^2\varphi_1}$$

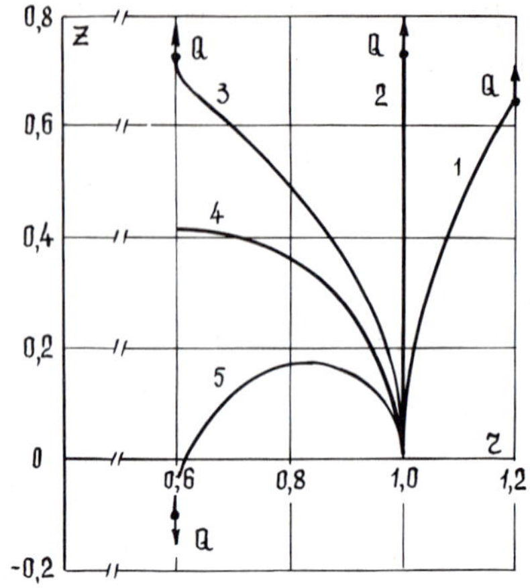

FIG. 20. Isotensoids for various loading conditions: 1, $p = 0$; $\varphi_1 = 45°$; 2, $p = 0$, $\varphi_1 = 0$; 3, $Q = \pi p a^2 r_0^2$, $\varphi_1 = 35°$; 4, $Q = 0$, $\varphi_1 = 36.9°$; 5, $Q = 0.5\pi p a^2 (1 - r_0^2)$, $\varphi_1 = 0$.

and its mass (M) is connected with the internal volume (V), specific strength of unidirectional material ($\gamma = \sigma_{\|}^{u}/\rho$) and bursting pressure (p) by the following formula:

$$M = k_s \frac{pV}{\gamma} .$$

Variations of k_s and $\bar{V} = V/a^3$ with the radius of the polar hole r_0 (see Fig. 19) are shown in Fig. 21.

Now, consider the second case corresponding to condition (5.9). Assuming $\varphi \equiv 0$, one is able to describe a class of composite shells of uniform strength

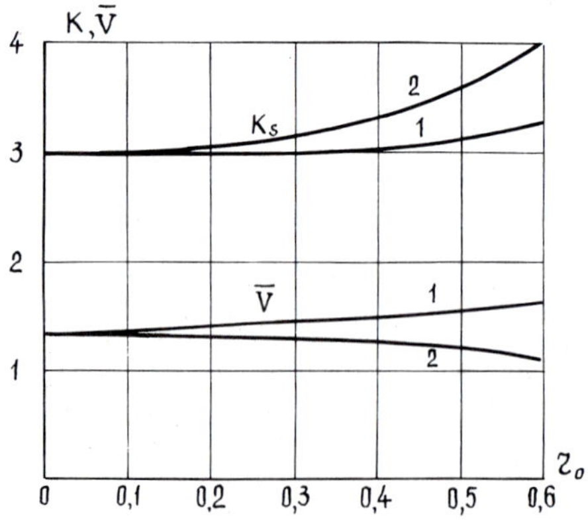

FIG. 21. Variations of coefficient k_s and internal volume of the shell ($\bar{V}$) with the radius of the closed (1) and open (2) polar hole.

and closed at the apex. Setting $p = \text{constant}$, $q = 0$, $Q = 0$ in equations (5.6) and (5.7) yields the following solutions:

$$z = \int_r^1 \frac{r^{2-n}\,dr}{(1 - r^{2(2-n)})^{1/2}}\,, \qquad h = h_1 r^{n-1}$$

specifying isotensoid with fibers oriented along the meridian if $n = 0$, spherical shell if $n = 1$ and cylindrical shell if $n = 2$. In the case of $0 \leqslant n < 2$, the integral can be expressed in terms of the β-function (B_x)

$$z = B_x(\tfrac{1}{2}, 1 - m)\,, \quad r = (1 - x)^m\,, \quad m = 1/[2(2 - n)]\,.$$

The results are presented graphically in Fig. 22. The thickness of the shell at the equator can be shown to be

$$h_1 = \frac{pa}{2}\,(F_{11} + nF_{12} + n^2F_{22})^{1/2}\,.$$

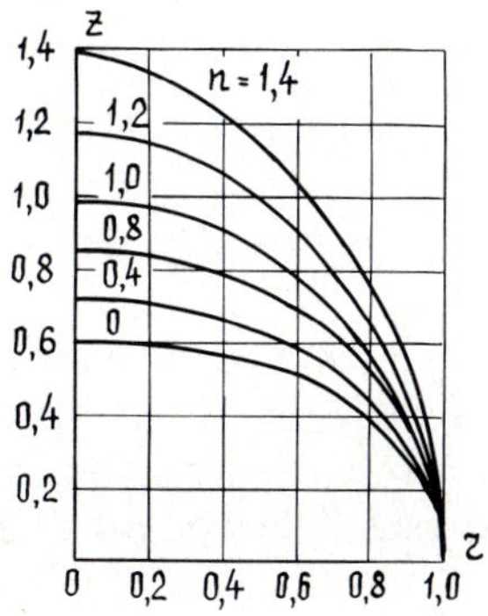

FIG. 22. Meridians of composite shells for a number of n-values.

Finally, let us note that the review of mechanics and optimum design of filament wound composite shells of revolution was given by OBRAZTSOV et al. (1977).

5.3. Spinning shells and disks

Uniformly stressed spinning filamentary disks and shells have been investigated by KAYSER (1965) and BOKOV et al. (1981). Consider a composite shell of revolution (Fig. 19) rotating at a constant angular velocity (ω) about its axis. Assuming in (5.6) and (5.7) that

$$p = -\frac{\rho h \omega^2 a r z'}{[1 + (z')^2]^{1/2}} \quad q = \frac{\rho h \omega^2 a r}{[1 + (z')^2]^{1/2}}\,,$$

the following set of equtions can be derived:

$$\frac{rz''}{z'[1+(z')]} + \frac{r^3hz'\lambda}{a\eta[1+(z')^2]^{1/2}} + f = 0\,, \tag{5.14}$$

$$\frac{\mathrm{d}}{\mathrm{d}r}\{rh[n+(1-n)\psi]\} - h[1-(1-n)\psi] + \lambda hr^2 = 0\,, \tag{5.15}$$

where

$$\lambda = \rho\omega^2a^2/\sigma\,, \quad \eta = Q/2\pi a^2\sigma\,. \tag{5.16}$$

For shells whose thickness is specified by (5.8) the first integral of (5.14) is as follows:

$$z' = \pm\frac{\eta}{\left(\frac{h_1^2\psi_1}{a^2\psi}[n+(1-n)\psi]^2 - \eta^2\right)^{1/2}}\,, \tag{5.17}$$

where

$$\eta = \frac{h_1}{a}[n+(1-n)\psi_1]\,, \tag{5.18}$$

taking into account that $z'(1)\rightarrow\pm\infty$. Eqs. (5.16) and (5.18) allow us to obtain the axial force

$$Q = 2\pi h_1 a\sigma[n+(1-n)\psi_1]\,, \tag{5.19}$$

which is caused by the interaction of the spinning shell with the driving shaft to which it is connected.

To simplify the analysis let us employ the netting theory of a composite shell assuming that the parameter n (5.3) is equal to zero. In this case, (5.15) to (5.17) and (5.19) reduce to

$$\frac{\mathrm{d}}{\mathrm{d}r}(r\sin\varphi) = \frac{\lambda r^2}{\sin\varphi}\,, \tag{5.20}$$

$$z' = \pm\frac{\cos\varphi_1}{(\cos^2\varphi - \cos^2\varphi_1)^{1/2}}\,, \tag{5.21}$$

$$Q = 2\pi h_1 a\sigma_{\parallel}^{\mathrm{u}}\cos^2\varphi_1\,, \quad \lambda = \rho\omega^2a^2/\sigma_{\parallel}^{\mathrm{u}}\,. \tag{5.22}$$

The + and − signs follow from the fact that the shell has a plane of axial symmetry. In what follows, only a part of its generator in the first quadrant of the coordinate system will be considered, hence the + sign will be ignored. The solution of equations (5.20) and (5.21) obtained subject to the boundary conditions $\varphi(1) = \varphi_1$, $z(1) = 0$ can be written in the form

$$\sin \varphi = r^{-1}[\sin^2 \varphi_1 - \tfrac{1}{2}\lambda(1 - r^4)]^{1/2} , \tag{5.23}$$

$$z = \frac{\cos \varphi_1}{\sqrt{2\lambda}} \left(\frac{\pi}{2} - \arcsin \frac{\lambda r^2 - \sin^2 \varphi_1}{\lambda - \sin^2 \varphi_1} \right) . \tag{5.24}$$

Eqs. (5.23) and (5.24), when analyzed, show that there exist the following six classes of uniformly stressed spinning filamentary systems illustrated in Fig. 23:

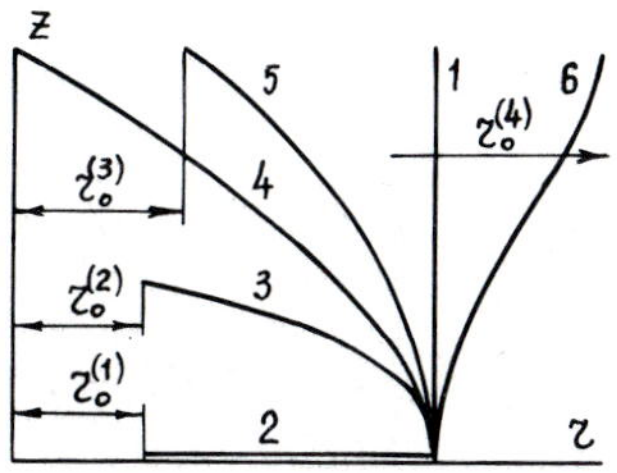

FIG. 23. Meridians of spinning composite shells.

(1) $\lambda = \sin^2 \varphi_1$, $r \equiv 1$, $\varphi \equiv \varphi_1$, $z' \to -\infty$.
In this case we have a cylindrical shell (straight line 1). In particular, for circumferential winding $Q = 0$, and the ultimate angular velocity is given by

$$\omega_u = \frac{1}{a} \left(\frac{\sigma_\parallel^u}{\rho} \right)^{1/2} . \tag{5.25}$$

(2) $1 < \lambda < 2$, $\varphi_1 = \pi/2$, $z \equiv 0$, $r_0^{(1)} \leqslant r \leqslant 1$,
where $r_0^{(1)} = [(2 - \lambda)/\lambda]^{1/2}$. The obtained solution defines a disk with a central hole of radius $r_0 = r_0^{(1)}$ (straight line 2) whose contour is free from stresses because the fibers are tangential to the contour, i.e. $\varphi(r = r_0^{(1)}) = \pi/2$. For a solid disk

$$r_0^{(1)} = 0 , \quad \lambda = 2 , \quad \sin \varphi = r , \quad \varphi_u = \frac{1}{a} \sqrt{\frac{2\sigma_\parallel^u}{\rho}} . \tag{5.26}$$

(3) $\lambda = m \sin^2 \varphi_1$, $1 < m < 2$, $r_0^{(2)} \leqslant r \leqslant 1$,
where $r_0^{(2)} = [(2 - m)/m]^{1/2}$. In this case the solution reduces to

$$\sin \varphi = \frac{\sin \varphi_1}{r} \left(1 - \frac{m}{2} (1 - r^4) \right)^{1/2} ,$$
$$z = \frac{\cot \varphi_1}{\sqrt{2m}} \left(\frac{\pi}{2} - \arcsin \frac{mr^2 - 1}{m - 1} \right) , \tag{5.27}$$

and the shape of the shell is represented by curve 3 in Fig. 23.

(4) $\lambda = 2\sin^2\varphi_1$, $0 \leq r \leq 1$.
The solution defines a shell closed at the apex (curve 4). Eqs. (5.27) allow us to obtain

$$\sin\varphi = r\sin\varphi_1 , \quad z = \tfrac{1}{2}\cot\varphi_1[\tfrac{1}{2}\pi - \arcsin(2r^2 - 1)] ,$$

$$\varphi_u = \frac{\sin\varphi_1}{a}\sqrt{\frac{2\sigma_{\parallel}^{u}}{\rho}} .$$

The curves $z(r)$ for the given case are shown in Fig. 24.

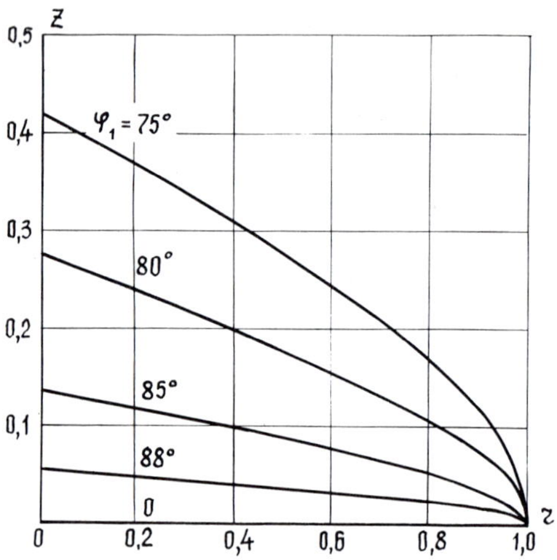

FIG. 24. Meridians of uniformly stressed spinning filamentary shells with $\lambda = 2\sin^2\varphi_1$ for a number of φ_1-values.

(5) $\lambda = m\sin^2\varphi_1$, $m > 2$, $r_0^{(3)} \leq r \leq 1$,
and

$$r_0^{(3)} = \left(\frac{m-2}{m}\right)^{1/4} , \quad \omega_u = \frac{\sin\varphi_1}{a}\left(\frac{m\sigma_{\parallel}^{u}}{\rho}\right)^{1/2} . \tag{5.28}$$

This case corresponds to a shell with a polar hole of radius $r_0 = r_0^{(3)}$ (curve 5) on the contour of which the fibers are tangential to the meridian, i.e. $\varphi(r = r_0^{(3)}) = 0$. If $\varphi_1 = \pi/2$, we have a disk connected with a shaft of radius $r_0 = r_0^{(3)}$. It follows from the second equation in (5.28) that an infinite angular velocity is available if $m \to \infty$. However, the first equation in (5.28) shows that in this case the radius of the rigid shaft $r_0^{(3)} \to 1$.

(6) $\lambda = m\sin^2\varphi_1$, $m < 1$, $1 \leq r \leq r_0^{(4)}$,
where $r_0^{(4)} = [(2-m)/m]^{1/2}$. The corresponding shape of the shell is represented by curve 6. In the case of $\varphi_1 = \pi/2$ we have a disk with an inner radius of $r = 1$ and outer radius of $r = r_0^{(4)}$ on the contour of which the fibers must be connected with a rigid ring.

One of the promising applications of fibrous composites is associated with the manufacture of inertial energy accumulators, i.e. flywheels. The high specific strength of composites allows a dramatic increase in the energy parameters to be anticipated as compared with flywheels made of conventional structural materials. Keeping in mind these applications, let us estimate the energy characteristics of spinning composite shells and disks. The kinetic energy and mass of a shell with a central hole of radius $r = r_0$ can be written as

$$E = km\omega^2 a^3 \int_{r_0}^{1} [1 + (z')^2]^{1/2} \frac{r^2\,dr}{\cos\varphi} ,$$

$$M = 2kma \int_{r_0}^{1} \frac{[1 + (z')^2]^{1/2}}{\cos\varphi}\, dr , \tag{5.29}$$

where m is the mass of roving per unit length and k denotes the number of rovings in any cross-section perpendicular to the rotation axis. Integrating (5.29) taking account of (5.23) and (5.24) yields

$$E = \frac{km}{\lambda}\omega^2 a^3 \left[\frac{\sin^2\varphi_1}{\sqrt{2\lambda}} F(\lambda, \varphi_1, r_0) + \Phi(\lambda, \varphi_1, r_0) \right] ,$$

$$M = \sqrt{\frac{2}{\lambda}}\, kmaF(\lambda, \varphi_1, r_0) , \tag{5.30}$$

where

$$F(\lambda, \varphi_1, r_0) = \frac{\pi}{2} - \arcsin \frac{\lambda r_0^2 - \sin^2\varphi_1}{\lambda - \sin^2\varphi_1} ,$$

$$\Phi(\lambda, \varphi_1, r_0) = \{(1 - r_0^2)[\tfrac{1}{2}\lambda(1 + r_0^2) - \sin^2\varphi_1]\}^{1/2} .$$

Let us introduce the relative energy

$$e = \frac{E}{M} = \frac{\omega^2 a^2}{2\lambda}\left(\sin^2\varphi_1 + \frac{\sqrt{2\lambda}\Phi}{F}\right) \tag{5.31}$$

and investigate some of the cases considered above. For a cylinder (case 1) with circumferential fibers ($\varphi_1 = \pi/2$) or a ring one can obtain with the help of (5.25)

$$E = \tfrac{1}{2}M\omega^2 a^2 = M\sigma_{\parallel}^{u}/2\rho .$$

Hence

$$e = E/M = \sigma_{\parallel}^{u}/2\rho . \tag{5.32}$$

For a disk (case 2), the same result follows from (5.16) and (5.31). Thus, the relative energy of a uniformly stressed fibrous disk is equal to that of a ring. It should be noted that the ultimate angular velocity of the solid disk (5.26) is $\sqrt{2}$ times the velocity of the ring (5.25).

For shells (cases 3, 4), (5.16) and (5.31) give

$$e = \frac{\sigma_{\parallel}^{u}}{2\rho} \sin^2 \varphi_1 . \tag{5.33}$$

It is obvious from (5.32) and (5.33) that the relative energy of a shell cannot exceed the corresponding energy of a ring or a disk. The closed shell (case 4) has the following absolute parameters (5.30):

$$E = \pi mk\omega^2 a^3/4 \sin \varphi_1 , \quad M = \pi mka/\sin \varphi_1 .$$

Finally, consider a shell whose angular velocity is theoretically unlimited (case 5). Eq. (5.31) in the general case allows us to obtain

$$e = \frac{\sigma_{\parallel}^{u}}{2\rho} \left[\sin^2 \varphi_1 + \frac{r_0^{(3)}\sqrt{2\lambda} \sin \varphi_1}{F(\lambda, \varphi_1, r_0^{(3)})} \right]$$

and for a shell with a relatively small polar hole ($r_0^{(3)} \ll 1$) the above takes the following approximate form:

$$e = \frac{\sigma_{\parallel}^{u}}{2\rho} \sin^2 \varphi_1 \left(1 + \frac{2r_0^{(3)}}{\pi} \right) .$$

It is interesting to investigate the possibility of increasing energy characteristics of shells under consideration by introduction of additional mass distributed along the equator. Therefore consider a shell connected at the equator with a ring having negligible stiffness and with mass per unit length denoted as m_0. In this case, the fiber orientation is specified as before by (5.23) and (5.24) can be generalized as

$$z = \frac{\theta \cos \varphi_1}{\sqrt{2\lambda}} F_m(\theta, \lambda, \varphi_1, r) , \tag{5.34}$$

where

$$\theta^2 = 1 - \frac{\lambda \bar{m}}{2 \cos^2 \varphi_1} , \quad \lambda = \frac{\rho \omega^2 a^2}{\sigma_{\parallel}^{u}} ,$$

$$\bar{m} = \frac{m_0}{h_1 \rho a} = \frac{2\pi m_0}{km} \cos \varphi_1 ,$$

$$F_m(\theta, \lambda, \varphi_1, r) = \arcsin \frac{\lambda + \theta^2 \cos^2 \varphi_1 - 1}{[(1 - \theta^2 \cos^2 \varphi_1)^2 - \lambda(2 \sin^2 \varphi_1 - \lambda)]^{1/2}}$$

$$- \arcsin \frac{\lambda r^2 + \theta^2 \cos^2 \varphi_1 - 1}{[(1 - \theta^2 \cos^2 \varphi_1)^2 - \lambda(2 \sin^2 \varphi_1 - \lambda)]^{1/2}} .$$

For $\varphi_1 = 80°$, the curves $z(r)$ (5.34) are shown in Fig. 25. The kinetic energy and mass of the shell are

$$E = \omega^2 a^3 \Big\{ \pi m_0 + \frac{km}{\lambda} \big[(r_0^2(1-\theta^2 \cos^2 \varphi_1) + \tfrac{1}{2}\lambda(1-r_0^4) - \sin^2 \varphi_1)^{1/2} - (1-\theta^2\cos^2\varphi_1 - \sin^2\varphi_1)^{1/2} + (2\lambda)^{-1/2}(1-\theta^2\cos^2\varphi_1)F_{\rm m}(\theta, \lambda, \varphi_1, r_0) \big] \Big\}, \tag{5.35}$$

$$M = a[2\pi m_0 + (2/\lambda)^{1/2} km F_{\rm m}(\theta, \lambda, \varphi_1, r_0)] .$$

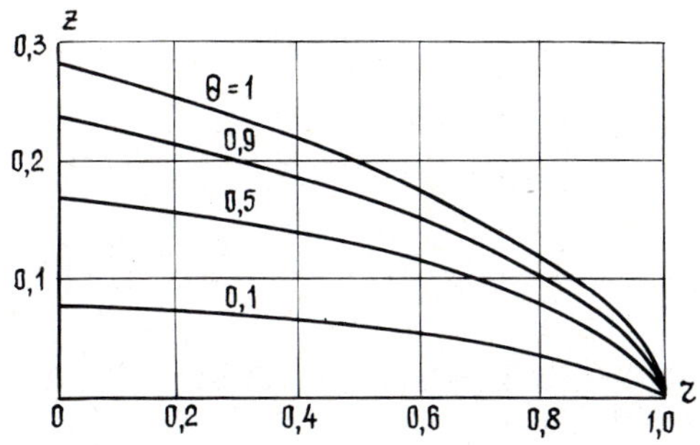

FIG. 25. Meridians of composite shells with additional mass for $\varphi_1 = 80°$ and a number of θ-values.

In the above r_0 is the radius of the polar hole. In the case of $\theta = 1$, which corresponds to a shell without additional mass, (5.35) reduce to (5.30). In the limiting case $\theta = 0$ we obtain a disk with additional mass. As an example, consider a solid disk for which $r_0 = 0$, $\lambda = 2\sin^2 \varphi_1$ and according to equations written above

$$\sin\varphi = r\sin\varphi_1 , \quad \sin^2\varphi_1 = 1/(1+\bar{m}) ,$$
$$\omega_{\rm u} = \frac{1}{a}\left(\frac{2\sigma_{\|}^{\rm u}}{\rho(1+\bar{m})}\right)^{1/2} . \tag{5.36}$$

Comparing (5.26) and (5.36) one can see that the additional circumferential mass reduces the ultimate value of angular velocity. It follows from the dependence of $\bar{e} = e/e(\bar{m} = 0)$ on $\bar{m}$ for the disk presented graphically in Fig. 26 that the additional mass does not allow one to increase the relative energy capacity of a uniformly stressed filamentary system. Now consider the shells satisfying condition (5.9). By putting $\psi = 1$ or $\varphi = 0$, (5.7) yields

$$\frac{\rm d}{{\rm d}r}(rh) - nh + \lambda h r^2 = 0 .$$

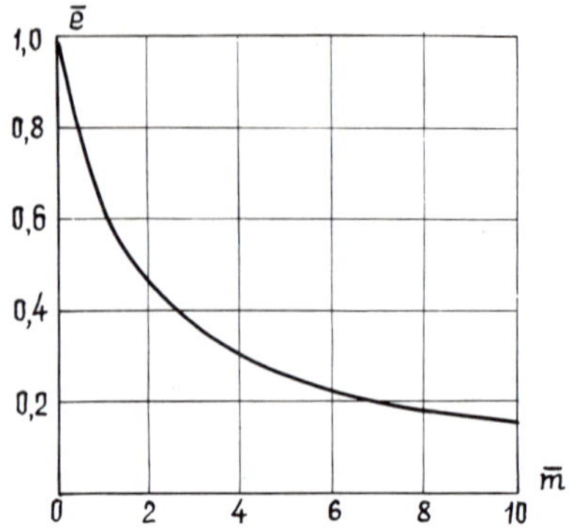

FIG. 26. Effect of additional mass on the relative characteristic energy of a spinning composite disk.

The solution to this equation is

$$h = h_1 r^{n-1} e^{\frac{1}{2}\lambda(1-r^2)} . \tag{5.37}$$

Taking into account the solution (5.37), (5.17) and (5.19) specifying the form of the shell and the axial force can be written as follows:

$$z' = \pm \frac{1}{(r^{2n} e^{\lambda(1-r^2)} - 1)^{1/2}} , \quad Q = 2\pi a h_1 \sigma .$$

The results are plotted in Figs. 27 and 28 for $n = 1$. The kinetic energy and

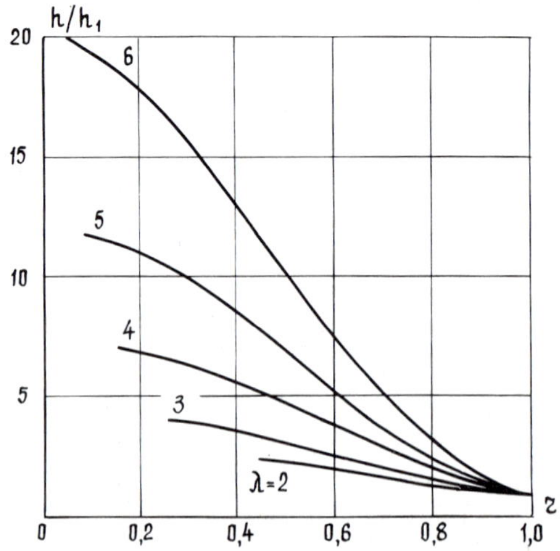

FIG. 27. Thickness variations for spinning shells made of orthotropic tapes for $\varphi = 0$ and a number of λ-values.

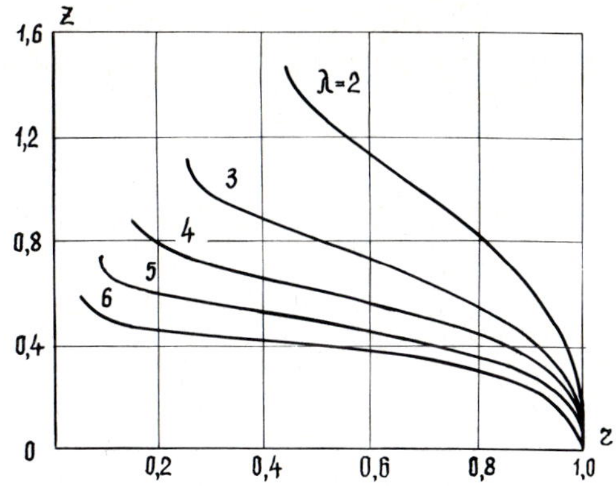

FIG. 28. Meridians of shells for a number of λ-values.

mass of the shell are

$$E = \pi\rho h_1 \omega^2 a^4 \int_0^1 \frac{r^{2(n+1)}\, e^{\lambda(1-r^2)}}{(r^{2n}\, e^{\lambda(1-r^2)} - 1)^{1/2}}\, dr$$

$$M = 2\pi\rho h_1 a^2 \int_0^1 \frac{r^{2n}\, e^{\lambda(1-r^2)}}{(r^{2n}\, e^{\lambda(1-r^2)} - 1)^{1/2}}\, dr\,.$$

Fig. 29 displays the variations of the relative mass and energy with the parameter λ for $n = 1$.

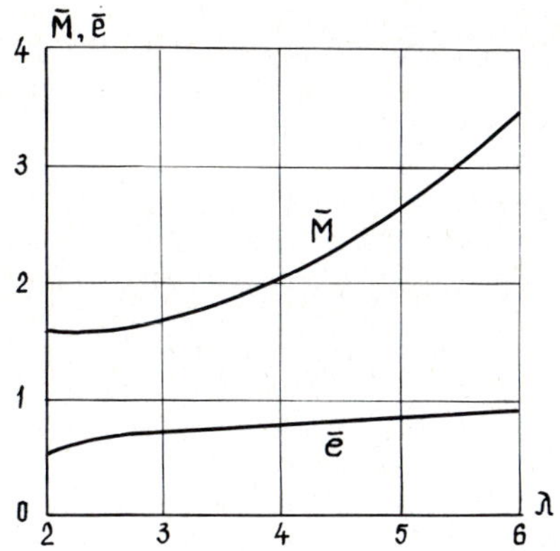

FIG. 29. Mass ($\bar{M} = M/2\pi\rho h_1 a^2$) and energy ($\bar{e} = E\rho/M\sigma$) of spinning composite shells for $n = 1$.

5.4. Optimum winding patterns for shells of revolution without strength constraints

The form and fiber orientation of shells considered above were mainly governed by strength conditions. If the level of active loads is relatively low, the strength conditions may be found to be inessential because the strength of fibrous composites is extremely high and the thickness of the shell cannot be smaller than that of an elementary ply. Therefore consider the problem of optimization of winding patterns in order to minimize the mass of the shell of revolution of an arbitrary form (see Fig. 19), regardless of strength constraints. The thickness of the filament wound shell is specified by the following relationship:

$$h = h_a a \cos \varphi_a / r \cos \varphi \ , \tag{5.38}$$

where $h_a = h(r = a)$, $\varphi_a = \varphi(r = a)$. It is assumed naturally that the minimum value of thickness (h_a) is given a priori and the problem is to find the winding pattern $\varphi(z)$ providing the minimum value of the volume of the material

$$v = 2\pi \int_0^{z_0} r(z)[1 + (r')^2]^{1/2} h \, \mathrm{d}z \ , \tag{5.39}$$

where $z_0 = z\ (r = r_0)$, r_0 is the radius of the polar hole (Fig. 19) and the prime, in this section, denotes differentiation with respect to z. The fibers must be tangent with the polar opening, hence $\varphi_0 = \varphi\ (r = r_0) = \pi/2$. Furthermore, that the following condition providing for equilibrium of the fibers on the surface of a mandrel during winding should be satisfied:

$$\left| \frac{(r \sin \varphi)'}{\sin^2 \varphi - \dfrac{rr'' \cos^2 \varphi}{1 + (r')^2}} \right| = \alpha k_{\mathrm{f}} \ . \tag{5.40}$$

In the above, k_{f} is the coefficient of friction and $|\alpha(z)| \leqslant 1$. The case of $\alpha = 0$ corresponds to winding along geodetic lines and if $\alpha = \pm 1$ (5.40) defines the lines of ultimate deviation from the geodetics. Thus, taking into account (5.38) to (5.40), we have the problem of finding $\varphi(z)$ and $\alpha(z)$ that provides us with

$$\min v = \int_0^{z_0} F(z, \theta) \, \mathrm{d}z \tag{5.41}$$

and satisfies the following constraints:

$$\begin{gathered} \theta' - r^{-1}[\alpha k_{\mathrm{f}} \Psi(z, \theta) - r'\theta] = 0 \ , \\ \theta_0 = 1 \ , \quad |\alpha(z)| \leqslant 1 \ , \end{gathered} \tag{5.42}$$

where

$$\theta = \sin \varphi \ , \quad \theta_0 = \sin \varphi_0 \ , \quad \theta_a = \sin \varphi_a$$

$$F(z, \theta) = 2\pi a h \left(\frac{1-\theta_a^2}{1-\theta^2} \left[1 + (r')^2\right] \right)^{1/2} ,$$

$$\Psi(z, \theta) = \theta^2 \left[1 + \frac{rr''}{1+(r')^2} \right] - \frac{rr''}{1+(r')^2} .$$

Let us introduce the Lagrangian multiplier $\lambda(z)$ and construct the augmented functional

$$L = \int_0^{z_0} \{F + \lambda[\theta' - r^{-1}(\alpha k_f \Psi - r'\theta)]\} \, \mathrm{d}z .$$

Stationarity of L according to Pontryagin's maximal principle (ALEKSEEV et al., 1979) in conjunction with the condition $\delta\theta_0 = 0$ leads to the variational equation and natural boundary condition for λ

$$\lambda' + \lambda f(z, \alpha, \theta) = \frac{\theta F}{1-\theta^2} , \quad \lambda_a = \frac{\theta_a}{1-\theta_a^2} \int_0^{z_0} F(z, \theta) \, \mathrm{d}z , \tag{5.43}$$

where

$$f(z, \alpha, \theta) = \frac{2}{r} \alpha k_f \theta \left[1 + \frac{rr''}{1+(r')^2} \right] - \frac{r'}{r} , \quad \lambda_a = \lambda(r = a)$$

and the following equation:

$$\max_{|\alpha| \leq 1} \left[\lambda \left(\frac{\alpha k_f}{r} \Psi - r'\theta \right) - F \right] = \lambda \left(\frac{\alpha_{op} k_f}{r} \Psi - r'\theta \right) ,$$

for the optimal control function $\alpha_{op}(z)$. The latter after some manipulations, allows us to conclude that for shells of zero or positive Gaussian curvature ($r'' \leq 0$), $\alpha(z) = \operatorname{sign} \lambda(z)$ if $\lambda \not\equiv 0$ and $-1 \leq \alpha \leq 1$ if $\lambda \equiv 0$. The solution of the boundary problem (5.43) can be written as

$$\lambda = \mathrm{e}^{-\int_0^z f \, \mathrm{d}z} \left(\frac{\theta_a}{1-\theta_a^2} \int_0^{z_0} F \, \mathrm{d}z + \int_0^z \frac{\theta F}{1-\theta^2} \mathrm{e}^{\int_0^z f \, \mathrm{d}z} \, \mathrm{d}z \right) . \tag{5.44}$$

General analysis of the solution shows that the function $\lambda(z)$ has only one switching point $z = z_c$, at which λ changes its sign, moreover, $\lambda \geq 0$ for $0 \leq z \leq z_c$ and $\lambda \leq 0$ for $z_c \leq z \leq z_0$. Thus, the optimal winding pattern is the solution to (5.12) in which one should take $\alpha = 1$ for $0 \leq z \leq z_c$ and $\alpha = -1$ for $z_c \leq z \leq z_0$. In general, this solution can be obtained only by numerical integration.

Determination of the coordinate z_c of the switching point from (5.44) involving the unknown function $\theta(z)$ seems to be a very cumbersome task. A simpler procedure for evaluating z_c may be based on plotting v (5.41) as a function of z_c and obtaining z_c as a coordinate of the point cooresponding to the minimum value of v.

As an example, consider a conical shell whose meridian $\bar{r}(\bar{z})$ is shown in Fig. 30 ($\bar{r} = r/r_0$; $\bar{z} = z/r_0$; $r_0 = 0.233$ m). This figure also gives a plot of the relative volume of the material $\bar{v} = v/v_g$ (v_g is the volume corresponding to winding along geodetic lines according to $\bar{r}\sin\varphi = 1$) as a function of the relative coordinate of the switching point $\bar{z}_c = z_c/r_0$. It can be seen that for the shell under consideration $\bar{z}_c = 1.54$. Eq. (5.42) for a conical shell with α = constant has a closed-form solution

$$\theta = \left[\bar{r} - \frac{\alpha k_f}{\bar{r}'}(\bar{r} - 1)\right]^{-1},$$

satisfying the boundary condition $\theta_0 = 1$. The optimal angle variation $\varphi(\bar{z})$ is shown in Fig. 30. The left-hand part of the curve (with respect to point A) corresponds to $\alpha = 1$ and the right-hand part to $\alpha = -1$. The optimal winding yields a 17.2% mass saving over the winding along geodetic lines.

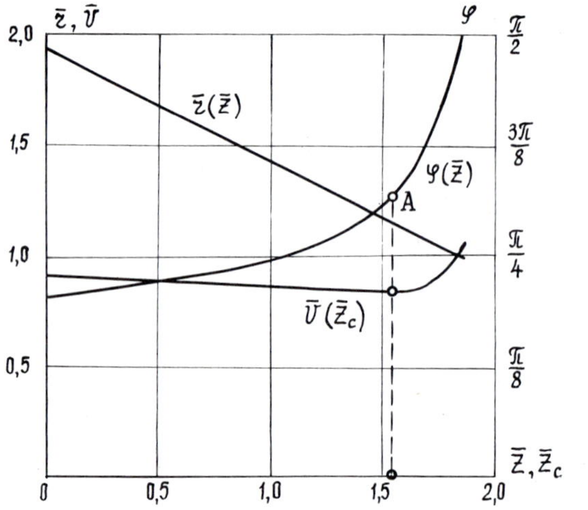

FIG. 30. Meridian $\bar{r}(\bar{z})$, fiber orientation $\varphi(\bar{z})$ and the volume of material $\bar{v}(\bar{z}_c)$ for an optimal conical shell.

6. Cylindrical shells

Cylindrical shells made of composite materials have gained extensive practical application in different fields of engineering. This section discusses some problems of optimum design of cylindrical shells subjected to uniform internal pressure (p) and axial load (N) which seem to be the most popular service conditions for structural elements under consideration.

Consider a composite cylindrical shell with a radius R, length l and thickness h. The overall wall thickness is composed of k layers, each layer having a thickness h_i and fiber orientation φ_i with respect to the longitudinal axis of the shell.

For a membrane stress state the conditions for minimum mass were discussed in Section 4.1. In particular, denoting longitudinal and circumferential stress resultants by N_1 and N_2, respectively, and assuming that $N_1 + N_2 > 0$ one can employ equations (4.32) and (4.34) which define the system of uniform strength. Since $N_{12} = 0$, it follows from the second equation in (4.34) that the ply angle distribution must be symmetric about the line on the surface parallel to the cylinder axis, i.e. the ith layer must consist of an even number of plies of the same material and thickness, but with alternating orientations of $+\varphi_i$ and $-\varphi_i$. Putting $N_1 = N$, $N_2 = pR$ (4.32) and (4.34) yield

$$h = \frac{pR + N}{1+n}(F_{11} + nF_{12} + n^2F_{22})^{1/2} ,$$

$$\sum_{i=1}^{k} \bar{h}_i \cos^2 \varphi_i = \frac{N - pRn}{(1-n)(pR+N)} .$$

Within the framework of the netting analysis one should set $n = 0$ and $F_{11} = (1/\sigma_{\parallel}^{u})^2$ in order to obtain

$$h = \frac{pR + N}{\sigma_{\parallel}^{u}} , \quad \sum_{i=1}^{k} \bar{h}_i \cos^2 \varphi_i = \frac{N}{pR+N} . \tag{6.1}$$

If a shell has the form of a cylindrical section of a pressure vessel, $N = 0.5pR$ and the equations in (6.1) reduce to

$$h = \frac{3pR}{2\sigma_{\parallel}^{u}} , \quad \sum_{i=1}^{k} \bar{h}_i \cos^2 \varphi_i = \frac{1}{3} . \tag{6.2}$$

For filament wound pressure vessels whose wall thickness h is composed of hoop windings of thickness h_{90} and helical windings of thickness h_{φ}, which are applied at an angle φ to the axis, the equations in (6.2) give

$$h_{\varphi} = \frac{pR}{2\sigma_{\parallel}^{u} \cos^2 \varphi} , \quad h_{90} = \frac{pR}{2\sigma_{\parallel}^{u}}\left(3 - \frac{1}{\cos^2 \varphi}\right) . \tag{6.3}$$

For a shell undergoing pure axial compression ($N_1 = -N$, $N_2 = N_{12} = 0$) and being simply supported at the edges $x_1 = 0$ and $x_1 = l$, the critical value of the axial normal stress resultant can be obtained as a result of minimization of the eigenvalue

$$N = \frac{1}{\lambda_m^2}\left(b_{33} - \frac{b_{11}b_{23}^2 + b_{22}b_{13}^2 - 2b_{12}b_{13}b_{23}}{b_{11}b_{22} - b_{12}^2}\right) , \tag{6.4}$$

with respect to $\lambda_m = \pi m/l$, $\lambda_n = n/R$. In (6.4), $b_{st} = a_{st} + c_{st}$, where the coefficients a_{st} are specified by formulas (4.67) and

$$c_{11} = 0\,, \quad c_{12} = \frac{\lambda_m \lambda_n}{R}\,(B_{12} + B_{33})\,,$$

$$c_{22} = \lambda_m^2\left(\frac{2B_{33}}{R} + \frac{D_{33}}{R^2}\right) + \lambda_n^2\left(\frac{2B_{22}}{R} + \frac{D_{22}}{R^2}\right),$$

$$c_{13} = \frac{\lambda_m}{R}\,A_{12}\,, \quad c_{23} = \frac{\lambda_n}{R}\left(\frac{B_{22}}{R} + \lambda_m^2(D_{12} + 2D_{33}) + \lambda_n^2 D_{22} + A_{22}\right),$$

$$c_{33} = \frac{A_{22}}{R^2} + \frac{2}{R}\,(B_{12}\lambda_m^2 + B_{22}\lambda_n^2)\,.$$

For a midsurface symmetrically laminated shell whose bending–stretching coupling stiffnesses (B_{mn}) are equal to zero and membrane (A_{mn}) and bending (D_{mn}) stiffnesses are defined by (4.64), the following approximate expression for the critical value can be obtained by means of the geometrical programming method:

$$N_c = \frac{2}{\eta R}\left(\frac{D_{11}}{A_{11}}\,(A_{11}A_{22} - A_{22}^2)\right)^{1/2}, \tag{6.5}$$

where η is a dimensionless factor given by

$$\eta = \left(\frac{A_{11}D_{22} + A_{22}D_{11} + \left(\dfrac{A_{11}A_{22} - A_{12}^2}{A_{33}} - 2A_{12}\right)(D_{11}D_{22})^{1/2}}{2A_{11}[D_{22} + (D_{12} + 2D_{33})(D_{22}/D_{11})^{1/2}]}\right)^{1/2},$$

or $\eta = 1$ whichever is smaller. The case of $\eta = 1$ corresponds to the axial symmetric mode of buckling.

For a homogeneous shell whose properties are constant through the thickness and stiffnesses are specified by equations (4.53) expression (6.5) reduces to

$$N_c = \frac{h^2(E_1E_2)^{1/2}}{R\eta[3(1 - \mu_{12}\mu_{21})]^{1/2}}, \tag{6.6}$$

where

$$\eta = \left(\frac{1 + \dfrac{1}{2}\left(\dfrac{E_2}{E_1}\right)^{1/2}\left(\dfrac{E_1}{G_{12}} - 2\mu_{21}\right)}{1 + \left(\dfrac{E_2}{E_1}\right)^{1/2}\left(\mu_{21} + \dfrac{2G_{12}}{E_2}\,(1 - \mu_{12}\mu_{21})\right)}\right)^{1/2}$$

or $\eta = 1$.

The half wave numbers are

$$m = \frac{l}{\pi R}\left(\frac{\dfrac{R}{2\eta h}\left(\dfrac{3E_2}{E_1}(1-\mu_{12}\mu_{21})\right)^{1/2}}{1+2\left(\dfrac{E_2}{E_1}\right)^{1/2}\left[\mu_{21}+\dfrac{2G_{12}}{E_1}(1-\mu_{12}\mu_{21})\right]}\right)^{1/2},$$

$$\frac{n}{m} = \frac{\pi R}{l}\left(\frac{E_1}{E_2}\right)^{1/4} \tag{6.7}$$

Note that m and n defined by (6.7) are not integers in general, so that the critical values (6.5) or 6.6) are lower than the exact ones.

Now, consider a cylindrical shell subjected to two mutually independent loads, namely: (1) to uniform internal pressure p and (2) to axial compressive stress resultant N. Let the shell be composed of helical and circumferential layers, the corresponding thicknesses (h_φ and h_{90}) being calculated for the first case of loading with aid of (6.3). The variation of the dimensionless buckling load $\bar{N}_c = N_c l^2/E_\| h^3$ with fiber orientation of the helical layer for a shell with $h/R = 0.0068$; $R/l = 0.375$; $E_\perp = 0.055E_\|$; $G_{\|\perp} = 0.04E_\|$ is shown in Fig. 31.

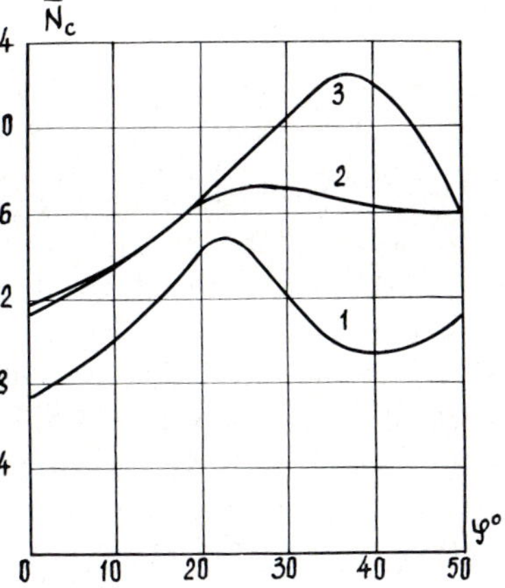

FIG. 31. Variation of the dimensionless buckling stress resultant with φ for a two-layer (1), three-layer (2) and quasi-homogeneous (3) cylindrical shell composed of circumferential and helical $\pm\varphi$ plies.

Curve 1 corresponds to the two-layer shell whose inner layer contains all the helical plies and outer layer consists of the hoop plies only. Curve 2 displays the variation of $\bar{N}_c$ with φ for the sandwich midsurface symmetrically laminated shell whose central layer is composed of helical plies and two identical outer layers that include all the hoop plies. Finally, curve 3 corresponds to the

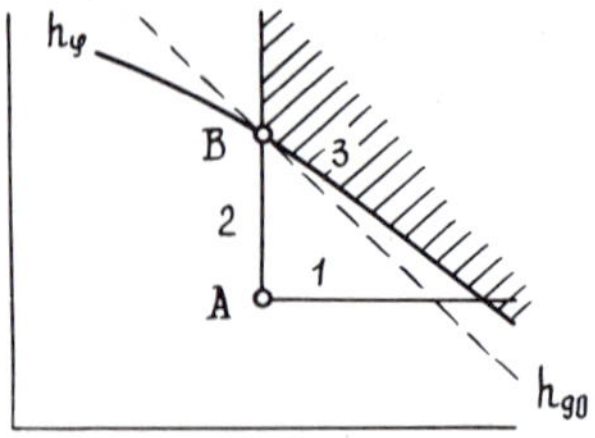

FIG. 32. Feasible area of thickness variation for a cylindrical shell subjected to internal pressure (solid lines 1, 2) and axial compression (curve 3) and the mass isoline (– – – –).

quasi-homogeneous shell composed of uniformly distributed laminate sets with ply angles of $\pm\varphi$ and 90°. It follows from the figure that the critical load is greatly influenced by the fiber orientation and stacking sequence of the layers.

A feasible area of thickness variation for a quasi-homogeneous shell with prescribed orientations of helical plies (φ) is shown in Fig. 32. Straight lines 1 and 2 display the strength constraints for helical and circumferential layers under internal pressure; curve 3 illustrates the stability condition under axial compression; the broken line represents the mass isoline ($h_\varphi + h_{90} =$ constant). Point A corresponds to the condition of simultaneous failure of helical and hoop fibers under internal pressure, and point B indicates the optimal thicknesses in view of the stability constraints.

7. Composite structural elements undergoing dynamic loading

7.1. Optimum design with natural frequency constraints

As was shown above, selection of appropriate material structural parameters enables the designer to control the strength and rigidity characteristics of structural elements, i.e. the load-carrying capacity and deformability of composite elements undergoing static loading. The same effects may be attained in the case of dynamic loading. The response of a shell undergoing time-dependent loading is governed by the relationship between its natural frequencies of vibration and the exciting frequency of external loads. One of the most unfavourable operating conditions is the resonance condition which occurs when the natural frequency coincides with the excitation frequency. As a result of this, high dynamic stresses occur in the structure and the reduction of these stresses is an important aspect of the design problem of load-carrying elements. Since the external loading is usually preassigned, the frequency separation problem may be resolved by an appropriate selection of design variables controlling the natural frequencies of the structure. In conformity with composite structural elements, these variables include in addition to conventional ones the material structural parameters, namely, fiber orientations (φ_i), coordinates of layer surfaces (t_i) and thickness of the layers (see Fig. 12). In this section, the formulation of the optimum design problem with natural frequency constraints is discussed and some examples for composite rectangular plates are considered.

Natural frequencies of a construction can be determined from the following characteristic equation:

$$|S - \lambda M| = 0\,, \tag{7.1}$$

where S denotes the stiffness matrix coefficients depending on membrane, bending–stretching and bending stiffnesses, which can be written as

$$\begin{aligned} A_{lr} &= h \sum_{i=1}^{k} C_{lr}^{(i)}(\gamma_i - \gamma_{i-l})\,, \\ B_{lr} &= \frac{h^2}{2} \sum_{i=1}^{k} C_{lr}^{(i)}(\gamma_i^2 - \gamma_{i-1}^2)\,, \\ D_{lr} &= \frac{h^3}{3} \sum_{i=1}^{k} C_{lr}^{(i)}(\gamma_i^3 - \gamma_{i-1}^3)\,. \end{aligned} \tag{7.2}$$

In the above, h is the overall thickness of a plate or a shell; $\gamma_i = t_i/h$ represents the coordinate of the ith layer (see Fig. 12); $C_{lr}^{(i)}$ ($lr = 11, 12, 22, 33$) are the stiffnesses specified by (2.11). Henceforth, the layers are restricted to be made of the same material, so one should take $E_{\parallel}^{(i)} = E_{\parallel}$, $E_{\perp}^{(i)} = E_{\perp}$, etc. in (2.11). The coefficients S also contain some geometrical parameters and half wave numbers m and n, which characterize the vibrational mode shapes. The intertia matrix coefficients denoted in (7.1) by M depend on the density (ρ) through the following inertial terms similar to (7.2):

$$\begin{aligned} A_{\rho} &= \rho h \sum_{i=1}^{k} (\gamma_i - \gamma_{i-1}) = \rho h\,, \\ B_{\rho} &= \tfrac{1}{2} \rho h^2 \sum_{i=1}^{k} (\gamma_i^2 - \gamma_{i-1}^2)\,, \\ D_{\rho} &= \tfrac{1}{3} \rho h^3 \sum_{i=1}^{k} (\gamma_i^3 - \gamma_{i-1}^3)\,. \end{aligned} \tag{7.3}$$

Eq. (7.1) involves the eigenvalues (λ) corresponding to various half wave numbers (m, n). Note that the fundamental frequency of free vibrations (ω) is given by $\omega^2 = \lambda$.

In order to obtain the natural frequencies for a structure with prescribed properties, (7.1) should be solved with respect to λ. Let us consider an inverse problem, i.e. the design problem of composite elements with prescribed frequencies. Assume the following objective function and optimality criterion:

$$\min_{\max} I(\varphi_i, \gamma_i, h) = \sum_{j=1}^{s} k_j |S(\varphi_i, \gamma_i, h) - \bar{\lambda}_j M(\gamma_i, h)|\,. \tag{7.4}$$

In the above s is the number of the preassigned eigen-frequencies $\bar{\lambda}_j = \bar{\omega}_j^2$ and k_j denote weight factors satisfying the condition

$$\sum_{j=1}^{s} k_j = 1 .$$

In the case of $I = 0$ $\bar{\omega}_j$ indicates the true natural frequencies of vibration of the structure under consideration. Minimization or maximization of I with respect to the design variables φ_i, γ_i, h corresponds respectively, to minimization or maximization of the deviation of the true frequencies (ω_j) from the prescribed ones $(\bar{\omega}_j)$. The design variables must satisfy the following geometric constraints:

$$0 \leqslant \varphi_i \leqslant \pi/2 , \quad \gamma_0 = 0 , \quad \gamma_k = 1 , \quad \gamma_i - \gamma_{i-1} \geqslant 0 , \quad h^- \leqslant h \leqslant h^+ , \qquad (7.5)$$

where h^- and h^+ are the lower and upper constraints on the thickness.

To illustrate the design of composite structural elements with natural frequency constraints, consider rectangular, simply supported plates. The numerical results discussed below were obtained with the random search technique.

First of all, let us note that the prescribed frequencies cannot be assumed as arbitrary without any preliminary estimates. For laminated, simply supported rectangular plates the following upper and lower bounds can be obtained:

$$\max_{\varphi} \omega_h \geqslant \omega \geqslant \min_{\varphi} \omega_h ,$$

where ω_h is the frequency of a homogeneous angle-ply $\pm\varphi$ plate.

As an example, consider a homogeneous, square angle-ply $\pm\varphi$ plate whose normalized fundamental frequencies are shown in Fig. 33. The problem is to

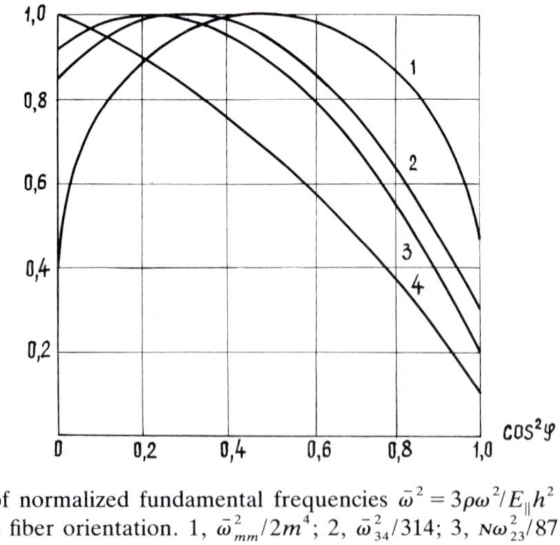

FIG. 33. Variation of normalized fundamental frequencies $\bar{\omega}^2 = 3\rho\omega^2/E_{\parallel}h^2$ of a square angle-ply plate with fiber orientation. 1, $\bar{\omega}_{mm}^2/2m^4$; 2, $\bar{\omega}_{34}^2/314$; 3, $N\omega_{23}^2/87$; 4, $\bar{\omega}_{12}^2/16$.

find an optimal angle φ_{op} providing the minimum overall deviation of the frequencies ω_{11}, ω_{23}, ω_{34} from their maximum values. Therefore, it is assumed that $\bar{\omega}_{11} = \max_\varphi \omega_{11}$, $\bar{\omega}_{23} = \max_\varphi \bar{\omega}_{23}$, $\bar{\omega}_{34} = \max_\varphi \omega_{34}$. Putting $k_1 = k_2 = k_3$ we have $\varphi_{\text{op}} = 56°$ and the deviations of the frequencies under consideration from their maximum values are as follows: $\Delta\omega_{11} = 10\%$, $\Delta\omega_{23} = 0.3\%$, $\Delta\omega_{34} = 0.6\%$. Let us change the weight factors, assuming that $k_1 = 0.9$, $k_2 = k_3 = 0.05$. In this case, we obtain $\varphi_{\text{op}} = 54.5°$ and $\Delta\omega_{11} = 5.5\%$, $\Delta\omega_{23} = 1.7\%$, $\Delta\omega_{34} = 0.0\%$. The deviations may be smoothed over by specification of the weight factors according to

$$k_j = \bar{\omega}_j^{-1} \Big/ \sum_{j=1}^{s} \bar{\omega}_j^{-1} . \tag{7.6}$$

For the above condition the result is $\varphi_{\text{op}} = 52°$, $\Delta\omega_{11} = 3.95\%$, $\Delta\omega_{23} = 2.3\%$, $\Delta\omega_{34} = 0.16\%$. Furthermore the weight factors will be specified by (7.6).

Consider an angle-ply, plate with an aspect ratio $a/b = 0.5$ and limiting thicknesses $h^-/b = 0.1$, $h^+/b = 0.12$, made of a glass–cloth, epoxy matrix composite with $E_\parallel = E_\perp = 42$ GPa, $G_{\parallel\perp} = 4.2$ GPa. The problem is to attain the minimum overall deviation of the frequencies ω_{mn} $(mn = 11, 12, 21)$ from the prescribed values $\bar{\omega}_{mn}$. The results of the optimal vibrational performance studies are given in Table 4, which presents the normalized frequencies $(\omega_{mn}/\omega_{11})$ and associated mode shapes (m, n) of the plate having some starting design variables (namely, $\varphi = 45°$, $h = h^-$), the prescribed frequencies $(\bar{\omega}_{mn}/\omega_{11})$ and the result of optimization $(\omega^0_{mn}/\omega_{11}$, $\varphi_{\text{op}} = 19.5°$, $h_{\text{op}} = h^+)$ according to criterion min I (7.4).

TABLE 4. Natural frequencies of angle-ply plates.

m	n	ω_{mn}/ω_{11} $\varphi = 45°, h = h^-$	$\bar{\omega}_{mn}/\omega_{11}$	$\omega^0_{mn}/\omega_{11}$ $\varphi_{\text{op}} = 19.5°, h_{\text{op}} = h^+$
1	1	1	1.16	1.22
1	2	1.29	1.55	1.42
2	1	2.23	2.58	2.48

To illustrate the frequency separation problem, i.e. the application of max I criterion, consider a two-layer plate with the same parameters as before and with the preassigned overall thickness $h = 0.1b$. The frequencies normalized by the first fundamental frequency (ω_{11}) and the associated mode shapes (m, n) are listed in Table 5, the superscripts (1), (2), (3) being used to indicate longitudinal and transversal in-plane vibrational modes and flexural vibrational modes, respectively. The table presents the frequencies $(\omega^{(j)}_{mn}/\omega^{(j)}_{11})$ of a cross-ply 0, 90° plate with the relative thickness of the first layer $\gamma_1 = 0.333$. The problem is to find the design variables φ_1, φ_2 and γ_1 providing the maximum overall deviation of the optimal flexural frequencies $(\omega^{(03)}_{mn}/\omega^{(3)}_{11}$, $mn = 11, 21, 31)$ from those corresponding to the cross-ply plate. The results of optimization are also given in Table 5.

TABLE 5. Natural frequencies of two-layer plates.

m	n	Cross-ply plate $\varphi_1 = 0$,	$\varphi_2 = 90°$,	$\varphi_1 = 0.333$	Optimal plate $\varphi_1 = 32°$,	$\varphi_2 = 17.6°$,	$\gamma_1 = 0.36$
		$\frac{\omega_{mn}^{(1)}}{\omega_{11}^{(1)}}$	$\frac{\omega_{mn}^{(2)}}{\omega_{11}^{(2)}}$	$\frac{\omega_{mn}^{(3)}}{\omega_{11}^{(3)}}$	$\frac{\omega_{mn}^{(01)}}{\omega_{11}^{(1)}}$	$\frac{\omega_{mn}^{(02)}}{\omega_{11}^{(2)}}$	$\frac{\omega_{mn}^{(03)}}{\omega_{mn}^{(3)}}$
1	1	1	1	1	1.22	0.90	1.15
1	2	1.46	1.98	2.17	2.13	1.73	2.08
2	1	1.69	1.04	1.36	1.74	1.04	1.43
3	1	2.45	1.11	1.65	2.34	1.20	1.72

The above analysis implies that the number of layers (k) has been prescribed. Let us generalize criterion (7.4) to other cases which are free from this restriction. With this in mind, let us introduce the following generalized design variables:

$$U_{sp} = \sum_{i=1}^{k} (\gamma_i^s - \gamma_{i-1}^s) \cos^{2p} \varphi_i \,, \tag{7.7}$$

where $s = 1, 2, 3$; $p = 0, 1, 2$. Taking into account expressions (7.7) and (2.11), the stiffnesses (7.2) and inertia terms (7.3) can be written as

$$A_{lr} = h \sum_{p=0}^{2} \alpha_{lr}^{(p)} U_{1p} \,, \quad B_{lr} = \frac{h^2}{2} \sum_{p=0}^{2} \alpha_{lr}^{(p)} U_{2p} \,, \quad D_{lr} = \frac{h^3}{3} \sum_{p=0}^{2} \alpha_{lr}^{(p)} U_{3p} \,, \tag{7.8}$$

$$A_\rho = \rho h U_{10} \,, \quad B_\rho = \tfrac{1}{2} \rho h^2 U_{20} \,, \quad D_\rho = \tfrac{1}{3} \rho h^3 U_{30} \,,$$

where

$$\alpha_{11}^{(0)} = \bar{E}_\perp \,, \quad \alpha_{12}^{(0)} = \bar{E}_\| \mu_{\|\perp} \,, \quad \alpha_{22}^{(0)} = \bar{E}_\| \,, \quad \alpha_{33}^{(0)} = G_{\|\perp} \,,$$

$$\alpha_{11}^{(1)} = 2(\bar{E}_\perp + \bar{E}_\| \mu_{\|\perp} + 2G_{\|\perp}) \,,$$

$$\alpha_{12}^{(1)} = \alpha_{33}^{(1)} = \alpha_{11}^{(2)} = \alpha_{22}^{(2)} = \bar{E}_\| + \bar{E}_\perp - 2\bar{E}_\| \mu_{\|\perp} - 4G_{\|\perp} \,,$$

$$\alpha_{22}^{(1)} = 2(\bar{E}_\| \mu_{\|\perp} - \bar{E}_\| + 2G_{\|\perp}) \,,$$

$$\alpha_{12}^{(2)} = \alpha_{33}^{(2)} = -\bar{E}_\| - \bar{E}_\perp + 2\bar{E}_\| \mu_{\|\perp} + 4G_{\|\perp} \,.$$

Note that (7.8) are valid only for a laminate whose layers are made of the same material. The generalized design variable must satisfy the following constraints:

$$U_{s0} = 1 \,, \quad 0 \leqslant U_{s2} \leqslant U_{s1} \leqslant 1 \,, \quad U_{s2} \geqslant U_{s1}^2 \,, \tag{7.9}$$

where $s = 1, 2, 3$. Thus, for a laminate consisting of an arbitrary number of layers, we have only six design variables U_{sp} ($s = 1, 2, 3$; $p = 1, 2$) and criterion (7.4) takes the form

$$\min_{\max} I(U_{sp}, h) = \sum_{j=1}^{t} k_j |S(U_{sp}, h) - \bar{\lambda}_j M(U_{sp}, h)| . \tag{7.10}$$

Parameters U_{sp} and h satisfying conditions (7.9) and the last inequality (7.5) can be obtained through the use of any of the various available techniques for numerical optimization. Let us assume that this procedure has been performed and the generalized design variables ($\bar{U}_{sp}$) as well as the overall thickness are known. The actual design variables φ_i and γ_i satisfying constraints (7.5) can be obtained then as a result of minimization of the following objective function:

$$Q(\varphi_i, \gamma_i) = \sum_{\nu=1}^{6} [U_\nu(\varphi_i, \gamma_i) - \bar{U}_\nu]^2 . \tag{7.11}$$

This problem can be resolved, for instance, by the penalty function technique. Fig. 34 gives a plot of Q (7.11) as a function of the number of layers for a square, fiberglass–epoxy composite plate symmetrically laminated with respect to its midplane. The problem was to provide the minimum overall deviation of the first three flexural frequencies from the corresponding prescribed values. The three variables h, U_{31}, U_{32} which characterize the problem under consideration were obtained with the aid of criterion (7.10) and the gradient optimization method. It can be seen from Fig. 34 that three layers provide the minimum value of Q, which practically does not decrease as the number of layers increases beyond $k = 3$.

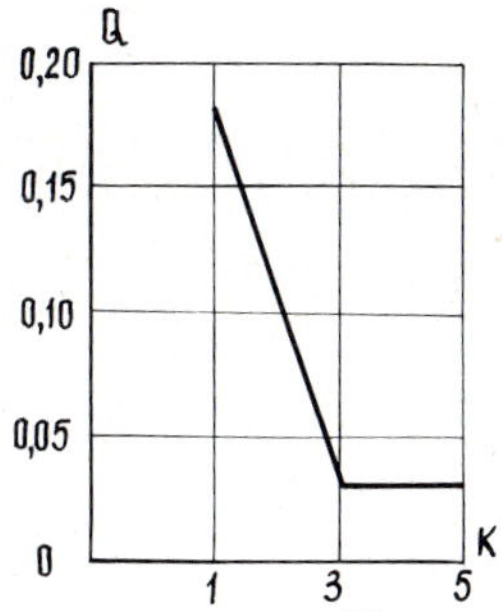

FIG. 34. Variation of objective function Q with the number of layers.

The techniques described above and based on criteria (7.4) or (7.10) and (7.11) allow us to implement frequency separation only in the average sense. The obtained results may be improved through the following considerations.

Let the equation of motion have the form

$$Lu = \lambda u \,, \tag{7.12}$$

where L is a linear differential operator whose coefficients depend on the design variables h, $\psi_i = \cos^2 \varphi_i$, γ_i $(i = 1, 2, 3 \ldots k)$; and where u designates displacement. Let λ_n, λ_m be some eigenvalues and u_n, u_m the associated eigen-functions, which are assumed to form a normalized orthogonal system, i.e.

$$(u_n, u_m) = \int u_n u_m \, \mathrm{d}x_1 \, \mathrm{d}x_2 = \delta_{nm} = \begin{cases} 1 & \text{if } m = n \\ 0 & \text{if } m \neq n \end{cases} , \tag{7.13}$$

where δ_{nm} is the Kronecker delta. For the nth mode of vibration, (7.12) can be written as

$$Lu_n = \lambda_n u_n \,. \tag{7.14}$$

Let us further introduce some increments in the design variables, namely, consider a structure whose variables are $h + \delta h$, $\psi_i + \delta\psi_i$, $\gamma_i + \delta\gamma_i$. Then in accordance with (7.4)

$$\delta L u_n + L \delta u_n = \delta\lambda_n u_n + \lambda_n \delta u_n \,.$$

Multiplying the above equation by u_m, integrating and using relations (7.13), yields

$$\delta\lambda_n = (\delta L u_n, u_n) + (L\delta u_n, u_n) - \lambda_n(\delta u_n, u_n) \,. \tag{7.15}$$

Expanding δu_n into a series in terms of the eigen-functions

$$\delta u_n = \sum_m a_{nm} u_m$$

one can obtain, with the aid of (7.13) and (7.14) the following relationships:

$$(L\delta u_n, u_n) = \sum_m a_{nm}(Lu_m, u_n) = \sum_m a_{nm}\lambda_m(u_m, u_n) = \lambda_n a_{nn} \,,$$

$$\lambda_n(\delta u_n, u_n) = \lambda_n \sum_m a_{nm}(u_m, u_n) = \lambda_n a_{nn} \,,$$

which allow (7.15) to be written as

$$\delta\lambda_n = (\delta L u_n, u_n) \,, \tag{7.16}$$

where

$$\delta L = \frac{\partial L}{\partial h} \delta h + \sum_{i=1}^{k} \frac{\partial L}{\partial \psi_i} \delta\psi_i + \sum_{i=1}^{k-1} \frac{\partial L}{\partial \gamma_i} \delta\gamma_i \,.$$

To illustrate the approach under consideration let us assume that we have a solution obtained on the basis of criteria (7.4) or (7.10) and (7.11); i.e. we have the design variables h, ψ_i, γ_i providing, for instance, the maximum deviation of the squares of natural frequencies λ_1 and λ_2 from the prescribed values. Let us further assume that λ_1 is quite satisfactory and λ_2 must be improved. Then (7.16) enables us to write the following conditions:

$$\Delta\lambda_1 = b_{11}\Delta h + \sum_{i=1}^{k} c_{1i}\Delta\psi_i + \sum_{i=1}^{k-1} d_{1i}\Delta\gamma_i = 0 ,$$
$$\Delta\lambda_2 = b_{21}\Delta h + \sum_{i=1}^{k} c_{2i}\Delta\psi_i + \sum_{i=1}^{k-1} d_{2i}\Delta\gamma_i \rightarrow \max , \tag{7.17}$$

which, in conjunction with the geometric constraints in (7.5), define the problem of linear programming. It should be noted that the increments in the design variables (Δh, $\Delta\psi_i$, $\Delta\gamma_i$) must be comparatively small. If the result is not satisfactory, the second step in the space of design variables is made according to expressions (7.17) and the procedure is repeated.

Fig. 35 displays the variation of flexural frequencies of a square midplane symmetrically laminated sandwich plate with the number of iteration steps (i), which were made in the space of design variables h, ψ_1, ψ_2, γ_1 in accordance with the following conditions:

$$\Delta\lambda_{11} = 0 , \quad \Delta\lambda_{21} \leqslant 0 , \quad \Delta\lambda_{12} \rightarrow \max .$$

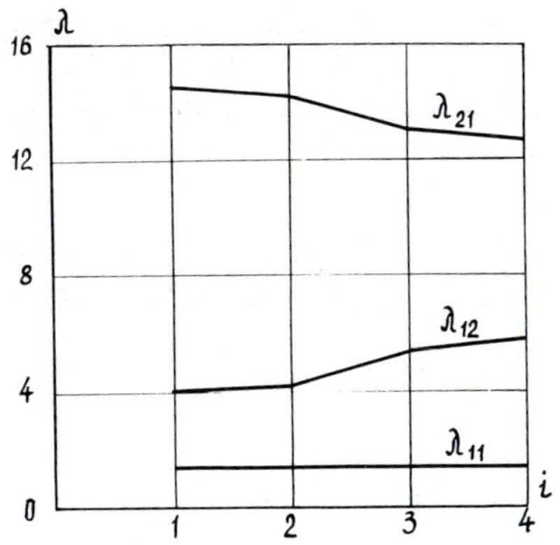

FIG. 35. Variation of eigenvalues $\lambda_{mn} = (a^2/\pi^2 h)(^3\rho/E_{\parallel})^{1/2}\omega_{mn}$ with the number of steps of an iterative process.

7.2. *Control of wave propagation in laminated composites*

As is known, as essential disadvantage of laminated composites is attributed to their relatively low interlaminar strength in tension perpendicular to the layers, which under some loading conditions may cause a special failure mode accompanied by delamination. In particular, short duration or impactlike forces acting in the direction normal to the layers generate elastic stress waves which (in spite of being compressive) can develop tensile stresses as the wave propagates through and is reflected from the boundaries of the layers. Composite materials have the unique feature that their mechanical properties can be varied in the material and hence the designer can change the directional distribution of stress waves by a judicious choice of the stacking sequence of the layer to avoid delamination.

Let us consider a one-dimensional wave propagation problem for the laminate which is shown in Fig. 36. The equation of motion along with the stress–strain relation, boundary and initial condition can be written as follows:

$$\frac{\partial}{\partial x}\left(E\,\frac{\partial u}{\partial x}\right)=\rho\,\frac{\partial^2 u}{\partial t^2}\,,\qquad \sigma=E\,\frac{\partial u}{\partial x}\,,\tag{7.18}$$

$$\sigma(x=0,\,t)=0\,,\quad \sigma(x=x_k,\,t)=-q(t)\,,\tag{7.19}$$

$$u(x,\,t=0)=0\,,\quad \frac{\partial u}{\partial t}\,(x,\,t=0)=0\,.$$

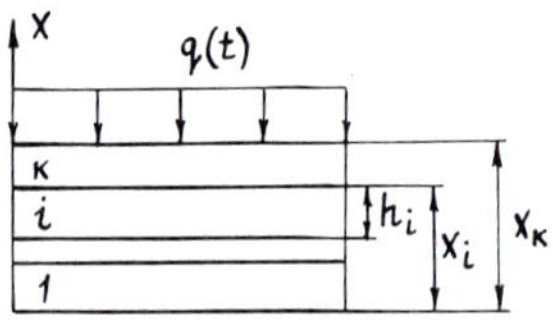

FIG. 36. A laminate subjected to time dependent loading.

In the above, $u(x, t)$ represents the displacement normal to the layers; $\sigma(x, t)$ is the normal stress at any plane parallel to the layers; and

$$E=\sum_{i=1}^{k}E_i(\theta_{i-1}-\theta_i)\,,\quad \rho=\sum_{i=1}^{k}\rho_i(\theta_{i-1}-\theta_i)\,,$$

where E_i and ρ_i denote the modulus of elasticity and the density of the ith layer; $\theta_i=\theta(x-x_i)$ is the Heaviside function ($\theta_i=0$ if $x<x_i$, $\theta_i=1$ if $x>x_i$, $\theta_0=1$, $\theta_k=0$). Let us apply the Laplace–Carson transformation with respect to the time variable (t) to (7.18) and boundary condition (7.19)

$$\frac{\mathrm{d}}{\mathrm{d}x}\left(E\,\frac{\mathrm{d}\bar u}{\mathrm{d}x}\right)=\rho p^2\bar u\,,\quad \bar\sigma=E\,\frac{\mathrm{d}\bar u}{\mathrm{d}x}\,,\tag{7.20}$$

$$\bar\sigma(x=0,\,p)=0\,,\quad \bar\sigma(x=x_k,\,p)=-\bar q(p)\,,\tag{7.21}$$

where, for instance,

$$\bar{u}(x, p) = p \int_0^\infty \mathrm{e}^{-pt} u(x, t)\, \mathrm{d}t .$$

Integration of (7.20) yields

$$\begin{aligned} \bar{\sigma} &= \int_0^x \rho p^2 \bar{u}\, \mathrm{d}x + \bar{\sigma}_0 , \\ \bar{u} &= \int_0^x \frac{\mathrm{d}x}{E} \int_0^x \rho p^2 \bar{u}\, \mathrm{d}x + \bar{\sigma}_0 \int_0^x \frac{\mathrm{d}x}{E} + \bar{u}_0 , \end{aligned} \tag{7.22}$$

where $\bar{\sigma}_0 = \bar{\sigma}\ (x = 0, p)$, $\bar{u}_0 = \bar{u}\ (x = 0, p)$. Assume that the solution $u(x, t)$ can be written in the form

$$u(x, t) = \sum_{i=1}^{k} u_i(x, t)(\theta_{i-1} - \theta_i) . \tag{7.23}$$

Substituting solution (7.23) into (7.22) and applying the Laplace–Carson transformation with respect to x gives

$$\sum_{i=1}^{k} \left(1 - \frac{\rho_i p^2}{E_i s^2}\right) s \int_{x_{i-1}}^{\infty} \bar{u}_i\, \mathrm{e}^{-sx}\, \mathrm{d}x - \sum_{i=1}^{k-1} \left(1 - \frac{\rho_i p^2}{E_i s^2}\right) s \int_{x_i}^{\infty} \bar{u}_i\, \mathrm{e}^{-sx}\, \mathrm{d}x$$

$$= \sum_{i=2}^{k} \frac{1}{E_i s}\, \mathrm{e}^{-sx_{i-1}} \sum_{j=1}^{i-1} \rho_i p^2 \int_{x_{j-1}}^{x_j} \bar{u}_j\, \mathrm{d}x - \sum_{i=1}^{k-1} \frac{1}{E_i s}\, \mathrm{e}^{-sx_i} \sum_{j=1}^{i} \rho_i p^2 \int_{x_j-1}^{x_j} \bar{u}_j\, \mathrm{d}x$$

$$+ \bar{\sigma}_0 \sum_{i=1}^{k} \frac{1}{E_i s}\, \mathrm{e}^{-sx_{i-1}} - \bar{\sigma}_0 \sum_{i=1}^{k-1} \frac{1}{E_i s}\, \mathrm{e}^{-sx_i} + \bar{u}_0 .$$

Taking into account the fact that the function e^{-sx_i} are linearly independent, we may obtain the following recurrent system of relations:

$$\begin{aligned} &\left(1 - \frac{\rho_1 p^2}{E_1 s^2}\right) v_1 = \frac{\bar{\sigma}_0}{E_1 s} + \bar{u}_0 , \\ &\left(1 - \frac{\rho_{n+1} p^2}{E_{n+1} s^2}\right) v_{n+1} = \left(1 - \frac{\rho_n p^2}{E_n s^2}\right) s \int_{x_n}^{\infty} \bar{u}_n\, \mathrm{e}^{-sx}\, \mathrm{d}x \\ &\quad + \frac{1}{s}\left(\frac{1}{E_{n+1}} - \frac{1}{E_n}\right) \sum_{j=1}^{n} \rho_i p^2 \int_{x_{j-1}}^{x_j} \bar{u}_j\, \mathrm{e}^{-sx}\, \mathrm{d}x + \frac{\bar{\sigma}_0}{s}\left(\frac{1}{E_{n+1}} - \frac{1}{E_n}\right) \mathrm{e}^{-sx_n} , \end{aligned} \tag{7.24}$$

where

$$v_n = s \int_{x_{n-1}}^{\infty} \bar{u}_n \, e^{-sx} \, dx \, .$$

Now, let us return to the x variable. The inversion of solution (7.24) has the form

$$\bar{u}_1(x, p) = \tfrac{1}{2} \sum_{\eta_1} \left(\bar{u}_0 + \frac{\bar{\sigma}_0 \eta_1}{E_1 \lambda_1} \right) \exp(\eta_1 \lambda_1 x) \, ,$$

$$\bar{u}_n(x, p) = 2^{-n} \sum_{\eta_1, \eta_2 \cdots \eta_n} \left(\bar{u}_0 + \frac{\bar{\sigma}_0}{E_1 \lambda_1} \right) \prod_{i=1}^{n-1} (1 \qquad (7.25)$$

$$+ \eta_i \eta_{i+1} \kappa_{i,i+1}) \exp\left(\sum_{i=1}^{n-1} \eta_i \lambda_i (x_i - x_{i-1}) + \eta_n \lambda_n (x - x_{n-1}) \right) ,$$

where

$$\lambda_i = p(\rho_i / E_i)^{1/2} , \quad \kappa_{i,j} = (\rho_i E_i / \rho_j E_j)^{1/2} , \quad \prod_{i=1}^{0} = 1 \, .$$

Eqs. (7.25) involve special parameters $\eta_i = \pm 1$, so that the sum on the right-hand side of the first equation in (7.25) contains two terms corresponding to $\eta_1 = \pm 1$ and the sum in the second equation contains $2n$ terms corresponding to $\eta_1 = \pm 1 \cdots \eta_n = \pm 1$. The Laplace–Carson transformation (with respect to t only) for the stresses in an arbitrary nth layer can be obtained with the aid of (7.18) and (7.25). Taking into account boundary conditions (7.21) we have

$$\bar{\sigma}_n(x, p) = 2^{k-n} \kappa_{n,k} \bar{q}(p) \Bigg\{ \sum_{\eta_1, \eta_2 \cdots \eta_n} \prod_{i=1}^{n-1} (1$$

$$+ \eta_i \eta_{i+1} \kappa_{i,i+1}) \exp\left[\prod_{i=1}^{n-1} \eta_i \lambda_i (x_i - x_{i-1}) \right.$$

$$\left. + \eta_n \lambda_n (x - x_{n-1}) \right] \Bigg\} \Bigg\{ \sum_{\eta_1, \eta_2 \cdots \eta_k} \eta_k \prod_{i=1}^{k-1} (1$$

$$+ \eta_i \eta_{i+1} \kappa_{i,i+1}) \exp\left[\prod_{i=1}^{k} \eta_i \lambda_i (x_i - x_{i-1}) \right] \Bigg\}^{-1} . \qquad (7.26)$$

Eqs. (7.25) and (7.26) permit the exact inversion of the Laplace–Carson transformation with respect to the time variable, however the final solutions $u_n(x, t)$, $\sigma_n(x, t)$ seem to be too lengthy to be written here.

As an example, consider a laminate composed of a composite bottom layer ($E_1 = 10$ GPa; $\rho_1 = 1400$ kg/m^3, $h_1 = 12 \times 10^{-3}$ m), lightweight central core ($E_2 = 0.28$ GPa; $\rho_2 = 250$ kg/m^3; $h_2 = 10^{-2}$ m) and a top metal layer ($E_3 = 72$ GPa; $\rho_3 = 2700$ kg/m^3; $h_3 = 10^{-3}$ m) subjected to a rectangular pulse of external pressure q (see Fig. 36). Fig. 37 displays the variation of normal stresses occurring in the section of the first layer, which is shown by the broken line, with the time variable. It can be seen from Fig. 37(a) that the compressive external loading develops tensile stresses in the composite layer of the three-layer material. One (Fig. 37(b)) or two (Fig. 37(c)) additional thin layers introduced into the core can eliminate the tensile stresses in the composite layer and thus prevent its delamination.

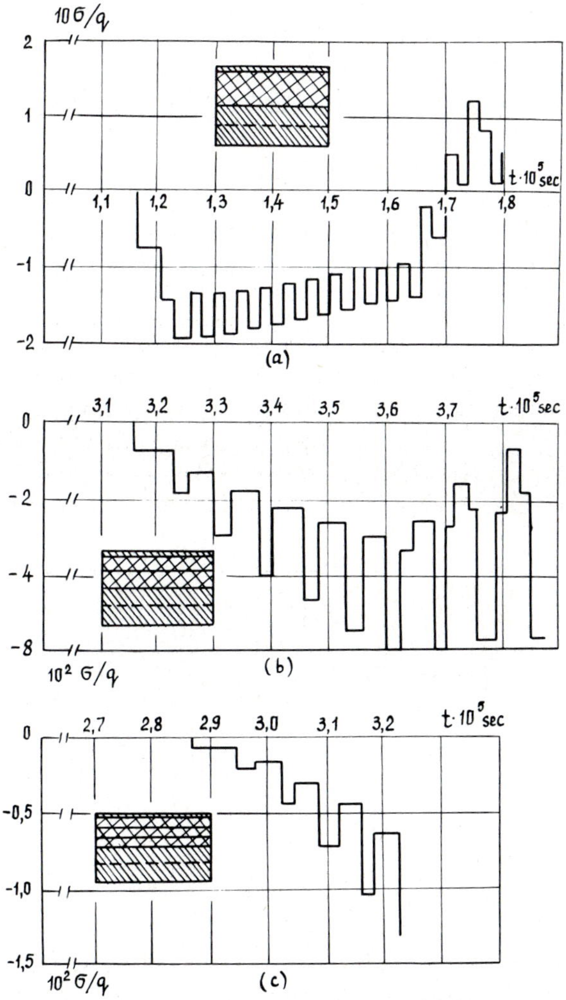

FIG. 37. Normal stresses occurring in (a) three-layer, (b) five-layer, (c) seven-layer laminates and acting in the section shown by the broken line.

8. Some elementary problems of spatial reinforcement of cylindrical shells

The problems considered above were concerned with the optimum design of composite structure consisting of a system of unidirectional layers, each layer being in the plane stress state, and all the fibers being parallel to the plane of the layer. In the presence of three-dimensional stress states, when the stresses between the layers and the in-plane stresses are of the same order of magnitude, the laminated structure is not efficient because of the low interlaminar strength which is largely controlled by the resin matrix properties.

Consider the problem of spatial reinforcement of a thick infinitely long cylindrical shell of inner radius a and outer radius b subjected to external pressure. The problem is described in terms of a cylindrical coordinate system z, θ, r where z is the axial coordinate, θ denotes the polar angle and r is the radius of an arbitrary point normalized by the inner radius a, i.e. $-\infty < z < \infty$, $0 \leqslant \theta \leqslant 2\pi$, $1 \leqslant r \leqslant \alpha$ $(\alpha = b/a)$. The shell is subjected to the lateral external pressure $k_1 p$ acting on the outer cylindrical surface and to the axial pressure $k_2 p$. Thus, putting $k_1 = k_2 = 1$ we have the case of overall external pressure; combinations $k_1 = 0$, $k_2 = 1$ and $k_1 = 1$, $k_2 = 0$ correspond to axial and lateral compression, respectively.

Let us specify the fiber orientation by $\pm\psi$ and $\pm\varphi$ where ψ is the angle between the fiber and the radius, and φ is the angle between the projection of the fiber on the cylindrical surface and the z axis. Note that at every point we have four, symmetric about the radius and the axis, systems of fibers with the orientations $\pm\psi$ and $\pm\varphi$. The case of $\psi = \pi/2$ corresponds to laminated cylindrical shells considered in Section 6.

Within the framework of the netting analysis which neglects the load-carrying capacity and rigidity of the matrix, one can obtain the following expressions specifying the stresses:

$$
\begin{aligned}
\sigma_z &= \sum_{i=1}^{4} \sigma_{\|}^{(i)} (l_z^i)^2 , \quad \sigma_r = \sum_{i=1}^{4} \sigma_{\|}^{(i)} (l_r^i)^2 , \\
\sigma_\theta &= \sum_{i=1}^{4} \sigma_{\|}^{(i)} (l_\theta^i)^2 , \quad \tau_{r\theta} = \sum_{i=1}^{4} \sigma_{\|}^{(i)} l_r^i l_\theta^i , \\
\tau_{rz} &= \sum_{i=1}^{4} \sigma_{\|}^{(i)} l_r^i l_z^i , \quad \tau_{\theta z} = \sum_{i=1}^{4} \sigma_{\|}^{(i)} l_\theta^i l_z^i ,
\end{aligned}
\tag{8.1}
$$

where

$$
l_z^i = \sin\psi_i \cos\varphi_i , \quad l_r^i = \cos\psi_i , \quad l_\theta^i = \sin\psi_i \sin\varphi_i .
$$

Assuming that the optimal system of fibers is the system of uniform strength, and putting in (8.1) $\sigma_{\|}^{(i)} = \sigma =$ constant, $\varphi_1 = -\varphi_2 = \varphi$ and $\psi_1 = -\psi_2 = \psi$, yields

$$
\begin{aligned}
&\sigma_z = \sigma \sin^2\psi \cos^2\varphi , \quad \sigma_r = \sigma \cos^2\psi , \quad \sigma_\theta = \sigma \sin^2\psi \sin^2\varphi , \\
&\tau_{r\theta} = \tau_{rz} = \tau_{\theta z} = 0 .
\end{aligned}
\tag{8.2}
$$

The averaged stresses in (8.2) must satisfy the following equilibrium equations

$$\frac{\mathrm{d}}{\mathrm{d}r}(r\sigma_r) - \sigma_\theta = 0\,, \quad \sigma_z = -\sigma_0 = -\frac{k_2 p \alpha^2}{\alpha^2 - 1} \tag{8.3}$$

and the boundary conditions

$$\sigma_r(r=1) = 0\,, \quad \sigma_r(r=\alpha) = -k_1 p\,. \tag{8.4}$$

Substitution of (8.2) into (8.3) gives

$$\frac{\mathrm{d}}{\mathrm{d}r}(r\cos^2\psi) - \sin^2\psi\sin^2\varphi = 0\,, \quad \sin^2\psi\cos^2\varphi = -\bar{\sigma}_0\,, \tag{8.5}$$

where $\bar{\sigma}_0 = \sigma_0/\sigma$. Taking into account the second equation in (8.2), one can find from the first condition in (8.5) that $\psi(r=1) = \pi/2$. The solution of (8.5) satisfying the above boundary condition has the following form:

$$\cos^2\psi = (1 + \bar{\sigma}_0)\,\frac{r^2 - 1}{2r^2}\,. \tag{8.6}$$

The second condition in (8.4) permits now to find the stress

$$\sigma = -\frac{p\alpha^2(2k_1 + k_2)}{\alpha^2 - 1}\,. \tag{8.7}$$

Taking into account (8.7), we finally obtain from the second equation in (8.5) and (8.6)

$$\cos^2\psi = \frac{k_1(r^2 - 1)}{r^2(2k_1 + k_2)}\,, \quad \cos^2\varphi = \frac{k_2 r^2}{k_1 + (k_1 + k_2)r^2}\,. \tag{8.8}$$

For pure axial compression ($k_1 = 0$) solution (8.8) gives $\psi = \pi/2$, $\varphi = 0$ and defines the system of axial fibers. In the case of lateral pressure ($k_2 = 0$) we have

$$\varphi = \frac{\pi}{2}\,, \quad \cos^2\psi = \tfrac{1}{2}(1 - r^{-2})\,.$$

For overall external pressure ($k_1 = k_2 = 1$)

$$\cos^2\psi = \frac{r^2 - 1}{3r^2}\,, \quad \cos^2\varphi = \frac{r^2}{1 + 2r^2}\,. \tag{8.9}$$

Eqs. (8.9) specify a spatial pattern of fiber orientation whose projection on the axial plane θ = constant has the form of a parabola defined by the following equation: $(z/a)^2 = r^2 - 1$.

For a thin shell of thickness h and radius R, upon putting $r \approx 1$, $\alpha \approx 1 + (h/R)$, we have the following result:

$$\psi = \tfrac{1}{2}\pi\ , \quad \cos^2 \varphi = \tfrac{1}{3},\ \sigma = -3pR/2h\ ,$$

which can also be obtained from (6.2).

References

ALEKSEEV, V.M., V.M. TIKHOMIROV and S.V. FOMIN (1979), *Optimum Control* (Nauka Press, Moscow) p. 430 (in Russian).

ALFUTOV, N.A. and P.A. ZINOV'EV (1982), Deformation and failure of fibrous composites with brittle polymeric matrix under plane stress, in: I.F. OBRAZTSOV and V.V. VASIL'EV, eds., *Mechanics of Composites* (Mir Publishers, Moscow) p. 166.

AZIKOV, N.S. and V.V. VASIL'EV (1982), On the reduction factor for compressed orthotropic rectangular plates, in: *Mech. Composite Materials* **75** (in Russian).

BANKS, W.M. and J.M. HARVEY (1979), Experimental study of stability problems in composite materials, in: *Stab. Probl. Eng. Struct. and Components*, Cardiff, 1978 (Applied Science Publishers, London) p. 1.

BOKOV, Yu.V., V.V. VASIL'EV and G.G. PORTNOV (1981), *Mech. Composite Materials* **5**, 846.

BRANDMAIER, H.E. (1970), *J. Comp. Mater.*, 422.

HANG, E.J. (1978), *Intern. J. for Num. Meth. in Eng.*, **12**, 299.

KAWATA, K. (1972), On the yielding-fracture criterion of angle-ply FW laminates, in: *Proc. Intern. Conf. Mechanical Behaviour Materials, Tokyo*, **5**, p. 146.

KAYSER, A. (1965), *AIAA J.* **7**, 1313.

KNOELL, A.C. and E.Y. ROBINSON (1975), Analysis of truss, beam, frame and membrane components, in: L.J. BROUTMAN and R.H. KROCK, eds., *Composite Materials* Vol. 7, *Structural Design and Analysis*, Part I, C.C. CHAMIS, ed., (Academic Press, New York, San Francisco, London) p. 99.

OBRAZTSOV, I.F., V.V. VASIL'EV and V.A. BUNAKOV (1977), *Optimum Design of Composite Shells of Revolution* (Mashinostroyeniye, Moscow) p. 144 (in Russian).

ROWLANDS, R.E. (1975), Flow and failure of biaxially loaded composites, in: C.T. HERAKOVICH, ed., *Inelastic Behaviour of Composite Materials*, AMD-Vol. 13, (American Society of Mechanical Engineers, New York) p. 97.

TARNOPOL'SKII, Yu.M. (1979), A delamination failure mode of composite bars in compression, in: G.C. SIH and V.P. TAMUŽS, eds. *Proc. 1st USA–USSR Symp. Fracture of Composite Materials*, Riga, 1978 (Sijthoff and Noordhoff, the Netherlands) p. 159.

TSAI, S.W. and V.D. AZZI (1966), *AIAA J.* **4**, 296.

TSAI, S.W. and H.T. HAHN (1975), Failure analysis of composite materials, in: C.T. HERAKOVICH, ed., *Inelastic Behaviour of Composite Materials*, AMD – Vol. 13, (American Society of Mechanical Engineers, New York) p. 73.

VASIL'EV, V.V. and A.N. ELPAT'EVSKII (1967), *Polymer Mech.* **5**, 915.

VICARIO, A.A. and R.H. TOLAND (1975), Failure criteria and failure analysis of composite structural components, in: L.J. BROUTMAN and R.H. KROCK, eds., *Composite Materials*, Vol. 7, *Structural Design and Analysis*, Part I, C.C. CHAMIS, ed. (Academic Press, New York, San Francisco, London) p. 51.

CHAPTER 2

Mechanics of Unsymmetric Laminates

M.W. Hyer

Department of Engineering Science and Mechanics
Virginia Polytechnic Institute and State University
Blacksburg, VA 24061
U.S.A.

Contents

HANDBOOK OF COMPOSITES, VOL. 2 – Structure and Design
Edited by C.T. HERAKOVICH and Y.M. TARNOPOL'SKII

1. Introduction

Generally laminates are constructed so that for every lamina with specific material properties, thickness, fiber orientation, and location on one side of the laminate's geometric midplane, there is a lamina with identical material properties, thickness, and fiber orientation an equal distance on the other side of the midplane. Such laminates are referred to as symmetric laminates. Symmetric laminates are characterized by the fact that the bending–stretching stiffnesses, commonly denoted by the B_{ij} matrix, are zero. If a laminate is constructed so that symmetry about the midplane is not maintained, even for one pair of laminae, the laminate is referred to as an unsymmetric laminate. One or more entries in the B_{ij} matrix will be nonzero. Such a laminate will exhibit bending–stretching coupling under the influence of mechanical loads. In addition, such a laminate will exhibit rather unusual characteristics under the influence of thermal loads. This chapter addresses the latter topic. Specifically the chapter deals with the fact that unsymmetrically laminated resin–matrix composites which are flat at their elevated cure temperature do not remain flat when their temperature is lowered. This characteristic might be expected but it is not widely understood nor is it utilized. It is the purpose of this chapter to present the theory necessary for understanding the deformation characteristics of unsymmetric laminates as a function of temperature. Particular attention will be given to the out-of-plane deformations. The theory and numerical results discussed are limited to unsymmetric cross-ply laminates. Even with these simple laminates, the basic characteristics of the thermally-induced deformations can be understood and the effects of geometric and material properties studied. The work is presented to support the notion that shape, as well as stiffness and strength, can be tailored for a particular application of composites. The potential for shape-tailoring is quite important if it is realized that most structures, particularly aircraft structures, call for components with simple or compound curvatures. A flat structure is not as efficient as a shell structure and finds less usage. This chapter begins by describing in quite general terms the curing procedure for resin–matrix composites. A brief description of the curing process is necessary in order to illustrate the problem being addressed and provide insight into the mechanics involved. Examples of cured unsymmetric laminates are presented. Following that section the theory is outlined which can be used to predict the deformations of cross-ply unsymmetric laminates when their temperature is lowered from the cure temperature. The section after that presents a variety of numerical results to illustrate the influence of temperature, material properties, and laminate geometry.

2. Curing and the out-of-plane shapes of unsymmetric laminates

Flat elevated-temperature cure laminates are usually made by using a hot press or autoclave. In either case the laminate is made by laying the preimpregnated material, one layer at a time, on a flat steel plate, commonly called a caul plate. Release cloth, bleeder plys and other layers are associated with the fabrication process but for this discussion they are not important. After the desired number of layers are stacked, one on top of the other, another caul plate is secured on top of the layers. With this arrangement the precured laminate is sandwiched between the two steel plates. Often the whole assembly is vacuum bagged. The curing process then begins by increasing the temperature and pressure of the sandwiched layers. The elevated temperature and pressure are maintained for a period of time, according to the material manufacturer's specifications, and the resin begins to crosslink. After sufficient time has elapsed, the resin has fully cured and the temperature is decreased slowly to room temperature. The pressure is released and the laminate is ready for removal from the hot press or autoclave. If the laminate is symmetric, the cooled laminate, when removed from between the caul plates, is flat. Due to differences in the thermal expansion and elastic properties of the different laminae there can be significant residual stresses in the laminate. If the residual stresses are large enough, they can cause cracking. Other than cracks, there is little evidence of residual effects. On the other hand, if the laminate is unsymmetric, removal from the caul plates results in very obvious residual effects. Specifically, if the laminate is unsymmetric, it will not be flat when it is removed. Depending on the stacking sequence, there can be significant out-of-plane deformations. Deformations many times the laminate thickness are possible. Curiously enough, despite the wide range of unsymmetric laminates possible, the shape of the out-of-plane deformation has been observed to be either saddle-like or cylindrical. This was discussed by Hyer (1981a). The principal curvatures of the saddle or cylinder may or may not be aligned with the edges of the laminae. Fig. 1 shows a $(0_4/90_4)_T$ graphite–epoxy laminate at room temperature. As can be seen, the cooled laminate is cylindrical. When flat at its elevated cure temperature the laminate in Fig. 1 was approximately 150 mm × 150 mm. The laminate in Fig. 1 has its principal curvatures aligned with the edges of the laminate. Fig. 2 shows an example of a cylindrical laminate with the principal curvatures not aligned with the edges of the originally flat laminate. When flat this laminate was 100 mm × 100 mm. The laminate in Fig. 2 has a stacking arrangement of $(0/60/90_2)_T$ (not a cross-ply). Fig. 3 shows a 200 mm × 50 mm $(+45_2/-45_2)_T$ laminate. This laminate has twist curvature in the coordinate system aligned with the edges of the originally flat laminate.

In contrast to these examples, Fig. 4 shows the room-temperature shape of a $(0_{10}/90_{10})_T$ laminate. The dimensions of the laminate when it is flat are $150 \times 150\ \text{mm}^2$. From the figure it is obvious the laminate is saddle shaped. The laminates in Figs. 1–4 were all made from a material which cures at 177°C. It is important to note that the only difference between the laminate of Fig. 1 and

FIG. 1. The room-temperature shape of a 150 mm × 150 mm $(0_4/90_4)_T$ graphite–epoxy laminate.

the laminate of Fig. 4 is the thickness. It is obvious from these two figures that the shape observed at room temperature is dependent on the geometry of the laminate. This is a very important characteristic of unsymmetric laminates.

Another important characteristic of unsymmetric laminates is associated with the cylindrical shape. For some cylindrical laminates it is possible to force the cylindrical shape into another cylindrical shape. This is accomplished by a

FIG. 2. The room-temperature shape of a 100 mm × 100 mm $(0/60/90_2)_T$ graphite–epoxy laminate.

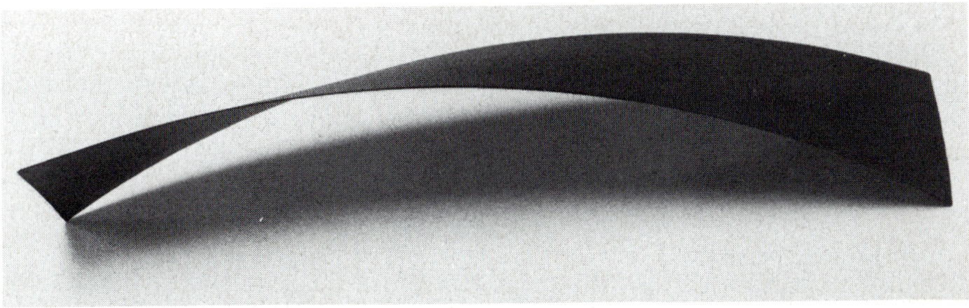

FIG. 3. The room temperature shape of a 200 mm × 50 mm $(+45_2/-45_2)_T$ graphite–epoxy laminate.

snap-through action. For other laminates, the cylindrical shape is unique, i.e., cannot be forced into another cylindrical shape. On the other hand, when the saddle shape appears, it is always unique. No other shape exists. As will be seen, it can be determined whether the shape is cylindrical or saddle, and whether the cylindrical shape is unique or whether it is possible for the unsymmetric laminate to have another cylindrical shape. In addition, it will be shown that while the laminate may have a cylindrical shape at room-temperature, it can be saddle shaped at a temperature closer to the cure temperature.

Figs. 1–4 show that there are unique characteristics associated with unsymmetric laminates. The behavior is not predicted by the method presented in the standard texts, for example, JONES (1975), AGARWAL and BROUTMAN (1980), TSAI and HAHN (1980). The method presented in the texts is usually referred to as classical lamination theory. The key to the behavior of unsymmetric laminates is the fact that the out-of-plane deflections associated with the shapes

FIG. 4. The room-temperature shape of a 150 mm × 150 mm $(0_{10}/90_{10})_T$ graphite–epoxy laminate.

are many, many times the thickness of the laminate. Any theory which does not account for geometrically nonlinear effects due to large deflections is in error. A theory which accounts for geometrically nonlinear effects and which explains this unusual behavior of unsymmetric laminates is developed in the next section.

3. Development of the theory

Fig. 5 depicts schematically the problem being studied. Fig. 5(a) shows a laminate flat and undeformed at its elevated cure temperature. The figure also shows the x_1–x_2–x_3 reference coordinates system used throughout the analysis. The origin of the coordinates system is at the geometric center of the volume of the undeformed laminate. As stated in the introduction the study here will be limited to cross-ply laminates. This means that in Fig. 5(a) the layers in the laminate will have the fibers aligned either with the x_1-axis or with the x_2-axis. Figs. 5(b)–(d) show the room-temperature shapes discussed in the previous section. Fig. 5(b) shows a schematic of the saddle configuration. Figs. 5(c) and 5(d) show cylindrical shapes. These two cylindrical shapes can be thought to represent the two shapes related through snap-through for certain laminates.

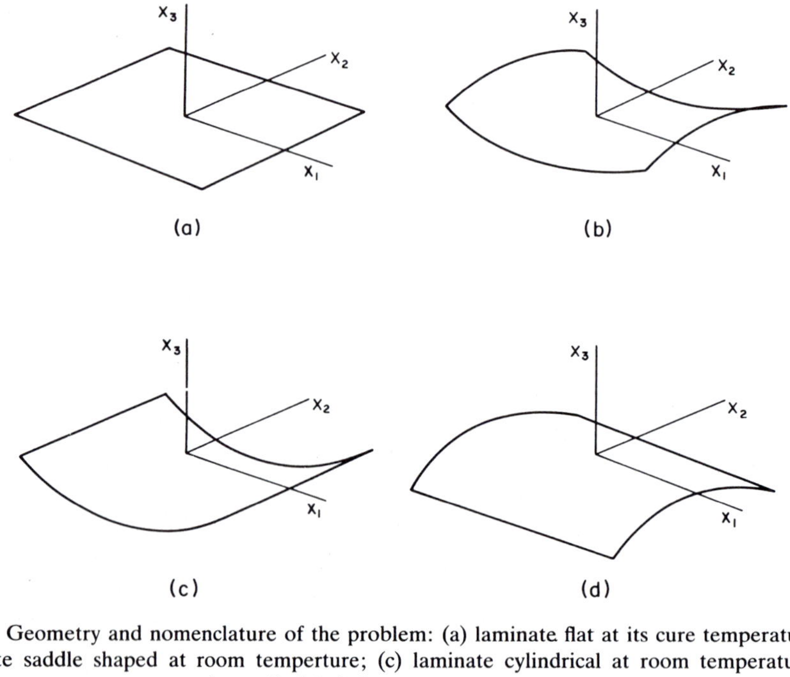

FIG. 5. Geometry and nomenclature of the problem: (a) laminate flat at its cure temperature; (b) laminate saddle shaped at room temperture; (c) laminate cylindrical at room temperature; (d) another cylindrical shape at room temperature.

On the other hand, one of the shapes can be thought to represent the unique cylindrical shape of other laminates. Since the laminates are cross-ply, the principal curvatures of the saddles or cylinders are aligned with the reference coordinate system. The laminate of Fig. 3 is cross-ply but not in a reference system aligned with the rectangular sides of the laminate. Such cases will not be considered here. The basic issue is to determine what fundamental mechanisms need to be included in an analysis of the thermo-mechanical cooling process so that the transition from the flat state of Fig. 5(a) to the deformed shape of Figs. 5(b), (c), or (d) can be accurately modeled. Of considerable importance is the fact that the geometric dimensions of a laminate have an influence on whether the flat laminate of Fig. 5(a) cools to the saddle shape given by 5(b) or whether it cools to the cylindrical shape given by Fig. 5(c). In addition, the snap-through to other cylindrical shape, Fig. 5(d), must be represented. Mathematically this latter point means multiple solutions to the governing equations (i.e. two equilibrium shapes at the same temperature). With multiple solutions the problem cannot be linear. It must be inherently nonlinear.

Here the theory governing the behavior of unsymmetric laminates will be developed from an energy viewpoint. The philosophy is simple: The shape (or shapes) observed for a given laminate at a given temperature is the shape that minimizes the total potential energy of the laminate. It will be assumed that the mechanical forces exerted on the laminate during the curing process do no net work on the laminate as the temperature changes from the cure temperature to room temperature. Thus the total potential energy to be considered is simply the strain energy in the cooled laminate.

The strain energy density, including the effects of thermal expansion, is given by

$$\omega = \tfrac{1}{2}\bar{C}_{ij}(e_i e_j - \alpha_i e_j \Delta T)\,, \quad i, j = 1\text{–}6\,. \tag{3.1}$$

Here $\bar{C}_{ij}$ is the stiffness matrix in the x_1–x_2–x_3 system. The e_i are components of Green's strain tensor in the x_1–x_2–x_3 system. The α_i are the material's linear coefficients of thermal expansion in the x_1–x_2–x_3 system. The temperature change is given by ΔT, ΔT positive corresponding to a temperature increase. Throughout the $\bar{C}_{ij}$ and α_i will be assumed to be independent of temperature. This is not a restriction on the analysis. Because of the large out-of-plane deformations associated with the temperature change, Green's strain tensor is used, as opposed to the strain tensor of linear elasticity. This assumption of the form of the strain tensor is the only deviation from classical lamination theory used in the present theory. It is assumed that Kirchhoff's hypothesis is valid and that each lamina is in a state of plane stress. Because of the latter assumption, the only components in eq. (3.1) that contribute to the strain energy density are e_1, e_2, and e_6. These three strain components are given by

$$e_1 = e_1^0 - x_3 \frac{\partial^2 u_3^0}{\partial x_1^2},$$
$$e_2 = e_2^0 - x_3 \frac{\partial^2 u_3^0}{\partial x_2^2}, \qquad (3.2)$$
$$e_6 = e_6^0 - x_3 \frac{\partial^2 u_3^0}{\partial x_2 \partial x_1},$$

with

$$e_1^0 = \frac{\partial u_1^0}{\partial x_1} + \frac{1}{2}\left(\frac{\partial u_3^0}{\partial x_1}\right)^2,$$
$$e_2^0 = \frac{\partial u_2^0}{\partial x_2} + \frac{1}{2}\left(\frac{\partial u_3^0}{\partial x_2}\right)^2, \qquad (3.3)$$
$$e_6^0 = \frac{1}{2}\left(\frac{\partial u_1^0}{\partial x_2} + \frac{\partial u_2^0}{\partial x_1} + \frac{\partial u_3^0}{\partial x_1}\frac{\partial u_3^0}{\partial x_2}\right).$$

The quantities e_i^0, $i = 1, 2, 6$, are the strains at the laminate midplane, u_1^0 is the displacement of a material point in the laminate in the x_1 direction, u_2^0 its displacement in the x_2 direction, and u_3^0 its displacement in the x_3 direction. The products of the strain gradients in eq. (3.3) represent the geometrically nonlinear effects due to large out-of-plane deformations (u_3^0).

Considering the plane stress state of the laminae and expanding eq. (3.1), the strain energy density becomes

$$\begin{aligned}\omega = {} & \tfrac{1}{2}C'_{11}e_1^2 + C'_{12}e_1e_2 + 2C'_{66}e_6^2 + \tfrac{1}{2}C'_{22}e_2^2 \\ & - (C'_{11}\alpha_1 + C'_{12}\alpha_2)e_1\Delta T - (C'_{12}\alpha_1 + C'_{22}\alpha_2)e_2\Delta T .\end{aligned} \qquad (3.4)$$

The C'-s are the reduced stiffnesses in the x_1–x_2–x_3 system. For cross-ply laminates C'_{16}, C'_{26}, and α_6 are zero.

Considering the entire laminate, the total potential energy is given by

$$W = \int_{x_1=-L_1/2}^{L_1/2} \int_{x_2=-L_2/2}^{L_2/2} \int_{x_3=-h/2}^{h/2} \omega \, \mathrm{d}x_1 \, \mathrm{d}x_2 \, \mathrm{d}x_3 \qquad (3.5)$$

The limits on the above volume integral reflect the fact that at the elevated cure temperature the flat laminate has length L_1 in the x_1 direction, L_2 in the x_2 direction and is of thickness h (in the x_3 direction). Recall that the origin of the coordinate system is located at the geometric center of the laminate.

To find the minimum value of total potential energy, eqs. (3.2) and (3.3) can be substituted into eq. (3.4) and those results substituted into eq. (3.5). Since the coordinate x_3 appears explicitly, integration with respect to x_3 will combine with the C'_{ij} to yield components of the extensional, coupling, and bending stiffnesses (A_{ij}, B_{ij}, and D_{ij}) of the laminate. In addition, integration with respect to x_3 results in effective thermal inplane forces, N_1^{T} and N_2^{T}, and moments, M_1^{T} and M_2^{T}. These effective thermal loads are given by

$$N_1^{\mathrm{T}} = \Delta T \int_{-h/2}^{h/2} (C'_{11}\alpha_1 + C'_{12}\alpha_2)\,\mathrm{d}x_3\,, \tag{3.6a}$$

$$N_2^{\mathrm{T}} = \Delta T \int_{-h/2}^{h/2} (C'_{12}\alpha_1 + C'_{22}\alpha_2)\,\mathrm{d}x_3\,, \tag{3.6b}$$

$$M_1^{\mathrm{T}} = \Delta T \int_{-h/2}^{h/2} (C'_{11}\alpha_1 + C'_{12}a_2)x_3\,\mathrm{d}x_3\,, \tag{3.6c}$$

$$M_2^{\mathrm{T}} = \Delta T \int_{-h/2}^{h/2} (C'_{12}\alpha_1 + C'_{22}\alpha_2)x_3\,\mathrm{d}x_3\,. \tag{3.6d}$$

The expression for the total potential energy, W, will then be of the form

$$W = \int_{x_1=-L_1/2}^{L_1/2} \int_{x_2=-L_2/2}^{L_2/2} \omega(A_{ij}, B_{ij}, D_{ij}, N_1^{\mathrm{T}}, N_2^{\mathrm{T}}, M_1^{\mathrm{T}}, M_2^{\mathrm{T}}, u_1^0, u_2^0, u_3^0, x_1, x_2)\,\mathrm{d}x_1\,\mathrm{d}x_2\,. \tag{3.7}$$

Seeking the stationary values of W with respect to variations in u_1^0, u_2^0, and u_3^0 leads to the differential equations and boundary conditions which govern the shape of the laminate at a given temperature. The equations will be nonlinear partial differential equations for u_1^0, u_2^0, and u_3^0 as functions of x_1 and x_2. These equations have not been derived. Except for selected cases it is felt that only approximate numerical solutions to the equations, e.g. finite-difference, finite-element, are possible. In addition, the first variation only provides information regarding equilibrium shapes. The equilibrium shapes may or may not exist in the laboratory. Since the problem is nonlinear, and since experimental observations indicated multiple shapes and a snap-through phenomenon, stability must be addressed. Obtaining solutions to the stability equations coupled with obtaining solutions to the equilibrium equations leads to a problem of significant proportions. The obtaining of numbers and conducting parameter studies becomes computer-intensive. Therefore, the approach here will be to seek approximate solutions in the sense of Ritz, starting with the energy expression of eq. (3.7), and minimizing the expression in the context of simple assumed displacement fields. These displacement fields will be valid for the laminate as a whole.

From Figs. 1–4 it is clear an assumed out-of-plane displacement of the form

$$u_3^0(x_1, x_2) = \tfrac{1}{2}(ax_1^2 + bx_2^2)\ , \tag{3.8}$$

with a and b being unknown and to-be-determined constants, represents the possible shapes observed. With this functional form for u_3^0, $a = -b$ represents the saddle shape of Fig. 5(b), $a > 0$, $b = 0$ represents the cylindrical shape of Fig. 5(c) and $a = 0$, $b < 0$ represents the cylinder of Fig. 5(d). Functional forms to assume for $u_1^0(x_1, x_2)$ and $u_2^0(x_1, x_2)$ are not so obvious from Figs. 1–4. However, Fig. 6 provides some insight into the choice of functional form for these other two displacement fields. Fig. 6 shows a cross-section in the x_1–x_3 plane of a deformed laminate. It is not important whether the cross-section is

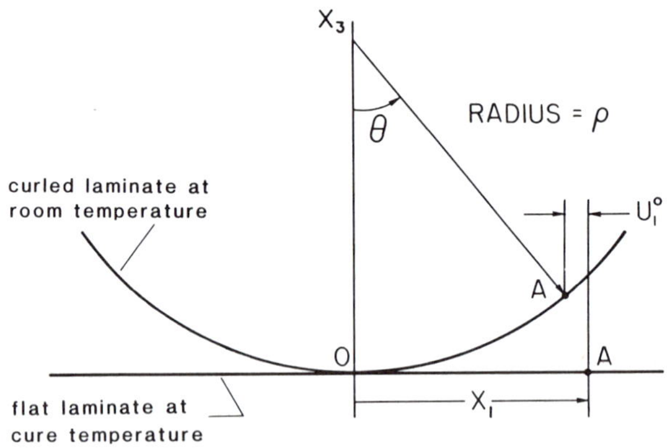

FIG. 6. Kinematics of a flat laminate deforming due to a temperature change.

associated with a cylinder or whether it is associated with a saddle. Shown in the figure is the cross-section when the laminate is flat and the cross-section when the laminate is at a lower temperature and curled out-of-plane. From the figure it is clear point A moves a significant amount in the negative x_1 direction due to the out-of-plane deformation. Specifically, point A moves an amount

$$u_1^0 = \rho \sin\theta - x_1\ , \tag{3.9}$$

x_1 being where point A was at the elevated cure temperature and ρ and θ geometric parameters indicated on Fig. 6. For small angular openings,

$$\sin\theta = \theta - \frac{\theta^3}{3!}\ . \tag{3.10}$$

If the inplane (arc-wise) strains are assumed small, the arc length from 0 to A is $\rho\theta$, or

$$\rho\theta = x_1 \tag{3.11}$$

Finally, from eq. (3.8),

$$1/\rho \cong a\ , \tag{3.12}$$

a being from eq. (3.8). Therefore, combining eqs. (3.9)–(3.12),

$$u_1^0 = -\tfrac{1}{6}a^2x_1^3\ . \tag{3.13}$$

Thermal effects, do however, contribute to the arc length 0A (Fig. 6) of the deformed laminate. As with classical lamination theory, it is assumed that the effect of thermal expansion (or contraction) in the x_1 direction is linear in x_1 and it superposes on the displacement due to curling, i.e.

$$u_1^0 = cx_1 - \tfrac{1}{6}a^2x_1^3\ , \tag{3.14}$$

c being an unknown constant.

Similar considerations in the x_2–x_3 plane lead to

$$u_2^0 = dx_2 - \tfrac{1}{6}b^2x_2^3\ , \tag{3.15}$$

d being a constant and b being from eq. (3.8). The quantities a, b, c and d will be determined as part of the minimization process.

Using eqs. (3.8), (3.14), and (3.15) in eq. (3.3) leads to

$$e_1^0 = c\ , \tag{3.16a}$$

$$e_2^0 = d\ , \tag{3.16b}$$

$$e_6^0 = abx_1x_2/2\ . \tag{3.16c}$$

For geometrically linear deformations of cross-ply laminates due to a temperature change, there are no midplane shearing strains, e_6^0. It can be reasoned that this is true for the large deformation geometrically nonlinear case. Eq. (3.16) contradicts this, i.e. $e_6^0 \neq 0$ in eq. (3.16), and so u_1^0 and u_2^0 of eqs. (3.14) and (3.15) must be modified to yield $e_6^0 = 0$. The final forms for u_1^0 and u_2^0 that do this are

$$\begin{aligned} u_1^0 &= cx_1 - \tfrac{1}{6}a^2x_1^3 - \tfrac{1}{4}abx_1x_2^2\ , \\ u_2^0 &= dx_2 - \tfrac{1}{6}b^2x_2^3 - \tfrac{1}{4}abx_2x_1^2\ . \end{aligned} \tag{3.17}$$

Eq. (3.17) along with eq. (3.8) form the assumed displacement fields for unsymmetric laminates deformed due to a temperature change relative to their cure temperature. These equations can be substituted into eqs. (3.2)–(3.4) and these in turn into eq. (3.5) and integration with respect to x_1 and x_2 performed in eq. (3.5). This substitution leads to an expression for W of the form

$$W = W(A_{ij}, B_{ij}, D_{ij}, N_1^T, N_2^T, M_1^T, M_2^T, L_1, L_2, a, b, c, d) . \tag{3.18}$$

Stationary values of W are then sought relative to variations in a, b, c and d. The integration with respect to x_1 and x_2 in eq. (3.5) is rather involved and results in a lengthy expression for W. However, the first variation of W with respect to a, b, c, and d results in an expression of the form

$$\delta W = f_1 \delta a + f_2 \delta b + f_3 \delta c + f_4 \delta d , \tag{3.19}$$

where

$$f_i = f_i(A_{ij}, B_{ij}, D_{ij}, N_1^T, N_2^T, M_1^T, M_2^T, L_1, L_2, a, b, c, d) , \quad i = 1, 4 .$$

For equilibrium

$$f_i = 0 , \quad i = 1, 4 . \tag{3.20}$$

Eqs. (3.20) constitute four algebraic equations for a, b, c, and d. The equations are nonlinear in a and b and linear in c and d. The equations are

$$\begin{aligned} f_1(a, b, c, d) = {} & -C_1 cb + C_2 ab^2 + 2C_3 ab - B_{11} c \\ & + D_{11} a - C_4 cb + 2C_5 ab^2 - C_6 db \\ & + D_{12} b - C_7 db + C_8 ab^2 + C_9 b^2 \\ & + (L_2^2/48) N_1^T b + M_1^T + (L_1^2/48) N_2^T b = 0 , \end{aligned} \tag{3.21a}$$

$$\begin{aligned} f_2(a, b, c, d) = {} & -C_1 ac + C_2 a^2 b + C_3 a^2 - C_4 ac + 2C_5 a^2 b + D_{12} a \\ & - C_6 da - C_7 da + C_8 a^2 b + 2C_9 ab - B_{22} d \\ & + D_{22} b + (L_2^2/48) N_1^T a + (L_1^2/48) N_2^T a + M_2^T = 0 , \end{aligned} \tag{3.21b}$$

$$f_3(a, b, c, d) = A_{11} c - C_1 ab - B_{11} a + A_{12} d - C_4 ab - N_1^T = 0 , \tag{3.21c}$$

$$f_4(a, b, c, d) = A_{12} c - C_6 ab - B_{22} b + A_{22} d - C_7 ab - N_2^T = 0 . \tag{3.21d}$$

In these equations

$$\begin{aligned} & C_1 = A_{11} L_2^2/48 , && C_2 = A_{11} L_2^4/1280 , \\ & C_3 = B_{11} L_2^2/48 , && C_4 = A_{12} L_1^2/48 , \\ & C_5 = A_{12} L_1^2 L_2^2/2304 , && C_6 = A_{12} L_2^2/48 , \\ & C_7 = A_{22} L_1^2/48 , && C_8 = A_{22} L_1^4/1280 , \\ & C_9 = B_{22} L_1^2/48 . \end{aligned} \tag{3.22}$$

It should be mentioned that when $L_1 = L_2 = 0$, the constants $C_1 - C_9$ are zero and eqs. (3.21) reduce to classical lamination theory, i.e., the geometrically linear case. Apparently, with the approach being used here, a laminate with zero sidelength is synonomus with classical lamination theory.

To obtain numerical results, eqs. (3.21) must be solved numerically. This can be done one of several ways. The four nonlinear algebraic equations can be solved using, for example, a Newton-type algorithm. Alternatively, it is possible to use the last two equations of eq. (3.21) and solve explicitly for c and d in terms of a and b. These expressions for c and d can be substituted into the first two equations to obtain two nonlinear equations for a and b. These two can be solved for by a and b by a Newton algorithm. As a final alternative, these two equations can be combined into a single equation for either a or b. HAMAMOTO and HYER (1985) used this approach. The alternative single equation for a is

$$\begin{aligned}(S^2U_1)a^5 + (S^2V_1)a^4 + (2SU_1U_2)a^3 + (STV_2 + 2SU_2V_1)a^2 \\ + (SV_2^2 - T^2U_2 + U_1U_2^2)a + (U_2^2V_1 - TU_2V_2) = 0\,,\end{aligned} \tag{3.23a}$$

with

$$b = -\frac{Ta + V_2}{Sa^2 + U_2}\,. \tag{3.23b}$$

Eqs. (3.23) use the following definitions:

$$\begin{aligned}
S &= \frac{A_{11}L_2^4 + A_{22}L_1^4}{2880}\,,\\
T &= D_{12} + \frac{A_{12}B_{11}B_{22}}{(A_{11}A_{22} - A_{12}^2)}\,,\\
U_1 &= D_{11} - \frac{A_{22}B_{11}^2}{(A_{11}A_{22} - A_{12}^2)}\,,\\
U_2 &= D_{22} - \frac{A_{11}B_{22}^2}{(A_{11}A_{22} - A_{12}^2)}\,,\\
V_1 &= M_1^{\mathrm{T}} - \frac{B_{11}(A_{22}N_1^{\mathrm{T}} - A_{12}N_2^{\mathrm{T}})}{(A_{11}A_{22} - A_{12}^2)}\,,\\
V_2 &= M_2^{\mathrm{T}} - \frac{B_{22}(A_{11}N_2^{\mathrm{T}} - A_{12}N_1^{\mathrm{T}})}{(A_{11}A_{22} - A_{12}^2)}\,.
\end{aligned} \tag{3.24}$$

With nonlinear algebraic equations, either eqs. (3.21) or eqs. (3.23), multiple roots and hence multiple configurations at a given ΔT can be expected. This correlates with the ability to predict multiple cylindrical shapes, as discussed in Section 2.

Given a specific laminate with a specific temperature change from cure, i.e., given A_{ij}, B_{ij}, D_{ij}, L_1, L_2, N_1^{T}, N_2^{T}, M_1^{T}, M_2^{T}, the out-of-plane deformations can be determined by solving for a and b. The inplane deformations c and d can also be determined, although here they are not of as much interest. As alluded to earlier, solutions to eqs. (3.21) are equilibrium solutions. At a given temperature one or more solutions may correspond to unstable equilibrium. Therefore the second variation of W must be examined for each solution. First, however, it is instructive to present numerical results regarding equilibrium shapes for specific laminates. The stability issues can thus be put into the context of specific shapes for specific laminates.

4. Numerical results

Numerical results for a specific laminate will illustrate clearly the important features of the relationship between laminate shape and the dimensions of the laminate, and between laminate shape and its temperature relative to cure. Recall in Section 2 that the only difference between the laminate in Fig. 1 and the laminate in Fig. 4 was the size of the laminate. This is the motivation for examining the influence of laminate size on its shape. On the other hand, raising and lowering the temperature of a laminate, relative to cure, is easily done and in fact is the method used by various investigators to determine the curing temperature of a laminate, PAGANO and HAHN (1977), HERAKOVICH et al. (1980). In addition, such an approach has been used to determine the influence of moisture and temperature of fiber-reinforced composite material, CROSSMAN et al. (1978), HYER and HAGAMAN (1979).

Fig. 7 illustrates the relationship between the size of a laminate and the shape of that laminate at room temperature (20°C). The results are from HAMAMATO and HYER (1985). The laminate considered is square, $L_1 = L_2 = L$, and the length of the side is on the horizontal axis. The values of a and b are the vertical axes of the two portions of the figure. The laminate is a $(0_4/90_4)_T$ laminate. Each lamina has the following properties

$$E_1 = 115\ \mathrm{GPa}\ ; \quad E_2 = 8\ \mathrm{GPa}\ ; \quad \nu_{12} = 0.28\ ,$$

$$\alpha_1 = -0.106 \times 10^{-6}/°\mathrm{C}\ ; \quad \alpha_2 = 25.6 \times 10^{-6}/°\mathrm{C}\ ,$$

$$\text{lamina thickness} = 0.175\ \mathrm{mm}\ ; \quad \text{cure temperature} = 121°\mathrm{C}\ .$$

For purposes of discussion, laminates made from laminae with these material properties will be referred to as type A laminates. The properties are representative of AS4/1908, a Hercules product, fabricated to 50% volume fraction fibers.

Immediately obvious from the figure are two factors. First, there is a symmetry to the solutions for the relation between a and b and sidelength. Second, there is the existence of three possible room-temperature equilibrium

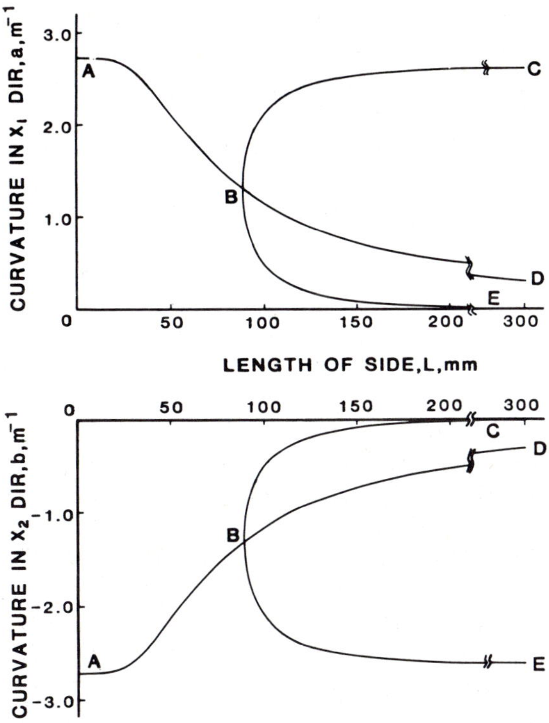

FIG. 7. Sidelength–curvature relation for square $(0_4/90_4)_T$ type A laminates at room temperature.

solutions if the laminate has a sidelength greater than some critical length. From the figure the critical length is 90 mm. For $L = 0$ there is only one solution. The solution is $b = -a$. This corresponds to a saddle shape. This, in fact, is the predicted shape if geometric nonlinearities are ignored in the analysis of shape prediction. This solution is denoted by point A on the figure. As the sidelength of the laminate increases from zero, say to 50 mm, the solutions are still single valued and the shape is still predicted to be a saddle. However, the saddle is one that is shallower than the linear prediction. As the sidelength increases even more, the saddle shape is still predicted to exist but it becomes even shallower. At some critical length the solution bifurcates. The bifurcation is denoted as point B on the figure. For sidelengths larger than this critical value, three equilibrium configurations are predicted to exist, each represented by a different solution branch on the figure. The branches are denoted as BC, BD and BE. Branch BD represents a continuation of the saddle shape. The other two solution branches represent a significant departure from a saddle shape. Ignoring for the moment the transition region near bifurcation ($75\text{ mm} \leq L \leq 150\text{ mm}$), branch BC represents a laminate with a large positive curvature in the x_1 direction and little or no curvature in the x_2 direction. By contrast, branch BE represents a laminate with little or no curvature in the x_1 direction and a large negative curvature in the x_2 direction. The shapes associated with these latter two branches can be considered cylindrical in nature and correspond with figs. 5(c) and 5(d). This variation in shape with laminate size corresponds well with the discussion in Section 2. For this 8 layer laminate, sidelengths greater than 90 mm will cause the laminate to cool to more than one equilibrium configuration. Obviously not all shapes occur simultaneously and some may correspond to unstable equilibrium configurations and will never be observed. This will be addressed later.

While the predictions of the dependence of laminate shape on size is interesting, the equations predict other interesting characteristics. Figs. 8–10 illustrate the dependence of type A laminate shape on temperature. In these figures the temperature–curvature relations of three laminates are shown, Fig. 8 illustrates the characteristics of a $125 \times 125\ \text{mm}^2$ laminate, Fig. 9 the characteristics of a $300 \times 300\ \text{mm}^2$, and Fig. 10 the characteristics of a $50 \times 50\ \text{mm}^2$ laminate. As before, there are two portions to each figure. Curvatures are on the vertical axes while the temperature of the laminate is on the horizontal axis. Room temperature (20°C) on the left end of the axis and the curing temperature (121°C) on the right end. The character of the shape can be traced with either increasing temperature of decreasing temperature. Only the latter situation will be discussed here.

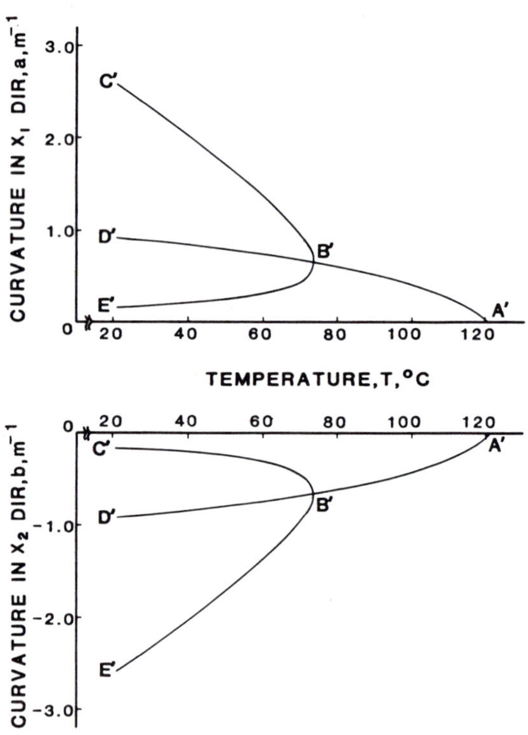

FIG. 8. Temperature–curvature relation for a 125 mm × 125 mm $(0_4/90_4)_T$ type A laminate.

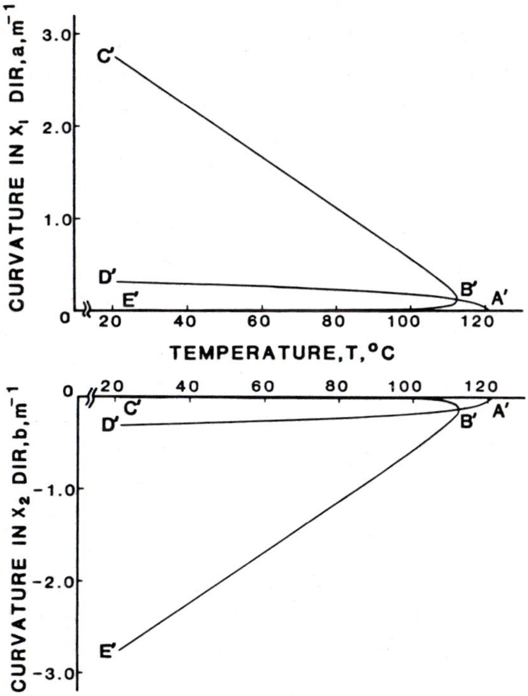

FIG. 9. Temperature–curvature relation for a 300 mm × 300 mm $(0_4/90_4)_T$ type A laminate.

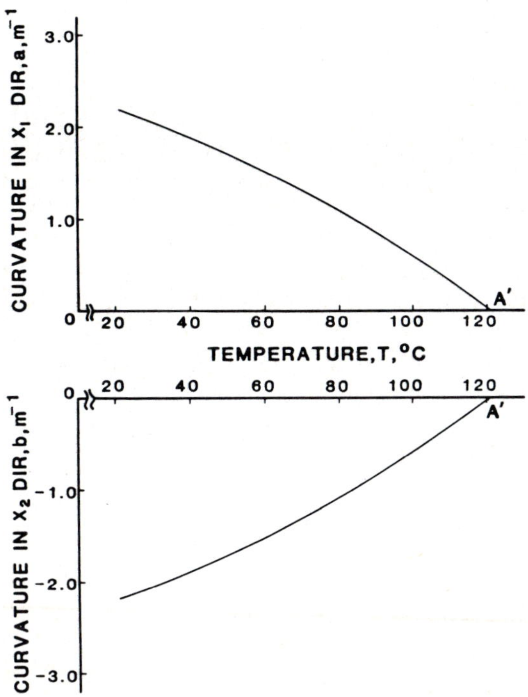

FIG. 10. Temperature–curvature relation for a 50 mm × 50 mm $(0_4/90_4)_T$ type A laminate.

Referring to Fig. 8, for the $125 \times 125\ \mathrm{mm}^2$ laminate at the cure temperature, point A′, the curvatures are zero. As the temperature decreases, curvatures develop. The curvatures are small but then are equal and opposite and so the laminate is saddle shaped. At roughly 77°C, point B′, the solution bifurcates. The equilibrium solutions can follow one of three paths. These paths are denoted as B′C′, B′D′, and B′E′. If the laminate follows path B′C′, the curvature in the x_1 direction continues to increase while the curvature in the x_2 direction decreases. Alternatively, if the laminate follows path B′E′, the curvature in the x_1 direction decreases while the curvature in the x_2 direction increases, in magnitude. As a third possibility, if the laminate follows path B′D′ as it cools, then it will remain saddle-shaped, the saddle becoming deeper and deeper as the temperature is lowered.

The temperature–curvature characteristics of the $300\ \mathrm{mm} \times 300\ \mathrm{mm}$ laminate, Fig. 9, are similar to the temperature–curvature characteristics of the $125\ \mathrm{mm} \times 125\ \mathrm{mm}$ laminate. For this larger laminate, however, the differences in the shapes represented by the different solutions branches are more distinct than they are for the $125 \times 125\ \mathrm{mm}^2$ laminate. As the temperature is lowered from the stress-free temperature, equal and opposite curvatures develop. However, at less than 10°C below the stress-free temperature, the solutions bifurcate. If as the temperature is further reduced and the laminate follows path B′C′, then the x_1-direction curvature continues to increase while the x_2-direction curvature virtually disappears. This corresponds to the cylindrical shape of Fig. 5(c). Alternatively, if the cooling path follows path B′E′, then the x_2-direction curvature increases in magnitude and the x_1-direction curvature disappears. This corresponds to the cylindrical shape of Fig. 5(d). Path B′D′ represents a laminate with equal and opposite curvatures, a saddle. For the $125 \times 125\ \mathrm{mm}^2$ laminate, the shapes represented by branches B′C′ and B′E′ are not quite cylindrical. The tendency is there but the effect is not as pronounced as it is with the $300 \times 300\ \mathrm{mm}^2$ laminate.

The small laminate, Fig. 10, shows a strikingly different temperature-curvature relation. For this laminate there is only one shape, a saddle, as the laminate cools from the cure temperature to room temperature. No bifurcation behavior occurs. In fact, the temperature–curvature relation is almost linear.

It should be mentioned that based on the description of the curing process earlier, the cooldown from cure occurs while the laminate is constrained between the caul plates. Therefore, the behavior shown in Figs. 8–10 does not occur. However, if after the laminate is removed from the caul plates it is heated to the cure temperature again and then cooled, the scenarios described in Figs. 8–10 are applicable.

With the numerical results for the $(0_4/90_4)_T$ laminate in hand, it is interesting to examine limiting value of a and b for large and small laminates. For a $(0_4/90_4)_T$ laminate $A_{22} = A_{11}$, $D_{22} = D_{11}$, $B_{22} = -B_{11}$. From eqs. (3.6), $N_2^{\mathrm{T}} = N_1^{\mathrm{T}}$, and $M_2^{\mathrm{T}} = -M_1^{\mathrm{T}}$. As a result, from eqs. (3.24), $U_2 = U_1$ and $V_2 = -V_1$. Eq. (3.23) provides the limiting information. For small laminates $L_1 \to 0$ and $L_2 \to 0$ and S of eq. (3.24) is zero. Eq. (3.23) gives

$$a = \left(\frac{24}{h} \frac{(Q_{11}Q_{22} - Q_{12}^2)(\alpha_1 - \alpha_2)\Delta T}{Q_{11}^2 + 14Q_{11}Q_{22} + Q_{22}^2 - 16Q_{12}^2} \right) \tag{4.1a}$$

and

$$b = -a \, . \tag{4.1b}$$

For large laminates $L_1 \to \infty$ and $L_2 \to \infty$. The first two terms in eq. (3.23) dominate and so

$$a = -\frac{U_1}{V_1} = \left(\frac{24}{h} \frac{(Q_{11}Q_{22} - Q_{12}^2)(\alpha_1 - \alpha_2)\Delta T}{Q_{11}^2 + 14Q_{11}Q_{22} + Q_{22}^2 - 16Q_{12}^2} \right) \left(\frac{Q_{11} + Q_{22} - 2Q_{12}}{Q_{11} + Q_{12}} \right) \tag{4.2a}$$

and

$$b = 0 \, . \tag{4.2b}$$

The first bracketed term in eq. (4.2a) is the linear solution of eq. (4.1a) and the second bracketed term is slightly larger than unity for any material. Therefore, the large-laminate limit for the nonzero curvature is asymptotic to a value slightly larger than the linear prediction. The other curvature for the large laminate is zero.

At this point it is tempting to examine the effects of other parameters, such as laminate material properties, laminate thickness, or laminate stacking sequence, on the predicted shape or shapes. However, the information presented in the previous figures is not really complete until the stability of the predicted shapes is examined. This will be done and then the effects of the other parameters on the shapes and the stability of the shapes will then be examined.

5. Stability considerations

For any particular solution to the four algebraic equations, eqs. (3.21), to correspond to stable equilibrium, the total potential energy must be a minimum for the particular solution. If the total potential energy is not a minimum, the equilibrium condition is not stable and that particular equilibrium configuration will not be physically realized. For this discretized system the stability of the equilibrium configurations for the laminate is determined by examining the following determinant of partial derivatives of the four functions f_i, $i = 1, 4$, given in eqs. (3.21):

$$\begin{bmatrix} \frac{\partial f_1}{\partial a} & \frac{\partial f_1}{\partial b} & \frac{\partial f_1}{\partial c} & \frac{\partial f_1}{\partial d} \\ \frac{\partial f_2}{\partial a} & \frac{\partial f_2}{\partial b} & \frac{\partial f_2}{\partial c} & \frac{\partial f_2}{\partial d} \\ \frac{\partial f_3}{\partial a} & \frac{\partial f_3}{\partial b} & \frac{\partial f_3}{\partial c} & \frac{\partial f_3}{\partial d} \\ \frac{\partial f_4}{\partial a} & \frac{\partial f_4}{\partial b} & \frac{\partial f_4}{\partial c} & \frac{\partial f_4}{\partial d} \end{bmatrix} .$$

SIMITSES (1976) discusses this. A particular equilibrium configuration of a laminate with specific material properties, stacking sequence, and sidelength, and at a specific temperature is stable if and only if the matrix is positive definite. (For the cases studied here, the determinant of the matrix was not zero.) If the matrix is not positive definite, then the equilibrium configuration is not stable. Each point on each branch of the previous figures must be checked.

Using this scheme it was found that for the $(0_4/90_4)_T$ laminates in Figs. 7–10, any equilibrium configuration corresponding to a saddle shape is unstable if cylindrical equilibrium configurations can exist under the same conditions. If the saddle shape is the only shape that is predicted to exist, then it is stable. All cylindrical configurations in the figures are stable. Therefore, in Fig. 7, all the saddle configurations of branch AB are stable and will be observed at room-temperature. On the other hand, none of the saddle configurations on Branch BD are stable. For laminates with sidelengths greater than 90 mm, a saddle configuration will never be observed. Either of the two cylindrical configurations will be observed. On Figs. 8 and 9, as the flat laminate is cooled from the curing temperature alone path A′B′, a stable saddle configuration will develop. As the laminate is cooled below the bifurcation point, point B′, the saddle will cease to exist because it is unstable. The laminate will deform into either the cylinder represented by path B′C′, or the cylinder represented by path B′E′. By contrast, when the smaller laminate is cooled, Fig. 10, it deforms into a stable saddle and remains that way for all temperatures below the cure temperature.

From the information presented in the analysis it is not possible to determine which of the two cylindrical configurations of Figs. 7, 8, and 9 will be observed. Since both cylindrical configurations are stable, they correspond to total potential energy minimums. These are local minimums in the sense that they were determined by examining variations of the configurations, i.e., variational displacements are by definition small. Therefore either configuration is likely. In reality, small imperfections in the laminate or small variations from a spatially uniform temperature will cause the laminate to favor one cylindrical configuration or the other.

With all the components available to determine the shapes of unsymmetric laminates and to determine whether or not the shapes will ever be observed, the effects of material properties, stacking sequence, and other variables are now examined.

6. Other numerical results

6.1. *Effect of material properties*

The effect of material properties on the shape response can be examined by considering laminates fabricated from lamina with the following material properties:

$$E_1 = 181\text{ GPa}\,;\quad E_2 = 10.3\text{ GPa}\,;\quad \nu_{12} = 0.28\,,$$
$$\alpha_1 = -0.106 \times 10^{-6}/°\text{C}\,;\quad \alpha_2 = 25.8 \times 10^{-6}/°\text{C}\,,$$
$$\text{lamina thickness} = 0.100\text{ mm}\,;\ \text{cure temperature} = 177°\text{C}\,.$$

Compared with the first material examined, this material has a higher cure temperature and is stiffer in both the fiber direction and perpendicular to the fibers. It is expected that this will influence the temperature–curvature and curvature–sidelength relations. Laminates made from lamina with these material properties will be referred to as type B laminates. The properties are representative of a material with T300 fibers and 5208 resin, with 60% volume fraction fibers, HYER (1981b).

Fig. 11 illustrates the sidelength–curvature relation at room-temperature for a $(0_4/90_4)_T$ laminate fabricated from this second material. Only the x_1-direction curvature is shown. The x_2-direction curve has the mirror image symmetry of Fig. 7. Compared to Fig. 7, the analogous figure for type A material, there are not many differences. The main differences are the magnitude of the large-laminate curvature, and the length corresponding to the bifurcation point. Compared with the type A material, the magnitude of the curvature, for a given sidelength, is slightly larger. On the other hand, the bifurcation sidelength is slightly smaller. A stability analysis indicates identical stability characteristics.

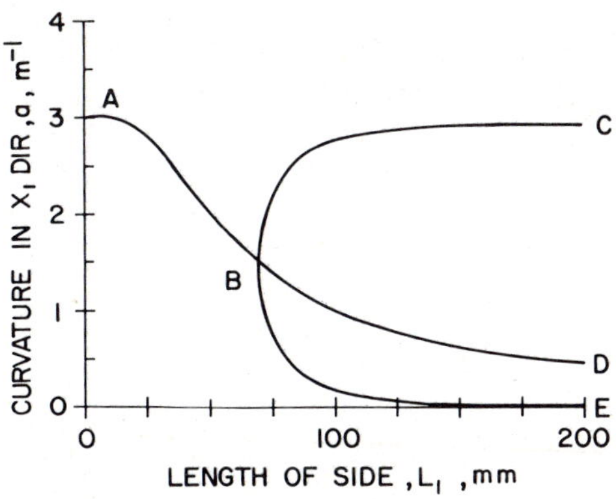

FIG. 11. Sidelength–curvature relation for square $(0_4/90_4)_T$ type B laminates at room temperature.

6.2. Effect of laminate thickness

Fig. 12 illustrates the shape characteristics of a $(0_2/90_2)_T$ laminate. The material considered is type B, HYER (1981b). Except for thickness, the laminate of Fig. 11 and the laminate of Fig. 12 are identical. To determine the effect of laminate thickness, this figure can be compared directly with Fig. 11.

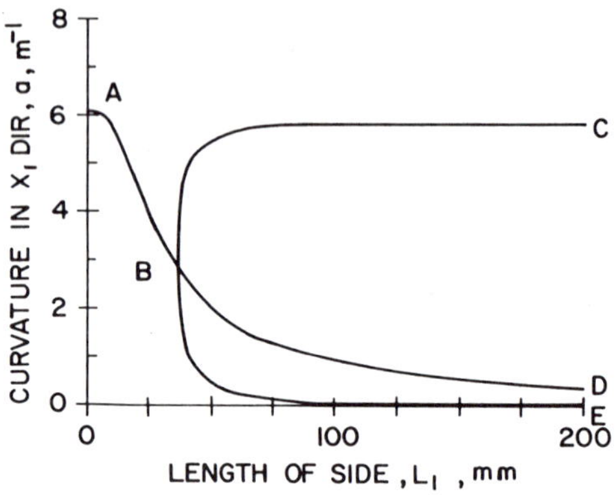

FIG. 12. Sidelength–curvature relation for square $(0_2/90_2)_T$ type B laminates at room temperature.

Two differences between the cases are immediately obvious. First the magnitude of the curvatures differ by a factor of two, the thinner laminate having larger curvatures. A 150 mm × 150 mm $(0_2/90_2)_T$ laminate, for example, has twice the out-of-plane deflection as a 150 mm × 150 mm^2 $(0_4/90_4)_T$ laminate. The second difference is the location of the bifurcation point. Bifurcation of the solutions for the $(0_2/90_2)_T$ laminate occurs when the sidelength reaches 36 mm. The thicker laminate can be twice as large before the saddle-shape configuration disappears and the laminate becomes cylindrical. If the governing equations are nondimensionalized, it can be shown that the laminates of the family $(0_n/90_n)_T$, $n = 1, 2, \ldots$, the transition from saddle to cylinder occurs at the value of L/h of 72.

6.3. Effect of laminate aspect ratio

So far the laminates discussed have all been square. The sidelength in the x_1 direction has been the same as the sidelength in the x_2 direction. Figs. 13–16 show the effect of two other aspect ratios on laminate shape, HYER (1981c). Here aspect ratio will be defined as the sidelength in the x_2 direction divided by the sidelength in the x_1 direction, L_2/L_1. Because of the symmetry of the a and b variation with sidelength, only the variation of a with sidelength is illustrated.

In addition, the sidelength in the x_1 direction, L_1, is used as the length variable.

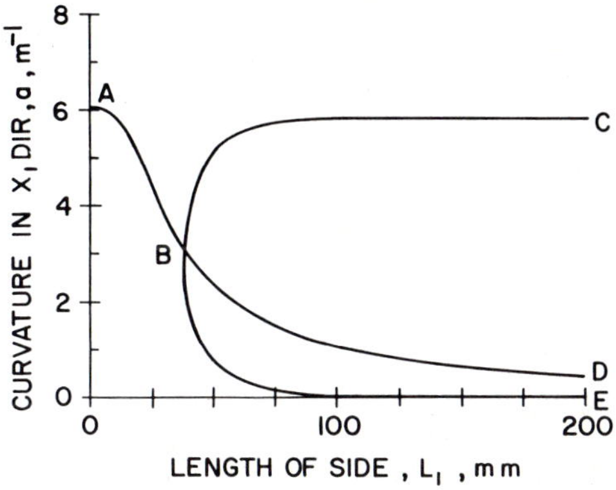

FIG. 13. Sidelength–curvature relation for $(0_2/90_2)_T$ type B laminates at room temperature, aspect ratio $L_2/L_1 = 0.25$.

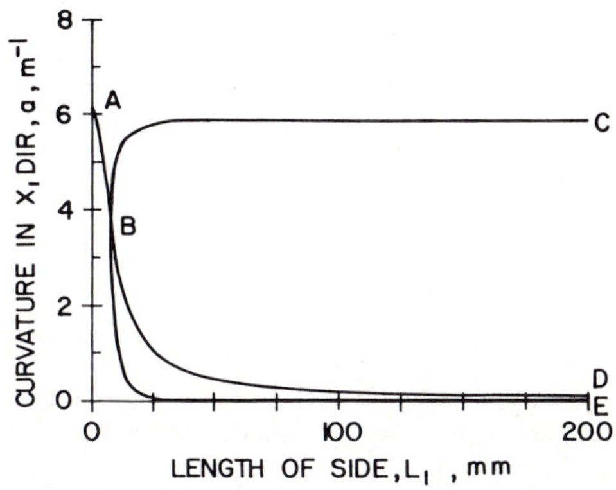

FIG. 14. Sidelength–curvature relation for $(0_2/90_2)_T$ type B laminates at room temperature, aspect ratio $L_2/L_1 = 4$.

Fig. 13 shows the relationship between the curvature in the x_1 direction and L_1 for a rectangular $(0_2/90_2)_T$ type B laminate with an aspect ratio of 0.25. Qualitatively the characteristics are similar to previous cases presented. There is a length range for which the solution is single-valued and a length range for which there are three solutions. Roughly, multiple solutions occur when L_1 is greater than 36 mm. All solution branches except the saddle branch BD represent stable equilibrium configurations. Except for the region around the bifurcation point, the sidelength–curvature relation for this laminate is very similar to the relation for a square $(0_2/90_2)_T$ laminate (Fig. 12).

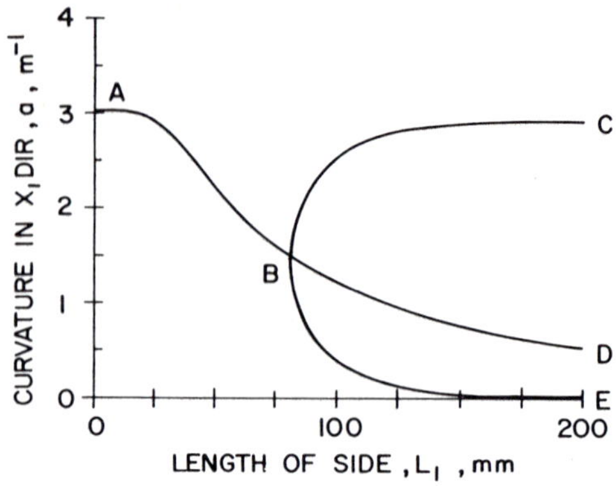

FIG. 15. Sidelength–curvature relation for $(0_4/90_4)_T$ type B laminates at room temperature, aspect ratio $L_2/L_1 = 0.25$.

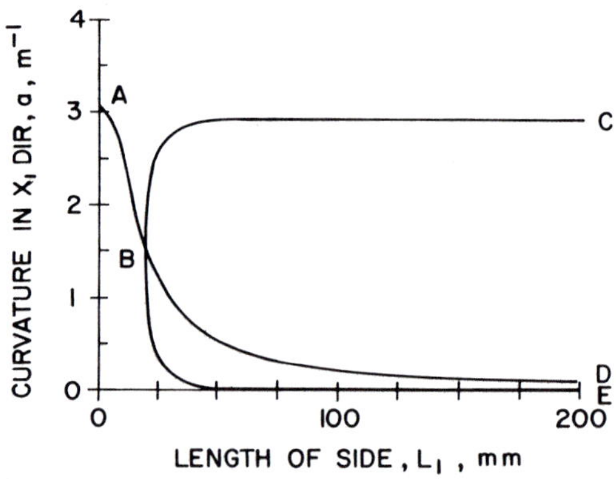

FIG. 16. Sidelength–curvature relation for $(0_4/90_4)_T$ type B laminates at room temperature, aspect ratio $L_2/L_1 = 4$.

Markedly different behavior is observed if the aspect ratio is changed so that the laminate is four times longer in the x_2 direction than in the x_1 direction. Fig. 14 shows the behavior of such a laminate. The figure indicates that the length range of single-valued solutions is much less for this laminate. Multiple solutions occur when L_1 exceeds 9 mm. The saddle configuration represented by branch BD is unstable and the curvature for large laminates, i.e., $L_1 >$ 25 mm, is the same as for square laminates or for laminates with an aspect ratio of 0.25.

Figs. 15 and 16 show the sidelength–curvature relation for the thicker $(0_4/90_4)_T$ type B laminates with aspect ratios of 0.25 and 4, respectively. Fig. 11 illustrated the behavior of square laminates of this thickness. By comparing Figs. 11, 15 and 16 it is clear that the square laminate and the laminate with an aspect ratio of 0.25 behave similarly. However, the laminate with an aspect ratio of 4 has larger sidelength range over which the multiple solutions occur.

From the results illustrated, it appears that for rectangular laminates, the location of the bifurcation point depends on the length of the longer side. In all cases shown, for the type B laminates, when the length of the longer side divided by the laminate's total thickness exceeds 72, multiple solutions occurred. Similar studies have not been conducted for the type A laminate but similar behavior is expected. The numerical value of the critical length-to-thickness ratio, however, is expected to be material-dependent.

6.4. *Effect of stacking sequence*

It is interesting to consider other stacking arrangements and study their shapes, HYER (1982). Here the discussion will be limited to square laminates. If just four-layer cross-ply laminates are considered, there are only four unique stacking arrangements. All other arrangements can be obtained from the four by a simple rotation of the laminate in its plane or by turning the laminate upside down. The four unique arrangements are: the $(0/0/90/90)_T$ previously discussed in Fig. 12; $(0/90/0/90)_T$; $(0/0/0/90)_T$; and $(0/90/0/0)_T$. The equilibrium configurations of these latter three laminates now will be discussed. For the discussion material properties of a type B laminate will be used.

Fig. 17 shows the sidelength–curvature relation for the $(0/90/0/90)_T$ laminate. The equilibrium configurations of this laminate are quite similar to the configurations of the $(0_2/90_2)_T$ of Fig. 12. However, some interesting differences can be observed. First, the out-of-plane deflections of the $(0/90/0/90)_T$ laminate are not as large as the deflections of the $(0_2/90_2)_T$ case. This is because the asymmetry in the laminate material properties for the $(0/90/0/90)_T$ is not as severe as the asymmetry in the $(0_2/90_2)_T$ case. Second, and directly related to the degree of asymmetry, this laminate will exhibit saddle-shape equilibrium configurations for sidelengths up to 87 mm. This is opposed to the 37 mm sidelength limit for the $(0_2/90_2)_T$ laminate. The level of asymmetry in the material properties of the laminate is reflected in the magnitude of the terms in the B_{ij} matrix and the magnitude of the effective thermal moments.

Figs. 18 and 19 show the sidelength–curvature relation for a $(0/0/0/90)_T$ and a $(0/0/90/0)_T$ laminate, respectively. These relationships are unlike any discussed so far. The most striking difference between the relationship for these laminates and the relationship for the other laminates discussed is the lack of a coalescence of the solution branches. Whereas the branching of solutions from a common point is referred to as bifurcation behavior, the disjoint behavior of the sidelength–curvature relationship for these laminates is referred to as limit point behavior. For unsymmetric laminates bifurcation behavior is generally associated with perfect or mathematically ideal conditions. Limit point behavior is associated with a lack of ideal conditions. The lack of idealism is

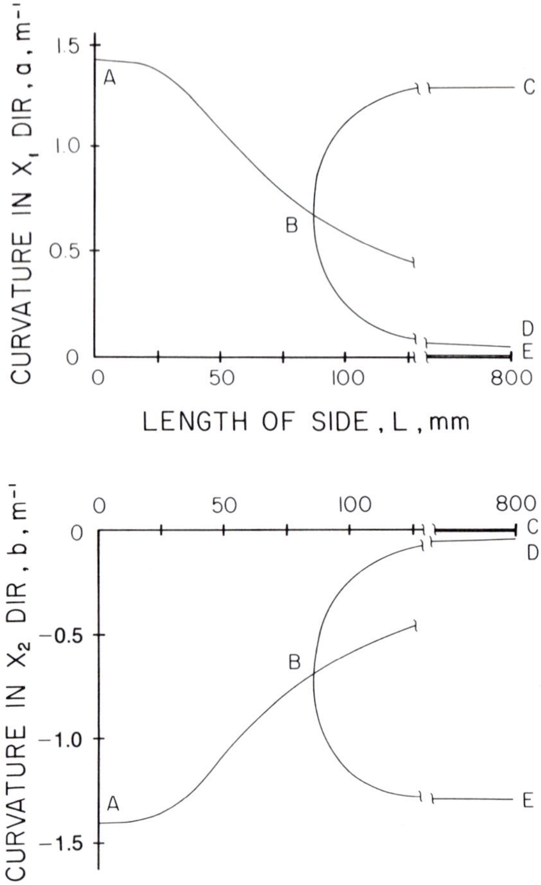

FIG. 17. Sidelength–curvature relation for $(0/90/0/90)_T$ type B laminates at room temperature.

reflected in the A, B and D matrices as well as in the effective thermal loads. For the ideal situation, $A_{22} = A_{11}$, $B_{22} = -B_{11}$, $D_{22} = D_{11}$, $N_1^{\mathrm{T}} = N_2^{\mathrm{T}}$, and $M_2^{\mathrm{T}} = -M_1^{\mathrm{T}}$. If one or more of these equalities is violated, bifurcation behavior will not occur.

Referring to Fig. 18, for a $(0/0/0/90)_T$ laminate of zero sidelength, i.e., the solution when geometric nonlinearities are ignored, the room-temperature shape is predicted to be a saddle. However, the saddle is not a perfect saddle, rather, the curvature in the x_1 direction is much smaller than the curvature in the x_2 direction. As the sidelength increases from zero, the smaller of the two curvatures begins to decrease rapidly. At a sidelength of 100 mm, the minor curvature has disappeared and the laminate is cylindrical, all the curvature being in the x_2 direction. As the sidelength increases to about 200 mm, multiple equilibrium configurations occur. One solution branch is the continuation of cylindrical branch AE. The other solution branches represent different equilibrium configurations. (Note that on the figure the curvatures in the x_2 direction

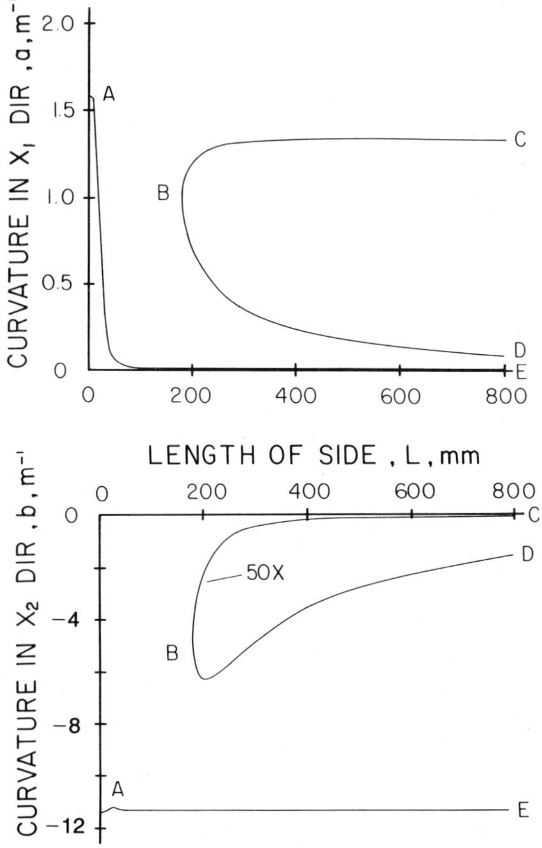

FIG. 18. Sidelength–curvature relation for square $(0/0/0/90)_T$ type B laminates at room temperature.

have been multiplied by 50 before plotting them.) At 400 mm, for example, branch BC represents a cylindrical configuration with no curvature in the x_2 direction. The curvature in the x_1 direction is roughly equal to the linear solution prediction. Branch BD represents a saddle configuration. Both curvatures are small but they are of opposite sign. It would be impossible, however, to detect the curvature in the x_2 direction for this particular case. Which of the three equilibrium will actually be observed is determined by examining stability.

A stability analysis predicts interesting results. Branch AE corresponds to stable equilibrium configurations. Branch BC also represents stable shapes. Branch BD corresponds to unstable configurations. Unlike the other laminates discussed, the existence of a unique and stable cylindrical shape is predicted. A 100 mm × 100 mm laminate, for example, will have a cylindrical shape and it cannot be forced into another cylindrical shape. Larger laminates, say

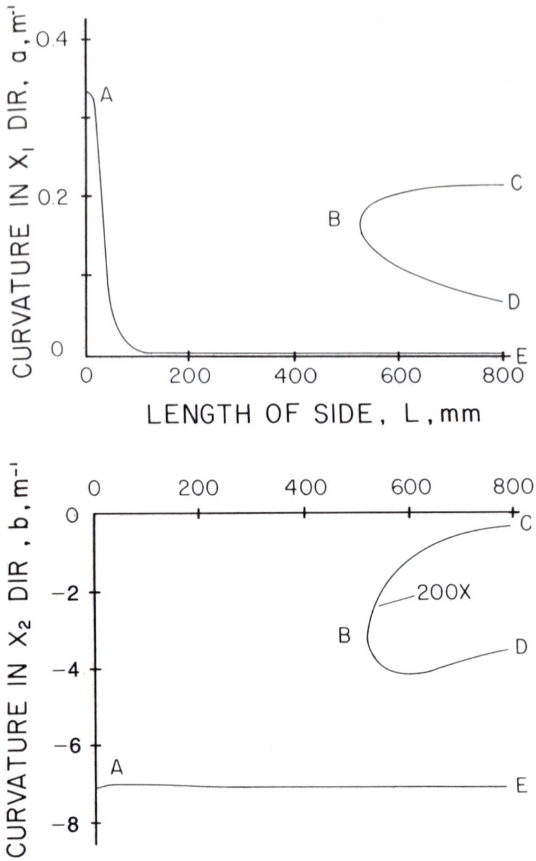

FIG. 19. Sidelength–curvature relation for square $(0/0/90/0)_T$ type B laminates at room temperature.

600 mm × 600 mm, can be made to have either of two cylindrical shapes. One of the shapes has the same curvature characteristics a 100 mm × 100 mm laminate would, i.e., branch AE. The other cylindrical shape would correspond to branch BC. The saddle configuration would never be observed for a 600 mm × 600 mm laminate.

The sidelength–curvature characteristics of a $(0/0/90/0)_T$ type B laminate, Fig. 19, are much like those of the $(0/0/0/90)_T$ laminate. Limit point behavior, instead of bifurcation behavior, is predicted. Compared to a $(0/0/0/90)_T$ laminate, the curvatures are smaller and the laminate can be made larger before multivalues solutions occur. Again, this can be attributed to the less severe degree of asymmetry for the $(0/0/90/0)_T$ can as compared to the $(0/0/0/90)_T$ case. This is clear if one considers that a $(0/0/0/0)_T$ has no asymmetry. If the lamina furthest from the laminate's geometric midplane is given a 90° orientation, the asymmetry in laminate material properties and thermal loads is greater than if the lamina closest to the geometric midplane is given a 90° orientation.

It should be noted that in the region of the sidelength–curvature relation where the smaller curvature is rapidly disappearing, the nonzero curvature (branch AE in the b portion of each figure) has a slight perturbation. The nonzero curvature decreases in magnitude very slightly as the length increases and then the curvature increases in magnitude again. This behavior would probably not be detected in an actual laminate.

7. Closure

This chapter has discussed an interesting phenomenon in the mechanics of composite materials. It has not been widely discussed before but the results explain behavior often observed in composite laminates. Though only simple laminates were considered, the same basic predictions are expected to occur for more complicated laminates.

Acknowledgements

The work discussed here was supported by the NASA-Virginia Tech Composites Program, Cooperative Agreements NCCl-15 and NAG-1-343, and Grant 81-0195 with the United States Air Force Office of Scientific Research, and Contract F49620-79-C-0038 with the United States Air Force Office of Scientific Research.

References

AGARWAL, B.D. and L.J. BROUTMAN (1980), *Analysis and Performance of Fiber Composites* (John Wiley and Sons, New York) pp. 145–156.

CROSSMAN, F.W., R.E. MAURI and W.J. WARREN (1978), Moisture-altered viscoelastic response of graphite–epoxy composites, in: J.R. Vinson, ed., *Advanced Composite Materials-Environmental Effects* (STP-658, American Society for Testing and Materials, Philadelphia, PA) pp. 205–220.

HAMAMOTO, A. and M.W. HYER (1987), *Int. J. of Solids and Structures* **23**, 919–935.

HERAKOVICH, C.T., J.G. DAVIS, Jr and J.S. MILLS (1980), Thermal microcracking in celion 6000/PMR-15 graphite-polyimide, in: D.P.H. HASSELMAN and R.A. HELLER, eds., *Thermal Stresses in Severe Environment* (Plenum Publishing Press, New York) pp. 649–665.

HYER, M.W. (1981a), *J. Comp. Mater.* **15**, 175–194.

HYER, M.W. (1981b), *J. Comp. Mater.* **15**, 296–340.

HYER, M.W. (1981c), An inherent instability in fiber-reinforced composite laminates, in: S.S. WANG and W.J. RENTON, eds., *Advances in Aerospace Structures and Materials-AD-01* (American Society of Mechanical Engineers, New York) pp. 239–246.

HYER, M.W. (1982), *J. Comp. Mater.* **16**, 318–340.

HYER, M.W. and J.A. HAGAMAN (1979), The effects of thermal cycling on the thermal deformations of graphite-polyimide, *Proc. Spring Meeting of the Society of Experimental Stress Analysis*, San Francisco, CA, paper no. R79-114 (Society of Experimental Mechanics, Brookfield Center, CT).

JONES, R.M. (1975), *Mechanics of Composite Materials* (McGraw-Hill, New York) pp. 147–156.

PAGANO, N.J. and H.T. HAHN (1977), Evaluations of composite curing stresses, in: J.G. DAVIS JR, ed., *Composite Materials: Testing and Design*, 4th *Conf.* (STP 617, American Society for Testing and Materials, Philadelphia, PA) pp. 317–329.

SIMITSES, G.J. (1976), *An Introduction to the Elastic Stability of Structures* (Prentice-Hall, Inc., Englewood Cliffs, NJ) pp. 8–14.

TSAI, S.W. and H.T. HAHN (1980), *Introduction to Composite Materials* (Technomic Publishing Co., Westport, CT) pp. 217–276.

CHAPTER 3

Nonlinear Analysis of Composite Structures

N.A. Alfutov and B.G. Popov

Moscow Higher Technical School
5, 2nd Bauman Street
107005 Moscow
U.S.S.R.

Contents

HANDBOOK OF COMPOSITES, VOL. 2 – Structure and Design
Edited by C.T. HERAKOVICH and Y.M. TARNOPOL'SKII

Preface

Advanced load-bearing structures are often made of fibrous composite materials which are composed of a comparatively compliant matrix reinforced with high-strength and high-modulus fibers. It is these composite materials that the typical elements of load-bearing structures–laminated shells and panels–are made from by continuous winding or layup techniques. This chapter is devoted to the design of such structures.

In the first section the simplest relationships for assessment of deformation properties of unidirectional and laminated composites are presented. A simple structural-phenomenological model of the composite material has been described which permits a clear notion to be obtained about specific features of the deformation process in the material. In this chapter the main attention has been focused on the second section, in which fundamentals of the variational-matrix method in the design of complex shell-type structures as related to laminated composite structures are outlined. Examples of the design of such structures are presented. In the Appendix, the basic variational formulations of static problems are enclosed, which can successfully be employed in practical designs.

List of Symbols

Section 1

$\boldsymbol{\Delta\sigma}'$ – a column vector of stress increments in the axes of a unidirectional layer;

$\boldsymbol{\Delta e}'$ – a column vector of strain increments in the axes of a unidirectional layer;

$\boldsymbol{C}'$ – a coupling matrix of stress increments with strain increments (a matrix of tangential moduli);

$E_{\parallel}$ – tangential modulus in uniaxial deformation of a layer along the fibers

$E_{\perp}$ – tangential modulus in uniaxial deformation of a layer across the fibers;

$G_{\parallel\perp}$ – tangential shear modulus in the plane of a layer;

$\nu_{\parallel\perp}$, $\nu_{\perp\parallel}$ – coefficients of the transverse strain of a layer;

$\boldsymbol{\Delta\sigma}$ – a column vector of stress increments in the axes of the structure;

$\boldsymbol{\Delta e}$ – a column vector of strain increments in the axes of the structure;

T_k – a matrix, connecting strains in axes of the k-th layer with strains in the axes of the structure;

α_k – reinforcement angle of the k-th layer;

Section 2

x_1, x_2, x_3 – a system of curvilinear orthogonal coordinates for a shell;
$\boldsymbol{e}$ – a column vector of total strains;
$\boldsymbol{\epsilon}$ – a column vector of linear components of strains;
$\boldsymbol{\eta}$ – a column vector of nonlinear components of strains;
A_1, A_2 – Lame's parameters for a coordinate surface;
k_1, k_2 – basic curvatures of a shell;
v_1, v_2, v_3 – displacements along coordinate lines at an arbitrary point of a shell;
u_1, u_2, w – displacements along coordinate lines of a point on the coordinate surface ($x_3 = 0$);
$\Delta\boldsymbol{\xi}$ – a column vector of the increments of generalized strains;
$\boldsymbol{D}$ – a matrix of reduced stiffness characteristics of a laminated shell;
$\boldsymbol{Q}$ – a column vector of the generalized running force factors for a laminated shell;
$\boldsymbol{X}$ – a column vector of amplitude values for the *m*th harmonics of the expansion of generalized displacements for a laminated shell of revolution;
$\boldsymbol{Y}$ – a column vector of derivatives of $\boldsymbol{X}$;
$\boldsymbol{\lambda}$ – a column vector of Lagrangian multipliers. In a physical sense, coefficients $\boldsymbol{\lambda}$ correspond to generalized force factors, conjugate with coefficients $\boldsymbol{X}$;
$\boldsymbol{C}_1, \boldsymbol{C}_2$ – matrices of a differential coupling conditions between $\boldsymbol{X}$ and $\boldsymbol{Y}$.

Appendix

x_1, x_2, x_3 – an orthogonal rectilinear system of coordinates;
$\boldsymbol{u}$ – a column vector for displacements;
$\boldsymbol{e}$ – a column vector of strains;
$\boldsymbol{\sigma}$ – a column vector of stresses;
$\boldsymbol{g}, \boldsymbol{p}$ – body and surface forces;
$\boldsymbol{C}$ – a matrix of elasticity coefficients;
$\boldsymbol{\phi}$ – a matrix of shape functions, approximating displacements;
$\boldsymbol{q}$ – a column vector of generalized (nodal) displacements;
$\boldsymbol{K}$ – a stiffness matrix;

General Symbols

$\{\cdot\}$ – coefficients of a column vector are shown in braces;
$[\cdot]$ – coefficients or blocks of the matrix are put into square brackets;
$\lceil\cdot\rfloor$ – non-zero coefficients of a diagonal matrix are put into French quotes;
$(\cdot)^{\mathrm{T}}$ – a superscript 'T' indicates transposition of a column vector or matrix;

$(\cdot)^{-1}$ – a superscript '-1' indicates conversion operation;

$(\hat{\cdot})$ – the upper sign '^' indicates the possible (virtual) values;

$(\cdot)_\tau$ – the lower index 'τ' shows that the given value has been determined for a balanced configuration of the system at the time instant τ;

$(\cdot)_{(m)}$ – the lower index in parentheses shows the number of iterations in the solution of nonlinear problems.

1. Simplest models for deformation of fibrous composites

1.1. Introduction

It is expedient in the calculations and rational design of composite structures that the stiffness and strength properties should be known. To investigate these properties the structural as well as phenomenological approaches are employed.

The structural approach makes it possible to evaluate the composite properties in terms of constituent properties and the matrix–fiber interface by using information about the material structure. As a rule, the structural approach is commonly employed in the stage of material development to appropriately select a rational relation between the constituent properties and the matrix–fiber interface. The difficulties in the development of valid structural models are quite considerable. The exact evaluation of the state of stress in composite constituents is necessary because the onset of composite fracture is usually associated with the local loss of integrity. Hence, the application of a variety of simple composite models becomes principally impossible in the nonlinear analysis of the deformation process. Serious difficulties require that the kinetics of material fracture should be taken into account, because the local loss of integrity occurs very often in the initial stages of loading. However, it does not necessarily result in bearing strength exhaustion. At present, the structural theories are far from perfect and, therefore, we will mainly be concerned with a phenomenological approach.

The phenomenological approach involves the use of a sufficiently simple mathematical model, taking specific features of composite deformation into account and using the results of a limited number of tests as initial data. The phenomenological approach involves the following possible mathematical models:

(1) models presupposing nonlinear-elastic behavior of the material (for instance, LOMAKIN, 1960; RABOTNOV and STEPANICHEV, 1968).

(2) models based on plasticity theory for anisotropic media (for instance, its simplest variant as treated by HILL, 1956).

(3) models based on generalization of the endochronic theory for an anisotropic material (ZINOV'EV and SARBAEV, 1982; SARBAEV, 1984).

Models of the first group are applicable only in the case of active deformation of the material. If we speak about theories with a smooth surface of plasticity, then the use of the models of the second group is not fully justified.

This may be accounted for by the fact that for a composite some superposition of plasticity surfaces must be considered, i.e. the concept of a plasticity surface is in need of some refinement in this case. The theories, presupposing the existence of angular singularities on the surface, are more acceptable, although generalizations for anisotropic materials are complex. The models of the third group have been developed comparatively recently and their study is rather promising.

1.2. Simplest structurally-phenomenological model of the nonlinearly-elastic behavior of a lamina

Due to the difficulties encountered in the experimental work of studying complex models, the researchers are often tempted to employ the simplest stress–strain relations in a composite. The case when the stress–strain relation is expressed analogously to the generalized Hooke's law is most often preferred. Coefficients in these relations are variable and can be obtained from the strain diagrams in uniaxial tension–compression in the direction of the orthotropic axes of a unidirectional lamina and in shear (PETIT and WADDOUPS, 1969).

Consider a plane state of stress, characteristic of thin-walled load-bearing structures such as plates and shells. In the solution of nonlinear problems of statics (see Subsection A.2 of the Appendix) the relation between stress and strain increments for a unidirectional lamina is expressed analogous to (A.20)

$$\Delta\boldsymbol{\sigma}' = \boldsymbol{C}'\Delta\boldsymbol{e}' , \tag{1.1}$$

where

$$\Delta\boldsymbol{\sigma}' = \{\Delta\sigma_1', \Delta\sigma_2', \Delta\sigma_6'\} , \quad \Delta\boldsymbol{e}' = \{\Delta e_1', \Delta e_2', \Delta e_6'\}$$

are column vectors of stress and strain increments; numeration of indices of vector components $\Delta\boldsymbol{\sigma}'$, $\Delta\boldsymbol{e}'$ corresponds to (A.1) and (A.2); a prime indicates that the value is related to a system of coordinates of a unidirectional lamina (Fig. 1). Coefficients of the matrix $\boldsymbol{C}'$ have the following structure

$$\boldsymbol{C}' = \begin{bmatrix} c_{11}' & c_{12}' & 0 \\ & c_{22}' & 0 \\ \text{sym.} & & c_{33}' \end{bmatrix} , \tag{1.2}$$

where

$$c_{11}' = \frac{E_\parallel}{1 - \nu_{\parallel\perp}\nu_{\perp\parallel}} \; (1, \parallel \rightleftarrows 2, \perp) , \quad c_{12}' = c_{11}'\nu_{\perp\parallel} , \quad c_{33}' = G_{\parallel\perp} ,$$

$E_\parallel$, $E_\perp$ and $G_{\parallel\perp}$ are tangential moduli determined from the strain diagrams under uniaxial loading along and perpendicular to the fibers and in-plane shear

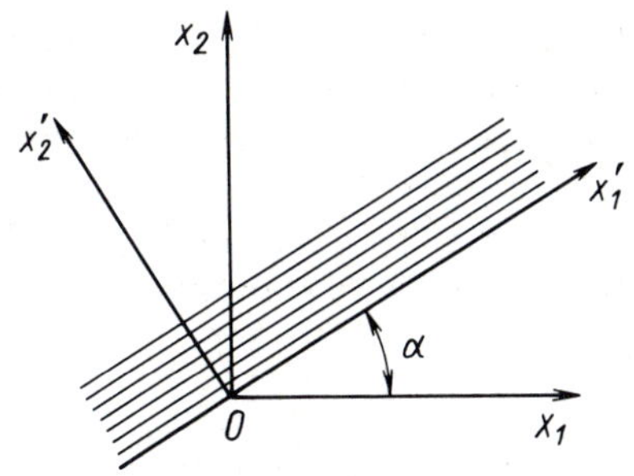

FIG. 1. Orientation of a unidirectional lamina.

for a unidirectional lamina; $\nu_{\|\perp}$ and $\nu_{\perp\|}$ are coefficients of transverse strain. At positive values of $E_{\|}$, $E_{\perp}$, assume that $\nu_{\|\perp} = \text{constant}$, $\nu_{\perp\|} = \nu_{\|\perp} E_{\perp}/E_{\|}$. We assume that $\nu_{\|\perp} = \nu_{\perp\|} = 0$ at $E_{\|} \leqslant 0$ or $E_{\perp} \leqslant 0$.

The technical elastic constants of a unidirectional material and strain diagrams in uniaxial tension–compression along and perpendicular to the fibers and in shear (Fig. 2), approximated in one way or another, have been used as initial data for the determination of properties of a unidirectional lamina.

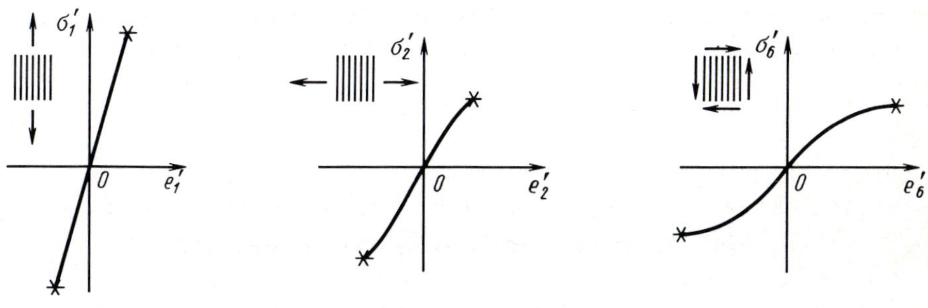

FIG. 2. Stress–strain diagrams of a unidirectional lamina.

It is assumed that in a complex state of stress the stiffness characteristics of a unidirectional material in the longitudinal and transversal directions depend only on the respective stresses. In this case, to evaluate the stresses and tangential moduli as functions of strains, the total deformations $\boldsymbol{e}'$ (generated in the complex state of stress) should not be used, but only the strain components $\boldsymbol{e}^*$ which are related to respective stresses. Then for strain increments $\Delta\boldsymbol{e}^* = \{\Delta e_1^*, \Delta e_2^*, \Delta e_6^*\}$ we have:

$$\Delta e_1^* = \frac{\Delta e_1' + \nu_{\perp\|}\Delta e_2'}{1 + \nu_{\|\perp}\nu_{\perp\|}} \ (1, \| \rightleftarrows 2, \perp)\,, \quad \Delta e_6^* = \Delta e_6'\,.$$

In the case of deformation of a unidirectional lamina in a multilayer material it is assumed that after having reached their ultimate values the stresses $\sigma_2' > 0$

and σ'_6 (marked by asterisks in Fig. 2) remain constant upon active loading whereas the respective tangential moduli are zero.

1.3. Modelling of fracture kinetics

The analysis of experimental results and analytical estimates (ALFUTOV et al., 1984; ZINOV'EV et al., 1982) permits us to assume in approximate calculations that the deformation of a unidirectional lamina in a laminated composite proceeds in accordance with the model as shown in Fig. 3.

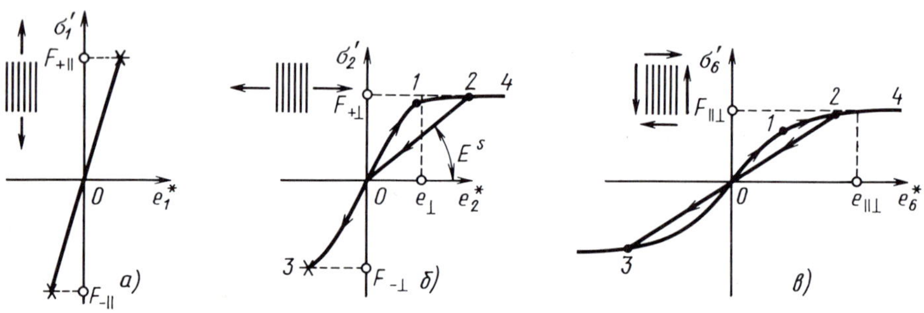

FIG. 3. Model stress–strain diagrams.

As a rule, deformation of a unidirectional material in the fiber direction is entirely linearly-elastic (Fig. 3(a)). Upon reaching the ultimate stresses $F_{+\|}$ or $F_{-\|}$ the lamina is considered to have failed.

Behavior of the lamina under loading in the direction orthogonal to fibers is more complex (Fig. 3(b)). Upon deformation the lamina is monolithic and elastic in the region 0–1. At the point 1 the process of intensive crack formation sets in which further develops in the region 1–2. The isolated lamina fails at the point 1. It is assumed that unloading from any point of the region 1–2 takes place at the modulus of load relief $E_\perp$, equal to the secant modulus of the diagram. Residual stresses are zero, this confirms the assumption of complete closing of cracks. Therefore, upon subsequent compression of a lamina (region 0–3) the initial modulus of the material is completely 'restored'. Repeated deformation of a unidirectional material at positive stresses σ'_2 occurs in the region 0–2 and further in 2–4. A scheme of algorithms for the evaluation of the tangential modulus of a lamina in the direction orthogonal to fibers, provided there are cracks in the matrix, is shown in Fig. 4. Here, σ'_2 and e^*_2 are stresses and strains in the transverse direction; $e_2^{\max}$ is the highest value of transverse strain, which has been attained throughout the loading history; $E^D_\perp$ is the tangential modulus; $E^S_\perp$ is the secant modulus (see Fig. 3(c)). If the stresses σ'_2 in compression reach their ultimate value $F_{-\perp}$, then the lamina is considered to have failed.

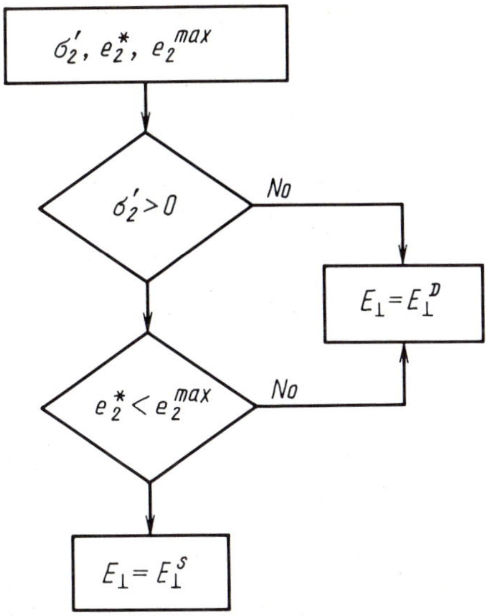

FIG. 4. A scheme of the algorithm for determination of $E_\perp$.

The behavior of a lamina in shear is in many respects analogous to that upon deformation in the direction orthogonal to the fibers (Fig. 3(d)). In the region 0–1 deformation of a lamina is elastic. The region 1–2 corresponds to the stage of crack formation in the matrix of a lamina. Unloading will take place (region 2–3) when the modulus of load relief is equal to the secant modulus. The process of shear deformation is believed to be independent of the sign of stress σ_6'. Repeated deformation proceeds along the path 3–0–2 and further in the region 2–4, where crack formation in the matrix recommences.

It is important to note that the process of crack formation in shear and deformation of a lamina in the transverse direction are interrelated. Let us assume that the appearance of a system of cracks, regardless of their origin, exert a simultaneous effect on the behavior of a lamina both in shear and upon deformation in the transverse direction.

In such a way, in accordance with the model of lamina behavior (ZINOV'EV et al., 1982), apart from its two natural states–the initial state (a monolithic material) and final (the material has failed), there is a group of intermediate states, characterized by the occurrence of cracks in the polymeric matrix. Within this group, we can distinguish groups of material states with open cracks (crack faces open) and closed cracks (crack faces closed). Transition from one group of states to another can be attributed to the sign of stress σ_2'. At positive stresses σ_2', cracks are open, at negative (compression)—closed.

1.4. Nonlinear deformation of a laminated composite

Let us treat uniform in-plane loading of a multilayer plate (Fig. 5). We shall associate the process of static loading with some conventional time τ. Assume

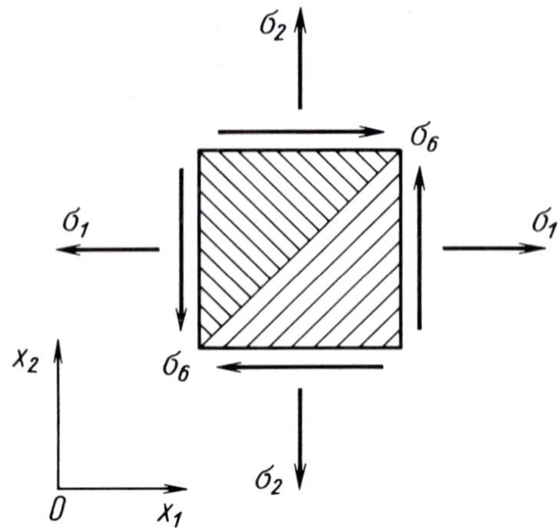

FIG. 5. Uniform loading of a laminated plate.

that during loading the variation of external average stresses $\boldsymbol{\sigma} = \{\sigma_1, \sigma_2, \sigma_6\}$ with a parameter τ is known. Then, by using a linearized formulation of the principle of virtual displacements (A.21) in the loading stage $(\tau, \tau + \Delta\tau)$ we obtain for the mth iteration that the increment of average strains of a multilayer material is

$$\Delta \boldsymbol{e}_{(m)} = \boldsymbol{C}_\tau^{-1}(\boldsymbol{\sigma}_{\tau+\Delta\tau} - \boldsymbol{\sigma}_{(m-1)}) ,$$

where

$$\Delta \boldsymbol{e}_{(m)} = \{\Delta e_{1(m)}, \Delta e_{2(m)}, \Delta e_{6(m)}\} ,$$

$$\boldsymbol{\sigma}_{(m-1)} = \frac{1}{h} \sum_{k=1}^{n} h_k \boldsymbol{T}_k^{\mathrm{T}} (\boldsymbol{\sigma}'_k)_{(m-1)} ,$$

$$\boldsymbol{C}_\tau = \frac{1}{h} \sum_{k=1}^{n} h_k \boldsymbol{T}_k^{\mathrm{T}} (\boldsymbol{C}'_k)_\tau \boldsymbol{T}_k ,$$

$$\boldsymbol{T}_k = \begin{bmatrix} c_k^2 & s_k^2 & s_k c_k \\ s_k^2 & c_k^2 & -s_k c_k \\ -2 s_k c_k & 2 s_k c_k & c_k^2 - s_k^2 \end{bmatrix} ;$$

$$s_k = \sin \alpha_k ; \quad c_k = \cos \alpha_k .$$

Here, n is the number corresponding to the given lamina in a laminated plate; h_k is the thickness of the kth layer; h is the total thickness; α_k is the layup angle of the kth layer; the index in parentheses indicates the number of iterations in the loading stage $(\tau, \tau + \Delta\tau)$. $(\boldsymbol{C}'_k)_\tau$ is the matrix relating stress and strain increments (see (1.1)). Of the kth layer, calculated for a balanced configuration at the time τ, a prime indicates that the respective coefficients have been obtained for the system of coordinates, related to the axis of elastic symmetry of the layer.

By employing the transformation matrix (1.3) for the mth iteration the strain increments in the kth layer are evaluated as

$$(\boldsymbol{\Delta e}'_k)_m = \boldsymbol{T}_k \boldsymbol{\Delta e}_{(m)} \quad k = 1, 2, \ldots, n \tag{1.4}$$

and the total strains in the kth layer as

$$(\boldsymbol{e}'_k)_{(m)} = (\boldsymbol{e}'_k)_{(m-1)} + (\boldsymbol{\Delta e}'_k)_{(m)} \quad k = 1, 2, \ldots, n \tag{1.5}$$

Analogously to (1.4), (1.5) and taking (1.2) into account the consistent strains $(\boldsymbol{\Delta e}^*_k)_{(m)}$ and $(\boldsymbol{e}^*_k)_{(m)}$ are calculated. With the help of the total consistent strains $(\boldsymbol{e}^*_k)_{(m)}$ the stresses in layers $(\boldsymbol{\sigma}'_k)_{(m)}$ and the average stresses in a laminated plate for the mth approximation are evaluated from the strain diagrams (Fig. 3)

$$\boldsymbol{\sigma}_{(m)} = \frac{1}{h} \sum_{k=1}^{n} h_k \boldsymbol{T}^{\mathrm{T}}_k (\boldsymbol{\sigma}'_k)_{(m)} \,.$$

The value of the relative residual can be used for checking the accuracy

$$\delta = \frac{\|\boldsymbol{\sigma}_{\tau+\Delta\tau} - \boldsymbol{\sigma}_{(m)}\|}{\|\boldsymbol{\sigma}_{\tau+\Delta\tau} - \boldsymbol{\sigma}_{\tau}\|} \,.$$

If δ is greater than the allowed error, the next iteration is accomplished. The number of iterations for refinement of the state of stress in layers must be controlled and if it exceeds some reasonable limit, the loading step must be reduced.

Upon passing over to the next loading step the matrix $(\boldsymbol{C}'_k)_{\tau+\Delta\tau}$ is corrected for each layer in accordance with the model described in Sections 1.2 and 1.3. Thereby, the positive definiteness of the matrix $\boldsymbol{C}_{\tau+\Delta\tau}$ is checked. For this reason it is enough to make sure that all the three diagonal coefficients of the matrix are positive.

Upon deformation, the variation of stiffness characteristics of the layers often results in the stiffness matrix of the composite becoming singular and thus lacking its inverse which is necessary for performing calculations according to formula (1.3). In this case, in accordance with the model, the composite attains the ability of boundless deformation under specified load. This situation can be explained by loss of stability in the process of deformation of a composite. After the matrix $\boldsymbol{C}_{\tau+\Delta\tau}$ has lost its positive definiteness further loading does not take place and the bearing strength of a composite is considered exhausted. As a rule, the multilayer composite fails, if, at least, in one of its layers, fracture occurs due to stresses σ_1' or $\sigma_2'<0$. As shown by experimental results, this has been confirmed, at least, for multilayer composites, composed of a few (up to 4–5) groups of layers of different orientation.

1.5. *Theoretical and experimental results*

The algorithm of the analysis of nonlinear deformation of a laminated composite, described in Section 1.4, permits us to model the loading process of a multilayer packet along various paths and to study its stiffness and strength properties. Thereby, it is possible to trace the fracture process in a packet starting from the loss of integrity of the layers right up to the moment when the stiffness matrix loses its positive definiteness and also to evaluate stresses in the individual layers. After having accomplished a series of calculations for various loading paths it is possible to obtain the strength surfaces of a laminated composite. Furthermore, the algorithm makes it possible to accomplish a parametric analysis with the aim of finding out rational reinforcement patterns. In doing so, the thicknesses and layup angles as well as the number of layers can be varied.

As a first example, the lines of the ultimate state in the 1st quadrant of a plane (σ_1, σ_2) for a fiberglass plastic of a quasi-isotropic structure [±30°/90°] are shown in Fig. 6. Dashed lines refer to a change of states in composite layers and solid lines refer to complete fracture of the material. Captions over the lines explain the reason for the change of states in the material. For instance, $\sigma_2^{30°} = F_{+\perp}$ stands for the onset of crack formation in the layers oriented at ±30° and caused by stresses σ_2'. Experimental results in Fig. 6 are indicated by circles (ALFUTOV and ZINOV'EV, 1982).

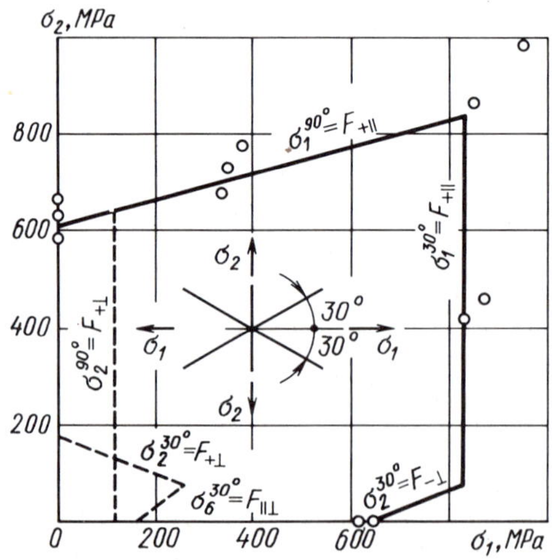

FIG. 6. Lines of ultimate states in a fiberglass plastic [±30°/90°].

The following initial data for a unidirectional fiberglass plastic have been used in the calculations $E_{\parallel} = 46.5 \times 10^3$ MPa; $E_{\perp} = 7 \times 10^3$ MPa; $G_{\parallel\perp} = 7 \times 10^3$ MPa; $\nu_{\parallel\perp} = 0.25$; $F_{+\parallel} = 1.6 \times 10^3$ MPa; $F_{-\parallel} = 0.5 \times 10^3$ MPa; $F_{+\perp} = 40$ MPa; $F_{-\perp} = 200$ MPa; $F_{\parallel\perp} = 60$ MPa. These initial data (if other characteristics are not indicated) will be used in all subsequent examples in this Section.

Depending on the structure of the material and the kind of state of stress, the character of stress–strain diagrams and material strength can be varied over a wide range. The specific features of nonlinear deformation of composites can be traced in angle-ply structures in uniaxial tension or compression. In an angle-ply composite, several types of bearing strength exhaustion are possible depending on the layup angle α (Fig. 7). Each of them is characterized by its own kind of stress–strain diagram.

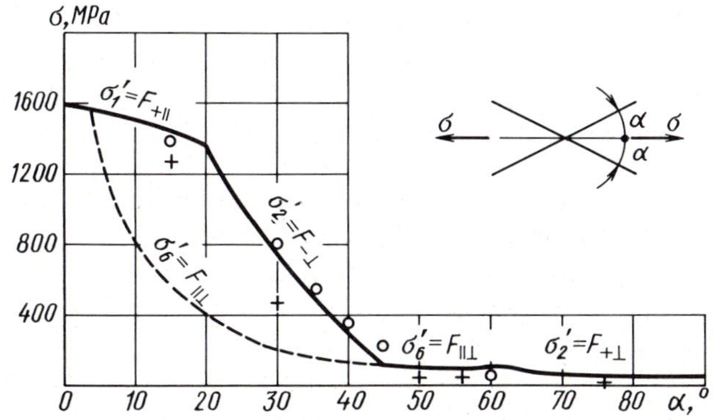

FIG. 7. Ultimate strengths of an angle-ply fiberglass plastic.

Solid lines in Fig. 7 refer to the calculated ultimate strength of a composite and the dotted line refers to the calculated integrity range, i.e. fulfilment of the conditions for the onset of crack formation in laminae. Captions over the lines indicate the reason for the change of state or fracture of the material. Experimental results have been obtained in the testing of cylindrical tubular fiberglass plastic specimens in axial tension (marked by crosses) and in tension in the circumferential direction (marked by circles).

Failure of the material is due to fiber breakage $\sigma_1' = F_{+\parallel}$ when the angle α varies within the range $0 \leqslant \alpha \leqslant 20°$. At angles less than 4° the strain diagram is linear right up to failure and fracture is the only change in the state of the material. Qualitatively, deformation within the limits $5° \leqslant \alpha \leqslant 20°$ is quite different. According to calculations, at first the process of matrix cracking sets in, since stresses σ_6' attain the value of the ultimate shear strength $F_{\parallel\perp}$. In the calculation, the modulus $G_{\parallel\perp}$ becomes zero and $E_{\perp}$ remains constant because the stresses σ_2' are compressive. There is a 'knee' in the stress–strain diagram for such structures.

Within the range of angles $20° < \alpha < 45°$ the process of bearing strength exhaustion has two stages. At first, at a certain stress level, when $\sigma_6' = F_{\parallel\perp}$, the material loses its integrity and as a result its stiffness changes. The failure of the material within this range of angles can be accounted for by the compression of unidirectional layers in the direction perpendicular to the reinforcement direction, i.e. at $\sigma_2' = F_{-\perp}$. Typical strain diagrams are presented in Fig. 8. First of all, they are characterized by strict nonlinearity. A sharp change in the transverse strain coefficient is also of interest. If in the initial stage of deformation the relation $-e_2/e_1$ is equal to 0.67, then as a result of crack formation in the matrix it amounts to 1.3. On the whole, the theoretical model describes the aforementioned phenomenon correctly. The severest divergence between the results of calculation and experiment is observed at the sites of 'a knee' in theoretical diagrams. This can be attributed to a piecewise-linear approximation of stress–strain curves for a unidirectional material, which was assumed in the construction of the model.

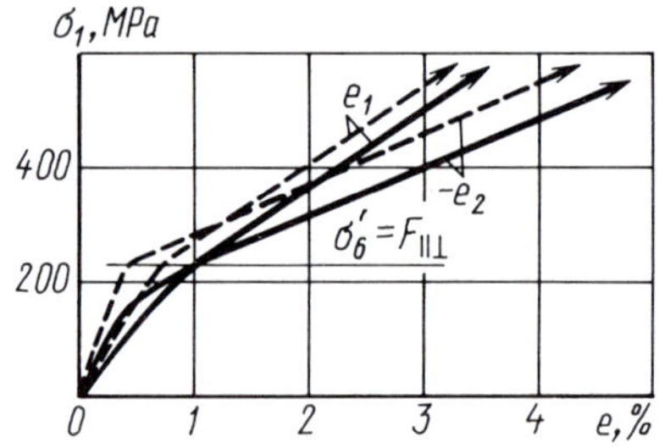

FIG. 8. Stress–strain diagrams of a fiberglass plastic [±30°].

The stress–strain diagrams of a composite reinforced at $\alpha = \pm 45°$ (Fig. 9) are outwardly similar to those for plastic metals. Thereby, the composite composed of layers, the ultimate strain of which in the unidirectional material does not exceed 0.6–3.5% (depending on the loading direction), undergoes deformation without failure up to the level of 9–10%. The effects of 'pseudo-plasticity' can successfully be employed in structural design.

In the final stage of deformation the observed increase in stiffness is caused by variation of the reinforcement angles. The theoretical curves in Fig. 9 were obtained in the case of 'deformational loading'.

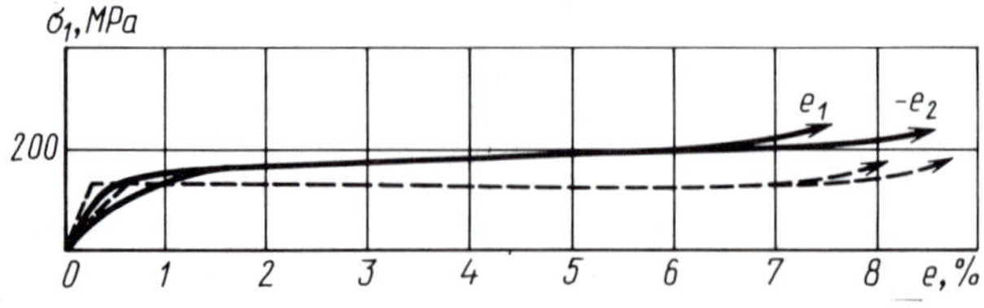

FIG. 9. Stress–strain diagrams of a fiberglass plastic [±45°].

With the angles range $45° < \alpha \leqslant 90°$, integrity disturbance leads to bearing strength exhaustion. Within the range of reinforcement patterns the materials are rather deformable (the above said pertains to the behavior of tubular specimens; plain specimens after integrity loss fail within this range of reinforcement angles). This type of bearing strength exhaustion in the material can be explained as a manifestation of instability of a deformation process.

Consider the deformation of multilayer composites in biaxial tension. In the discussion of features of such materials under loading the concept of 'a balanced' structure of reinforcement is of great significance. A structure is considered to be balanced if under the effect of a specified system of loads: σ_1, $\sigma_2 = k\sigma_1$, the material undergoes uniform tension $e_1 = e_2 = e_1' = e_2' = e$. It is easy to show that by neglecting the stiffness characteristics of a lamina in shear and perpendicular to the fibers we shall obtain uniform tension of the material at $\sigma_2 = k\sigma_1$ of angle-ply composite with a layup angle $\alpha = \operatorname{arctg}\sqrt{k}$.

The following conditions (ALFUTOV et al., 1984) can serve as a qualitative orienting point in forecasting the behavior of angle-ply reinforced materials in a plane state of stress at $\sigma_6 = 0$, $k > 0$.
Within the range $90° > \alpha \geqslant 45°$

$$k > \operatorname{tg}^2 \alpha \quad \text{at} \quad \sigma_1 > 0\,,$$

$$k < \operatorname{tg}^2 \alpha \quad \text{at} \quad \sigma_1 < 0\,.$$

Within the range of $45° > \alpha > 0°$

$$k < \operatorname{tg}^2 \alpha \quad \text{at} \quad \sigma_1 > 0\,,$$

$$k < \operatorname{tg}^2 \alpha \quad \text{at} \quad \sigma_1 < 0\,.$$

In satisfying the inequalities after the onset of crack formation in the matrix, the cracks are closed and the bearing strength of the system preserved.

In accordance with the inequalities, the reinforcement angles at which integrity loss does not result in bearing strength exhaustion for $k = 2$ are within the range of $45° \leqslant \alpha < 55°$. An equilibrium layup angle is close to 55°.

Quantitative data about the behavior of angle-ply reinforced materials in biaxial tension are presented in Fig. 10. Circles in Fig. 10 indicate the

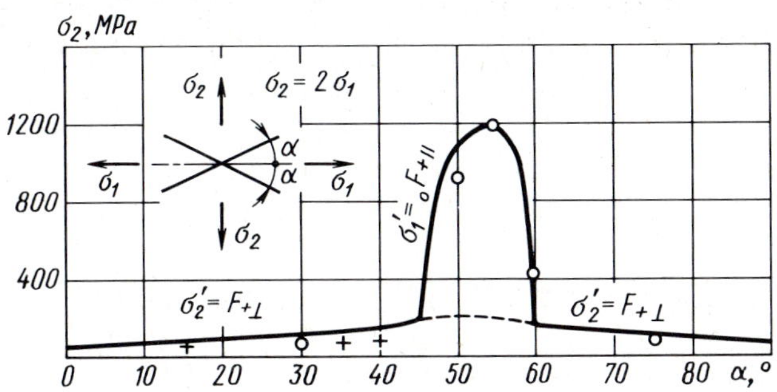

FIG. 10. Strength of an angle-ply fiberglass plastic in biaxial tension.

experimental results, obtained at $\sigma_2/\sigma_1 = 2$, crosses—at $\sigma_2/\sigma_1 = \frac{1}{2}$. In the latter case, the actual reinforcement angle in Fig. 10 has been substituted by a complementary angle $90° - \alpha$. Theoretical and experimental stress–strain diagrams for the material at $\alpha = 55°$ (Fig. 11) have two characteristic parts which are in good agreement. After the loss of integrity in the matrix (in the calculation, $\sigma_2' = F_{+\perp}$, $E_\perp = 0$) stress–strain diagrams are practically parallel straight lines. This testifies to the absence of increments of shear strains $\Delta e_6'$ in laminae, proportional to a change in the difference $e_2 - e_1$. Failure is accompanied by fiber breakage in tension.

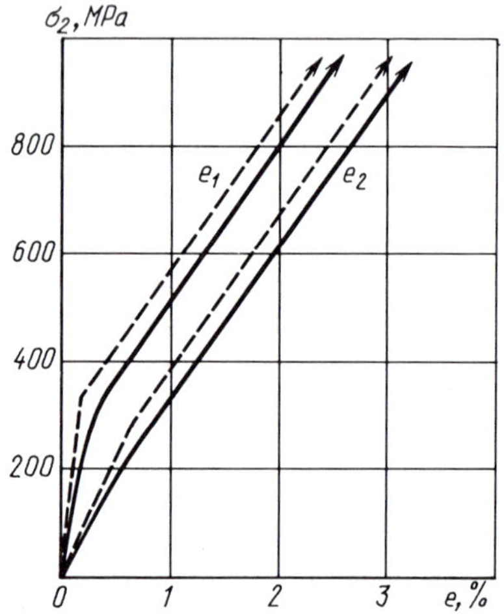

FIG. 11. Stress–strain diagram of a fiberglass plastic [±55°].

The stress–strain diagrams of fiberglass plastic with reinforcement angles $\alpha = \pm 50°$ have a more complex and quite different shape (Fig. 12). Up to the stress level of $\sigma_2 = 200$ MPa, the deformation proceeds, on the whole, analogously to that of the structure with $\alpha = \pm 55°$, but after the process of matrix failure has set in it sharply changes its character. The strain e_1 starts to decrease and then alters its sign. In the calculation, attaining of ultimate stresses $\sigma_2' = F_{+\perp}$ turns out to be the reason for such a sharp change in the material behavior. Just after the onset of crack formation in the matrix unloading at stresses σ_2' occurs and modulus of load relief $E_\perp$ is equal to the initial one. At the same time, the tangential shear modulus $G_{\parallel\perp}$ reduces to zero. The increase in shear compliance of the material permits the reinforcement angles to approach those for a balanced structure of $\alpha = \pm 55°$. The theoretical curves in Fig. 12 describe qualitatively the correct character of material deformation, but the abrupt decrease in the tangential modulus $G_{\parallel\perp}$ from the initial value to zero turns out to be excessively severe.

Let us note that alteration of the sign for increments of individual components of strain and stress tensors under constantly increasing load is charac-

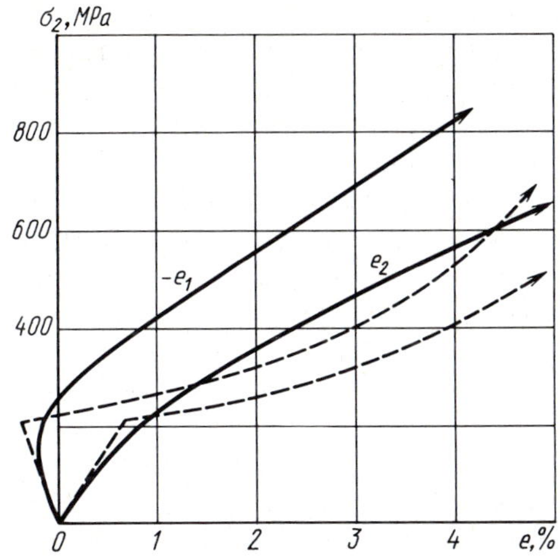

FIG. 12. Stress–strain diagram of a fiberglass plastic [±50°].

teristic in many cases. Upon changing the average stresses $\boldsymbol{\sigma}$ in accordance with the rules of a 'simple loading' when all the components of $\boldsymbol{\sigma}$ vary proportionally to some parameter, the laminae in the composite can be under the conditions of 'complex loading'.

For structures with layup angles within the ranges of $0° < \alpha < 45°$ and $60° < \alpha < 90°$ the matrix discontinuity leads to bearing strength exhaustion in the material. Further deformation of such materials is unstable and the increase in strains occurs at a constant load. This result has also been confirmed experimentally.

2. Laminated shells

2.1. Introduction

With the advent of advanced composites for load-bearing structures of a shell type the possibilities have arisen in the process of manufacture to achieve the necessary distributions of stiffness and strength characteristics through proper selection of materials, layer thicknesses and reinforcement directions. However, specific features of anisotropic and deformation characteristics are closely connected with the structure of composites as well as with the technology of manufacture in the first stages of transition to new materials and have interfered with the prediction of mechanical characteristics of such structures. At first, this can be accounted for by the fact that the rich industrial, experimental and theoretical experience which has been accumulated on structures made of isotropic materials turned out to be insufficient for the evaluation of reliability of new items. The urgent necessity of filling up this gap served as an impetus for the fast development of new theoretical and experimental research. Surveys of the principal directions and research results can be found in the works of AMBARTSUMYAN (1974), BOLOTIN and NOVICHKOV (1980), GRIGOL'UK and KOGAN (1972).

We will now be concerned with two approaches to obtaining approximate resolvent relations for laminated shells. The first approach presupposes an independent kinematic description of the deformation of isolated layers and their subsequent incorporation into a multilayer packet by taking the continuity conditions of displacement distribution into account. Three kinematic models of layers in shells, which allow both the stiff layers and transversally compliant layers to be approximately described, are presented. The laborious procedure of procuring resolvent systems of differential equations for shells of revolution has been automatized with the help of a variational-matrix method (see Subsection A.4 of the Appendix). The second approach involves unified hypotheses for the entire multilayer packet and, as a rule, is formulated on the basis of functionals of mixed type. In this case, the order of resolvent equations does not depend on the number of layers and is determined from initial assumptions. Such an approach is treated here in terms of the simplest model, permitting us to take the transverse shear strains in a laminated shell into account.

The suggested methods are intended for computer aided design.

2.2. *Strain–displacement relationships and geometric characteristics*

Let us associate the orthogonal system of curvilinear coordinates in Fig. 13 with a shell. As x_1-, x_2-lines we shall choose lines of curvature and direct the

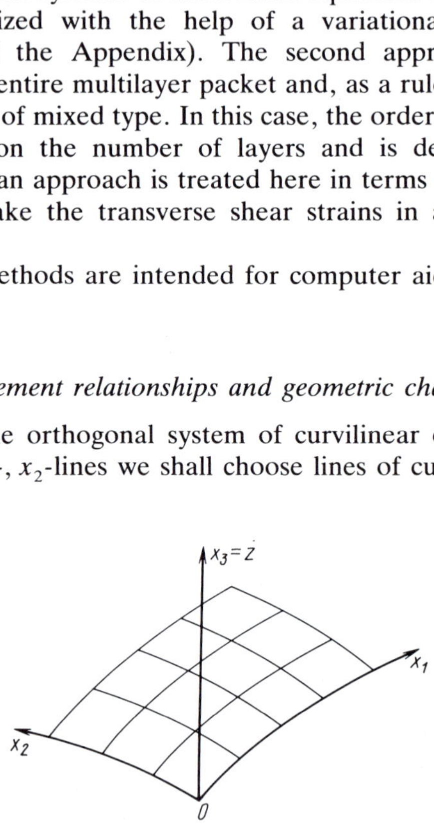

FIG. 13. Orthogonal system of curvilinear coordinates.

$x_3 = z$ axis along the outer normal. Let us designate displacements along the coordinate lines x_1, x_2, x_3 by v_1, v_2, v_3, respectively. Express the total deformations $\boldsymbol{e}$ as linear $\boldsymbol{\epsilon}$ and nonlinear $\boldsymbol{\eta}$ components, $\boldsymbol{e} = \boldsymbol{\epsilon} + \boldsymbol{\eta}$. The linear strain components $\boldsymbol{\epsilon} = \{\epsilon_1, \epsilon_2, \epsilon_3, \epsilon_4, \epsilon_5, \epsilon_6\}$ will be determined as

$$\epsilon_1 = \frac{1}{1+k_1 z}\left(\frac{1}{A_1} v_{1,1} + \varphi_{12} v_2 + k_1 v_3\right)(1 \rightleftarrows 2)\,,$$

$$\epsilon_3 = v_{3,3}\,,$$

$$\epsilon_4 = \frac{1}{1+k_2 z}\left(\frac{1}{A_2} v_{3,2} - k_2 v_2\right) + v_{2,3}(4, 2 \rightleftarrows 5, 1)\,, \tag{2.1}$$

$$\epsilon_6 = \frac{1}{1+k_2 z}\left(\frac{1}{A_2} v_{1,2} - \varphi_{21} v_2\right) + \frac{1}{1+k_1 z}\left(\frac{1}{A_1} v_{2,1} - \varphi_{12} v_1\right),$$

where

$$\varphi_{12} = \frac{1}{A_1 A_2}\frac{\partial A_1}{\partial x_2}(1 \rightleftarrows 2)\,, \quad (\cdot)_{,i} = \frac{\partial}{\partial x_i}(\cdot)\,,$$

A_1 and A_2 are Lame's parameters of the coordinate surface $z = 0$; k_1 and k_2 are the principal curvatures of a shell; $(4, 2 \rightleftarrows 5, 1)$ means that index 4 changes to 5 after substitution of the indices $(1 \rightleftarrows 2)$.

For nonlinear strain components $\boldsymbol{\eta} = \{\eta_1, \eta_2, \eta_3, \eta_4, \eta_5, \eta_6\}$ let us confine ourselves to a quadratic representation

$$\eta_1 = \frac{1}{2}\frac{1}{(1+k_1 z)^2}\left[\left(\frac{1}{A_1} v_{1,1} + \varphi_{12} v_2 + k_1 v_3\right)^2 + \left(\frac{1}{A_1} v_{2,1} - \varphi_{12} v_1\right)^2 + \left(\frac{1}{A_1} v_{3,1} - k_1 v_1\right)^2\right](1 \rightleftarrows 2)\,,$$

$$\eta_3 = \tfrac{1}{2}(v_{1,3}^2 + v_{2,3}^2 + v_{3,3}^2)\,,$$

$$\eta_4 = \frac{1}{1+k_2 z}\left[v_{2,3}\left(\frac{1}{A_2} v_{2,2} + \varphi_{21} v_1 + k_2 v_2\right) + v_{1,3}\left(\frac{1}{A_2} v_{1,2} - \varphi_{21} v_2\right) + v_{3,3}\left(\frac{1}{A_2} v_{3,2} - k_2 v_2\right)\right](4, 2 \rightleftarrows 5, 1)\,, \tag{2.2}$$

$$\eta_6 = \frac{1}{(1+k_1 z)(1+k_2 z)}\left[\left(\frac{1}{A_1} v_{1,1} + \varphi_{12} v_2 + k_1 v_3\right)\left(\frac{1}{A_2} v_{1,2} - \varphi_{21} v_2\right) + \left(\frac{1}{A_1} v_{2,1} - \varphi_{12} v_1\right)\left(\frac{1}{A_2} v_{2,2} + \varphi_{21} v_1 + k_2 v_3\right) + \left(\frac{1}{A_1} v_{3,1} - k_1 v_1\right)\left(\frac{1}{A_2} v_{3,2} - k_2 v_2\right)\right].$$

Let us decide upon the calculation of the necessary geometric characteristics entering (2.1) and (2.2). Let us limit ourselves to the treatment of shells of revolution. For a shell, obtained by rotation of a plane curve, prescribed by parametric functions $r(x)$, $t(x)$ (Fig. 14), the geometric characteristics are determined by the following relationships

$$A_1 = [(r')^2 + (t')^2]^{1/2}, \quad A_2 = r,$$

$$k_1 = \frac{r''t' - r't''}{A_1^3}, \quad k_2 = -\frac{t'}{A_1 A_2},$$

$$\sin\theta = -\frac{t'}{A_1}, \quad \cos\theta = \frac{r'}{A_1},$$

$$\varphi_{12} = 0, \quad \varphi_{21} = \frac{r'}{A_1 A_2}, \quad (\cdot)' = \frac{\mathrm{d}}{\mathrm{d}x}(\cdot).$$

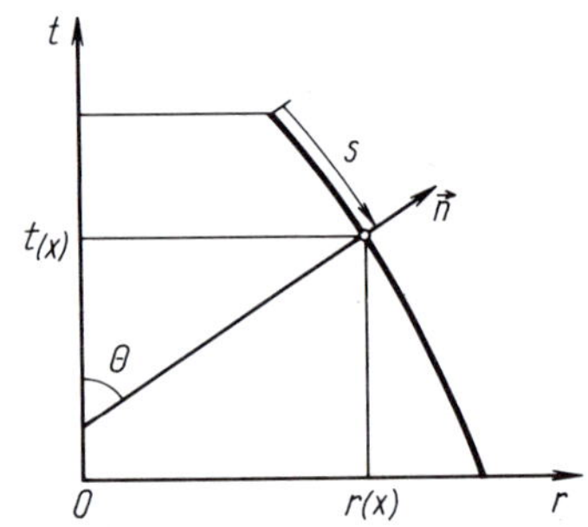

FIG. 14. The shape of a meridian in a shell of revolution.

One of the values t, r, θ, s is usually chosen as a parameter x (t is a coordinate along the rotation axis, r is the radius of a parallel, θ is the angle between the axis of rotation and a normal).

In a particular case, we shall have for a cone (Fig. 15)

$$x = s, \quad A_1 = 1, \quad A_2 = r_0 + s\cos\theta_0, \quad k_1 = 0,$$

$$k_2 = \frac{\sin\theta_0}{A_2}, \quad \sin\theta = \sin\theta_0, \quad \cos\theta = \cos\theta_0, \quad \varphi_{21} = \frac{\cos\theta_0}{A_2}.$$

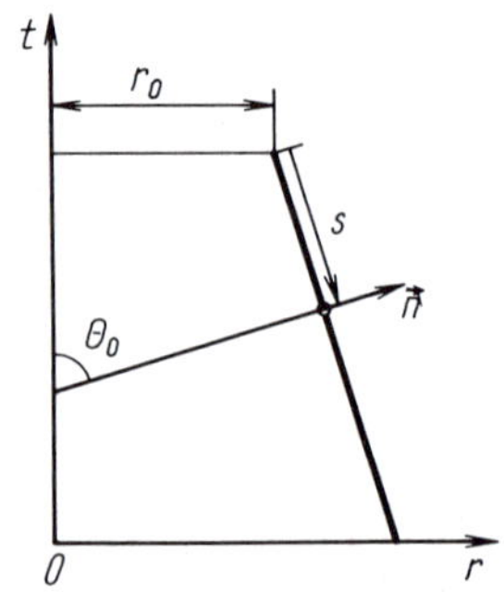

FIG. 15. Cone.

For the case of an elliptic torus (Fig. 16) we shall have

$$x = \theta\,,\quad A_1 = \frac{a^2b^2}{c^3}\,,\quad A_2 = d + \frac{a^2 \sin x}{c}\,,$$

$$k_1 = \frac{1}{A_1}\,,\quad k_2 = \frac{\sin x}{A_2}\,,\quad \sin\theta = \sin x\,,$$

$$\cos\theta = \cos x\,,\quad \varphi_{21} = \frac{\cos x}{A_2}\,,$$

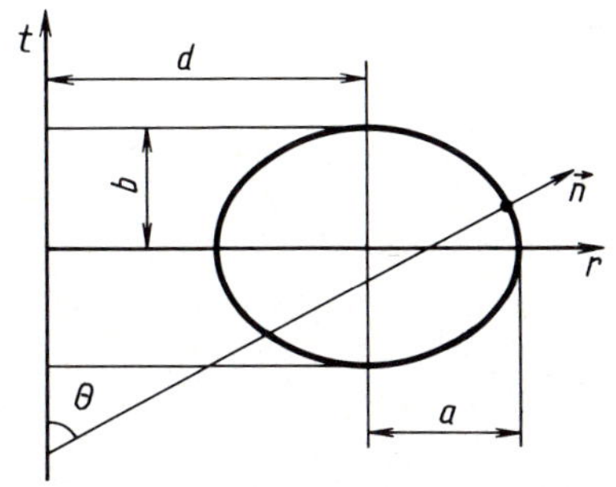

FIG. 16. Elliptic torus.

where

$$c = (a^2 \sin^2 x + b^2 \cos^2 x)^{1/2}\,.$$

In the case of specifying the shell element by coordinates r_1, r_2, t_1, t_2 and angles of inclination of a meridian φ_1, φ_2 (STRICKLIN et al., 1966) (Fig. 17) the

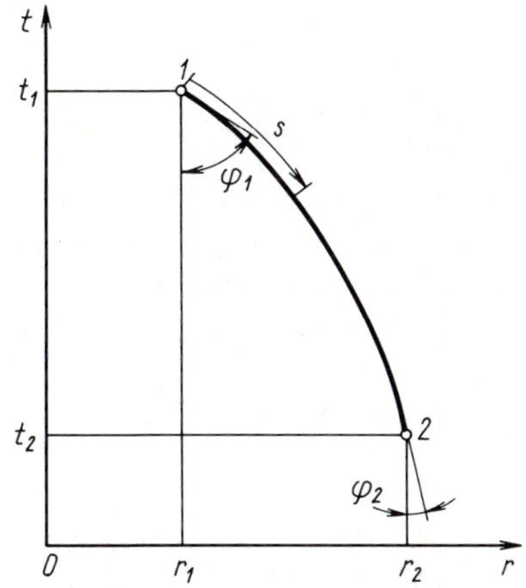

FIG. 17. An element of the shell of revolution.

geometric characteristics of a shell are determined as follows

$$x = s\,,\quad A_1 = 1\,,\quad A_2 = r\,,\quad k_1 = a_1 - 2a_2 s\,,$$

$$k_2 = \frac{\cos\varphi}{r}\,,\quad \sin\theta = \cos\varphi\,,\quad \cos\theta = \sin\varphi\,,\quad \varphi_{21} = \frac{\sin\varphi}{r}\,,$$

where

$$\varphi = a_0 + a_1 s + a_2 s^2\,,$$

$$r = c_0 + c_1 s + c_2 s^2 + c_3 s^3 + c_4 s^4 + c_5 s^5\,,$$

$$c_0 = r_1\,,$$

$$c_1 = b_1 \cos\varphi_c + (1 - \tfrac{1}{2} b_1^2) \sin\varphi_c\,,$$

$$c_2 = \tfrac{1}{2}(a_1 \cos\varphi_c - b_1 a_1 \sin\varphi_c)\,,$$

$$c_3 = \tfrac{1}{3}(a_2 \cos\varphi_c - b_1 a_2 \sin\varphi_c - \tfrac{1}{2} a_1^2 \sin\varphi_c)\,,$$

$$c_4 = -\tfrac{1}{4} a_1 a_2 \sin\varphi_c\,,\quad c_5 = -\tfrac{1}{10} a_2^2 \sin\varphi_c\,,\quad \operatorname{tg}\varphi_c = \frac{r_2 - r_1}{t_1 - t_2}\,,$$

$$a_0 = \varphi_1\,,\quad a_1 = \frac{6\varphi_c - 4\varphi_1 - 2\varphi_2}{L}\,,\quad a_2 = \frac{3\varphi_1 + 3\varphi_2 - 6\varphi_c}{L^2}\,,$$

$$L = l\left(1 + \frac{2b_1^2 - b_1 b_2 + 2b_2^2}{30}\right),\quad l = [(r_2 - r_1)^2 + (t_2 - t_1)^2]^{1/2}\,,$$

$$b_1 = \varphi_1 - \varphi_c\,,\quad b_2 = \varphi_2 - \varphi_c\,.$$

It should be noted that for such an approximation of the geometry of a general type of shell of revolution the curvature of a meridian k_1 can abruptly change on passing over from one element to another. This, upon rare division into elements, introduces a certain error in the geometric characteristics of a shell.

2.3. *Deformation of thin laminated shells*

For the description of deformation of thin shells, Kirchhoff–Love's kinematic hypothesis is assumed to be valid. According to this hypothesis the rectilinear elements of a shell, normal until deformation to the mid-surface, remain rectilinear and normal to the deformed mid-surface and maintain their length. With the help of expressions (2.1) and (2.2) this hypothesis can be written as

$$\epsilon_i + \eta_i = 0 \quad (i = 3, 4, 5)\,. \tag{2.3}$$

According to NOVOZHILOV (1948), an approximate solution of a system of nonlinear equations (2.3) can be obtained by means of the expansion of displacements v_i $(i = 1, 2, 3)$ into a power series relative to the argument z.

Upon retaining the first two terms $v_i = u_i + z\psi_i$ and under the further assumption that it is possible to neglect elongations, shears and squares of angles of rotation for a normal compared to unity, the following distribution of displacements through the thickness will be the solution of (2.3)

$$v_1 = u_1 + z\vartheta_1(1 \rightleftarrows 2)\,,$$
$$v_3 = w\,, \tag{2.4}$$

where

$$\vartheta_1 = -\frac{1}{A_1} w_{,1} + k_1 u_1 (1 \rightleftarrows 2)\,, \tag{2.5}$$

are angles of rotation of a normal.

For strain distributions through the thickness we shall obtain the following linear

$$\epsilon_1 = \varepsilon_1 + z\kappa_1(1 \rightleftarrows 2)\,,$$
$$\epsilon_6 = \gamma_{12} + z\kappa_{12}\,, \tag{2.6}$$

and nonlinear components

$$\eta_1 = \tfrac{1}{2}\vartheta_1^2(1 \rightleftarrows 2)\,,$$
$$\eta_6 = \vartheta_1\vartheta_2\,, \tag{2.7}$$

where

$$\varepsilon_1 = \frac{1}{A_1} u_{1,1} + \varphi_{12}u_2 + k_1 w(1 \rightleftarrows 2)\,,$$
$$\gamma_{12} = \omega_1 + \omega_2 \tag{2.8}$$
$$\omega_1 = \frac{1}{A_1} u_{2,1} - \varphi_{12}u_1(1 \rightleftarrows 2)\,,$$

determine the linear strains of the coordinate surface; and

$$\kappa_1 = \frac{1}{A_1}\vartheta_{1,1} + \varphi_{12}\vartheta_2 - \underline{k_1\varepsilon_1}(1 \rightleftarrows 2)\,,$$
$$\kappa_{12} = \tau_1 + \tau_2 + \underline{k_1\omega_2} + \underline{k_2\omega_1}\,, \tag{2.9}$$
$$\tau_1 = \frac{1}{A_1}\vartheta_{2,1} - \varphi_{12}\vartheta_1(1 \rightleftarrows 2)\,,$$

determine the change of curvature parameters. In the design of thin shells the underlined terms in (2.5) are often neglected.

To obtain resolvent equations for the problem of nonlinear deformation of a laminated shell, let us employ the linearized formulation of the principle of virtual displacements and consider each term entering (A.21) separately.

For transformation of the first term we shall designate increments of generalized strains by the column vector $\Delta\boldsymbol{\xi}$

$$\Delta\boldsymbol{\xi} = \{\Delta\epsilon_1, \Delta\epsilon_2, \Delta\epsilon_6, \Delta\kappa_1, \Delta\kappa_2, \Delta\kappa_{12}\}\ . \tag{2.10}$$

We shall determine the components $\Delta\boldsymbol{\xi}$ according to (2.6) and (2.9) as follows

$$\begin{aligned}
\Delta\epsilon_1 &= \Delta\varepsilon_1 + \vartheta_{1\tau}\Delta\vartheta(1\rightleftarrows 2)\ ,\\
\Delta\epsilon_6 &= \Delta\gamma_{12} + \vartheta_{1\tau}\Delta\vartheta_2 + \vartheta_{2\tau}\Delta\vartheta_1\ ,\\
\Delta\kappa_1 &= \frac{1}{A_1}\,\Delta\vartheta_{1,1} + \varphi_{12}\Delta\vartheta_2(1\rightleftarrows 2)\ ,\\
\Delta\kappa_{12} &= \Delta\tau_1 + \Delta\tau_2 + k_1\Delta\omega_1 + k_2\Delta\omega_2\ ,
\end{aligned}$$

where the index τ indicates that the given value has been estimated for an equilibrium configuration τ (see (A.3)). In the solution of linear problems, the values with the index τ are assumed to be zero. Components of the vector $\Delta\boldsymbol{\xi}$ are linearly related to increments of displacements $\Delta\boldsymbol{u} = \{\Delta u_1, \Delta u_2, \Delta w\}$. After integration through the thickness of a laminated shell the first term in (A.21) is expressed as

$$\int_V \Delta\hat{\boldsymbol{\epsilon}}^{\mathrm{T}}_{(m)}\boldsymbol{C}_\tau\Delta\boldsymbol{\epsilon}_{(m)}\,\mathrm{d}V = \int_S \Delta\hat{\boldsymbol{\xi}}^{\mathrm{T}}_{(m)}\boldsymbol{D}_\tau\Delta\boldsymbol{\xi}_{(m)}\,\mathrm{d}S\ ,$$

where $\Delta\boldsymbol{\xi}_{(m)} = \boldsymbol{L}_\tau\Delta\boldsymbol{u}_{(m)}$; $\mathrm{d}S = A_1A_2\,\mathrm{d}x_1\,\mathrm{d}x_2$; S is the surface of a shell, the matrix coefficients of differential operators $\boldsymbol{L}_\tau$ are determined from relationships (2.5), (2.8) and (2.9); $\boldsymbol{D}_\tau$ is a matrix of reduced stiffness characteristics for a thin laminated shell

$$\boldsymbol{D}_\tau = \begin{bmatrix} \boldsymbol{D}_{11} & \boldsymbol{D}_{12} \\ \text{sym.} & \boldsymbol{D}_{22} \end{bmatrix},$$

and the matrices $\boldsymbol{D}_{ij}$ are calculated as

$$(\boldsymbol{D}_{11}; \boldsymbol{D}_{12}; \boldsymbol{D}_{22}) = \sum_{k=1}^{n}\int_{z_{k-1}}^{z_k} \boldsymbol{C}^k_\tau(1; z; z^2)\,\mathrm{d}z\ . \tag{2.11}$$

Here n is the number of layers; z_{k-1}, z_k are z coordinates of the inner and outer surfaces of the kth layer; $\boldsymbol{C}^k_\tau$ is a matrix of tangential moduli of the kth layer, evaluated in the x_1 and x_2 axes

$$C_\tau^k = T_k^{\mathrm{T}}(C_k')_\tau T_k \ .$$

The matrix T_k has been given earlier (see (1.3)).

For a laminated composite $(C_k')_\tau$ is the matrix relating stress and strain increment for the kth layer, evaluated for the configuration τ (see Section 1.4).

The second term in the condition (A.21) will be written taking (2.7) into account, and then by integrating across the thickness we shall obtain

$$\int_V \Delta\hat{\boldsymbol{\eta}}_{(m)}\boldsymbol{\sigma}_\tau \,\mathrm{d}V = \int_S \Delta\hat{\boldsymbol{\vartheta}}^{\mathrm{T}}_{(m)}\boldsymbol{T}_\tau \Delta\boldsymbol{\vartheta}_{(m)} \,\mathrm{d}S \ ,$$

where $\Delta\boldsymbol{\vartheta} = \boldsymbol{J}\Delta\boldsymbol{u}$; $\Delta\boldsymbol{\vartheta} = \{\Delta\vartheta_1, \Delta\vartheta_2\}$;

$$\boldsymbol{T}_\tau = \begin{bmatrix} T_1 & T_6 \\ T_6 & T_2 \end{bmatrix}, \quad T_i = \int_h \sigma_i \,\mathrm{d}z \ (i = 1, 2, 6) \ .$$

The matrix $\boldsymbol{T}_\tau$ characterizes the state of stress in an equilibrium configuration τ. Components of a column vector of increments in angles of rotation are linearly related to the increments of displacements through a matrix $\boldsymbol{J}$, the coefficients of which are determined by relationship (2.5).

We shall write the third term in (A.21) by using the vector of increments of generalized strains

$$\int_v \Delta\hat{\boldsymbol{\epsilon}}^{(\mathrm{T})}_{(m)}\boldsymbol{\sigma}_{(m-1)} \,\mathrm{d}v = \int_S \Delta\hat{\boldsymbol{\xi}}^{\mathrm{T}}_{(m)}\boldsymbol{Q}_{(m-1)} \,\mathrm{d}S \ ,$$

where

$$\boldsymbol{Q}_{(m-1)} = \{\boldsymbol{N}, \boldsymbol{M}\}_{(m-1)} \ ,$$

$$\boldsymbol{N}_{(m-1)} = \int_h \boldsymbol{\sigma}_{(m-1)} \,\mathrm{d}z \ , \quad \boldsymbol{M}_{(m-1)} = \int_h \boldsymbol{\sigma}_{(m-1)} z \,\mathrm{d}z \ .$$

In the case of conservative surface loads $\boldsymbol{p}_{\tau+\Delta\tau}$ the work of external forces will be defined by the relationship

$$f_{(m)} = \int_S \Delta\hat{\boldsymbol{u}}^{\mathrm{T}}_{(m)}\boldsymbol{p}_{\tau+\Delta\tau} \,\mathrm{d}S \ .$$

In its final form, the variational formulation of the solution of the nonlinear problem of statics for a laminated shell in the mth approximation at the loading stage $\tau + \Delta\tau$ will be expressed as follows

$$\int_S (\Delta\hat{\boldsymbol{\xi}}^{\mathrm{T}}_{(m)}\boldsymbol{D}_\tau \Delta\boldsymbol{\xi}_{(m)} + \Delta\hat{\boldsymbol{\vartheta}}^{\mathrm{T}}_{(m)}\boldsymbol{T}_\tau \Delta\boldsymbol{\vartheta}_{(m)} + \Delta\hat{\boldsymbol{\xi}}^{\mathrm{T}}_{(m)}\boldsymbol{Q}_{(m-1)} - \Delta\hat{\boldsymbol{u}}^{\mathrm{T}}_{(m)}\boldsymbol{p}_{\tau+\Delta\tau})\,\mathrm{d}S = 0\,, \tag{2.12}$$

where

$$\Delta\boldsymbol{\xi}_{(m)} = \boldsymbol{L}_\tau \Delta\boldsymbol{u}_{(m)}\,, \quad \Delta\boldsymbol{\vartheta}_{(m)} = \boldsymbol{J}\Delta\boldsymbol{u}_{(m)}\,,$$

i.e. such displacements $\Delta\boldsymbol{u}_{(m)}$ are to be found which ensure fulfilment of the condition (2.12) for any virtual $\Delta\hat{\boldsymbol{u}}_{(m)}$. Formulation of the problem in the form of (2.12) permits us to derive resolvent differential equations, or with the help of direct methods to obtain an approximate solution, and accomplish the next step of the iteration process $\boldsymbol{u}_{(m)} = \boldsymbol{u}_{(m-1)} + \Delta\boldsymbol{u}_{(m)}$. After the completion of the iteration process, the coefficients of matrices $\boldsymbol{D}_\tau$, $\boldsymbol{L}_\tau$, $\boldsymbol{T}_\tau$ are specified and the new load increment defined. A general scheme of solving the nonlinear problem is presented in Fig. A.3 of the Appendix.

2.4. Transverse shear strains

In the cases when the relative thickness of a laminated shell (Fig. 18) is considerable and/or the material in some layers has lower transverse shear stiffness, the theory for shells based on Kirchhoff–Love's hypotheses introduces essential errors. For the design of shells a variety of refined theories based on other hypotheses have been developed. Let us consider the simplest

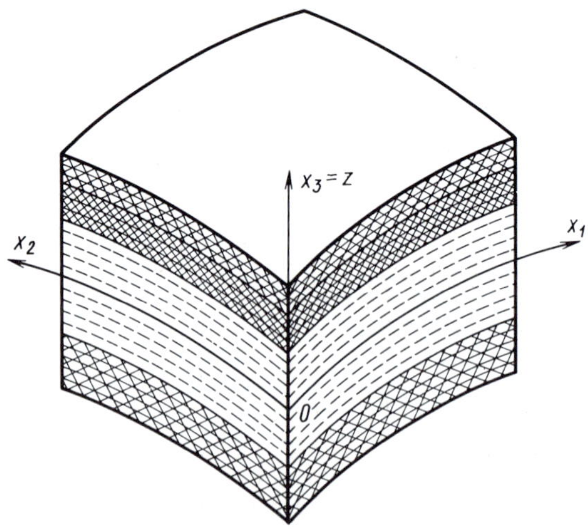

FIG. 18. An element of the laminated shell.

design method based on the hypothesis of linear through-the-thickness distribution of tangential displacements for the independent approximation of transverse shear stresses. In Fig. 19, the distribution of tangential displacements in the plane $(0zx_1)$ is presented. Here, ψ_1 is the angle of rotation of a section; ϑ_1 is the angle of rotation of a normal; and u_1 is the displacement of the coordinate surface. In Fig. 20 an illustrative distribution of transverse shear stresses is presented.

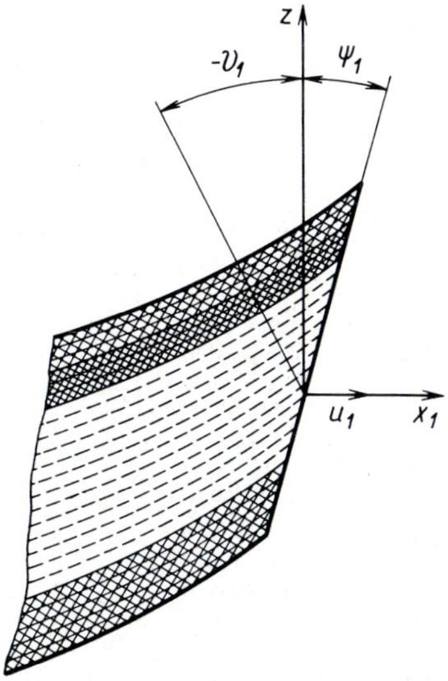

FIG. 19. Deformation of a multi-layer packet.

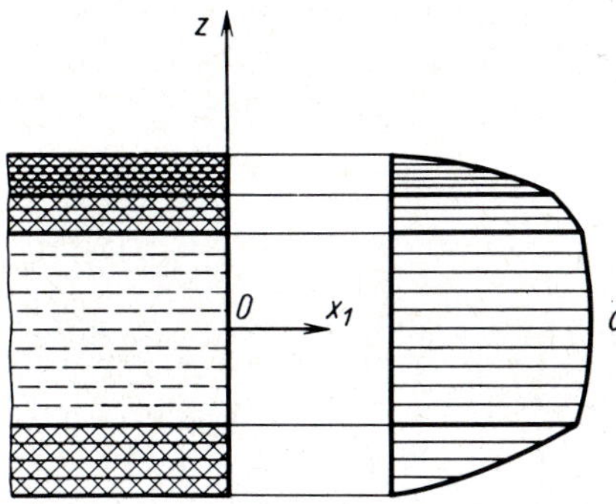

FIG. 20. Transverse shear stress distribution.

In accordance with the assumed kinetic hypotheses, let us take the next approximation in the form

$$v_1 = u_1 + z\psi_1 (1 \rightleftarrows 2) ;$$

$$v_3 = w .$$

By neglecting the change of metrics through the thickness of a shell, we shall get the following distribution of the respective linear strains from the general expressions (2.1)

$$\epsilon_1 = \varepsilon_1 + z\kappa_1 (1 \rightleftarrows 2) ,$$
$$\epsilon_6 = \gamma_{12} + z\kappa_{12} ,$$
$$\epsilon_4 = \gamma_{23} (4, 2 \rightleftarrows 5, 1) ,$$

where ε_1 and γ_{12} are determined from the relations (2.8);

$$\kappa_1 = \frac{1}{A_1} \psi_{1,1} - \varphi_{12}\psi_2 (1 \rightleftarrows 2) ,$$

$$\kappa_{12} = \frac{1}{A_2} \psi_{1,2} - \varphi_{21}\psi_2 + \frac{1}{A_1} \psi_{2,1} - \varphi_{12}\psi_1 ,$$

$$\gamma_{13} = \psi_1 - \vartheta_1 (1 \rightleftarrows 2) ,$$

where ϑ_1 is the angle of rotation of a normal (2.5). For nonlinear strain components we shall take into account only η_1, η_2, η_6 defined by the angle of rotation of the normal (2.7).

Besides kinematic hypotheses, we shall also employ an independent approximation of transverse shear stresses $\boldsymbol{\tau}$ which is specified as follows:

$$\boldsymbol{\tau} = \boldsymbol{fd} , \tag{2.13}$$

where $\boldsymbol{\tau} = \{\sigma_5, \sigma_4\}$ is a column vector of tangential stresses in transverse shear; $\boldsymbol{d}$ is a vector of indefinite coefficients; and $\boldsymbol{f} = \lceil f_{11}, f_{22} \rfloor$ is a diagonal matrix of approximating functions (RYABOV and RASSKAZOV, 1974)

$$f_{11} = \tfrac{1}{2} \int_{-h/2}^{z} (z - t_1)(c_{11} + c_{12} + 2c_{33}) \, \mathrm{d}z \; (1 \rightleftarrows 2) ,$$

where the parameter t_1 has been taken from the condition $f_{11}(h/2) = 0$; c_{11}, c_{12}, c_{33} are the coefficients of the stiffness matrix of an orthotropic laminated composite (in the $0x_1x_2$ plane) which are assumed to be known functions of the argument z and the loading history. By employing the variational formulation of Reissner's principle, analogous to (A.29), let us write

$$\int_h \hat{\boldsymbol{\tau}}^{\mathrm{T}} (\boldsymbol{\gamma} - \boldsymbol{C}_\gamma^{-1} \boldsymbol{\tau}) \, \mathrm{d}z = 0 , \tag{2.14}$$

where $\boldsymbol{\gamma} = \{\gamma_{13}, \gamma_{23}\}$; $\boldsymbol{C}_\gamma = \lceil G_{13}, G_{23} \rfloor$ is a diagonal matrix of transverse shear moduli. The integral condition (2.14) allows us to associate the indefinite coefficients of the vector $\boldsymbol{d}$ (see 2.13) with the transverse shear strains $\boldsymbol{\gamma}$

$$\boldsymbol{d} = \boldsymbol{H}_*^{-1}\boldsymbol{G}_*\boldsymbol{\gamma} , \tag{2.15}$$

where

$$\boldsymbol{H}_* = \int_h \boldsymbol{f}^{\mathrm{T}}\boldsymbol{C}_{\gamma}^{-1}\boldsymbol{f}\,\mathrm{d}z , \quad \boldsymbol{G}_* = \int_h \boldsymbol{f}^{\mathrm{T}}\,\mathrm{d}z .$$

Let us introduce a vector of independent generalized variables comprising, with the displacements of the coordinate surface u_1, u_2, w, the angles of rotation of cross sections ψ_1, ψ_2. Then, the increments of the generalized variables are given by

$$\Delta\boldsymbol{u} = \{\Delta u_1, \Delta u_2, \Delta w, \Delta\psi_1, \Delta\psi_2\} . \tag{2.16}$$

We shall express the vector of linear components of the generalized strains $\Delta\boldsymbol{\xi}$ by a set of the following components

$$\Delta\boldsymbol{\xi} = \{\Delta\epsilon_1, \Delta\epsilon_2, \Delta\epsilon_6, \Delta\kappa_1, \Delta\kappa_2, \Delta\kappa_{12}, \Delta\gamma_{13}, \Delta\gamma_{23}\} , \tag{2.17}$$

where

$$\Delta\epsilon_1 = \Delta\varepsilon_1 + \vartheta_{1\tau}\Delta\vartheta_1 \ (1 \rightleftarrows 2) ,$$

$$\Delta\epsilon_6 = \Delta\gamma_{12} + \vartheta_{1\tau}\Delta\vartheta_2 + \vartheta_{2\tau}\Delta\vartheta_1 ,$$

$$\Delta\kappa_1 = \frac{1}{A_1}\Delta\psi_{1,1} - \varphi_{12}\Delta\psi_2 \ (1 \rightleftarrows 2) ,$$

$$\Delta\kappa_{12} = \frac{1}{A_2}\Delta\psi_{1,2} - \varphi_{21}\Delta\psi_2 + \frac{1}{A_1}\Delta\psi_{2,1} - \varphi_{12}\Delta\psi_1 ,$$

$$\Delta\gamma_{13} = \Delta\psi_1 - \Delta\vartheta_1 \ (1 \rightleftarrows 2) .$$

Subsequently, the variational formulation of the solution of a nonlinear problem of statics for a laminated shell, taking into account the transverse shear strains for the mth approximation in the loading stage $\tau + \Delta\tau$ will be expressed analogously to (2.12). In the above, the vector representation of $\Delta\boldsymbol{u}$, $\Delta\boldsymbol{\xi}$ is to be understood in accordance with (2.16) and (2.17). The matrix of elasticity relations in the increments $\boldsymbol{D}_\tau$ will have the following structure:

$$\boldsymbol{D}_\tau = \begin{bmatrix} \boldsymbol{D}_{11} & \boldsymbol{D}_{12} & 0 \\ & \boldsymbol{D}_{22} & 0 \\ \text{sym.} & & \boldsymbol{D}_{33} \end{bmatrix} , \tag{2.18}$$

where the matrices $\boldsymbol{D}_{11}, \boldsymbol{D}_{12}, \boldsymbol{D}_{22}$ are defined by relationships (2.11), whereas the matrix of given stiffness characteristics in transverse shear, $\boldsymbol{D}_{33}$, is determined by taking into account (2.15)

$$\boldsymbol{D}_{33} = \boldsymbol{G}_*^{\mathrm{T}} \boldsymbol{H}_*^{-1} \boldsymbol{G}_* \,.$$

In the case of a uniform isotropic elastic material, the selected approximation of transverse stresses (2.13) represents a quadratic parabola and the non-zero coefficients of the diagonal matrix $\boldsymbol{D}_{33}$ will be equal to $5Gh/6$, where G is the shear modulus of the isotropic material.

2.5. Statics of laminated shells of revolution

Shells of revolution are among the most widely spread wound laminated composite structures. The mathematical analysis of deformation of laminated shells of revolution can be reduced to systems of ordinary differential equations. At present, effective numerical methods for the solution of such systems have been worked out by BIDERMAN (1977), GRIGORENKO (1973), MYACHENKOV and GRIGOR'EV (1981). The most convenient way for computer aided integration involves expressing the resolvent differential equations through a system of first order differential equations (or a canonical system). The variational-matrix method of obtaining such systems is shown in its general form in the Appendix (see (A.4)). Let us consider application of this method to laminated shells of revolution (ALFUTOV and POPOV, 1977; 1983; POPOV, 1979; 1982; POPOV and RAMAN, 1984).

Consider a shell, composed of n layers having differential mechanical characteristics. In the construction of the model of deformation for an isolated layer let us employ the kinematic approach, i.e. by specifying the laws of displacement variation along the normal coordinate z. Such a method offers the simplest way of reducing, in an approximate fashion the three-dimensional problem of elasticity theory to a two-dimensional one. Let us perform expansion of the displacements v, v_2, v_3 of the layer directed along coordinate lines x_1, x_2, z (Fig. 18) into a power series relative to the argument z. In the solution of a wide range of problems in the shell theory we can retain only a small number of the first terms of expansion. It is more convenient for the selected model of layer deformation that the displacement approximation should not be expressed through indefinite coefficients of a polynomial but through displacements and angles of rotation of characteristic (base) surfaces of the layer. In this case, the functions of arguments x_1, x_2 will have an obvious geometric sense.

Let us point out the several most widely used models of shells. Upon retaining the first expansion terms only, i.e. under the assumption that the tangential and normal displacements are constant through the thickness, we shall get the model of a momentless layer.

Upon retaining two terms of expansion for the tangential displacements v_1, v_2 and the first term for the normal displacement v_3 we shall obtain a refined model of a shell with uniform transverse shear strain distribution through the thickness (a generalization of S.P. Timoshenko's model of beams to shells)

$$\begin{aligned} v_1 &= u_1^{i-1}(1-\zeta) + u_1^i\zeta \ (1 \rightleftarrows 2)\,, \\ v_3 &= u_3^i = u_3^{i-1}\,, \end{aligned} \qquad (2.19)$$

where $\zeta = (z - z_{i-1})/h_i$; $h_i = z_i - z_{i-1}$ is the layer thickness; and i is the number of the base surface.

Under an additional condition of the absence of transverse shear strains we shall obtain the model of deformation corresponding to the Kirchhoff–Love hypotheses

$$\begin{aligned} v_1 &= u_1^{i-1} + (z - z_{i-1})\vartheta_1^{i-1} \ (1 \rightleftarrows 2)\,, \\ v_3 &= u_3^{i-1}\,, \end{aligned} \qquad (2.20)$$

where

$$\vartheta_1^{i-1} = -\frac{1}{H_1^{i-1}}\, u_{3,1}^{i-1} + k_1^{i-1}u_1^{i-1} \ (1 \rightleftarrows 2)\,,$$

$$H_1^{i-1} = A_1(1 + k_1 z_{i-1})\,, \quad (\cdot)_{,1} = \frac{\partial}{\partial x_1}(\cdot) \ (1 \rightleftarrows 2)\,,$$

k_1^{i-1}, k_2^{i-1} are curvatures of a base surface; and A_1, A_2 are Lame's parameters for a coordinate surface.

To take into account the deformation of transverse compression along with transverse shear strains in a transversally compliant layer in the first appproximation, the first two terms of the expansion should be retained in approximations of all components of displacement vectors

$$v_k = u_k^{i-1}(1-\zeta) + u_k^i\zeta \ (k = 1, 2, 3)\,. \qquad (2.21)$$

For substantiation of the classification of the layer models in terms of relations of elastic and geometric characteristics we can use energetic estimates (BOLOTIN and NOVICHKOV, 1980).

After expanding the displacement into a Fourier series relative to the angular coordinate x_2 we can write down the expression for deformation of an isolated layer. Here let us temporarily treat each layer as being isolated from other layers. Without taking nonlinear components into account, let us employ (2.1) and express formally the strain–displacement relationships for the ith layer in the following matrix form:

$$\boldsymbol{e}_i = \boldsymbol{L}_1^i\boldsymbol{X}_i + \boldsymbol{L}_2^i\boldsymbol{Y}_i \ (i = 1, 2, \ldots, n)\,, \qquad (2.22)$$

where the displacements and their derivatives involved in the essential (geometric) conditions of conjugation enter the vector $\boldsymbol{X}_i$. The components of the vector $\boldsymbol{Y}_i$ are the derivatives of the components of the vector $\boldsymbol{X}_i$ (or their combinations), which are necessary for constructing (2.22). A characteristic feature of deformations expressed in the form (2.22) is that the matrices $\boldsymbol{L}_1^i$, $\boldsymbol{L}_2^i$ do not contain differential operators, since differential operators relative to the coordinate x_2 are substituted by $\pm m$ (m is the number of expansion harmonics in the circumferential direction). The necessary differential operators relative to the coordinate x_1 enter the components of the vector $\boldsymbol{Y}_i$.

By employing the kinematic model of deformation of the ith layer, the distribution of displacements through the thickness can generally be expressed as follows:

$$\boldsymbol{v}_i = \boldsymbol{F}_1^i \boldsymbol{X}_i + \boldsymbol{F}_2^i \boldsymbol{Y}_i \,, \tag{2.23}$$

where $\boldsymbol{v}_i = \{v_1^i, v_2^i, v_3^i\}$ is a vector of displacements and $\boldsymbol{F}_1^i$ and $\boldsymbol{F}_2^i$ are matrices of the displacement distribution. By displacements, deformations and forces we shall understand their coefficients of expansion for the mth harmonic.

For the sake of clarity, the components of vector $\boldsymbol{e}_i$, $\boldsymbol{X}_i$, $\boldsymbol{Y}_i$ in (2.22) are given below for the deformation models under consideration.

For the model (2.19)

$$\boldsymbol{e}_i = \{\epsilon_1^i, \epsilon_2^i, \epsilon_6^i, \epsilon_5^i, \epsilon_4^i\} \,,$$

$$\boldsymbol{X}_i = \{u_1^{i-1}, u_1^i, u_2^{i-1}, u_2^i, u_3^{i-1}\} \,,$$

$$\boldsymbol{Y}_i = \{u_{1,1}^{i-1}, u_{1,1}^i, u_{2,1}^{i-1}, u_{2,1}^i, u_{3,1}^{i-1}\} \,.$$

For the model of layer deformation based on the Kirchhoff–Love hypotheses (2.20), the above vectors are now given by

$$\boldsymbol{e}_i = \{\epsilon_1^i, \epsilon_2^i, \epsilon_6^i\} \,,$$

$$\boldsymbol{X}_i = \{u_1^{i-1}, u_2^{i-1}, u_3^{i-1}, \vartheta_1^{i-1}\} \,,$$

$$\boldsymbol{Y}_i = \{u_{1,1}^{i-1}, u_{2,1}^{i-1}, \vartheta_{1,1}^{i-1}\} \,.$$

For a transversally compliant layer with the displacement approximation expressed in the form (2.21), the components of vectors $\boldsymbol{e}_i$, $\boldsymbol{X}_i$, $\boldsymbol{Y}_i$ will be given by

$$\boldsymbol{e}_i = \{\epsilon_1^i, \epsilon_2^i, \epsilon_3^i, \epsilon_6^i, \epsilon_5^i, \epsilon_4^i\} \,,$$

$$\boldsymbol{X}_i = \{u_1^{i-1}, u_2^{i-1}, u_3^{i-1}, u_1^i, u_2^i, u_3^i\} \,,$$

$$\boldsymbol{Y}_i = \{u_{1,1}^{i-1}, u_{2,1}^{i-1}, \vartheta_1^{i-1}, u_{1,1}^i, u_{2,1}^i, \vartheta_1^i\} \,.$$

For a transverally compliant layer with the displacement approximation expressed in the form (2.21), the components of vectors $\boldsymbol{e}_i$, $\boldsymbol{X}_i$, $\boldsymbol{Y}_i$ will be given by

$$\boldsymbol{e}_i = \{\epsilon_1^i, \epsilon_2^i, \epsilon_3^i, \epsilon_6^i, \epsilon_5^i, \epsilon_4^i\} ,$$
$$\boldsymbol{X}_i = \{u_1^{i-1}, u_2^{i-1}, u_3^{i-1}, u_1^i, u_2^i, u_3^i\} ,$$
$$\boldsymbol{Y}_i = \{u_{1,1}^{i-1}, u_{2,1}^{i-1}, \vartheta_1^{i-1}, u_{1,1}^i, u_{2,1}^i, \vartheta_1^i\} .$$

Construction of the matrices $\boldsymbol{L}_1^i, \boldsymbol{L}_2^i$ (2.22), which define the relationship between deformations and generalized displacements of a layer $\boldsymbol{X}_i$ and their derivatives $\boldsymbol{Y}_i$ (see (2.21)–(2.23)) is accomplished through substitution of the respective variants of displacement approximations into the relationship (2.1).

Prior to variational derivation of the resolvent system of differential equations it is necessary to make the transition from the generalized degrees of freedom for the individual layers to the generalized displacements for a multilayer packet as a whole. For this reason let us incorporate all the linearly independent generalized displacements $\boldsymbol{X}_i$ of individual layers into the vector $\boldsymbol{X}$ employing the condition of displacement continuity between adjacent layers. Furthermore, analogous incorporation of the vectors of derivatives $\boldsymbol{Y}_i$ into the vector $\boldsymbol{Y}$ we shall disregard those components, which already enter the vector $\boldsymbol{X}$ (for instance, the angles of rotation ϑ_1^{i-1} for the Kirchhoff–Love model are already included in the vector $\boldsymbol{X}_i$, for the model of a transversally compliant layer, on the other hand, they will be included in the vector $\boldsymbol{Y}_i$). The vectors $\boldsymbol{X}_i$ and $\boldsymbol{Y}_i$ for an isolated layer will be expressed in terms of the formulated generalized degrees of freedom for a multilayer packet $\boldsymbol{X}$ and their derivatives $\boldsymbol{Y}$ as follows

$$\boldsymbol{X}_i = \boldsymbol{K}_0^i \boldsymbol{X} , \quad \boldsymbol{Y}_i = \boldsymbol{K}_1^i \boldsymbol{X} + \boldsymbol{K}_2^i \boldsymbol{Y} . \tag{2.24}$$

The coefficient of the matrix $\boldsymbol{K}_k^i$ $(k = 0, 1, 2)$ depends on the actual structure of a multilayer packet. The relationships (2.24) permit us to determine $\boldsymbol{X}_i$ and $\boldsymbol{Y}_i$ for a particular layer from the respective vectors for a packet as a whole.

As has been noted earlier, the components of the vector $\boldsymbol{Y}$ represent the first derivatives of the components of generalized displacements vector $\boldsymbol{X}$ on their combinations. Generally, this differential coupling can be written as

$$\frac{1}{A_1} \boldsymbol{X}_{,1} - \boldsymbol{C}_1 \boldsymbol{X} - \boldsymbol{C}_2 \boldsymbol{Y} = 0 . \tag{2.25}$$

Now, we shall take into account an additional condition of integration in (2.25).

To obtain the resolvent equations let us employ the variational formulation of the principle of virtual displacements (A.5), which for a laminated shell of revolution we write down by taking the elasticity relations and the coupling condition (2.25) into account, and obtain

$$\int_{x_1^0}^{x_1^k}\int_0^{2\pi}\sum_{i=1}^{n}\int_{h_i}\hat{\boldsymbol{e}}_i^{\mathrm{T}}\boldsymbol{C}_i\boldsymbol{e}_iH_1^iH_2^i\,\mathrm{d}z\,\mathrm{d}x_2\,\mathrm{d}x_1$$

$$-\int_{x_1^0}^{x_1^k}\int_0^{2\pi}\sum_{i=0,\,n}\hat{\boldsymbol{u}}_i^{\mathrm{T}}\boldsymbol{p}_iH_1^iH_2^i\,\mathrm{d}x_2\,\mathrm{d}x_1$$

$$-\int_0^{2\pi}\sum_{i=1}^{n}\int_{h_i}\hat{\boldsymbol{v}}_i^{\mathrm{T}}\boldsymbol{\sigma}_b^iH_2^i\,\mathrm{d}z\,\mathrm{d}x_2\Big|_{x_1^0}^{x_1^k}$$

$$+\int_{x_1^0}^{x_1^k}\int_0^{2\pi}\Big(\frac{1}{A_1}\hat{\boldsymbol{X}}_{,1}-\boldsymbol{C}_1\hat{\boldsymbol{X}}-\boldsymbol{C}_2\hat{\boldsymbol{Y}})^{\mathrm{T}}\boldsymbol{\lambda}A_1A_2\,\mathrm{d}x_2\,\mathrm{d}x_1=0\,,$$

where x_1^0 and x_1^k are the initial and final coordinates x_1 of the shell under consideration; h_i is the thickness of the ith layer; $\hat{\boldsymbol{e}}_i$ is the vector of virtual strains of the ith layer; $\boldsymbol{C}_i$ is the matrix of coefficients of elasticity of the ith layer; $\boldsymbol{p}_i(i=0,n)$ are the vectors of surface loads acting on the inner and outer facing; $\boldsymbol{\lambda}$ is the vector of Lagrangian multipliers; $\boldsymbol{\sigma}_b^i=\{\sigma_1^i,\sigma_6^i,\sigma_5^i\}_b$ is the vector of end section stresses of the ith layer; and $\hat{\boldsymbol{v}}_i$ is the vector of virtual displacements of the ith layer.

After integrating out the x_2, z coordinates in expression (2.25), by taking (2.22) and (2.24) into account, we obtain the following formulation:

$$\begin{aligned}&\int_{x_1^0}^{x_1^k}(\hat{\boldsymbol{X}}^{\mathrm{T}}(\boldsymbol{S}_{11}\boldsymbol{X}+\boldsymbol{S}_{12}\boldsymbol{Y}-\boldsymbol{N}_1)+\hat{\boldsymbol{Y}}^{\mathrm{T}}(\boldsymbol{S}_{21}\boldsymbol{X}+\boldsymbol{S}_{22}\boldsymbol{Y}-\boldsymbol{N}_2)\\&\quad+\Big(\frac{1}{A_1}\hat{\boldsymbol{X}}_{,1}-\boldsymbol{C}_1\hat{\boldsymbol{X}}-\boldsymbol{C}_2\hat{\boldsymbol{Y}})^{\mathrm{T}}\boldsymbol{\lambda}\Big)A_1\,\mathrm{d}x_1\\&\quad-[\hat{\boldsymbol{X}}^{\mathrm{T}}\boldsymbol{\lambda}_1+\hat{\boldsymbol{Y}}^{\mathrm{T}}\boldsymbol{\lambda}_2]_{x_1^0}^{x_1^k}=0\,,\end{aligned}\tag{2.26}$$

where

$$\boldsymbol{S}_{jk}=A_2\sum_{i=1}^{n}\sum_{l=j}^{2}\sum_{m=k}^{2}(\boldsymbol{K}_{l+j-2}^i)^{\mathrm{T}}\boldsymbol{S}_{lm}^i\boldsymbol{K}_{m+k-2}^i\qquad j,k=1,2\,,$$

$$\boldsymbol{S}_{lm}^i=\int_{h_i}(\boldsymbol{L}_l^i)^{\mathrm{T}}\boldsymbol{C}_i\boldsymbol{L}_m^id_i\,\mathrm{d}z\qquad l,m=1,2\,,$$

$$d_i=(1+z_ik_1)(1+z_ik_2)\qquad i=1,2,\ldots,n\,,$$

$$N_k = A_2 \sum_{i=0,n} \sum_{m=k}^{2} (K^i_{k+m-2})^{\mathrm{T}} P^i_m d_i \quad k = 1, 2 ,$$

$$P^i_m = (F^i_m(z_i))^{\mathrm{T}} p_i \quad m = 1, 2 ,$$

$$\lambda_k = \sum_{i=1}^{n} \sum_{m=k}^{2} (K^i_{k+m-2})^{\mathrm{T}} \int_{h_i} (F^i_m)^{\mathrm{T}} \sigma^i_b H^i_2 \, \mathrm{d}z \quad k = 1, 2 .$$

Employing integration by parts, we do away with the derivatives $\hat{X}_{,1}$ in (2.26). By setting the vector multipliers for varying $\hat{X}$ and $\hat{Y}$ in the integrand equal to zero we obtain the following Euler's equations

$$S_{11}X + S_{12}Y - N_1 - C_1^{\mathrm{T}}\lambda - \frac{1}{A_1}\lambda_{,1} = 0 , \tag{2.27}$$

$$S_{21}X + S_{22}Y - N_2 - C_2^{\mathrm{T}}\lambda = 0 , \tag{2.28}$$

together with the boundary conditions at the end sections of the shell

$$\hat{X}^{\mathrm{T}}(\lambda - \lambda_1)|_{x_1^0}^{x_1^k} = 0 , \tag{2.29}$$

as well as an additional condition for the contour loads

$$\lambda_2 = 0 \text{ at } x_1 = x_1^0 \text{ and } x_1 = x_1^k .$$

It follows from the boundary conditions (2.29) that either the components of the vector X or the respective components of the vector λ should be specified at the end sections of a shell. In the physical sense, the vector of Lagrangian multipliers λ represents the vector of the coefficients of expansion of the generalized force factors in a Fourier series (for the mth harmonic) multiplied by Lame's coefficient A_2.

To derive a canonical system of differential equations let us eliminate the vector Y by (2.28), then the differential equations (2.25) and (2.27) will be of the form

$$\frac{1}{A_1}\begin{Bmatrix} X \\ \lambda \end{Bmatrix}_{,1} = \begin{bmatrix} A_{11} & A_{12} \\ A_{21} & A_{22} \end{bmatrix}\begin{Bmatrix} X \\ \lambda \end{Bmatrix} + \begin{Bmatrix} H_1 \\ H_2 \end{Bmatrix} , \tag{2.30}$$

where

$$\begin{gathered} A_{11} = C_1 + bS_{21} , \quad A_{12} = bC_2^{\mathrm{T}} , \\ A_{21} = S_{11} - dS_{21} , \quad A_{22} = -A_{11}^{\mathrm{T}} , \\ H_1 = bN_1 , \quad H_2 = -N_1 + dN_2 , \\ b = C_2 S_{22}^{-1} , \quad d = S_{21}^{\mathrm{T}} S_{22}^{-1} . \end{gathered} \tag{2.31}$$

The system of differential equations (2.30) obtained for the mth harmonic of the expansion consists of the components of the generalized displacements vector $\boldsymbol{X}$ and generalized forces $\boldsymbol{\lambda}$ which are unknown functions of the argument x_1. To calculate the matrix blocks of a resolvent system $\boldsymbol{A}_{11}$, $\boldsymbol{A}_{12}$, $\boldsymbol{A}_{21}$, $\boldsymbol{A}_{22}$ it is necessary to perform the operations specified by (2.31). We must have the following data: kinematic models of deformation of isolated layers $\boldsymbol{L}_1^i$, $\boldsymbol{L}_2^i$, $\boldsymbol{F}_1^i$, $\boldsymbol{F}_2^i$ (see (2.22) and (2.23)); elasticity relations $\boldsymbol{C}_i$; representation of the structure of a multilayer packet (matrices $\boldsymbol{K}_0^i$, $\boldsymbol{K}_1^i$, $\boldsymbol{K}_2^i$ see (2.24)); as well as coupling matrices $\boldsymbol{C}_1$, $\boldsymbol{C}_2$ (see (2.25)).

The suggested variational-matrix method permits us to use computer techniques for obtaining resolvent equations for laminated shells of revolution in terms of complex models of deformation.

2.6. *Stability and vibrations of laminated shells of revolution*

Let us consider the way of obtaining, by the variational-matrix method, a canonical system of differential equations for solution of stability and vibration problems. In deriving the resolvent equations let us assume that the shell is stressed but not deformed in the initial non-disturbed state. The initial state of stress is determined by solving the linear problem of statics. In deriving the equations of motion in the vicinity of the initial state let us take into account the initial state of stress. In deformation relations, besides the linear components, we take into account the nonlinear terms associated with additional angles of rotation of the normals. In solving the problem we shall treat only an axisymmetric initial state of stress. Let us assume that neither the value nor the direction of the external loads acting on the structure changes during the motion of the system. The system as a whole, including the external loads and the coupling conditions, will be considered as conservative. We shall investigate the motion of the system relative to its initial state without considering the damping characteristics. Just as in the previous section let us employ the kinematic approach to describe the models of deformation of isolated layers. We shall represent the distribution of additional displacements through the layer thickness analogously to (2.23)

$$\Delta \boldsymbol{v}_i = \boldsymbol{F}_1^i \boldsymbol{X}_i + \boldsymbol{F}_2^i \boldsymbol{Y}_i \,. \tag{2.32}$$

Let us relate the distribution of additional deformations specified by the first order infinitesimals to the degrees of freedom of the layer $\boldsymbol{X}_i$ and their derivatives $\boldsymbol{Y}_i$ analogously to (2.22)

$$\Delta \boldsymbol{\epsilon}_i = \boldsymbol{L}_1^i \boldsymbol{X}_i + \boldsymbol{L}_2^i \boldsymbol{Y}_i \,. \tag{2.33}$$

Furthermore, for the ith layer, we shall express the angles of rotation of the normal $\Delta\vartheta_1^i$, $\Delta\vartheta_2^i$ (see (2.5)) in terms of $\boldsymbol{X}_i$ and $\boldsymbol{Y}_i$. In general, this coupling can also be written in a matrix form

$$\Delta\boldsymbol{\vartheta}_i = \boldsymbol{R}_1^i\boldsymbol{X}_i + \boldsymbol{R}_2^i\boldsymbol{Y}_i \,, \tag{2.34}$$

where $\Delta\boldsymbol{\vartheta}_i = \{\Delta\vartheta_1^i, \Delta\vartheta_2^i\}$ and $\boldsymbol{R}_1^i, \boldsymbol{R}_2^i$ are matrices whose coefficients are filled for particular $\boldsymbol{X}_i$ and $\boldsymbol{Y}_i$ by using relationships (2.5).

Assembling layers into a unified multilayer packet is performed with the help of matrices $\boldsymbol{K}_0^i, \boldsymbol{K}_1^i, \boldsymbol{K}_2^i$ (2.24). Consider the obvious differential coupling (2.25) between the degrees of freedom of a multilayer packet $\boldsymbol{X}$ and their derivatives $\boldsymbol{Y}$.

Let us consider the harmonic vibrations of a laminated shell relative to the initial state of stress. To obtain the resolvent equations we shall employ a linearized formulation of the principle of virtual displacements (A.26), which is supplemented by the work of inertial forces

$$\int_V (\Delta\hat{\boldsymbol{\epsilon}}^{\mathrm{T}}\boldsymbol{C}\Delta\boldsymbol{\epsilon} + \boldsymbol{\Pi}\Delta\hat{\boldsymbol{\eta}}^{\mathrm{T}}\boldsymbol{\sigma}_0 - \omega^2\rho\Delta\hat{\boldsymbol{V}}^{\mathrm{T}}\Delta\boldsymbol{V})\,\mathrm{d}V = 0 \,, \tag{2.35}$$

where $\boldsymbol{\Pi}$ is a parameter of loading; $\Delta\boldsymbol{\eta}$ are additional deformations specified by the second order infinitesimals; ω is circular frequency; ρ is specific density of the material.

Let us write down the variational formulation (2.35) for a laminated shell of revolution with the condition of coupling taken into account (2.25)

$$\begin{aligned}
&\int_{x_1^0}^{x_1^k}\int_0^{2\pi}\sum_{i=1}^{n}\int_{h_i} (\Delta\hat{\boldsymbol{\epsilon}}_i^{\mathrm{T}}\boldsymbol{C}_i\Delta\boldsymbol{\epsilon}_i + \boldsymbol{\Pi}\Delta\hat{\boldsymbol{\eta}}_i^{\mathrm{T}}\boldsymbol{\sigma}_0^i \\
&\qquad - \omega^2 p_i\Delta\hat{\boldsymbol{V}}_i^{\mathrm{T}}\Delta\boldsymbol{V}_i)H_1^iH_2^i\,\mathrm{d}z\,\mathrm{d}x_2\,\mathrm{d}x_1 \\
&\int_{x_1^0}^{x_1^k}\int_0^{2\pi}\Big(\frac{1}{A_1}\hat{\boldsymbol{X}}_{,1} - \boldsymbol{C}_1\hat{\boldsymbol{X}} - \boldsymbol{C}_2\hat{\boldsymbol{Y}}\Big)^{\mathrm{T}}\boldsymbol{\lambda}A_1A_2\,\mathrm{d}x_1\,\mathrm{d}x_2 = 0\,.
\end{aligned} \tag{2.36}$$

For additional deformations specified by the second order infinitesimals let us confine ourselves to the values, associated with the angles of rotation of the normals and assume that

$$\begin{aligned}
&\Delta\boldsymbol{\eta}_i = \{\Delta\eta_1^i, \Delta\eta_2^i, \Delta\eta_6^i\}\,, \\
&\Delta\eta_1^i = \tfrac{1}{2}(\Delta\vartheta_1^i)^2\ (1\rightleftarrows 2)\,, \quad \Delta\eta_6^i = \Delta\vartheta_1^i\Delta\vartheta_2^i\,.
\end{aligned} \tag{2.37}$$

Taking (2.32)–(2.34) and (2.37) into account, and performing the necessary integrations in (2.36), we obtain the following variational formulation of the problem

$$\int_{x_1^0}^{x_1^k} \left(\hat{\boldsymbol{X}}^{\mathrm{T}}\left(\boldsymbol{S}_{11}\boldsymbol{X} + \boldsymbol{S}_{12}\boldsymbol{Y} - \boldsymbol{C}_1^{\mathrm{T}}\boldsymbol{\lambda} - \frac{1}{A_1}\boldsymbol{\lambda}_{,1}\right)\right.$$
$$\left. + \hat{\boldsymbol{Y}}^{\mathrm{T}}(\boldsymbol{S}_{21}\boldsymbol{X} + \boldsymbol{S}_{22}\boldsymbol{Y} - \boldsymbol{C}_2^{\mathrm{T}}\boldsymbol{\lambda})\right) A_1\,\mathrm{d}x_1 - \boldsymbol{X}^{\mathrm{T}}\boldsymbol{\lambda}\Big|_{x_1^0}^{x_1^k} = 0\,, \tag{2.38}$$

where

$$\boldsymbol{S}_{jk} = A_2 \sum_{i=1}^{n} \sum_{l=j}^{2} \sum_{m=k}^{2} (\boldsymbol{K}^i_{l+j-2})^{\mathrm{T}} \boldsymbol{S}^i_{lm} \boldsymbol{K}^i_{m+k-2} \quad j, k = 1, 2\,,$$
$$\boldsymbol{S}^i_{lm} = \int_{h_i} ((\boldsymbol{L}^i_l)^{\mathrm{T}} \boldsymbol{C}_i \boldsymbol{L}^i_m + \boldsymbol{\Pi}(\boldsymbol{R}^i_l)^{\mathrm{T}} \boldsymbol{T}^i_0 \boldsymbol{R}^i_m$$
$$- \omega^2 \rho_i (\boldsymbol{F}^i_l)^{\mathrm{T}} \boldsymbol{F}^i_m) d_i\,\mathrm{d}z \quad l, m = 1, 2\,, \tag{2.39}$$

$T^i_0 = \lceil \sigma^0_{1i}, \sigma^0_{2i} \rfloor$ is a diagonal matrix of initial stresses (axisymmetric state of stress).

By using Euler equations in (2.38) together with the condition of coupling (2.25) we derive the desired canonical system

$$\frac{1}{A_1}\begin{Bmatrix} \boldsymbol{X} \\ \boldsymbol{\lambda} \end{Bmatrix}_{,1} = \begin{bmatrix} \boldsymbol{A}_{11} & \boldsymbol{A}_{12} \\ \boldsymbol{A}_{21} & \boldsymbol{A}_{22} \end{bmatrix} \begin{Bmatrix} \boldsymbol{X} \\ \boldsymbol{\lambda} \end{Bmatrix} \tag{2.40}$$

along with the homogeneous boundary conditions $\boldsymbol{X}^{\mathrm{T}}\boldsymbol{\lambda}|_{x_1^0}^{x_1^k} = 0$, which define the associated eigenvalue problem. The matrix blocks $\boldsymbol{A}_{ij}$ of the system (2.40) are calculated from relationships (2.31), whereas the matrices $\boldsymbol{S}^i_{lm}$—by taking the initial stresses and inertial forces into account according to (2.39).

The derived resolvent system of differential equations allows one, for the mth harmonic of expansion, to solve the following problems: to determine the critical loads ($\boldsymbol{\Pi}_{\mathrm{cr}}$) at $\omega = 0$, $\boldsymbol{T}_0 \neq 0$, the natural frequencies of vibration at $\boldsymbol{T}_0 = 0$ and frequencies of vibration at $\boldsymbol{\Pi} = 1$, $\boldsymbol{T}_0 \neq 0$ in the presence of preloading

2.7. Axisymmetric nonlinear problems in statics of thin laminated shells

In a number of applications of thin-walled composite structures large displacements are possible which should be taken into account in the design for geometric nonlinearity. Also, structural changes in individual layers associated with crack formation caused by high shear and transverse stresses become imminent. Generation of initial cracks in a laminated composite results in the redistribution of stresses in the individual layers and specifies the composites physical nonlinearity. For this reason, there is a need to perform the calculation with a more detailed consideration of the internal mechanisms of deformation.

Consider a thin laminated shell of revolution under the action of axisymmetric loads. Let us obtain the initial matrices for the solution of a physically and geometrically nonlinear problem of deformation for such a shell by the method of finite elements (ZINOV'EV et al., 1980). We shall employ the method of step-by-step loading and perform integration for the previous equilibrium configuration (see Appendix A.2). For convenience let us pass over to the following designations

$$x = x_1\,, \quad A = A_1\,, \quad B = A_2\,, \quad \psi = \varphi_{21}\,,$$

$$u_{(z)} = v_1\,, \quad w = v_3\,, \quad u = u_1\,.$$

For the description of the kinematic model of deformation we shall use Kirchhoff–Love hypotheses. Consequently, the distribution of the through-the-thickness displacements will be defined by the expressions

$$u_{(z)} = u + z\vartheta_1\,,$$

$$w_{(z)} = w\,,$$

whereas the distribution of the remaining deformations will be given by

$$e_1 = \varepsilon_1 + z\kappa_1 + \tfrac{1}{2}\vartheta_1^2\,,$$

$$e_2 = \varepsilon_2 + z\kappa_2\,,$$

where

$$\vartheta_1 = -\frac{1}{A}\frac{\mathrm{d}w}{\mathrm{d}x} + k_1 u\,,$$

$$\varepsilon_1 = \frac{1}{A}\frac{\mathrm{d}u}{\mathrm{d}x} + k_1 w\,, \quad \varepsilon_2 = \psi u + k_2 w\,,$$

$$\kappa_1 = \frac{1}{A}\frac{\mathrm{d}\vartheta}{\mathrm{d}x}\,, \quad \kappa_2 = \psi\vartheta_1\,, \quad \psi = \frac{1}{AB}\frac{\mathrm{d}B}{\mathrm{d}x}$$

and u and, w are the tangential and normal displacements of the coordinate surface $z = 0$ of the shell configuration at the time instant τ. The principal curvatures k_1, k_2 and Lame's parameters are also assumed to be known for the previous equilibrium state.

Within each individual finite element let us approximate the displacements u and w by linear and cubic polynomials

$$u = a_1 + a_2 x\,,$$
$$w = a_3 + a_4 x + a_5 x^2 + a_6 x^3\,. \tag{2.41}$$

Let us express the indefinite coefficients

$$\boldsymbol{a} = \{a_1, a_2, a_3, a_4, a_5, a_6\}$$

in terms of the generalized displacements of the finite elements and sections

$$\boldsymbol{q} = \{u^1, w^1, \vartheta_1^1, u^2, w^2, \vartheta_1^2\}$$

(Fig. 21) as follows

$$\boldsymbol{a} = \boldsymbol{c}\boldsymbol{q}\ , \tag{2.42}$$

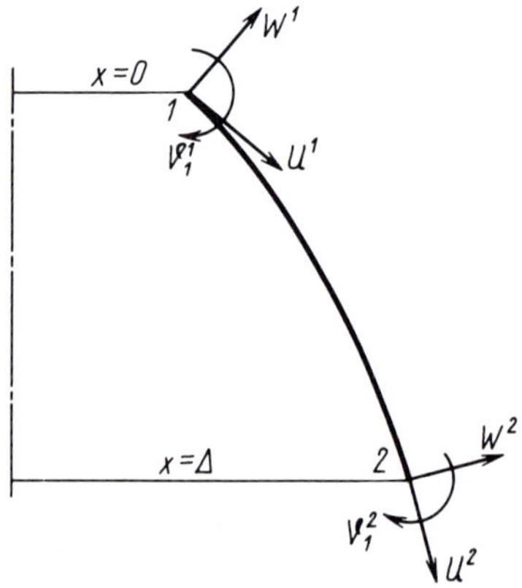

FIG. 21. Generalized displacements for a finite element of a thin shell upon axisymmetric deformation.

where

$$\boldsymbol{C} = \begin{bmatrix} 1 & 0 & 0 & 0 & 0 & 0 \\ -1/\Delta & 0 & 0 & 1/\Delta & 0 & 0 \\ 0 & 1 & 0 & 0 & 0 & 0 \\ A^0 k_1^0 & 0 & -A^0 & 0 & 0 & 0 \\ -\dfrac{4A^0k_1^0}{\Delta} & -\dfrac{3}{\Delta^2} & \dfrac{4A^0}{\Delta} & \dfrac{A^\Delta k_1^\Delta}{\Delta} & \dfrac{3}{\Delta^2} & \dfrac{A^\Delta}{\Delta} \\ \dfrac{3A^0k_1^0}{\Delta^2} & \dfrac{2}{\Delta^3} & -\dfrac{3A^0}{\Delta^2} & \dfrac{A^\Delta k_1^\Delta}{\Delta^2} & -\dfrac{2}{\Delta^3} & -\dfrac{A^\Delta}{\Delta^2} \end{bmatrix} .$$

Here, the upper indices '0' and 'Δ' are associated with the $x = 0$ and $x = \Delta$, respectively. The parameter Δ characterizing the length of the element is determined for the configuration τ (see (A.2)).

Approximation of the displacements given by (2.41), taking (2.42) into account, allows one to write down the distribution of the kinematic displacement characteristics within the boundaries of one element in the following form:

$$\boldsymbol{u} = \boldsymbol{\phi} \boldsymbol{q} \,,$$

where

$$\boldsymbol{u} = \{u, w\} \,, \quad \boldsymbol{\phi} = \boldsymbol{N} \boldsymbol{c}$$

and the angle of rotation of a normal as:

$$\boldsymbol{\vartheta} = \boldsymbol{J} \boldsymbol{q} \,,$$

where

$$\boldsymbol{\vartheta} = \{\vartheta_1\} \,, \quad \boldsymbol{J} = \boldsymbol{N}_\vartheta \boldsymbol{c} \,,$$

and the linear deformations and changes of curvatures are related in the following manner:

$$\boldsymbol{\xi} = \boldsymbol{B} \boldsymbol{q} \,,$$

where

$$\boldsymbol{\xi} = \{\varepsilon_1, \varepsilon_2, \kappa_1, \kappa_2\} \,; \quad \boldsymbol{B} = \boldsymbol{B}_{\mathrm{a}} \boldsymbol{c} \,.$$

The matrices $\boldsymbol{N}$, $\boldsymbol{N}_\vartheta$ and $\boldsymbol{B}_{\mathrm{a}}$ are presented below.
The matrix of displacements

$$\boldsymbol{N} = \begin{bmatrix} 1 & x & 0 & 0 & 0 & 0 \\ 0 & 0 & 1 & x & x^2 & x^3 \end{bmatrix} .$$

The matrix of angles of rotation

$$\boldsymbol{N}_\vartheta = [k_1 \;\; k_1 x \;\; 0 \;\; -1/A \;\; -2x/A \;\; -3x^2/A] \,.$$

The matrix of linear deformations and changes of curvatures.

$$\boldsymbol{B}_{\mathrm{a}} = \begin{bmatrix} 0 & 1/A & k_1 & k_1 x & k_1 x^2 & k_1 x^3 \\ \psi & \psi_x & k_2 & k_2 x & k_2 x^2 & k_2 x^3 \\ k_1'/A & (k_1' x + k_1)/A & 0 & A'/A^3 & 2A'x/A & 3A'x^2/A^3 - 6x/A^2 \\ \psi k_1 & \psi k_1 x & 0 & -\psi/A & -2\psi x/A & -\psi x^2/A \end{bmatrix} .$$

Here $(\cdot)' = (\cdot)_{,x}$.

Such approximations are also applied to the increments of aforementioned values for the mth iteration of the calculation process within the loading range $(\tau, \tau + \Delta\tau)$.

The above obtained matrices $\boldsymbol{\phi}$, $\boldsymbol{J}$ and $\boldsymbol{B}$ provide a relationship between the increments of reaction $\Delta \boldsymbol{t}^e_{(m)}$ and the increments of generalized displacements $\Delta \boldsymbol{q}^e_{(m)}$ of an element e for the mth iteration. For this reason, one should employ the formulation of the equilibrium condition analogous to (2.12). After integrating we have

$$\Delta \boldsymbol{t}^e_{(m)} = (\boldsymbol{K}^e_\tau + \boldsymbol{S}^e_\tau)\Delta \boldsymbol{q}^e_{(m)} + \boldsymbol{F}^e_{(m-1)} - \boldsymbol{P}_{\tau+\Delta\tau} , \tag{2.43}$$

where

$$\boldsymbol{K}^e_\tau = \int_0^\Delta \boldsymbol{B}^{\mathrm{T}} \boldsymbol{D}_\tau \boldsymbol{B} AB \,\mathrm{d}x , \tag{2.44}$$

$$\boldsymbol{S}^e_\tau = \int_0^\Delta \boldsymbol{J}^{\mathrm{T}} \boldsymbol{T}_\tau \boldsymbol{J} AB \,\mathrm{d}x , \tag{2.45}$$

$$\boldsymbol{F}^e_{(m-1)} = \int_0^\Delta \boldsymbol{B}^{\mathrm{T}} \boldsymbol{Q}_{(m-1)} AB \,\mathrm{d}x , \tag{2.46}$$

$$\boldsymbol{P}^e_{\tau+\Delta\tau} = \int_0^\Delta \boldsymbol{\phi}^{\mathrm{T}} \boldsymbol{p}_{\tau+\Delta\tau} AB \,\mathrm{d}x . \tag{2.47}$$

In computing the stiffness matrix $\boldsymbol{K}^e_\tau$, symmetric matrix $\boldsymbol{D}_\tau$ is involved which establishes the relationship between the increments of fractional forces and moments with the increments of linear deformations and changes of curvatures. This matrix has the following structure:

$$\boldsymbol{D}_\tau = \begin{bmatrix} \boldsymbol{D}_{11} & \boldsymbol{D}_{12} \\ \text{sym.} & \boldsymbol{D}_{22} \end{bmatrix} .$$

The matrix blocks $\boldsymbol{D}_{ij}$ are calculated as follows

$$(\boldsymbol{D}_{11}, \boldsymbol{D}_{12}, \boldsymbol{D}_{22}) = \sum_{k=1}^{n} \int_{z_{k-1}}^{z_k} \boldsymbol{C}^k_\tau (1, z, z^2) \,\mathrm{d}z ,$$

where n is the number of layers in a laminated shell; z_{k-1} and z_k are the coordinates z of the inner and outer surfaces of the kth layer; $\boldsymbol{C}^k_\tau$ is a matrix of tangential moduli of the kth layer; transformed for the x_1, x_2 axes of the shell

$$\boldsymbol{C}^k_\tau = \underset{(3\times 2)}{\boldsymbol{T}}{}^{\mathrm{T}}_k (\boldsymbol{C}'_k)_\tau \underset{(3\times 2)}{\boldsymbol{T}}{}_k .$$

Here, the matrix $\underset{(3\times 2)}{T}{}_k$ necessary for recalculation of stiffness characteristics of the kth layer upon transition from the system of coordinates of a layer to that of a shell in accordance with (1.3) has the components

$$\underset{(3\times 2)}{T}{}_k = \begin{bmatrix} c_k^2 & s_k^2 \\ s_k^2 & c_k^2 \\ -2s_k c_k & 2s_k c_k \end{bmatrix}, \quad \begin{matrix} c_k = \cos \alpha_k \,, \\ s_k^2 = \sin \alpha_k \,, \end{matrix}$$

where α_k is the layup angle of the kth layer which is read off from the meridian of the shell. The coefficients of the matrix C_k' characterizing the stiffness of a unidirectional composite have been treated in detail in Sections 1.2 and 1.3.

The matrix T_τ used in computation of the matrix of consistent stresses S_τ (2.45) contains only one coefficient which is equal to the linear meridian force T_1^τ at the time instant τ.

The vector of inner linear force factors $Q_{(m-1)}$ which is necessary for computation of the vector $F_{(m-1)}$ (2.46) has the following components:

$$Q_{(m-1)} = \{T_1, T_2, M_1, M_2\}_{(m-1)} \,,$$

i.e. the overall linear meridian and circumferential forces T_1, T_2 and moments M_1, M_2 which had been computed during the previous $(m-1)$th iteration

$$T_{1(m-1)} = \sum_{k=1}^{n} \int_{z_{k-1}}^{z_k} \sigma^k_{1(m-1)} \, dz \; (1 \rightleftarrows 2) \,,$$

$$M_{1(m-1)} = \sum_{k=1}^{n} \int_{z_{k-1}}^{z_k} \sigma^k_{1(m-1)} z \, dz \; (1 \rightleftarrows 2) \,.$$

The overall tangential and normal loads $P_{\tau+\Delta\tau}$ acting at the stage of loading $\tau+\Delta\tau$ serve as components of the vector of external surface forces, with the help of which the consistent nodal external forces are calculated.

By carrying out the standard procedures of assembling individual elements and taking the boundary conditions into account, we obtain a linear system of algebraic equations for evaluation of the increments of nodal displacements

$$K_{\Sigma\tau} \Delta q_{(m)} = P_{\tau+\Delta\tau} - F_{(m-1)} \,, \tag{2.48}$$

where $K_{\Sigma\tau} = K_\tau + S_\tau$. As a result of solving the above system, the displacements $\Delta q_{(m)}$ are calculated thus allowing us to refine the previous solution $q_{(m)} = q_{(m-1)} + \Delta q_{(m)}$ where at $q_0 = 0$, $m = 1$.

Consider briefly the design algorithm. To describe the geometry of a general type of a laminated shell of revolution it is convenient to specify coordinates of points for individual elements on the meridian profiles (Fig. 17). Such a way of describing each finite element is still more convenient for it allows one to trace the geometry of the coordinate surface of a shell in the process of deformation. To describe stiffness characteristics of isolated layers the structural-phenomenological model of deformation for the composite material with a brittle matrix can be successfully employed (Sections 1.2, 1.3).

Let us consider the common sequence of operations in solving the problem. For a known geometry of a structure and internal force factors and stiffness characteristics which have been evaluated in the previous equilibrium state at the time instant τ a succession of problems (3.48) is solved and increments of nodal displacements found at successive external forces $\boldsymbol{P}_{\tau+\Delta\tau}$. These increments correspond to the increments of parameters in the strained state which are summed up with the corresponding parameters found at the previous stage of loading.

Further states of the layers in a shell are analyzed in accordance with the model of deformation, their stiffness is evaluated and correction of reinforcement angles made by taking shear deformations in an isolated layer into account (structural nonlinearity) (ZINOV'EV et al., 1980).

New coordinates of a profile of the deformed shell are also calculated from the known increments of displacements. For the new configuration, consistent with time $\tau + \Delta\tau$, the possible residual $\boldsymbol{\chi}^{\tau+\Delta\tau}_{(m-1)}$ attributed to the errors in the previous iteration process and transition to new geometry is calculated

$$\boldsymbol{\chi}^{\tau+\Delta\tau}_{(m-1)} = \boldsymbol{P}_{\tau+\Delta\tau} - \boldsymbol{F}_{(m-1)} ,$$

where m is the number of the last iteration. In the case when the norm of the residual vector is high and the condition

$$\|\boldsymbol{\chi}^{\tau+\Delta\tau}_{(m-1)}\| \leqslant \mu \|\boldsymbol{P}_{\tau+2\Delta\tau} - \boldsymbol{P}_{\tau+\Delta\tau}\| \qquad (2.49)$$

is not satisfied the norm of increments for the vector of external nodal forces is shown in the right part of the inequality; μ is the allowable relative error ($\mu = 0.1–0.01$)), the previous solution is reiterated

$$\boldsymbol{K}_{\Sigma\tau+\Delta\tau}\Delta\boldsymbol{q}_{(m)} = \boldsymbol{\chi}^{\tau+\Delta\tau}_{(m-1)} ,$$

until the condition (2.49) is satisfied. Subsequently, the new matrix $\boldsymbol{K}_{\Sigma\tau+\Delta\tau}$ is found, the next portion of load applied and the new iteration process sets in which is analogous to (2.48).

In the case when upon passing over to the new configuration the residuals are negligible and the condition (2.49) is satisfied the next portion of the load is applied immediately and the 'old' stiffness matrix used in the iteration process.

2.8. *A triangular finite element of a laminated shell*

In Section 2.4 a refined model of deformation for a laminated shell has been treated which allows one, in the first approximation, to take transverse shear strains and anisotropy of elastic properties of individual layers into account. On the basis of this model we shall obtain stiffness characteristics of the finite element for a laminated shell of double curvature (BIKOV and POPOV, 1984). For simplicity, let us consider a linear problem of statics. In this case the initial variational formulation will be expressed as follows: it is expedient that such generalized displacements $\boldsymbol{u}$ should be evaluated for which

$$\int_S (\hat{\boldsymbol{\xi}}^{\mathrm{T}} \boldsymbol{D} \boldsymbol{\xi} - \hat{\boldsymbol{u}}^{\mathrm{T}} \boldsymbol{p})\, \mathrm{d}S = 0 \tag{2.50}$$

at any admissible $\hat{\boldsymbol{u}}$. Here, S is the area of the shell under consideration; $\mathrm{d}S = A_1 A_2\, \mathrm{d}x_1\, \mathrm{d}x_2$. By generalized deformations $\boldsymbol{\xi}$ we understand that they can be expressed through generalized displacements $\boldsymbol{\xi} = \boldsymbol{L}\boldsymbol{u}$ where

$$\boldsymbol{u} = \{u_1, u_2, w, \psi_1, \psi_2\} \tag{2.51}$$

are the components of the vector $\boldsymbol{u}$, i.e. displacements of the coordinate surface and angles of rotation of sections (see Fig. 19). The components of the vector $\boldsymbol{\xi}$ are given by the following:

$$\boldsymbol{\xi} = \{\varepsilon_1, \varepsilon_2, \gamma_{12}, \kappa_1, \kappa_2, \kappa_{12}, \gamma_{13}, \gamma_{23}\} \tag{2.52}$$

which are evaluated in terms of displacements and angles of rotation of sections from the following relationships

$$\begin{aligned}
\varepsilon_1 &= \frac{1}{A_1} u_{1,1} + \varphi_{12} u_2 + k_1 w \ (1 \rightleftarrows 2)\,, \\
\gamma_{12} &= \omega_1 + \omega_2\,, \\
\omega_1 &= \frac{1}{A_1} u_{2,1} - \varphi_{12} u_1 \ (1 \rightleftarrows 2)\,, \\
\kappa_1 &= \frac{1}{A_1} \psi_{1,1} - \varphi_{12} \psi_2 \ (1 \rightleftarrows 2)\,, \\
\kappa_{12} &= \frac{1}{A_2} \psi_{1,2} - \varphi_{21} \psi_2 + \frac{1}{A_1} \psi_{2,1} - \varphi_{12} \psi_1\,, \\
\gamma_{13} &= \psi_1 + \frac{1}{A_1} w_{,1} - k_1 u_1 \ (1 \rightleftarrows 2)\,.
\end{aligned} \tag{2.53}$$

The matrix of elasticity relationships $\boldsymbol{D}$ in (4.50) is determined analogously to $\boldsymbol{D}_\tau$ (2.18), whereas the coefficients of the matrix $\boldsymbol{L}$ are constituted in accordance with (2.53).

In order to have the capability to independently approximate the deformations $\boldsymbol{\xi}$ and displacements $\boldsymbol{u}$, we shall employ Hellinger–Reissner's variational formulation (see Section A.3) instead of (2.50) which will take the form

$$\int_S \hat{\boldsymbol{\xi}}^{\mathrm{T}} \boldsymbol{D}(\boldsymbol{L}\boldsymbol{u} - \boldsymbol{\xi})\, \mathrm{d}S = 0\,, \tag{2.54}$$

$$\int_S (\boldsymbol{L}\hat{\boldsymbol{u}})^{\mathrm{T}} \boldsymbol{D}\boldsymbol{\xi}\, \mathrm{d}S - f(\hat{\boldsymbol{u}}) = 0\,, \tag{2.55}$$

where

$$f(\hat{\boldsymbol{u}}) = \int_S \hat{\boldsymbol{u}}^{\mathrm{T}} \boldsymbol{p}\, \mathrm{d}S\,.$$

Let us choose approximations of displacements for the finite element of a shell in the form

$$\boldsymbol{u} = \boldsymbol{\phi}\boldsymbol{q} \tag{2.56}$$

and of deformations

$$\boldsymbol{\xi} = \boldsymbol{\omega}\boldsymbol{\alpha}\,, \tag{2.57}$$

where $\boldsymbol{\phi}$, $\boldsymbol{\omega}$ are the shape matrices; $\boldsymbol{q}$ is the vector of generalized nodal displacements of the element; and $\boldsymbol{\alpha}$ is the vector of approximation coefficients for deformations. The conditions (2.54) and (2.50), by taking (2.46) and (2.57) into account, allow us to obtain the following equations

$$\boldsymbol{G}\boldsymbol{q} = \boldsymbol{H}\boldsymbol{\alpha}\,, \tag{2.58}$$

$$\boldsymbol{G}^{\mathrm{T}}\boldsymbol{\alpha} = \boldsymbol{P} \tag{2.59}$$

where

$$\boldsymbol{G} = \int_S \boldsymbol{\omega}^{\mathrm{T}} \boldsymbol{D}\boldsymbol{B}\, \mathrm{d}S\,, \quad \boldsymbol{H} = \int_S \boldsymbol{\omega}^{\mathrm{T}} \boldsymbol{D}\boldsymbol{\omega}\, \mathrm{d}S\,,$$
$$\boldsymbol{P} = \int_S \boldsymbol{\phi}^{\mathrm{T}} \boldsymbol{p}\, \mathrm{d}S\,, \quad \boldsymbol{B} = \boldsymbol{L}\boldsymbol{\phi}\,. \tag{2.60}$$

By eliminating coefficients $\boldsymbol{\alpha}$ from (2.58) we shall rewrite (2.59) in the form

$$\boldsymbol{K}\boldsymbol{q} = \boldsymbol{P}\,, \tag{2.61}$$

where

$$\boldsymbol{K} = \boldsymbol{G}^{\mathrm{T}}\boldsymbol{H}^{-1}\boldsymbol{G} \tag{2.62}$$

defines the stiffness matrix of the finite element.

It can be concluded from the analysis of expressions for deformations (2.53) that in solving the problem by the finite element technique it is necessary to ensure smoothness of conjugation C^0 at the boundaries of the elements from five components of the vector $\boldsymbol{u}$ (2.51). These components will be selected as the degrees of freedom at the nodes.

Consider a triangular finite element for a laminated shell with six nodes (Fig. 22). The total number of generalized nodal displacements of the element $n_q = 5 \times 6 = 30$. Consequently, the respective stiffness matrix $\boldsymbol{K}$ of the finite element will have the dimension (30×30). Any component of the displacement vector $\boldsymbol{u}$ can be approximated by a perfect square polynomial of arguments x_1, x_2. For instance, $u_1 = c_1 + c_2x_1 + c_3x_2 + c_4x_1^2 + c_5x_1x_2 + c_6x_2^2$. Since we have six sets of nodal displacements, the unknown coefficients c_1 and c_6 can be uniquely expressed in terms of six nodal displacements u_1^1 and u_1^6. The smoothness of conjugation of elements in terms of displacements will be realized upon passing over from one element to another. This can easily be demonstrated, since along the boundary of the element, for instance, 1–4–2 (Fig. 22) the distribution of displacements will be a quadratic function of s and the coefficients of this function will be uniquely defined by displacements of the three nodes 1, 4, 2.

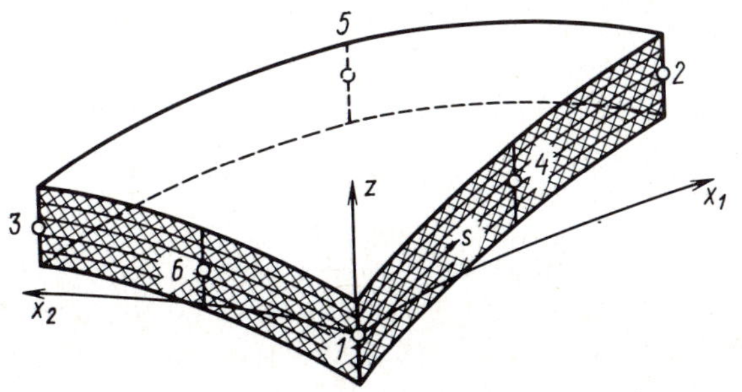

FIG. 22. A triangular finite element of a laminated shell with six nodes.

Displacements of the finite element of a shell as a rigid body involve three displacements and three angles of rotation, therefore the total number of displacements as a rigid body is $n_r = 6$. The stiffness matrix of the finite element $\boldsymbol{K}$ must be singular. Therefore, from the known nodal forces $\boldsymbol{P}$ from (2.61), the nodal displacements of an individual non-fixed element must be evaluated with the accuracy of the rigid body displacements and the rank of the matrix must be equal to

$$\operatorname{rank} \boldsymbol{K} = n_q - n_r \tag{2.63}$$

In our case, rank $\boldsymbol{K} = 30 – 6 = 24$. Fulfilment of the condition (2.63) can be attained by properly selecting the approximation of deformation $\boldsymbol{\xi}$. Let us represent the stiffness matrix of the element in the form (2.62), then rank $\boldsymbol{K} =$

min(rank G, rank H^{-1}). Since for a positive-defined matrix H rank H^{-1} = rank $H = n_\alpha$ (where n_α is the dimension of the vector α in approximation (2.47)), whereas for the matrix G rank $G = \min(n_q - n_*, n_\alpha)$, where $n_* \leqslant n_r$, n_α can be selected such that

$$n_\alpha = n_q - n_r \tag{2.64}$$

so that the condition (2.63) might be fulfilled. In our case, we have obtained $n_\alpha = 24$. Since in accordance with (2.52) we have eight components of the vector ξ, we can single out three constants for each component and approximate it by a perfect linear polynomial.

In such a way, the use of a mixed-type functional makes it possible, in terms of the necessary stiffness properties of the matrix (2.63), to select the shape of a finite element and the order of approximation of displacements and deformations which would uniquely correspond to the model of deformation for a shell.

Let us treat the selected approximations in more detail. The vector of generalized nodal displacements for an element will be presented in the form

$$q = \{q_1, q_2, \ldots, q_6\},$$

where q_i is the vector of generalized displacements of the ith node

$$q_i = \{u_1^i, u_2^i, w^i, \psi_1^i, \psi_2^i\} \quad (i = 1, 2, \ldots, 6).$$

We shall determine the matrix of the shape function ϕ (5 × 30) (see (2.56)) as

$$\phi = [\phi_{q_1}, \phi_{q_2}, \ldots, \phi_{q_6}],$$

where $\phi_{q_i} = \phi_i E$ $(i = 1, 2, \ldots, 6)$; E (5 × 5) is the unit matrix; ϕ_i are the quadratic shape functions. To represent ϕ_i it is convenient to employ natural coordinates

$$\phi_1 = L_1(2L_1 - 1) \quad \overset{\longrightarrow}{\underset{\longleftarrow}{(1, 2, 3)}},$$

$$\ldots\ldots\ldots\ldots\ldots$$

$$\phi_4 = 4L_1L_2 \quad \overset{\longrightarrow}{\underset{\longleftarrow}{(1, 2, 3)}}.$$

$$\ldots\ldots\ldots$$

The functions ϕ_2, ϕ_3 can be obtained from ϕ_1, and ϕ_5, ϕ_6 from ϕ_4 by cyclic rearrangement of indices 1, 2, 3. Numeration of shape functions corresponds to numeration of nodes of the element (Fig. 22). The natural coordinates L_i are determined in terms of the coordinates x_1 and x_2 as follows:

$$L_i = \frac{1}{2S_*}(a_i + b_i x_1 + c_i x_2) \quad (i = 1, 2, 3),$$

where

$$a_1 = x^{(2)}y^{(3)} - x^{(3)}y^{(2)}\,, \quad b_1 = y^{(2)} - y^{(3)}\,, \quad c_1 = x^{(3)} - x^{(2)} \quad \overrightarrow{\underleftarrow{(1,2,3)}}\,,$$

$$S_* = \tfrac{1}{2}\,(b_1c_2 - b_2c_1)\,,$$

$x^{(i)}$ and $y^{(i)}$ are coordinates x_1 and x_2 of the ith node.

It is convenient to represent the matrix ω (8×24), approximating generalized deformations (2.57), in a block diagonal form

$$\boldsymbol{\omega} = \lceil \boldsymbol{f}_1, \boldsymbol{f}_2, \ldots, \boldsymbol{f}_8 \rfloor\,,$$

where $\boldsymbol{f}_i = [L_1\; L_2\; L_3]$ is the row matrix (1×3).

By employing the formulas of differentiation

$$\phi_{1,1} = \frac{b_1}{2S_*}\,(4L_1 - 1)\,, \quad \phi_{4,1} = \frac{2}{S_*}\,(b_1L_2 + b_2L_1) \quad \overrightarrow{\underleftarrow{(1,2,3)}}\,,$$

$$\phi_{1,2} = \frac{c_1}{2S_*}\,(4L_1 - 1)\,, \quad \phi_{4,2} = \frac{2}{S_*}\,(c_1L_2 + c_2L_1) \quad \overrightarrow{\underleftarrow{(1,2,3)}}\,,$$

it is possible to construct the matrix $\boldsymbol{B}$ (see (2.50))

$$\boldsymbol{B} = [\boldsymbol{B}_1\boldsymbol{B}_2 \cdots \boldsymbol{B}_6]\,,$$

where the matrix blocks $\boldsymbol{B}_i$ (8×5) are constructed in accordance with (2.53)

$$\boldsymbol{B}_i = \begin{bmatrix} \bar{\phi}_{i,1} & \varphi_{12}\phi_i & k_1\phi_i & 0 & 0 \\ \varphi_{21}\phi_i & \bar{\phi}_{i,2} & k_2\phi_i & 0 & 0 \\ \bar{\phi}_{i,2} - \varphi_{12}\phi_i & \bar{\phi}_{i,1} - \varphi_{21}\phi_i & 0 & 0 & 0 \\ 0 & 0 & 0 & \bar{\phi}_{i,1} & \varphi_{12}\phi_i \\ 0 & 0 & 0 & \varphi_{21}\phi_i & \bar{\phi}_{i,2} \\ 0 & 0 & 0 & \bar{\phi}_{i,2} - \varphi_{12}\phi_i & \bar{\phi}_{i,1} - \varphi_{21}\varphi_i \\ -k_1\phi_i & 0 & \bar{\phi}_{i,1} & \phi_i & 0 \\ 0 & -k_2\phi_i & \bar{\phi}_{i,2} & 0 & \phi_i \end{bmatrix},$$

$$\bar{\phi}_{i,k} = \frac{1}{A_k}\,\phi_{i,k}\;(k = 1, 2)\,, \quad \varphi_{12} = \frac{1}{A_1A_2}\,A_{1,2}\;(1 \rightleftarrows 2)\,.$$

To calculate the integrals (2.60) for a triangular element let us use the following transformation

$$\int_S F(x_1, x_2)\,\mathrm{d}S = \int_0^1 \int_0^{1-L_1} F(L_1, L_2, L_3)\,\det J\,\mathrm{d}L_1\,\mathrm{d}L_2\,,$$

where $\det J = 2S_*$, along with the quadrature formulas of numerical integration. For the matrix $\boldsymbol{G}$ we shall have

$$\boldsymbol{G} = \sum_{i=1}^{M} \boldsymbol{\omega}_i^{\mathrm{T}} \boldsymbol{D}_i \boldsymbol{B}_i w_i^* ,$$

where $w_i^* = S_* A_{1i} A_{2i} w_i$; w_i is the weight coefficient of numerical integration; i is the number of a given integration point; and M is the number of integration points.

The procedure of direct calculation of the matrix $\boldsymbol{H}$ (24×24) and its subsequent inversion require high machine time expenditures. To avoid this (if stiffness characteristics are constant throughout the element) it is possible to employ the block structure of matrices and write

$$\boldsymbol{H}^{-1} = [\bar{\boldsymbol{H}}_{ij}] \quad (i, j = 1, 2, \ldots, 8) ,$$

where the blocks $\bar{\boldsymbol{H}}_{ij}$ (3×3) are calculated in terms of coefficients of the compliance matrix $\boldsymbol{D}^{-1} = [\bar{d}_{ij}]$ and the symmetric matrix $\boldsymbol{N}^{-1}$ (3×3)

$$\bar{\boldsymbol{H}}_{ij} = d_{ij} \boldsymbol{N}^{-1} ,$$

where

$$\boldsymbol{N} = \sum_{k=1}^{M} \boldsymbol{f}_k^{\mathrm{T}} \boldsymbol{f}_k , \quad \boldsymbol{f}_k = [L_1, L_2, L_3]$$

is the row matrix with coefficients calculated at the kth point of integration.

After the stiffness matrices of the elements have been constructed and the vectors of nodal forces found, the standard algorithms of finite element method are employed to derive a system of resolvent algebraic equations and its solution. Subsequently, the necessary deformations and stresses in the layers of a shell can be determined from the nodal displacements of the appropriate elements.

2.9. Numerical results

2.9.1. *State of stress in a sandwich structure* (ALFUTOV et al., 1984)

Let us consider a structure loaded with external uniform pressure which consists of two sandwich shells of conical and cylindrical shape and two rings—that of the end section and an intermediate one (Fig. 23). The lower edge of the shell is rigidly restrained. The geometrical dimensions of the shells are assumed to be $R_1 = 450$ mm, $R_2 = 600$ mm, $L = 800$ mm, $\theta = 70°$. Each load-bearing layer of a sandwich packet consists of five layers of a unidirectional composite material, laid up according to the following scheme [0°/90°/0°/90°/0°] (the layup angles have been recorded from the direction of the meridian). The thickness of a unidirectional layer is assumed to be 0.12 mm, and the elastic characteristics are as follows: $E_{\parallel} = E$; $E_{\perp} = 0.067E$; $G_{\parallel\perp} =$

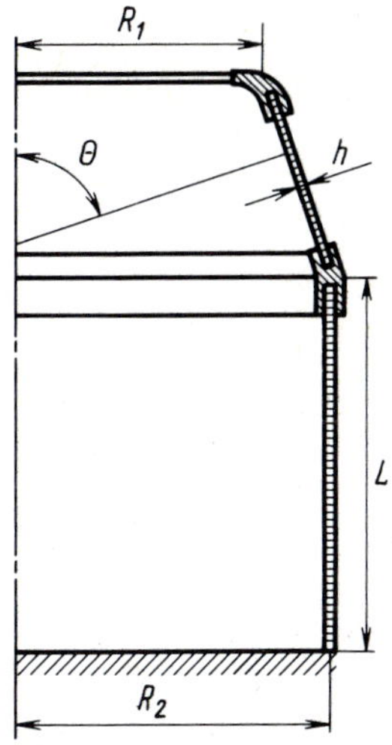

FIG. 23. A sandwich shell of revolution.

$0.046E$; $\nu_{\parallel\perp} = 0.25$. The isotropic core has the modulus of elasticity $0.146 \times 10^{-3}E$ and Poisson's ratio 0.22. Thickness of the core layer is 20 mm and the modulus of elasticity of the ring material is $0.51E$. The geometrical characteristics of the end section (and intermediate) rings are assumed to be $J_1 = 3.84 \times 10^4$ (4.92×10^4) mm^4; $J_3 = 3.21 \times 10^3$ (6.07×10^4) mm^4; $J_{13} = -2.66(0)$ mm^4; $J_k = 7 \times 10^4$ (1.1×10^5) mm^4; $F = 6.48 \times 10^2$ (8.1×10^2) mm^2.

In the calculations involving the face layers, Kirchhoff–Love's hypotheses were used and the transverse shear strains in the core taken into account. Graphs of variation of normal deflection, transverse shear strains and longitudinal stresses in facings along the length of a structure are presented in Figs. 24–27. The curves 1, 4 refer to the inner and outer surfaces of a sandwich

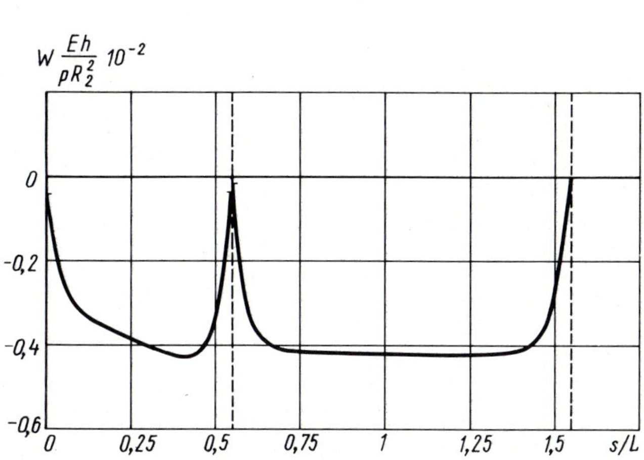

FIG. 24. Normal deflection.

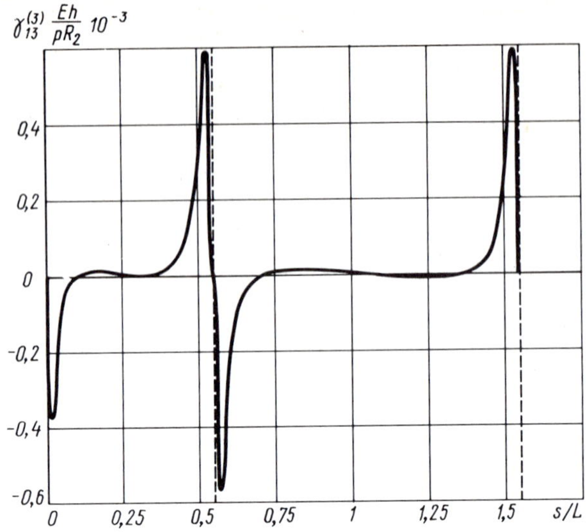

FIG. 25. Transverse shear strain.

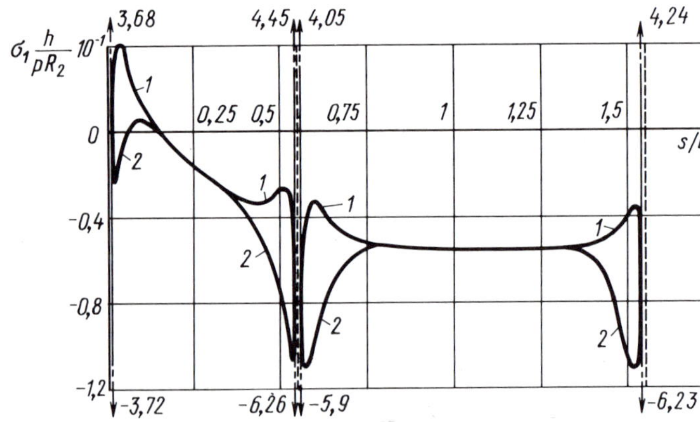

FIG. 26. Longitudinal stresses in inner facing.

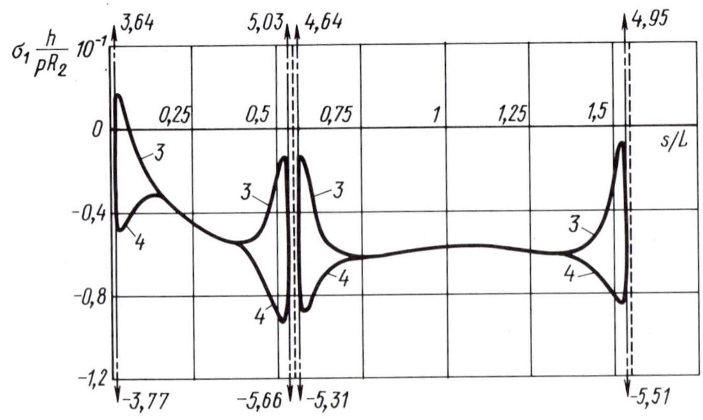

FIG. 27. Longitudinal stresses in outer facing.

packet, curves 2, 3—to the surfaces of the inner and outer facings which are contiguous with the core. As is evident from the graphs, pronounced edge effects occur in the vicinity of rings and restraint.

2.9.2. *Tension of a fiberglass shell*

A cylindrical shell rigidly restrained at the edges is loaded with tensile forces (Fig. 28). The shell was made by winding a fiberglass composite at the angle

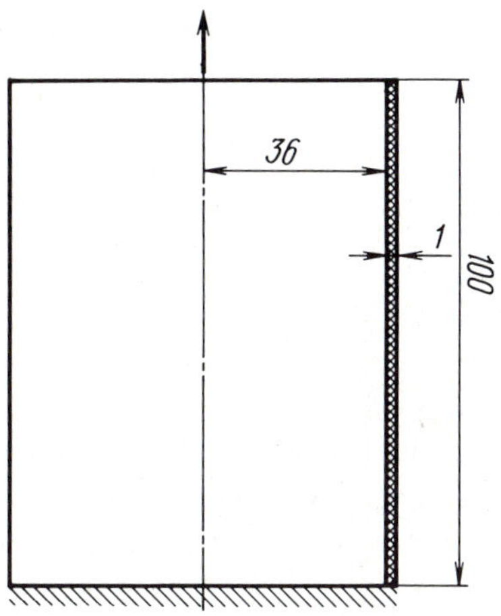

FIG. 28. Tension of a fiberglass plastic shell [±40°].

±40° to the generatrix. The unidirectional material had the following characteristics: $E_{\|} = 5.4 \times 10^4$ MPa; $E_{\perp} = 1.2 \times 10^4$ MPa; $G_{\|\perp} = 0.5 \times 10^4$ MPa; $\nu_{\|\perp} = 0.28$; $F_{+\perp} = F_{-\perp} = 40$ MPa. It can be seen on the stress–strain diagram of an angle-ply fiberglass composite reinforced at ±40° (Fig. 29) that at $\varepsilon = 0.6\%$ when the matrix starts to fail, fracture occurs, the tangential modulus E diminishes by an order of magnitude and then as the level of deformation increases this modulus increases somewhat due to the decrease in the reinforcement angle. The distribution of the radial displacements w along the generatrix at different values of increments of the total length of a shell Δ is given in Fig.

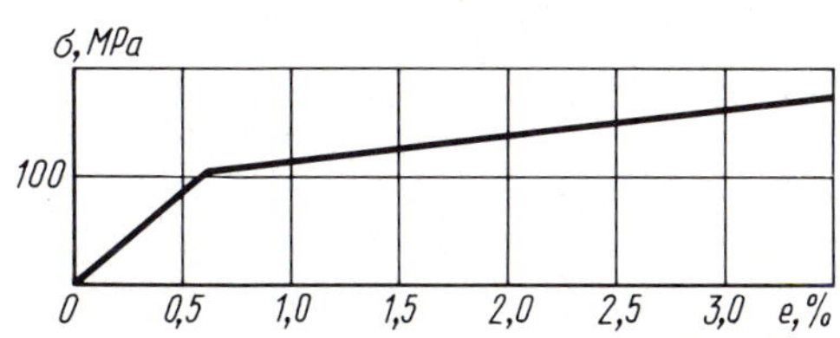

FIG. 29. Stress–strain diagram of a fiberglass plastic [±40°].

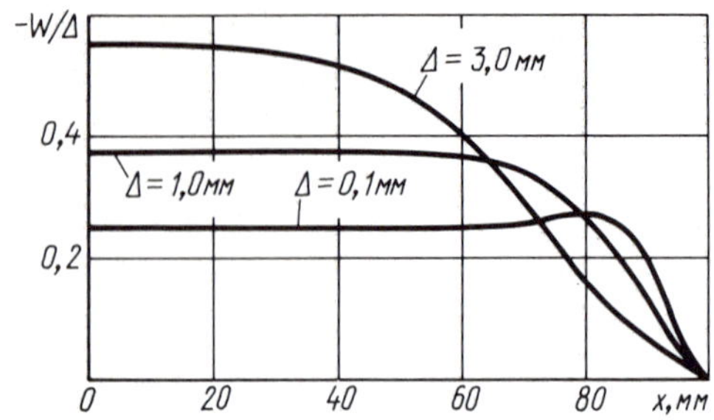

FIG. 30. Distribution of radial displacements.

30. As can be seen, the character of deformation has essentially changed with the increase in Δ within the limits of 0.1 to 3 mm, edge disturbance due to restraint has levelled out and the area of its action increased. All three types of nonlinear response have manifested themselves in this problem, namely geometrical, physical and structural.

2.9.3. Design of a sandwich panel

The sandwich panel shown in Fig. 31 was loaded with external uniform pressure p (Fig. 31). Stiffening diaphragms were placed along the contour of the panel and the edges were simply supported. Geometrical dimensions of the panel were assumed to be $R = 2.0$ m, $r = 0.25$ m, $L = 2.0$ m, $a = 1.0$ m, $\varphi_1 = 30°$, $\varphi_2 = 60°$. Each load-bearing layer of a sandwich packet consisted of five layers of a unidirectional composite material laid up according to the following scheme [90°/0°/90°/0°/90°], the layup angles were read off from the rectilinear

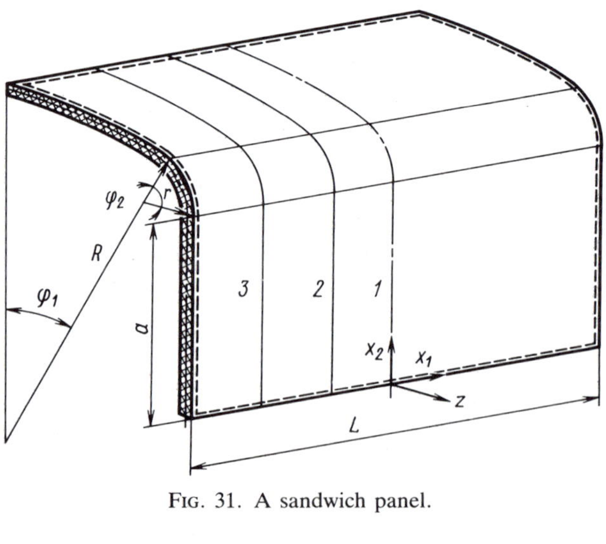

FIG. 31. A sandwich panel.

generatrix. The properties of the unidirectional layer and the core were the same as for problem 2.9.1. The thickness of the unidirectional layer was assumed to be 0.2×10^{-2} m, whereas that of the core—0.5×10^{-1} m.

Calculation of the deformed state was carried out with the finite element method. The triangular element for a laminated shell, which was described in Section 2.8, was used. One half of the panel divided into 48 finite elements was considered with allowance for symmetry of the problem. In Figs. 32–35, the graphs of variation of normal deflection, bending moment, shear and tangential forces have been presented. The curves of normal deflection (Fig. 32) are shown for sections 1, 2, 3 of Fig. 31. The curves of variation of force factors (Figs. 33–35) have been drawn for the plane of symmetry of the panel. As can be seen from the graphs, the greatest deflections and bending moments occur in the panel center, the highest gradients of force factors take place in the cylindrical part of the panel of small radius.

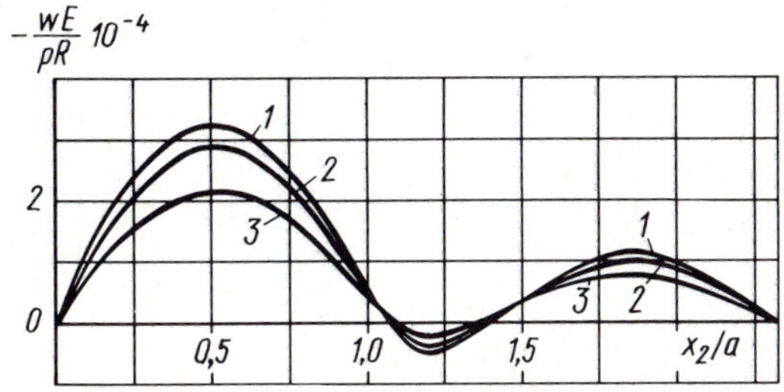

FIG. 32. Distribution of normal displacements.

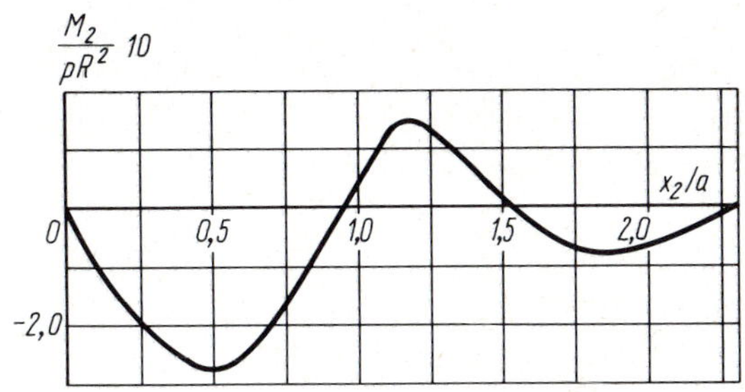

FIG. 33. Distribution of bending moment.

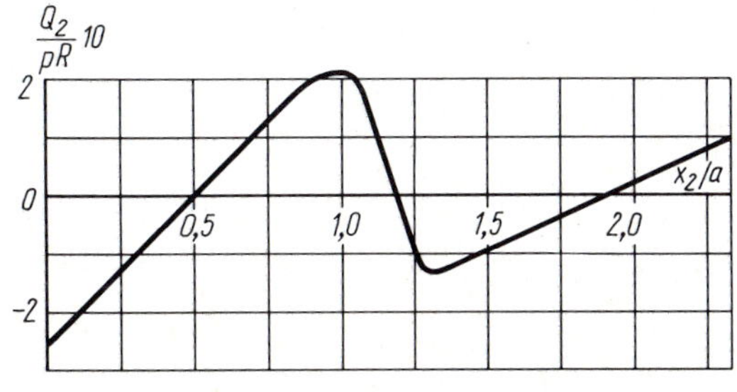

FIG. 34. Distribution of shearing force.

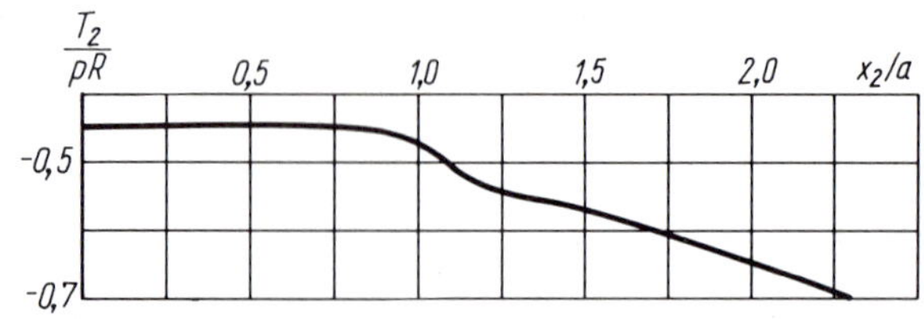

FIG. 35. Distribution of consistent tangential force.

Appendix. Variational formulations of problems in the mechanics of solids

It is convenient to obtain the basic equations and boundary conditions, which are employed in the design of laminated structures in terms of complex models of deformation, on the basis of variational formulations. In the Appendix, different variational formulations of the problems of statics and stability of a rigid deformable body have been considered from a common standpoint of the principle of virtual displacements. For the general case, it has been shown—that by employing these formulations—it is possible to obtain resolvent differential equations, boundary conditions and approximate solutions. Special attention is devoted to two types of mixed variational formulations. The first type corresponds to a mixed variational principle of Reissner, the second—to problems of conventional extremum with restrictions to differential equations. For one-dimensional problems, the variational-matrix method of deriving a canonical system of the Hamilton–Jacobi type has been obtained. For two-dimensional problems, the design methods of Rayleigh–Ritz type and finite element procedures which take into account peculiarities introduced by mixed variational formulation have been treated. The vector-matrix notation, used in the chapter, allows one to trace general stages in solving a wide range of questions, to write down equations and to make the process of preparation of programs for computer aided design substantially easier.

A.1. Principle of virtual displacements

Consider a mechanical system consisting of an arbitrarily deformable solid body and applied distributed body and surface forces $\boldsymbol{g}$ and $\boldsymbol{p}$ (Fig. A.1); the surface of the body S_u is fastened in space by means of some ideal bonds eliminating its displacements as a rigid body. Let us assume that the system is in the state of equilibrium. Denote the real displacements characterizing transition of body points from the initial unloaded state into the state of equilibrium by $\boldsymbol{u} = \{u_1, u_2, u_3\}$. The real state of stress is designated by a column vector $\boldsymbol{\sigma}$ whose components are the normal and tangential stresses in the orthogonal system of coordinates

$$\boldsymbol{\sigma} = \{\sigma_1, \sigma_2, \sigma_3, \sigma_4, \sigma_5, \sigma_6\} ,$$

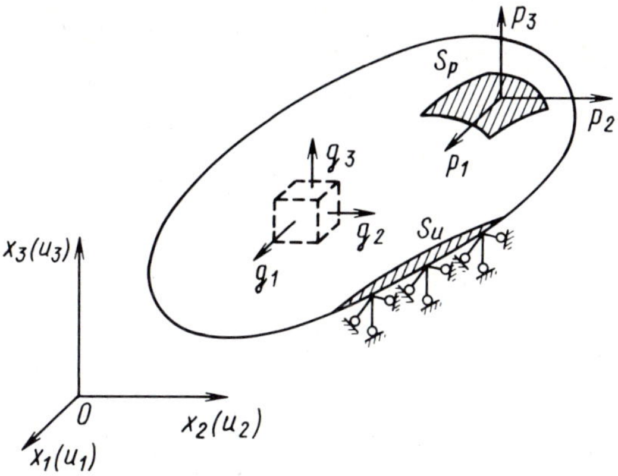

FIG. A.1. A deformable body and external forces.

where

$$\sigma_m = \begin{cases} \sigma_{ii}, & m = i \\ \sigma_{ij}, & m = 9 - i - j \quad (i \neq j), \end{cases} \tag{A.1}$$
$$i, j = 1, 2, 3 .$$

Let us denote the strains in the body caused by real displacements by the column vector $\boldsymbol{e}$, whose components are the relative elongations and shear angles in the orthogonal system of coordinates

$$\boldsymbol{e} = \{e_1, e_2, e_3, e_4, e_5, e_6\} ,$$

where

$$e_m = \begin{cases} e_{ii}, & m = i \\ e_{ij} + e_{ji}, & m = 9 - i - j \quad (i \neq j) \end{cases} \tag{A.2}$$
$$i, j = 1, 2, 3 .$$

the strains in the body are considered to be sufficiently small and the volume and surface of the body are identified with its volume V and surface S in the initial undeformed state.

Along with real displacements, we small mentally apply the virtual displacements $\hat{\boldsymbol{u}}$ to the body. Components of the vector $\hat{\boldsymbol{u}}$ can be treated as small changes in real displacements. Assume that functions $\hat{\boldsymbol{u}}$ are continuous throughout the volume of the body, have continuous derivatives of the necessary order and reduce to zero on the surface S_u, where the respective displacements are disallowed. The enforced virtual displacements are compatible with the internal and external deformation fields of the system, i.e. they are kinematically (or geometrically) permissible. We shall denote the space of such functions by $\boldsymbol{U}$.

In accordance with the principle of virtual displacements, the sum of work performed by external and internal forces on virtual displacements relative to the state of equilibrium is zero. In the case of dead (conservative) external forces, their work on virtual displacements is defined by the linear functional

$$f(\hat{\boldsymbol{u}}) = \int_V \hat{\boldsymbol{u}}^{\mathrm{T}} \boldsymbol{g} \, \mathrm{d}V + \int_{S_p} \hat{\boldsymbol{u}}^{\mathrm{T}} \boldsymbol{p} \, \mathrm{d}S , \tag{A.3}$$

where T is the sign of transposition. In calculating the sum of work of all internal forces let us observe that the deformations $\boldsymbol{e}$ are generalized displacements at stresses $\boldsymbol{\sigma}$. Therefore, apply additional deformations $\hat{\boldsymbol{e}}$ to the body due to virtual displacements $\hat{\boldsymbol{u}}$, it is necessary that the following work should be done

$$a = \int_V \hat{\boldsymbol{e}}^{\mathrm{T}} \boldsymbol{\sigma} \, \mathrm{d}V . \tag{A.4}$$

In accordance with the principle of virtual displacements let us write down that $a - f = 0$. Ultimately, by the use of expressions (A.1) and (A.2) we obtain a variational equation representing a mathematical formulation of the principle of virtual displacements for a deformable solid body:

$$\int_V \hat{\boldsymbol{e}}^{\mathrm{T}} \boldsymbol{\sigma} \, \mathrm{d}V - \int_V \hat{\boldsymbol{u}}^{\mathrm{T}} \boldsymbol{g} \, \mathrm{d}V - \int_{S_p} \hat{\boldsymbol{u}}^{\mathrm{T}} \boldsymbol{p} \, \mathrm{d}S = 0 . \tag{A.5}$$

The variational equation (A.5) allows one to obtain differential equilibrium equations together with all the variants of the boundary conditions. For this reason it is expedient to establish a more concrete relationship between deformations and displacements in order to carry out integration by parts operations. Such a formal way of procuring equilibrium equations and boundary conditions turns out to be rather convenient for laminated composite shells in deriving the resolvent equations of different variants of approximate theories based on these or alternative kinematic hypotheses. In the general case of linear problems, the relationship between deformations and displacements is prescribed by the linear relationship

$$\boldsymbol{e} = \boldsymbol{L} \boldsymbol{u} , \tag{A.6}$$

where $\boldsymbol{L}$ is a matrix of linear differential operators. The relationship between virtual deformations and displacements is determined analogously as $\hat{\boldsymbol{e}} = \boldsymbol{L}\hat{\boldsymbol{u}}$. Thus the equilibrium condition (A.5) is written down as

$$\int_V (\boldsymbol{L}\hat{\boldsymbol{u}})^{\mathrm{T}} \boldsymbol{\sigma} \, \mathrm{d}V - \int_V \hat{\boldsymbol{u}}^{\mathrm{T}} \boldsymbol{g} \, \mathrm{d}V - \int_{S_p} \hat{\boldsymbol{u}}^{\mathrm{T}} \boldsymbol{p} \, \mathrm{d}S = 0 . \tag{A.7}$$

To obtain equilibrium equations and boundary conditions let us do away with the differential operators $\boldsymbol{L}$ at virtual displacements $\hat{\boldsymbol{u}}$. For this reason we shall integrate the first term in (A.7) by parts and obtain the result

$$\int_V \hat{\boldsymbol{u}}^{\mathrm{T}}(\boldsymbol{L}^*\boldsymbol{\sigma} - \boldsymbol{g})\,\mathrm{d}V + \int_{S_p} \hat{\boldsymbol{u}}^{\mathrm{T}}(\boldsymbol{L}^{\nu}\boldsymbol{\sigma} - \boldsymbol{p})\,\mathrm{d}S = 0 \quad \forall \hat{\boldsymbol{u}} \in \boldsymbol{U}\,, \tag{A.8}$$

where $\boldsymbol{L}^*$ is the matrix of conjugate differential operators, $\boldsymbol{L}^{\nu}$ is the matrix of direction cosines.

Since the requirement (A.8) must be fulfilled for any $\hat{\boldsymbol{u}}$ belonging to $\boldsymbol{U}$, we obtain the following equilibrium equations

$$\boldsymbol{L}^*\boldsymbol{\sigma} - \boldsymbol{g} = 0 \quad \text{in } V\,, \tag{A.9}$$

$$\boldsymbol{L}^{\nu}\boldsymbol{\sigma} - \boldsymbol{p} = 0 \quad \text{on } S_p\,. \tag{A.10}$$

For a linearly-elastic body, the stress–strain relation is defined by Hooke's law

$$\boldsymbol{\sigma} = \boldsymbol{C}\boldsymbol{e} \tag{A.11}$$

where $\boldsymbol{C}$ is the symmetric positively definite matrix of elasticity coefficients. Then, by taking (A.6) and (A.11) into account, the formulation of the principle of virtual displacements (A.7) poses the following problem: determining such displacements $\boldsymbol{u}$ from among a great variety of functions $\boldsymbol{U}$, for which the following relationship is valid

$$a(\hat{\boldsymbol{u}}, \boldsymbol{u}) = f(\hat{\boldsymbol{u}}) \quad \forall \hat{\boldsymbol{u}} \in \boldsymbol{U}\,, \tag{A.12}$$

where

$$a(\hat{\boldsymbol{u}}, \boldsymbol{u}) = \int_V (\boldsymbol{L}\hat{\boldsymbol{u}})^{\mathrm{T}}\boldsymbol{C}\boldsymbol{L}\boldsymbol{u}\,\mathrm{d}V \tag{A.13}$$

is a continuous and positively definite bilinear form. It is easy to observe that the variational formulation of the problem (A.12) is equivalent to the problem of determining the minimum of the functional

$$J(\boldsymbol{u}) = \tfrac{1}{2}a(\boldsymbol{u}, \boldsymbol{u}) - f(\boldsymbol{u})$$

representing the total potential energy of the system.

In solving the problem (A.12) by the Rayleigh–Ritz method or by a finite element procedure the field of displacements $\boldsymbol{u}$ is approximated to the shape

$$\boldsymbol{u} = \boldsymbol{\phi}\boldsymbol{q}\,, \tag{A.14}$$

where $\boldsymbol{\phi}$ is the matrix of shape functions; $\boldsymbol{q}$ is the vector of generalized displacements. Thus the variational formulation of the problem results in the system of algebraic linear equations

$$\boldsymbol{K}\boldsymbol{q} = \boldsymbol{P} , \tag{A.15}$$

where

$$\boldsymbol{K} = \int_V \boldsymbol{B}^{\mathrm{T}}\boldsymbol{C}\boldsymbol{B}\, \mathrm{d}V ;$$
$$\boldsymbol{B} = \boldsymbol{L}\boldsymbol{\phi} , \quad \boldsymbol{P} = \int_V \boldsymbol{\phi}^{\mathrm{T}}\boldsymbol{g}\, \mathrm{d}V + \int_{S_p} \boldsymbol{\phi}^{\mathrm{T}}\boldsymbol{p}\, \mathrm{d}S . \tag{A.16}$$

The matrix $\boldsymbol{K}$ (A.16) is called the stiffness matrix. After the system (A.15) has been solved, the deformations $\boldsymbol{e} = \boldsymbol{B}\boldsymbol{q}$ and stresses $\boldsymbol{\sigma} = \boldsymbol{C}\boldsymbol{B}\boldsymbol{q}$ are evaluated.

A.2. Linearized formulation of the principle of virtual displacements for nonlinear problems

Let us obtain the formulation of the problem of the equilibrium of a physically nonlinear system with large displacements. We can mentally associate the process of static loading with some reference time τ. Let us further assume that the exact increments in external loads are known within the time interval $(\tau, \tau + \Delta\tau)$.

Let the object under consideration have the initial undeformed configuration at the initial time instant $\tau = 0$. Now, let us trace the change of this configuration in the process of loading. In Fig. A.2 the coordinates of the body point in the process of deformation at time instants $\tau = 0, \tau, \tau + \Delta\tau$ are conventionally shown by positions 0, 1, 2. Let us apply a portion of the load at the instants

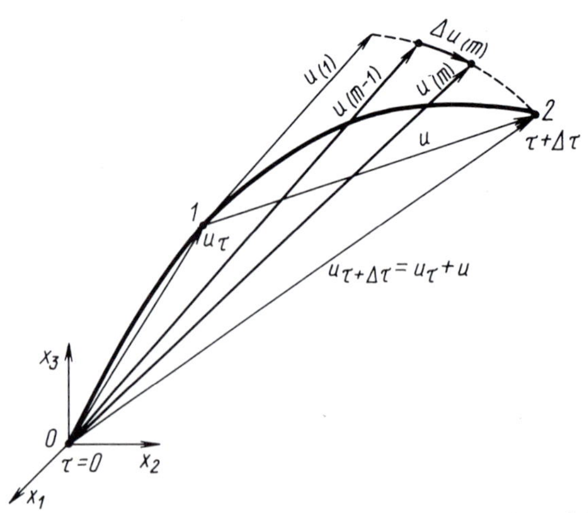

FIG. A.2. Displacement of a point upon nonlinear deformation of the system.

$\tau, \tau+\Delta\tau$. Denote the real increments of displacements which are to be evaluated by $\boldsymbol{u}$. The problem being nonlinear, the displacements cannot be evaluated during one step but through a process of iteration. Such an iteration process can be accomplished by employing linearized formulation of the principle of virtual displacements which must be written down for an equilibrium state of the body at the time instant $\tau+\Delta\tau$.

Assume that the mth iteration results in a state of equilibrium. In this case, we can write down

$$\int_V \Delta\hat{\boldsymbol{e}}^{\mathrm{T}}_{(m)}\boldsymbol{\sigma}_{(m)}\,\mathrm{d}V - f_{(m)} = 0\,, \tag{A.17}$$

where

$$f_{(m)} = \int_V \Delta\hat{\boldsymbol{u}}^{\mathrm{T}}_{(m)}\boldsymbol{g}_{\tau+\Delta\tau}\,\mathrm{d}V + \int_{S_p} \Delta\hat{\boldsymbol{u}}^{\mathrm{T}}_{(m)}\boldsymbol{p}_{\tau+\Delta\tau}\,\mathrm{d}S$$

represents the work of external forces. The total stresses at the mth iteration are evaluated as follows

$$\boldsymbol{\sigma}_{(m)} = \boldsymbol{\sigma}_{\tau} + \sum_{k=1}^{m-1}\Delta\boldsymbol{\sigma}_{(k)} + \Delta\boldsymbol{\sigma}_{(m)} \tag{A.18}$$

and the deformations $\Delta\boldsymbol{e}_{(m)}$ contain linear and nonlinear terms

$$\Delta\boldsymbol{e}_{(m)} = \Delta\boldsymbol{\epsilon}_{(m)} + \Delta\boldsymbol{\eta}_{(m)}\,. \tag{A.19}$$

The indices τ and $\tau+\Delta\tau$ pertain to the respective configurations.

The portions of the load $\Delta\boldsymbol{g}, \Delta\boldsymbol{p}$ must be sufficiently small so that the relationship between stress and strain increments might remain linear, i.e.

$$\Delta\boldsymbol{\sigma}_{(m)} = \boldsymbol{C}_{\tau}\Delta\boldsymbol{e}_{(m)}\,, \tag{A.20}$$

where $\boldsymbol{C}_{\tau}$ is the matrix of tangential moduli, evaluated for the configuration τ.

Taking (A.18)–(A.20) into consideration, let us write down the principle of virtual displacements (A.17) by retaining the terms of the first and second order in the infinitesimals. Then we have

$$\int_V (\Delta\hat{\boldsymbol{\epsilon}}^{\mathrm{T}}_{(m)}\boldsymbol{C}_{\tau}\Delta\boldsymbol{\epsilon}_{(m)} + \Delta\hat{\boldsymbol{\eta}}^{\mathrm{T}}_{(m)}\boldsymbol{\sigma}_{\tau} + \Delta\hat{\boldsymbol{\epsilon}}^{\mathrm{T}}_{(m)}\boldsymbol{\sigma}_{(m-1)})\,\mathrm{d}V - f_{(m)} = 0\,. \tag{A.21}$$

The formulation of the equilibrium condition, expressed in the form (A.21), allows one to obtain the resolvent equation for the evaluation of displacements $\Delta\boldsymbol{u}_{(m)}$ and accomplish the iteration $\boldsymbol{u}_{(m)} = \boldsymbol{u}_{(m-1)} + \Delta\boldsymbol{u}_{(m)}$.

There are two variants of integration (A.21) which are most convenient from the viewpoint of numerical realization. In the first method, integration is accomplished for the initial undeformed volume, i.e. for the configuration $\tau = 0$; in the second—for the configuration of the previous equilibrium state, corresponding to the time instant τ. Thereby in a Cartesian coordinate system, to calculate the linear components of strain increments in integration for the initial volume the following expressions must be used for components of the vector $\boldsymbol{\Delta\epsilon}$, which are ordered analogously to (A.2)

$$\Delta\epsilon_{ij} = \tfrac{1}{2}\left(\frac{\partial \Delta u_i}{\partial x_j} + \frac{\partial \Delta u_j}{\partial x_i} + \frac{\partial u_k^\tau}{\partial x_i}\frac{\partial \Delta u_k}{\partial x_j} + \frac{\partial u_k^\tau}{\partial x_j}\frac{\partial \Delta u_k}{\partial x_i}\right), \tag{A.22}$$

where i, j, k are equal to the values $1, 2, 3$ (index of iteration m is omitted). For the second method, since the configuration τ includes displacements $\boldsymbol{u}_\tau$, the following relationships must be employed in the calculation of components $\boldsymbol{\Delta\epsilon}$

$$\Delta\epsilon_{ij} = \tfrac{1}{2}\left(\frac{\partial \Delta u_i}{\partial x_j} + \frac{\partial \Delta u_j}{\partial x_i}\right) \quad i, j = 1, 2, 3\,. \tag{A.23}$$

The nonlinear components of the strain increments $\boldsymbol{\Delta\eta}$ are estimated for the first and second method as follows:

$$\Delta\eta_{ij} = \tfrac{1}{2}\,\frac{\partial \Delta u_k}{\partial x_i}\,\frac{\partial \Delta u_k}{\partial x_j} \quad i, j, k = 1, 2, 3\,. \tag{A.24}$$

With regard to the physical relationships (A.20) between stress and strain increments, let us assume that the matrix of tangential moduli $\boldsymbol{C}_\tau$, calculated for an equilibrium configuration τ maintains constant its components for iterations during the loading step. Furthermore, let us suppose deformations to be sufficiently small so that in relationships (A.20) there might be no distinction between matrices $\boldsymbol{C}_\tau$ for the two aforementioned methods of integration.

Such a method of solving the nonlinear problems is called Newton's step-by-step method with intermediate iterative correction—Newton's modified method.

In Fig. A.3 the process of calculation is conventionally shown. By p we understand the load, by u-displacements. As can be seen from the figure, in the loading step $(\tau, \tau + \Delta\tau)$ the stiffness of the system, characterized by tangent of the angle α_τ, remains constant.

In solving physically and geometrically nonlinear problems of statics by the finite element method, the linearized formulation of the equilibrium condition (A.21) allows one to procure a system of equations relative to increments of generalized nodal displacements on the mth iteration (BATHE, 1982)

$$(\boldsymbol{K}_\tau + \boldsymbol{S}_\tau)\Delta\boldsymbol{q}_{(m)} = \boldsymbol{P}_{\tau+\Delta\tau} - \boldsymbol{F}_{(m-1)}\,, \tag{A.25}$$

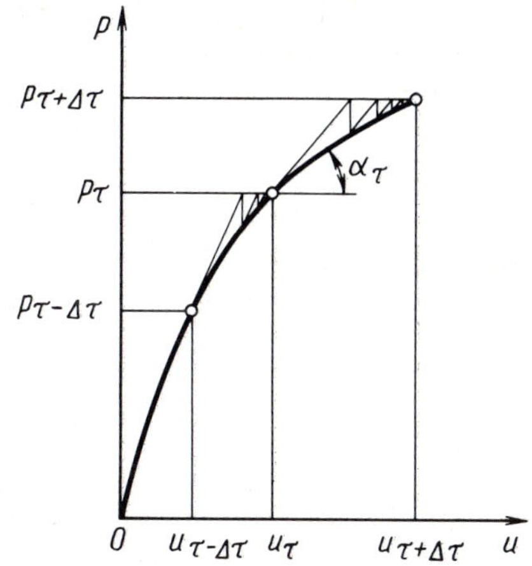

FIG. A.3. A scheme of the step-by-step procedure for solving a nonlinear problem.

where

$$K_\tau = \int_V B^T C_\tau B \, dV , \quad B = \mathscr{L}\phi ,$$

$$S_\tau = \int_V R^T T_\tau R \, dV , \quad R = \nabla\Phi ,$$

$$P_{\tau+\Delta\tau} = \int_V \phi^T g_{\tau+\Delta\tau} \, dV + \int_{S_p} \phi^T p_{\tau+\Delta\tau} \, dS ,$$

$$F_{(m-1)} = \int_V B^T \sigma_{(m-1)} \, dV .$$

Here, ϕ is the matrix of shape functions, relating the displacements $\Delta u_{(m)}$ to the generalized nodal degrees of freedom $\Delta q_{(m)}$; the matrices $\mathscr{L}$ and ∇ are defined by the relationships (A.22) or (A.23), (A.24); T_τ is the matrix where coefficients characterize the state of stress in the body at the time instant τ.

In solving the problems of static stability of conservative systems (ALFUTOV, 1978) the equilibrium state adjacent to the initial one is being studied. In this case, the variational formulation (A.21) is transformed as follows:

$$\int_V (\Delta\hat{\epsilon}^T C \Delta\epsilon + \Pi \Delta\hat{\eta}^T \sigma_0) \, dV = 0 . \qquad (A.26)$$

where Π is the loading parameter; σ_0 are initial stresses. Thereby, as a rule, the initial state is assumed to be stressed but undeformed.

A.3. Variational formulations of mixed type

Let us write down the principle of virtual displacements (A.5) in the following form:

$$\int_V (\hat{\boldsymbol{e}}^{\mathrm{T}}\boldsymbol{\sigma} - \hat{\boldsymbol{e}}^{\mathrm{T}}\boldsymbol{\sigma} + \hat{\boldsymbol{e}}^{\mathrm{T}}\boldsymbol{\sigma})\,\mathrm{d}V - f(\hat{\boldsymbol{u}}) = 0\,, \tag{A.27}$$

where $f(\hat{\boldsymbol{u}})$ is the work of external forces on virtual displacements (A.3). Also, we shall employ Hooke's law (A.11) and express the virtual strains in terms of virtual stresses $\hat{\boldsymbol{e}} = \boldsymbol{C}^{-1}\hat{\boldsymbol{\sigma}}$, where $\boldsymbol{C}^{-1}$ is the symmetric positive definite matrix of compliance coefficients. Thus the first two terms in the condition (A.27) can be transformed to $\hat{\boldsymbol{e}}^{\mathrm{T}}\boldsymbol{\sigma} - \hat{\boldsymbol{e}}^{\mathrm{T}}\boldsymbol{\sigma} = \hat{\boldsymbol{\sigma}}^{\mathrm{T}}(\boldsymbol{e} - \boldsymbol{C}^{-1}\boldsymbol{\sigma})$. As a result, the initial variational formulation of the equilibrium condition will take the following form upon taking (A.6) into account

$$\int_V \hat{\boldsymbol{\sigma}}^{\mathrm{T}}(\boldsymbol{L}\boldsymbol{u} - \boldsymbol{C}^{-1}\boldsymbol{\sigma})\,\mathrm{d}V + \int_V (\boldsymbol{L}\hat{\boldsymbol{u}})^{\mathrm{T}}\boldsymbol{\sigma}\,\mathrm{d}V - f(\hat{\boldsymbol{u}}) = 0\,. \tag{A.28}$$

The above variational principle (A.28) is known as Hellinger–Reissner's principle. Displacements as well as stresses in (A.28) can be varied independently. In view of the arbitrary character of $\hat{\boldsymbol{\sigma}}$ and $\hat{\boldsymbol{u}}$, the condition (A.28) can be separated into two parts

$$\int_V \hat{\boldsymbol{\sigma}}^{\mathrm{T}}(\boldsymbol{L}\boldsymbol{u} - \boldsymbol{C}^{-1}\boldsymbol{\sigma})\,\mathrm{d}V = 0\,, \tag{A.29}$$

$$\int_V (\boldsymbol{L}\hat{\boldsymbol{u}})^{\mathrm{T}}\boldsymbol{\sigma}\,\mathrm{d}V - f(\hat{\boldsymbol{u}}) = 0\,. \tag{A.30}$$

Let us introduce a variety of allowable functions of stresses $\boldsymbol{T}$ and displacements $\boldsymbol{U}$. Then, the evaluation of the deformed state is reduced to the following task: such functions $\boldsymbol{\sigma} \in \boldsymbol{T}$ and $\boldsymbol{u} \in \boldsymbol{U}$ must be found for which the variational equations are statisfied:

$$b(\boldsymbol{u}, \hat{\boldsymbol{\sigma}}) - c(\hat{\boldsymbol{\sigma}}, \boldsymbol{\sigma}) = 0 \quad \forall \hat{\boldsymbol{\sigma}} \in \boldsymbol{T}\,, \tag{A.31}$$

$$b(\hat{\boldsymbol{u}}, \boldsymbol{\sigma}) - f(\hat{\boldsymbol{u}}) = 0 \quad \forall \hat{\boldsymbol{u}} \in \boldsymbol{U}\,, \tag{A.32}$$

where

$$b(\boldsymbol{u}, \hat{\boldsymbol{\sigma}}) = \int_V \hat{\boldsymbol{\sigma}}^{\mathrm{T}}\boldsymbol{L}\boldsymbol{u}\,\mathrm{d}V\,,$$

$$b(\hat{\boldsymbol{u}}, \boldsymbol{\sigma}) = \int_V (\boldsymbol{L}\hat{\boldsymbol{u}})^{\mathrm{T}}\boldsymbol{\sigma}\,\mathrm{d}V\,,$$

$$c(\hat{\boldsymbol{\sigma}}, \boldsymbol{\sigma}) = \int_V \hat{\boldsymbol{\sigma}}^{\mathrm{T}}\boldsymbol{C}^{-1}\boldsymbol{\sigma}\,\mathrm{d}V\,.$$

It is easy to note that the problems (A.31), (A.32) correspond to the problem of a saddle point of the functional $R\,(\boldsymbol{u}, \boldsymbol{\sigma})$

$$R(\boldsymbol{u}, \boldsymbol{\sigma}) = b(\boldsymbol{u}, \boldsymbol{\sigma}) - \tfrac{1}{2}c(\boldsymbol{\sigma}, \boldsymbol{\sigma}) - f(\boldsymbol{u}) .$$

In a physical sense, the condition (A.31) requires fulfilment of the elasticity relationships, whereas (A.32) requires fulfilment of equilibrium relations. For rough approximation of stresses and displacements, the elasticity relationships $\boldsymbol{Lu} = \boldsymbol{C}^{-1}\boldsymbol{\sigma}$ and equilibrium equations $\boldsymbol{L}^*\boldsymbol{\sigma} = \boldsymbol{g}$ might not be fully realized, however, the condition (A.31) allows one to obtain integral relationships of elasticity which can be further employed in integral equilibrium equations (A.32).

In a number of cases it is convenient to use, instead of (A.29) and (A.30), the variational formulation of Hellinger–Reissner's principle in terms of deformations and displacements (ALFUTOV et al., 1983; ODEN and CAREY, 1983)

$$\int_V \hat{\boldsymbol{e}}^{\mathrm{T}}\boldsymbol{C}(\boldsymbol{Lu} - \boldsymbol{e})\,\mathrm{d}V = 0 \quad \forall \hat{\boldsymbol{e}} \in \boldsymbol{E} ,$$

$$\int_V (\boldsymbol{L}\hat{\boldsymbol{u}})^{\mathrm{T}}\boldsymbol{Ce}\,\mathrm{d}V - f(\hat{\boldsymbol{u}}) = 0 \quad \forall \hat{\boldsymbol{u}} \in \boldsymbol{U}$$

or

$$d(\boldsymbol{u}, \hat{\boldsymbol{e}}) = c(\hat{\boldsymbol{e}}, \boldsymbol{e}) \quad \forall \hat{\boldsymbol{e}} \in \boldsymbol{E} , \tag{A.33}$$

$$d(\hat{\boldsymbol{u}}, \boldsymbol{e}) = f(\hat{\boldsymbol{u}}) \quad \forall \hat{\boldsymbol{u}} \in \boldsymbol{U} , \tag{A.34}$$

where

$$d(\hat{\boldsymbol{u}}, \boldsymbol{e}) = \int_V (\boldsymbol{L}\hat{\boldsymbol{u}})^{\mathrm{T}}\boldsymbol{Ce}\,\mathrm{d}V ,$$

$$c(\hat{\boldsymbol{e}}, \boldsymbol{e}) = \int_V \hat{\boldsymbol{e}}^{\mathrm{T}}\boldsymbol{Ce}\,\mathrm{d}V ,$$

$\boldsymbol{E}$ is a space of functions of virtual deformations.

In solving the problems by the finite element method and employing (A.33) and (A.34) it is possible to correct the rank of the stiffness matrix of an element (POPOV, 1984). This procedure is rather effective for the curvilinear elements of shell type (LEE and PIAN, 1978; BIKOV and POPOV, 1984) as well as in taking into account transverse shear and compression strains in multilayer elements. Approximation of the field of displacements within one element is accomplished analogously to (A.14), when, as a rule, polynomials are used as shape functions, which ensure geometrical isotropy throughout the element, and the necessary smoothness of conjugation on the boundaries between elements. In accordance with (A.14), the field of deformations in the finite element is evaluated as $\boldsymbol{e} = \boldsymbol{Bq}$ in the case of solution by displacement method, where $\boldsymbol{B} = \boldsymbol{L\phi}$, whereas the respective stiffness matrix of the element is calculated analogously to (A.16).

Let n_q be the number of generalized nodal degrees of freedom in the element (equal to dimensionality of the vector $\boldsymbol{q}$), and n_r the dimensionality of the operator kernel $\boldsymbol{L}$ (equal to the number of independent rigid displacements of the element). Thus the rank of the stiffness matrix $\boldsymbol{K}$ must be equal to

$$\text{rank}\,\boldsymbol{K} = n_q - n_r \,. \tag{A.35}$$

If the shape functions entering the matrix $\boldsymbol{\phi}$ (see (A.14)) do not contain functions describing exactly the rigid displacements of elements, the condition (A.35) will not be fulfilled. In such elements false deformations will be generated at rigid displacements and convergence of solution will be low.

By employing the variational formulation of the problem of a mixed form (A.33) and (A.34), the dimensionality of independent approximation of the field of deformation

$$\boldsymbol{e} = \boldsymbol{\omega\alpha} \tag{A.36}$$

can be specified as follows

$$n_\alpha = n_q - n_r \,, \tag{A.37}$$

where n_α is the dimensionality of the vector $\boldsymbol{\alpha}$. Fulfilment of (A.37) upon reasonable approximation of (A.36) ensures the necessary rank of the stiffness matrix (A.35), since in accordance with (A.37), (A.34), (A.14), (A.36)

$$\begin{aligned} \boldsymbol{G}^{\mathrm{T}}\boldsymbol{\alpha} &= \boldsymbol{P} \,, \\ \boldsymbol{Gq} &= \boldsymbol{H\alpha} \,, \end{aligned} \tag{A.38}$$

where

$$\boldsymbol{G} = \int_{V_e} \boldsymbol{\omega}^{\mathrm{T}}\boldsymbol{CB}\,\mathrm{d}V \,, \quad \boldsymbol{H} = \int_{V_e} \boldsymbol{\omega}^{\mathrm{T}}\boldsymbol{C\omega}\,\mathrm{d}V \,,$$

V_e is volume of the element; $\boldsymbol{P}$ the external load applied to the nodes. Then by eliminating the coefficients $\boldsymbol{\alpha}$ from (A.38), we shall obtain $\boldsymbol{Kq} = \boldsymbol{P}$, where

$$\boldsymbol{K} = \boldsymbol{G}^{\mathrm{T}}\boldsymbol{H}^{-1}\boldsymbol{G} \tag{A.39}$$

is the stiffness matrix of the element, obtained from an independent approximation of displacement and deformation fields. Since at best rank $\boldsymbol{G} =$

$\min(n_q - n_r, n_\alpha)$ and rank $\boldsymbol{H} = n_\alpha$, then by selecting the dimensionality n_α as in (A.37), the condition (A.35) will be fulfilled for the stiffness matrix (A.39).

In such a way, the independent approximation of deformation (A.36) for displacements (A.14) allows one to maintain in integral form one of the principal properties of the operator $\boldsymbol{L}$—its kernel, and apply the necessary rigid displacements to the element. Hereby, it should be remembered that within the element, the condition of deformation consistency is disturbed. However, if the Babushka–Brezzi's requirements are fulfilled for making approximations in finite elements, the solution convergence will be ensured.

In constructing finite elements by independent approximation of deformations in the element, the geometrical isotropy of displacement and deformation fields and fulfilment of the requirement (A.37) must be ensured. The most critical is disturbance of the requirement (A.37) at $n_\alpha < n_q - n_r$. In this case, the stiffness matrix will contain extra zero eigenvalues and the finite element will be transformed into a mechanism.

A.4. Derivation of canonical systems for one-dimensional problems

In computer aided design the most appropriate form of presentation of resolvent differential equations for one-dimensional problems are systems of first-order differential equations of canonical systems. Standard integration programs as well as different design procedures, ensuring sufficient accuracy of solutions for edge problems (BIDERMAN, 1977; VALISHVILI, 1976; GRIGORENKO, 1973; MYACHENKOV and GRIGOR'EV, 1981) have been evolved. Problems of rods, beams as well as of round plates and shells of revolution under axisymmetric loading belong to one-dimensional problems. In some cases of two- and three-dimensional problems in the theory of elasticity, the method of separation of variables can be employed and the solution obtained in Fourier series or by Kantorovich method. The problems, for which it is possible in one way or another to pass over from equations in partial derivatives to usual equations, will also belong to one-dimensional ones. The variational-matrix method has been suggested below for procuring the system of first-order differential equations for one-dimensional systems, based on a mixed variational formulation (ALFUTOV and POPOV, 1977; POPOV, 1979; 1982; POPOV and RAMAN, 1984). In the derivation, the matrix notation is used which allows us to formally describe the model of deformation, to perform computations compactly and prepare programs for computer aided design.

The kinematic components in the section of a one-dimensional system will be characterized by the vector of generalized displacements $\boldsymbol{X}$.

The distribution of displacements and deformation through the section is determined from the initial hypotheses of deformation and, in a general case, can be expressed as

$$\boldsymbol{u} = \boldsymbol{F}_1\boldsymbol{X} + \boldsymbol{F}_2\boldsymbol{Y} , \tag{A.40}$$

$$\boldsymbol{e} = \boldsymbol{L}_1\boldsymbol{X} + \boldsymbol{L}_2\boldsymbol{Y} , \tag{A.41}$$

where $\boldsymbol{Y}$ is the vector of derivatives, the coefficients of which are the first derivatives $\mathrm{d}/\mathrm{d}s$ (conventionally, it is assumed that $s = x_1$) of components of the vector of generalized displacements $\boldsymbol{X}$ or their combination with displacements, necessary for recording the distribution of displacements and deformations; $\boldsymbol{F}_1$, $\boldsymbol{F}_2$, $\boldsymbol{L}_1$, $\boldsymbol{L}_2$ are matrices containing specified functions of arguments x_2, x_3 approximating the solution for the section. The expressions (A.40) and (A.41) are peculiar in the sense that the approximated matrix coefficients do not contain differential operators $\mathrm{d}/\mathrm{d}s$ because all the necessary first derivatives enter the components of the vector $\boldsymbol{Y}$. For this reason, a differential relationship can be established between the vectors $\boldsymbol{X}$ and $\boldsymbol{Y}$ which may generally be conveniently expressed as

$$\frac{\mathrm{d}}{\mathrm{d}s}\boldsymbol{X} - \boldsymbol{C}_1\boldsymbol{X} - \boldsymbol{C}_2\boldsymbol{Y} = 0\,. \tag{A.42}$$

The differential coupling equation (A.4) is considered to be an additional requirement in deriving the resolvent equations.

Let us employ the principle of virtual displacements (A.5). For a one-dimensional system we shall express it in the following form

$$\int_{s_1}^{s_2}\int_A (\hat{\boldsymbol{e}}^{\mathrm{T}}\boldsymbol{\sigma} - \hat{\boldsymbol{u}}^{\mathrm{T}}\boldsymbol{g})\,\mathrm{d}A\,\mathrm{d}s - \hat{\boldsymbol{X}}^{\mathrm{T}}\bar{\boldsymbol{\lambda}}\big|_{s_1}^{s_2} = 0\,, \tag{A.43}$$

where A is the cross sectional area; s_1 and s_2 are the initial and final coordinates s of the one-dimensional system; $\bar{\boldsymbol{\lambda}}(s_1)$, $\bar{\boldsymbol{\lambda}}(s_2)$ are external force factors applied to the system at $s = s_1$ and $s = s_2$.

By the use of Lagrangian multipliers $\boldsymbol{\lambda}$ let us supplement (A.43) with the condition of coupling for possible $\hat{\boldsymbol{X}}$ and $\hat{\boldsymbol{Y}}$ (see (A.42)) and integrate by parts to obtain

$$\int_{s_1}^{s_2} (\hat{\boldsymbol{X}}^{\mathrm{T}}\boldsymbol{\psi}_1 + \hat{\boldsymbol{Y}}_2^{\mathrm{T}}\boldsymbol{\psi}_2)\,\mathrm{d}s + \hat{\boldsymbol{X}}^{\mathrm{T}}\boldsymbol{\psi}_3\big|_{s_1}^{s_2} = 0\,, \tag{A.44}$$

where the vectors $\boldsymbol{\psi}_i$ $(i = 1, 2, 3)$ are denoted by

$$\begin{gathered}\boldsymbol{\psi}_1 = \int_A (\boldsymbol{L}_1^{\mathrm{T}}\boldsymbol{\sigma} - \boldsymbol{F}_1^{\mathrm{T}}\boldsymbol{g})\,\mathrm{d}A - \boldsymbol{C}_1^{\mathrm{T}}\boldsymbol{\lambda} - \frac{\mathrm{d}}{\mathrm{d}s}\boldsymbol{\lambda}\,,\\ \boldsymbol{\psi}_2 = \int_A (\boldsymbol{L}_2^{\mathrm{T}}\boldsymbol{\sigma} - \boldsymbol{F}_2^{\mathrm{T}}\boldsymbol{g})\,\mathrm{d}A - \boldsymbol{C}_2^{\mathrm{T}}\boldsymbol{\lambda}\,, \quad \boldsymbol{\psi}_3 = (\boldsymbol{\lambda} - \bar{\boldsymbol{\lambda}})\big|_{s_1}^{s_2}\,.\end{gathered} \tag{A.45}$$

In a physical sense, components of the vector of indefinite Lagrangian multipliers $\boldsymbol{\lambda}$ correspond to generalized forces in the section. Since components of the vectors $\hat{\boldsymbol{X}}$ and $\hat{\boldsymbol{Y}}$ are arbitrary functions, then for the equilibrium state in accordance with (A.44) the following requirements must be fulfilled

$$\boldsymbol{\psi}_1 = 0\,, \quad \boldsymbol{\psi}_2 = 0\,. \tag{A.46}$$

For specified force boundary conditions at the sides of a one-dimensional system it is expedient that $\boldsymbol{\psi}_3(s_1) = 0$, $\boldsymbol{\psi}_3(s_2) = 0$ at $s = s_1$, $s = s_2$. The kinematic boundary conditions are specified by the vectors of generalized displacements $\boldsymbol{X}$.

For a linearly elastic one-dimensional system, in accordance with (A.46), we shall obtain one differential and one algebraic equation by employing Hooke's law $\boldsymbol{\sigma} = \boldsymbol{C}(\boldsymbol{L}_1\boldsymbol{X} + \boldsymbol{L}_2\boldsymbol{Y})$

$$\boldsymbol{S}_{11}\boldsymbol{X} + \boldsymbol{S}_{12}\boldsymbol{Y} - \boldsymbol{N}_1 - \boldsymbol{C}_1^{\mathrm{T}}\boldsymbol{\lambda} - \frac{\mathrm{d}}{\mathrm{d}s}\boldsymbol{\lambda} = 0\,, \tag{A.47}$$

$$\boldsymbol{S}_{21}\boldsymbol{X} + \boldsymbol{S}_{22}\boldsymbol{Y} - \boldsymbol{N}_2 - \boldsymbol{C}_2^{\mathrm{T}}\boldsymbol{\lambda} = 0\,, \tag{A.48}$$

where

$$\boldsymbol{S}_{ij} = \int_A \boldsymbol{L}_i\boldsymbol{C}\boldsymbol{L}_j \,\mathrm{d}A \quad (i, j = 1, 2)\,, \tag{A.49}$$

$$\boldsymbol{N}_i = \int_A \boldsymbol{F}_i^{\mathrm{T}}\boldsymbol{g}\,\mathrm{d}A \quad (i = 1, 2)\,. \tag{A.50}$$

Here, the matrices $\boldsymbol{F}_1$, $\boldsymbol{F}_2$, $\boldsymbol{L}_1$, $\boldsymbol{L}_2$ are specified by initial approximations of displacements and deformations (A.40) and (A.41). The algebraic equation (A.48) allows us to express the vector $\boldsymbol{Y}$ through the vectors of generalized displacements $\boldsymbol{X}$ and force factors $\boldsymbol{\lambda}$

$$\boldsymbol{Y} = \boldsymbol{S}_{22}^{-1}(\boldsymbol{N}_2 - \boldsymbol{C}_2^{\mathrm{T}}\boldsymbol{\lambda} - \boldsymbol{S}_{21}\boldsymbol{X})\,. \tag{A.51}$$

Let us employ (A.51) and eliminate $\boldsymbol{Y}$ from the differential equations (A.42) and (A.47) to procure the desired system of first-order differential equations and express it in the unit form as

$$\frac{\mathrm{d}}{\mathrm{d}s}\begin{Bmatrix}\boldsymbol{X}\\ \boldsymbol{\lambda}\end{Bmatrix} = \begin{bmatrix}\boldsymbol{A}_{11} & \boldsymbol{A}_{12}\\ \boldsymbol{A}_{21} & \boldsymbol{A}_{22}\end{bmatrix}\begin{Bmatrix}\boldsymbol{X}\\ \boldsymbol{\lambda}\end{Bmatrix} + \begin{Bmatrix}\boldsymbol{H}_1\\ \boldsymbol{H}_2\end{Bmatrix}, \tag{A.52}$$

where

$$\begin{aligned}
&\boldsymbol{A}_{11} = \boldsymbol{C}_1 - \boldsymbol{b}\boldsymbol{S}_{21}\,, \quad \boldsymbol{A}_{12} = \boldsymbol{b}\boldsymbol{C}_2^{\mathrm{T}}\,,\\
&\boldsymbol{A}_{21} = \boldsymbol{S}_{11} - \boldsymbol{d}\boldsymbol{s}_{21}\,, \quad \boldsymbol{A}_{22} = -\boldsymbol{A}_{11}^{\mathrm{T}}\,,\\
&\boldsymbol{H}_1 = \boldsymbol{b}\boldsymbol{N}_2\,, \quad \boldsymbol{H}_2 = -\boldsymbol{N}_1 + \boldsymbol{d}\boldsymbol{N}_2\,,\\
&\boldsymbol{b} = \boldsymbol{C}_2\boldsymbol{S}_{22}^{-1}\,, \quad \boldsymbol{d} = \boldsymbol{S}_{21}^{\mathrm{T}}\boldsymbol{S}_{22}^{-1}\,.
\end{aligned} \tag{A.53}$$

The system of differential equations (A.52) contains components of the vector of generalized displacements $\boldsymbol{X}$ and force factors $\boldsymbol{\lambda}$ as unknown functions of the argument s. Matrix blocks of the resolvent system and the vector of free terms have been obtained by a formal variational-matrix method. To calculate them in accordance with (A.53) we need to have the laws of distribution of displacements and deformations through the section (matrices $\boldsymbol{F}_1, \boldsymbol{F}_2$, see (A.40)) and $\boldsymbol{L}_1, \boldsymbol{L}_2$ (see (A.41)); elasticity relationships (matrix $\boldsymbol{C}$); matrices of coupling $\boldsymbol{C}_1$ and $\boldsymbol{C}_2$ (see (A.42)) and the vector of external distributed loads $\boldsymbol{g}$ as the initial data. The obtained relationships (A.49), (A.50) and (A.53), defining the algorithm for deriving canonical systems, are common for a wide class of one-dimensional systems. The suggested algorithm can be sufficiently easily generalized for solving nonlinear problems, problems of dynamics and stability.

References

ALFUTOV, N.A. and B.G. POPOV (1977), *Proc. USSR Academy of Sciences, Mechanics of Solids* **3** (in Russian).

ALFUTOV, N.A. (1978), *Design of Fundamentals for Stability of Elastic Systems* (Mashinostroyeniye, Moscow) (in Russian).

ALFUTOV, N.A. and P.A. ZINOV'EV (1982), Deformation and failure of fibrous composites with a brittle polymer matrix under plane stress, in: *Mechanics of Composites* (Mir Publishers, Moscow) Ch. 7.

ALFUTOV, N.A., B.G. POPOV and E.V. BIKOV (1983), *Proc. of Higher Education Institutions, Machine Building* **9** (in Russian).

ALFUTOV, N.A. and B.G. POPOV (1983), The use of a variational-matrix approach to the solution of stability problems for laminated shells, in: *Problems of Stability and Ultimate Load-Carrying Capacity of Structures* (Leningrad Civil Engineering Institute, Leningrad) p. 4 (in Russian).

ALFUTOV, N.A., P.A. ZINOV'EV and B.G. POPOV (1984), *Design of Laminated Plates and Shells Made of Composite Materials* (Mashinostroyeniye, Moscow) (in Russian).

AMBARTSUMYAN, S.A. (1974), *General Theory of Anisotropic Shells* (Nauka Press, Moscow) (in Russian).

BATHE, K.J. (1982), *The Finite Element Procedures in Engineering Analysis* (Prentice-Hall, Englewood Cliffs, N.J.).

BIDERMAN, V.L. (1977), *Mechanics of Thin-Walled Structures. Statics* (Mashinostroyeniye, Moscow) (in Russian).

BIKOV, E.V. and B.G. POPOV (1984), *Proc. of Higher Education Institutions, Machine Building* **10** (in Russian).

BOLOTIN, V.V. and YU.N. NOVICHKOV (1980), *Mechanics of Multilayered Structures* (Mashinostroyeniye, Moscow) (in Russian).

GRIGOL'UK, E.I. and F.A. KOGAN (1972), *Appl. Mech.* **6** (in Russian).

GRIGORENKO, YU.M. (1973), *Isotropic and Anisotropic Laminated Shells of Revolution of Variable Stiffness* (Naukova Dumka, Kiev) (in Russian).

HILL, R. (1956), *Mathematical Theory of Plasticity* (State Publishing House for Technical Literature, Moscow) (in Russian).

LEE, S.W. and T.H.H. PIAN (1978), *AIAA J.* **16**(1).

LOMAKIN, V.A. (1960), *Proc. USSR Academy of Sciences, Mechanics and Machine Building* **4** (in Russian).

Myachenkov, V.I. and I.V. Grigor'ev (1981), *Calculation of Composite Shell-Type Structures on Electronic Computer* (Mashinostroyeniye, Moscow) (in Russian).

Novozhilov, V.V. (1948), *Fundamentals of the Nonlinear Theory of Elasticity* (Gostekhizdat, Press, Leningrad, Moscow) (in Russian).

Oden, J.T. and G.F. Carey (1983), *Finite Elements. Mathematical Aspects* (Prentice-Hall, Englewood Cliffs, N.J.).

Petit, P.H. and M.E. Waddoups (1969), *J. Comp. Mater.* **3**.

Popov, B.G. (1979), *Proc. of Higher Education Institutions, Machine Building* **9** (in Russian).

Popov, B.G. (1982), *Proc. of Higher Education Institutions, Machine Building* **9** (in Russian).

Popov, B.G. (1984), *Proc. of Higher Education Institutions, Machine Building* **12** (in Russian).

Popov, B.G. and E.V. Raman (1984), *Appl. Mech.* **20**(7) (in Russian).

Rabotnov, Yu.N. and E.I. Stepanitchev (1968), *Proc. USSR Academy of Sciences, Mechanics of Solids* **1** (in Russian).

Ryabov, A.F. and A.O. Rasskazov (1974), *Appl. Mech.* **10**(2) (in Russian).

Sarbaev, B.S. (1984), *Proc. of Higher Education Institutions, Machine Building* **4** (in Russian).

Stricklin, I.A., D.R. Navaratha and T.H.H. Pian (1966), *AIAA J.*, 2069.

Valishvili, N.V. (1976), *Calculation Methods of Shells of Revolution by Using Electronic Computer* (Mashinostroyeniye, Moscow) (in Russian).

Zinov'ev, P.A., A.I. Tarakanov and B.Ya. Fomin (1980), *Proc. 12th All-Union Conf. on Theory of Shells and Plates* (Erevan State University Press, Erevan) p. 157 (in Russian).

Zinov'ev, P.A., A.I. Tarakanov and B.Ya. Fomin (1982), Deformation and fracture of composite material in biaxial tension, in: *The Use of Plastics in Machine Building* (Moscow Higher Technical School, Moscow) p. 33 (in Russian).

Zinov'ev, P.A. and B.S. Sarbaev (1982), *Proc. of Higher Education Institutions, Machine Building* **7** (in Russian).

CHAPTER 4

Free Edge Effects in Laminated Composites

C.T. Herakovich

Civil Engineering Department
University of Virginia
Charlottesville, VA 22903
U.S.A.

Contents

HANDBOOK OF COMPOSITES, VOL. 2 – Structure and Design
Edited by C.T. HERAKOVICH and Y.M. TARNOPOL'SKII

1. Introduction

1.1. The free edge problem

A typical composite laminate with free edges is shown in Fig. 1. For the sake of simplicity of notation in this chapter, we shall consider the laminate to be in the x–y plane. The interlaminar stresses are then the z-components σ_z, τ_{yz} and τ_{xz}. Interlaminar normal and shear stresses, some of which exhibit singular behavior, are present in a boundary layer region along the free edges of laminated composite materials. These interlaminar stresses are of critical concern to designers because they can lead to delamination type failures at loads well below those corresponding to inplane failure. Interlaminar stresses may be present as a result of mechanical, thermal or moisture loading.

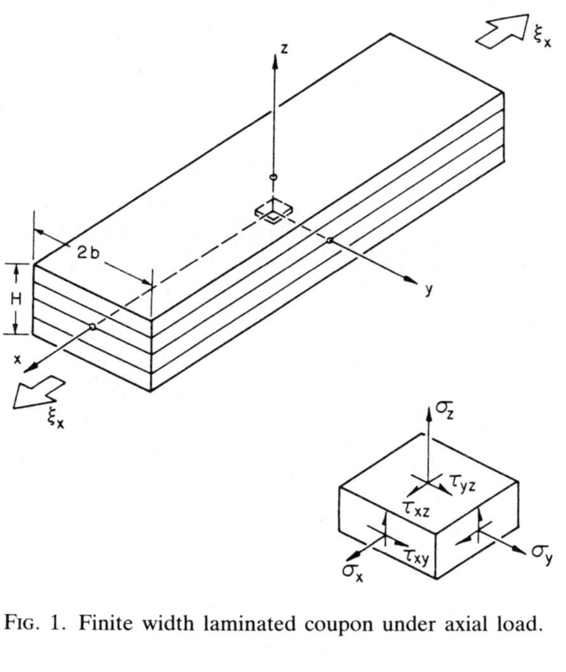

FIG. 1. Finite width laminated coupon under axial load.

Interlaminar stresses are not unique to anisotropic materials. They may be present (and usually are) in any layered material with free edges. These stresses between layers are caused by the mismatch in material properties of bonded adjacent layers and the presence of gradients in the inplane stresses.

Stress free boundaries present the most severe stress gradients and hence the most severe interlaminar stresses. For isotropic layers, the mismatch in properties is usually small and interlaminar stresses are ignored in most applications.

In contrast to isotropic materials, the properties of orthotropic composite laminae made of high modulus fibers in a less stiff matrix exhibit a very broad range of properties as a function of fiber orientation. For example, Poisson's ratio of a typical unidirectional graphite–epoxy composite lamina varies from a low of 0.02 to a high of 0.49. The mismatch in Poisson's ratios leads to interlaminar shear and interlaminar normal stresses.

Further, off-axis laminae exhibit axial shear coupling which can be expressed in terms of the coefficients of mutual influence (the ratio of induced shear strain to applied normal strain). The inplane coefficient of mutual influence can range from a low of −2.2 to a high of +2.2. This potentially large mismatch in coefficient of mutual influence, including opposite signs, can lead to very large interlaminar shear stresses. Other material properties whose mismatch can result in interlaminar stresses include the coefficients of thermal and hygroscopic expansion.

Interlaminar stresses near free edges can be controlled to some extent through the choice of materials, fiber orientations, stacking sequence and layer thicknesses. However, they can be eliminated totally only through the use of homogeneous materials. They must be considered whenever laminated composite materials are used in the design of a structure which has free edges. They are present at all free edges of laminated materials including holes and cutouts.

Edge reinforcement techniques can and have been used to suppress the deleterious effects of interlaminar stresses, but only with limited success and at the expense of additional cost. These techniques do not eliminate the interlaminar stresses. They only provide a restraint against delamination. A thorough understanding of the mechanics of the free edge problem is indepensible to the designer of composite structures. With a basic understanding of the problem, materials (and structures) can be 'designed' to minimize these deleterious effects while still meeting all other structural design requirements.

This chapter will attempt to provide the bases for a fundamental understanding of the mechanics of free edge effects as well as guidelines for optimum design in the presence of free edges.

1.2. Historical review

The existence of interlaminar stresses in a boundary layer region along free edges of laminated materials has been known for some time. The first publication dealing with the singular behavior of stresses at the interesection of a boundary and bonded dissimilar materials is that of Bogy (BOGY, 1968). He considered the problem of two wedges made of different isotropic, homogeneous elastic materials which are bonded together along their common plane. Other studies of isotropic layers have also appeared in the literature (HEIN, 1968; HESS, 1969).

The earliest work dealing with anisotropic materials appears to be a study of anisotropic layers separated by isotropic shear layers (PUPPO and EVENSEN, 1970). The first complete three-dimensional analysis of interlaminar stresses in fibrous composites involved the application of finite difference techniques to the governing elasticity equations (PIPES and PAGANO, 1970). They obtained solutions to the problem of a finite width laminated coupon under tensile axial load (Fig. 1). Their results showed the existence of interlaminar stresses in a boundary layer region along the free edge of the coupon. Results were presented for a variety of fiber orientations and laminate stacking sequences. It was shown that the width of the boundary layer is approximately equal to the thickness of the laminate, that the interlaminar normal stress σ_z and the interlaminar shear stress τ_{zx} can exhibit singular behavior at the intersection of the layer interfaces with the free edge, and that the sign and magnitude of the interlaminar stresses are functions of the laminate configuration (material, fiber orientations, layer thicknesses and order of stacking). The relationship between interlaminar stresses and delamination was discussed in this and subsequent papers (PAGANO and PIPES, 1971; PAGANO and PIPES, 1973; PAGANO, 1974; PAGANO, 1978; SONI and PAGANO, 1983).

The free edge problem in laminated composites has been studied on a continuing basis ever since the original work of Pipes and Pagano. Their finite difference solution was followed quickly by finite element solutions (ISAKSON and LEVY, 1971; RYBICKI, 1971). Isakson and Levy considered the problem of an isotropic shear layer in the absence of interlaminar normal stresses whereas Rybicki did a three-dimensional finite element analysis of the full three-dimensional problem considered by Pipes and Pagano.

It was also recognized that the tensile coupon problem could be formulated as a two-dimensional finite element problem because of the independence of the stress and strain states on the axial coordinate. Initially, cross-ply laminates were studied with two-dimensional finite elements (FOYE and BAKER, 1971; HERAKOVICH, 1976). It was then recognized that general laminates could also be analyzed using two-dimensional finite element models (RENIERI and HERAKOVICH 1976; HERAKOVICH et al., 1976). Subsequently, many results have been published using the finite element technique. In addition to those of Herakovich and co-workers (HERAKOVICH and O'BRIEN, 1979; HERAKOVICH, NAGARKAR and O'BRIEN, 1979; HERAKOVICH and NAGARKAR, 1980) a number of papers have been published by WANG and CROSSMAN (1977). They used the same formulation as Herakovich et al., but with a finer finite element mesh. The finer mesh provided more detail in the neighborhood of the singularity.

The finite element solutions are valuable in that they are an efficient means for obtaining results for a variety of laminate configurations and geometries. They have been obtained for thermal and moisture loading as well as mechanical loading (HERAKOVICH, 1976; FARLEY and HERAKOVICH 1978; CROSSMAN and WANG, 1978; HERAKOVICH and NAGARKAR, 1980). Most finite element results have been obtained using nonsingular elements; however, Wang and co-workers have obtained results with singular elements (WANG and YUAN, 1983).

Analytical solutions have been obtained using perturbation methods (TANG,

1975; HSU and HERAKOVICH, 1977) and complex stress potentials and eigenfunction series (WANG, 1983; WANG and CHOI, 1982). The free edge problem has also been investigated experimentally. Significant works are those of PIPES and DANIEL (1971), OPLINGER et al. (1974), and HERAKOVICH et al. (CZARNEK et al., 1983; HERAKOVICH et al., 1984). The experimental investigations have verified the existence of a boundary layer and large displacement gradients (strains) along free edges. It has been shown that the free edge shear strains in angle-ply laminates are finite and do not exceed nine times the applied axial strain. Also, the boundary layer width has been shown to be approximately equal to the laminate thickness.

2. Problem formulation

2.1. The basic problem

The simplest problem to consider for the study of edge effects in composites is the laminated tensile cupon of Fig. 1. The n ply laminate has total thickness, h, half thickness, H, and finite width, $2b$. For simplicity of exposition, we shall consider only laminates which are symmetric about the midplane and balanced with a $+\theta$ ply for every $-\theta$ ply. Individual layers of the laminate are unidirectional laminae of fibrous composites oriented at an angle θ_i with respect to the global x axis of the coupon. It is assumed that all layers are perfectly bonded together and behave elastically under the application of load. Tensile load is applied at the ends of a long coupon. Away from the ends, the states of stress and strain are independent of the axial coordinate x. The problem is quasi three-dimensional in that all six components of stress and strain may be non-zero, but the analysis can be restricted to a generic two-dimensional cross-section. In interior regions away from edges, all layers are in a state of plane stress and the intralaminar stresses σ_x, σ_y and τ_{xy} can be determined from classical lamination theory.

As depicted in Fig. 2(a), individual layers of a laminate subjected to a given axial strain would exhibit unequal lateral contractions due to differing Poisson's ratios if they were not bonded together. For perfectly bonded laminae, displacement continuity at layer interfaces requires the existence of non-zero interlaminar shear stresses τ_{yz} (Fig. 2(b)). These shear stresses are equilibrated by the inplane laminate stresses σ_y. Moment equilibrium requires the existence of non-zero interlaminar normal stresses σ_z (Fig. 2(c)).

Unidirectional off-axis laminae also experience inplane shear strain when subjected to axial strain. The ratio of induced shear strain to applied axial strain is called the coefficient of mutual influence $\eta_{xy,x}$ (LEKHNITSKII, 1963).

$$\eta_{xy,x} = \gamma_{xy}/\varepsilon_x \,. \tag{2.1}$$

Fig. 3(a) depicts the differing shear strains of $+\theta$ and $-\theta$ laminae when

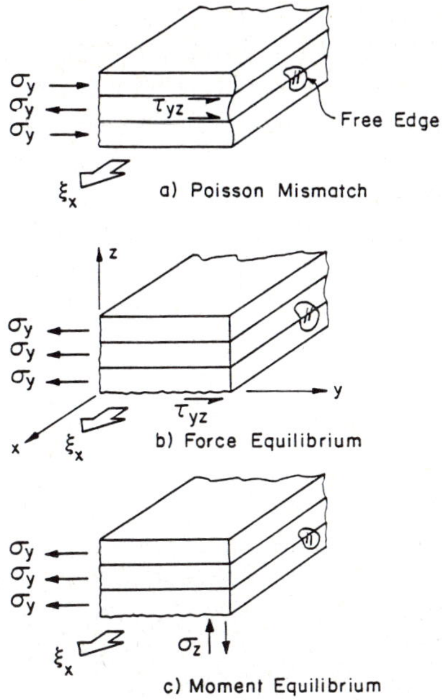

FIG. 2. Effects of Poisson contractions.

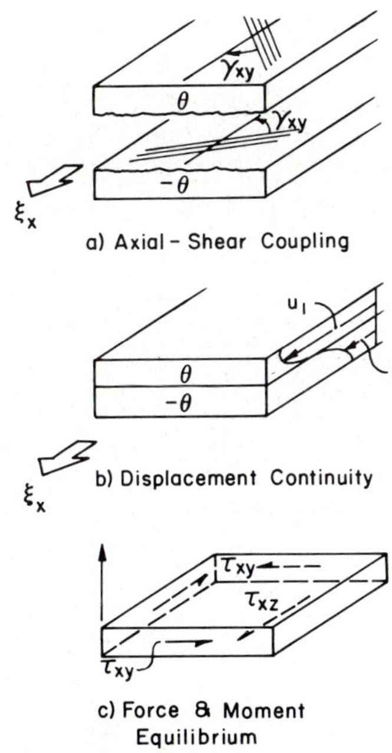

FIG. 3. Effects of axial shear coupling.

subjected to the same axial strain. The shear strains are of opposite sign and thus the layers have opposing shear displacements. Displacement continuity at the free edge of such a laminate is maintained exclusively by the interlaminar shear stresses τ_{zx} (Fig. 3(b)) because the boundary condition requires that τ_{yz} be zero at the edge. Both force equilibrium in the x direction and moment equilibrium about the z axis are maintained by the distribution of interlaminar shear stresses τ_{zx} over the interface and the constant intralaminar shear stresses τ_{xy} (Fig. 3(c)).

Figs. 2 and 3 demonstrate the necessity of non-zero interlaminar stresses when there is material mismatch at free edges. As will be shown, elasticity solutions (analytical and numerical) show that these interlaminar stresses are concentrated in a narrow boundary layer region along the free edge.

2.2. Equilibrium considerations

Equilibrium of a group of layers of a typical laminate clearly shows the necessity for non-zero interlaminar stresses when free edges are present. Fig. 4(a) shows the non-self-equilibrating y-component stresses acting on a group of layers above the interface z_k. Force equilibrium in the y-direction can be

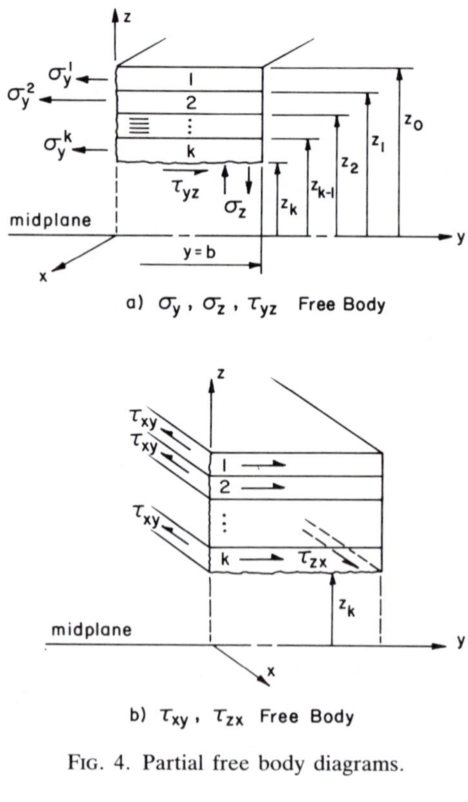

FIG. 4. Partial free body diagrams.

written

$$\sum_{i=1}^{k} \sigma_y^i h_i = \int_0^b \tau_{yz}^k \, dy \, , \tag{2.2}$$

where h_i is the thickness of the ith layer.

Classical lamination theory (CLT) is valid in interior regions away from the edge, hence, σ_y^i is constant in each layer along the centerline $y = 0$. If $\sum \sigma_y^i h_i$ is non-zero for all layers above the interface of interest, then τ_{yz} must be non-zero over some portion $0 < y/b < 1.0$ of that interface. However, the free edge boundary conditions require that $\tau_{yz} = 0$ for all z at $y/b = 1.0$. Hence, this interlaminar shear stress is zero in interior regions away from free edges, is non-zero near the edge, but returns to zero at the edge. The same statement can be made for any arbitrary z-coordinate. Interlaminar stresses are not restricted to the interface between layers.

Moment equilibrium about the x-axis for the same free body diagram requires that the interlaminar normal stress σ_z be related to the laminate stresses σ_y^i through

$$\sum_{i=1}^{k} \sigma_y^i h_i (Z_i - z_k) = \int_0^b \sigma_z^k y \, dy \, , \tag{2.3}$$

where Z_i is the midply coordinate.

Further, z-force equilibrium requires that the total force associated with the distribution of σ_z along any plane $z = z_k'$ be zero. Thus, the distribution of σ_z must, at most, be equivalent to a couple of non-zero moment, but zero force. Clearly, the sign of this moment is a function of the distribution of σ_y along $y = 0$. Hence, the interlaminar normal stress may vary through the thickness of the laminate. It may be positive, negative or zero at any point, but its distribution over any plane z = constant must be equivalent to a couple.

Force equilibrium of the free body diagram in Fig. 4(b) requires that the unbalanced stresses τ_{xy} along the centerline be equilibrated by non-zero stresses τ_{zx} along the interface. This can be expressed as

$$\sum_{i=1}^{k} \tau_{xy}^i h_i = \int_0^b \tau_{zx}^k \, dy \, . \tag{2.4}$$

From (2.4), non-zero interlaminar shear stresses τ_{zx} must exist over some region of the interface if $\sum \tau_{xy}^i h_i$ is non-zero.

The above equilibrium considerations show that the existence of non-zero interlaminar shear and normal stresses over some portion of any plane $z = z_k$ can be verified using laminate analysis. Elasticity theory must be used to determine the actual distribution of the interlaminar stresses.

2.3. Mismatch considerations

The presence of interlaminar stresses is solely the result of the mismatch in material properties from layer to layer. Homogeneous materials do not exhibit interlaminar stresses under the application of uniaxial loading. The two properties of primary interest for the problem under consideration are Poisson's ratio and the coefficient of mutual influence. Poisson's ratio ν_{xy} is defined as the negative ratio of the induced lateral strain ε_y to the applied axial strain ε_x

$$\nu_{xy} = -\varepsilon_y/\varepsilon_x \,. \tag{2.5}$$

The coefficient of mutual influence was defined in (2.1) to be the ratio of induced inplane shear strain γ_{xy} to the applied axial strain. The equation is repeated here for completeness.

$$\eta_{xy,x} = \gamma_{xy}/\varepsilon_x \,. \tag{2.6}$$

The variation of these material properties with the fiber orientation for a unidirectional off-axis graphite–epoxy lamina is shown in Fig. 5. It is evident from the figure that Poisson's ratio is an even function of theta ranging from 0.37 to 0.02. However, the coefficient of mutual influence is an odd function of theta ranging from −2.17 to +2.17.

A more dramatic representation of the mismatch characteristics is presented in Fig. 6. This figure shows the mismatch in Poisson's ratios and coefficients of

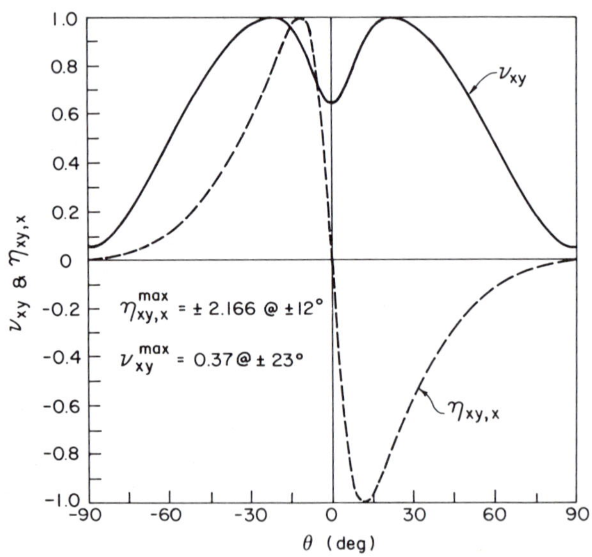

FIG. 5. Poisson's ratio and axial shear coefficient.

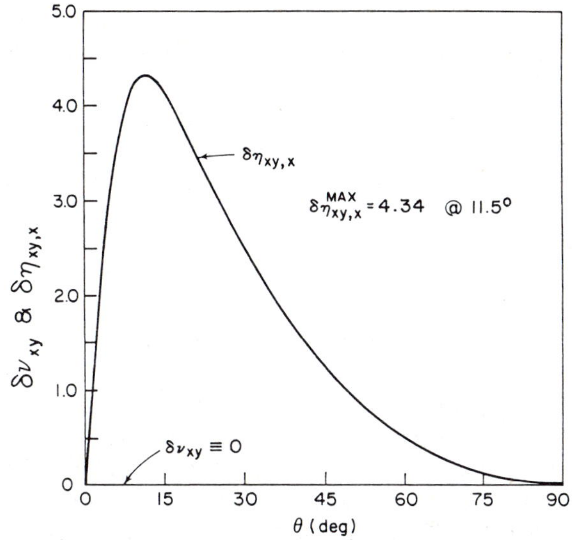

FIG. 6. Mismatch in Poisson's ratio and axial shear coefficient for $+\theta/-\theta$ adjacent layers.

mutual influence for adjacent $+\theta$ and $-\theta$ layers. The mismatch in Poisson's ratios for adjacent $+\theta/-\theta$ layers is identically zero for all angles because Poisson's ratio is an even function of the fiber orientation. It can also be shown that the mismatch in Poisson's ratio for $\theta/90$ and $\theta/0$ adjacent layer combinations has a maximum value of 0.34 for a 22/90 grouping.

The curve for the mismatch in coefficients of mutual influence for $+\theta/-\theta$ layers exhibits much higher values (more than ten times greater) than the maximum Poisson mismatch for the laminates considered. The mismatch in coefficients of mutual influence shows a strong local maximum at 11.5° for the $+\theta/-\theta$ combination. This mismatch in $\eta_{xy,x}$ for $\theta/90$ and $\theta/0$ interfaces is exactly half that of the $+\theta/-\theta$ configuration. This is because $\eta_{xy,x}$ of the 0 and 90 layers is identically zero and $\eta_{xy,x}$ is an odd function of θ.

Several important design guidelines can be drawn from Fig. 6.

- For angle-ply laminates, Poisson mismatch is zero and hence the interlaminar normal stress σ_z and interlaminar shear stress τ_{yz} are expected to be zero for all y and all z.
- The largest interlaminar shear stresses τ_{zx} are expected for adjacent $+11.5/-11.5$ layers where the mutual influence mismatch is greatest.
- Interspersing 0 or 90 layers between $+\theta$ and $-\theta$ layers reduces the mutual influence mismatch by a factor of 2.0. This should result in a corresponding reduction in interlaminar shear stresses τ_{zx}.
- There is no mismatch of mutual influence at 0/90 interfaces and hence cross-ply laminates should be free of interlaminar shear stresses τ_{zx}.

2.4. Elasticity formulation

The elasticity formulation for the balanced, symmetric, finite width tensile coupon problem can be developed starting from the condition of x-independence of stresses and strains. Functional analysis then leads to the most general displacement field

$$\begin{aligned} u(x, y, z) &= x\xi_x + U(y, z) , \\ v(x, y, z) &= V(y, z) , \\ w(x, y, z) &= W(y, z) , \end{aligned} \tag{2.7}$$

where ξ_x is the applied axial strain and $U(y, z)$, $V(y, z)$ and $W(y, z)$ are functions representing the unknown displacement fields. Given the applied axial strain, all displacements are functions of y and z only and hence the analysis may be limited to a generic cross-sectional plane (Fig. 7). Further, since the laminate is symmetric about its midplane, the analysis may be limited to a symmetric half-plane.

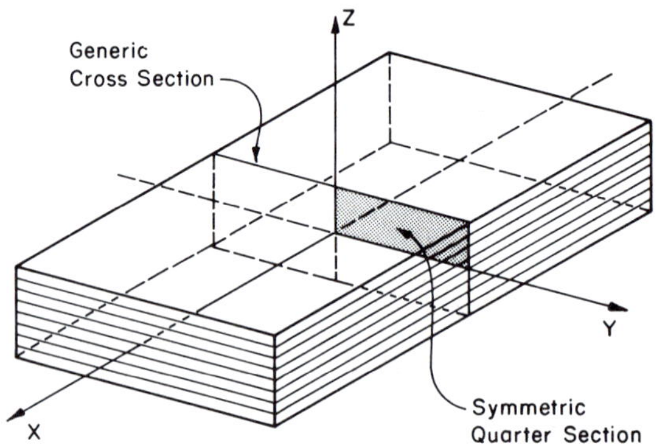

FIG. 7. Generic cross-sectional plane of coupon.

Several investigations (DANA, 1974; GRIFFIN et al., 1980) have shown that for thin, balanced, symmetric laminates, the analysis may be limited further to a symmetric quarter section of the laminate without substantial sacrifice in accuracy. Thus, the problem is reduced to the determination of the U, V, and W displacement fields in a quarter section of a laminate (Fig. 7) which is subjected to uniform axial strain.

The boundary conditions on the quarter section are:

– traction boundary conditions:

$$\begin{aligned} \sigma_y = \tau_{xy} = \tau_{yz} = 0 \quad &\text{along } y = b , \\ \sigma_z = \tau_{yz} = \tau_{zx} = 0 \quad &\text{along } z = H , \end{aligned}$$

– displacement boundary conditions:

$$W(y,0)=0\,,$$
$$U(0,z)=V(0,z)=0\,.$$

The formal solution to this problem is the set of functions U, V, and W which satisfy the displacement equilibrium equations, the compatibility conditions, the appropriate boundary conditions, and the interfacial stress and displacement continuity conditions. The displacement equilibrium equations are 2nd order, linear partial differential equations in terms of the unknown functions U, V and W. The coefficients are constant within each layer. If C_{ij} are the material stiffness coefficients for a given layer, the displacement equilibrium equations for that layer may be written:

$$\begin{aligned} &C_{66}U_{,yy}+C_{55}U_{,zz}+C_{26}V_{,yy}+C_{45}V_{,zz}+(C_{36}+C_{45})W_{,yz}=0\,,\\ &C_{26}U_{,yy}+C_{45}U_{,zz}+C_{22}V_{,yy}+C_{44}V_{,zz}+(C_{23}+C_{44})W_{,yz}=0\,,\\ &(C_{45}+C_{36})U_{,yz}+(C_{44}+C_{23})V_{,yz}+C_{44}W_{,yy}+C_{33}W_{,zz}=0\,. \end{aligned} \tag{2.8}$$

In (2.8), commas denote partial differentiation with respect to the indicated variables.

As discussed previously, results for this problem have been obtained numerically and analytically. The analytical solution of WANG (1983) is formulated in terms of Lekhnitskii's stress potentials $F(y,z)$ and $\Psi(y,z)$ (LEKHNITSKII, 1963). The solution is expressed in terms of the complex roots of the characteristic equation. The interested reader is referred to WANG (1983) for details of the mathematical solution.

3. Experimental results

3.1. Axial displacements in angle-ply laminates

Moire fringe patterns showing the axial displacements along the face and free edge of three angle-ply graphite–epoxy tensile coupons are shown in Fig. 8 (CZARNEK et al., 1983). Side by side, face and edge fringe patterns. On both the face and the edge of the coupons, the fringe patterns are independent of the axial coordinate. The displacements are uniform on the face of the specimen with the exception of a boundary layer region along the edges. The nonuniform displacements in the boundary layer region and on the edges have the same form along the entire length of the coupon. It is evident from the face patterns that the nonuniform axial displacements in the boundary layer region are antisymmetric about the coupon centerline for these angle-ply laminates. The experimental results clearly verify the assumption that the strains (fringe gradients) are independent of the axial coordinate. These experimental results also show that the boundary layer width is approximately equal to the laminate thickness for all three laminates.

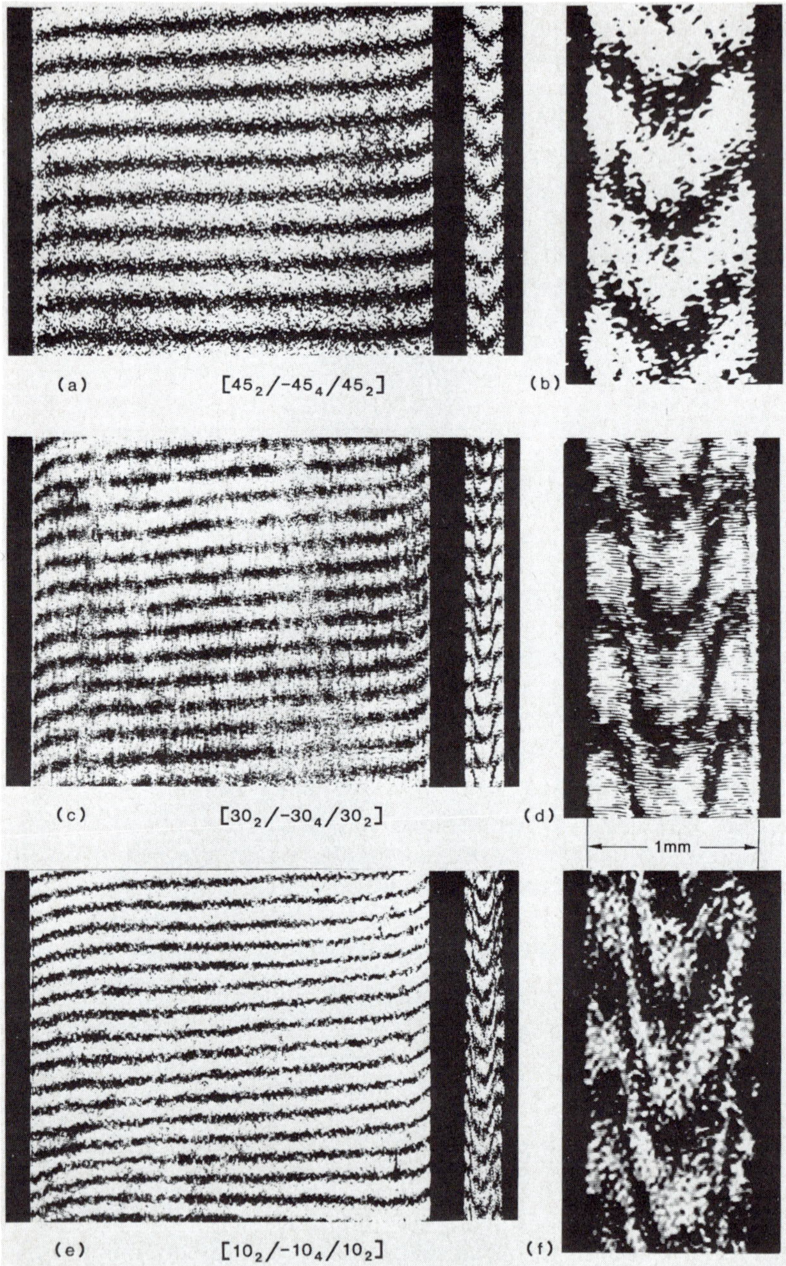

FIG. 8. Face and edge Moire fringe patterns.

3.2. Shear strains in angle-ply laminates

On the specimen edges, the fringe patterns show large gradients indicating very high strains (Fig. 9). The shear strains γ_{zx} on the specimen edges are proportional to the gradient of the fringe lines with respect to the through the thickness z coordinate. The maximum slope of the fringes is at the $+\theta/-\theta$ interface of the laminate and is finite. The influence of fiber orientation is very evident in the figure. The maximum strains for the 10° and 30° coupons are several times larger than that of the 45° coupon. The 30° specimen exhibits strains slightly larger than the 10° specimen.

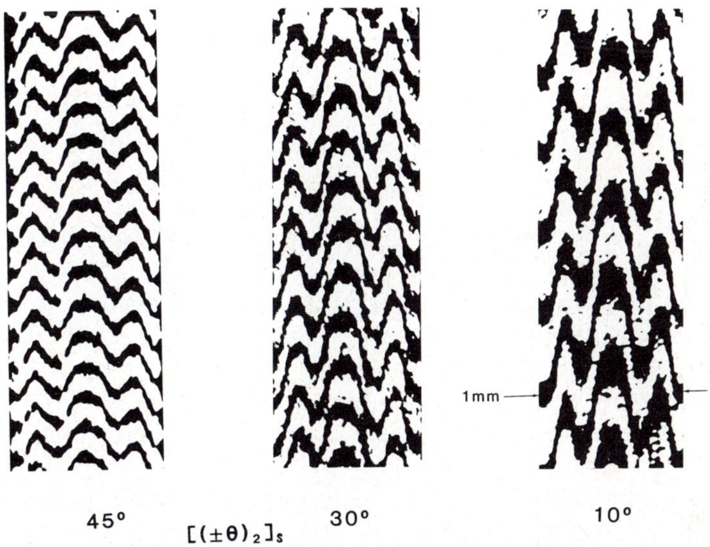

FIG. 9. Moire fringes along edges of $[(\pm\theta)_2]_s$ laminates.

Fig. 10 shows a comparison of edge shear strains obtained from the moire fringe patterns with those obtained using finite elements. The finite element results were obtained for 5° increments of fiber orientations in order to better identify the critical angle. Since the finite element results are a function of the type and size of element used, a study was performed to determine the 'best' finite element representation. As indicated in the figure, the best results were obtained with a four node, isoparametric finite element representation. The figure shows reasonably good correlation between theory and experiment using the same finite element grid for all laminates considered. The finite element results indicate that the critical angle is 15°. This result compares well with the mismatch considerations discussed previously.

The experimental results indicate that the ratio of maximum shear strain to applied axial strain varies with fiber orientation, is finite and does not exceed 9.0 for the most critical angle.

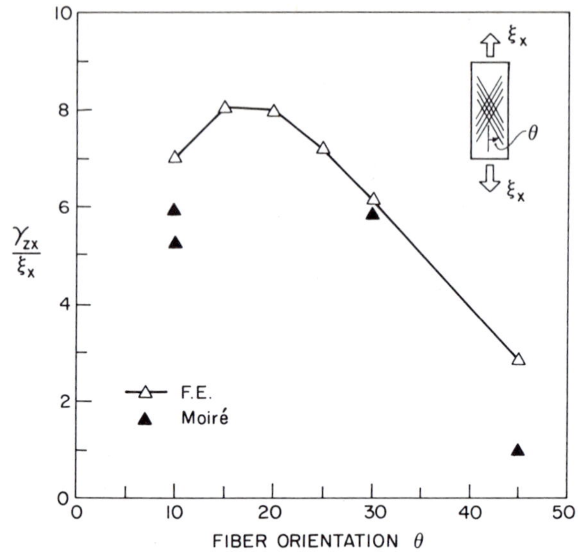

FIG. 10. Theoretical experimental correlation for edge shear strains.

A disconcerting observation is that the finite element results predicted that the shear strain γ_{zx} is larger for the [10] laminate than for [30] laminate. The experimental results indicate that the opposite is true. The differences are small and could be attributed to accuracy of the input properties to the finite element program as well as to accuracy in the analysis of the maximum slopes from the experimental data.

These comparisons show that the finite element method can be applied to the free edges problem with confidence. It is not necessary to use a singular element, but care should be used in selecting the element size (NAGARKAR and HERAKOVICH, 1979).

4. Numerical predictions

As discussed previously, theoretical results have been obtained by numerous investigators using a variety of methods. Comparisons between analytical and numerical solutions have generally shown good agreement. Differences that have been observed are usually associated with the accuracy of a given solution in the vicinity of the mathematical singularity which is present in the idealized problem of two distinct layers. (As shown in the previous section, interlaminar shear strains in real laminates are not singular.)

The results which have been published in the literature tend to address specific cases rather than present a complete set of results for the laminates of general interest. For this survey article, it has been decided to present a set of results for cross-ply, angle-ply and quasi-isotropic laminates which demonstrate various facets of the free edge problem. All the results are for a typical graphite–epoxy whose properties are given in Table 1.

The results presented here were obtained using a four-node isoparametric finite element representation (BUCZEK et al., 1983). It was decided to use finite elements because of the availability of the computer program and associated graphics routines, the comparison between finite element and analytical solutions demonstrated by WANG (1983), and the comparison between experiment and finite element results demonstrated by HERAKOVICH et al. (1984).

The same finite element mesh (Fig. 11) was used for all laminates. It consisted of 356 elements, 353 nodes and 1028 degrees of freedom. A high

TABLE 1. Typical properties for graphite epoxy.

$E_1 = 132$ GPa	(19.2×10^6 psi)
$E_2 = 10.8$ GPa	(1.56×10^6 psi)
$E_3 = 10.8$ GPa	(1.56×10^6 psi)
$G_{23} = 3.38$ GPa	(0.49×10^6 psi)
$G_{13} = 5.65$ GPa	(0.82×10^6 psi)
$G_{12} = 5.65$ GPa	(0.82×10^6 psi)
$\nu_{23} = 0.49$	
$\nu_{13} = 0.24$	
$\nu_{12} = 0.24$	

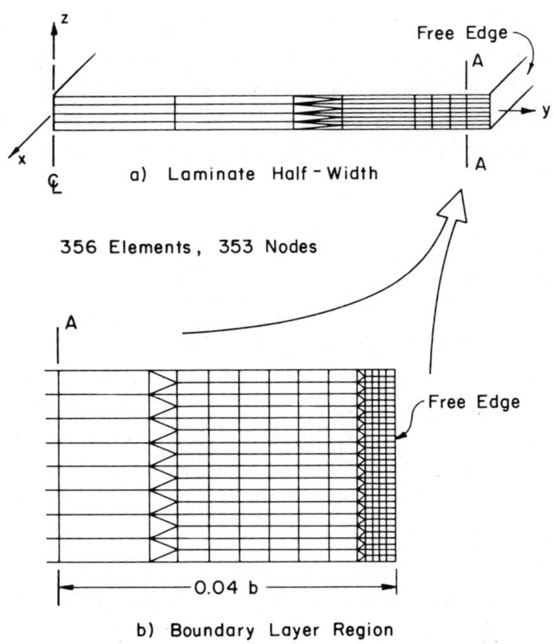

FIG. 11. Finite element mesh for finite width coupon.

density of smaller elements was used in the boundary layer region. There are 32 elements through the thickness at the edge. Thus, the number of elements per layer ranged from 8 to 24 depending upon the laminate configuration. The triangular transition elements were obtained by double numbering one of the nodes. All results are for an applied axial strain of 0.1%.

The influence of fiber orientation and stacking sequence on edge effects will be demonstrated through deformed finite element grids and interlaminar stress distributions along layer interfaces and through the thickness near the free edge. Failure analysis as predicted by the tensor polynomial failure criterion will be discussed and compared with experimental results.

4.1. Cross-ply laminates

Cross-ply laminates are the simplest laminates to consider because there are no off-axis plies. Hence $\eta_{xy,x}$ is identically zero for all piles and there is no coupling between axial and shear strains. The interlaminar shear stress τ_{zx} is identically zero. This class of laminates also exhibits exact symmetry about the x–z plane and hence the analysis of one-quarter of the cross-section introduces no approximation into the solution, independent of thickness. Six configurations have been chosen to demonstrate the influence of edge effects in cross-ply laminates. They are: $[0_2/90_2]_s$; $[90_2/0_2]_s$; $[0/90_3]_s$; $[90/0_3]_s$; $[(0/90)_2]_s$; $[(90/0)_2]_s$.

4.1.1. Free edge deformations

Deformed finite element meshes for six cross-ply laminates are shown in Fig. 12. All deformed plots presented in this chapter have been scaled such that the maximum V displacement equals the maximum W displacement. This was done to enhance the deformations so that local effects could be distinguished easily. Only the boundary layer portion of each laminate is shown. Figs. 12(a) through 12(d) clearly show the influence of stacking sequence on laminates with equal numbers of 0 and 90 plies, but different stacking sequences. The mismatch in Poisson's ratio is most evident in the $[(90/0)_2]_s$ laminate where the alternating layers provide a distinct contrast in lateral contraction. The influence of percent 0° plies is demonstrated in Figs. 12(e) and 12(f). Figures such as these are very helpful for understanding edge effects. Close study of the deformed meshes provides insight into the expected distribution of interlaminar stresses. The interlaminar normal stress is expected to be tensile in regions where the mesh exhibits relative expansion and compressive in regions where the relative deformation indicates compaction. Based upon the deformed mesh shapes, it is anticipated that, on the average, the laminates with the 0 ply on the outside exhibit interlaminar tension while those with the 90 ply on the outside exhibit interlaminar compression. Comparison of the deformed meshes with the stress distributions presented in the following sections will show the correlation between deformed shape and stress distribution.

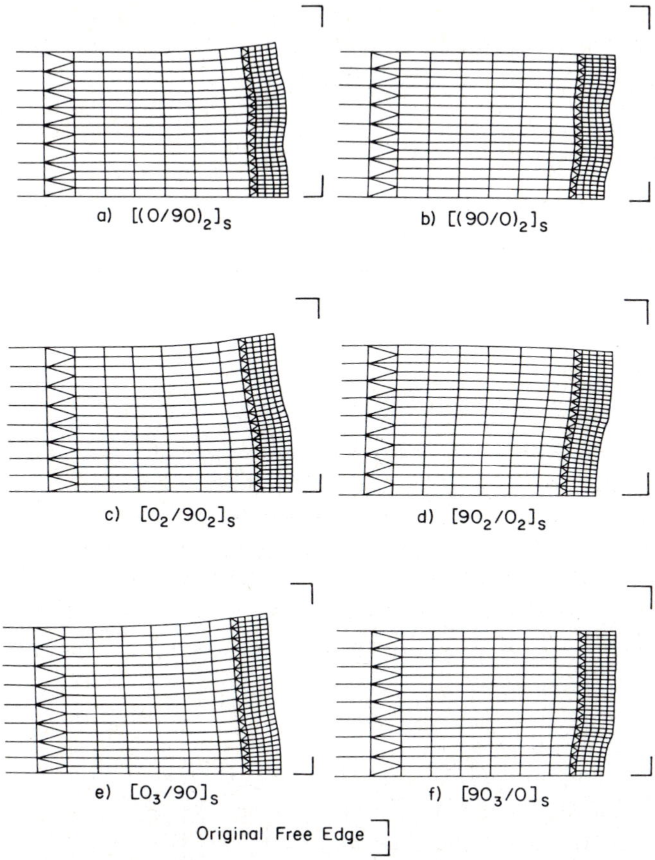

FIG. 12. Deformed shapes of cross-ply laminates.

4.1.2. *Interlaminar stress distributions*

The distributions of σ_z along the midplane of $[90_2/0_2]_s$ and $[0_2/90_2]_s$ laminates are shown in Figs. 13 and 14. It is evident from the curves that σ_z exhibit singular behavior at the free edge, corresponds to a moment, and reverses sign when the stacking sequence is reversed. However, reversing the stacking sequence does not result in a mirror image of the stress distribution. There is a 20% difference in the magnitude of the maximum σ_z between the two stacking sequences. The width of the boundary layer varies from 4% to 20% of the laminate width depending upon the definition chosen. The smaller value corresponds to the point where σ_z changes sign and the larger value roughly corresponds to the point where all interlaminar stresses approach zero. In terms of laminate thickness, the boundary layer width ranges from $0.5h$ to $2.5h$ depending upon the definition.

As the deformed shapes of Fig. 12 indicate, the laminate with the 0 layer on the outside has tensile interlaminar normal stress at the free edge whereas the

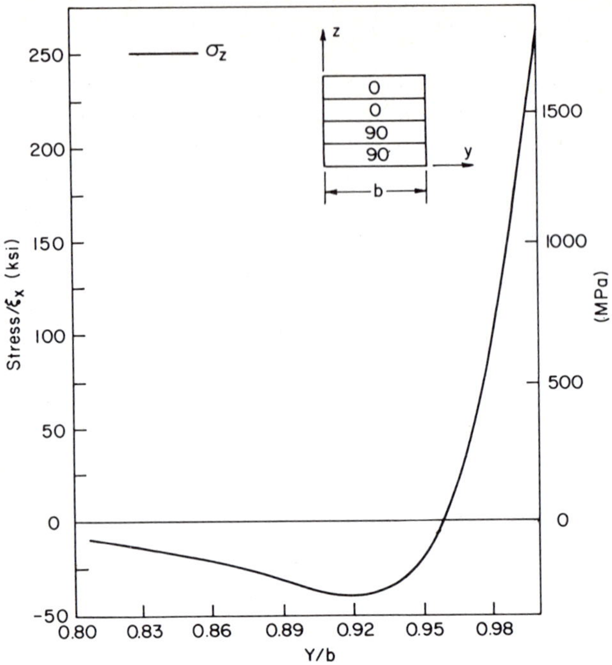

FIG. 13. Interlaminar stresses along the midplane of a $[0_2/90_2]_s$ laminate.

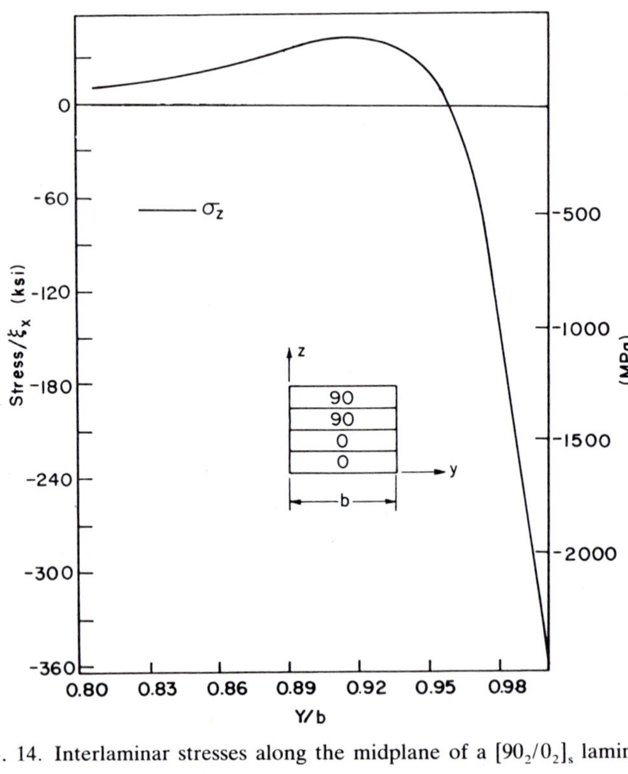

FIG. 14. Interlaminar stresses along the midplane of a $[90_2/0_2]_s$ laminate.

laminate with the 90 ply on the outside has compressive interlaminar normal stress. The interlaminar shear stresses are identically zero along the midplane, consistent with the symmetry requirements of the problem.

Figs. 15 and 16 shows the interlaminar stress distributions along the 0/90 and 90/0 interfaces of the same two laminates. It is evident from this figure that there are differences between the two laminates. The distribution of σ_z is quite different in the two laminates. In the $[0_2/90_2]_s$ laminate, the distribution of σ_z along the 0/90 interface is very similar to the distribution along the midplane. The behavior is singular and tensile at the free edge. However, in the $[90_2/0_2]_s$ laminate the interlaminar normal stress distribution along the 90/0 interface reverses sign just prior to the free edge and appears to be approaching a tensile singular value. Thus, both laminates exhibit singular tensile behavior.

The distribution of interlaminar shear stress τ_{yz} also provides some unexpected results. For the $[90_2/0_2]_s$ laminate, the distribution is as expected reaching a maximum value a small distance in from the free edge and then tending toward zero as the free edge is approached. However, for the $[0_2/90_2]_s$ laminate, the magnitude continues to increase as the free edge is approached. This result shows a limitation of displacement based finite element solutions. They do not satisfy stress free boundary conditions exactly. The sign of the interlaminar shear stress is opposite in the two laminates as it should be based upon lamination theory and equilibrium (2.2).

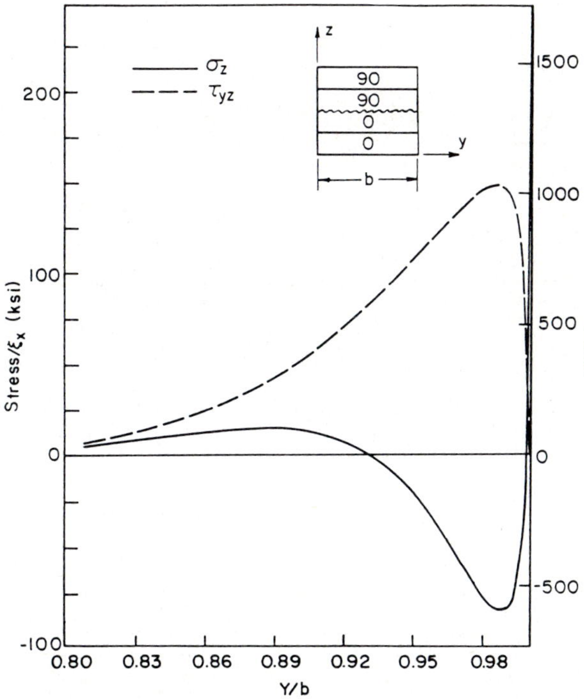

FIG. 15. Interlaminar stresses along the 90/0 interface of a $[90_2/0_2]_s$ laminate.

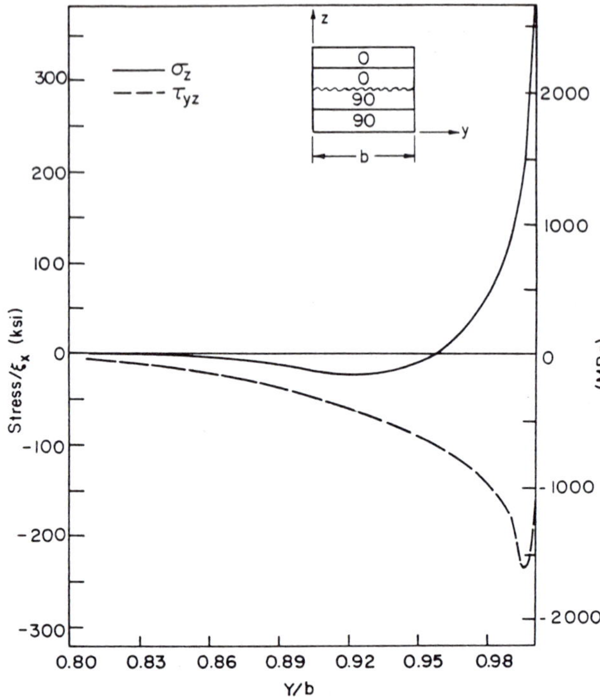

FIG. 16. Interlaminar streses along the 0/90 interface of a $[0_2/90_2]_s$ laminate.

4.1.3. Through the thickness stress distributions

The through the thickness distributions of σ_z at the free edge of $[0_3/90]_s$ and $[90_2/0_2]_s$ laminates are shown in Figs. 17 and 18, respectively. These two figures demonstrate the influence of stacking sequence and percent 0° plies on the interlaminar normal stress.

The singular behavior at the intersection of the interface and the free edge is evident in both figures in the form of stress concentrations and steep gradients. Although the interlaminar normal stress is tensile at the interface in both figures, the interfacial distributions of Figs. 15 and 16 show that the $[90_2/0_2]_s$ laminate exhibits a very steep gradient including a change in sign very near the free edge. In contrast, the interlaminar normal stress in the $[0_2/90_2]_s$ laminate is positive and increases monotonically over the last 4% of the width.

Changing the location of the 90° plies from exterior to interior changes the sign of the normal stress at the midplane where there is no singularity and the stress distribution is well-behaved. The maximum interlaminar normal stress occurs at the midplane for the $[90_2/0_2]_s$ laminate, but near the interface in the $[0_3/90]_s$ laminate. The latter laminate represents the worst case of the six considered because the largest interlaminar normal stresses are predicted for this case.

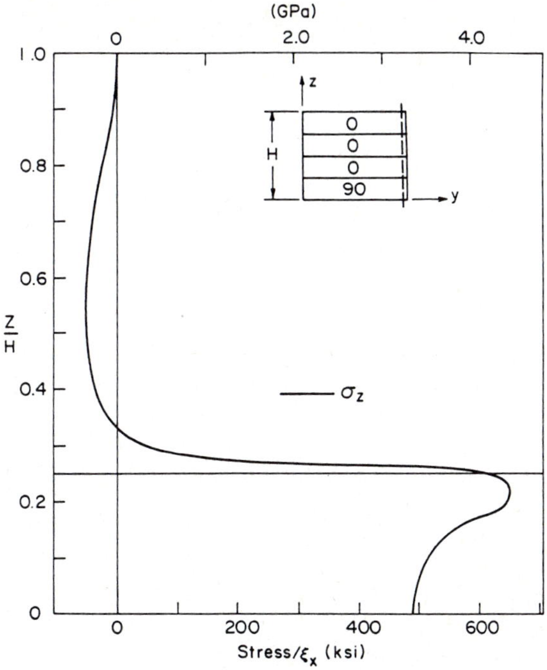

FIG. 17. Through the thickness distribution of interlaminar normal stress in a $[0_3/90]_s$ laminate.

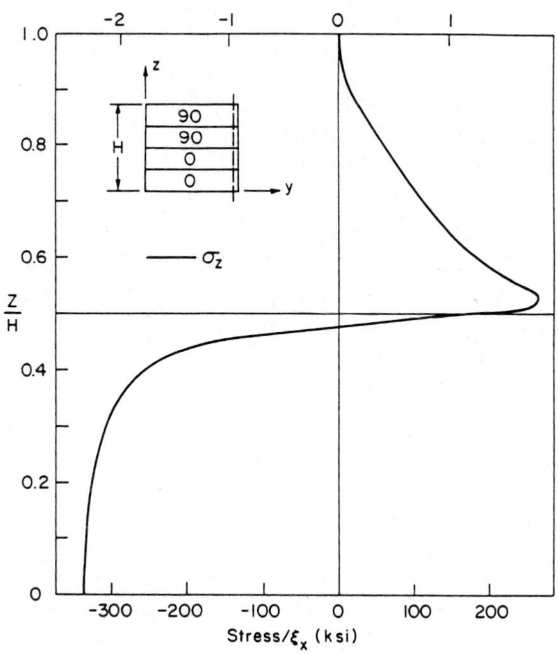

FIG. 18. Through the thickness distribution of interlaminar normal stress in a $[90_2/0_2]_s$ laminate.

Through the thickness distributions for the other cross-ply laminates show results similar to those in Figs. 17 and 18. In general, the magnitudes of the stresses are lower when the 0° and 90° plies are interspersed rather than clustered.

4.2. *Angle-ply laminates*

Balanced, symmetric laminates with fibers oriented at $+\theta$ and $-\theta$ to the loading direction provide fundamental insight into the influence of fiber orientation on interlaminar stresses. Changing the stacking sequence of these laminates provides additional insight into the effect of layer thickness.

The mismatch considerations discussed previously for angle-ply laminates indicate that the interlaminar normal stress should be small or zero whereas the interlaminar shear stress τ_{zx} is expected to be quite large. Further, it was shown that the interlaminar shear stress τ_{zx} is expected to be maximum in the 10–15 deg. fiber range. Four fiber orientations have been chosen to demonstrate interlaminar stresses in angle-ply laminates: 10, 30, 45 and 60. Two stacking sequences were investigated: $[\theta_2/-\theta_2]_s$ and $[(\theta/-\theta)_2]_s$.

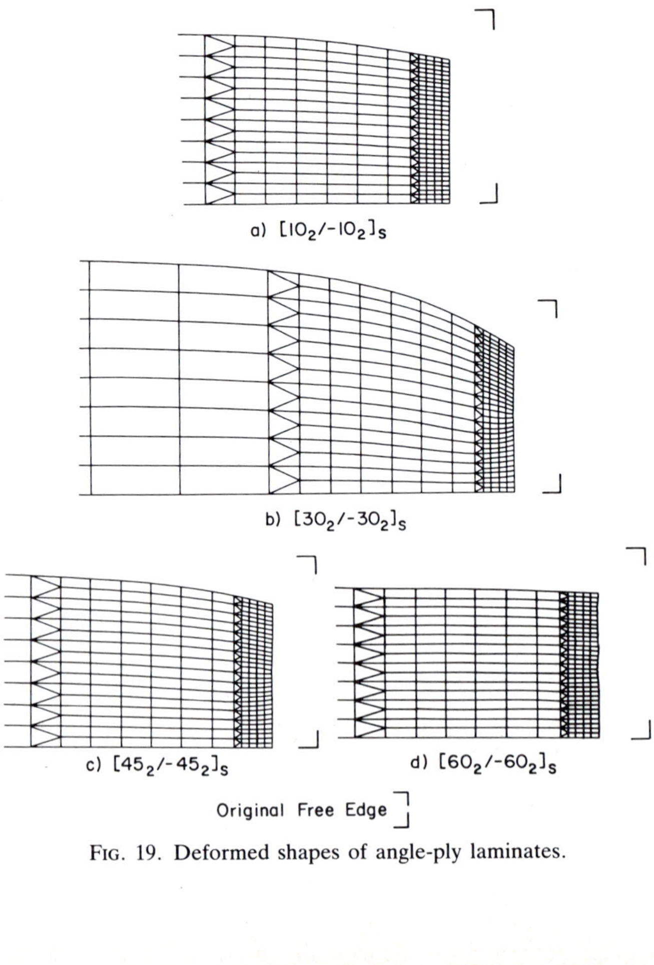

FIG. 19. Deformed shapes of angle-ply laminates.

4.2.1. Free edge deformations

The most striking feature of the deformed meshes for angle-ply laminates in Fig. 19 is that the edges are essentially straight. This is in sharp contrast to the deformed shapes for the cross-ply laminates and is a direct result of the equality of Poisson's ratio in $+\theta$ and $-\theta$ layers (Fig. 5). Another interesting feature of these deformed shapes is that the thickness of the central region (away from the edges) of the 30° laminates increases with the application of load. This corresponds to a negative through the thickness effective laminate Poisson's ratio. This unusual phenomenon has been explained in the literature (HERAKOVICH, 1984). It is the result of the laminated material behaving as a structure with compressive internal intralaminar stresses. The compressive intralaminar stresses result in expansion of the individual layers in the thickness direction. Such behavior is not limited to pure angle-ply laminates.

4.2.2. Interlaminar stress distributions

Typical distributions of interlaminar stresses along the interfaces of 15° angle-ply laminates are presented in Figs. 20 and 21. These figures show that the interlaminar shear stress τ_{zx} exhibits singular behavior near the free edge. The other two components of interlaminar stress are essentially zero along the entire interface. The width of the boundary layer ranges from 10 to 20% of the

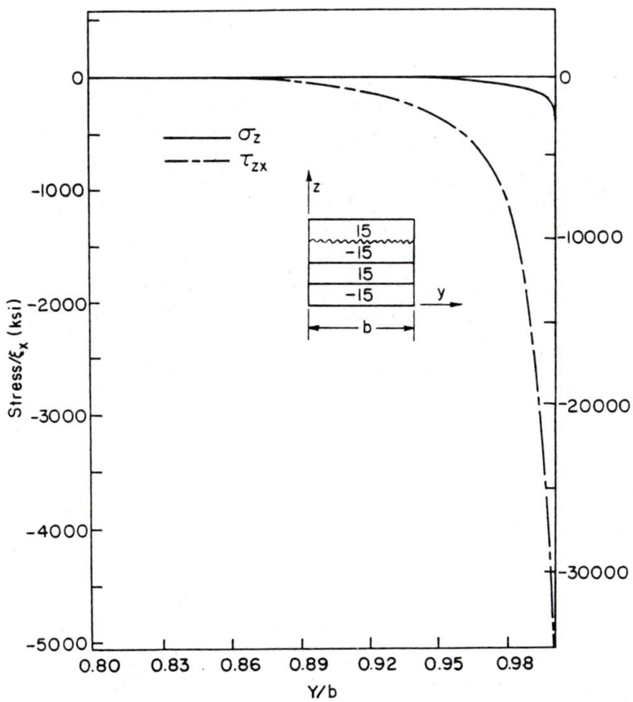

FIG. 20. Interlaminar stresses along the 15/−15 interface of a $[(\pm 15)_2]_s$ laminate.

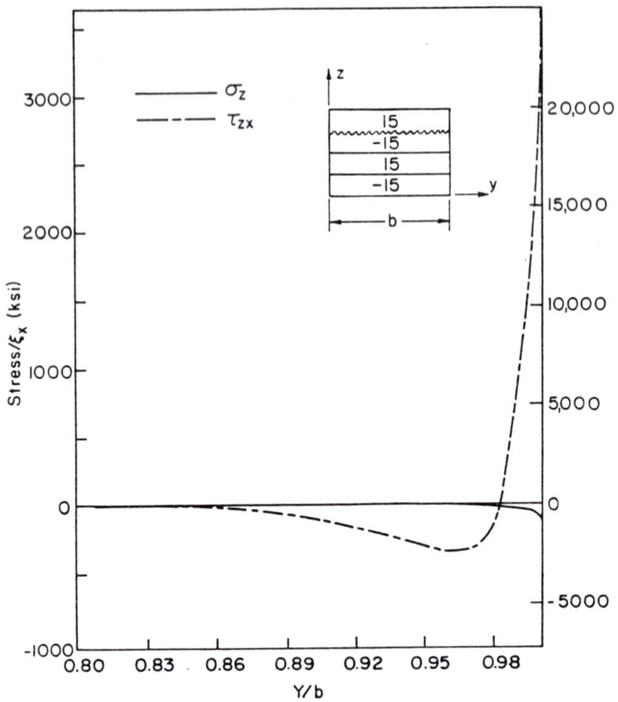

FIG. 21. Interlaminar stresses along the 15/−15 interface of a $[(\pm 15)_2]_s$ laminate.

laminate width depending upon the stacking sequence and the definition of boundary layer. The boundary layer width is smaller in the alternating $+\theta/-\theta$ configuration.

The maximum magnitude of the interlaminar shear stress is a function of fiber orientation. Fig. 22 shows a plot of the maximum shear stress τ_{zx} in

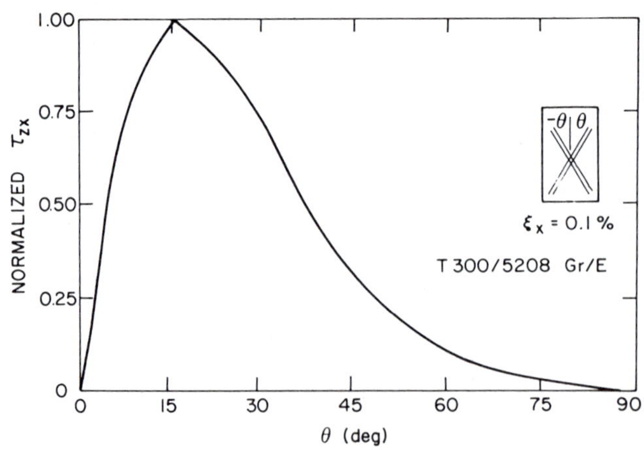

FIG. 22. Maximum interlaminar shear stress τ_{zx} in angle-ply laminates.

angle-ply laminates for 5° increments of fiber orientation. The maximum value is for $\theta = 15°$. This is the same value predicted by the mismatch considerations discussed previously. These results support the application of the mismatch concept for design considerations.

4.2.3. Through the thickness stress distributions

Figs. 23 and 24 show through the thickness distributions of interlaminar stresses near the free edge of alternating $[(\pm 15)_2]_s$ and clustered $[15_2/-15_2]_s$ laminates. The results show that the stacking sequence has a dramatic effect on the through the thickness distributions. Alternating the layers causes the sign of τ_{zx} to alternate between plus and minus. The clustered laminates exhibit uniform sign through the thickness and higher maximum stresses. The maximum interlaminar stress in the clustered laminates is approximately 20% higher than the maximum values in the alternating laminates. The higher stresses in the clustered laminates leads to delamination at lower applied strains. This will be discussed in more detail in a later section on edge damage. These results clearly show that the alternating stacking sequence is preferred when free edges are present. It should be noted that this conclusion is independent of the loading direction since shear strength is independent of sign. The stress distributions for the 15° laminates are typical of the distributions for all other angle-ply laminates.

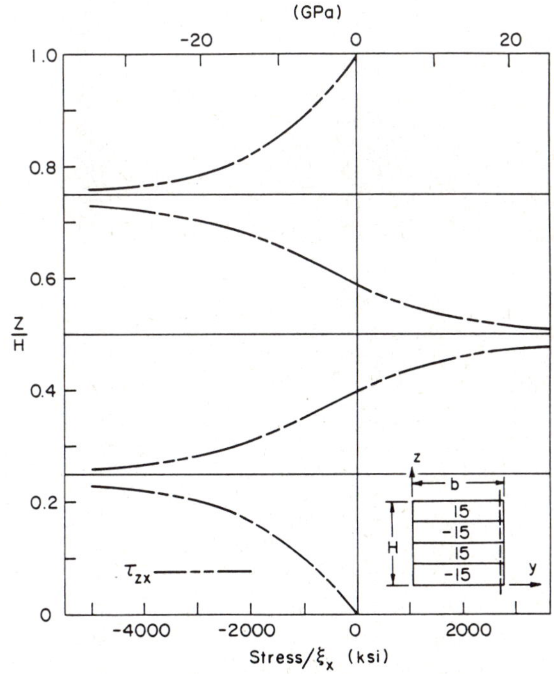

FIG. 23. Through the thickness distribution of interlaminar stresses in a $[(\pm 15)_2]_s$ laminate.

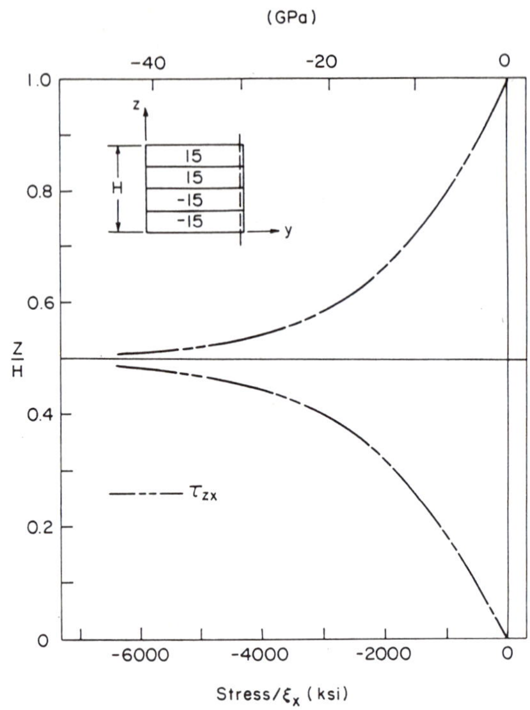

FIG. 24. Through the thickness distribution of interlaminar stresses in a $[15_2/-15_2]_s$ laminate.

4.3. *Quasi-isotropic laminates*

There are twelve unique configurations of symmetric quasi-isotropic laminates made with equal percentages of 0, 90, +45 and −45 laminae orientations if the +45 and −45 layers are assumed interchangeable. These laminates have been used extensively because the inplane elastic properties are isotropic. Quasi-isotropic laminates of the type described can be divided into two groups. Those with adjacent ±45 layers and those with interspersed ±45 layers. The twelve unique configurations are presented in Table 2.

TABLE 2. Quasi-isotropic laminates.

Interspersed ± 45 laminates	Adjacent ± 45 laminates
$[90/45/0/-45]_s$	$[90/0/\pm 45]_s$
$[0/-45/90/45]_s$	$[90/\pm 45/0]_s$
$[45/90/0/-45]_s$	$[0/90/\pm 45]_s$
$[45/90/-45/0]_s$	$[0/\pm 45/90]_s$
$[45/0/90/-45]_s$	$[\pm 45/90/0]_s$
$[45/0/-45/90]_s$	$[\pm 45/0/90]_s$

4.3.1. *Free edge deformations*

The deformed shapes of these twelve laminates are shown in Figs. 25 and 26. It is clear from these figures that stacking sequence can have a dramatic effect on the deformed shape. These figures show that some laminates exhibit mainly contraction along the free edge (e.g., $[90/45/0/-45]_s$) while others exhibit mainly expansion (e.g., $[\pm 45/0/90]_s$). (As will be shown, there is direct correspondence between the through the thickness distribution of interlaminar normal stress and the deformed shape.) The effect of large differences in Poisson's ratios from layer to layer is apparent in the form of ragged edges. This is in sharp contrast to the angle-ply laminates which are free of Poisson mismatch and exhibit essentially straight edges.

It should be noted that all of these quasi-isotropic laminates have identical through the thickness contraction in the central region away from the edge. Differences in the figures are the result of the normalization scheme used for this presentation.

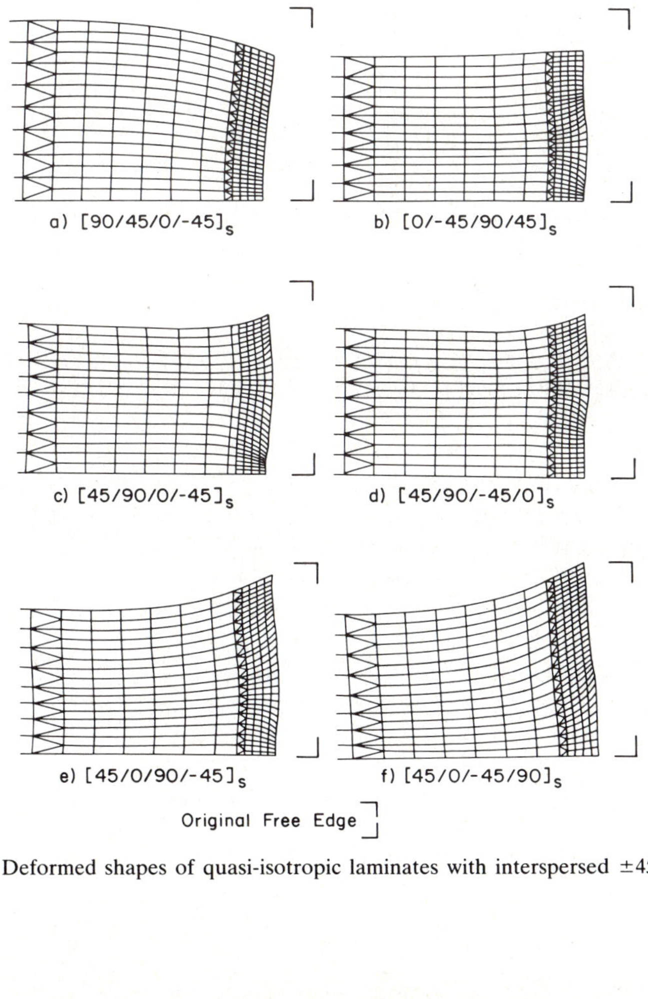

FIG. 25. Deformed shapes of quasi-isotropic laminates with interspersed ±45 layers.

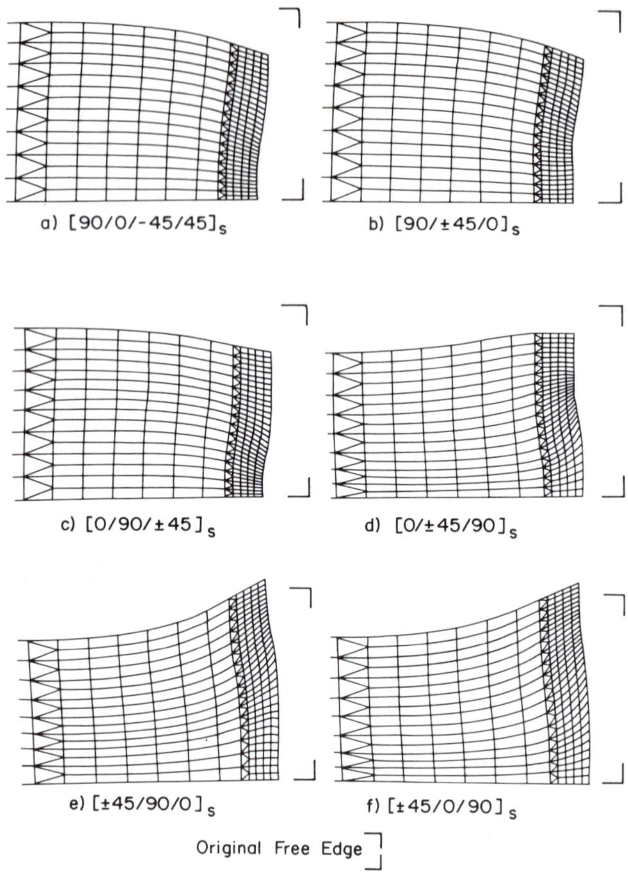

FIG. 26. Deformed shapes of quasi-isotropic laminates with adjacent ±45 layers.

4.3.2. Interlaminar stress distributions

Three quasi-isotropic laminates have been chosen to exhibit typical through the thickness and interlaminar distributions of interlaminar stresses in quasi-isotropic laminates. The laminates chosen for presentation are: $[90/45/0/-45]_s$, $[0/90/\pm45]_s$, and $[\pm45/0/90]_s$. These three laminates were chosen to illustrate tensile and compressive interlaminar normal stresses as well as adjacent and interspersed +45/−45 configurations. The stress distributions are shown in Figs. 27, 28 and 29. Several important features are evident in all three figures. The distribution of interlaminar normal stress is equivalent to a couple with non-zero moment, but zero force. This is consistent with the equilibrium requirements discussed earlier. Further, σ_z exhibits singular behavior at the intersection of layer interfaces and free edges. The interlaminar normal stress may be positive or negative depending on stacking sequence.

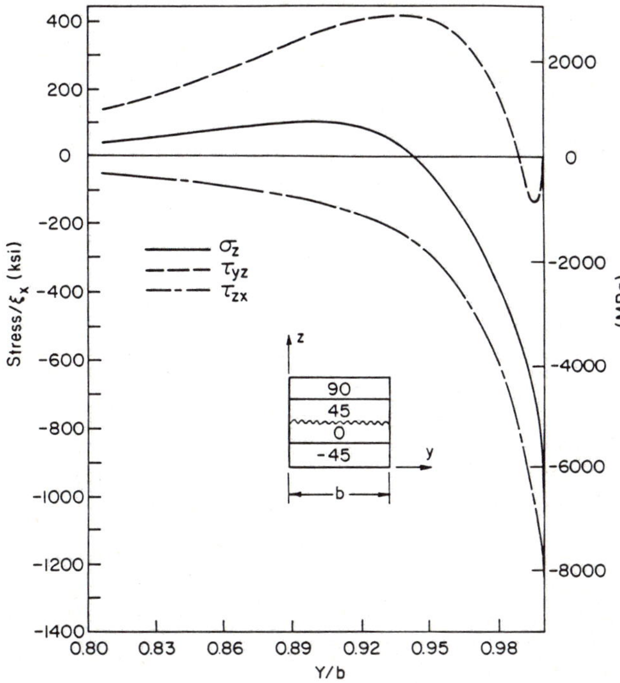

FIG. 27. Interlaminar stress along the 45/0 interface of a $[90/45/0/-45]_s$ laminate.

The results in Figs. 28 and 29 show the effect of interchanging the position of the 0/90 and ±45 combinations. The sign of the interlaminar normal stress, at the midplane, reverses with this change in stacking sequence. Since tensile interlaminar normal stresses lead to delamination, the least preferred laminate is the one with the largest tensile stress.

Figure 27 shows that the interlaminar shear stress τ_{yz} rises in the boundary layer, but approaches zero at the boundary in order to satisfy the stress free boundary condition. The interlaminar shear stress τ_{yz} exhibits strong gradients near the free edge, but its behavior is not singular. Both components of interlaminar shear stress are identically zero along the midplane because of symmetry.

The distribution of interlaminar shear stress τ_{zx} is very similar throughout the laminate, but the magnitude is very dependent on location. The behavior is basically singular with the strength of the singularity dependent upon the mismatch in coefficient of mutual influence of the adjacent layers (Fig. 6). (This component of shear stress is not constrained by any boundary condition at the free edge.) The maximum stress occurs at the ±45 interface where the mismatch is largest, and the stress is zero at the midplane where there is no mismatch. The sign of this shear stress may be positive or negative depending upon location and stacking sequence.

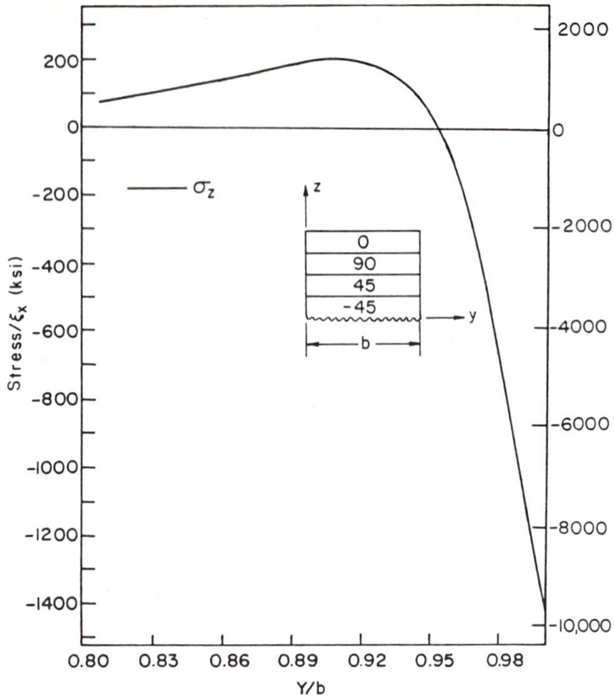

FIG. 28. Interlaminar stresses along the midplane of a $[0/90/\pm45]_s$ laminate.

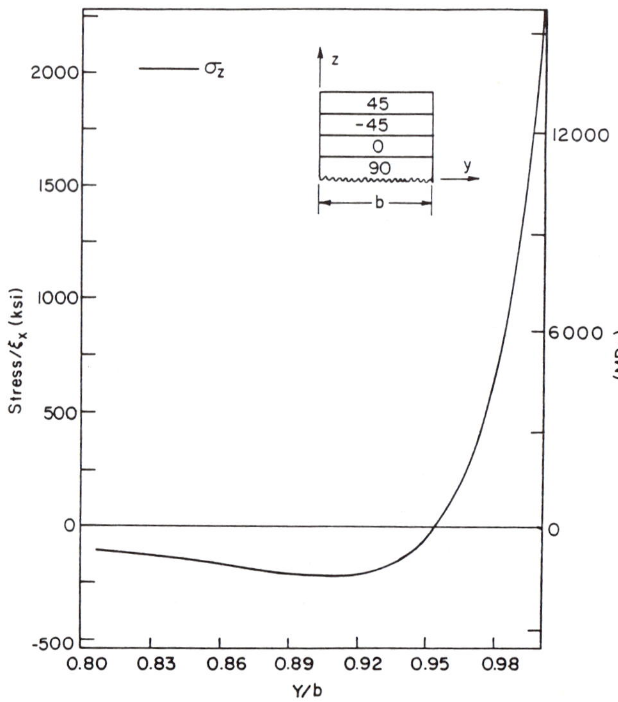

FIG. 29. Interlaminar stresses along the midplane of a $[\pm45/0/90]_s$ laminate.

4.3.3. Through the thickness stress distributions

The variation of interlaminar stresses through the thickness and near the free edge of the same three quasi-isotropic laminates are shown in Figs. 30, 31 and 32. These results clearly show the influence of stacking sequence on the interlaminar normal stress σ_z and the interlaminar shear stress τ_{zx}.

The maximum magnitude of interlaminar normal stress generally occurs at the first interface from the midplane. This is a consequence of the larger moment developed by σ_y as the midplane is approached. As indicated previously and as shown in the figures, the sign of this stress may be positive or negative depending upon stacking sequence. The location of the maximum interlaminar normal stress is not associated with any particular set of adjacent layers. It is primarily dependent upon the moment developed by the through the thickness distribution of intralaminar stresses σ_y. Since the magnitude of intralaminar stresses is independent of stacking sequence, the moment (and thereby the maximum σ_z) is primarily dependent on stacking sequence.

Figs. 33 and 34 show the magnitude and location of the maximum interlaminar normal and shear stresses for each of the twelve laminates considered along with the intralaminar stresses σ_y and the interfacial moments. The correspondence between maximum value of σ_z and interfacial moment is evident.

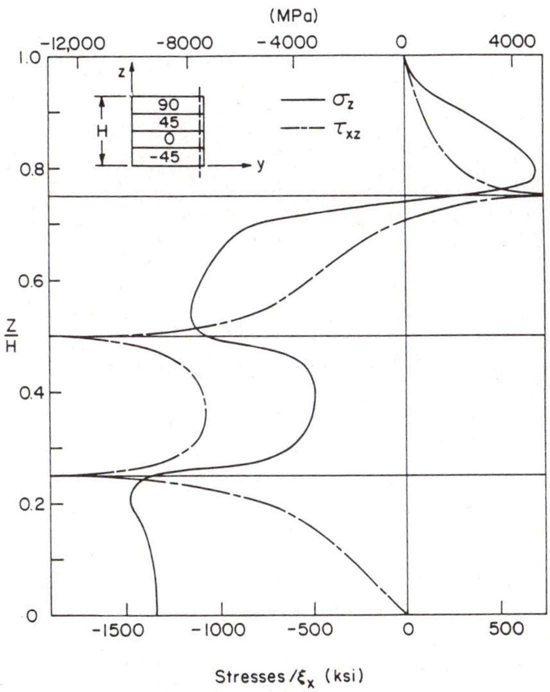

FIG. 30. Through the thickness distribution of interlaminar stresses in a $[90/45/0/-45]_s$ laminate.

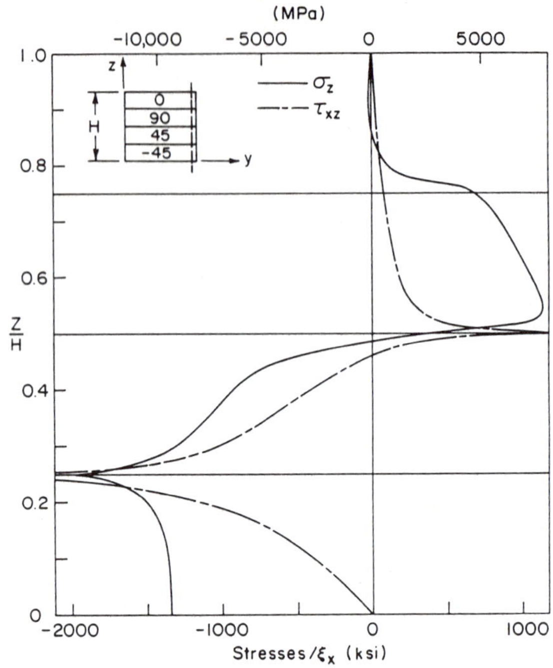

FIG. 31. Through the thickness distribution of interlaminar stresses in a $[0/90/\pm45]_s$ laminate.

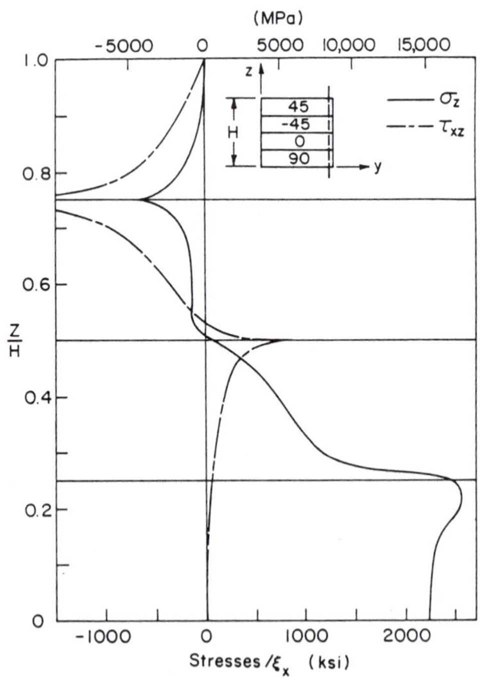

FIG. 32. Through the thickness distribution of interlaminar stresses in a $[\pm45/0/90]_s$ laminate.

Laminate Stress σ_y,(MPa)	Laminate	Interface Moment $\left(\frac{N-m}{m}\right)$	Finite Element Maximum Stresses (MPa)
-36.5 18.6 -0.7 18.6	90 45 0 -45	-1.47 -3.69 -5.16 -5.92	σ_z = -9.4 τ_{zx} = -9.5
-0.7 18.6 -36.5 18.6	0 -45 90 45	-0.04 0.67 0.67 -0.09	σ_z = 8.6 τ_{zx} = 9.1
18.6 -36.5 -0.7 18.6	45 90 0 -45	0.76 0.76 -0.71 -1.47	σ_z = 9.1 τ_{zx} = 8.1
18.6 -36.5 18.6 -0.7	45 90 -45 0	0.76 0.76 0.09 0.09	σ_z = 9.5 τ_{zx} = -9.0
18.6 -0.7 -36.5 18.6	45 0 90 -45	0.76 2.23 2.23 1.47	σ_z = 10.5 τ_{zx} = -8.0
18.6 -0.7 18.6 -36.5	45 0 -45 90	0.76 2.23 4.41 5.92	σ_z = 14.3 τ_{zx} = -8.3

o - Maximum $|\tau_{zx}|$

□ - Maximum $|\sigma_z|$

ϵ_x = 0.1%

FIG. 33. Stresses and interlaminar moments in interspersed ±45 quasi-isotropic laminates.

The distributions of τ_{zx} indicate that if adjacent ±45 layers are present in the laminate, then the maximum interlaminar shear stress occurs at this interface. This was true for all six quasi-isotropic laminates with adjacent ±45 layers. Further, the magnitude of the maximum interlaminar shear stress τ_{zx} in the six laminates with interspersed ±45 layers is less than all maximum interlaminar edge values found in the six laminates with adjacent ±45 layers (Figs. 33 and 34). Both of these results are further evidence of the importance of mismatch considerations for design. The interlaminar stress distributions also show that the magnitude of maximum interlaminar shear stress is equal to or greater than the magnitude of the maximum interlaminar normal stress. Hence, delamination may be the result of interlaminar shear as well as interlaminar normal stress.

Laminate Stress σ_y, (MPa)	Laminate	Interface Moment $\left(\frac{\text{N-m}}{\text{m}}\right)$	Finite Element Maximum Stresses (MPa)
-36.5 → -0.7 → 18.6 ← 18.6 ←	90 0 -45 45	-1.47 -4.45 -6.68 -7.43	σ_z = -11.3 τ_{zx} = 12.4
-36.5 → 18.6 ← 18.6 ← -0.7 →	90 45 -45 0	-1.47 -3.65 -4.36 -4.36	σ_z = -10.2 τ_{zx} = -12.7
-0.7 → -36.5 → 18.6 ← 18.6 ←	0 90 45 -45	-0.04 -1.51 -3.78 -4.54	σ_z = -10.5 τ_{zx} = -12.7
-0.7 → 18.6 ← 18.6 ← -36.5 →	0 45 -45 90	-0.04 0.67 2.89 4.36	σ_z = 13.8 τ_{zx} = -11.4
18.6 ← 18.6 ← -36.5 → -0.7 →	45 -45 90 0	0.76 2.98 4.54 4.54	σ_z = 12.4 τ_{zx} = -10.6
18.6 ← 18.6 ← -0.7 → -36.5 →	45 -45 0 90	0.76 2.98 1.51 7.43	σ_z = 15.0 τ_{zx} = -9.9

o - Maximum $|\tau_{zx}|$

□ - Maximum $|\sigma_z|$

ϵ_x = 0.1 %

FIG. 34. Stresses and interlaminar moments in adjacent ±45 quasi-isotropic laminates.

5. Failure of angle-ply laminates

The preceding sections have discussed the distribution of interlaminar stresses in great detail. These stresses are of concern because they cause delaminations at free edges. In this section we will present results for edge damage in several angle-ply laminates. Angle-ply laminates have been chosen in order to limit the discussion to the influence of fiber orientation. This will keep the discussion within bounds for this presentation. A more exhaustive treatment would be necessary in order to include the influence of stacking sequence. An example of the influence of edge effects on ultimate stress and mode of failure will also be demonstrated.

5.1. Edge damage

The availability of stress distributions throughout the laminates makes it possible to predict the location and mode of initial failure with the aid of a failure criterion. The tensor polynomial failure criterion (TSAI and WU, 1971) has been chosen for use in the study. The location and mode of failure were predicted with the theory and then compared to experimental results. The experimental results were obtained by taking edge replicas at selected intervals during the load history of tensile tests on finite width coupons. The edge replicas show the type of damage (in the form of cracks) present along the free edge.

The tensor polynomial failure criterion hypothesizes that failure occurs when the following condition is first satisfied:

$$F_i \sigma_i + F_{ij} \sigma_i \sigma_j = 1 . \tag{5.1}$$

Table 3 presents results showing individual terms of the tensor polynomial at failure for a series of angle-ply laminates. Fig. 35 presents results showing the distribution of the tensor polynomial function along the interface of $+\theta/-\theta$ laminates. It is clear from the figure that failure is predicted to initiate at the free edge. Further, it is predicted that edge effects are much more critical for laminates with fiber orientations near 15°. For low fiber angles (<37°) edge effects totally dominate the initiation of failure. As the fiber orientation is increased, edge effects become less important.

TABLE 3. Tensor polynomial for angle-ply laminates.

θ	$F_{11}\sigma_1^2$	$F_2\sigma_2$	$F_{22}\sigma_2^2$	$F_3\sigma_3$	$F_{44}\tau_{23}^2$	$F_{55}\tau_{13}^2$	$F_{66}\tau_{12}^2$
10	0.020	−0.061	0.001	−0.056	0.044	1.04	0.004
15	0.015	−0.109	0.003	−0.112	0.115	1.07	0.010
30	0.012	−0.251	0.015	−0.375	0.584	0.867	0.104
37.5	0.001	0.511	0.060	0.184	0	0	0.240
45	0	0.588	0.080	0.144	0	0	0.185
60	0	0.724	0.151	0.062	0	0	0.092
75	0	0.810	0.152	0.014	0	0	0.025

The results in Table 3 show that failure is dominated by interlaminar shear stresses for small fiber angles, but shifts to a transverse tensile failure mode with increasing angle. These results are consistent with the mismatch considerations presented previously. Where there is large mismatch in the coefficient of mutual influence, interlaminar shear stresses are high resulting in delamination. Where this mismatch is small, interlaminar shear stress are small, and thus the initial mode of failure is transverse tension.

These predictions are consistent with experimental results as shown by Figs. 36–39 (HERAKOVICH and KLANG, 1984). These figures show edge replicas taken

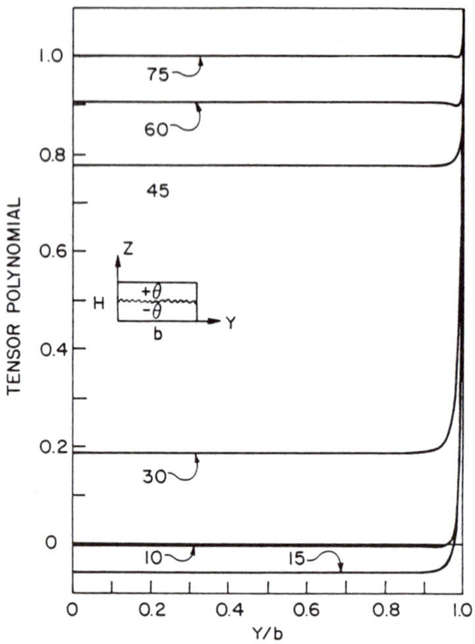

FIG. 35. Tensor polynomial along interface of angle-ply laminates.

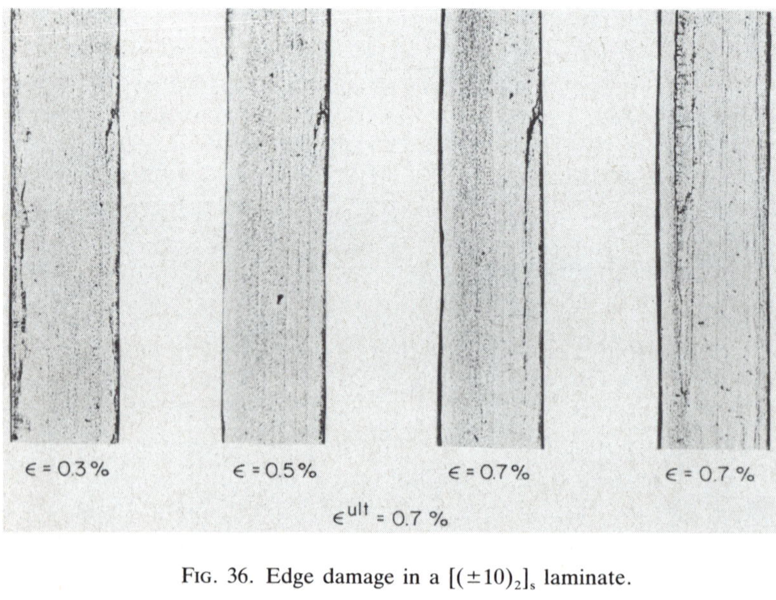

FIG. 36. Edge damage in a $[(\pm 10)_2]_s$ laminate.

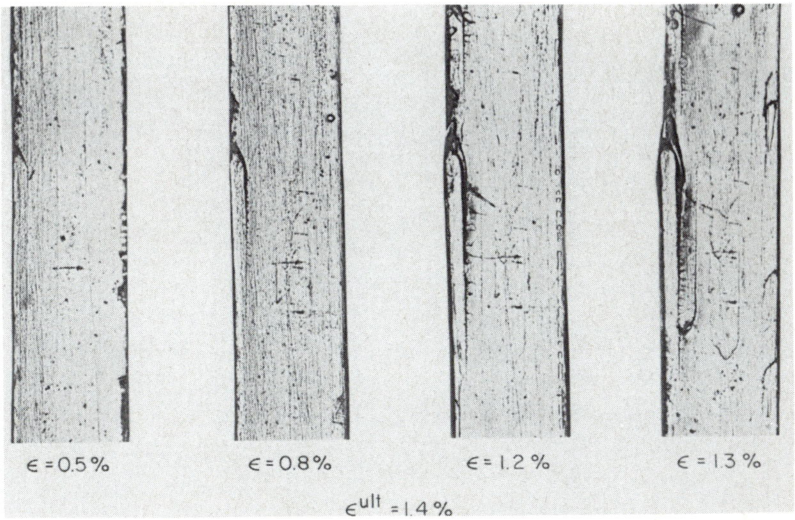

FIG. 37. Edge damage in a $[(\pm 30)_2]_s$ laminate.

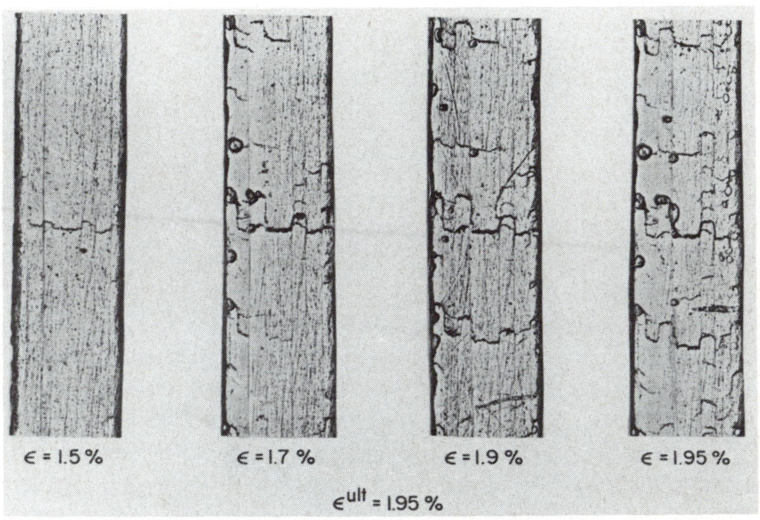

FIG. 38. Edge damage in a $[(\pm 45)_2]_s$ laminate.

just prior to complete failure of the laminates. It can be seen that the mode of failure is delamination for small fiber angles and shifts to transverse cracking as the fiber angle is increased.

The results for angle-ply laminates are summarized in Fig. 40 which shows a comparison of the mismatch in coefficient of mutual influence, the maximum interlaminar shear stress, and the variation in tensor polynomial, all with respect to fiber orientation. All three curves exhibit strong peaks for $\theta = 15°$.

FIG. 39. Edge damage in a $[(\pm 60)_2]_s$ laminate.

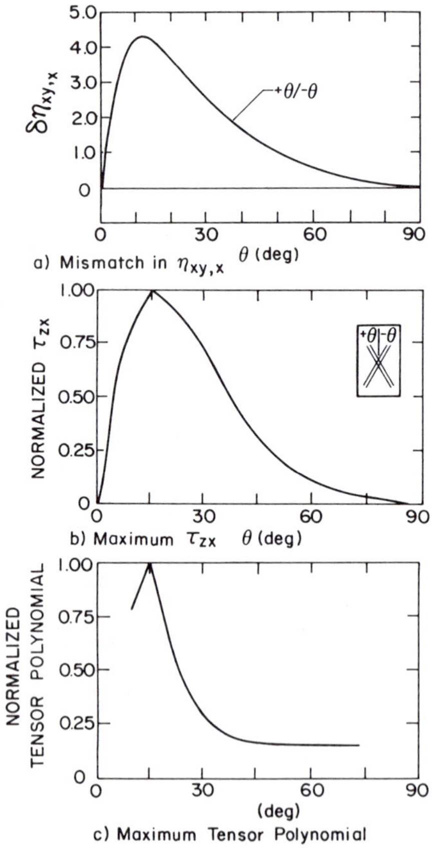

FIG. 40. Summary of edge effects in angle-ply laminates.

5.2. Failure modes

Figure 41 shows the difference in failure modes for alternating and clustered 10 and 30 angle-ply laminates. It is evident from the figure that the mode of failure is completely different depending upon stacking sequence (layer thickness) (HERAKOVICH, 1982). The alternating laminates fail due to a combination of fiber breakage and matrix cracking or fiber/matrix debonding. Failure of the clustered laminates is entirely the result of matrix failure. There is no fiber failure in the clustered laminates.

The failure surface of the alternating laminates is a single surface parallel to the fiber direction in one of the layers. In contrast, the failure surface of the clustered laminates consists of delaminations at the $+\theta/-\theta$ interface and matrix failure in the planes of the fibers. This significant difference in failure modes is solely the result of the interlaminar stresses. The ultimate stress of the alternating laminates shown here is 20-30% higher than that for the clustered laminates (HERAKOVICH, 1982). The effect of stacking sequence diminishes for large fiber orientations because the mismatch in coefficient of mutual influences also diminishes.

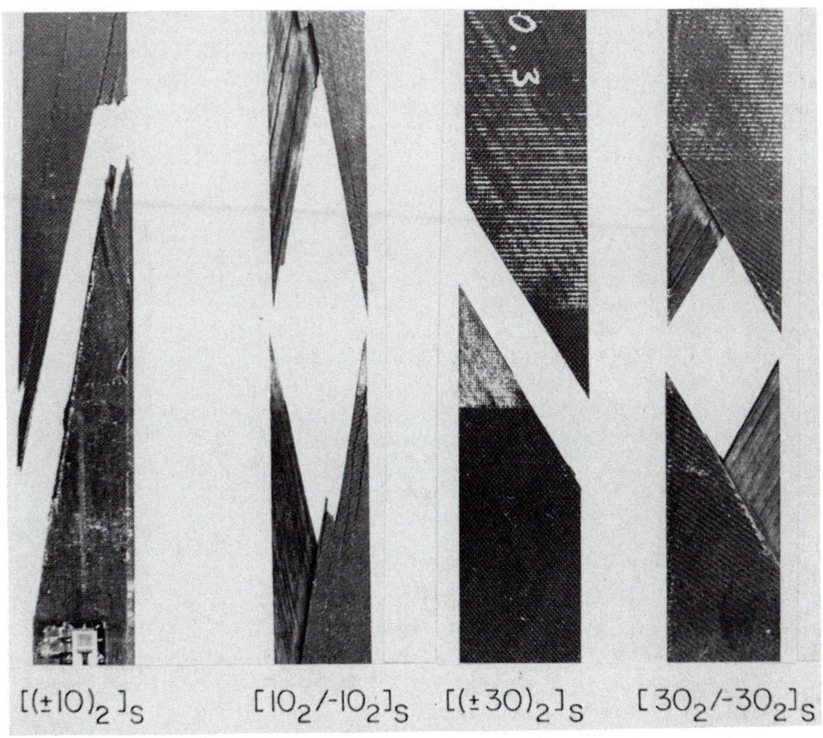

FIG. 41. Failure modes for angle-ply laminates.

6. Summary

Theoretical and experimental results have been presented showing the importance of edge effects in laminated composite materials. It has been shown that the interlaminar normal stress and one component of interlaminar shear stress exhibit singular behavior at the free edge. The fiber orientation corresponding to the maximum interlaminar shear stress is directly related to the mismatch in coefficient of mutual influence of adjacent layers. The maximum interlaminar normal stress is a function of the laminate stacking sequence. The mode of initial failure is variable depending upon the fiber orientations and stacking sequence.

The results presented here are for mechanical loading and straight edges. However, the same basic principles apply to curved edges and other types of loading such as thermal and moisture.

Acknowledgements

The author is indebted to NASA Langley Research Center for the continuous support of research on mechanics of composite materials beginning in 1971, the past eleven years of which has been provided under the NASA Virginia Tech Composites Program. Thanks are also due to Ms. Elizabeth Strauss for numerical work associated with many of the figures presented in this chapter.

References

BOGY, D.B. (1968), Edge-bonded dissimilar orthogonal elastic wedges under normal and shear loading, *J. Applied Mechanics* **35**, 460.

BUCZEK, M.B., M.A. GREGORY, and C.T. HERAKOVICH, (1983), CLFE2D – A generalized plane strain finite element program for laminated composites subjected to mechanical and hygrothermal loading, VPI-E-83-40, Virginia Polytechnic Institute.

CROSSMAN, F.W. and A.S.D. WANG, (1978), Stress field induced by transient moisture sorption in finite-width composite laminates, *J. Composite Materials* **12**, 2.

CZARNEK, R., D. POST, and C.T. HERAKOVICH, (1983), Edge effects in composites by moire interferometry, *Experimental Techniques* **7**, 18.

DANA, J.R. (1974), Three dimensional finite element analysis of thick laminated composites – Including interlaminar and boundary effects near circular holes, Ph.D. Dissertation, Virginia Polytechnic Institute.

FARLEY, G.L. and C.T. HERAKOVICH, (1978), Influence of two-dimensional hygrothermal gradients on interlaminar stresses near free edges, Environmental Effects on Advanced Composite Materials, ASTM STP 658, Am. Soc. for Testing and Materials, p. 143.

FOYE, R.L. and D.J. BAKER, (1971), Design/analysis methods for advanced composite structures, AFML-TR-70-299, Vol. 1.

GRIFFIN, O.H., M.P. KAMAT, and C.T. HERAKOVICH, (1980), Three dimensional inelastic finite element analysis of laminated composites, VPI-E-80-28, Virginia Polytechnic Institute.

HEIN, V.L. (1986), Residual stresses in a two-material wedge and a finite cylinder, Ph.D. Thesis, Lehigh University.

HERAKOVICH, C.T. (1976), On thermal edge effects in composite laminates, *International Journal of Mechanical Sciences* **18**, 129.

HERAKOVICH, C.T. (1982), Influence of layer thickness on the strength of angle-ply laminates, *J. Composite Materials* **16**, 216.

HERAKOVICH, C.T. (1984), Composite laminates with negative through-the-thickness Poisson's ratios, *J. Composite Materials* **18**, 447.

HERAKOVICH, C.T. and E.C. KLANG, (1984), Theoretical/experimental correlation for edge cracking in graphite/polyimide laminates, *Int. Symposium on Composites: Materials and Engineering*, Univ. Delaware.

HERAKOVICH, C.T. and A.P. NAGARKAR, (1980), On failure modes in finite width angle-ply laminates, *Proceedings ICCM-III* (BUNSELL ET AL., ed.) p. 425.

HERAKOVICH, C.T., A. NAGARKAR, and D.A. O'BRIEN, (1979), Failure analysis of composite laminates with free edges, *Modern Developments in Composite Materials and Structures* (J.R. VINSON, ed.) ASME, p. 53.

HERAKOVICH, C.T. and D.A. O'BRIEN, (1979), Failure analysis of an idealized composite damage zone, *Proceedings 29th MFPG Symposium, Advanced Composites: Design and Applications*, p. 242.

HERAKOVICH, C.T., D. POST, M.B. BUCZEK, and R. CZARNEK, (1984), Free edge strain concentrations in real composite laminates: Experimental–theoretical correlation, CCMS-84-10, VPI-E-84-24, Virginia Polytechnic Institute.

HERAKOVICH, C.T., G.D. RENIERI, and H.F. BRINSON, (1976), Finite element analysis of mechanical and thermal edge effects in composite laminates, *Army Symposium on Solid Mechanics, Composite Materials: The Influence of Mechanics of Failure on Design* (Cape Cod, MA) p. 237.

HESS, M.S. (1969), The end problem for a laminated elastic strip – II. Differential expansion stresses, *J. Composite Materials* **3**, 630.

HSU, P.W. and C.T. HERAKOVICH, (1977), Edge effects in angle-ply composite laminates, *J. Composite Materials* **11**, 422.

ISAKSON, G. and A. Levy, (1971), Finite-element analysis of interlaminar shear in fibrous composites, *J. Composite Materials* **5**, 273.

LEKHNITSKII, S.G. (1963), *Theory of Elasticity of an Anisotropic Elastic Body* (Holden-Day, San Francisco).

NAGARKAR, A.P. and C.T. HERAKOVICH, (1979), Nonlinear temperature dependent failure analysis of finite width composite laminates, VPI-E-79-36, Virginia Polytechnic Institute.

OPLINGER, D.W., B.S. PARKER, and F.P. CHIANG, (1974), Edge-effect studies in fiber-reinforced laminates, *Experimental Mechanics* **14**, 347.

PAGANO, N.J. (1974), On the calculation of interlaminar normal stress in composite laminates, *J. Composite Materials* **8**, 65.

PAGANO, N.J. (1978), Free edge stress fields in composite laminates, *Int. J. Solids and Structures* **14**, 401.

PAGANO, N.J. and R.B. PIPES, (1971), The influence of stacking sequence on laminate strength, *J. Composite Materials* **5**, 50.

PAGANO, N.J. and R.B. Pipes, (1973), Some observations on the interlaminar strength of composite laminates, *Int. J. Mech. Sci.* **15**, 679.

PIPES, R.B. and I.M. DANIEL, (1971), Moire analysis of the interlaminar shear edge effect in laminated composites, *J. Comp. Matls.* **5**, 255.

PIPES, R.B. and N.J. PAGANO, (1970), Interlaminar stresses in composite laminates under uniform axial extension, *J. Composite Materials* **4**, 538.

PUPPO, A.H. and H.A. EVENSEN, (1970), Interlaminar shear in laminated composites under plane stress, *J. Composite Materials* **4**, 204.

RENIERI, G.D. and C.T. HERAKOVICH, (1976), Nonlinear analysis of laminated fibrous composites, VPI-E-76-10, also NASA CR-148317, Virginia Polytechnic Institute.

RYBICKI, E.F. (1971), Approximate three-dimensional solutions for symmetric laminates under inplane loading, *J. Composite Materials* **5**, 345.

SONI, S.R. and N.J. PAGANO, (1983), Elastic response of composite laminates, *1982 IUTAM*

Symposium, Mechanics of Composite Materials: Recent Advances (HASHIN and HERAKOVICH, eds.) p. 227.

TANG, S. (1975), A boundary layer theory – Part I: Laminated composites in plane stress, *J. Composite Materials* **9**, 33.

TSAI, S.W. and E.M. WU, (1971), A general theory of strength for anisotropic materials, *J. Composite Matls.* **5**, 58.

WANG, A.S.D. and F.W. CROSSMAN, (1977), Some new results on edge effect in symmetric composite laminates, *J. Composite Materials* **11**, 92.

WANG, S.S. (1983), Elasticity solutions for a class of composite laminate problems with stress singularities, *IUTAM Symposium, Mechanics of Composite Materials: Recent Advances* (HASHIN and HERAKOVICH, eds.) p. 259.

WANG, S.S., and I. CHOI, (1982), Boundary-layer effects in composite laminates: Part I – Free-edge stress singularities, *J. of Applied Mechanics* **49**, 541.

WANG, S.S. and F.G. YUAN, (1983), A singular hybrid finite element analysis of boundary-layer stresses in composite laminates, *Int. J. Solids and Structures* **19**(9), 825.

CHAPTER 5

Service Induced Damage in Composite Structures

K.L. Reifsnider

Department of Engineering Science and Mechanics
Virginia Polytechnic Institute and State University
Blacksburg, VA 24061
U.S.A.

Contents

HANDBOOK OF COMPOSITES, VOL. 2 – Structure and Design
Edited by C.T. HERAKOVICH and Y.M. TARNOPOL'SKII

1. Introduction

This chapter will deal with the manner in which the characteristics of composite materials influence the performance of structures made from those materials. We will be concerned with material behavior over periods of time and under load histories associated with common service conditions associated with structures made from composite materials. Because there is only limited treatment of the subject here, emphasis will be placed on those aspects which are most generic, most general, and—by judgment of the author—most important.

A designer must ultimately answer three questions concerning any material he uses: How stiff is the material? How strong is the material? How long will it last? Those questions must be answered not only at the time of manufacture, but also following service of the structure. Indeed, the strength, stiffness and life of composite materials in service are functions of the past history of loading and the expected loading history at any point in time, if the internal nature of the material changes during these loadings.

Damage development causes such changes. In fact, the (finite) life of any material component loaded at levels below the quasi-static ultimate strength is defined and determined by damage development, since it is that development which ultimately reduces strength to the level of load application causing failure which defines the life of the component.

This chapter will deal with damage development associated with long-term, primarily cyclic loading. We will begin with a discussion of damage development, presented by example, and continue with a discussion of the mechanics associated with the subject. Finally, a treatment of life prediction and representation will be offered.

2. Damage development

The advent of the development and use of composite materials for a variety of major primary engineering structures and an even greater variety of secondary ones has ushered in a new era in the design and construction of structural components. While it is the engineering properties of composite materials (such as their strength to weight ratios and stiffness to weight ratios) that have been most widely heralded, it is the feature which allows designers to create a material with strength and stiffness of their choosing as part of the design process which is at the center of this new era from the design standpoint. This feature is the basis for the creation of entirely new designs which have structural components that are tailored for performance, even to the extent of choosing different components of tensor stiffness and strength, and coupling coefficients so as to generate deformations under loading which change the shape of the engineering components in a manner that increases their functionality and efficiency as in the case of certain aereolastic tailoring operations. These are exciting concepts, indeed, and they suggest degrees of freedom in applications and design which we have only begun to suspect or exploit.

However revolutionary composite materials may be, however, they are not without reality; few materials last forever and no proper designs should do so. In realistic engineering situations, and especially in unexpected circumstances, the strength, stiffness and life of composite materials are degraded by service induced damage. We will be concerned in this chapter with states of stress and states of material that are peculiar to high-modulus fibrous composite laminates which endure loading histories known to cause changes in laminate strength, stiffness and life due to micro-failure events collectively called 'damage'. In this section we shall be concerned with the description of damage events that can be expected and which have been observed in engineering structures subjected to quasi-static, cyclic, and dynamic loading, especially over long periods of time. The experimental data that will be presented have been carefully chosen so as to represent generic behavior (in the opinion of the author) whenever possible.

It is well established that the micro-events which reduce the strength and stiffness and determine the life of composite laminates, commonly referred to as damage, are complex, various, and intricately related to a variety of failure modes under different circumstances. The literature contains a profuse record of observations which should be consulted for specific and special interests. Starting points for such an effort include a number of Special Technical Publications published by the American Society for Testing and Materials (DANIEL, 1982; REIFSNIDER, 1982; LAURAITIS, 1981; FONG, 1979; TSAI, 1979) and a myriad of papers (O'BRIEN, 1982; REIFSNIDER and HIGHSMITH, 1981; MANDELL et al., 1981; BADER et al., 1979; REIFSNIDER et al., 1979). Although important new details are still being added to these observations, and many important matters have not yet been settled, at least in a relative sense we can say that the investigators concerned with this field have a fairly clear picture of what is happening; but it is much less certain how those events occur and why they occur in the order and combinations observed.

Since we are concerned here with generic features and not specific details, we will attempt to present a discussion which centers around evidence of behavior which is general. To that end, we consider examples of damage development in laminates having stacking sequences which are cross-plies, such as $[0, 90_2]_s$, those having only 0 and 45° plies such as $[0, \pm 45, 0]_s$, and stacking sequences which are quasi-isotropic such as $[0, 90, \pm 45]_s$. We will consider development of damage in graphite epoxy and glass epoxy systems specifically, although many of these results apply to metal matrix composites as well. Many of the detailed experimental results that will appear below are taken from work by JAMISON (1982) and REIFSNIDER (1982). We will begin our discussion with a description of damage development in situations where tensile load excursions dominate the cyclic load spectra. This subject is usually discussed in the literature as ‘tension–tension fatigue’, abbreviated as the T–T case for our purposes. While the addition of compression load excursions adds several different failure modes and possibly some specialized types of crack coupling and growth, virtually all of the damage modes that are observed in tension or compression loading are seen to occur in various degrees under cyclic tensile loading alone. There are at least two major reasons for this important observation. First, although only tensile loading may be applied in a single axial direction to a given specimen, if the specimen consists of several plies having different orientations, the individual plies will in general be subjected to two-dimensional stress, and unless their Poisson contractions are the same, some of the plies will be subjected to transverse compression upon axial tensile loading. Hence, although a one-dimensional tensile load is applied to the global laminate, the stresses in the individual plies will in general be two-dimensional and may include compressive components. A second reason for the generality of tensile damage modes has to do with the complex stress states that develop near the edges and boundaries of laminates. It is well established that interlaminar stresses form between the plies of a laminate at a free edge due to the requirements of equilibrium on material elements at the edge and the presence of biaxial stresses in the individual plies in the interior of a laminate (cf. PIPES and PAGANO, 1970; 1971). While it is true that reversing the direction of the applied load reverses the sign of the three-dimensional stresses at the edge of a laminate, due to the complexity of that stress state there is nothing inherently unique about the damage created by a tensile or compressive loading. Of course, it is possible for local of global instabilities to contribute to damage growth and perhaps even its initiation in certain cases when global compressive stress states dominate the behavior. More will be said of this in our discussion of compressive loading.

2.1. *Quasi-isotropic laminates*

One of the most common types of stacking sequences used in composite structures is a quasi-isotropic laminate having repeated units of the type $[0, 90, \pm 45]_s$ or perhaps variations of that stacking sequence. For the purpose of our discussion we assume that a reasonable fatigue life is 100 000 cycles for a given example application. The maximum stress amplitude for such a life in T–T loading would be about 0.70 times the ultimate strength of the laminate in quasi-static loading. If such a laminate is loaded at that maximum stress amplitude at a frequency of about 10 cycles per second in a load-controlled test, one can expect the stiffness of the specimen to change as shown in Fig. 1. The shape of this curve is characteristic in the sense that there is an initial drop of stiffness in the early part of the life followed by a period of less rapid stiffness change during a large segment of the middle portion of the life and a

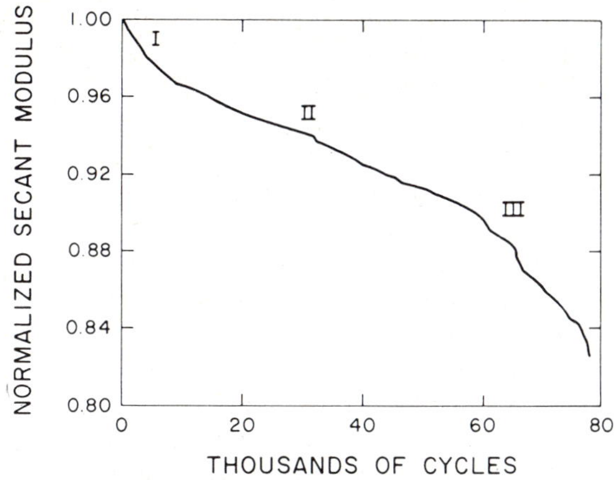

FIG. 1. Reduction in normalized laminate stiffness with applied cycles of loading for a quasi-isotropic graphite epoxy laminate.

rapid and rapidly accelerating degradation of stiffness in the late part of the life just prior to fracture. The amounts of change indicated by the figure are also typical, with a 4–5% change in the first region and greater changes thereafter with maximum changes of 18–20% commonly measured before fracture. From a structural standpoint, this stiffness change is an extremely important characteristic of damage development in composite laminates.

We can illustrate the nature of this stiffness change by referring to Fig. 2. Let us suppose that the maximum cyclic stress at which failure occurs after many cycles of loading is σ_a. If the strain to failure is the same as that for quasi-static failure, or greater (as the literature suggests—HAHN and KIM, 1976, O'BRIEN and REIFSNIDER, 1977), the modulus will have reduced from E_0 to E_1 as shown in Fig. 2. This would suggest that for a constant amplitude fatigue test wherein the maximum cyclic stress was held at a value of, say, 70% of the ultimate

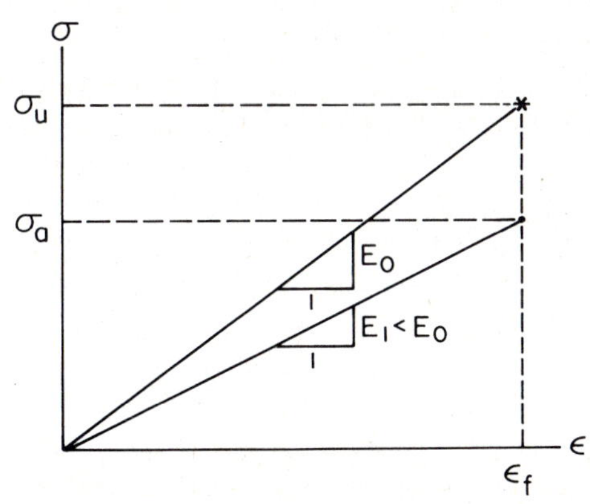

FIG. 2. Schematic diagram of stiffness change concept for a fixed level of strain to failure.

strength that a corresponding reduction in stiffness of the order of 30% should occur. Whether the laminate is fiber dominated or not makes no difference in this argument if the specimen does eventually fail. It does make a difference how carefully the modulus is measured and monitored. A major part of the stiffness change may occur in the late stages of fatigue life, as suggested in Fig. 1. If stiffness is not monitored continuously or measured during that short period, a large part of the change may not be recorded. Also, a large part of the stiffness change may occur quite early in the test. If stiffness is not carefully measured at the beginning of a test or in the early cycles when test parameters are being established and cyclic amplitudes first reached this stiffness change may also be missed. Of course, the integrity and accuracy of the stiffness measurement device over long periods of time usually associated with cyclic loading must be assured, and proper measurement methods must be used. However, when measurements are properly made, large changes such as those shown in Fig. 2 are routinely recorded. These changes in stiffness are, therefore, a parameter which can be used to nondestructively monitor and assess the severity of damage development during fatigue loading of composite laminates.

Fig. 3 shows a sequence of replicas of the central region of a single $[0, 90, \pm 45]_s$ specimen at various stages of its fatigue life. Replicas were made with the specimens in the loading fixture under static load while cyclic loading was temporarily interrupted. A 25.5 mm long strip of cellulose acetate tape was affixed to the specimen edge, a small amount of acetone was injected between the tape and the specimen, and light pressure was applied to the acetate tape. The acetone-softened tape flowed into the edge surface details and upon drying provided a high fidelity impression of the edge topology. As the replicas show, transverse cracks develop in the matrix of the off-axis plies and are clearly in evidence quite early in the test, when as little as 2% stiffness reduction is observed, Fig. 5(a). Matrix cracking usually occurs through the thickness of plies that are inclined in orientation to the principal load axis. These cracks usually form along fiber directions and, for coupons, usually transverse the entire width of the specimen. In the simplest sense, the cracks form when the normal stress perpendicular to the fibers in a given ply exceeds the traverse strength of that ply. During fatigue loading, and during service loading for composite structures, the number of such cracks in a given ply frequently saturates. An illustration of this is presented in Fig. 4 which shows the spacing between cracks in a $-45°$ ply in a $[0, 90, \pm 45]_s$ AS3501-5 laminate as a function of quasi-static load level and cycles of loading at about two-thirds of the ultimate strength ($R = 0.1$). It can be seen that cracks develop quite early in the life and quickly stabilize to a very nearly constant pattern with a fixed spacing. It is surprising however that the same behavior occurs for quasi-static loading, in the sense that crack development occurs over a large range of load and quickly stabilizes into a pattern which has the same spacing as the fatigue crack pattern. In fact, the two patterns are essentially identical regular crack arrays in each ply regardless of load history. Similar behavior is observed for

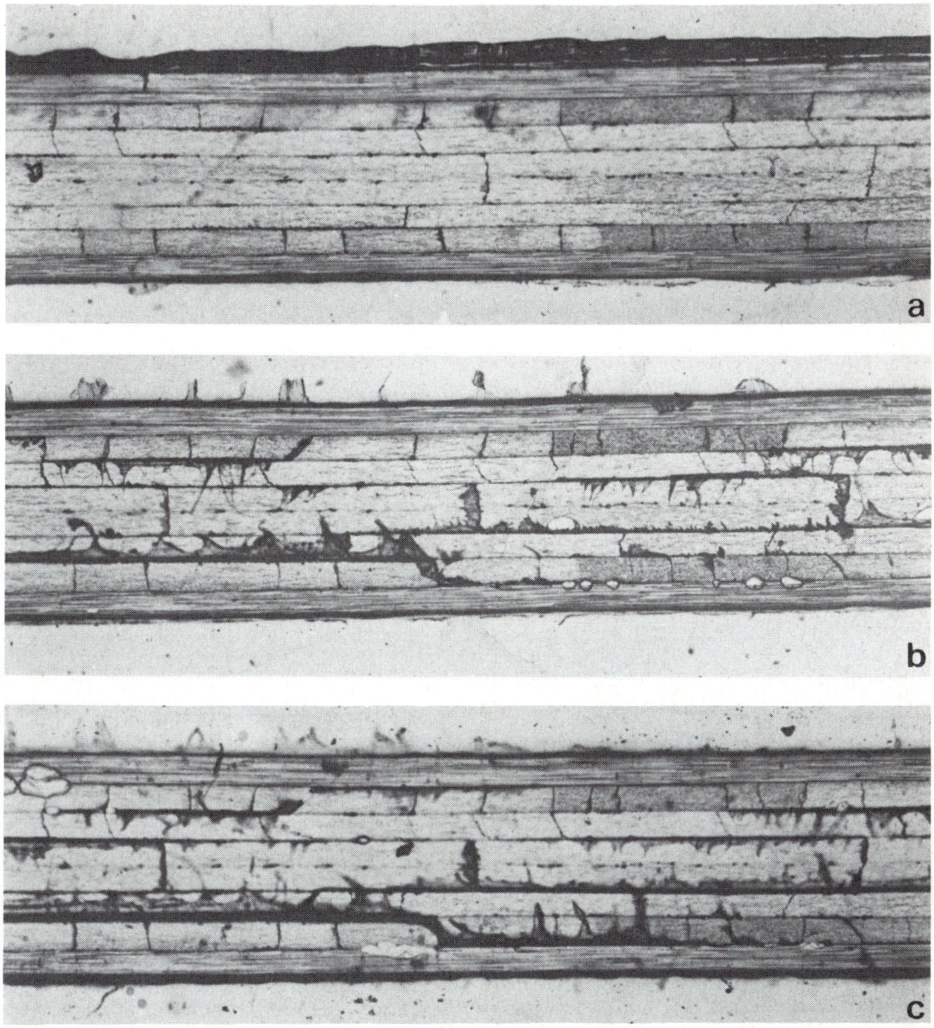

FIG. 3. Edge replicas of progressively damaged $[0, 90 \pm 45]_s$ graphite epoxy specimens at: (a) 2% stiffness reduction; (b) 8% stiffness reduction; (c) 15% stiffness reduction (magnification 13×).

other laminates. The author has named these crack patterns ‘characteristic damage states’ (CDS) for matrix cracking in laminates having off-axis plies (REIFSNIDER, 1977). The CDS is a laminate property, i.e., it is completely defined by the properties of the individual plies, their thickness, and the stacking sequence of the variously oriented plies. The CDS is independent of extensive variables such as load history and environment (except as the ply properties are altered) and internal affairs such as residual or moisture related stresses. A more thorough discussion of the CDS can be found in O’BRIEN (1980); REIFSNIDER and HIGHSMITH (1981); REIFSNIDER (1977); REIFSNIDER and

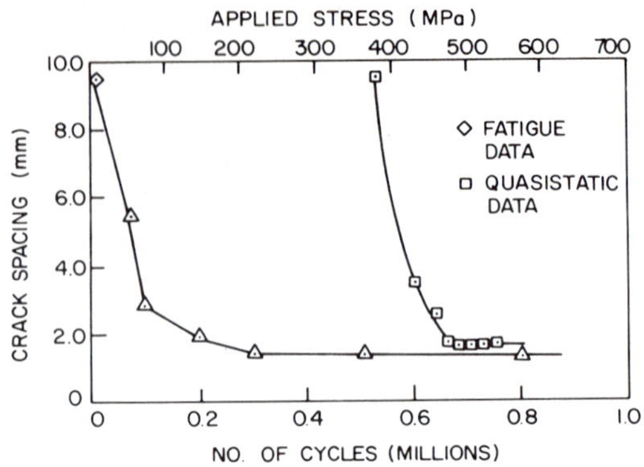

FIG. 4. Spacing of cracks in the −45° plies of $[0, 90, \pm 45]_s$ Gr/Ep laminates as a function of quasi-static and fatigue loading.

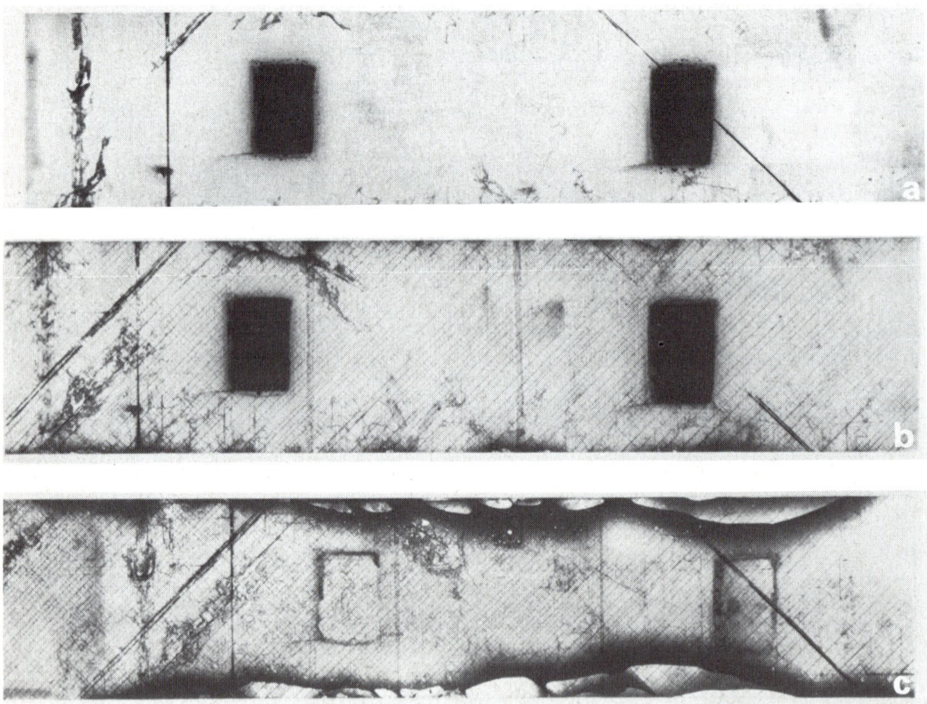

FIG. 5. Sequence of penetrant-enhanced X-ray radiographs of a $[0, 90, \pm 45]_s$ graphite epoxy specimen: (a) undamaged, (b) at 6% stiffness reduction; (c) at 18% stiffness reduction (actual size).

TALUG (1980); KRIZ (1980); and REIFSNIDER et al. (1979). While the CDS is common to both quasi-static and cyclic loading, it frequently does not develop completely in all plies of a a laminate for the quasi-static case. It should be mentioned that these cracks do not reduce the residual strength of a laminate as such, since cracks occur prior to failure in a quasi-static test of strength.

Hence, we have seen that the early stiffness reduction seen in a coupon test and to be expected in the first 20% or so of the life of a composite structural component is associated primarily with matrix cracking in the off-axis plies of the laminates involved. At increasing cycles beyond the 4% stiffness reduction in Fig. 1, edge delaminations were observed at the 90/+45 interfaces as shown in Figs. 3(b) and (c). The appearance of incipient delamination at this point is a precursor to the edge delamination-dominated damage progression which appears to characterize the middle region of the stiffness change curve. At 8% stiffness reduction (Fig. 3(b)) these delaminations had coupled to the 0/90° interface and growth significantly. Long delaminations were seen to have formed at the +45/−45° interfaces. At this interface the largest interlaminar shear stress (τ_{xz}) in the laminate occurs. Such a stress state would be expected to provide a strong driving force for the production and extension of the delaminations at that interface. At 15% stiffness reduction (Fig. 3(c)) the longitudinal delamination growth along the edge was essentially complete, with existing delaminations growing across the width.

Edge delamination is a very common damage mode in composite laminates which have free edges at locations such as cut-outs and joints. This type of delamination is one of the most widely discussed and thoroughly described damage modes in composite laminates (O'BRIEN, 1982).

Edge delamination for the quasi-isotropic laminates under discussion can be followed by X-ray radiography as indicated in Fig. 5. The radiographs in that figure were obtained by interrupting the cyclic test at specified stiffness reduction levels and removing the specimen from the test machine for radiographic examination. A solution of zinc iodide in ethyl alcohol and water was applied to the edges to enhance the definition of the delaminations. The specimen was then placed in a small load frame and loaded to about 1300 N to obtain sufficient contrast for X-ray observation. Radiographs were made at operating conditions of 25 kV and 25 mA. Exposure times were approximately 0.5 min. Fig. 5(a) is a radiograph of an as-fabricated undamaged specimen. The two dark rectangular images are extensometer mounting tabs; the dark regular horizontal and diagonal lines are ply seams resulting from fabrication; the irregular dark detail is porosity. Fig. 5(b) is the same specimen retreated with penetrant and X-rayed after 32 000 cycles, which induced a 6% stiffness decrease. Dark lines in the +45, −45, and 90° directions indicate transverse cracks (the two horizontal lines between the extensometer mounting tabs are ink lines drawn on the surface to serve as fiducial marks for edge replication). The −45° cracks appear darker in the X-ray radiograph because they involve double −45° ply thickness. Matrix cracks in the other plies are clearly resolved. Dark regions near the edge of Fig. 5(b) indicate the initiation of delamination.

Fig. 5(c) shows the same specimen after 78 000 cycles with an 18% stiffness reduction. The transverse cracks have achieved a saturation spacing and the delaminations have extended across nearly half of the width of the specimen in some regions. This situation is typical for a specimen which is quite near to final failure during cyclic loading. The specimen shown in Fig. 5 is 25 mm wide. The delamination in a structure would not be constrained by such a small width, and delaminations would normally propagate over much larger distances. Delaminations do reduce the strength and stiffness of coupon specimens such as those shown in Fig. 5, and can reduce the stiffness and strength of composite structures. More will be said of this in the next section.

The complexity of damage that can be expected in composite laminate materials commonly used for structural components is very great indeed. Fig. 6 shows an enlargement of the information shown in Fig. 5(c) and indicates that in addition to major matrix cracks that form through the thickness of off-axis plies under tensile loading, there is another entire substructure of micro-cracks with very limited extent which appear to form at the interface between off-axis plies, transverse to the major matrix cracks in the ply adjacent to the one in which the cracks form. In fact, although these micro-cracks are widely distributed in a given ply, they tend to cluster in zones centered around transverse cracks in the adjacent plies. This is shown schematically in Fig. 7. Several points should be noted by the reader who is concerned about the presence of

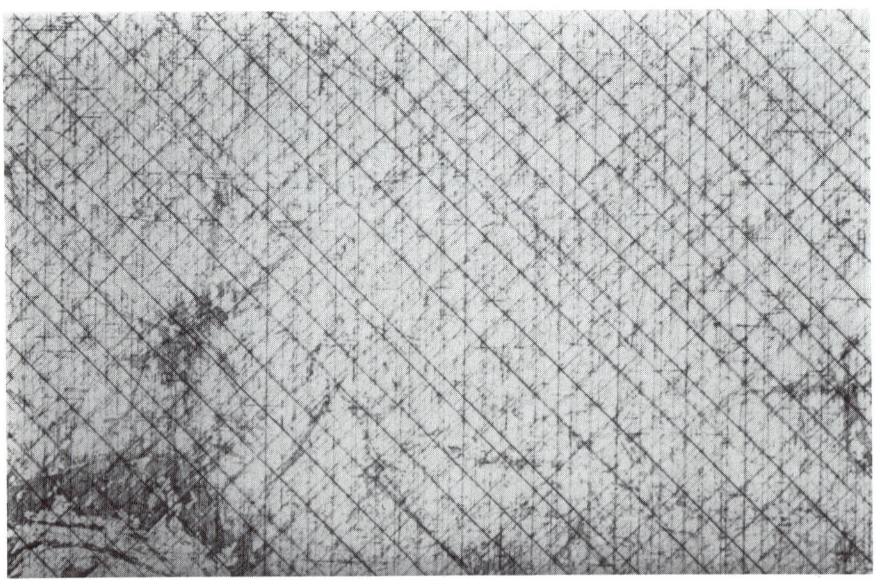

FIG. 6. Detail of Fig. 5(c) showing the extent and distribution of microdamage (magnification 6.5×).

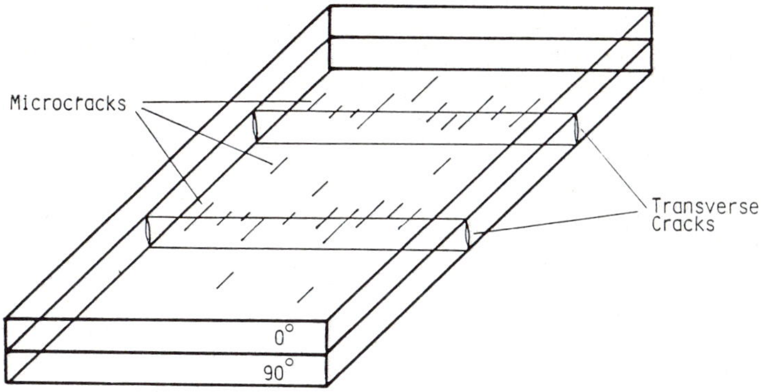

FIG. 7. Schematic diagram of a primary transverse matrix crack and secondary longitudinal matrix crack formation in a $[0, 90]_s$ laminate.

these cracks in composite structures. First, as we have indicated earlier and as we will explain to greater extent in the next section, the presence of these cracks does not, per se, reduce the residual strength of a laminate or a structural component. Second, over long periods of cyclic loading, the significant damage caused by matrix cracks usually has to do with the fact they are nucleation points for a variety of other damage including fiber fractures and delaminations. And third, because of these facts, and despite the fact that matrix cracks are virtually ubiquitous in the polymeric matrix materials commonly used for structural composites and also are found in a variety of metal matrix systems, it is desirable to choose when possible a composite system which minimizes the occurrence of matrix cracks for a given composite structure to minimize the reductions in stiffness and strength (and possibly life) when long-term cyclic loads are to be sustained by the structure.

2.2. *Cross-ply laminates*

We continue our discussion with a description of damage development in $[0, 90_2]_s$ laminates which illustrate another generic extreme of the damage development process. This type of laminate, commonly called a cross-ply laminate, is not as commonly used as the quasi-isotropic stacking sequence but does appear in a variety of applications especially those having cylindrical shapes. Fig. 8 shows a typical stiffness-cycles curve for a cross-ply graphite epoxy laminate loaded in tension–tension fatigue with $\sigma_{max} = 0.7 \times S_{ult}$, with the stress ratio, $R = 0.1$, at a frequency of 10 Hz. As for the quasi-isotropic specimens, the shape of the curve suggests three regions of interest: an initial, rapid stiffness reduction of 2–3%; and intermediate region in which an additional 1–2% of stiffness reduction occurs in approximately linear fashion; and a final region in which stiffness reduction occurs in abrupt steps ending in specimen fracture. Edge replicas made at various stages of the fatigue life for a typical cross-ply specimen are shown in Fig. 9. Approximately one-half of the

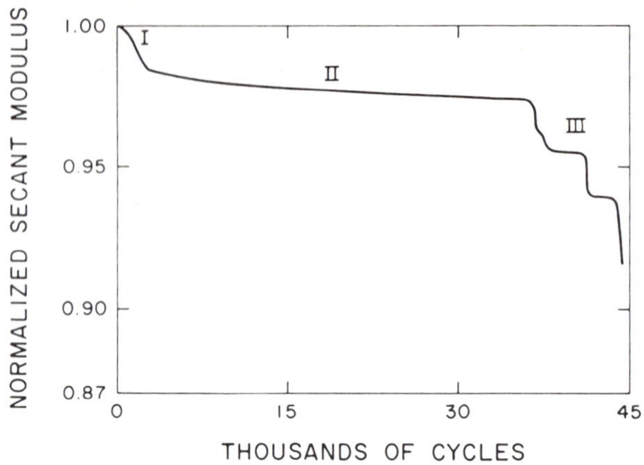

FIG. 8. Change in normalized stiffness with applied cycles for a $[0, 90_2]_s$ graphite epoxy specimen.

FIG. 9. Edge replicas of a progressively damaged $[0, 90_2]_s$ Gr/Ep specimen at (a) 1 cycle; (b) after 1% stiffness reduction; (c) after 3% additional stiffness change (magnification 25×).

transverse cracks which ultimately form to produce the characteristic damage state are seen to be formed in the first half-cycle. This contribution to the overall stiffness change was not reflected in the dynamic modulus measure of stiffness shown in Fig. 8, nor was the dynamic modulus measurement shown in that figure influenced by cracks formed in the initial loading to the mean level of fatigue cycling. Combined, these effects reduce the stiffness by approximately 4.5% measured statically for specimens of this type. In the middle region of Fig. 9, incipient edge delamination begins to form. At 3% stiffness reduction (Fig. 9(c)) one interface has completely delaminated along the length of the specimen and delamination has begun at the other interface near the edge of the specimen.

In the intermediate stiffness reduction region of Fig. 8, examination of penetrant-enhanced X-ray radiographs at this stage show the development of cracks along the fiber direction in the 0° plies, hereafter referred to as longitudinal splits. Fig. 10 shows the development of longitudinal splitting and transverse cracks in one example specimen. The longitudinal splits are observed to grow in both number and length with number of applied cycles. For this particular laminate, the region where these splits cross the major matrix cracks in the transverse 90° plies is of particular interest. Fig. 11 shows a detailed enlargement of one of the radiographs shown in Fig. 10. Several of the regions where longitudinal splits and transverse cracks cross are also regions where dark shaded areas have formed in the X-ray radiograph. One such region is indicated by an arrow in Fig. 11. A stereoscopic X-ray radiographic system was used to determine that these shaded regions are in fact delaminations at one or other of the 0/90° interfaces. By use of the stereoscopic X-ray method, each delamination was associated with the longitudinal split which bisects it. In order to verify the presence of the delamination and the longitudinal and transverse cracks that intersect at that position, several specimens were sectioned destructively in a serial fashion in order to establish the location and nature of the local details at these points of interest. It has been firmly established that these local delaminations are associated with the peculiar situation created by the coincidence of major matrix cracks which cross at an interface as is the case for the longitudinal splits and transverse matrix cracks in the cross-ply laminate under discussion. This type of damage localization is a serious consequence of matrix crack formation which is brought about by long-term cyclic loading and the complex stress fields associated with the progressive localization process. It should be emphasized, however, that this extreme situation where the local stresses cause relatively large regions of delamination has only been observed in the case of cross-ply laminates for graphite epoxy materials. While the nature of the stress fields that produce these local delaminations is certainly generic, the extent and severity of the result of those stress fields is highly dependent upon the material properties of the composite system involved and the stacking sequence of the laminates to be used.

So far we have mentioned only two major types of damage, matrix cracks which extend through the thickness of each ply and run parallel to the fibers in

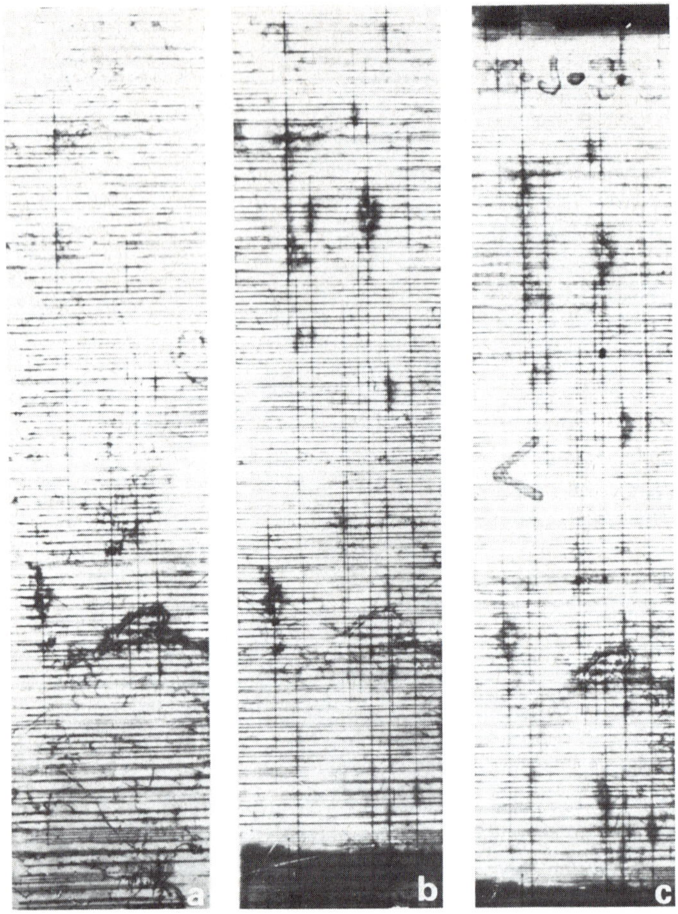

FIG. 10. Sequence of penetrant-enhanced X-ray radiographs of a $[0, 90_2]_s$ graphite epoxy specimen: (a) after 100 k cycles; (b) after 500 k cycles; (c) after 1 M cycles (magnification 0.55×).

those plies, and delaminations which form between the plies of a laminate either at the edge where complex stress states associated with a boundary cause initiation or in the interior where complex stress states associated with matrix cracks and intersecting matrix cracks cause delamination initiation. However, laminates ultimately fail because fibers break. Although most of what we consider to be 'the fatigue effect' in composite materials is associated with matrix damage and ply separation, the performance and degradation of the load bearing (usually 0°) plies is influenced by the damage that forms in the plies around it. The most prominent aspect of that influence is the relationship between transverse matrix crack formation and the fracture of fibers in plies which are adjacent to the plies in which the matrix cracks form. In order to understand this aspect of damage development, a number of specimens

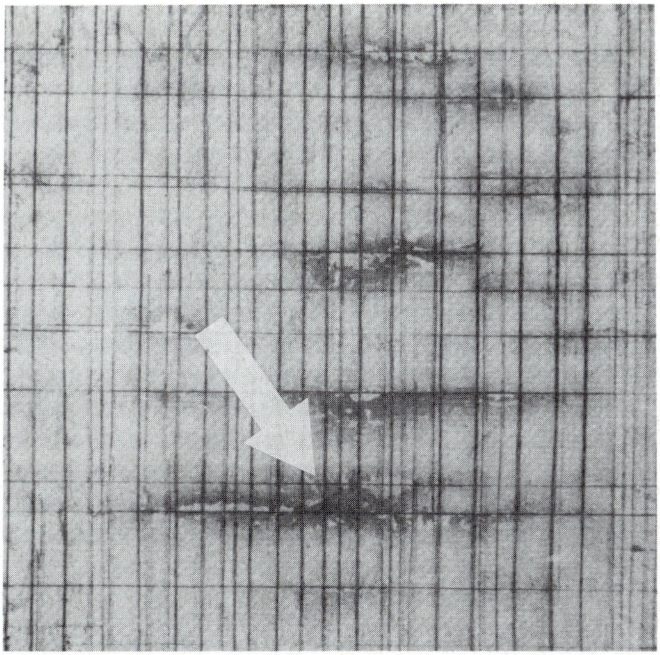

FIG. 11. Detail of a radiograph of a fatigue-damaged $[0, 90_2]_s$ Gr/Ep specimen showing delamination associated with longitudinal splitting (magnification 5×).

including several of the $[0, 90_2]_s$ specimens under discussion were subjected to fatigue loading and were then deplied and the surfaces of the individual plies were examined in a transmission electron microscope. (The method of deply was essentially identical to that of FREEMAN, 1982.) In this process, the damaged laminate is heated to a temperature sufficient to pyrolize the resin matrix, about 418°C in the case of the graphite epoxy material involved in the present tests. With the interlaminar binding medium thus eliminated, plies can be separated for individual examination. In order to establish the exact location of cracks in a 90° ply relative to positions on the face of the adjacent 0° plies, a solution of gold chloride in diethyl ether was applied to the fatigue damaged specimens in the same manner that the X-ray enhancing penetrant was applied. The ether carried deposits of the gold chloride into all of the damaged zones and was then purged prior to the pyrolization step by moderate heating. For the $[0, 90_2]_s$ specimens this produced internal gold chloride 'traces' of transverse cracks, delaminations, and longitudinal splits in the 0° plies which remained on the face of the adjacent plies after ply separation providing an excellent fiducial marker of those locations. Coupons cut from deplied 0° laminae were examined in a scanning electron microscope to determine the relationship between fiber fractures in the 0° plies and damage in the adjacent plies. Figs. 12 and 13 show examples of those results. The bright, dust-like particles in Fig. 12 are gold chloride particles which are associated with matrix

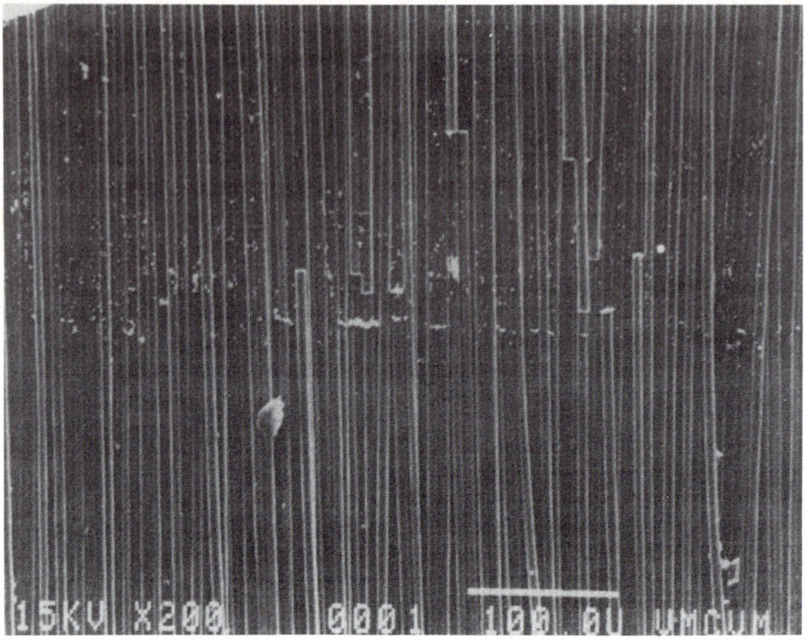

FIG. 12. Scanning electron micrograph of a 0° ply taken from a fatigue-damaged $[0, 90_2]_s$ Gr/Ep specimen showing a fiber fracture zone (magnification 200×).

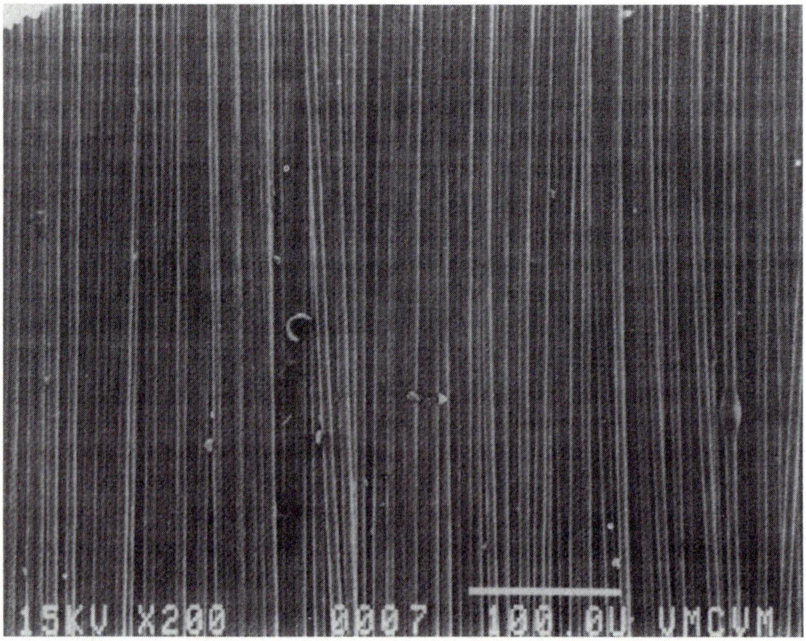

FIG. 13. Scanning electron micrograph of a 0° ply taken from a fatigue-damaged $[0, 90_2]_s$ Gr/Ep specimen showing a fiber fracture-free zone (magnification 200×).

cracks in the adjacent 90° plies as described earlier. Fiber breaks in the 0° ply are seen to lie within a band which is associated with those particles that mark the trace of the adjacent 90° transverse cracks. In regions where no gold chloride is seen, few fiber breaks are found (c.f. Fig. 13). This result was seen to be repeated for coupons taken from various points along the length and across the width of the specimen, i.e., the bands of fiber fractures with gold chloride fiducial marks occurred across the width of the specimen in periodic arrays that were associated with the characteristic damage state of matrix cracking in the 90° plies of the laminate. An attempt was also made to establish the nature of the preferential fiber fracture patterns through-the-thickness of the 0° plies in regions of concentrated fiber fracture. To this end, several fiber layers were carefully removed with transfer tape from the deplied 0° coupon, which has previously been examined. The density of fiber breaks in the next layer through-the-thickness of the 0° ply was less than the density of breaks in the fiber layer nearest to the 0/90° interface in the same coupon. In addition, the zones of fiber breakage were somewhat broader in the sublayers of the 0° ply. Hence, it appears that fiber fractures are induced by matrix cracks and that the number of fiber failures so induced drops off rather quickly with distance from the crack tip in the plane of the 0° plies and through-the-thickness of the 0° plies, i.e., as a function of distance from the matrix crack tip in the adjacent 90° plies.

The structural significance of this result derives from the fact that fiber failures diminish the strength and life of composite laminates and composite structures. Although matrix cracks do not diminish those quantities by their occurrence alone, they do induce significant localized damage including fiber fracture during cyclic loading which does contribute significantly to the degradation of the laminate properties. It should also be mentioned that these fiber fractures begin quite early in the fatigue life, usually within the first third of the total life of a specimen or component. The number of fiber fractures is also significantly increased by cyclic loading, in general (JAMISON, 1982).

2.3. Other angle-ply laminates

The third generic type of laminate that we will address for demonstration purposes is the $[0, \pm 45]_s$ laminate cycled at the same relative amplitudes (about 65% of the ultimate tensile strength) which produce a life of a few hundred thousand cycles. Fig. 14 shows the comparative stiffness changes for the two laminates just discussed and the one under present consideration. Curve C of that figure indicates a stiffness change due to matrix cracking (in the 45° plies in this case) occurs quite early in the life of the specimen as was observed for the other two laminates, but is larger in magnitude for this situation. After that initial change, there is a 'stable' damage development region where very little stiffness change is observed and very little change in properties of the laminate occur. Perhaps the most distinctive aspect of the stiffness change curve for this laminate is the very abrupt drop is stiffness quite near the end of the life of the specimen which corresponds to a very rapid coalescence, localization, and

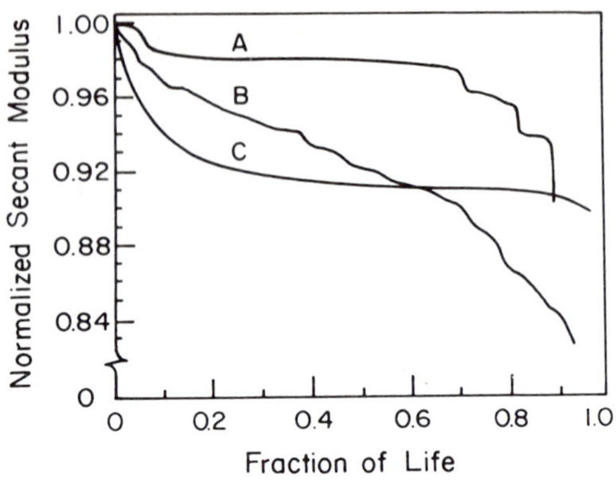

FIG. 14. Typical graphs of normalized stiffness change versus cycles of loading for: (A) $[0, 90_2]_s$; (B) $[0, 45, 90]_s$, (C) $[0, 45]_s$ graphite epoxy laminates.

increased rate of damage development. That sequence of events happens so rapidly that stiffness changes are very difficult to record over that small period of time. No edge delamination was observed in these specimens. Although initial stiffness reductions due to matrix cracking in this type of laminate is relatively large, the laminate demonstrates excellent stability during cyclic loading.

We have examined various modes of damage development in composite laminates that can be expected during cyclic loading of composite structures at load levels which correspond to a life of several hundred thousand cycles to approximately one million cycles. Fig. 15 illustrates a schematic representation of the relationships of the changes in properties, damage development, and life of such laminates and composite structures. As indicated in that figure, although initial matrix cracking produces stiffness changes and a rapid increase in the damage curve, residual strength changes are small since that damage mode would also be present in a quasi-static test of strength. The residual strength is also changed by only small amounts during the relatively stable (slow rate of development) damage development region throughout a large portion of the middle section of the life diagram. As coalescence, localization, damage growth, and the rate of damage initiation increase near the end of the life of the specimen, the residual strength begins a rather rapid decrease until it reaches the stress amplitude at which the load is applied at which time failure occurs defining the life of the specimen. As mentioned earlier, the presence of compressive load excursions during cyclic loading may introduce load or global instability (buckling) as a mode of damage development and failure. This type of situation is extremely geometry dependent and very difficult to characterize in a generic way. It should be mentioned, however, that damage development is very much more rapid when both tension and compression load excursions are present in a laminate in a composite structure. Regions of uniform strain in a laminate will develop matrix crack patterns if the strain level or cyclic

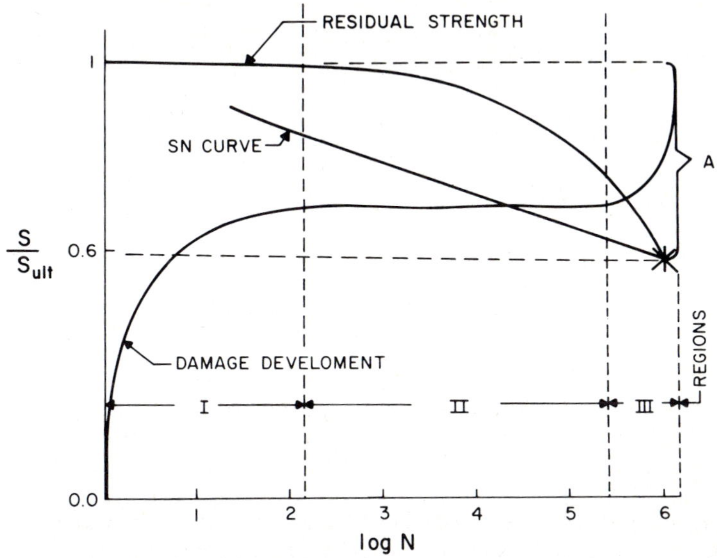

FIG. 15. Schematic diagram of a damage development curve superposed on the residual strength and life curves of a typical composite laminate.

amplitude is large enough to cause a single crack, as mentioned above. This characteristic damage state, or CDS, is characteristic of the laminate properties, the ply properties, the ply thicknesses, and the stacking sequence of the plies in a specific laminate.

These patterns can be predicted from any valid mechanics representation of the local stresses and some philosophy such as that suggested by REIFSNIDER (1977). An example of such a predicted and observed pattern is reproduced in Fig. 16. Some of the fundamental aspects of the internal stress redistribution associated with these matrix cracks can be extracted from simple laminate stress analysis and physical observations. Fig. 17(a) portrays a generic situation

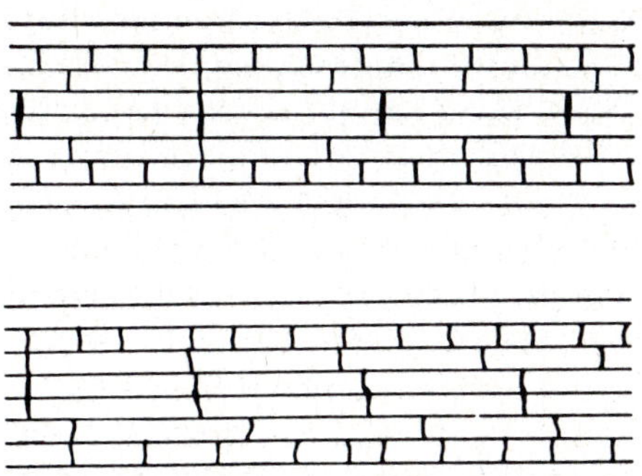

FIG. 16. Predicted (top) and observed (bottom) characteristic damage states for matrix cracking in a $[0, 90, \pm 45]_s$ graphite epoxy laminate.

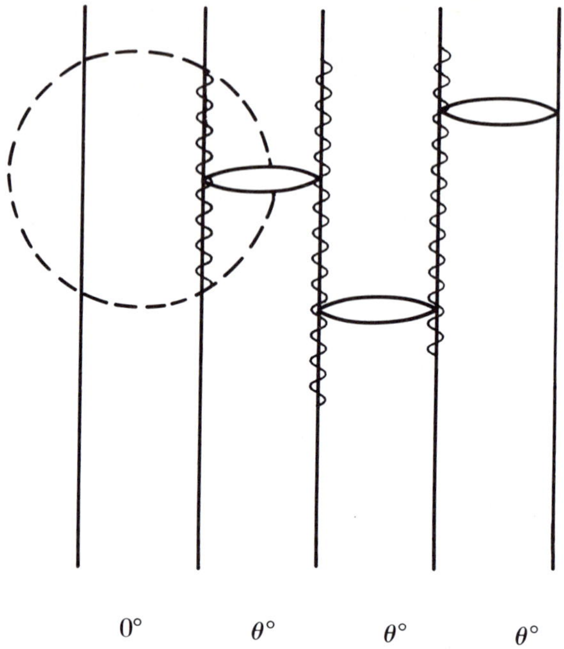

FIG. 17. Sketch of local damage and regions of influence for matrix cracking in off-axis plies and subsequent induced local delamination.

wherein off-axis plies having orientations of + or $-\theta^0$ have sustained matrix cracking causing redistribution of stresses in the regions near those cracks as suggested by the dotted circle. The basic nature of that stress redistribution can be established by considering the quasi-static strength of such a laminate since such cracks also form before fracture under quasi-static loading. The most common method of calculating an engineering value of static strength for laminates not dominated by edge delamination is to calculate the ply stresses using laminate analysis, invoke some failure criterion to predict first-ply failure (usually matrix cracking), reduce the shear and in-plane modulus of the broken ply, recalculate the ply stresses, test for second-ply failure, etc. until 'last-ply failure' is predicted. This scheme, commonly referred to as the ply discount method, has been widely used over a period of at least 15 years and is known to provide good engineering estimates of laminate strength when edge effects do not dominate the failure process. Table 1 shows the stresses in individual plies of an example laminate before and after matrix cracks form in the 90° and + and −45° plies. As cracks develop in the off-axis plies the axial stress in the 0° plies increases by about 14%. From a structural standpoint, it is important to note that the stiffness of the laminate changes by almost exactly the same amount. This is true in a great number of cases examined by the authors and suggests that the residual properties of the laminate which are determined by

TABLE 1.

Laminate condition	Engineering Moduli (GPa)				Ply Stresses (MPa for 1000 MPa applied) 0			90°			+45°			−45°		
	E_L	E_T	G_{LT}	ν_{LT}	σ_x	σ_y	τ_{xy}	σ_x	σ_y	τ_{xy}	σ_x	σ_y	τ_{xy}	σ_x	σ_y	τ_{xy}
Undamaged [0, 90, ±45]$_s$ laminate (T300–5208)	54.4	54.5	20.5	0.32	2632	−2.3	0	167	−797	0	600	400	418	600	400	−418
90° plies cracked	52.4	54.6	19.5	0.31	2735	0.1	0	0	−852	0	632	426	442	632	426	−442
90°, ±45° plies cracked	47.8	50.0	19.0	0.33	2993	−4.7	0	0	−1000	0	503	503	503	503	503	−503
Delaminated to [0, 90] and	77.2	77.2	4.1	0.04	3131	59	0	217	−59	0						
[±45]$_s$ sublaminates (laminate stiffness $E_L \approx 46.1$)	15	15	37.2	0.81							325	0	136	325	0	−136
All off-axis plies delaminated and cracked	35.8	—	—	—	4000	0	0	0	0	0	0	0	0	0	0	0

the 0° plies and the stress state in those plies can be implied directly from measuring stiffness changes caused by damage development during quasi-static or fatigue loading. If matrix cracking were the only damage mode present, this would indeed be so. However, these stress redistributions and the stiffness reductions that cause them are not uniform. They exist only near the matrix cracks in the off-axis plies as implied by Fig. 17. The first direct proof of that to our knowledge was provided by HIGHSMITH (1981) and JAMISON et al. (1981) in the Materials Response Laboratory who measured local strains in the 0° ply caused by the stress concentrations associated with matrix cracks in adjacent plies (HIGHSMITH, 1981; JAMISON et al., 1981). These local stresses frequently cause the initiation growth of delaminations in local regions in the interior of the laminate. We can illustrate this by considering the interlaminar stresses near a matrix crack in a $[0, \pm 45, 90]_s$ graphite epoxy laminate. Fig. 18 shows distributions of interlaminar stresses calculated using the quasi three-dimensional finite difference analysis generated by TALUG and REIFSNIDER (1978). The stresses shown are generated at comparable amplitudes (such as the $R = -1$ case). While this phenomenon is not clearly understood at this time, it is clear that the presence of matrix cracks induced by tensile loading greatly enhance the initiation and propagation of delamination during the compressive part of the load cycle. As a consequence, it is common to find that the life of a specimen at a maximum stress amplitude of 60% of the ultimate strength for $R = 0.1$ is about the same as the life of a specimen subjected to a stress amplitude of 45% of the static ultimate strength for an R value of -1 for common types of graphite epoxy laminates for example. This type of damage development is still under study.

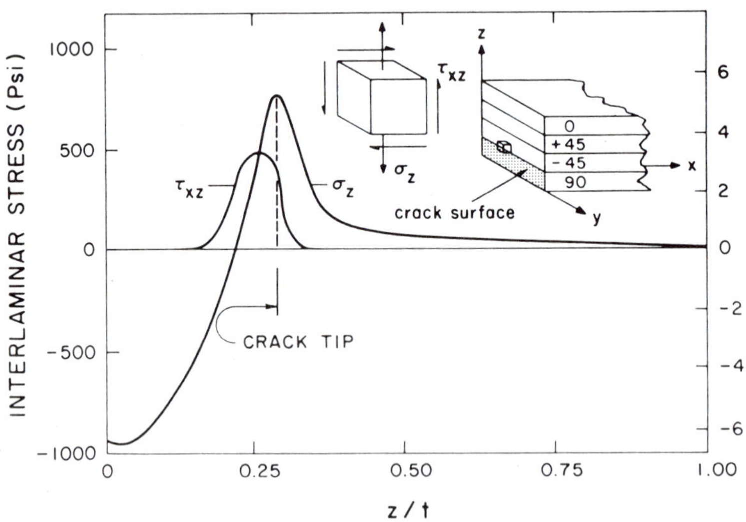

FIG. 18. Calculated distributions of interlaminar shear and normal stress caused by matrix cracking for a quasi-isotropic laminate.

3. Damage mechanics

In the previous section we have examined the damage modes that can be expected to occur in composite laminates of the type commonly used for engineering structures subjected to quasi-static and cyclic load histories of the type one would expect to occur during service of those structures. In this section, we will attempt to provide a summary discussion of the micro-mechanics associated with the damage modes discussed earlier, in the hope of providing some understanding of the consequences of damage development and also of providing a background for our discussion in the next section of life prediction and representation.

The subject of micro-mechanics is extensive. Even now the literature on this subject is profuse and there is much activity in the field at this time. Many of the topics that we will discuss here in a very brief way are well developed or under vigorous development in the literature. The reader with specific interest in particular subjects of this type is advised to survey the literature carefully and heed the most recent work. The objective of our discussion here will be to concentrate on principles and to develop the underlying concepts associated with damage development and the consequences of that development during cyclic loading in composite laminates.

3.1. Delamination

We will begin with a discussion of delamination. Delamination growth is a major concern to designers of composite structures, and is the major damage mode in many situations. Edges, notches, and other geometric discontinuities are sites of local interlaminar normal and shear stresses, and are therefore sites of potential delamination initiation. The interlaminar stresses may be tensile, compressive or shear and may vary in magnitude depending on the stacking sequence and mode of loading. Tensile interlaminar stresses are thought to be the most common and effective initiators of delamination. It should be noted that while uniaxial tensile loading of an unnotched $[0, 90, \pm 45]_s$ laminate produces small tensile interlaminar normal stresses at the edge of the 0° plies and compressive interlaminar normal stresses at the edge of 90 and 45° plies, if a compressive load is applied to the laminate the sign of the interlaminar normal stresses reverses. In fact, such a laminate will generally not delaminate during tensile cyclic loading but delamination will initiate and grow when compressive load excursions are added to the cyclic history. Under cyclic loading, delaminations will also initiate at interfaces where the interlaminar shear stresses are high. Delamination of those interfaces at holes is associated with a threshold value of shear stress (STINCHCOMB et al., 1981; RATWANI and KAHN, 1979). Interlaminar shear stresses greater than a threshold stress value initiate delaminations which reduce compressive strength and fatigue life. There is still a considerable amount of uncertainty about the prediction of delamination initiation. Two approaches are common. One approach holds that delamination can be predicted by determining the stress level, i.e., if one determines all the three-dimensional stress components from a suitable analysis and uses those stresses in an appropriate failure theory based on stress level concepts, delamination can be accurately and successfully predicted. A second approach is based on energy release rate, and claims that a critical value of energy release (a single parameter) can be associated with the resistance of a laminate to delamination initiation. Both approaches appear to have elements of truth since each has been used successfully in certain circumstances, but neither is without failures at this point.

There are several difficulties associated with theses endeavors. First, as mentioned earlier, it is difficult to discuss this subject in a generic way since the particulars of each situation are highly dependent upon geometry and stacking sequence and are, therefore, generally quite different from case to case. Second, both of the major approaches mentioned above require an extremely meticulous and accurate account of the micro-stresses in the neighborhood of the delamination. While great strides have been made in the analysis of such stresses, the development of techniques which are capable of determining three-dimensional micro-stresses is by no means complete. Nevertheless, in well controlled and isolated conditions, good correlations with experimental behavior have been observed using both approaches. The interested reader is referred to STARNES and WILLIAMS (1982); RAMKUMAR (1983); WHITCOMB (1981a); WHITCOMB (1981b); WHITCOMB (1983); KIM and SONI (1983); O'BRIEN (1981); O'BRIEN (1983); O'BRIEN (1982); and HERAKOVICH et al. (1979) and HERAKOVICH (1981, 1982). Because delamination growth is planar and self-similar, it is the one type of damage in composite materials that is ideally suited for study and description using principles of fracture mechanics. As the delamination grows along an interface, the growth rate ($\mathrm{d}a/\mathrm{d}N$) is related to the strain energy release rate ($\mathcal{G}$) by the power law shown below (WILKINS et al., 1982; O'BRIEN, 1981).

$$\mathrm{d}a/\mathrm{d}N = A\mathcal{G}^m \,. \tag{3.1}$$

The quantities A and m are assumed to be material constants. In general practice, $\mathcal{G}$ is the total strain energy release rate with components due to Mode I, Mode II and Mode III deformation.

The growth rates of interfacial cracks and delaminations are particularly sensitive to the mode of cyclic loading. For example, tensile cyclic loading of 48-ply, quasi-isotropic graphite epoxy laminates at a maximum cyclic strain of 4500 $\mu\varepsilon$ and $R = 0.1$ has been shown to produce characteristic matrix cracking and limited crack coupling after one million cycles (LIECHTI, 1982). Under those conditions the residual strength was 77–92% of the static tensile strength. Compressive cyclic loading at the same strain level but with $R = 10$ produced no damage and no strength degradation after one million cycles. However, specimens loaded in tension–compression fatigue at ± 4500 $\mu\varepsilon$ ($R = -1$) were heavily damaged by 50 000 cycles and suffered tensile strength degradations of 10–25%. Matrix cracks and interfacial cracks initiated and grew during the tensile portion of the loading wave form; however the compressive portion of the wave form caused damage mode interaction, accelerated the growth rates of the existing damage modes, and produced interfacial cracks which extended along the length of the specimen. As mentioned earlier, this type of synergistic interaction of damage modes in combined tension–compression loading when edges are present (and edge delamination is a possible damage mode) is the most serious potential damage development situation that the designer of composite structures must be concerned with, at least in the sense that the strain amplitudes which cause failure in less than a million cycles are quite low compared to the ultimate strain to failure in quasi-static loading (about 10 000 $\mu\varepsilon$ in this case).

3.2. Matrix cracking

The other major damage mechanics subject to be dealt with in this section is associated with the stresses and stress redistributions brought about by matrix cracking. These are important primarily because of their influence on the residual strength and life of composite laminates and structures.

We are concerned here with matrix cracks which extend completely through the thickness of the cracked ply in a laminate and which follow the direction of the fibers in that ply across the entire width of a laboratory specimen or over distances which are several times thickness 90/−45° ply interface when cracks form in the double 90° plies. These stresses occur in the interior of the laminate and vanish when no cracks are present. It is these stresses which frequently cause local delamination to initiate along the ply interfaces near matrix cracks as illustrated in the former section and shown in Fig. 11. It is important to remember that these stresses are associated with the presence of matrix cracks regardless of the global applied stress as long as the crack is in an opening mode. In that situation, it is also important to note that local stresses develop in the σ_y direction, parallel to the matrix crack tip as shown in Fig. 19. For a crack opening situation, these stresses are tensile and tend to cause the crossed-crack patterns described in Fig. 6. These stresses are suggested from equations for Mode I crack opening stresses as shown in Fig. 19. There has been no confirmation that these equations represent actual stresses or strain fields near matrix cracks. However, the general nature of the Poisson effect along the crack front direction is reflected in the σ_y component shown in the diagram, and does provide a 'driving force' for the type of extensive cross cracking observed in many cases.

Although the degree and extent of the stress redistribution associated with the damage mechanics discussed above varies according to the material system and stacking sequence concerned, the generic nature of the mechanics treatments discussed is general.

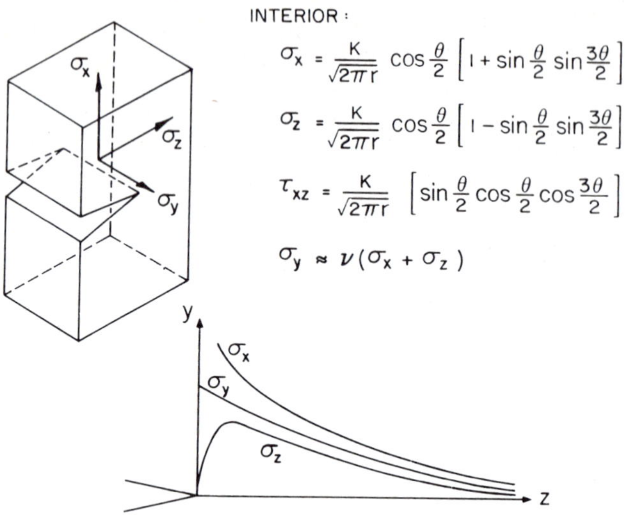

FIG. 19. Linear elastic fracture mechanics stress components for an ideal crack in Mode I opening.

4. Life prediction and representation

In order to illustrate the consequences of the damage development discussed above, we will develop in this section a mechanistic cumulative damage relationship based on those damage concepts and details. For the purpose of our accumulation model we adopt the fatigue behavior of the unidirectional material as a basic 'critical' element. This is done in the context of the realization that strength and life considerations must involve phenomenological characterizations, since rational derivations of those concepts has not been achieved. In our case, we assume that the phenomenological characterization necessary to describe cumulative damage, i.e., the basic data that must be developed which characterizes the performance of the material to be used in our laminates, is the *S–N* relationship for unidirectional material. This first basic premise is motivated by two important factors. First, the 0° plies are indeed the 'critical elements' in the fiber dominated laminates that we wish to consider in the sense that those plies are the last to fail during cyclic and quasi-static loading and are, therefore, the critical elements in the determination of strength and life. And second, these critical elements (the 0° plies) do degrade under cyclic loading. It should be mentioned that it does appear that some generalization of the characterization of fatigue performance of unidirectional material appears to be possible. To the author's knowledge, Mandell (MANDELL et al., 1981) was the first to suggest such a generalization, and noted that for a large range of both unidirectional and multiaxial laminates made from glass epoxy material a normalized fatigue life equation such as

$$S/S_u = 1 - 0.1 \log n \tag{4.1}$$

could be used to characterize the data with reasonable engineering accuracy. In that equation, S is the applied stress amplitude, S_u is the static ultimate strength, and n is the number of cycles. Similar generalizations can be made for other material systems. While this type of characterization is certainly no more than a baseline generalization, the simplicity and universality of these representations which serve as the phenomenological characterization necessary in a philosophy of damage accumulation is certainly an attractive feature.

The second basic element of our damage accumulation philosophy is a representation of the damage that develops in the plies of the laminate which have orientations that are not coincident with the principal loading direction, as described in former sections. These plies can be considered to be 'subcritical elements' in the sense that their failure does not cause failure of the laminate but causes stress redistribution which influences the rate of degradation of the critical elements, the 0° plies in our case. Hence, as those damage modes develop, the internal stress which causes the degradation of the 0° plies as represented by S in eq. (4.1) for example, increases as the number of cycles of load increases due to the fact that other off-axis plies are failing and redistributing the load they were holding into the remaining unbroken plies. For our purposes, then, we adopt an extension of the 'ply discount scheme' discussed earlier in the sense that we adjust the stress state and the rate of degradation of the 0° plies as the degradation of the off-axis plies during fatigue loading occurs. However, the details of the application of such a philosophy are considerably more complex than the discount scheme. For example, the life of the 0° plies calculated by solving eq. (4.1) for N becomes a continuous function of the number of applied cycles since the local stress in the 0° plies as damage develops is a function of cycles also. In addition, the stress field in the 0° plies is not unidirectional, a fact which must be considered when computing the strength of the laminate determined from the stress state in those plies. In order to accommodate these features, a generalized summation equation is postulated as shown in eq. (4.2).

$$\Delta S(n) = \int_0^{\gamma} (1 - F_{\mathrm{L}}(n))\, i \left(\frac{n}{N(n)}\right)^{i-1} \mathrm{d}\left(\frac{n}{N(n)}\right). \qquad (4.2)$$

This equation states that the change in residual strength of the specimen is equal to the integral of a quantity which describes the degradation of the 0° plies in terms of the local failure function, F_{L}, the current lifetime of the plies, N, and the number of applied cycles, n. A parameter i is introduced to account for nonlinearity in the degradation of residual strength. While the development of this equation is beyond the scope of the present paper, two forms of the equation will be mentioned for illustration purposes. If the stress in the 0° plies were unidirectional and were not dependent upon the number of applied cycles (that is, were not influenced by the redistribution of stress as damage develops in the off-axis plies) and if the degradation of residual strength were a linear function so that the parameter i in eq. (4.2) became equal to 1, then eq. (4.2) could be integrated by quadratures and expressed as eq. (4.3) below.

$$(S_r(n) - S_u) = (1 - S_a)n/N \,. \tag{4.3}$$

(It should be noted that the quantities in eq. (4.2) were normalized by the quasi-static tensile strength.) It is apparent that eq. (4.3) is a simple straight line anchored at the end points by the static ultimate strength which corresponds to $n = 0$ and the stress amplitude which corresponds to the life of the specimen when $n = N$. This linear relationship is in fact the expression postulated by BROUTMAN and SAHU (1969). Hence, for the situation when the applied stress amplitude is constant, the stress in the critical elements is unidirectional and constant, the failure of the unidirectional plies is defined by simple maximum stress criterion, and the variation of residual strength is linear between the static ultimate strength and the constant applied stress amplitude, eq. (4.2) reduces to the simple identity stated in eq. (4.3).

A second illustration of the meaning of the terms in eq. (4.2) is provided by Fig. 20. That figure illustrates the interpretation of the terms in eq. (4.2) for the instance when tension–tension loading is to be described. Since the quantities in the integrand are functions of the number of applied cycles, it is necessary to compute the normalized change in residual strength by performing the integration numerically while incrementing the number of applied cycles, n. The two basic elements of our cumulative damage model are accounted for in the computations which enter the equation. The damage in the subcritical

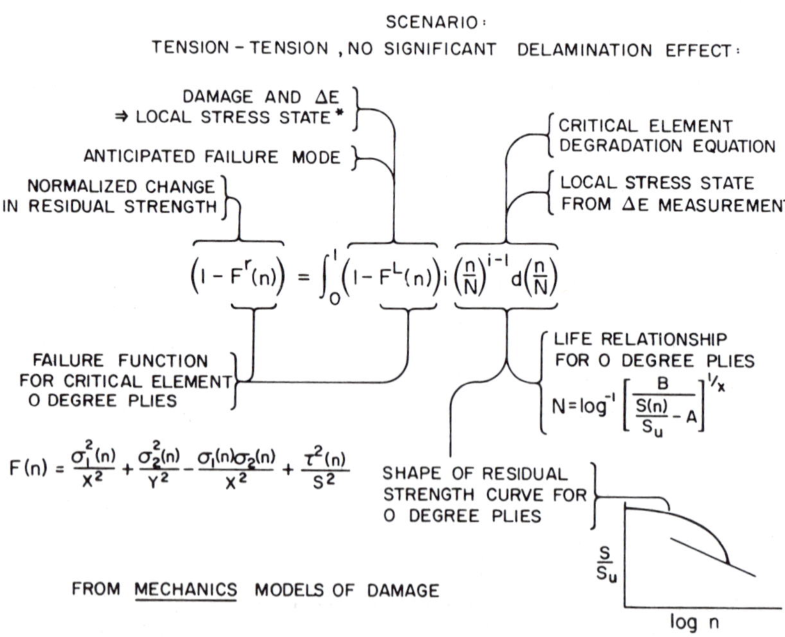

FIG. 20. Definition of terms in the degradation equation for tension–tension cyclic loading.

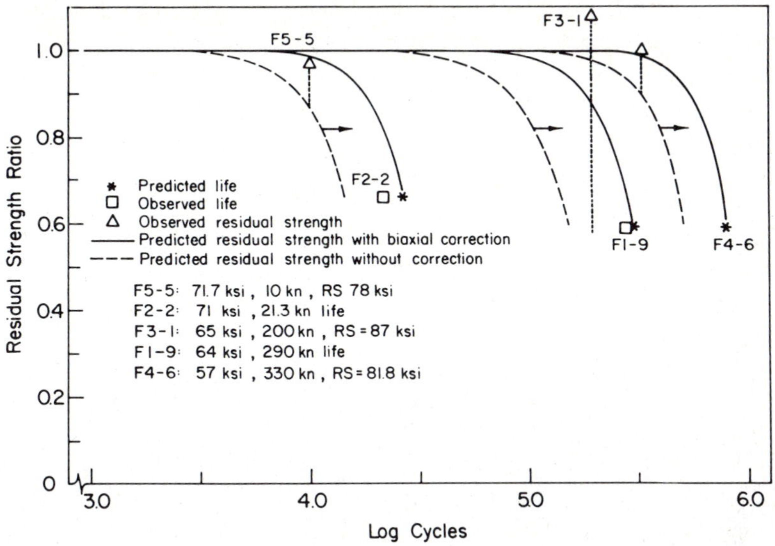

FIG. 21. Predicted and observed residual strength and life for five specimens under cyclic tensile loading, showing the importance of considering the local biaxiality of stress in the model.

elements enters through the variation in the local stress state which is used to calculate the local failure function, F_L, and also influences the current life, N, by changing the local fiber direction normal stress in the 0° plies, S, which is entered into the phenomenological S–N curve of the 0° plies. The degradation of the critical elements, 0° plies in this case, is represented by the phenomenological S–N curve which is determined from experimental data. This model is able to account for sequence effects in block loading, the effects of unknown load histories, biaxial stress states, manufacturing variations, and the influence of different laminate type. The model has been applied to tension–tension loading, tension–compression loading, and compression–compression sequences (REIFSNIDER and STINCHCOMB, 1984). An example of the predictions of that equation is shown in Fig. 21. The residual strength curves are predicted by the model as is the life of the specimens by the coincidence of the residual strength prediction and the applied stress level. While this mechanistic model is only the first step in a very long journey to the achievement of the representation and prediction of life of composite structures through the mechanistic representation of damage development in composite laminates, it is at least an encouraging step in that direction.

5. Closure

It is the practice in engineering to avoid problems that we do not understand. Such is frequently the case for damage development in composite materials. It is regarded as prudent for the designer to prescribe design strain levels which are sufficiently small so that damage development is thought to be precluded. If all materials used in structures were ideal, and if all loading histories were completely and rigidly established, then perhaps one could be sure that damage would be avoided. However, reality requires that something be known of damage development even if it is to be avoided, and the basic (and critical) question of 'how long will this structure last?' cannot be answered or even properly addressed without some knowledge and understanding of damage development in composite materials. As new material systems and applications are developed, and old material systems and applications are further exploited, a more and more precise understanding of damage development will be required. What is needed is a philosophy and rational analysis that can be used to describe the explain damage development in composite laminates so that we can anticipate their residual strength, stiffness, and life under engineering conditions. In a general sense, this requires a replacement for the single crack problem (and fracture mechanics) in homogeneous isotropic materials. The discovery and development of such a philosophy is the ultimate goal of our efforts to study damage. Without such a development the practical application of composite materials will be a disorganized effort which produces a random array of limited successes and costly failures. If such a development is achieved, it will provide a foundation for rational exploitation of the widely recognized advantages of composite materials for many engineering purposes.

References

BADER, M.G., J.E. BAILEY, P.T. CURTIS and A. PARVIZ (1979), The mechanisms of initiation and development of damage in multi-axial fibre-reinforced plastic laminates, in: *Proc. 3rd Int. Conf. on Mechanical Behavior of Materials*, Cambridge.

BROUTMAN, L.J. and S. SAHU (1969), Progressive damage of a glass-reinforced plastic during fatigue, *24th Annual Technical Conf.*, *Reinforced Plastics/Composites Dir.*, SPI.

DANIEL, I.M., Ed. (1982), *Composite Materials: Testing and Design (Sixth Conference)*, STP-787 (American Society for Testing and Materials, Philadelphia).

FONG, J.T., Ed. (1979), *Fatigue Mechanisms*, STP-675 (American Society for Testing and Materials, Philadelphia).

FREEMAN, S.M. (1982), *Composite Materials: Testing and Design (Sixth Conference)*, STP 787 (American Society for Testing and Materials, Philadelphia).

HAHN, H.T. and R.Y. KIM (1976), *J. Comp. Mater.* **10**, 156.

HERAKOVICH, C.T. (1981), *J. Comp. Mater.* **15**, 336–348.

HERAKOVICH, C.T. (1982), *J. Comp. Mater.* **16**, 216–227.

HERAKOVICH, C.T., A. NAGARKAR and D.H. O'BRIEN (1979), *Modern Developments in Composite Materials and Structures*, J.R. VINSON, Ed., ASME, pp. 53–66.

HIGHSMITH, A.A. (1981), Stiffness reduction from transverse cracking in fiber reinforced composite laminates, Master's Thesis, College of Engineering, Virginia Polytechnic Institute and State University, Blacksburg, VA.

JAMISON, R.D., A.L. HIGHSMITH and K.L. REIFSNIDER (1981), *Comp. Technol. Rev.*, **3**, No. 4, 158.

JAMISON, R.D. (1982), Damage development and failure of fiber-reinforced composite materials, Ph.D. Dissertation, College of Engineering, Virginia Polytechnic Institute and State University, Blacksburg, VA.

KIM, R.Y. and S.R. SONI (1983), Experimental and analytical studies of the onset of delamination in laminated composites, UDR-TR-83-40, Univ. of Dayton Research Inst., Dayton, OH.

KRIZ, R.D. and W.W. STINCHCOMB (1980), Effects of moisture, residual thermal curing stresses and mechanical load on the damage development in quasi-isotropic laminates, in: *Damage in Composite Materials*, STP-775 (American Society for Testing and Materials, Philadelphia) pp. 63–80.

LAURAITIS, K.N., Ed. (1981), *Fatigue of Fibrous Composite Materials*, STP-723 (American Society for Testing and Materials, Philadelphia).

LIECHTI, K.M. (1982), Cumulative damage model for advanced composite materials, Final Report, Phase I, Contract No. F33615-81-C-5049, AFML, WPAFB, OH.

MANDELL, J.F., D.D. HUANG and F.J. MCGARRY (1981), *Comp. Technol. Rev.* **3**, No. 3, 96.

O'BRIEN, T.K. (1980), Stiffness change as a nondestructive damage measurement, in: W.W. STINCHCOMB, Ed., *Mechanics of Nondestructive Testing* (Plenum Press, NY).

O'BRIEN, T.K. (1981), Characterization of delamination onset and growth in a composite laminate, NASA Technical Memorandum 81940, NASA Langley Research Center, Hampton, VA.

O'BRIEN, T.K. (1982), Characterization of delamination onset and growth in a composite laminate, *Damage in Composite Materials*, K.L. REIFSNIDER, Ed. (American Society for Testing and Materials, Philadelphia) pp. 140–167.

O'BRIEN, T.K. (1983), Mixed-mode strain–energy release rate effects on edge delamination of composites, NASA Technical Memorandum 84592, NASA Langley Research Center, Hampton, VA.

O'BRIEN, T.K. and K.L. REIFSNIDER (1977), *J. Testing Eval.* **5**, No. 5, 384.

PIPES, R.B. and N.J. PAGANO (1970), *J. Comp. Mater.*, Oct., 538.

PIPES, R.B. and N.J. PAGANO (1971), *J. Comp. Mater.*, Apr., 255.

RAMKUMAR, R.L. (1983), Performance of a quantitative study of instability-related delamination growth, NASA Contractor Report 166046, NASA Langley Research Center, Hampton, VA.

RATWANI, M.M. and H.P. KAN (1979), Compression fatigue analysis of fiber composites, Contract Report, NADC-78049-60.

REIFSNIDER, K.L. (1977), Some fundamental aspects of the fatigue and fracture response of composite materials, in: *Proc. 14th Annual Meeting, Soc. of Engrg. Sci.*, Lehigh Univ.

REIFSNIDER, K.L., Ed. (1982), *Damage in Composite Materials*: Basic Mechanisms, Accumulation, Tolerance, and Characterization, STP-775 (American Society for Testing and Materials, Philadelphia).

REIFSNIDER, K.L. and A.L. HIGHSMITH (1981), Characteristic damage states: a new approach to representing fatigue damage in composite laminates, in: *Materials: Experimentation and Design in Fatigue* (Westbury House, Guildford, Surrey) pp. 246–260.

REIFSNIDER, K.L. and R.D. JAMISON (1982), *Int. J. Fatigue*, Oct., 187.

REIFSNIDER, K.L. and W.W. STINCHCOMB (March 1984), Cumulative damage model, Final Report, Phase II, General Systems.

REIFSNIDER, K.L., E.G. HENNEKE, II and W.W. STINCHCOMB (1979), Defect-property relationships in composite materials, AFML-TR-76-81, Wright-Patterson AFB, OH.

REIFSNIDER, K.L. and A. TALUG (1980), *Int. J. Fatigue*, **3**. No. 1, 3.

STARNES, J.H., Jr and J.G. WILLIAMS (1982), Failure characteristics of graphite epoxy structural components loaded in compression, NASA Technical Memorandum 84552, (NASA Langley Research Center, Hampton, VA).

STINCHCOMB, W.W. and K.L. REIFSNIDER (1983), Damage accumulation concepts for fatigue loaded composite laminates, *Conf. on Mechanics of Damage in Composites*, Winter Annual Meeting, ASME, Boston, MA.

STINCHCOMB, W.W., N.F. BLACK, K.L. REIFSNIDER and E.G. HENNEKE (1981), Damage development mechanisms in notched composite laminates under compressive fatigue loading, Contract Report, NADC-79055-60.

TALUG, A. and K.L. REIFSNIDER (1978), Analysis of stress fields in composite laminates with interior cracks, VPI-E-78-23, College of Engineering, Virginia Polytechnic Institute and State University, Blacksburg, VA.

TSAI, S.W., Ed. (1979), *Composite Materials: Testing and Design (Fifth Conference)*, STP-674 (American Society for Testing and Materials, Philadelphia).

WILKINS, J.D., J.R. EISENMANN, R.A. CAMIN, W.S. MARGOLIS and R.A. BENSON (1982), *Characterizing Delamination Growth in Graphite Epoxy*, STP-775 (American Society for Testing and Materials, Philadelphia).

WHITCOMB, J.D. (1981a), *Finite Element Analysis of Instability-related Delamination Growth*, NASA Technical Memorandum 81964, (NASA Langley Research Center, Hampton, VA).

WHITCOMB, J.D. (1981b), *Approximate Analysis of Postbuckled Through-width Delaminations*, NASA Technical Memorandum 83147, (NASA Langley Research Center, Hampton, VA).

WHITCOMB, J.D. (1983), *Strain Energy Release Rate Analysis of Cyclic Delamination Growth in Compressively Loaded Laminates*, NASA Technical Memorandum 84598, (NASA Langley Research Center, Hampton, VA).

CHAPTER 6

Reliability of Composite Structures

V.V. Bolotin

USSR Academy of Sciences
Institute of Mechanical Engineering
Griboyedova 4
101830 Moscow
U.S.S.R.

Contents

HANDBOOK OF COMPOSITES, VOL. 2 – Structure and Design
Edited by C.T. HERAKOVICH and Y.M. TARNOPOL'SKII

List of Symbols

$R(t)$, R – function of reliability, reliability index
$H(t)$, H – hazard function, hazard index
$\boldsymbol{P}(A)$ – probability of event A
$F_x(x)$, $p(x)$ – cumulative distribution function and probability density, respectively, of a random value x
$\boldsymbol{E}[x]$, $\boldsymbol{D}[x]$ – mathematical expectation and dispersion of a random value x
v_x – coefficient of variation
x, y, z – rectangular Cartesian coordinates (the axis z is along the fibers)
ρ – characteristic size of the crack cross-section
λ_e, λ_p – transfer (non-effective) lengths of fibers in the vicinity of a breakage for elastic and elastic–plastic matrices, respectively
l – characteristic size of a crack, e.g. the radius of a brush-like transverse crack in a unidirectional composite
l_*, l_{**} – characteristic sizes of newly-formed and critical cracks, respectively
λ – half-length of a debonded fiber's segment or half-length of a brush-like crack
V – volume of a specimen or a structural component
v_f – fiber volume ratio in composite
σ – nominal normal stress in a unidirectional composite in tension or compression along fibers
σ_f – nominal normal stress in fibers
r – ultimate tensile strength of fibers assumed as a random value
τ_m – shear stress in a matrix or fiber–matrix interface
t – time
t_*, t_{**} – time up to macrocrack nucleation and up to crack instability, respectively
t^* – time up to failure in the loss of integrity
T – design lifetime of a structure or a structural component
φ – damage measure of primary fiber segments
τ – time up to failure of a primary element
ψ – damage measure of composite
ψ_*, ψ^* – values of damage measure corresponding to the macrocrack nucleation and the loss of integrity, respectively
$P(\mathrm{D} \mid l)$ – probability to detect a crack of size l located in the working field of an instrument

$P_D(l)$ – probability to detect a crack of size $\geq l$ from a given random set of cracks

μ – mathematical expectation of the number of cracks in unit volume

ν – mathematical expectation of a number of undetected cracks in unit volume

$H_k(l)$ – probability of non-detection of k cracks with sizes $\geq l$

1. Theory of structural reliability

1.1. The subjects of the theory of structural reliability

The theory of structural reliability covers the following topics:

(1) the statistical analysis of loads and actions on structures, such as external forces, temperature, humidity and other environmental conditions of loads and actions;

(2) the statistical analysis of the mechanical properties of structural materials, as well as structural components and joints;

(3) the development of probabilistic models of load–structure interaction; the statement and solution of problems of the theory of structural reliability using both analytical and numerical methods; the planning of laboratory experiments and in-service structural monitoring;

(4) the development of the philosophy and advanced methods of codified design of structures, decision making based on statistical data and probabilistic models both at the design and operation stages.

Apart from the analytical and numerical difficulties which are natural for all developing theories using probabilistic methods, the theory of structural reliability meets specific conceptual and methodological difficulties. The first group of difficulties arise from the necessity to apply results of the theory to unique or small-scale structures. This relates more to the philosophical interpretation of the concepts of the theory of probability than to engineering applications. The aim of the theory of structural reliability is to give engineers a set of tools to arrive at rational and scientifically based decisions, without going into sophisticated arguments about the nature of probabilistic concepts. The second group of difficulties is more significant for applications. These difficulties arise from uncertainties about statistical information concerning materials, structural components and joints as well as loading and environmental conditions. In fact, each time one comes to practical applications, one suffers from a lack of sufficient information. The situation is improving, although not so rapidly as is desirable from the viewpoint of the theory of structural reliability. Still, there is significant progress in the accumulation of statistical data, even relating to seismic loading, i.e. in a topic that recently appeared almost beyond hope. One of the applications of the theory of structural reliability is to indicate rational ways of gathering, treatment and use of statistical information (BENJAMIN and CORNELL, 1970). This feedback between theory and application supports the development of the theory to a large extent.

1.2. Historical remarks. Reliability functions

The theory of structural reliability, which is presently recognized as an important branch of engineering sciences, has a long and rather vague history. Elements of the theory can be traced in the conventional safety factor approach using the ratio $n = r/q$ of design strength r to design load q. An understanding of the statistical nature of safety factors came later, in the first third of this century. A number of authors introduced independently the reliability index R equal to the probability that the characteristic strength r is larger than the characteristic load q. Thus $R = \boldsymbol{P}\{r > q\}$ where $\boldsymbol{P}\{\cdot\}$ is the probability of a random event specified within into the brackets. After the Second World War, this approach developed rapidly. Meanwhile, elementary reliability concepts were introduced into civil engineering codes. The safety factor was split into parts corresponding to various aspects of load–structure interaction, and each part was given a certain statistical interpretation. Further progress was achieved in the 1950s when the time factor was introduced into reliability considerations. A survey of earlier attempts was presented by BOLOTIN (1961). Year by year the contemporary point of view regarding structural failure was being accepted. Now, structural failure is interpreted as an excursion of the random vector $\boldsymbol{v}$ from an admissible region Ω in the quality space V. Vector $\boldsymbol{v}$ describes the current state and operational quality of a structure. The function of time t given below

$$R(t) = \boldsymbol{P}\{\boldsymbol{v}(\tau) \in \Omega\ ; \quad \tau \in [t_0, t]\} \tag{1.1}$$

is called the reliability function. Function $R(t)$ is one of the leading means of characterizing reliability. Using $R(t)$, other reliability, longevity and safety factors can be calculated.

There are a large number of ways to generalize eq. (1.1) (BOLOTIN, 1971). For example, allow the properties of a structure to be given by vector $\boldsymbol{r}$, and the loading conditions by vector $\boldsymbol{s}$. In place of $R(t)$ therefore, the conditional reliability function has to be introduced, e.g.

$$R(t \,|\, \boldsymbol{r}, \boldsymbol{s}) = \boldsymbol{P}\{\boldsymbol{v}(\tau \,|\, \boldsymbol{r}, \boldsymbol{s}) \in \Omega(\boldsymbol{r})\ ; \quad \tau \in [t_0, t]\}\ . \tag{1.2}$$

Here $\boldsymbol{v}(t \,|\, \boldsymbol{r}, \boldsymbol{s})$ is the conditional process in the quality space V. The admissible region Ω, generally, depends on $\boldsymbol{r}$. The unconditional reliability function is given below

$$R(t) = \int_{D(\boldsymbol{r},\boldsymbol{s})} R(t \,|\, \boldsymbol{r}, \boldsymbol{s}) p(\boldsymbol{r}, \boldsymbol{s})\, \mathrm{d}\boldsymbol{r}\, \mathrm{d}\boldsymbol{s}\ , \tag{1.3}$$

where $p(\boldsymbol{r}, \boldsymbol{s})$ is a joint probability density function of vectors $\boldsymbol{r}$ and $\boldsymbol{s}$. Integration in (1.3) is applied over the definition domain $D(\boldsymbol{r}, \boldsymbol{s})$ of these vectors.

Other generalizations have been stimulated by the attempts to develop methods of predicting the reliability of individual operating systems (BOLOTIN, 1977). The aposteriori reliability function $R(t\,|\,t_k)$ enters into these considerations. This function is equal to the probability that the vector $\boldsymbol{v}$ remains in the region Ω during the time interval $(t_k, t]$ under the condition that certain information is known about the behavior of the structure and loading conditions at $t \in [t_0, t_k]$:

$$R(\,|\,t_k) = \boldsymbol{P}\{\boldsymbol{v}(\tau\,|\,t_k) \in \Omega(t_k)\,;\quad \tau \in (t_k, t]\}\,. \tag{1.4}$$

In the beginning of the 1950s, the theory of system reliability began to develop very rapidly with special applications to large systems composed of mass produced elements. The main features of the theory are: the assumption of logical interaction between failures of elements without consideration of the physical, mechanical, etc. Processes leading to the failures, and the wide use of statistical data relating to the elements. The first approaches were based on the so-called exponential law of failures $R(t) = \mathrm{e}^{-\lambda t}$ where λ is the failure rate, assumed to be constant. Simplified models of interactions were used, e.g. the ones based on the assumption of independent failures of elements resulting in equations of the type $R = R_1 \cdots R_n$. Later, more sophisticated models were developed taking into account ageing of elements, redundancy, etc. (BARLOW and PROSCHAN, 1965; GNEDENKO et al., 1965).

In principle, all the fundamental concepts of the theory of system reliability can be formulated in terms of the probabilities given by (1.1)–(1.4). Using multidimensional quality spaces, we describe the logical interaction of elements and subsystems by means of unions and intersections of partial regions. On the other hand, an analogue of the quality vector (t) can be introduced into the

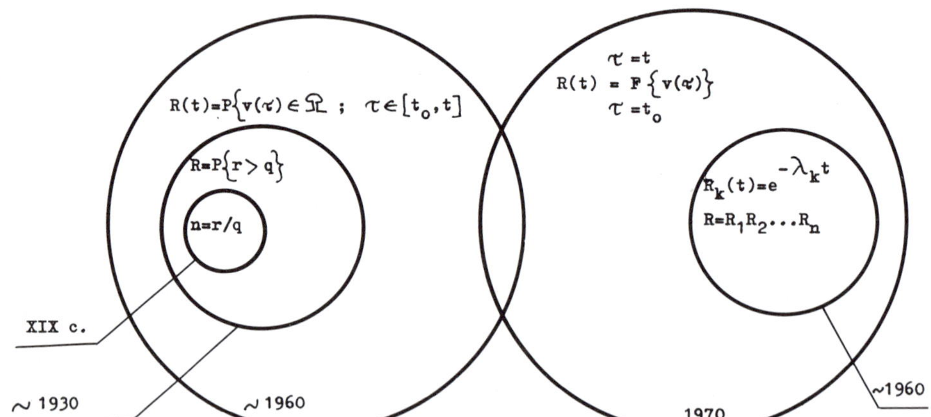

(a) (b)

FIG. 1. Illustration of the history of the reliability theory; (a) structural reliability, (b) system reliability; date milestones are stated approximately.

theory of system reliability. Then the reliability function $R(t)$ can be interpreted as a functional $F\{\boldsymbol{v}(\tau)\}$ of history of the process $\boldsymbol{v}(t)$ in the interval $[t_0, t]$ (GNEDENKO et al., 1965). Now we can begin an extensive study of the physical, mechanical and chemical phenomena that result in failures. Therefore, both approaches to the concept of reliability are in the state of convergence. This is illustrated in Fig. 1, the left-hand side of which corresponds to the theory of structural reliability while the right-hand side corresponds to the theory of system reliability.

1.3. Failures of structures and structural components

One of the tendencies of contemporary reliability theory is to emphasize the difference between various kinds of failures taking into account their operational, economical and social consequences. We consider three groups of failures here. The first group consists of failures of secondary components and does not result in long or dangerous stoppages in the operation of a structure. It is convenient to describe these failures in the framework of system reliability theory using as a principal parameter the failure rate $\lambda(t) = -R'(t)/R(t)$ (Fig. 2(a)).

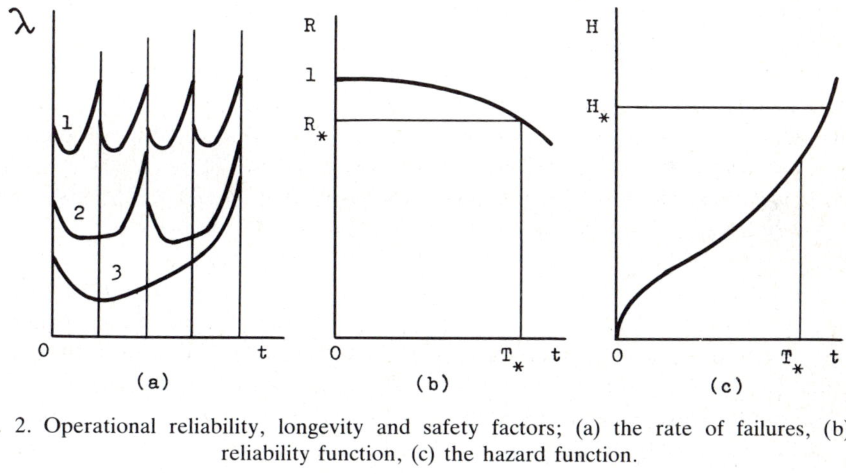

FIG. 2. Operational reliability, longevity and safety factors; (a) the rate of failures, (b) the reliability function, (c) the hazard function.

Failures of the second group are limit states of the fundamental components of a structure. These types of failure impose limitations on the lifetime of a structure as a whole and lead to its significant physical deterioration. Further, we use the term 'lifetime' or even briefly 'time' with respect to the fundamental components of a structure or to a structure as a whole. The life T is measured by the units of physical calendar or operating time.

Lifetime consumption takes place, as a rule, under normal operating, loading and environmental conditions. Mechanical, physical, etc. processes leading to the limit state of a structure are strongly irreversible (if, of course, no repair and replacement of components are taken into account). These processes develop rather slowly. Being random by nature, they are relatively predictable in the probabilistic sense. The main characteristics of longevity is the cumulative distribution function $F_T(T)$ of the lifetime T (BOLOTIN, 1984a). The function $F_T(T) = 1 - R(T)$ is connected with the corresponding reliability function $R(t)$. The latter is to be calculated with respect to that admissible region Ω the boundary of which corresponds to the limit states of a structure. One of the most important features of the longevity prediction is that the design probability R_* corresponding to the design lifetime T_* is not necessarily very close to unity (Fig. 2(b)).

Accidents that result in danger to the life and health of people, or in large economic and normal losses, form the third group of failures. These failures are, as a rule, results of very rare natural events and/or very rare unfavourable combinations of loads and actions and/or human errors at the design, manufacturing and operation stages. The probability of such accidents ought to be very small, even though it is impossible to eliminate them in principle. The main probabilistic characteristics of the third group of failures is the hazard function $H(t)$ equal to the probability of occurrence of accidents in the given time interval (Fig. 2(c)). The hazard function is connected with the probabilities $R(t)$ defined by (1.1)

$$H(t) = 1 - R(t) . \tag{1.5}$$

The region Ω corresponds to the safe state of a structure. Design values of the hazard function for the end of the design life are usually $H_* = 10^{-3} \cdots 10^{-6}$. Hence, contrary to the problems of the longevity prediction, the safety prediction requires one to deal with very rare random events.

The presented classification is rather conditional. In particular, it is difficult to draw a distinct line between the three groups. Mutual effects of failures related to the various groups are of significance, too. For example, when the rate of operational failures increases, it results in approaching the limit state of a structure as a whole. The life of a structure can be limited by safety requirements, etc. This is illustrated in Fig. 2(c), where the value T_{**} corresponds to the restriction $H(t) \leq H_*$ imposed on the hazard function.

1.4. *Elementary models of structural failures*

In the following, a brief survey of models mostly suitable for the reliability analysis of structures and their components is presented. So-called elementary models have been used at the earlier stage. Loading is supposed to be of a single act nature, and a deterministic relation between vectors $\boldsymbol{v}$, $\boldsymbol{q}$ and $\boldsymbol{r}$ is assumed. Then the reliability factor is given by

$$R = \int_{\Omega} p(\boldsymbol{v})\, \mathrm{d}\boldsymbol{v} \tag{1.6}$$

where $p(\boldsymbol{v})$ is the probability density of the quality vector $\boldsymbol{v}$. The integration domain Ω coincides with the admissible region. Eq. (1.6) is in fact a very special case of (1.1). Using the function $\boldsymbol{v}(\boldsymbol{r}, \boldsymbol{q})$, (1.6) becomes

$$R = \iint_{\Omega'} p(\boldsymbol{r}, \boldsymbol{q})\, \mathrm{d}\boldsymbol{r}\, \mathrm{d}\boldsymbol{q} \tag{1.7}$$

with the joint probability density $p(\boldsymbol{r}, \boldsymbol{q})$ and the integration domain $\Omega' = \{\boldsymbol{r}, \boldsymbol{q} : \boldsymbol{v}(\boldsymbol{r}, \boldsymbol{q}) \in \Omega\}$.

Let the load and resistance vectors be one-dimensional, and the domain Ω defined by the inequality $\boldsymbol{v} = \boldsymbol{r} - \boldsymbol{q} > 0$. Then (1.7) results in

$$R = \iint_{r-q>0} p(r, q)\, \mathrm{d}r\, \mathrm{d}q\,. \tag{1.8}$$

A common graphical interpretation of this model is drawn in Fig. 3. The load and strength parameters are assumed to be independent, i.e. $p(r, q) = p_r(r)p_q(q)$. Together with the assumption that the values q and r are normally distributed, the elementary models are used widely to substantiate the codified design in civil engineering. Some applications to composite structures are given by JONES (1975).

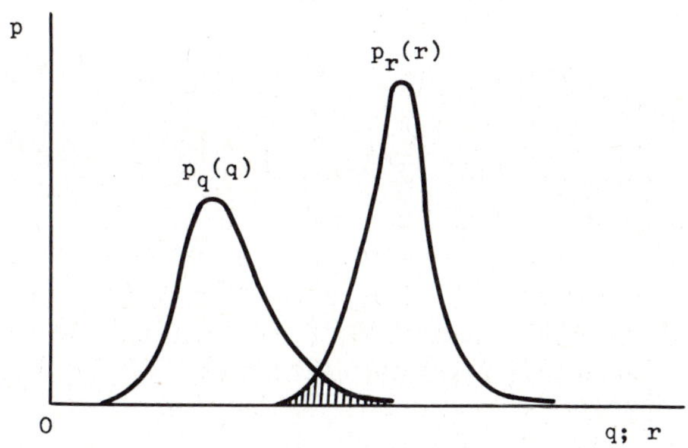

FIG. 3. Elementary model of structural failure in terms of probability density functions $p_q(q)$ of the load parameter q and $p_r(r)$ of the strength parameter r.

1.5. *Cumulative models*

The next type of model used especially for life prediction, originated from the well-known Palmgren–Miner rule for the fatigue life prediction. We call these models cumulative because they, generally speaking, describe irreversible damage accumulation in structures. Let the region Ω occupy a part of the first octant of the space V adjoining the origin (Fig. 4). Then, all components of the vector $\boldsymbol{v}$ are non-negative. Let the exterior normal to the limit surface be

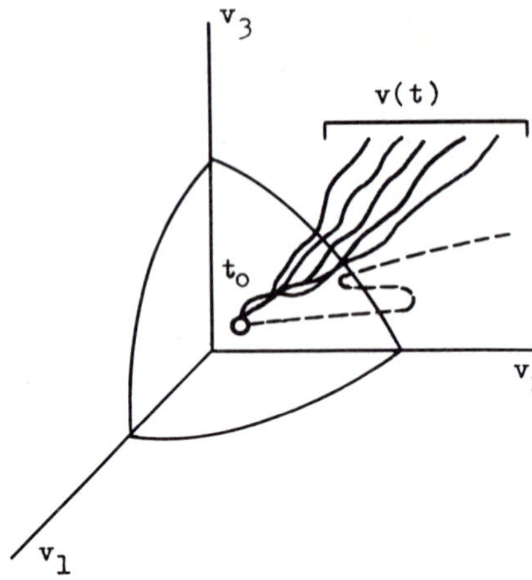

FIG. 4. Cumulative model of structural failure. The dashed line corresponds to the case when the cumulative model is inapplicable.

positive. More precisely, all of its projections on the coordinate axes are non-negative. If the components of the n-dimensional process $\boldsymbol{v}(t)$ in the considered time interval satisfy the conditions.

$$v_k(t_2) \geqslant v_k(t_1) \quad (t_2 > t_1;\, k = 1, \ldots, n)\,, \tag{1.9}$$

and $\boldsymbol{v}(t_0) \in \Omega_0$, the reliability function (1.1) reduces to

$$R(t) = \boldsymbol{P}\{\boldsymbol{v}(t) \in \Omega\}\,. \tag{1.10}$$

Another approach to introduce cumulative models of failure is based on the use of a norm $\|\boldsymbol{v}\|$ in the space V. Let the condition $\boldsymbol{v} \in \Omega$ be equivalent to $\|\boldsymbol{v}\| < 1$. It is always possible to assume this if the region Ω is convex. We will call the process $\boldsymbol{v}(t)$ cumulative with respect to the norm $\|\boldsymbol{v}\|$ if in the considered time interval

$$\|\boldsymbol{v}(t_2)\| \geqslant \|\boldsymbol{v}(t_1)\| \quad (t_2 > t_1)\,. \tag{1.11}$$

The reliability function (1.10) takes the form

$$R(t) = \boldsymbol{P}\{\|\boldsymbol{v}(t)\| < 1\}\,. \tag{1.12}$$

Evidently, (1.9) and (1.11) put essential restrictions on the equation governing damage accumulation in a structure. For example, consider the equation

$$\mathrm{d}\boldsymbol{v}/\mathrm{d}t = \boldsymbol{f}[\boldsymbol{v}, \boldsymbol{q}(t\,|\,\boldsymbol{s}), \boldsymbol{r}]\,. \tag{1.13}$$

Each component of the right-hand side is to be non-negative to provide the component-wise cumulative property defined by (1.9).

The difference between the reliability functions (1.1) and (1.10) is as follows. Eq. (1.1) takes into account the possibility that the vector $\boldsymbol{v}(t)$ leaves the admissible region at a certain moment $t_0 < \tau < t$ and then returns at some point in time. Perhaps the vector $\boldsymbol{v}(t)$ repeats this excursion many times in the interval $(t_0, t]$. This behavior is illustrated in Fig. 4 by the dashed line. Failure corresponds to the first excursion out of the region. Thus, generally, to estimate the reliability function given by (1.1), we have to find the solution of the problem of the first excursion of a random process. In the one-dimensional case it means that this is a problem of the first crossing by a scalar random process of a certain level (TIKHONOV, 1970).

1.6. *Poissonian models*

Since structural failures are rather rare events, the simplified approach based on a special assumption regarding the distribution of times prior to the first excursion is valid. Let $\boldsymbol{v}(t_0) \in \Omega$, then $R(t_0) = 1$. Assuming the stream of excursions to be Poissonian, we obtain

$$R(t) = \exp\left[-\int_{t_0}^{t} \lambda(\tau)\,\mathrm{d}\tau\right]. \tag{1.14}$$

Here $\lambda(t)$ is the mathematical expectation of the number of excursions per unit time. Mostly, it may be identified with the failure rate of the theory of system reliability.

We call models of failures described by (1.14) Poissonian. Using the second, third, etc. moments of the number of excursions, we obtain lower and upper estimates of the function $R(t)$. The simplest lower estimate is

$$R(t) \geqslant 1 - \int_{t_0}^{t} \lambda(\tau)\,\mathrm{d}\tau . \tag{1.15}$$

It is evident that the right-hand side of (1.15) coincides with the two first terms of the power expansion of the exponent in (1.14).

Poissonian models of failures are most suitable for structures subjected to vibration. Details, applications and references can be found in the books by BOLOTIN (1961, 1971, 1979). To calculate $\lambda(t)$ and the joint probability density $p(\boldsymbol{v}, \dot{\boldsymbol{v}}; t)$ of the process, $\boldsymbol{v}(t)$ and its derivative $\dot{\boldsymbol{v}}(t)$ must be known at each considered time t. Then

$$\lambda(t) = \int_{\Gamma} \mathrm{d}\Gamma \int_{\dot{v}_n > 0} p(\boldsymbol{v}_\Gamma, \dot{\boldsymbol{v}}; t)\dot{v}_n\,\mathrm{d}\boldsymbol{v} . \tag{1.16}$$

Here Γ is the boundary surface of the admissible region Ω (see Fig. 4). The surface Γ is supposed to be smooth (or at least piece-wise smooth if the process $\boldsymbol{v}(t)$ is sufficiently well-mixed). Values of the vector $\boldsymbol{v}$ on the surface Γ are denoted by $\boldsymbol{v}_\Gamma$ and those of the normal component of the derivative $\dot{\boldsymbol{v}}$ are denoted by $\dot{v}_n$. If the process $\boldsymbol{v}(t)$ is one-dimensional, and the admissible region is $\Omega = \{v : v_*^- < v < v_*^+\}$, as is shown in Fig. 5, (1.16) should be replaced by

$$\lambda(t) = \int_0^\infty [p(v_*^-, \dot{v}; t) + p(v_*^+, \dot{v}; t)]\dot{v}\, d\dot{v} . \qquad (1.17)$$

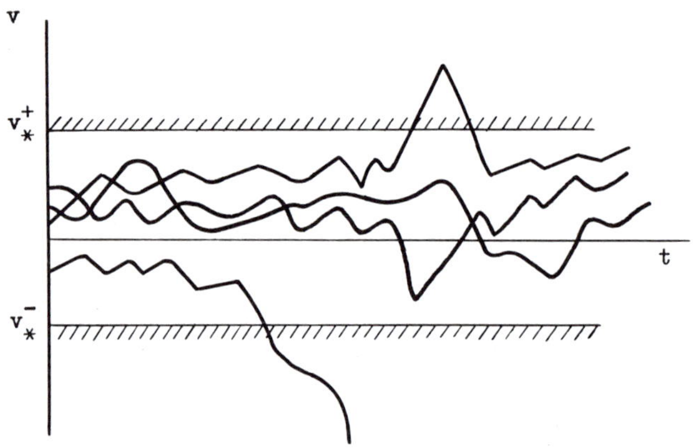

FIG. 5. Poissonian model of structural failure for the one-dimensional process $\boldsymbol{v}(t)$; the admissible region is $v_*^- < v < v_*^+$.

More or less consistent models of structures and their environments, as a rule, are complicated to such a degree that analytical methods become almost helpless. Generally, numerical simulation, i.e. the Monte Carlo method is the most universal tool to treat problems expressed in the probabilistic form. But since very small (or very close to unity) probabilities are involved in the reliability analysis, the Monte Carlo method requires very large samples and hence very large amounts of computer time. Obstacles can be partially overcome if the Monte Carlo method is used together with the limit theorems of the theory of probability. These theorems are of asymptotic nature, and some of them relate to the properties of tails of distributions. The problem is to select suitable asymptotic distributions and to estimate their parameters when only rather small samples are available.

1.7. Composite structures from the reliability viewpoint

In principle, there should be no difference between the reliability approaches to structures of composites than to those of conventional materials. The loading analysis, and to a large degree the analysis of load–structure interaction, are similar in both cases. The main distinction occurs in the study of mechanical properties, interpretation and analytical presentation of experimental data.

Properties of conventional structural materials such as common steels, aluminium alloys, concrete, etc., have been studied in detail, including the statistical scatter of mechanical properties which is of essential significance for reliability considerations. Contrary to conventional materials, the properties of composites are not well-known. Hundreds of new composites are introduced into the industry each year. Composites have a wide range of mechanical properties which depend largely on the choice of components, on the structure of a composite, and on the technology of fabrication. Often, a composite intended for a particular set of structures must be specially produced. The amount of such a composite available for test specimens is limited, and the structure or its components can only be simulated by small-scale models. Moreover, one may say that a composite as a structural material is often created together with a structure. It is difficult to separate the properties of the composite from those of the structure as a whole. The testing of small specimens does not generate enough information to predict the properties of a composite incorporated into a structure. Due to the higher sensitivity of composites to secondary effects, small specimens display properties which may differ substantially from those in a structure. Transfer of testing results from small specimens and even from small-scaled models to real-size structures is complicated since the scale effect in composites has not been studied properly yet. Anyway, the scale effect is not the same as that in common brittle solids, and it is rather non-predictable. Additional difficulties arise when we consider the behavior of composites with time. Fatigue, ageing and other deterioration effects in composite structures are now at the first stage of investigation. Therefore, extrapolation of results of short-term tests to the design life of a structure is rather questionable.

The above listed reasons emphasize the importance of theoretical methods to predict the mechanical properties of composites in comparison with those of conventional materials. For example, an experienced engineer would look sceptically at the attempts of a theoretician to predict the mechanical properties of a common polycrystalline material, say, carbon steel, assuming the knowledge of the properties of crystallites and their statistical distribution. Moreover it is simpler to test samples of standard specimens or, perhaps, to consult a handbook. The situation changes substantially when composite materials are involved. As a rule, the properties of components are more or less known, and the properties of a composite have to be predicted.

1.8. Mechanics of composites and structural reliability

From the viewpoint of the theory of structural reliability, the most important problems of the mechanics of composites are the following:

(1) the prediction of mechanical properties of composites taking into account the stochastic nature of the properties of components, random imperfections of the composite structure, random distribution of defects, random factors entering into the fabrication process;
(2) the prediction of the scale effect in fracture, both for short-term and persistant loading, in order to develop a basis for the extrapolation of results obtained from the testing of specimens and small-scale models to real-size structures;
(3) the prediction of the statistical distribution of the lifetime of structural components under persistant quasistatic, cyclic as well as combined loading.

It is comparatively easy to predict the so-called bulk parameters such as elastic moduli (MALMEISTER et al., 1980). Being the result of the averaging of sufficiently large volumes of a composite, they are comparatively non-sensitive to small-scale imperfections. Hence the scatter of bulk parameters as well as the related scale effect are, as a rule, small and do not enter into the reliability analysis. On the other hand, stochastic effects are of significance for those phenomena which are governed mostly by local properties including random flaws and imperfections in the structure of a composite. Among these are, for instance, the tensile strength of unidirectional fibrous composites in the fiber direction the local compression strength of layered composites with comparatively pliable matrices, especially with regard to surface buckling, etc.

Most of the problems of the reliability of composite structures relate in the same degree to fracture mechanics. At present, the latter is developing basically as the mechanics of cracked solids. The cracks are usually represented as mathematical cuts in a body the material of which is assumed to be homogeneous or at least piece-wise homogeneous. Composites are by nature essentially nonhomogeneous. The strength of a composite is affected by certain parameters having the dimension of length but not entering into the classical equations of the theories of elasticity and plasticity. Among them are characteristic sizes of grains, fibers, etc. For example, in the case of unidirectional fibrous composites at least two scale parameters are present $\lambda_1 = \rho$, $\lambda_2 = \rho (E_f / E_m)^{1/2}$. Here ρ is the characteristic size of the fiber cross section, e.g. a radius of the section, and E_f and E_m are characteristic moduli of elasticity of the fibers and matrix respectively. The scale parameter λ_1 characterizes the structure in a section across the fibers. The scale parameter λ_2 is of the order of magnitude of the transfer (ineffective) length of the fibers (ROSEN, 1965). From another viewpoint, λ_2 may be interpreted as a characteristic length of the boundary effect caused by the interaction between the fiber bending and shear of the matrix (BOLOTIN, 1965). Another example is a laminated composite. The thickness of layers and the characteristic length of the boundary effect enter into consideration at the lamina level. If each layer is of fibrous structure, scale parameters typical of fibrous composites should be added, etc.

Thus, the recognized models and analytical results of classical fracture mechanics are of limited applicability to composites because of their essentially nonhomogeneous structure. This fact is now commonly recognized. Due to the nonhomogeneity, composites show a rather wide variety of modes of fracture. Some of them, such as the accumulation of dispersed damage and the loss of integrity have not yet been investigated on a large scale. The interaction between the various modes of damage as well as the interaction between macroscopic cracks and dispersed damage in a composite are of interest. Thus, unified models of fracture arise in the mechanics of composites in a natural way, generated by the intrinsic features of the fracture of composites, although analogous models, generally, are of interest for conventional materials, too.

A new class of models has developed especially for laminated composites. Delaminations as a result of initial fabrication flaws, residual and thermal stresses, local surface impacts, etc. are typical if the matrix or the interlayer boundary are not sufficiently strong. Behavior of delaminations under loading, generally, differs from that of a common crack in a quasi-isotropic solid. In particular, delaminations can grow, both in stable and unstable ways, under compression. Interaction between fracture of the interlaminar boundary and elastic instability is an issue which is new to fracture mechanics (BOLOTIN, 1984a).

The intrinsic inhomogeneity of most composites and imperfections introduced by the fabrication processes are not the only sources of the scatter of mechanical properties. Often, at the design stage, suppliers of a composite and/or components are not yet known. Therefore, the scatter in the properties of the composites or their components of the same trademark but delivered by various suppliers has to be taken into account. The situation is the same as for conventional materials. To take into consideration this part of scatter, one must use the concept of conditional reliability functions given by (1.2) and (1.3).

Although material properties form an essential part of the reliability theory developed for composite structures, some special problems also arise in the load–structure interaction area. Generally, high-performance composites, have a wide range of mechanical properties. High strength in the direction of reinforcement is accompanied by low resistance against shear and tension orthogonal to that direction. Hence, composite structures are sensitive to rather low loads and actions the character of which might not coincide with the assumed design. Since loading conditions during the life of a structure are rather unstable, even small random deviations of loads and actions can be of significance.

2. Statistical fracture mechanics of composites

2.1. Introductory remarks

In this section a survey of the modes of fracture is presented with special attention to unidirectional fibrous composites. These composites are chosen as a basic subject of the study because they are most commonly used for high-performance structures. Moreover, laminated structures are very often composed of fibrous layers, and the material of each layer may be considered as a unidirectional fibrous composite. The behavior of laminated structures under loading has a number of specific features. Interaction between the layers, edge effects, local instability phenomena, etc. are the subjects of the mechanics of laminated structures (AMBARTSUMYAN, 1974; BOLOTIN and NOVICHKOV, 1978). If the properties of composites forming all the layers are studied, considerable scope will be offered for the study of the fracture of the laminated structure as a whole (PROTASOV et al., 1980; OBRAZTSOV and VASIL'EV, 1982).

The process of the fracture of conventional quasi-isotropic materials consists of two stages. The first stage is the accumulation of dispersed damage. This stage can amount to a significant part of the structure's lifetime preceding the stage of macroscopic cracking. If there were no initial or specially planted macrocracks in a body, nucleation of the first macrocrack is the result of dispersed damage accumulation. The process of damage accumulation continues after the macrocrack begins to grow. Thus, these processes interact.

High-performance composites display a larger variety of fracture modes. A scheme illustrating eventual modes and their connections is given in Fig. 6. To simplify the presentation, a specimen in uniform uniaxial tension is considered,

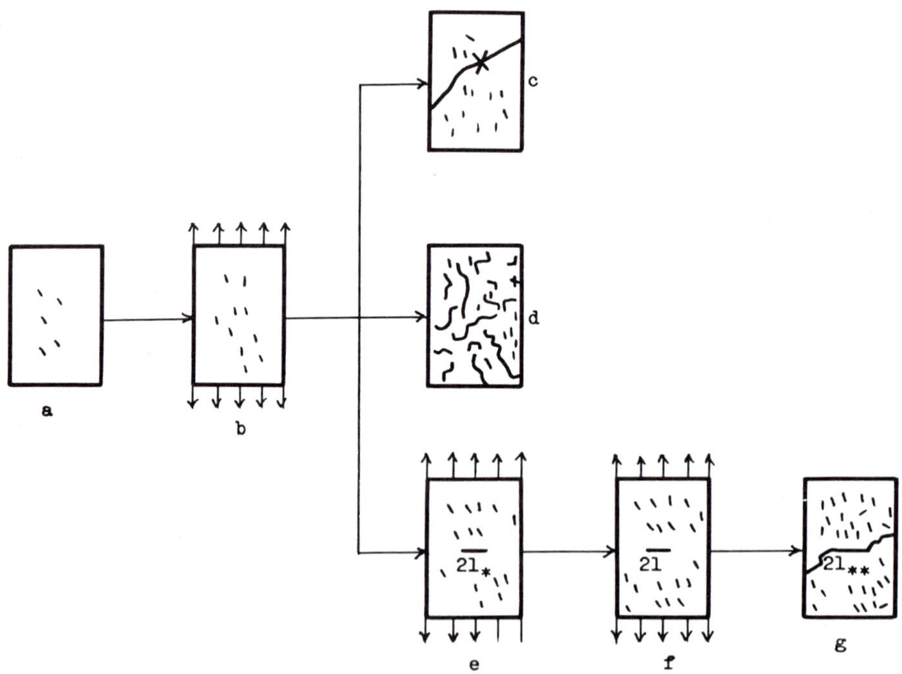

FIG. 6. Modes of failure of composites; (a) initial dispersed damage, (b) damage accumulation under loading, (c) brittle fracture, (d) loss of integrity, (e) macrocrack initiation, (f) stable macrocrack growth, (g) final failure in macrocrack instability.

and all properties of the composite are assumed to be statistically homogeneous in the volume V of a specimen. The crack size l is supposed to be small compared to the characteristic dimensions of the specimen so that its influence on the nominal stress field may be considered negligible.

2.2. *A unified model of fracture*

We will treat a specimen or a structural component as a statistical sample of interacting primary elements with random properties. The grains of filler with the adjacent parts of the matrix present an example of primary elements. Their characteristic size ρ is supposed to be very small compared to the specimen size. Thus the number N of these elements in the volume V is very large. In the case of unidirectional fibrous composites, segments of fibers with the adjacent matrix can be considered as primary elements. Assume the length of these elements to be equal to $2\lambda_e$ where λ_e is the elastic transfer length. Typically, all fiber sections are assumed to be circular with radius ρ, and the fiber and matrix isotropic and linear elastic with a fiber Young's modulus E_f, and matrix shear modulus G_m. Then

$$\lambda_e \sim \rho(E_f/2G_m)^{1/2} . \tag{2.1}$$

In (2.1) we use the sign of equality with the order of magnitude. More exact equations contain multipliers which are of the order of unity when common fiber volume ratios are considered such as $v_f \sim (v_f^{-1/2} - 1)^{1/2} \sim 1$. Generally, the exact and rigorous development of probabilistic models is rather complicated and results in completely non-transparent equations. On the other hand, most fundamental results of linear fracture mechanics are of very simple form. We will simplify as much as possible the analytical part of calculations, omitting multipliers whose order is equal to unity, as well as small additional terms. For example, consider the bulk tensile sress σ for a unidirectional fibrous composite. This value is related to the fiber and matrix stresses, σ_f and σ_m respectively by the 'mixture rule' as follows:

$$\sigma \approx v_f\sigma_f + (1 - v_f)\sigma_m .$$

When $v_f \sim 1$, and the matrix is sufficiently pliable, let us say, $E_m \ll E_f$ for elastic materials, the order of magnitude estimate is $\sigma \sim \sigma_f$.

Limit theorems of the theory of probability and asymptotic estimates will be used for simplification, too. Some estimates will be calculated with the probability of the order of unity, assuming, e.g. that $1 - e^{-1} = 0.632 \ldots \sim 1$. Although this rough approach and so-called physical level of rigor are rather vulnerable from a mathematically strict point of view the final results appear to be quite reasonable.

A number of microstructural parameters can be considered whose magnitudes can be determined only indirectly, by the statistical treatment of the data of standard tests. Therefore, there exists certain freedom in the interpretation of final equations. We may consider them as 'exact' provided that the involved parameters are properly chosen. For this reason, the sign of approximate equality and even of exact equality may be used in equations developed analytically, even if initial equations contain the sign of equality within an order of magnitude.

Dispersed damage accumulation is characterized by a multi-dimensional process $\psi(t)$ called, in what follows, just the damage vector. The multi-dimensional nature of damage is rooted in a variety of damage mechanisms for most composites. In the case of unidirectional fibrous composites one should distinguish at least two types of damage. The first type is formed by single breaks of fibers. This damage may be characterized by the ratio of the number of breaks in a certain volume to the total number of primary elements in this volume. The second type relates to matrix damage. The ratio of the sum of the lengths of damaged segments of the fiber–matrix interface to the total length of fibers can be suggested as the second damage measure. If the matrix is elasto-plastic, one should distinguish segments of the interface which are deformed plastically, and those that are completely debonded. Then the third damage measure enters into consideration, etc.

At the initial time t_0, initial dispersed damage is present in the specimen and is characterized by the damage vector ψ_0. The origin of this damage are fabrication flaws, as well as breaks induced between the time of fabrication and actual service (Fig. 6(a)). Breaks and damage are assumed to be nonhealing and therefore the process $\psi(t)$ is irreversible. Damage is accumulated gradually under loading, but the specimen as a whole is not considered to have failed (Fig. 6(b)). At a certain moment in time, the damage density reaches some critical level. The nature of the process varies qualitatively.

There are three possibilities for the subsequent development. Firstly, brittle fracture of the specimen can occur due to the rapid propagation of a crack from one of the damage centers (Fig. 6(c)). Secondly, a system of multiple cracks can be formed which develops rapidly and almost simultaneously over the whole volume of the specimen. This cracking is distributed almost uniformly over the volume if the nominal stress field and mechanical properties are homogeneous. The second type of failure is called the loss of integrity (Fig. 6(d)). The external appearance of specimens which have failed by brittle fracture and by loss of integrity may be almost similar. But in fact, the loss of integrity is accompanied by multiple cracking which can be observed by performing a detailed study of the structure of the failed specimen. On the other hand, a specimen that fails by brittle fracture remains relatively undamaged in the whole volume. The source of brittle fracture is always localized.

If neither brittle fracture nor loss of integrity is possible due to the properties of a composite and environmental conditions, then a third possibility can occur (Fig. 6(e)). Suppose that at time t_*, a stable macroscopic crack initiates with an initial size of l_*. The choice of this size is more or less arbitrary. For example, if a composite has a fine microstructure, it is expedient to assume the size l_* equal to the lower detection threshold of common non-destructive methods of cracks detection. Alternatively a collection of several adjacent broken fibers can be treated as an initial macroscopic crack in a unidirectional fibrous composite. For most composites the numbers $n_* = 6$ or 7 may be assumed (see Section 3).

The macrocrack grows in a stable manner under further loading, as shown in Fig. 6(f), until the moment t_{**} at which the crack size reaches the critical value l_{**}. At this point the crack becomes unstable and total failure of the specimen (Fig. 6(g)) occurs.

One should distinguish the fracture which appears naturally due to macrocrack initiation and development from the fracture of a specimen with a given initial notch such as that used in experimental fracture mechanics. In the latter case the damage accumulation begins in a virgin composite with the stress concentration at the tips of the notch. However, after a period of loading time, damage accumulates both near the tips of the crack and in the whole volume of the specimen. From the viewpoint of structural reliability, natural initiation and propagation of cracks are of primary interest.

2.3. *Dispersed damage accumulation*

Following the paper by BOLOTIN (1976), the behavior of each primary element is governed by a kinetic equation with respect to the damage measure of the element. Generally, the damage measure φ of elements is multidimensionally similar to that of the specimen as a whole. A typical kinetic equation for an element under long-acting quasistatic loading is

$$\mathrm{d}\boldsymbol{\varphi}/\mathrm{d}t = f(\boldsymbol{\varphi}, \boldsymbol{q}, \boldsymbol{r}, \boldsymbol{\psi}) . \tag{2.2}$$

The right-hand side contains a non-negative function of the damage vector $\boldsymbol{\varphi}$, the loading vector $\boldsymbol{q}$ and the resistance vector $\boldsymbol{r}$ of the element under consideration. Generally, the damage vector of the composite $\boldsymbol{\psi}$, enters into (2.2) as well. Resistance $\boldsymbol{r}$ is considered a random vector with a given probability density. For an undamaged element $\boldsymbol{\varphi} = 0$, and for a completely ruptured one $\|\boldsymbol{\varphi}\| = 1$. In the one-dimensional case we assume simply that the rupture of a primary element occurs when $\boldsymbol{\varphi} = 1$.

If the loading is cyclic, (2.2) should be replaced by a finite-difference analogue

$$\boldsymbol{\varphi}_n - \boldsymbol{\varphi}_{n-1} = f_1(\boldsymbol{\varphi}_{n-1}, \boldsymbol{q}_n, \boldsymbol{r}_n, \quad \boldsymbol{\psi}_{n-1}) \quad (n = 1, 2, \ldots) . \tag{2.3}$$

Here n is the number of cycles, and the right-hand side satisfies the same conditions as those in (2.2). If one considers a process of cyclic loading with slowly varying amplitudes, the number n may be treated as a continuous parameter. Then (2.3) is to be replaced by (2.2) using instead of t the 'slow' time n. Processes which include both quasistatic and cyclic components of loading sometimes enter into reliability analysis. In those cases the additiveness of the damage vector φ may be supposed, and the resulting kinetic equation is the union of (2.2) and (2.3). Later on, unless specified otherwise, persistant quasistatic loading is considered, and the damage measures of both primary elements and structural components are assumed to be scalar quantities.

At first, consider the stage of damage accumulation. Let the damage density be very small. Thus the primary elements are deformed and fractured almost independently. The damage takes place in the weakest elements. The cumulative distribution function of the lifetime τ of an arbitrarily chosen element $F_\tau(\tau)$ is the solution of the inverse boundary-value problem of (2.2) subject to boundary conditions $\varphi(0) = \varphi_0$, $\varphi(\tau) = 1$. The damage to the composite is characterized by the ratio $\psi = m/N$, where m is the number of ruptured elements. Strictly speaking, $\psi(t)$ is the step function

$$\psi(t) = (1/N) \sum_{k=1}^{N} \eta(t - \tau_k) , \tag{2.4}$$

where $\eta(t)$ is the Heaviside function, and τ_k are the lifetimes of individual elements. Since $m = \psi N \gg 1$, it is possible to use a smoothed approximation of the process $\psi(t)$ replacing the function given by (2.4) by an appropriate, continuous function.

The probability that not more than m elements from the sample of N elements will be ruptured in the time interval $[t_0, t]$ is given by

$$P_N^m(t) = \sum_{k=1}^{m} C_N^k F_\tau^k(t)[1 - F_\tau(t)]^{N-k} , \tag{2.5}$$

where C_N^k are binominal coefficients. The cumulative distribution function $F_\tau(t)$ is to be determined in the same time interval. Since N is very large and $m = \psi N$ is sufficiently large, the Laplace–Moivre limit theorem is application. The asymptotic estimate for the cumulative distribution function $F_\psi(\psi; t)$ of the magnitudes of the process $\delta(t)$ is

$$F_\psi(\psi; t) \approx \Phi\left(\frac{[\psi - F_\tau(t)]N^{1/2}}{\{F_\tau(t)[1 - F_\tau(t)]\}^{1/2}} \right) . \tag{2.6}$$

The standard cumulative function of the Gauss distribution is introduced in (2.6), i.e.

$$\Phi(z) = (2\pi)^{-1/2} \int_{-\infty}^{z} \exp(-u^2/2)\, du . \tag{2.7}$$

The mathematical expectation $\boldsymbol{E}[\psi(t)]$ and the dispersion (variance) $\boldsymbol{D}[\psi(t)]$ corresponding to (2.6) are

$$\boldsymbol{E}[\psi(t)] = F_\tau(t) ; \quad \boldsymbol{D}[\psi(t)] = F_\tau(t)[1 - F_\tau(t)]/N . \tag{2.8}$$

Further simplification is possible if the coefficient of variation of the process $\psi(t)$ is small as compared to unity. Then (2.6) gives

$$F_{\psi}(\psi; t) \approx \eta[\psi - F_{\tau}(t)]\,. \tag{2.9}$$

This means that under the assumed circumstances $\psi(t) \approx F_{\tau}(t)$, and the process $\psi(t)$ is almost deterministic.

2.4. Loss of integrity

The damage accumulation process terminates either in brittle fracture or in loss of integrity. Also, a macroscopic crack can be initiated. In all these cases, the character of the subsequent process is determined by the damage measure ψ.

At first, let us consider the case of loss of integrity. The moment t^* at which loss of integrity occurs is determined from the condition

$$\psi(t^*) = \psi^*\,. \tag{2.10}$$

The critical level of damage ψ^* enters the right-hand side of the above equation. To calculate ψ^* it is necessary to find the solution of the stability problem for a set of microcracks and flaws formed as a result of damage accumulation. The value of ψ^* can also be found from experiments. For high-performance fibrous composites, $\psi^* \sim 10^{-1}$ and even less. Analytical attempts to evaluate ψ^* for composites are based on combinatorial considerations. Let the number n^* be calculated for adjacent elements which results in the probability of the order of unity in subsequent association of multiple damage centers. Then $\psi^* \sim (n^*)^{-1}$.

The critical value ψ^* entering into (2.10), generally depends on the stress level at the moment t^*. To account for that, a simple model can be used based on Daniels 'fiber bundle' model (ROSEN, 1965; FREUDENTHAL, 1968). Let $F_r(r)$ be the cumulative distribution function of the short-term strength of primary elements, and $\sigma(\psi)$ be a characteristic stress in neighboring elements calculated by taking into account load redistribution and stress concentration at the microstructural level. The value of ψ^* is defined as one of the roots of the equation

$$\psi = F_r[\sigma(\psi)]\,. \tag{2.11}$$

The root corresponding to the unstable branch of the equation $\sigma = \sigma(\psi)$, i.e. to that with $\mathrm{d}\sigma/\mathrm{d}\psi \leq 0$, should be chosen. If the load is constant, the obtained value of ψ^* should be put into (2.10). Otherwise, (2.10) should be replaced by (2.11) and both sides of the equation should be taken as functions of time.

2.5. Macrocrack initiation. Brittle fracture

Unlike the loss of integrity, which is an almost deterministic event under deterministic loading especially if (2.9) is applicable, brittle fracture and macrocrack initiation are essentially random events. The ways to estimate the probabilities of these events are very similar, and we consider more closely the macrocrack initiation. We treat a macrocrack nucleus as an aggregate of n_* adjacent primary elements which rupture almost simultaneously. Assuming that $n_* \ll \psi(t_*)N$, the cumulative distribution function $F_*(t_*)$ at the time instant t_* of the macrocrack initiation is given by:

$$F_*(t_*) \approx 1 - [1 - \psi^{n_*-1}(t_*)]^{\psi(t_*)N} . \tag{2.12}$$

The asymptotic distribution in (2.12) belongs to the Weibull-type class:

$$F_*(t_*) \approx 1 - \exp[-N\psi^{n_*}(t_*)] . \tag{2.13}$$

For comparison with results obtained both from experiments and various deterministic models, the semi-deterministic estimate for the damage measure ψ_* responsible for crack initiation is useful:

$$\psi_* \sim N^{-1/n_*} . \tag{2.14}$$

Clearly, this estimate corresponds to the probability $1 - e^{-1} \sim 1$. Eqs. (2.13) and (2.14) indicate that the damage level at the instant of macrocrack initiation is very sensitive to the size of a specimen. For example, if $N = 10^6$, $n_* = 6$, (2.14) gives $\psi_* \sim 10^{-1}$. If $N = 10^{12}$, one obtains $\psi_* \sim 10^{-2}$.

The assumptions leading to (2.13) are rather strong. In fact, various ways for macrocrack nucleation should be considered taking into account the time lags between ruptures of primary elements, the difference in stress concentration factors, etc. Not only the cumulative distribution function $F_\tau(t)$ for a single rupture, but also a conditional cumulative distribution function such as $F_\tau^{(2)}(t \mid \tau)$, $F_\tau^{(3)}(t \mid \tau_1, \tau_2)$, etc. for the second, third, etc. ruptures of neighboring elements ought to be introduced. For example, the cumulative distribution function for the time prior to the second rupture is

$$\int_{t_0}^{t} F_\tau^{(2)}(t \mid \tau)\, \mathrm{d}F_\tau(\tau) .$$

A number of calculations with application to unidirectional fibrous composites has been given by TAMUZS (1982), SMITH et al. (1983), and other authors.

A similar approach is valid for finding the probability of occurrence of brittle fracture. Let fracture occur at the instant t^{**} when the first aggregate of breaks originates consisting of n^{**} ruptured primary elements. Then the cumulative distribution function $F^{**}(t^{**})$ will be defined by (2.12) and (2.13) where the number n_* is replaced by n^{**}. Weibull distribution for t^{**} follows from (2.13) subject to the condition that the damage measure is a power function of time.

To obtain the Weibull distribution for the ultimate nominal stress σ^{**}, it is sufficient to suppose that the damage measure is proportional to a power of the nominal stress.

The mode of fracture (macrocrack initiation, loss of integrity, or brittle fracture) depends on the relationship between the critical levels ψ_*, ψ^* and ψ^{**}, and on the number of primary elements in the considered volume. Comparing (2.10) and (2.14) we find that if $N^{-1/n_*} \ll \min\{\psi^*, \psi^{**}\}$, macrocrack initiation is the more probable mode. If the number N is not very large, the probabilities of macrocrack initiation and brittle fracture are also small, and loss of integrity becomes more probable.

2.6. *Stable macrocrack growth*

To obtain an equation describing the growth of a stable macroscopic crack one needs to analyze the damage and rupture of primary elements located near the front of the crack. Since crack propagation and damage accumulation are taking place within the damaged composite, both processes should be studied together. The analysis of the stochastic problem is complicated in large measure since the behavior of a large number of elements needs to be taken into account.

The simplest stochastic model is based on the assumption that the stress concentration is negligible in all the primary elements except in those located at the tip of the crack at a given instant of time. Let the right-hand side of (2.2) be independent of φ and ψ. The increment of the lifetime of an element starting from the moment when the element first comes into contact with the crack tip is approximately evaluated as

$$\Delta\tau \approx \left[1 - \int_0^t f(\sigma, r)\,\mathrm{d}\tau\right][f(\kappa\sigma, r)]^{-1}\,. \tag{2.15}$$

Here σ is the nominal stress, and κ is the stress concentration factor which, generally, depends on the crack size l. The integral on the right-hand side takes into account the damage of an element before it approaches the tip of the crack. If the term in the first brackets is negative, one has to set $\Delta\tau \equiv 0$.

For the averaged rate of crack growth, a semi-deterministic equation has been proposed by BOLOTIN (1981a)

$$\mathrm{d}l/\mathrm{d}t \approx 2\rho/\boldsymbol{E}[\Delta\tau]\,. \tag{2.16}$$

This equation is based on the assumption that the number of primary elements at the tip of the crack is large, and that they form a sufficiently large representative sample to consider crack propagation as a kind of a natural averaging process of the individual element fracture.

Linear fracture mechanics does not incorporate dispersed damage in the study of macrocrack propagation. To obtain results in aggreement with those of linear fracture mechanics one has to omit the integral on the right-hand side of (2.15). Employing (2.15) and (2.16) we obtain:

$$dl/dt \approx 2\rho \left\{ \int_0^\infty [f(\kappa\sigma, r)]^{-1} p_r(r)\, dr \right\}^{-1}, \tag{2.17}$$

where $p_r(r)$ is the probability density of the strength parameter.

2.7. Macrocrack instability

The crack grows in a stable fashion until the time t_{**} when the size l reaches its critical value l_{**}. We can estimate l_{**} within the framework of the given theory. The simplest approach is to find l_{**} from the condition that almost every primary element coming into contact with the crack tip is immediately ruptured with the probability of the order of unity. Since the short-term strength distribution is given by the function $F_r(r)$, the following equation for the evaluation of l_{**} holds:

$$F_r[\kappa(l)\sigma] \sim 1\,. \tag{2.18}$$

When the crack size l increases by one order of magnitude of ρ, Eq. (2.18) will again be satisfied and the crack will subsequently propagate. Thus, (2.18) may be interpreted as a critical relation which allows us to separate stable cracks from the unstable ones.

Another approach is based on the use of (2.15). If t_c is the characteristic time of rapid fracture, the value l_{**} can be found from the condition

$$\boldsymbol{E}[\Delta\tau] \sim t_c\,. \tag{2.19}$$

The more exact analysis has to take into account scatter in the properties of the primary elements. A lower bound of the cumulative distribution function $F(l_{**}; \sigma)$ can be obtained from the condition that at the given nominal stress level all the primary elements located at the crack tip will be ruptured almost immediately. Let a crack be penny-shaped with a radius of l. Then the total number of these elements will be approximately $\sim \pi l/\rho$. Hence

$$F(l_{**}; \sigma) \approx \{F_r[\kappa(l_{**})\sigma]\}^{\pi l_{**}/\rho}\,. \tag{2.20}$$

Thus far the influence of dispersed damage on macrocrack stability has not been taken into consideration. The damage measure ψ is approximately equal to the probability that an arbitrary primary element coming into contact with the tip of the crack is already ruptured. Instead of (2.18), we obtain the following equation for determining l_{**}:

$$F_r[\kappa(l)\sigma] + \psi \sim 1\,. \tag{2.21}$$

Eqs. (2.15)–(2.21) are valid for cracks propagating under long-acting quasi-static loading. Therefore, the considered cracks are to be interpreted as cracks under static fatigue conditions. Cracks under cyclic and combined fatigue conditions however, are of primary interest from the applications point of view. To make the results applicable to cyclic fatigue cracks, we should use finite-difference equations similar to (2.3). If a loading process with constant or slowly varying amplitudes is considered, the equations of static fatigue remain valid after a small modification. To this end t is replaced by the cycle number n considered as a continuous argument. Other parameters and functions entering the equations are also to be properly re-interpreted. Further details can be found in the book by BOLOTIN (1984a).

2.8. A model of a granular composite

The presented theory partially covers the damage and fracture of composites with granular, e.g. spheroidal, fillers. There is similarity between the behavior of granular composites and conventional polycrystalline materials for which linear fracture mechanics has been developed (RABOTNOV, 1979). Therefore, applications of the presented theory to granular composites can be used for comparison with the well-known results of linear fracture mechanics. The most important among the latter are the Griffith–Irwin criterion of stability of a macroscopic crack and the Paris–Erdogan equation for the rate of fatigue crack growth.

Let the material be reinforced with spheroidal grains whose characteristic size is of the order of ρ. All grains belong to the same statistical set. The scatter of grain properties is the result of their different orientation with respect to the field of nominal stress as well as of random flaws present in the grains. If the grain–matrix interface is of sufficiently high strength, it is natural to treat the grain as a primary element of the composite. Rupture of a primary element occurs because of cracks crossing the entire grain or because of ultimate plastic damage of the latter. A macroscopic crack consists of a bonded set of ruptured grains, and the crack clearance in an unstressed specimen is of the order of ρ. In contrast with linear fracture mechanics where a crack is considered as a mathematical cut, a non-zero radius of curvature ρ_1 of the crack-tip is introduced. Usually, but not necessarily, $\rho_1 \sim \rho$.

If the grain–matrix interface is weak, most of the damage will be accumulated in the matrix. Thus, segments of the interface should be considered as primary elements. A macrocrack goes through the matrix, and the radius of curvature of the crack-tip $\rho_1 \sim h$. Here $2h$ is the characteristic thickness of the matrix layer. Generally, a combination of both mechanisms is present with $h \leqslant \rho_1 \leqslant \rho$.

As a specific illustration we take (2.2) in the following form

$$\mathrm{d}\varphi/\mathrm{d}t = \begin{cases} 0 & (\sigma < r_0) \\ (1/t_c)[(\sigma - r_c)/r]^m & (\sigma \geqslant r_0)\,. \end{cases} \tag{2.22}$$

Here t_c is the characteristic time of a short-term test and the exponent $m > 0$ characterizes grain properties related to delayed fracture. A threshold value of the strength $r_0 \geqslant 0$ enters into (2.22).

We suppose that the only random variable in (2.22) is the characteristic grain strength r. Let r have the following cumulative distribution function:

$$F_r(r) = \begin{cases} 0 & (r < r_0) \\ [(r - r_0)/r_c]^\alpha & (r_0 \leqslant r \leqslant r_c) \\ 1 & (r > r_0) \end{cases} . \tag{2.23}$$

Eq. (2.23) seems, at first glance, to be artificial. Yet, in fact, it is consistent with the right-hand side of (2.22) reflecting both the structure of granular composites and polycrystalline materials. The lower bound of the strength r_0 corresponds to grains or crystallites oriented in the least favourable way relative to the field of nominal stresses. The upper bound r_c corresponds to grains or crystallites with the most favourable orientation. The exponent α in (2.23) has a deterministic positive value.

2.9. *Analytical treatment*

Consider the basic case when $\sigma = \text{constant}$. An approximate equation for the damage measure follows from (2.9), (2.22) and (2.23):

$$\psi(t) \approx \begin{cases} 0 & (\sigma < r_0) \\ [(\sigma - r_0)/r_c]^\alpha [(t - t_0)/t_c]^\beta & (r_0 \leqslant \sigma \leqslant r_c) \\ [(t - t_0)/t_c]^\beta & (\sigma > r_c) \end{cases} \tag{2.24}$$

The exponents α and β are related by

$$\beta = \alpha / m . \tag{2.25}$$

To find the cumulative distribution function of time up to macrocrack initiation, (2.24) is to be substituted into (2.17). This results in a three-parameter Weibull distribution

$$F_*(t_*) = 1 - \exp\{-(M/M_0)[(\sigma - r_0)/r_c]^{\alpha n_*}[(t_* - t_0)/t_c]^{\beta n_*} , \tag{2.26}$$

which is in agreement with common stochastic models employed in fracture mechanics (BOLOTIN, 1961; FREUDENTHAL, 1968). Here M is a measure of the volume of a specimen proportional to the number N of primary elements, and M_0 is a reference, e.g. unit volume. Eq. (2.26) includes the scale effect of macrocrack initiation. Let us consider, for example, the ratio of mathematical expectations of the times t_1 and t_2 for the appearance of the first macrocrack at a given stress level. If $r_0 \ll r_c$, $t_0 = 0$, then

$$\boldsymbol{E}[t_1]/\boldsymbol{E}[t_2] \approx (M_2/M_1)^{1/\beta n_*} . \tag{2.27}$$

here M_1 and M_2 are measures of specimens proportional to their volumes V_1 and V_2 respectively. Let a macrocrack initiate from the grains located at the specimen's surface only. This is, as a rule, the case for industrial structural materials. Then, $M_2/M_1 = (V_2/V_1)^{2/3}$ for geometrically similar specimens.

It is significant that the exponent βn_* characterizing both the scatter and scale effects, is equal to the corresponding exponent for the grain damage law multiplied by the number of grains n_* forming the nucleus of a macrocrack. The larger the number of grains n_*, the smaller the scatter of times prior to crack initiation as well as the scale effect.

Quite a similar approach is applicable in the case of brittle fracture. To obtain the cumulative distribution function $F^{**}(t^{**})$ of lifetimes prior to brittle fracture, it is sufficient to replace n_* by n^{**} in the right-hand side of (2.26). Here we briefly remark that (2.26) also gives the distribution of ultimate stresses for specimens that fail in brittle fracture. Interpreting σ as the strength at some given time t, and treating t as a parameter, (2.26) results in a three-parameter Weibull distribution of σ. For short-term tests we should replace $t_* - t_0$ by t_c = constant where t_c is the characteristic time from (2.22).

Turning to the stage of macrocrack growth, it is of interest to obtain equations similar in form to those of fracture mechanics (RABOTNOV, 1979). Let us consider (2.17). The function $f(\sigma, r)$ is given by the right-hand side of (2.22). The probability density is $p_r(r) = F_r'(r)$ where $F_r(r)$ is given by (2.23). The stress concentration factor $\kappa(l)$ enters into (2.17). Let the material be linear elasticly up to final fracture, the nominal stress field uniaxial tension orthogonal to the crack plane, and the crack penny-shaped with the radius l. The stress distribution near the crack-tip can be obtained from the solution of Sack's problem. The characteristic radius of curvature of the crack-tip is ρ_1. Then the estimate holds

$$\kappa \sim (l/\rho_1)^{1/2} . \tag{2.28}$$

This estimate is valid in a number of other cases, e.g. for internal or edge plane cracks if the size l is properly chosen.

Substituting (2.22), (2.23) and (2.28) into (2.17) we obtain

$$\mathrm{d}l/\mathrm{d}t = 2\rho(m + \alpha)[(l/\rho_1)^{1/2}\sigma - r_0]^m/(t_c \alpha r_c^m) . \tag{2.29}$$

Eq. (2.29) is valid when $(l/\rho_1) \geqslant r_0$. Otherwise $\mathrm{d}l/\mathrm{d}t \equiv 0$.

2.10. Comparison with linear fracture mechanics

The right-hand side of (2.29) contains a combination of the nominal stress σ and the crack size l while forms the stress intensity factor

$$K = \sigma(\pi l)^{1/2} . \tag{2.30}$$

Introducing notations

$$\begin{aligned} r_c(\pi\rho_1)^{-1/2} = K_c\,, \quad r_0(\pi\rho_1)^{-1/2} = K_0\,, \\ c = (2\rho/t_c)[(m+\alpha)/\alpha]\,, \end{aligned} \tag{2.31}$$

Eq. (2.29) can be re-written as follows

$$\mathrm{d}l/\mathrm{d}t = c[(K-K_0)/K_c]^m\,. \tag{2.32}$$

It will be shown below that K_c is of the order of magnitude of the critical value of the stress intensity factor. In other words, K_c is equivalent to the fracture toughness used in linear fracture mechanics. The parameter K_0 holds for the threshold stress intensity factor. Eq. (2.32) is valid when $K_0 \leqslant K < K_c$. When $K \geqslant K_0$ and time t is replaced by the cycle number n, (2.32) results in the well-known Paris–Erdogan equation for the cyclic fatigue crack propagation. In this case the range ΔK enters the right-hand side of the equation in place of K.

In contrast with the Paris–Erdogan equation and a number of its analogues (PANASYUK et al., 1977) which are of semi-empirical nature, (2.32) is derived analytically from the probabilistic model which takes into account the microstructural scale. The macroscopic parameters in (2.32) are connected with the parameters of the microstructure by (2.31). Incidentally, the form of the right-hand side of (2.32) is a direct consequence of the initial analytical equations, in particular, (2.32) and (2.33). Choosing other initial equations, a variety of fatigue growth equations can be obtained, e.g. such as those listed in the book by PANASYUK et al. (1977).

In order to find the critical size l_{**} (2.18)–(2.21) are to be used. Let us take, at first, (2.18) assuming that approximation of the order of unity corresponds to the quantile $1 - e^{-1} = 0.632\ldots$. Eq. (2.18) thus results in $l_{**}/\rho_1 \sim (r_c/\sigma)^2$. Hence the Griffith–Irwin criterion follows

$$K_{**} = \sigma(\pi l_{**})^{1/2} \sim K_c\,, \tag{2.33}$$

where K_c is determined by the first formula of (2.31). This confirms the way of interpreting K_c mentioned in the discussion of (2.32).

The second way to estimate the critical value K_{**} is by using (2.19). The latter is equivalent to $\mathrm{d}l/\mathrm{d}t \sim 2\rho/t_c$. With the account of (2.31) and (2.32) it follows that $K_{**} \sim K_0 + [\alpha/(m+\alpha)]^{1/m}(K_c - K_0)$. As $K_0 \ll K_c$, $\alpha \sim m$ and $m \gg 1$, the determined value of K_{**} is of the same order as K_c. The appropriate cumulative distribution function $F(l_{**};\sigma)$ is given by (2.20). In terms of the stress intensity factors it takes the form

$$F(K_{**}) \approx \begin{cases} 0 & K_{**} < K_0)\,, \\ [(K_{**} - K_0)/K_c]^{\pi l_{**}\alpha/\rho_1} & (K_0 \leqslant K_{**} \leqslant K_c)\,, \\ 1 & (K_{**} > K_c)\,. \end{cases}$$

Thus, the distribution of the values of K_{**} in the interval $[K_0, K_c]$ depends substantially on the exponent $\pi l_{**}\alpha/\rho_1$. Since the critical crack length l_{**} also enters into the function $F(K_{**})$ explicitly, one should consider the obtained distribution as relating to the random value l_{**}. But $\pi l_{**}\alpha/\rho_1 \gg 1$, and therefore the expected value of the critical values intensity factor is close to that defined by (2.33).

To study the influence of dispersed damage on the critical value of the stress intensity factor, the approximate condition given by (2.21) may be used. For the model considered here, (2.21) takes the following form: $K_{**} \approx K_0 + K_c(1-\psi)^{1/\alpha}$. Assume $\sigma = \text{constant}$. Then the damage measure is given by (2.24). Under the additional assumption $K_0 \ll K_c$ the result is $K_{**} \approx K_c[1 - (\sigma/r_c)^\alpha]^{1/\alpha}$. This equation is similar to those discussed in nonlinear fracture mechanics when the nominal stress is not small compared to the ultimate stress for an undamaged specimen. In this context, the ultimate stress may be identified with the upper strength parameter r_c. Further details concerning polycrystalline materials can be found in the book by BOLOTIN (1984a).

3. Unidirectional fibrous composites in tension

3.1. Composites with elastic matrices

Some aspects regarding probabilistic models of damage and fracture of unidirectional fibrous composites were discussed in Section 2 in connection with the unified approach to damage and fracture phenomena. In this section, a unidirectional fibrous composite subjected to tension along the fibers is considered. Fibers are assumed to be linear-elastic until fracture. In the first stage of study the matrix will also be considered linear elastic and sufficiently ductile so that matrix rupture does not occur before the failure of the specimen as a whole. The definition of a primary element for this case was given by (2.1). The total number of primary elements in the volume V can be evaluated as $N \sim V/(2\pi\rho^2\lambda_e)$.

The main feature of the probabilistic model is the scale effect of fiber strength. The tensile strength of a fiber segment of length λ is a random variable whose distribution depends on λ. Commonly, the Weibull distribution is used to describe random properties of fibers in tension. Here, the following cumulative distribution function (ARGON, 1974) is assumed:

$$F_r(r; \lambda) = 1 - \exp\{-(\lambda/\rho)[(r - r_0)/r_c]^\alpha\} . \tag{3.1}$$

The same notation is used as in (2.1) and (2.23). The scale effect is introduced into (3.1) through the ratio λ/ρ, where ρ is the characteristic size of the fibers' cross section. In particular, fibers of circular section are considered henceforth. Since only sufficiently long fibers can be tested in tension, the value of r_c has to be estimated by extrapolation. Another approach to evaluate r_c is to study the statistics of large samples of fibers or the corresponding scale effect (BOLOTIN, 1961).

3.2. Dispersed damage accumulation

Let damage in primary elements be governed by (2.22). Assume that $\sigma =$ constant. The damage measure calculated by means of (2.9), (2.25) and (3.1) is

$$\psi(t) = 1 - \exp\{-(\lambda_e/\rho)[(\sigma - r_0)/r_c]^\alpha[(t - t_0)/t_c]^\beta\} . \tag{3.2}$$

Critical values ψ^*, ψ^{**} and ψ_* are usually small compared to unity. Consequently, the exponential function in (3.2) may be expanded in power series. Retaining the first two terms of expansion, the power law of damage accumulation can be expressed in terms of both time and nominal stresses as follows:

$$\psi(t) \approx (\lambda_e/\rho)[(\sigma - r_0)/r_c]^\alpha[(t - t_0)/t_c]^\beta \tag{3.3}$$

A result similar to (5.2) is obtained for $\sigma = r_0 + \dot{\sigma}(t - t_0)$ and $\dot{\sigma} = \text{constant} > 0$

$$\psi(t) = 1 - \exp\{-(\lambda_e/\rho)[\dot{\sigma}(t - t_0)/r_c]^\alpha[(t - t_0)/(m + 1)t_c]^\beta . \tag{3.4}$$

Eqs. (3.2) and (3.4) are illustrated in Fig. 7. The following data were used: $m = 4$, $\lambda_e/\rho = 10^2$, $r_0 = 0$. The heavy lines in Fig. 7 show the damage accumulation at $\sigma =$ constant for the case $\sigma/r_c = 10^{-1}$. The values of α are indicated near

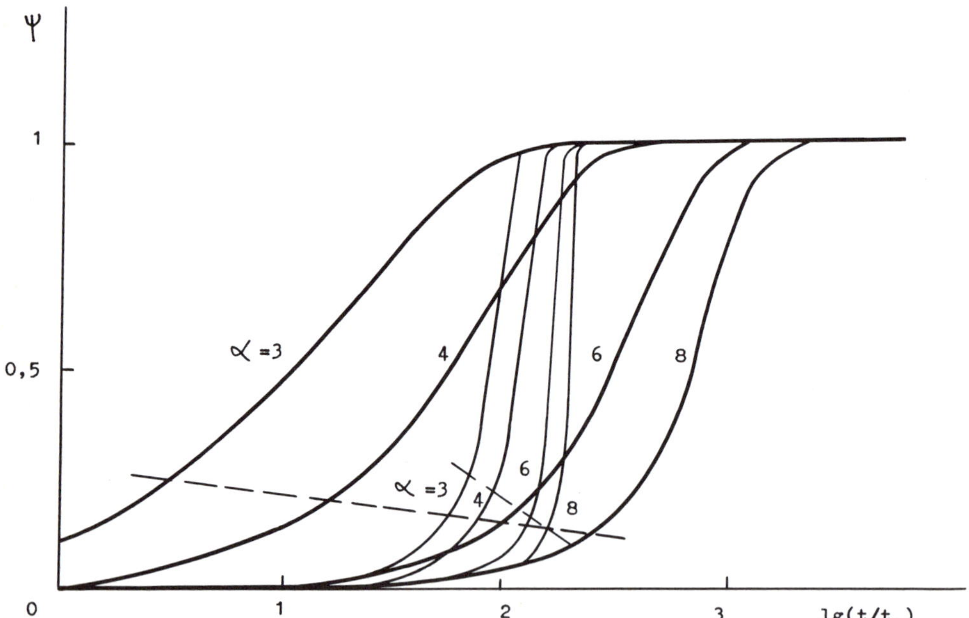

FIG. 7. Damage accumulation in a unidirectional fibrous composite in tension for $m = 4$, $\lambda_e/\rho = 10^2$, $r_0 = 0$ and various α; heavy lines at $\sigma/r_c = 10^{-1}$; light lines at $\dot{\sigma}t_c/r_c = 10^{-3}$.

the lines, and according to (2.25) $\beta = \alpha/m$. The light lines correspond to the case $\sigma = \dot{\sigma}(t - t_0)$ with $\dot{\sigma}t_c/r_c = 10^{-3}$.

Re-distribution of stresses in fibers due to dispersed damage should be, generally, introduced into the constitutive equations of damage in primary elements. This effect is taken into account by (2.2), but omitted in the special case presented by (2.22). Consider now the equation whose right-hand side contains the damage measure ψ:

$$\mathrm{d}\varphi/\mathrm{d}t = (1/t_c)(\sigma/r)^m(1-\psi)^{-m\gamma} . \tag{3.5}$$

For $\gamma = 0$ (3.5) takes the form of (2.22) at $r_0 = 0$. For $\gamma > 0$ (3.5) takes into account overloading of the remaining primary elements in the process of damage accumulation. Let σ = constant. In the first approximation, we replace $\psi(t)$ in the right-hand side of (3.5) by the cumulative distribution function $F_\tau(t)$. Integrating (3.5) subject to the initial conditions $\varphi(0) = 0$ and equating the result to unity, we obtain

$$\varphi(\tau\,|\,r) = (\tau/t_c)(\sigma/r)^m \int_0^\tau [1 - F_\tau(\tau_1)]^{-m\gamma}\, \mathrm{d}\tau_1 .$$

The lifetime τ is a random variable related explicitly to another random variable r. The latter is defined by (3.1). Hence the integral equation with respect to $F_\tau(t)$ follows

$$F_\tau(t) = 1 - \exp\left\{-\left[(1/t_\sigma)\int_0^t [1 - F_\tau(\tau)]^{-m\gamma}\, \mathrm{d}\tau\right]^\beta\right\},$$

where the notation $t_\sigma = t_c(\rho/\lambda_e)^{1/\beta}(r_c/\sigma)^m$ is used. This equation is equivalent to a certain differential equation. Its solution may be presented in the form containing the complementary function $u(t) = 1 - F_\tau(t)$:

$$t = (t_\sigma/\beta)\int_u^1 v^{m\gamma-1}(-\ln v)^{-(\beta-1)/\beta}\, \mathrm{d}v .$$

Explicit formulas for the damage measure can be obtained for $\beta = 1$:

$$\psi(t) = \begin{cases} 1 - \exp(-t/t_\sigma) & (\gamma = 0) , \\ 1 - (1 - m\gamma t/t_\sigma)^{1/m} & (\gamma > 0,\ m\gamma tr < t\sigma) , \\ 1 & (\gamma > 0,\ m\gamma t > t\sigma) . \end{cases}$$

When $\gamma = 0$, we arrive at a very special case of the model governed by (2.22). Acceleration of the damage accumulation process takes place for $\gamma > 0$, and at certain t the measure ψ becomes equal to unity. This is illustrated in Fig. 8 drawn for $\sigma = 0.1\ r_c$, $\lambda_e/\rho = 10^2$, $m = \alpha = 4$ and various values of γ.

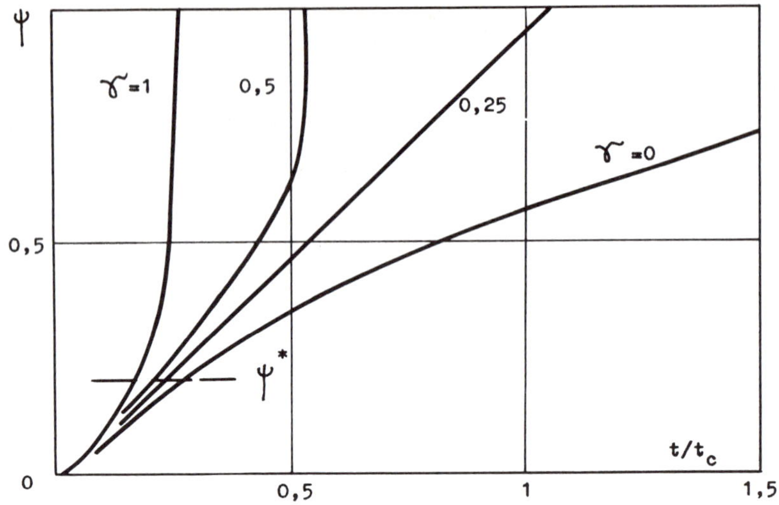

FIG. 8. Damage accumulation in tension with account of stress re-distribution for $m = \alpha = 4$, $\lambda_e/\rho = 10^2$, $r_0 = 0$, $\sigma/r_c = 10^{-1}$ and various γ.

Although the functions in Figs. 7 and 8 have been plotted up to $\psi = 1$, it ought to be noted that the models become unsuitable even at relatively low damage level. Actually, the interaction between primary elements is not assumed at the present stage. Approximate upper boundaries of the applicability of the curves $\psi(t)$ are shown in Figs. 7 and 8 with broken lines.

3.3. Critical damage measure

The simplest estimate of the critical value of the damage measure corresponding to loss of integrity is $\psi^* \sim (n^*)^{-1}$. Here n^* is the number of damaged neighboring primary elements whose association results in multiple damage over the whole volume of a specimen. For a hexagonal array of fibers a reasonable number for n^* is 7. This means that if one element from each bundle of seven elements is ruptured, rupture should cover the entire volume. Taking also into consideration the skew located elements, we obtain $n^* = 20$. Hence

$$1/20 \leqslant \psi^* \leqslant 1/7\,. \tag{3.6}$$

Quite another approach is based on (2.11). Stresses acting in fibers of homogeneously damaged composite are of the order of $\sigma(1-\psi)^{-\gamma}$ where $\gamma > 0$. This function should be substituted into (2.11) instead of $\sigma(\psi)$. Using the short-term fiber strength distribution given by (3.1), we obtain

$$\psi = 1 - \exp\{-(\lambda_e/\rho)[\sigma(1-\psi)^{-\gamma} - r_0]^{\alpha}/r_c^{\alpha}\}\,. \tag{3.7}$$

Solution of (3.7) is presented in Fig. 9 where the critical damage level ψ^* is plotted versus the non-dimensional nominal stress σ/r_c. Fig. 9(a) is drawn for $r_0 = 0$, Fig. 9(b) for $r_0 = 0.1r_c$, and $\lambda_e/\rho = 10^2$, $\gamma = 1$ in both cases. Magnitudes of the exponent α are shown near the curves. Only the descending branches of the curves have physical meaning. From the figures we see that the higher is the stress level, the smaller is the critical value ψ^*. The maximum of the function $\sigma(\psi)$ corresponds to the 'fiber bundle' model (ROSEN, 1965). At $r_0 \ll r_c$ (3.7) can be solved explicitly for σ whose magnitude is approximately proportional to $\psi^{1/\alpha}(1-\psi)^{-\gamma}$. The maximum of σ is located near the value

$$\psi^* \approx (1 + \alpha\gamma)^{-1} . \tag{3.8}$$

Eq. (3.8) is plotted in Figs. 7–9 with broken lines.

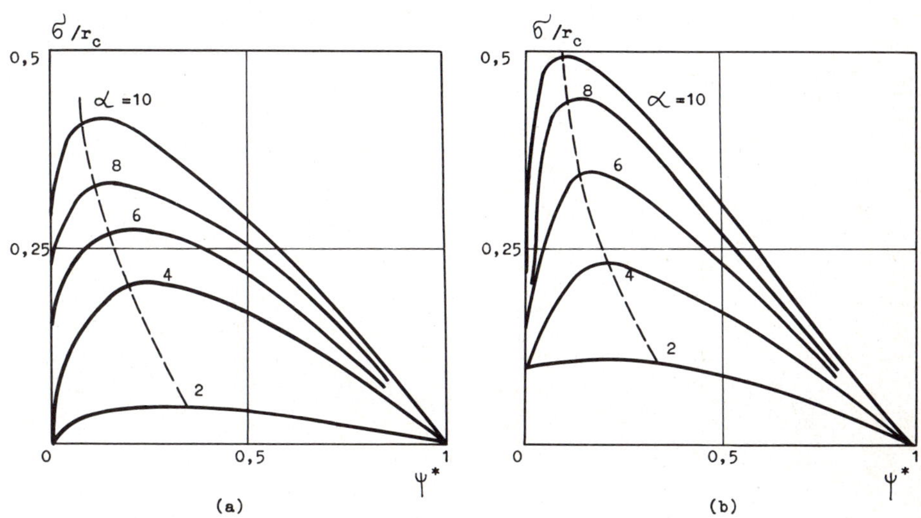

FIG. 9. Critical damage level ψ^* corresponding to loss of integrity in tension versus non-dimensional nominal stress σ/r_c for $\lambda_e/\rho = 10^2$, $\gamma = 1$; (a) $r_0 = 0$; (b) $r_0 = 0.1r_c$.

The nature of (3.6) and (3.8) is quite different. The first equation is based on intuitive topological notions related to those of the percolation theory (CHELIDZE, 1983). The second one contains a fiber strength parameter which, in fact, is the power exponent of (3.1). Nevertheless, numerical results are of the same order of magnitude when industrial fibrous composites are considered. Actually, for $\alpha = 6$ which is a rather low value for common glass fibers, and $\gamma = 1$, (3.8) gives $\psi^* \approx \frac{1}{7}$. This is in fact the upper bound in (3.6). KOP'YEV and OVCHINSKIY (1977), TAMUSZ (1982) and a number of other authors studied dispersed damage accumulation by direct computational simulation of samples of interacting primary elements. The results of such simulation are, generally, in agreement with the presented analytical estimates. However, when the stress level is low, the values of ψ^* are rather high as is shown in Fig. 9. In such a situation other mechanisms of failure becomes dominant, namely, brittle fracture and macrocrack initiation.

3.4. *Composites with elastic–plastic matrices*

Now, let us discuss a composite whose matrix material is ideally elastic–plastic. The matrix ultimate shear stress τ_m^* may be interpreted either as the yield stress of the matrix, or the friction stress on the damaged fiber–matrix interface, or, generally, as some limit stress including all non-elastic phenomena in the matrix occurring in the vicinity of a ruptured fiber. Evidently, a single damage measure ψ is insufficient in the case of non-elastic matrix. This measure is denoted below by ψ_1 to distinguish it from the newly introduced damage measures.

If the fiber tensile stresses are rather low, the transfer length is defined by (2.1). In the case of sufficiently large tensile stresses, the matrix is deformed near a broken fiber non-elastically. The characteristic transfer length of the stress from a broken fiber was introduced by KELLY (1973) as

$$\lambda_{\mathrm{p}} \sim \rho\sigma_{\mathrm{f}}/(2\tau_{\mathrm{m}}^*) . \tag{3.9}$$

Thus, the transfer length depends on the nominal stress in the fibers. For simplicity, we assume that the transition from elastic to non-elastic transfer length occurs at $\sigma \sim \sigma_{\mathrm{f}} \sim \sigma_{\mathrm{Y}}$, where the stress σ_{Y} satisfies the condition $\lambda_{\mathrm{e}} = \lambda_{\mathrm{p}}$. Then

$$\sigma_{\mathrm{Y}} = 2\tau_{\mathrm{m}}^*(E_{\mathrm{f}}/2G_{\mathrm{m}})^{1/2} . \tag{3.10}$$

When the case $\sigma = \text{constant} > \sigma_{\mathrm{Y}}$ is discussed, it is natural to take the length of a primary element equal to $2\lambda_{\mathrm{p}}$. However, for variable stresses it is expedient to assume the total number of primary elements to be constant. For this reason, the length $2\lambda_{\mathrm{e}}$ is taken as the characteristic length of the primary elements also in the case of non-elastic deformation of the matrix.

We introduce a second damage measure ψ_2 as the ratio of the total length of all fiber segments whose interface is deformed non-elastically to the total length of the fibers. This measure is related to the measure ψ_1 in the following way:

$$\psi_2(t) = \int_0^t \Lambda(t, \tau)\, \mathrm{d}\psi_1(\tau) . \tag{3.11}$$

The kernel in (3.11) is

$$\Lambda(t, \tau) = \begin{cases} 0 & \sigma(t, \tau) < \sigma_{\mathrm{Y}} , \\ [\sigma(t, \tau)/\sigma_{\mathrm{Y}}) - 1 & \sigma(t, \tau) \geq \sigma_{\mathrm{Y}} \end{cases} , \tag{3.12}$$

where the notation $\sigma(t, \tau) = \sup \sigma(\tau_1)$ at $\tau_1 \in [\tau, t]$ is used, and σ_Y is defined by (3.10). If $\sigma(t)$ is a quasimonotonously increasing function, (3.12) results in

$$\psi_2(t) = \begin{cases} 0 & t < t_Y, \\ \psi_1(t)\{[\sigma(t)/\sigma_Y] - 1\} & t \geq t_Y \end{cases} . \tag{3.13}$$

It is sometimes useful to introduce the third measure $\psi_3 = \psi_1 + \psi_2$ which is equal to the ratio of the sum of all transfer lengths to the total length of the fibers. The condition $\psi_1 = 1$ signifies the rupture of all primary elements, and $\psi_3 = 1$ signifies total delamination of a specimen. As in the case of elastic matrix, values of ψ_1 and ψ_2 are small compared to unity because even at a rather low damage level the process makes a transition to a new stage of failure.

Results of computation based on (3.13) are plotted in Fig. 10. The case $\dot{\sigma}$ = constant was considered, and the following numerical data were used: $m = \alpha = 4$, $\lambda_e/\rho = 10^2$, $\sigma_Y = 0.2r_c$. Heavy lines correspond to ψ_1, light lines to ψ_2, and the process $\sigma(t)$ is represented by dashed lines. The following values of $\dot{\sigma}t_c/r_c$ are chosen 10^{-3}, 10^{-4} and 10^{-5} were chosen for the curves 1, 2 and 3 respectively.

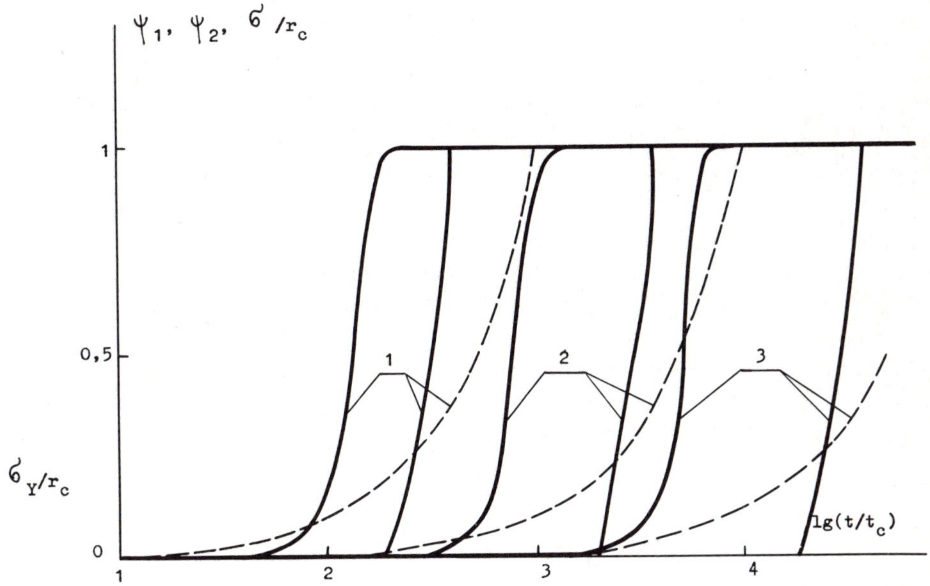

FIG. 10. Damage accumulation in unidirectional fibrous composite in tension for $m = \alpha = 4$, $\lambda_e/\rho = 10^2$, $r_0 = 0$, $\sigma_Y = 0.2r_c$; heavy lines: ψ_1, light lines: ψ_2, dashed lines: σ; (1) $\dot{\sigma}t_c/r_c = 10^{-3}$; (2) 10^{-4}; (3) 10^{-5}.

3.5. Short-term tensile strength

Consider some applications of the presented theory, apart from those related directly to the reliability analysis. First, approximate estimates for the short-term ultimate strength may be obtained. In fact, putting in (3.3) $t = t_c$ and $\psi_1(t_c) = \psi_1^*$ we obtain

$$\sigma^* \sim r_c(\psi_1)^{1/\alpha}(2G_m/E_f)^{-1/2}, \tag{3.14}$$

where $\sigma \sim \sigma_f$, $r_0 \ll r_c$ is assumed. Even at $\psi_1 \ll 1$ one may assume $(\psi_1)^{1/\alpha} \sim 1$ if α is sufficiently large. This means that (3.14) is analogous to the well-known equation by ROSEN (1965), differing from the latter by a numerical multiplier of the order of unity. In the case of the elastic–plastic matrix (3.3) and (3.13) should be used together. Let the short-term fracture occur when $\psi_3(t_c) = \psi_3^*$. Then at $\sigma \sim \sigma_f$, $r_0 \ll r_c$, the ultimate strength is estimated as

$$\sigma^* \sim r_c(\psi_3)^{1/(\alpha+1)}(2\tau_m^*/r_c)^{1/(\alpha+1)}. \tag{3.15}$$

Depending on the fiber and matrix properties, either (3.14) or (3.15) gives the actual strength of a composite. Although (3.14) and (3.15) are valid under certain assumptions and are only order of magnitude estimates, they are nevertheless useful since they illustrate explicitly the effect of various factors on the ultimate tensile strength of a composite. The fiber volume ratio v_f does not enter in (3.14) and (3.15) because of the initial assumption that $v_f \sim (v_f^{-1/2} - 1)^{1/2} \sim 1$. If necessary, one could take this factor into consideration, and in particular, make a distinction between the nominal tensile stress in a composite, σ, and that in fibers, σ_f. For details see the paper by BOLOTIN (1981b).

3.6. Acoustic emission of dispersed damage

Another application of the theory of damage accumulation is related to non-destructive testing and especially to the acoustic emission technique. The latter has good prospects for predicting the reliability and residual lifetime parameters of operating structures. A number of pieces of information is used in the acoustic emission technique. For example, when each significant damage event is counted, the summed number of counts $N(t)$, the count rate $\dot{N}(t)$, and the level of emission $L(t)$ proportional to the logarithm of the count rate correlate with the damage of a structure. Acoustic signatures and their Fourier transforms may be used for more detailed information.

To distinguish between significant insignificant damage, e.g. minor cracking of the interface or matrix, a proper detection threshold should be chosen. Suppose that the chosen threshold allows us to register all the damage events beginning with ruptures of single fibers. Then, at the dispersed damage accumulation stage, the number of counts is related to the damage measure

$\psi(t)$, as was suggested by BOLOTIN and LOBANOV (1983). For example, let damage be homogeneously distributed in volume V, and all signals emitted from this volume and corresponding to fiber ruptures be counted. The total number of counts at time t is

$$N(t) \sim \nu_0 V \psi(t) . \tag{3.16}$$

Here ν_0 is the number of primary elements per unit volume. Under constant loading, (3.3) and (3.16) yield $N(t) \sim \text{constant} \times \sigma^{\alpha} t^{\beta}$. Under loading with constant velocity, $N(t) \sim \text{constant} \times t^{\alpha+\beta}$, etc. this is in agreement with the empirical 'power law' used in discussions dealing with the acoustic emission technique.

Fig. 7 may be interpreted as an illustration of (3.16). The value $\psi(t)$ is in fact equal to the normalized summed number of counts. The energy release from a single fiber breakage is proportional to σ^2. Hence the logarithmic level of acoustic emission $L = \text{constant} \times \lg(\dot{N}\sigma^2)$. This quantity is plotted in Fig. 11. The same numerical data were used as in Fig. 7. Only the initial segments of the plotted curves have physical meaning using the same reasoning as in the previous discussion.

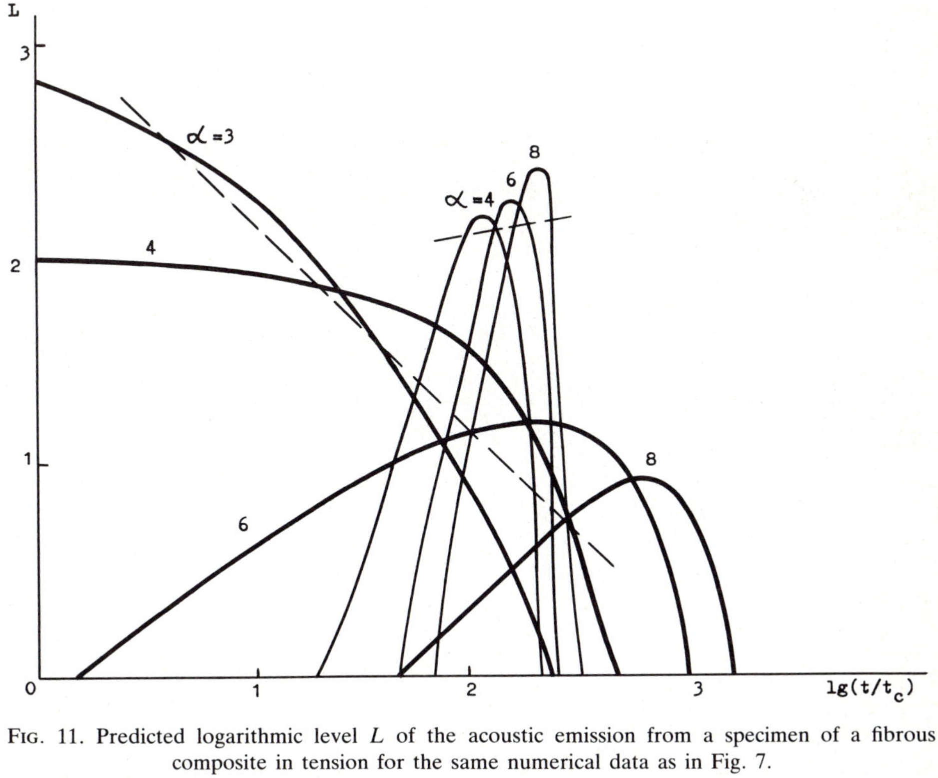

FIG. 11. Predicted logarithmic level L of the acoustic emission from a specimen of a fibrous composite in tension for the same numerical data as in Fig. 7.

If no time effects are included, $N(t) \sim \nu_0 V F_r[s(t)]$. Here $F_r(r)$ is a cumulative distribution function similar to that of (3.1), and $s(t) = \sup \sigma(\tau)$, where $\tau \in [t_0, t]$. Since damage is irreversible, this model describes the Kaiser effect well known in the acoustic emission technique.

It is easy to generalize the model for spatial inhomogeneity of stress and strength fields. Attenuation of signals which depends on the distance between the source and the transducer, as well as on dispersion, dissipation and multiple diffraction of signals by inhomogeneities can be taken into consideration, too. Suppose only those signals $u(x, t)$ coming from the point x are recorded by the transducer put in the origin $x = 0$, whose level is above the threshold u_{th}. Instead of (3.16) we thus obtain

$$N(t) \sim \nu_0 \int_{V_{\text{E}}(t)} F_\tau(\tau \mid x)\, \mathrm{d}V . \tag{3.17}$$

Here, $F_\tau(\tau \mid x)$ is the cumulative distribution function of the lifetime of a primary element at the point x under the given loading history and the domain of integration is $V_{\text{E}}(t) = \{x : u(x, t) \rangle u_{\text{th}}\}$. Fig. 12 illustrates the effect of the signal attenuation and the detection threshold. A long thin specimen is considered and the attenuation law is taken in the form $u = u_0 F(a/x)^{1/2} \exp(-\varepsilon x)$. Here u_0 is a characteristic signal magnitude, a is a characteristic length, and ε is an effective damping factor taking into consideration all the dispersion, dissipation and multiple diffraction phenomena. The nominal stress is $\sigma = 10^{-3} r_{\text{c}}(t - t_0)/t_{\text{c}}$. Heavy lines correspond to the case when

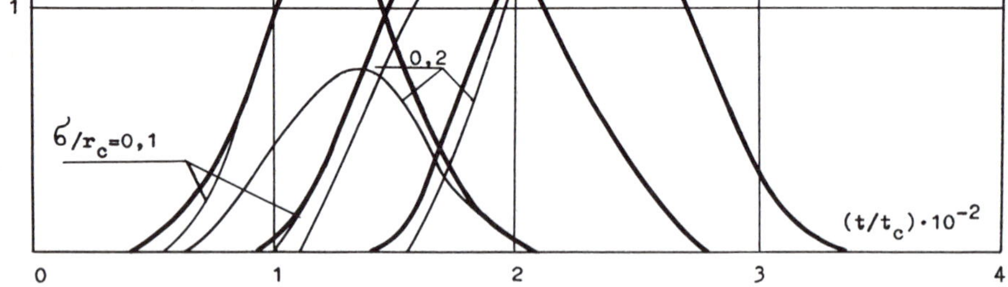

FIG. 12. Predicted non-dimensional count rate $\dot{N}/N_0$ of acoustic emission from a specimen in tension taking into account the attenuation and detection thresholds corresponding to nominal stresses $\sigma = 0$ (no threshold is present), $\sigma = 0.1 r_{\text{c}}$, and $0.2 r_{\text{c}}$. Other numerical data are the same as in Fig. 7.

all the signals are registered. Light lines correspond to the signals with a detection threshold. The threshold level is nominally chosen according to the amount of energy released from fibers located close to the transducer and subjected to a certain nominal stress. In Fig. 12 the following threshold stresses were chosen: $0.1r_c$ and $0.2r_c$. When $\alpha = 4$ the attenuation effect is rather considerable, but at $\alpha = 8$ it becomes negligible.

3.7. A model of a transverse crack

Whereas cracks, in unidirectional fibrous composites oriented along the fibers are more or less similar to those studied in linear fracture mechanics, transverse cracks require a special consideration. Transverse cracking is accompanied by significant delamination and debonding of fibers from the matrix. There are two consequences of this. First, the stress concentration at the front of a crack becomes less significant than for common materials where the stress concentration is the main mechanism of crack growth. This is, incidentally, the reason for using the expression 'front of a crack' instead of 'tip of a crack' or 'crack-tip' when transverse cracks in fibrous or laminated composites are considered. Secondly, when a transverse crack grows, the length of delamination increases, and an effective fiber length on which fiber strength depends also increases. Due to the significant scale effect of fiber strength, delamination seems to become the dominant mechanism instead of stress concentration. These conclusions are in agreement with observations of cracking of unidirectional fibrous composites subjected to tension along the fibers. Typically, such fractured specimens have a brush-like appearance because of the significant debonding of fibers in the vicinity of final breaks. Even in cases when an artificially initiated transverse cut or notch is planted in a specimen, the appearance of cracking becomes brush-like after the crack proceeds some distance under the influence of the load. The notion that delamination of fibers and fiber strength scale effect form the main mechanism of transverse cracking in unidirectional fibrous composites was suggested by BOLOTIN (1981a, 1981b).

An internal transverse crack in a composite is, in fact, three-dimensional. Consider the analogue of a penny-shaped crack. The domain occupied by the crack may be treated schematically as a circular cylinder with radius l and length 2λ (Fig. 13). To determine λ, the same approach is used as that in estimating the transfer length of a fiber near a single rupture. If the matrix is linear elastic, we obtain an estimate

$$\lambda \sim (l\rho)^{1/2}(E_f/2G_m)^{1/2}, \tag{3.18}$$

analogous to (2.1). In the case of ideal elastic–plastic matrix an estimate analogous to (3.9) is valid, i.e.

$$\lambda \sim l(\sigma_f/2\tau_m^*). \tag{3.19}$$

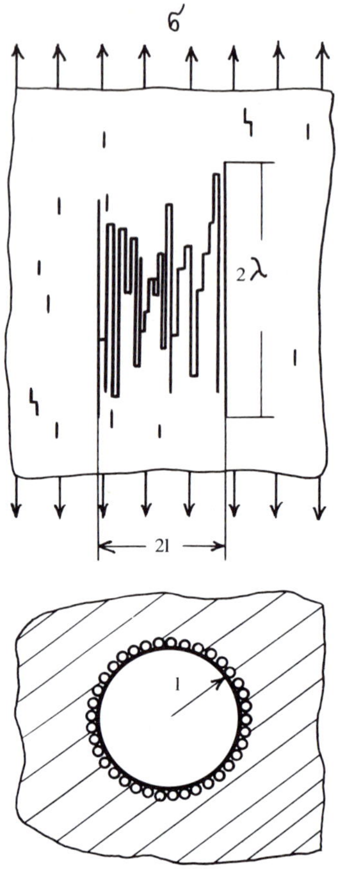

FIG. 13. A model of a transverse crack in a unidirectional fibrous composite in tension along the fibers direction.

Transition from (3.18) to (3.19) takes place at $\sigma \sim \sigma_f \sim \sigma_Y(l)$ where

$$\sigma_Y(l) = 2\tau_m^*(\rho/l)^{1/2}(E_f/2G_m)^{1/2} . \tag{3.20}$$

Eqs. (3.18) and (3.19) give, in fact, the transfer length of a bundle of fibers with a certain radius. When $l \sim \rho$, (3.18)–(3.20) reduce to (2.1), (3.9) and (3.10) respectively.

3.8. Nucleation and growth of cracks

The nucleus of a transverse macroscopic crack contains, by definition a certain amount of ruptured fibers. Let us denote this number by n_*. Then, the initial size of the nucleus is approximately, $l_* \sim n_*^{1/2}\rho$. According to (3.18) the transfer length of the nucleus is proportional to $n_*^{1/4}$. Alternatively, if (3.19) holds, then it is proportional to $n_*^{1/2}$. In both cases, the transfer length is larger than twice the length of a primary element, and that should be taken into account when the probabilities of macrocrack nucleation are estimated. This conclusion is of significance with respect to failure due to loss of integrity, too. Details can be found in the paper by BOLOTIN (1981a).

The process of stable growth of a transverse crack is governed by (2.15)–(2.17). The new point is that the transfer length λ enters into the probability density function $p_r(r)$ and the stress concentration factor κ. Since λ depends on the crack size l, the latter enters into the right-hand side of (2.17). This means that (2.17) becomes rather complicated. Very simple estimates of the right-hand side may be obtained for $r_0 \ll r_c$. Substituting (2.22), (3.1) and (3.18) into (2.17) we find

$$\mathrm{d}l/\mathrm{d}t \approx (2\rho/t_c)(l/\rho)^{1/(2\beta)}(\kappa\sigma/r_c)^m(E_f/2G_m)^{1/(2\beta)}\Gamma(1+1/\beta)\,, \qquad (3.21)$$

where $\Gamma(\cdot)$ is a gamma-function. Eq. (3.21) is valid when $\sigma < \sigma_Y(l)$. Otherwise, (3.19) should be used instead of (3.18). This results in

$$\mathrm{d}l/\mathrm{d}t \approx (2\rho/t_c)(l/\rho)^{1/\beta}(\kappa\sigma/r_c)^m(\sigma/2\pi_m^*)^{1/\beta}\Gamma(1+1/\beta)\,. \qquad (3.22)$$

Eqs. (3.21) and (3.22) are of the same structure as the semi-empirical equations for fatigue crack growth. In terms of static fatigue, the latter equations can be written as

$$\mathrm{d}l/\mathrm{d}t = f(K)\,, \qquad (3.23)$$

where the right-hand side is a function of the stress intensity factor. Similarly to (2.30) $K = \text{constant} \times \sigma l^{1/2}$. In the Paris–Erdogan equation it is assumed that $f(K) = \text{constant} \times K^m$.

It is easy to see that both (3.21) and (3.22) reduce to the form of (3.23) only when $\alpha + 1$. The function $f(K)$ in (3.23) is then a power function with the exponent $1/\beta$ if the matrix is deformed elastically and $2/\beta$ for the elastic–plastic matrix. The case $\alpha = 1$ corresponds to the exponential distribution of fiber strength, as follows from (3.1). As was mentioned, the parameter α takes the values of 5 and higher for industrial fibers. The value of α incidentally, characterizes the scatter of fibers strength. At $\alpha \gg 1$, an asymptotic estimate $v_\sigma \sim (\pi/\alpha)6^{-1/2}$ for the coefficient of variation v_σ of the tensile strength. It is possible to increase the value α up to 10 and even more with a well controlled fabrication process. It is of interest to note that the value $\alpha = 1$ is not a random result. A hypothetic fibrous composite whose fibers strength is distributed exponentially approaches in a large measure to conventional granular and polycrystalline materials for which linear fracture mechanics has been developed.

3.9. *Acoustic emission from a growing crack*

Some of the presented results have been discussed by BOLOTIN and LOBANOV (1983) in connection with the acoustic emission technique. Actually, transverse cracks emit acoustic signals whose magnitude is larger than that emitted from a single broken fiber. The energy released from a single broken fiber imbedded in an elastic matrix is estimated as $U_e \sim \rho^2 \lambda_e \sigma^2 / E_f$. When a fiber is located in front of the crack, the energy release is $U \sim \rho^2 \lambda \sigma^2 / E_f$. Magnitudes of acoustic signal are proportional to the square root of the amounts of released energy. Under the same circumstances, we have, following the estimates $u \sim$ constant $\times\, 6\rho^{3/2}$ for single fibers and $u \sim$ constant $\times\, \sigma(\rho^2 l)^{1/2}$ for fibers in front of a crack. Here (2.1) and (3.18) have been used. The magnitudes of signals related to a propagating brush-like crack are $(l/\rho)^{1/2}$ times larger than those of signals related to dispersed damage. Thus, propagating brush-like cracks can be, in principle, distinguished with an appropriate choice of the detection threshold.

The same conclusion follows when the matrix is ruptured plastically. The energy released from a single broken fiber is $U_p \sim \rho^2 \lambda_p \sigma^2 / E_f$ or, taking account of (3.9), $U_p \sim \rho^3 \sigma^3 / (E_f \tau_m^*)$. Using (3.19), we find that the energy released from a fiber in front of a crack $U \sim \rho^2 l \sigma^3 / (E_f \tau_m^*)$. Evidently, magnitudes of the signals are in the same ratio as in the case of the elastic matrix. Other outputs and inputs of energy are not taken into account in this discussion. They are supposed to be small in comparison with the amount of energy released from broken fibers. For example, the energy of the attached matrix segments is comparatively small since from the beginning it was assumed that $G_m \ll E_f$, $\tau_m^* \ll \sigma_f$.

3.10. *Critical sizes of transverse cracks*

A crack becomes unstable when every fiber in front of it can be ruptured with a probability of the order of unity. In this case, (2.18) takes the form

$$F_r(\kappa \sigma_f; \lambda) \sim 1 . \tag{3.24}$$

The critical size l_{**} enters into (3.24) through λ which is defined by (3.18) or (3.19) depending on the magnitude of the stress σ in comparison with that given by (3.20). With the accuracy given by the relationships $r_0 \ll r_c$, $1 - e^{-1} \sim 1$, we obtain the following estimate of l_{**} for the case of elastic matrix

$$l_{**} \sim \rho (r_c / \kappa \sigma)^{2\alpha} (2 G_m / E_f) . \tag{3.25}$$

This estimate is valid if $\sigma < \sigma_Y(l_{**})$. Otherwise,

$$l_{**} \sim \rho (r_c / \kappa \sigma)^{\alpha} (2 \tau_m^* / \sigma) . \tag{3.26}$$

If $\alpha > 1$, an explicit formula for $\sigma_Y(l_{**})$ follows from the equality condition of the right-hand sides in (3.25) and (3.26):

$$\sigma_Y(l_{**}) = (r_c/\kappa) r_c G_m / (\kappa \tau_m^* E_f)^{\alpha - 1} . \tag{3.27}$$

It should be mentioned that the Griffith–Irwin criterion, $K = \text{constant}$ follows from (3.25) and (3.26) only for $\alpha = 1$. This conclusion is of physical significance and it is similar to that obtained in connection with (3.21) and (3.22). The interpretation given there remains applicable to (3.25) and (3.26).

3.11. Generalization

Until now the simplest case of quasistatic loading along the fibers has been discussed. This case is the basic one, and may be considered as an initial point from which extension to more general cases of crack initiation, nominal stress–strain state, and loading conditions can be carried out. To apply the theory to cyclic fatigue, one should change the kinetic equation of damage of the primary elements by using a finite-difference equation with the number of cycles as an argument. This argument will enter into further analysis. Therefore, final results such as analogues to (3.2)–(3.4), (3.21), (3.22), etc., will contain the number of cycles as an argument. In the case of cyclic loading with slowly varying parameters, the final equations are of the same form as in the case of static fatigue with t replaced by n. However, all the initial equations relate to a single loading cycle. Consequently, numerical values of parameters entering these equations differ generally from those for the case of static fatigue.

Let a specimen contain a cut inclined at the angle θ to the fiber direction and be subjected to tension along the fibers (Fig. 14(a)). This case can be approximately reduced to the basic one with the characteristic crack size $l_0 \sin \theta$. After delamination of the first group of fibers located close to the cut tip occurs, the situation differs only slightly from the case of a brush-like crack.

A more general case is shown in Fig. 14(b). A specimen of a unidirectional fibrous composite is subjected to a plane stress state with nominal stresses σ_x, σ_z and τ_{xz}. Fiber stresses estimated within an order of magnitude are $\sigma_f \sim \sigma_z$. For matrix stresses the estimates $\sigma_m \sim \sigma_x$, $\tau_m \sim \tau_{xz}$ may be used. If the matrix is deformed elastically, which is the case in the presence of very small stresses σ_x and τ_{xz}, most of the previous results remain unchanged, beginning with (2.1) for the elastic transfer length. In fact, this parameter defines the size of primary elements subjected to tensile stress with the order of magnitude of σ_z.

The situation becomes more complicated if the matrix is deformed non-elastically. For the limit state of the matrix we take the Mohr criterion

$$|\tau| - k\sigma_x = \tau_m^* , \tag{3.28}$$

where $k \geqslant 0$ is an empirical constant, and τ_m^* is the ultimate shear stress of the

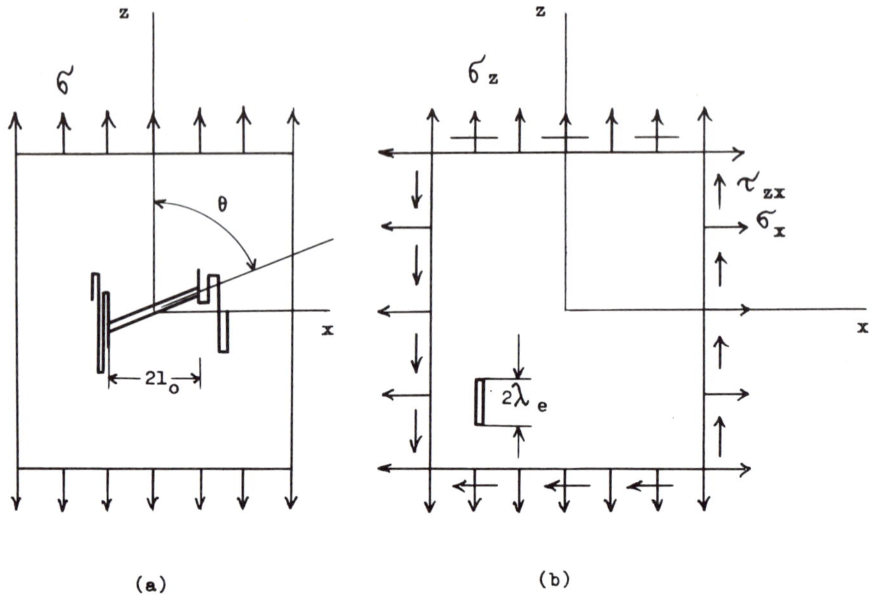

FIG. 14. Loading of a specimen of a unidirectional fibrous composite with initial cut; (a) inclined cut, (b) plane stress state.

matrix. In order to study dispersed damage accumulation, macrocrack initiation and other kinds of failures, one must estimate the transfer length from a ruptured fiber under the conditions of plane stress state. Ineffective segments λ_1 and λ_2 located on opposite ends of a broken fiber generally are not equal. The equations of load transfer have the form $2\pi\rho\lambda_{1,2}|\Delta\tau_{1,2}| = \pi\rho^2\sigma_z$ where the shear stress increments $\Delta\tau_{1,2}$ are determined from (3.28). The latter gives $|\tau_{xz} + \Delta\tau_{1,2}| - k\sigma_x = \tau^*_{\mathrm{m}}$. For the length of the damage fiber–matrix interface $2\lambda_{\mathrm{p}} = \lambda_1 + \lambda_2$ we obtain an estimate

$$2\lambda_{\mathrm{p}} \sim \rho\sigma_z(\tau^*_{\mathrm{m}} + k\sigma_x)[(t^*_{\mathrm{m}} + k\sigma_x)^2 - \tau^2_{xz}]^{-1} . \tag{3.29}$$

In the case when $\sigma_x = \tau_{xz} = 0$, (3.29) reduces to (3.9) which, in fact, is thus seen to be a very special case of (3.29). Setting $\lambda_{\mathrm{p}} = \lambda_{\mathrm{e}}$ in (3.29), we obtain a relationship between the components of the stress states of elastically and inelastically deformed matrices:

$$\sigma_z(\tau^*_{\mathrm{m}} + k\sigma_x) \approx 2(\tau^*_{\mathrm{m}} + k\sigma_x)^2 - \tau^2_{xz}(E_{\mathrm{f}}/2G_{\mathrm{m}})^{1/2} .$$

Eqs. (3.18) and (3.19) for the characteristic length of a crack can be generalized in the same way.

4. Applications to the reliability analysis

4.1. Reliability with respect to loss of integrity

The probabilistic model of fracture presented in the previous sections can be applied to the reliability analysis with a minor additional development. In fact, the reliability function defined by (1.1) is a complementary cumulative distribution function of time until failure. For example, if a failure is associated with the initiation of a macroscopic crack, then $R(t) = 1 - F_*(t)$. Here $F_*(t_*)$ is the cumulative distribution function of time until crack initiation. The corresponding rate of failures is $\lambda(t) = F'(t) \times [1 - F_*(t)]^{-1}$, etc.

Some special points should be discussed in more detail. Among these are: (1) nonstationary and random loading in general; (2) non-homogeneity of the stress–strain field as well as of the stress–damage due to the geometry of a structural component and/or to the character of load–structure interaction. As a rule, both issues enter into the reliability analysis, jointly.

An approach to the estimation of reliability factors when failure is the result of loss of integrity has been discussed by BOLOTIN (1976). Consider, at first, the simplest case. Let the damage vector be one-dimensional, the damage and strength fields homogeneous, and the critical damage level ψ^* entering into (2.10) independent of the stress level at a given t. The damage process $\psi(t)$ is by definition a cumulative one in the sense of (1.9) or (1.11). Then, following (1.10), the reliability function is

$$R(t) = \boldsymbol{P}\{\psi(t) < \psi^*\} . \tag{4.1}$$

There are three sources for the stochastic nature of the process $\psi(t)$. First, the scatter of damage arises from the random sampling of ruptured primary elements, as is shown by (1.5) and its asymptotic approximation based on Laplace–Moivre theorem. Second, the loading process $\boldsymbol{q}(t)$, is generally random. Finally, the strength parameters given by the vector $\boldsymbol{r}$ are random, too. In principle, all the three factors are taken into account by (2.5), (2.6) and (2.9). If (4.1) is valid, then

$$R(t) = F_\psi(\psi^*; t) . \tag{4.2}$$

The cumulative distribution function $F_\psi(\psi)$ entering in (2.5), (2.6) and (2.9) is a solution of (2.2) which, in the case of random loading, becomes a stochastic differential equation. The solution depends parametrically on the vector $\boldsymbol{s}$ characterizing loading and environment conditions as a whole, as well as on the strength vector $\boldsymbol{r}$. Hence, the life distribution of primary elements is a conditional one in the sense of (1.2) and (1.3). Let us denote conditional cumulative distribution functions by $F_\tau(\tau \mid \boldsymbol{r}, \boldsymbol{s})$, $F_\psi(\psi; t \mid \boldsymbol{r}, \boldsymbol{s})$, etc. The unconditional reliability function $R(t)$ is a result of the convolution of distributions of all the involved random values and processes. Particularly, in place of (4.2) we obtain

$$R(t) = \iint\limits_{D(r,s)} F_{\psi}(\psi^*; t \mid \boldsymbol{r}, \boldsymbol{s}) p(\boldsymbol{r}, \boldsymbol{s}) \, \mathrm{d}\boldsymbol{r} \, \mathrm{d}\boldsymbol{s} \,, \tag{4.3}$$

where the notation of (1.3) is used.

For the more general case of the vector process $\boldsymbol{\psi}(t)$, the inequality $\psi(t) < \psi^*$ in (4.1) is to be replaced by the condition $\boldsymbol{\psi}(t) \in \Omega$ or $\|\boldsymbol{\psi}(t)\| < 1$, and the reliability function is defined by (1.10) or (1.12), respectively. A scalar process $\psi(t)$ is discussed later since the methods of generalization are more or less apparent. A number of problems concerning multidimensional processes will be considered in Section 6 with application to structural components containing sets of cracks and crack-like defects.

4.2. *Scale effect in the loss of integrity*

First, let discuss just the scatter and scale effects due to random sampling of ruptured primary elements. Eqs. (2.6) and (4.2) result in

$$R(t) \approx \Phi\left(\frac{[\psi^* - F_{\tau}(t)] N^{1/2}}{\{F_{\tau}(t)[1 - F_{\tau}(s)]\}^{1/2}} \right) . \tag{4.4}$$

The reliability function and other reliability indicators depend substantially on the number N, i.e. on the dimensions of a specimen or a structural component. If N is sufficiently large, the right-hand side of (2.6) closely approaches that of (2.9). This means that under the assumed circumstances, the process $\psi(t)$ is almost deterministic. Hence, significant scatter is absent when the loss of integrity of a sufficiently large volume is concerned.

For illustration, consider the model of the unidirectional fibrous composite discussed in Section 3. The function $F(t)$ coincides with the damage measure $\psi(t)$ given by (3.2) and (3.3), etc. The reliability function $R(t)$ is plotted in Fig. 15 for various numbers of primary elements. Numerical data employed here are: $\alpha = 6$, $\beta = 1$, $r_0 = 0$, $\psi^* = 0.2$ and $\sigma = 0.5 r_c$ When the number N increases, the curves closely resemble the step-like curve corresponding to (2.9) with $\psi = \psi^*$. The mathematical expectation $\boldsymbol{E}[T]$ and coefficient of variation v_T of the life T are plotted in Fig. 16 as a function of the number N. The critical damage level ψ^* is taken as a parameter. When the number N increases, the mathematical expectation of the lifetime asymptotically approaches a certain value independent of N (Fig. 16(a)). The coefficient of variation of the lifetime approaches zero (Fig. 16(b)) on the other hand. The number of primary elements in real structural components fabricated with fibrous composites is much larger than that considered in Fig. 16. Hence, (2.9) presents a rather good approximation. This means that $R(t) \approx 1$ at $F_{\tau}(t) < \psi^*$, and $R(t) \approx 0$

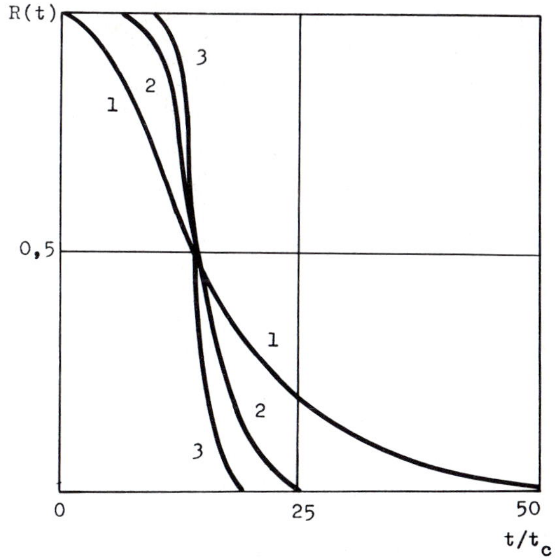

FIG. 15. Reliability function $R(t)$ for $\alpha = 6$, $\beta = 1$, $\psi^* = 0.2$, and $\sigma = 0.5 r_0$ for various numbers of primary elements N.

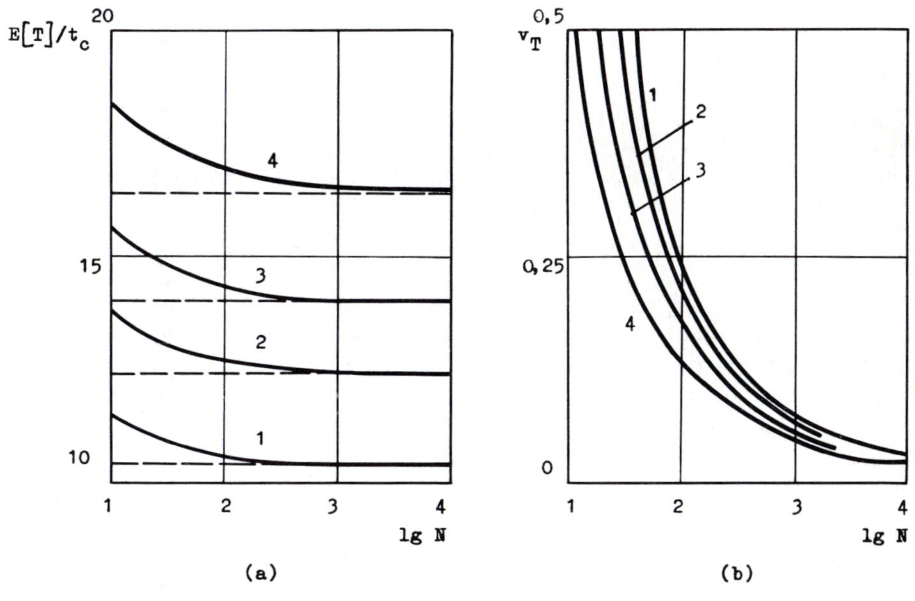

FIG. 16. Parameters of lifetime distribution as a function of the number of primary elements N for various critical damage levels ψ^* for numerical data as in Fig. 15; (a) mathematical expectation $E[T]$, (b) coefficient of variation v_T.

otherwise (this conclusion relates only to the conditional reliability function, i.e. under the given magnitudes of $\boldsymbol{r}$ and $\boldsymbol{s}$).

The scale effect of tensile strength is illustrated in Fig. 17. A qualitative relationship is shown there between the short-term ultimate stress σ and the volume V of specimens which are supposed to be geometrically similar. Three levels of reliability are considered, namely, $R = 0.99$, 0.5 and 0.01. Fig. 17(a) corresponds to the scale effect defined by (4.4). The scale effect at higher levels of reliability has an opposite tendency in comparison with the common scale effect in brittle fracture. When V increases substantially, the scale effect becomes negligible.

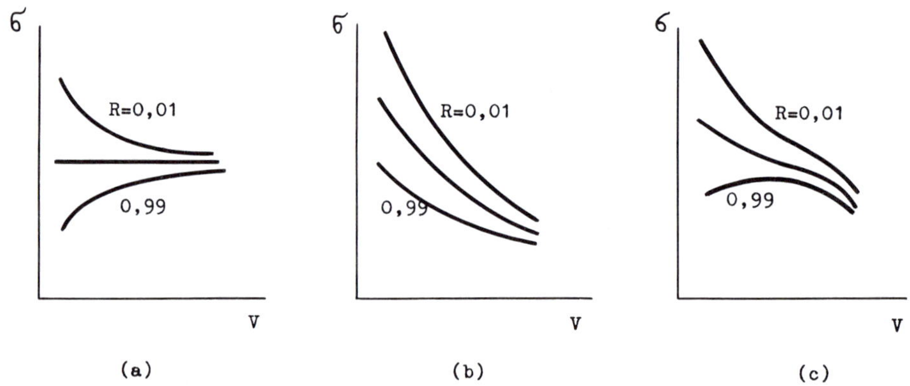

FIG. 17. Ultimate tensile strength σ versus the volume V of specimens for various reliability levels R; (a) critical volume V_c is proportional to V, (b) and (c) critical volume V_c is constant.

It was proposed by BOLOTIN (1976) that the reliability of large-scale structures failing by multiple cracking is of a mixed nature. Local failures are governed by the same law as the ones of small laboratory specimens. Failure of a structure occurs when just one of critical volumes with the order $V_c \ll V$ fails by loss of integrity. Hence, critical volumes of a structure interact in the same way as in the case of brittle fracture. The reliability function of a structure as a whole is equal to the product of partial reliability functions, and each of the latter is determined analogously to (4.4). This type of interaction is illustrated in terms of the scale effect in Figs. 17(b) and (c). The median line descends when the volume V increases because of the 'brittle' character of interaction between partial critical volumes. The curves corresponding to high-level reliability may be descending (Fig. 17(b)) and non-monotonic (Fig. 17(c)). With regard to Fig. 17(a), it can be interpreted as relating to the case when the critical volume V_c is proportional to V (Fig. 18(a)). The scale effect illustrated in Figs. 17(b) and (c) is closer to the situation when the critical volume is constant (Fig. 18(b)).

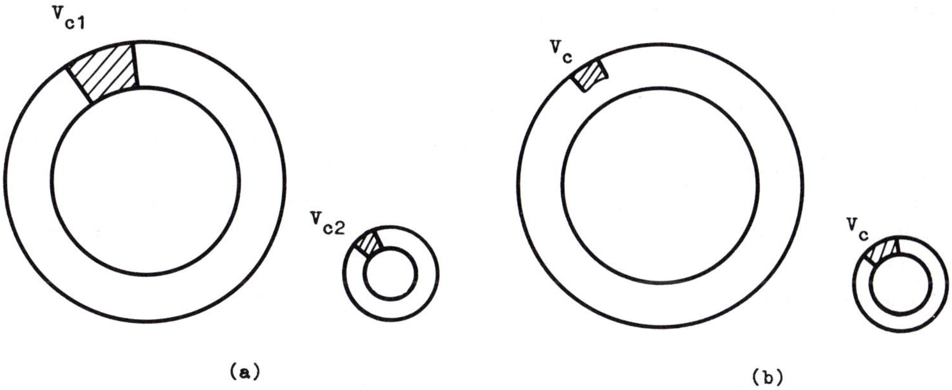

FIG. 18. Geometrically similar ring specimens; (a) critical volumes and volumes of specimens are in the ratio $V_{c1}/V_{c2} = V_1/V_2$, (b) critical volume V_c is independent of the volume of a specimen.

The order of the volume V_c responsible for the failure of a structure can be estimated only by a rather detailed experimental study. Samples of specimens and models of various dimensions including, perhaps, natural-scale prototypes ought to be tested to predict the scale effect with a high confidence level.

4.3. Loading as a stochastic process

Until now, loading was assumed to be deterministic. For a structure with a long lifetime, loading should be treated as a stochastic process. Characteristic correlation times of loading are, as a rule, small in comparison to the design life of a structure. Memory of the loading process is rather short, and this process may be considered as well-mixed. There are two ways to use this assumption. First, the technique of the theory of Markov processes is applicable. Second, the central limit theorem of the theory of probability extended to integrals of random functions may be used. Both ways were discussed in a number of papers concerning structural reliability (see survey in the book by BOLOTIN, 1984a). Under equivalent assumptions, the final results of the above methods coincide.

The loading process with the nominal fiber stress $\sigma_f \sim \sigma$ is considered. Let $\sigma(t)$ be a stochastic process and (2.2) is given in the special form

$$\mathrm{d}\varphi/\mathrm{d}t = f_\varphi(\varphi) f_\sigma(\sigma \mid r) . \tag{4.5}$$

The right-hand side contains two deterministic functions. The function $f_\varphi(\varphi)$ is positive, and function $f_\sigma(\sigma \mid r)$ non-negative. It is essential that the right-hand side of (4.5) be a product of functions separately depending on the damage measure φ and the nominal stress σ. Integrating (4.5) subject to the initial condition $\varphi = \varphi_0$ at $t = t_0$ we arrive at the equation $U(\varphi) = u(t \mid r)$ where

$$U(\varphi)=\int\limits_{\varphi_0}^{\varphi}[f_\varphi(\varphi)]^{-1}\,\mathrm{d}\varphi\,,\quad u(t\,|\,r)=\int\limits_{t_0}^{t}f_\sigma[\sigma(\tau)]\,\mathrm{d}\tau\,. \tag{4.6}$$

Functions $f_\varphi(\varphi)$ and $f_\sigma(\sigma\,|\,r)$ are supposed to provide the existence of both integrals in (4.6).

Function $u(t\,|\,r)$ is equal to the integral of a random function. The integral may be considered as a limit of a sum of almost independent random variables if the process $\sigma(t)$ is well-mixed in the interval $[t_0, t]$. Hence, the asymptotically normal distribution of the function $u(t\,|\,r)$ follows. Rigorous conditions are given by a theorem similar to the central limit theorem (IBRAHIMOV and LINNIK, 1965).

Taking into account that the function $U(\varphi)$ is monotonically increasing, an asymptotic estimate follows for the cumulative distribution function (BOLOTIN, 1977)

$$F_\varphi(\varphi;t\,|\,r)\approx\Phi\left(\frac{U(\varphi)-\boldsymbol{E}[u(t\,|\,r)]}{\{\boldsymbol{D}[u(t\,|\,r)]\}^{1/2}}\right). \tag{4.7}$$

Here $\Phi(\cdot)$ is again the standard cumulative function of a Gauss distribution defined by (2.7). The mathematical expectation and dispersion of the process $u(t\,|\,r)$ are given by

$$\begin{gathered}\boldsymbol{E}[\varphi(t\,|\,r)]=\int\limits_{t_0}^{t}\boldsymbol{E}\{f_\sigma[\sigma(\tau\,|\,r)]\}\,\mathrm{d}\tau\,,\\ \boldsymbol{D}[u(t\,|\,r)]=\int\limits_{t_0}^{t}\int\limits_{t_0}^{t}K_{f_\sigma}(\tau_1,\tau_2)\,\mathrm{d}\tau_1\,\mathrm{d}\tau_2\,.\end{gathered} \tag{4.8}$$

The covariance function $K_{f_\sigma}(\sigma_1, \sigma_2)$ enters into the second equation. To calculate the covariance function, a two-moment probability density function of the process $\sigma(t)$ must be given.

Suppose that the loading process is stationary. Then, (4.8) takes the form $\boldsymbol{E}[u(t\,|\,r)]=\mu(r)(t-t_0)$, $\boldsymbol{D}[u(t\,|\,r)]\approx\nu(r)(t-t_0)$ with the notation

$$\mu(r)=\boldsymbol{E}\{f_\sigma[\sigma(t)]\}\,,\quad \nu=2\int\limits_0^\infty K_{f_\sigma}(\tau)\,\mathrm{d}\tau\,.$$

Here the covariance function $K_{f_\sigma}(\tau)$ which depends on the time shift $\tau=\tau_1-\tau_2$ is introduced. Eq. (4.7) thus yields

$$F_\varphi(\varphi;t\,|\,r)\approx\Phi\left\{\frac{U(\varphi)-\mu(r)(t-t_0)}{[\nu(r)(t-t_0)]^{1/2}}\right\}. \tag{4.9}$$

Similar equations can be obtained using the technique of the theory of Markov processes. In such a case, the random function $\sigma(t)$ should be approximated by an appropriate 'white noise'.

To calculate the conditional cumulative function $F_\tau(\tau \mid r)$ of the lifetime of primary elements, one must set $\varphi = 1$, $t = \tau$ in (4.7) and (4.9). Then, $F_\tau(\tau \mid r) = 1 - F_\varphi(1, \tau \mid r)$. For example, it follows from (4.9) that at $t_0 = 0$

$$F_\tau(\tau \mid r) \approx 1 - \Phi\left\{\frac{1 - \mu(r)}{[\nu(r)\tau]^{1/2}}\right\}. \tag{4.10}$$

It is taken into account that the function $f_\varphi(\varphi)$ entering (4.5) can always be chosen from the condition $U(1) = 1$.

The unconditional cumulative function of the lifetime of primary elements is given by

$$F_\tau(\tau) = 1 - \int_0^\infty F_\varphi(1; \tau \mid r) p_r(r)\, \mathrm{d}r\,, \tag{4.11}$$

where the probability density $p_r(r)$ of the strength parameter r is introduced. The function defined by (4.11) is to be substituted into (4.4), etc.

Eq. (4.11) includes two sources of randomness: the scatter of fiber strength and the scatter of loading. If the strength parameters were deterministic, $F_\tau(\tau) \equiv F_\tau(\tau \mid r)$. On the other hand, when the scatter of loading is negligible, (4.9) degenerates into $F_\varphi(\varphi, t \mid r) = \eta\{t - E[u(\tau \mid r)]\}$ with Heaviside function $\eta(\cdot)$. Subsequently, $F_\tau(\tau) = F_r[r_0(\tau)]$ where $r = r_0(\tau)$ is the equation $E[u(\tau \mid r)] = 1$ solved with respect to r. The latter case was discussed in Subsection 3.2. Generally, the scatter of fiber strength is more important, especially when the considered time intervals are very large in comparison to the characteristic correlation time τ_c. In the case of (4.10), the coefficient of variation of the conditional lifetime is of the order of $\nu(r)/\mu(r)$. It is easy to show that in this special case $\nu(r)$ is of the order of τ_c, and $\mu(r)$ is of the order of the characteristic lifetime. Further details and more complicated examples can be found in the book by BOLOTIN (1984a).

One should emphasize that the stochastic character of loading discussed above is to be understood in a limited sense. Only the scatter between different samples of the same random process was taken into account. A large group of random factors, such as scatter of general loading and environment conditions, lack of information and uncertainties should be included into the reliability analysis separately, as was done in (1.2) and (1.3) by means of the vector $\boldsymbol{s}$.

4.4. Failure as the first excursion event

It has been assumed until now that the critical damage measure ψ^* is a given deterministic variable. If it is a random variable, no additional difficulties arise since the scatter in ψ^* may be included in the vector $\boldsymbol{r}$ characterizing all the stochastic material and structural properties. Consequently (4.3) is also applicable in a broader sense.

The situation changes substantially when the critical damage measure depends on the nominal stress $\sigma(t)$ at a considered instant of time. In fact, this dependence can be very strong as was shown, for example, in Fig. 9. Loss of integrity occurs when the inequality $\psi^*[\sigma(t)] - \psi(t) > 0$ is violated. Hence, the reliability function is

$$R(t) = \boldsymbol{P}\{\psi^*[\sigma(\tau)] - \psi(\tau) > 0\,; \quad \tau \in [t_0, t]\}\,. \tag{4.12}$$

Compared to (4.1), (4.12) relates the reliability function to the notion of a first excursion of a random process. For example, using (1.1) we may introduce a quality process as $v(t) = \psi(t) - \psi^*[\sigma(t)]$. Then, the admissible region is given by $\Omega = \{v : v < 0\}$. The damage process $\psi(t)$ is quasi-monotonically increasing and, therefore, a cumulative one. If the level of loading is constant or quasi-monotonically increasing, then the critical damage measure $\psi^*[\sigma(t)]$ will be constant or quasi-monotonically decreasing. The difference $\psi(t) - \psi^*[\sigma(t)]$ will be quasi-monotonically increasing. This case is illustrated in Fig. 19(a). Since no return to the admissible region is possible under these conditions, (4.12) may be replaced by the much simpler (4.1).

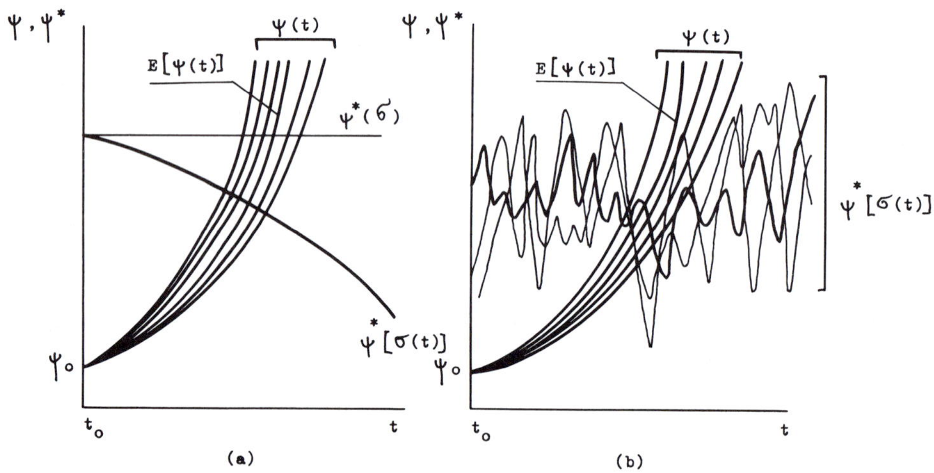

FIG. 19. Failure in the loss of integrity as the first crossing of processes $\psi(t)$ and $\psi^*[\sigma(t)]$; (a) nominal stresses σ are constant or quasi-monotonically increasing in time, (b) $\sigma(t)$ is an oscillation process.

If the loading process is non-monotonic, the critical damage measure is non-monotonic, as well. Consequently (4.1) is inapplicable, and we must consider the problem of the first excursion in its whole scope (Fig. 19(b)). Additional complications arise because stochastic factors from various sources enter into the quality process $v(t) = \psi(t) - \psi^*[\sigma(t)]$ jointly. An approach to overcome this difficulty has been suggested by BOLOTIN (1980). This approach has been developed especially for random vibrations and cyclic fatigue problems. Let the process $\sigma(t)$ be an oscillatory one, and its characteristic time (e.g. characteristic duration of a single cycle) very small in comparison to the lifetime of a structural component. We may then distinguish between the 'slow' time when dealing with the process $\psi(t)$, and the 'rapid' time during which the process $\sigma(t)$ and, therefore, $\psi^*[\sigma(t)]$ are evolving. Instead of upward crossings of the zero level by the process $\psi(t) - \psi^*[\sigma(t)]$, we consider downward crossings of the level $\psi(t)$ by the process $\psi[\sigma(t)]$. We subsequently apply the Poissonian model of failure (see Subsection 1.6). The mathematical expectation $\lambda(t)$ of the number of downward crossings per time unit is to be calculated in two steps. On the first step, we find the conditional mathematical expectation $\lambda(t | \psi)$ of the number of downward crossings of the given level ψ. Since the process $\psi(t)$ is very slow, time t may be treated as a parameter at this stage of calculation. Using (1.17), we obtain

$$\lambda(t | \psi) = \int_0^\infty p[\psi(t), \dot{\chi}; t] \dot{\chi} \, \mathrm{d}\dot{\chi} \, . \tag{4.13}$$

Here, $p(\chi, \dot{\chi}; t)$ is the joint probability density of the process $\chi(t) \equiv \psi^*(t)$ and its derivative $\dot{\chi}(t)$ depends parametrically on t. To find this function, the corresponding probability density $p_\sigma(\sigma, \dot{\sigma}; t)$ of the process $\sigma(t)$ and the relationship $\psi^*(\sigma)$ must be known. The second step is the averaging of the conditional mathematical expectation $\lambda(t | \psi)$ over all values of the slow process $\psi(t)$, Then

$$\lambda(t) \approx \int_0^1 \lambda(t | \psi) p_\psi(\psi; t) \, \mathrm{d}\psi \, , \tag{4.14}$$

where $p_\psi(\psi; t)$ is the corresponding probability density defined, for example, by (1.6) or (1.9).

4.5. Nonhomogeneity of stress and/or strength fields

The notion of the critical volume allows one to include nonhomogeneities into the reliability analysis. Suppose that the stress and/or strength parameters vary slowly in the volume V depending on the reference vector $\boldsymbol{x}$. Assume that at $V_c \ll V$ variation of the fields within each critical volume V_c is negligible. Then, (4.4) and analogous equations may be applied to each critical volume separately considering the fields within each volume to be dependent of x. Denote the total number of critical volumes by N_c. Ordering critical volumes $V_1, V_2, \ldots,$ we calculate their reliability functions $R_1(t), R_2(t), \ldots,$ respectively. The reliability function corresponding to the volume V is

$$R(t) = \prod_{k=1}^{N_c} R_k(t) . \tag{4.15}$$

In fact, (4.15) was used in the qualitative discussion of the scale effect in conjunction with Figs. 17(b) and (c).

Structures should be highly-reliable, and failure of even one of the critical volumes ought to be a rare event. To obtain an asymptotic estimate of the reliability function, we suppose that the failed volumes form a Poissonian set among the total number of samples in the volume V. It is expedient to discuss the problem in terms of hazard functions introduced by (1.5). The hazard function $H(t) = 1 - R(t)$ is equal to the probability that even just one of the critical volumes has failed. This probability depends only on the mathematical expectation $\boldsymbol{E}[N(t)]$ of the number of failure critical volumes, if the hypothesis of Poisson distribution is accepted:

$$H(t) = 1 - \exp\{-\boldsymbol{E}[N(t)]\} . \tag{4.16}$$

It is easy to see that

$$\boldsymbol{E}[N(t)] = \frac{1}{N_c} \sum_{k=1}^{N_c} H_k(t) , \tag{4.17}$$

where $H_k(t) = 1 - R_k(t)$. As $V_c \ll V$, the sum in (4.17) may be approximated by an appropriate integral over the volume V. Eqs. (4.16) and (4.17) result in

$$R(t) \approx \exp\left[-\int_V H_c(t, x)\, dV/V_c(x)\right] . \tag{4.18}$$

Here $H_c(t, x)$ and $V_c(x)$ are hazard functions and critical volumes, respectively, considered as functions of the continuous vector argument $\boldsymbol{x}$.

The right-hand side of (4.18) is the complementary function of a far-reaching generalization of the Weibull distribution. If all the fields under consideration are homogeneous and all the critical volumes equal, (4.18) takes the form

$$R(t) \approx \exp[-(V/V_c)H_c(t)] , \tag{4.19}$$

which is closer to the Weibull distribution that the general equation. Eq. (4.19) shows, in a very explicit form, the scale effect of reliability. The calculations leading from (4.15) to (4.18) and (4.19) are analogous to those in the statistical theory of brittle fracture (BOLOTIN, 1961).

Only the lower tails of the hazard functions carry a significant contribution to the reliability functions defined by (4.18) and (4.19). When (4.4) and analogous equations containing the Gauss distribution function are considered, it is expedient to replace the function $\Phi(z)$ with its asymptotic expansion

$$1-\Phi(z)=\frac{1}{(2\pi)^{1/2}u}\exp\left(-\frac{u^2}{2}\right)\left(1-\frac{1}{u^2}+\frac{1\times 3}{u^4}-\cdots\right). \tag{4.20}$$

Since for highly-reliable structures $u^2 \ll 1$, the first term of the expansion gives good numerical accuracy. Generally, asymptotic approximations are of use even when using computers since they provide better numerical accuracy for highly-reliable structures than the 'exact' solutions.

4.6. Reliability with respect to brittle fracture

The cumulative distribution function of the time up to brittle fracture for the homogeneous case is obtained from (2.13) by replacing n^* with n^{**}. In terms of the reliability functions we obtain

$$R(t)\approx \exp[-N\psi^{n^{**}}(t)]\,. \tag{4.21}$$

Here $\psi(t)$ is considered a deterministic process, i.e. (2.9) is used. In more general cases, (4.21) defines a conditional reliability function which should be further treated according to (1.3).

Let us apply (4.21) to a unidirectional fibrous composite under quasistatic tension having a nominal stress $\sigma(t)$. Rupture of fibers is described by (2.25) and (2.34). The process $\psi(t)$ is determined by (3.2)–(3.4), etc. depending on the type of loading. For example, (3.3) is valid for $\sigma=$ constant and sufficiently small values of ψ. Substituting (3.3) in (4.21), the reliability function takes the form

$$R(t)\approx \exp\{-(V/V_0)[(\sigma-r_0)/r_c]^{\alpha n^{**}}[(t-t_0)/t_c]^{\beta n^{**}}\}\,. \tag{4.22}$$

The volume V_0 is chosen from condition $V/V_0=N(\lambda_e/\rho)^{n^{**}}$. Eq. (4.22) shows that both the tensile strength of a specimen and the times up to brittle fracture follow the Weibull distributions. The shape parameters of these distributions are αn^{**} and βn^{**}, respectively. Thus the shape parameters differ from those of the initial distribution of fiber strength and lifetime by the factor n^{**}. The latter is equal to the number of the ruptured neighboring fibers responsible for brittle fracture of the whole specimen.

Eq. (4.22) is similar to (2.26) which defines the cumulative distribution of time up to the macrocrack initiation in a granular composite. The conclusions concerning that case remain valid in the case of brittle fracture, too. In particular, the relationships between the mathematical expectations of tensile strength (for the same duration of loading), and the times to fracture (for the same loading level) are

$$\begin{aligned} \boldsymbol{E}[\sigma_1]/\boldsymbol{E}[\sigma_2]&\approx (V_2/V_1)^{1/\alpha n^{**}}\,,\\ \boldsymbol{E}[t_1]/\boldsymbol{E}[t_2]&\approx (V_2/V_1)^{1/\beta n^{**}}\,. \end{aligned} \tag{4.23}$$

Here V_1 and V_2 are volumes of the compared specimens, and condition $r_0 \ll r_c$ is used. The exponent m is larger than unity, let us say, $m = 4$ and can be even higher. It follows from (2.25) that $\beta < \alpha$ at $m > 1$. Therefore, the scale effect of longevity is more significant than that of tensile strength.

Similar conclusions follow when statistical scatter is studied. The coefficient of variation of tensile strength and that of time to brittle fracture at $r_0 \ll r_c$ are

$$
\begin{aligned}
v_\sigma &\approx \{[\Gamma(1 + 2/\alpha n^{**})/\Gamma^2(1 + 1/\alpha n^{**})] - 1\}^{1/2}, \\
v_t &\approx \{[\Gamma(1 + 2/\beta n^{**})/\Gamma^2(1 + 1/\beta n^{**})] - 1\}^{1/2}.
\end{aligned}
\qquad (4.24)
$$

The first variable depends on αn^{**} only, and the second on βn^{**}. As α and β are connected by (2.25), (4.24) may be interpreted as a parametric relationship between v_σ and v_t. The exponent m from (2.25) enters into this relationship as a parameter. The relationship between v_σ and v_t, as it follows from (4.24), is shown in Fig. 20. Values of m are indicated near the curves. The scatter of longevity is more significant than the scatter of tensile strength for all $m > 1$. This difference grows with an increase in the value of m. This fact, of course, is well-known from common fatigue experiments (both static and cyclic), and can be seen directly from Woehler diagrams.

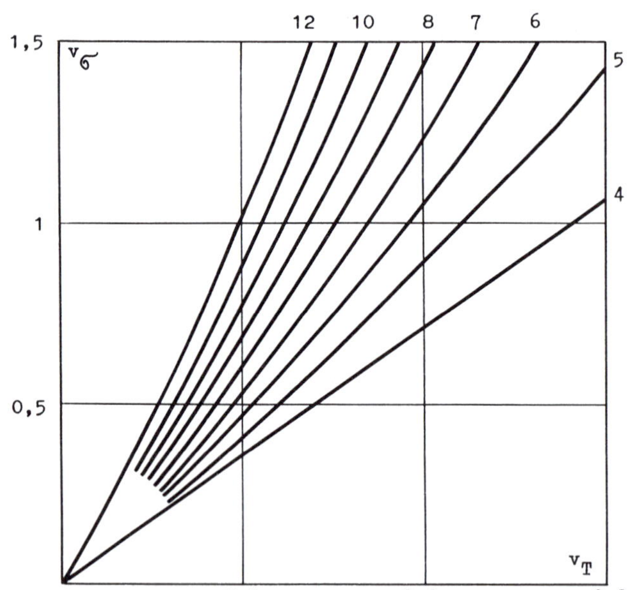

FIG. 20. Relationship between coefficients of variation of lifetime v_T and ultimate stress v_σ for various m.

The scale effect and statistical scatter for fibrous composites are not as strong as for the corresponding fibers. This follows directly from (4.22)–(4.24). In fact, $n^{**} > 1$, and typically $n^{**} = 4$ and even more (SMITH et al., 1983). For example, for $\alpha = m = n^{**} = 4$, the coefficient of variation of fiber tensile strength $v_r = 0.28$, and that for a sufficiently large specimen of composite $v_\sigma = 0.08$.

Analogous results are valid for macrocrack initiation when this event is considered a structural failure. However, if the assumed size of the nucleus corresponds to a stable macrocrack (and apparently this case is more typical), subsequent propagation of the macrocrack should be included.

4.7. *Reliability with respect to crack growth*

Composite structures, generally, contain a rather large number of cracks and crack-like defects whose evolution eventually leads to failure. Therefore, a single crack is quite an idealization. The reliability analysis in the presence of a random set of cracks will be developed in Section 6. Comparatively simple cases are discussed in the Subsection concerning a single crack in a specimen of a structural component in order to identify the typical statements of reliability problems with respect to crack growth.

Let the nominal stress–strain state and material properties be homogeneous in the volume V and loading a one-dimensional process which, for instance, is specified by the nominal tensile stress $\sigma(t)$. A crack is described by means of the characteristic size l, let us say, the half-length of a Griffith crack or the radius of a cylinder corresponding to brush-like cracked zone in a unidirectional fibrous composite.

To statements of the problem are to be distinguished from the standpoint of reliability. If the limit size l_{**} of a crack has a given deterministic value, the reliability function is

$$R(t) = \boldsymbol{P}\{l(t) < l_{**}\} . \tag{4.25}$$

This definition is similar to that of (4.1). It reflects the fact that the crack growth is irreversible and, therefore, a cumulative process. From the general point of view, (4.25) corresponds to (1.10) and (1.12).

If failure is the result of the crack's instability in the sense of linear fracture mechanics, (4.25) should be replaced by the following

$$R(t) = \boldsymbol{P}\{l(\tau) < l_{**}[\sigma(\tau)] \, ; \quad \tau \in [t_0, t]\} . \tag{4.26}$$

The critical size l_{**} depends on the loading level at time t. Under random loading $l_{**}(t)$ is a stochastic process. The definition of the reliability function given by (4.26) is similar to that given by (4.12) or, speaking in terms of general notions, to that given by (1.1). Thus, when the criterion of stability is stated, we arrive at the problem of the intersection of two random processes.

One of those is the crack growth process $l(t)$, and the second is the process of the variation of the critical size $l_{**}[\sigma(t)]$ with time. To transform this problem into that of the first excursion, it is sufficient to introduce the process $v(t) = l(t) - l_{**}[\sigma(t)]$ with the admissible values located on the negative semi-axis. The condition of reliability is also definable in terms of stresses. The admissible region obtained from (1.1) corresponds to the condition $\sigma(t) < \sigma_{**}[l(t)]$, where σ_{**} is the critical stress depending on the size of the crack at the considered moment. The critical stress characterizes the residual strength of the cracked structural component.

Analogous problems were discussed in Subsections 4.1–4.4 where failures were associated with the loss of integrity. Approximate approaches developed in those Subsections may be applied to estimate the reliability function defined by (4.26). In fact, when the stress level is constant or increasing in time, the difference $v(t) = l(t) - l_{**}[\sigma(t)]$ is a cumulative process, and (4.26) may be replaced by (4.25). When the stress level is nonmonotonic, an oscillating function of time for instance, it is expedient to estimate $R(t)$ by considering downward crossings of the slowly varying level $l(t)$ by the rapid process $l_{**}[\sigma(t)]$ and, reciprocally, upward crossings of the slowly varying level $\sigma_{**}[l(t)]$ by the rapid process $\sigma(t)$ (Fig. 21). The central limit theorem generalized to integrals of random functions is of use when the loading process is sufficiently well-mixed.

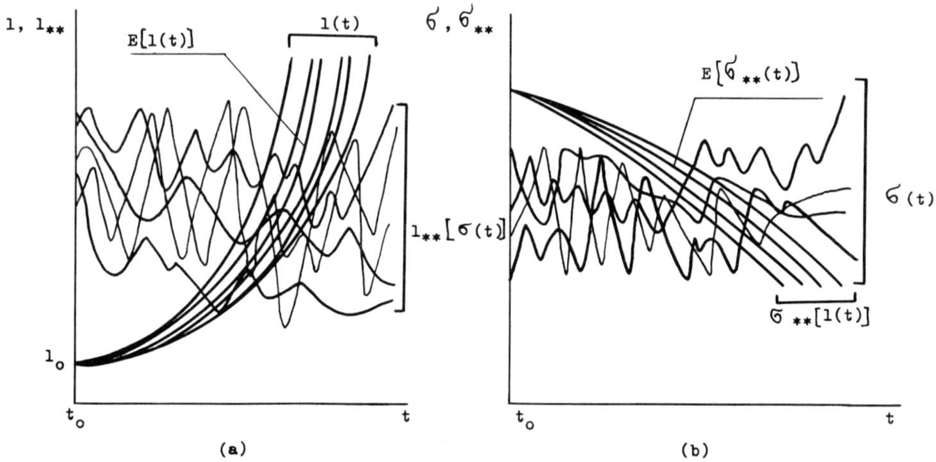

FIG. 21. Crack instability as the first crossing event of two stochastic processes; (a) in terms of crack sizes, (b) in terms of stresses.

Equations governing crack growth should be known in order to calculate the probabilities on the right-hand side of (4.25) and (4.26). Semi-empirical equations of fracture mechanics are of use when conventional structural materials are concerned. Special analytical and experimental studies are, as a rule, required to develop equations of crack in composites. As examples of such equations, (3.21) and (3.22) describing the propagation of a brush-like transverse crack in unidirectional fibrous composites can be mentioned. Other examples have been presented by BOLOTIN (1984b) concerning the growth of delaminations in composites of laminated construction. The general form of all one-parameter equations is the same. In the case of static fatigue one obtains

$$\mathrm{d}l/\mathrm{d}t = f(\sigma, l)\,. \tag{4.27}$$

A finite-difference analogue of (4.27) is used in the case of cyclic fatigue, i.e.

$$l_n - l_{n-1} = f_1(\sigma_n, l_{n-1})\,, \quad n = 1, 2, \ldots\,. \tag{4.28}$$

Here, n is the number of cycles of the standard segments of the loading process. Eqs. (4.27) and (4.28) are of the same structure as (2.2) and (2.3), respectively, and their right-hand sides satisfy analogous conditions. In particular, (4.28) may be approximated by (4.27) upon replacing t by the continualized argument n when the right-hand side of (4.28) is a slowly varying function of n.

Generally, dispersed damage should be taken into account since the processes of crack growth and damage accumulation interact. The right-hand side of (4.27) will depend in such a case on the damage vector $\boldsymbol{\psi}$. This means that the equation of crack growth has to be treated jointly with the equation of dispersed damage accumulation. This idea has been developed by BOLOTIN (1984b) with special applications to crack propagation in laminated composites. It was shown that, under certain conditions, (4.27) may be interpreted to include dispersed damage in an inexplicit form. From this point on, we will limit our attention in this Subsection to (4.27) since development of results for other cases is more or less evident.

One of the central points of the discussion is the formulation of initial conditions. If the initial size l_0 of a crack is given deterministically, the initial condition is $l(t_0) = l_0$. This condition is typical for laboratory fatigue studies of using notched specimens. Nevertheless, the deterministic initial condition can also be met in engineering practice, e.g. when the size of a crack is determined by inspection, thus allowing one to predict the residual strength or the residual lifetime of a structural component. When the loading may be assumed to be deterministic, the problem of crack growth becomes completely deterministic, too. The solution of (4.27) satisfying the initial condition is thus

$$l = L(t \,|\, l_0, t_0)\,. \tag{4.29}$$

Here $L(t \,|\, l_0, t_0)$ is a deterministic function of time. The initial data are given to the right of the vertical line. This notation is useful since conditional probabilities will enter into subsequent analysis.

When both $l(t)$ and l_{**} are strictly deterministic, the reliability function defined by (4.25) is either equal to unity or to zero. More precisely,

$$R(t \mid l_0, t_0) = \eta[l_{**} - L(t \mid l_0, t_0)] , \tag{4.30}$$

where $\eta(\cdot)$ is again the Heaviside function.

The next step in the course of generalization is to assume that initial crack sizes are randomly distributed with a given probability density $p_0(l_0)$. Using the solution in the form of (4.29), the probability density $p(l; t)$ of crack sizes for $t > t_0$ is easily calculated. The relationship between the two probability density functions is illustrated in Fig. 22. In terms of reliability functions we obtain

$$R(t) = \int_{l_*}^{l_{**}} R(t \mid l_0, t_0) p_0(l_0) \, dl_0 . \tag{4.31}$$

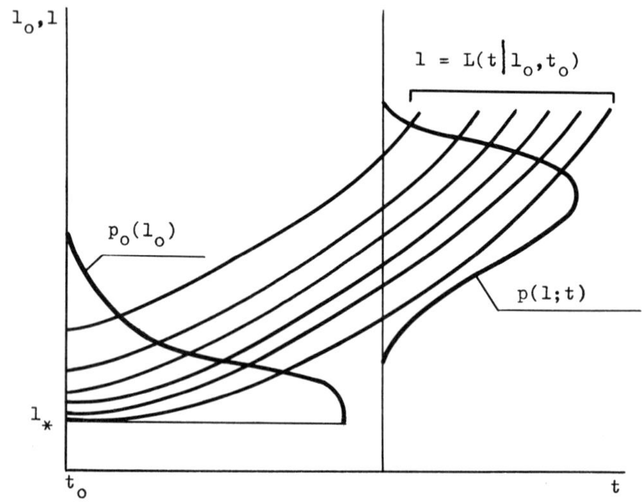

FIG. 22. Stable growth of a random crack with probability densities $p_0(l_0)$ at $t = t_0$, $p(l; t)$ at $t > t_0$; $L(t \mid l_0, t_0)$ is a deterministic solution of the equation of crack growth for $l(t_0) = l_0$.

The right-hand side of (4.31) may be re-written, using the cumulative distribution function $F_0(l_0)$, as follows: $R(t) = F_0[L_0(l_{**}, t, t_0)] - E_0(l_*)$. Here, $l_0 = L_0(l, t, t_0)$ is the solution of (4.29) with respect to l_0.

When no initial crack is present at $t = t_0$, the entire life of a structural component can be divided into two stages. The major part of the components life relates, as a rule, to the stage during which dispersed damage accumulates, and only a minor part is ascribed to macrocrack development. For conventional metallic materials, such as steel, the first stage takes up 80% and more of the time to final rupture. Both stages have to be taken into account in the reliability analysis if macrocracks are supposed to be nucleated during the service of a structure. In terms of (4.27) this means that initial conditions are stochastic even if the initial size l_* is supposed to have a given deterministic value. Indeed, the initial condition is $l(t_*) = l_*$ where t_* is a random time for the crack's initiation. The cumulative distribution function $F_*(t_*)$ of times for macrocrack initiation discussed in Subsections 2.5 and 3.8 becomes of significance since this function enters into stochastic initial conditions. This is illustrated in Fig. 23 in terms of the probability densities $p_*(t) = F'_*(t_*)$ and $p_{**}(t_*) = F'_{**}(t_{**})$.

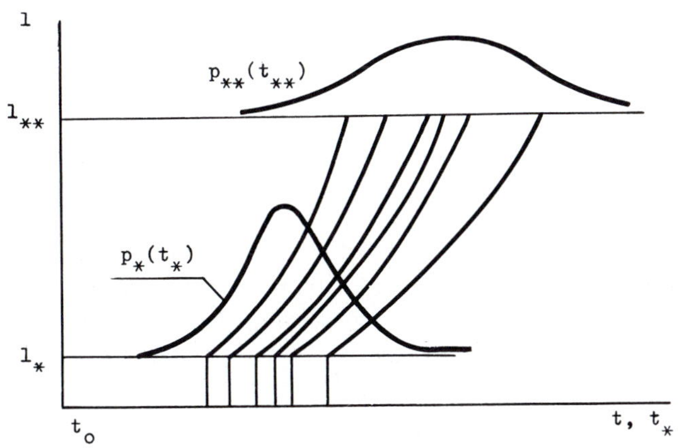

FIG. 23. Stable growth of cracks randomly initiated under loading; $p_*(t_*)$ and $p_{**}(t_{**})$ are probability densities of times up to crack initiation and crack instability, respectively; the critical size l_{**} is assumed constant.

The conditional reliability function for the case of a crack initiated at time t_* is given by (4.30) upon replacing l_0 by l_*, and t_0 by t_*. The unconditional reliability function is

$$R(t) = = \int_{t_0}^{t} R(t \mid l_*, t_*) p_*(t_*) \, \mathrm{d}t_* \,. \tag{4.32}$$

Similarly to (4.31), it follows from (4.32) that $R(t) = F_*[T_0(l_{**}, l, t)] - F_*(t_0)$. Here $t_0 = T_0(l, l_0, t)$ is the solution of (4.29) with respect to t_0. Since at $t = t_0$ the initial macrocrack is absent, $F_*(t_0) \equiv 0$.

Eq. (4.32) is illustrated in Fig. 24. The complementary function $1 - F_*(t_*)$ is, in fact, the reliability function for the condition that failure is associated with the initiation of a macroscopic crack. The instant of crack instability is denoted by t_{**}, and the difference $t_{**} - t_*$ is equal to the life of a specimen or structural component in the presence of a stable crack.

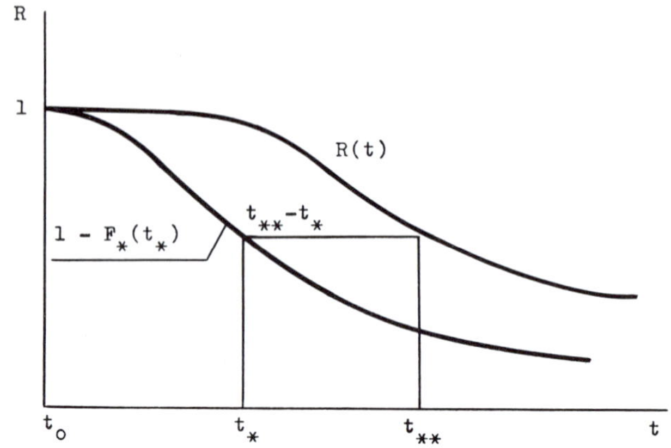

FIG. 24. Relationship between the reliability function $R(t)$ with respect to crack instability, and the cumulative distribution function $F_*(t_*)$ of times up to crack initiation; $t_{**} - t_*$ is equal to the duration of the crack stable growth stage.

In the general case, both possibilities should be included in the study, namely: one that a crack of a random size l_0 is present at $t = t_0$; and that a crack of a given size l_* is initiated at $t = t_* > t_0$. If these events are incompatible, we obtain the resulting reliability function using (4.31) and (4.32). However, this assumption appears to be too strong since macrocracks of both origins can be simultaneously present in a structural component. In such a situation, the concept of a single crack becomes invalid, and we must turn to random sets of cracks (see Section 6).

The case of deterministic loading has been discussed up to this point. If loading is stochastic, the solution of (4.27) should be presented in the form of the conditional cumulative distribution function $F_l(l, t \mid l_0, t_0)$. Here l is the size of a crack for $t > t_0$ under the condition that $l(t_0) = l_0$. Since $R(t \mid l_0, t_0) = 1 - F_l(l_{**}, t \mid l_0, t_0)$, (4.31) has to be replaced by

$$R(t) = 1 - \int_{l_*}^{l_{**}} F_l(l, t \mid l_0, t_0) p_0(l_0) \, \mathrm{d}l_0 \, . \tag{4.33}$$

Similarly, instead of (4.32) we take

$$R(t) = 1 - \int_{t_*}^{t} F_l(l_{**}, t \mid l_*, t_*) p_*(t_*) \, \mathrm{d}t_* \, . \tag{4.34}$$

From the application point of view, well-mixed loading processes are of interest. As in Subsection 4.4, we define the process $\sigma(t)$ to be well-mixed if its correlation time τ_c is small in comparison to all considered $t - t_0$. The central limit theorem is also applicable for this case, and, furthermore, it is possible to obtain a semi-deterministic approximation of the conditional process $l(t \mid l_0, t_0)$.

Let (4.27) be of the form

$$\mathrm{d}l/\mathrm{d}t = f_l(l) f_\sigma(\sigma) . \tag{4.35}$$

Eq. (4.35) contains, as special cases, equations of crack growth in composites such as (3.20) and (3.22), etc, and is analogous to (4.5) describing the damage accumulation process. Functions on the right-hand side of (4.35) should satisfy the same conditions as those of (4.5). The solution of (4.35) subject to the initial condition $l(t_0) = l_0$ is $U(l \mid l_0) = u(t \mid t_0)$. Here, similarly to (4.6)

$$U(l \mid l_0) = \int_{l_0}^{l} [f_l(l)]^{-1} \, \mathrm{d}l \; ; \quad u(t \mid t_0) = \int_{t_0}^{t} f_\sigma[\sigma(\tau)] \, \mathrm{d}\tau . \tag{4.36}$$

Under known conditions (IBRAHIMOV and LINNIK, 1965) integrals of well-mixed random functions may be assumed to have asymptotically normal distributions. It follows then, that

$$f_l(l, t \mid l_0, t_0) \approx \Phi\left(\frac{U(l \mid l_0) - \boldsymbol{E}[u(t \mid t_0)]}{\{\boldsymbol{D}[u(t \mid t_0)]\}^{1/2}} \right) \tag{4.37}$$

with notations similar to those used in (4.7) and (4.8).

The coefficient of variation of the function $u(t \mid t_0)$ decreases when $t - t_0$ inçreases. For example, if $\sigma(\tau)$ is a stationary ergodic process, then $\tau_c =$ constant, and the coefficient of variation is of the order of constant $\times [\tau_c/(t - t_0)]^{1/2}$. Hence, further approximation follows

$$P(l, t \mid l_0, t_0) \approx \eta\{U(l \mid l_0) - \boldsymbol{E}[u(t \mid t_0)]\} . \tag{4.38}$$

In other words, the scatter of samples of the conditional process $l(t \mid l_0, t_0)$ is assumed to be very small even if the scatter of loading samples is significant. Eq. (4.38) is equivalent to $l(t \mid l_0, t_0) \approx U^{-1}\{\boldsymbol{E}[u(t \mid t_0)]\}$ where $U^{-1}\,\{\cdot\}$ is the inverse function corresponding to the function $U(l)$ defined by (4.36). This semi-deterministic approximation is of use when other significant sources of randomness are involved, for example, when loading is a rather scattered oscillating process. The critical size $l_{**}[\sigma(\tau)]$ entering (4.26) depends on the current stress $\sigma(\tau)$, which is being a scattered oscillating process, too. This was illustrated in Fig. 21.

Reliability functions introduced by (4.25) and (4.26), etc., correspond to a structural component with given properties and to a loading process with given probabilistic parameters. These functions are, therefore, conditional ones in the sense of (1.2). In order to take into account the uncertainties and statistical scatter of mechanical properties, loading and environmental conditions, an additional group of random parameters should be included in study by means of the vectors $\boldsymbol{r}$ and $\boldsymbol{s}$ entering (1.2) and (1.3). This remark relates, obviously, to all the problems discussed in this Section.

5. Unidirectional fibrous composites in compression

5.1. Modes of failure in compression

Local buckling of fibers is usually assumed to be the main mechanism of fracture of unidirectional fibrous composites subjected to compression along the fiber direction. This assumption, when applied to the case of a sufficiently pliable matrix, implies that the failure stresses are of the order of the shear modulus of the matrix. The simplest equation for the critical stress, when the shear (coherent) mode of local instability is supposed, is

$$\sigma_*^0 \sim (1 - v_f)^{-1} G_m . \tag{5.1}$$

Assuming an anti-coherent mode of local buckling we obtain

$$\sigma_*^{tr} \sim [(1 - v_f)^{-l} v_f E_f E_m^{tr}]^{1/2} , \tag{5.2}$$

where E_m^{tr} is the transverse Young's modulus of the matrix taking the fiber–matrix interaction and Poisson's effect into account.

If the fiber volume ratio v_f is not sufficiently small, the critical stress estimated by (5.1) is smaller than that given by (5.2). In such a case, the shear mode should take place (Fig. 25(a)). However, this conclusion does not seem sufficiently convincing. Critical stresses for very short columns are close to the value of the bulk shear modulus, and, therefore, are of the same order as the critical stress of fiber buckling in the coherent mode (CHOU and KELLY, 1980). A more detailed study by PIGGOT and HARRIS (1980) shows that compression strength depends on the fiber volume ratio in a different way from that predicted by (5.1). In particular, when v_f is large, and $E_m \ll E_f$, the compressive strength approaches that predicted by the common 'rule-of-mixtures'. The

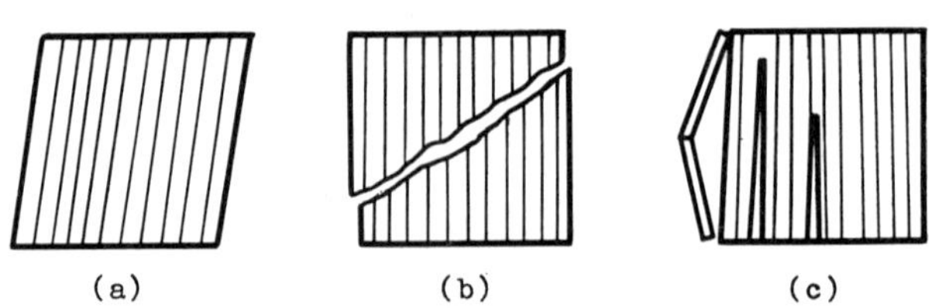

FIG. 25. Modes of failure of unidirectional fibrous composites in compression; (a) shear instability, (b) skew rupture, (c) delamination and splitting.

failure of a specimen with low-strength fibers is accompanied by the rupture of fibers in compression. The patterns of failure become similar to those observed in the testing of wood specimens (Fig. 25(b)). Shearing at angles of the order of 45° frequently occurs as well as delamination resulting in splitting of the specimen (Fig. 25(c)).

The situation becomes more complicated when thermal compression stresses are involved. Additional effects appear in laminated composites due to anisotropy and inhomogeneity of mechanical and thermal properties, especially when the properties vary substantially from layer to layer.

In order to increase the reproducibility of compression tests, a number of special testing methods have been proposed. The survey of these methods is presented by TARNOPOL'SKI and KINCIS (1981). Nevertheless, there are reasons for the statistical scatter of experimental results due to the intrinsic nature of fracture of composites in compression. The probabilistic model proposed by BOLOTIN (1982) is based on the assumption that buckling of fibers is responsible for failure in compression. Contrary to conventional deterministic models dealing with initially straight fibers, a composite with imperfect fibers is considered. It is supposed that the axes of fibers deviate initially from the ideal straight form. These deviations are the result of imperfections in the fabrication process, and they are assumed to be random functions of the coordinates along fibers. Buckling of fibers occurs from the beginning of loading. Bending stresses in the fibers and shear stresses in the matrix or fiber–matrix interface increase when the buckling process develops. At a certain level, bending stresses result in the rupture of the fibers, and shear stresses result in damage to the matrix or the interface. To judge the load-carrying capacity and reliability of a composite in compression, it is necessary to know the density of fiber ruptures as well as the portion of the total fiber length with respect to the failed interface. The probabilistic distribution of the damage length is also of interest, since the maximum damage length characterizes to a certain degree the damage to the composite as a whole, and can be detrimental from the standpoint of integrity. The following analysis is of a qualitative nature only. Estimates of orders of magnitude will be widely used as well as other simplifications introduced in an ad hoc fashion.

5.2. Buckling of fibers with initial imperfections

Let us consider a specimen or a structural component subjected to compression along the fiber direction. Assume that every fiber has small deviations $u_0(z)$ and $v_0(z)$ from the perfect straight form. These deviations represent samples of homogeneous ergodic random functions. The character of correlation between the imperfections of adjacent fibers is of special interest. Consider two extreme cases: first, when the fibers are almost equidistant, and, secondly, when no correlation exists between the imperfections of adjacent fibers. The first case is typical of composites fabricated by winding with yarns, bundles or tapes containing a large number of fibers. During compression of these composites the shear (coherent) patterns of buckling dominate. The second case is more typical for composites wound with single fibers and having comparatively low fiber volume fractions. Since adjacent fibers are deformed almost independently, the patterns of failure are close to those corresponding to out-of-phase fiber buckling.

Let us express the random functions $u_0(z)$ and $v_0(z)$ in the form

$$u_0(z) = \int_{-\infty}^{\infty} U_0(k)\,\mathrm{e}^{ikz}\,\mathrm{d}k\,, \quad v_0(z) = \int_{-\infty}^{\infty} V_0(k)\,\mathrm{e}^{ikz}\,\mathrm{d}k\,. \tag{5.3}$$

Generalized random functions $U_0(k)$ and $V_0(k)$ of the wave number k are connected with spectral densities $S_{u_0}(k)$ and $S_{v_0}(k)$ as follows:

$$\langle U_0^*(k)U_0(k')\rangle = S_{u_0}(k)\delta(k-k'),\ \langle V_0^*(k)V_0(k')\rangle = S_{v_0}(k)\delta(k-k')\,.$$

The asterisk indicates complex conjugation, and $\delta(\cdot)$ is the Dirac delta function. It is natural to assume that $u_0(z)$ and $v_0(z)$ are mutually independent. Then $\langle U_0^*(k)V_0(k')\rangle \equiv 0$. The functions $u_0(z)$ and $v_0(z)$ are at least twice differentiable so that the integrals

$$\varphi_0^2 = \int_{-\infty}^{\infty} S_{u_0}(k)k^2\,\mathrm{d}k\,, \quad \kappa_0^2 = \int_{-\infty}^{\infty} S_{u_0}(k)k^4\,\mathrm{d}k \tag{5.4}$$

and similar integrals with $S_{v_0}(k)$ exist. The first integral in (5.4) is the mean square angle of inclination of the fiber axis to the z axis in the y–z plane, and the second corresponds to the mean square curvature. The ratio $\lambda_0 = 2\pi\varphi_0/\kappa_0$ is interpreted as a characteristic wave length of the initial imperfections.

For simplicity we consider a composite whose structure is isotropic in the x–y plane. Then $u_0(z)$ and $v_0(z)$ are samples of the same random function, and $S_{u_0}(k) \equiv S_{v_0}(k)$. Additional transverse displacements of fibers in homogeneous axial compression represent samples of homogeneous ergodic functions. Denoting the latter by $u(z)$ and $v(z)$, we take, similarly to (5.3),

$$u(z) = \int_{-\infty}^{\infty} U(k)\,\mathrm{e}^{ikz}\,\mathrm{d}k\,, \tag{5.5}$$

where $U(k)$ satisfies

$$\langle U^*(k)U(k')\rangle = S_u(k)\delta(k-k')\,, \tag{5.6}$$

and $S_u(k)$ is the spectral density of the function $u(z)$. Similar equations are valid for $v(z)$.

For a linear elastic composite and small transverse displacements (5.3)–(5.6) result in

$$U(k) = H(k)U_0(k)\,,$$

where the function $H(k)$ depends on the model of the composite, including the assumed mechanism of deformation of a fiber embedded in a deformable matrix. Later on, a model which has been suggested primarily for laminated composites with random initial imperfections (BOLOTIN, 1966) will be employed. If the shear mode of buckling is considered,

$$H(k) \approx \sigma[\tfrac{1}{4}k^2\rho^2 v_f E_f + (1 - v_f)^{-1} G_m - \sigma]^{-1} . \tag{5.7}$$

Here, ρ is the radius of the fiber cross-section, assuming circular fibers, and σ is the nominal compression stress in the fibers. The first term inside the brackets of (5.7) accounts for the bending stiffness of the fibers, the second accounts for the shear stiffness of the matrix, and the third accounts for the effect of compression on fiber bending. Local buckling takes place when $H(k) \to 0$. The critical stress

$$\sigma_*(k) = \tfrac{1}{4}k^2\rho^2 v_f E_f + (1 - v_f)^{-1} G_m \tag{5.8}$$

depends on the wave number k. The minimum value of $\sigma_*(k)$ is attained when $k \to 0$. This value corresponds to buckling in very long waves, i.e. to the shear mode of instability. We thus obtain the critical stress given by (5.1).

When adjacent fibers are deformed almost independently, the matrix plays the role of an elastic foundation with a normal reaction force. Then, instead of (5.7)

$$H(k) \approx [\tfrac{1}{4}k^2\rho^2 v_f E_f + (1 - v_f)^{-1} E_m^{tr}/(k^2\rho^2) - \sigma]^{-1} . \tag{5.9}$$

Setting $H(k)$ to zero we find the critical stress to be

$$\sigma_*(k) = \tfrac{1}{4}k^2\rho^2 v_f E_f + (1 - v_f)^{-1} E_m^{tr}/(k^2\rho^2) . \tag{5.10}$$

Minimizing $\sigma_*(k)$ with respect to k we obtain (5.2). This shows that there is agreement between the proposed probabilistic model for composites with initial imperfections and conventional models for composites with perfect fibers.

Eq. (5.7), taking (5.8)–(5.10) into account, assumes the form

$$U(k) = \sigma[\sigma_*(k) - \sigma]^{-1} U_0(k) . \tag{5.11}$$

The critical stress $\sigma_*(k)$ is obtained from (5.8) or (5.10) depending on the mode of fiber buckling.

5.3. Reliability with respect to rupture of fibers

Normal stresses in fibers due to the combination of axial compression and bending in the y–z plane are estimated as $\sigma_1 \approx \sigma + E_f\rho(\partial^2 u/\partial z^2)$. It should be recalled that fibers with circular cross section are considered. The function $\sigma_1(z)$ is stationary ergodic. Similarly to (5.3) and (5.5), we present it in the form

$$\sigma_1(z) = \sigma + \int_{-\infty}^{\infty} \Sigma(k)\, \mathrm{e}^{-ikz}\, \mathrm{d}k$$

with the generalized random function $\Sigma(k)$ of the wave number k. It is obvious that the relationship $\Sigma(k) = E_f \rho k^2 U(k)$ holds, where $U(k)$ is the generalized random function entering in (5.5). Taking into account (5.6) and (5.11) the spectral density of the function $\sigma_1(z)$ is connected with that of the function $u_0(k)$ as follows

$$S_{\sigma_1}(k) = (k^2 \rho E_f \sigma)^2 [\sigma_*(k) - \sigma]^{-2} S_{u_0}(k) . \qquad (5.12)$$

Eq. (5.12) is valid for $\sigma < \sigma_*(k)$. Since k is distributed along the entire real axis, we should have $\sigma < \sigma_*^0$, where σ_*^0 is the critical stress calculated from (5.1) or (5.2). The maximum of $S_{\sigma_1}(k)$ is attained when $k \to \infty$. This means that initial imperfections with very short waves are most unfavourable from the point of view of fiber bending.

Consider quasistatic monotonic loading in compression with the bulk nominal normal stress σ. With an order of magnitude estimate the corresponding nominal stress in fibers is $\sigma_f \approx \sigma$. Denote the tensile strength of fibers as 35_+ and the compression strength as r_-. If these parameters are deterministic, the admissible region is $\Omega = \{\sigma_1 : -r_+ < \sigma_1 < r_-\}$. The signs are chosen in this Section with the convention that compression stresses are positive. Failure is associated with the first excursion of the function $\sigma_1(z)$ outside of the corresponding band (Fig. 26(a)). To estimate the reliability with respect to this kind of failure, we apply the Poissonian model of failure (Subsection 2.3). First, let us calculate the mathematical expectation ν_σ of the number of excursions per unit length. Assume that the initial imperfections and, therefore, the fiber stresses are normal functions of z. Using (1.17) we obtain

$$\nu_\sigma = \frac{1}{\lambda_\sigma} \exp\left(- \frac{(r_- - \sigma)^2}{2D[\sigma_1]} \right) + \frac{1}{\lambda_\sigma} \exp\left(- \frac{(r_+ + \sigma)^2}{2D[\sigma_1]} \right), \qquad (5.13)$$

where λ_σ is a characteristic length of the function $\sigma_1(z)$ introduced as follows:

$$\lambda_\sigma = 2\pi (D[\dot{\sigma}_1]/D[\sigma_1])^{1/2} . \qquad (5.14)$$

Dispersions entering (5.14) are expressed through the spectral density given by (5.12) as

$$D[\sigma_1] = \int_{-\infty}^{\infty} S_{\sigma_1}(k)\, \mathrm{d}k , \quad D[\dot{\sigma}_1] = \int_{-\infty}^{\infty} S_{\sigma_1}(k) k^2\, \mathrm{d}k . \qquad (5.15)$$

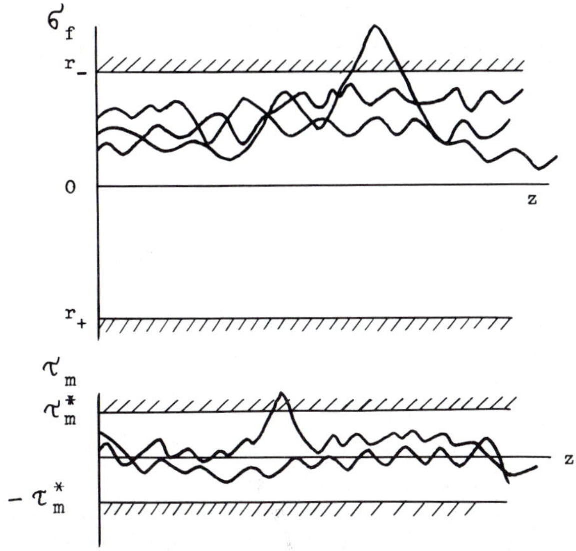

FIG. 26. Failures in compression as the first excursion events; (top) rupture of fibers, (bottom) damage of matrix.

Compared with (1.14)–(1.17), the mathematical expectations of the number of excursions are denoted by ν (since λ is already used for the parameters of length).

The probability that not a single excursion occurs out of the admissible region along the total length L of all fibers in a specimen or structural component is $R = \exp(-\nu_\sigma L)$ according to the Poissonian model of failure. Here, L is the total length of all fibers in the stressed volume V, which is proportional to V. The value L here plays the same role as the time t in (1.14).

Fiber strength has scatter both in compression and tension. Therefore, we should consider the introduced reliability factor R as a conditional one. Suppose that the joint probability density function $p_r(r_+, r_-)$ is given. Then the unconditional reliability factor is

$$r = \int_0^\infty \int_0^\infty \exp[-\nu_\sigma(r_+, r_-)L] p_r(r_+, r_-) \, \mathrm{d}r_+ \, \mathrm{d}r_- \, . \tag{5.16}$$

The approach resulting in (5.16) appears to be very conservative since the rupture of a single fiber does not mean total failure. Similarly as in the case of tension, a more realistic statement of the problem is based on the concept of dispersed damage. We return to the treatment of a composite as a sample of primary elements. The value ν_σ determined by (5.13) may be interpreted as an averaged number of ruptures per unit length. The number of primary elements per the same length is $(2\lambda_e)^{-1}$. Hence, the mean ratio of the number of ruptured elements to the total number of all elements in a certain volume is $2\nu_\sigma \lambda_e$. This value may be interpreted as a conditional damage measure of a composite with imperfect fibers. The unconditional damage measure ψ_f should be calculated in the same way as was accomplished in (5.16). Hence,

$$\psi_f = 2\lambda_e \int_0^\infty \int_0^\infty \nu_\sigma(r_+, r_-) p_r(r_+, r_-)\, dr_+\, dr_- . \tag{5.17}$$

All the primary elements are ruptured at $\psi_f = 1$. However, a failure occurs much earlier when the damage density approaches a certain critical level $\psi^* < 1$. Then, the coalescence of centers of damage takes place throughout the entire stressed volume which is equivalent to failure by loss of integrity. The reliability factor is equal to unity at $\psi_f < \psi^*$ and to zero at $\psi_f > \psi^*$. Thus, neither the statistical scatter nor the scale effect is associated with this model of failure as opposed to the weakest link model resulting in (5.16). However, the reliability parameters defined by (5.16) and (5.17) do not take into consideration the scatter of bulk material properties as well as that of loading conditions.

Only buckling of fibers in the y–z plane has been treated so far. In fact however, buckling occurs simultaneously in two planes. The maximum normal stress in the fibers in this case is $\sigma_1 = \sigma \pm E_f\rho[(\partial^2 u/\partial z^2)^2 + (\partial^2 v/\partial z^2)^2]^{1/2}$. If $u(z)$ and $v(z)$ are normal random functions, there are two ways to estimate the reliability factors. First, one can consider excursions of a two-dimensional normal process out of a circular ring area defined by the condition $-r_+ < \sigma_1 < r_-$. The mean number of excursions is determined with on the basis of (1.16). Secondly, an equivalent one-dimensional problem of the first excursion can be considered. In fact, the magnitudes $|\sigma_1 - \sigma|$ follow the Rayleigh distribution when both $u(z)$ and $v(z)$ are independent with identical Gauss distributions. Estimates of the reliability factors remain, however, of the same order of magnitude as those relating to bending in the y–z plane.

5.4. *Reliability with respect to damage of the matrix*

Tangential stresses in the matrix due to bending of the fibers may be estimated as follows: $\tau \sim (1 - v_f)^{-1} G_m(\partial u/\partial z)$. Representing the random function $\tau(z)$ in a fashion similar to (5.3)

$$\tau(z) = \int_{-\infty}^{\infty} T(k)\, e^{ikz}\, dk$$

we find that the generalized random function $T(k) = ik(1 - v_f)^{-1} G_m U(k)$. The spectral density of the tangential stress is determined by an equation similar to (5.12):

$$S_\tau(k) = (k\sigma_*^0 \sigma)^2 [\sigma_*(k) - \sigma]^{-2} S_{u_0}(k) . \tag{5.18}$$

Here σ_*^0 is the critical stress defined by (5.1) and not by (5.2). It is easy to show that $S_\tau(k)$ attains its maximum at $k^2 = 4(\sigma_*^0 - \sigma)/(\rho^2 v_f E_f)$. In terms of the wave length it is equivalent to $\lambda \approx 2\lambda_e(1-\sigma/\sigma_*^0)^{-1/2}$. If $\sigma \ll \sigma_*^0$ then $\lambda = 2\pi/k \approx 2\lambda_e$. Thus, the characteristic length of the most unfavourable initial imperfections responsible for tangential stresses is of the order of the length of the primary elements (in other words, of the characteristic length of the boundary effect for the bending of fibers embedded in the matrix).

When the nominal stress σ increases, tangential stresses in the matrix also increase. Let failure occur when, even at just one point of the interface, the critical level of tangential stresses τ_m^* is reached (Fig. 26(b)). The admissible region is $\Omega = \{\tau : |\tau| < \tau_m^*\}$. The mathematical expectation of the number of excursions out of this region per unit length of the fibers is

$$\nu_\tau = (2/\lambda_\tau)\exp(-k_\tau^2/2)\,. \tag{5.19}$$

Eq. (5.19) is similar to (5.13). The characteristic length of fluctuations of the function $\tau(z)$ is denoted by λ_τ. Analogously to (5.14) $\lambda_\tau = 2\pi(\boldsymbol{D}[\dot{\tau}]/\boldsymbol{D}[\tau])^{1/2}$. Dispersions $\boldsymbol{D}[\dot{\tau}]$ and $\boldsymbol{D}[\tau]$ are connected with the spectral density, defined by (5.18) in a similar fashion as in (5.15). According to the Poissonian model of failure the reliability factor is $R = \exp(-\nu_\tau L)$.

Let us discuss another statement of the reliability problem with respect to matrix damage. We allow the condition $|\tau| < \tau_m^*$ to be violated, but require that the length λ of damaged segments of the fiber–matrix interface should not reach a limit level λ_*. The admissible region is $\Omega = \{\lambda : \lambda < \lambda_*\}$. Assume for simplification that the function $\tau(z)$ remains normally distributed despite nonlinearities resulting from violating the condition $|\tau| < \tau_m^*$. The distribution of the length of segments of a normal process whose magnitudes exceed a given level is needed in order to estimate the required reliability factor. Using the approximate solution presented by TICHONOV (1970) we obtain the probability density function of the length of damaged segments $p_\lambda(\lambda) \approx (\pi^2 k_\tau^2 \lambda/\lambda_\tau^2)$ $\exp[-\pi^2 k_\tau^2/(2\lambda_\tau^2)$. Here $k_\tau^2 = (\tau_m^*)^2/\boldsymbol{D}[\tau]$. Using (1.17) we find the mathematical expectation of the number of excursions out of the region $|\tau| < \tau_m^*$ whose length λ is larger than the limit level λ_*

$$\nu_\lambda = (2/\lambda_\tau)\exp[-(k^2/2)(1+\pi^2\lambda_*^2/\lambda_\tau^2)]\,. \tag{5.20}$$

The reliability factor is $R = \exp(-\nu_\lambda L)$. When $\lambda_* \ll \lambda_\tau$, the second term in parentheses is small compared to unity and (5.20) approaches (5.19). This is quite natural because the condition $\lambda_* \ll \lambda_\tau$ means that the limiting damaged length is very short. The difference between (5.19) and (5.20) becomes significant, on the other hand, when the limiting length λ_* is large compared to the characteristic length of the process $\tau(z)$.

An alternative approach to reliability with respect to matrix damage is as follows. We introduce a damage measure ψ_m equal to the ratio of the total length of the damaged segments of the fiber–matrix interface to the total length of all the fibers in the considered volume. Assuming that $\tau(z)$ remains a Gaussian random function even if a significant part of the interface is damaged, we obtain

$$\psi_m = 2[1 - \Phi(k_\tau)] , \tag{5.21}$$

where $\Phi(\cdot)$ is defined by (2.7), and k_τ has the same meaning as in (5.19) and (5.20). If $k_\tau \gg 1$, the asymptotic expansion presented by (4.20) may be used. Then approximately, $\psi_m \approx (2/\pi)^{1/2} k_\tau^{-1} \exp(-k_\tau^2/2)$. It is obvious that the latter approximation corresponds to sufficiently small damage.

A numerical example is given in Fig. 27. Initial imperfections of the fibers are assumed to be narrow-band random functions of z whose spectral densities are concentrated near the wave numbers $\pm k_0$. The damage measure ψ_m is plotted in Fig. 27 versus the nominal stress ratio σ/σ_*^0 for $\tau_m^*/\sigma_*^0 = 10^{-2}$ and $\sigma_*(k_0) \approx \sigma_*^0$. Three values of the mean square angle of initial imperfections defined by the first equation in (5.4) are considered. Diagrams are drawn both for compression (positive half-axis) and tension. Matrix damage increases in tension, too, although the damage measure approaches a certain level $\psi_m < 1$ with increasing tensile stresses.

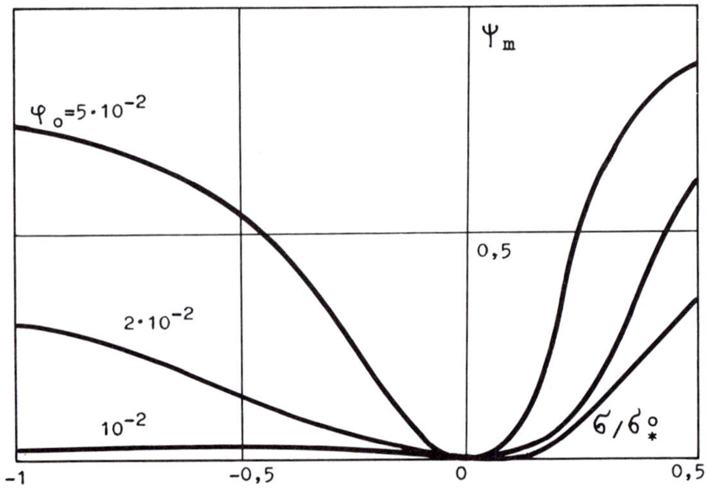

FIG. 27. Damage measure of the matrix ψ_m versus nondimensional nominal stress σ/σ_*^0 for $\tau_m^*/\sigma_*^0 = 10^{-2}$, $\sigma_*(k_0) \approx \sigma_*^0$, and various values of φ_0; compression stresses are assumed positive.

6. Reliability in the presence of sets of cracks

6.1. Cracks and crack-like defects in composite structures

Composite structures, especially the large-scale ones, contain a number of crack-like defects of various origins and statistical properties. We arrive, therefore, at a more general and realistic problem of predicting the reliability and longevity of a structure in the presence of sets of cracks.

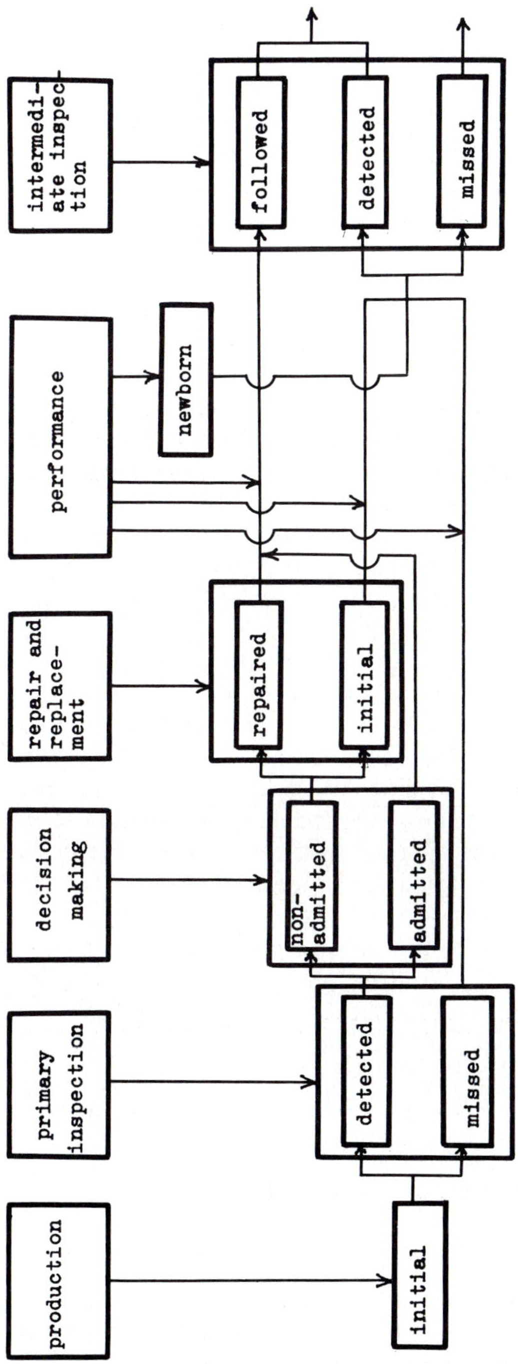

FIG. 28. Random sets of cracks in a structure.

Stated in its entire scope, the problem is very complicated. A simplified model has been suggested by BOLOTIN (1984a). The model is based on the assumption that all cracks and crack-like defects are non-interacting and forming Poissonian random sets distributed in the volume of a structural component. Furthermore, the concept of Poissonian random sets is followed through all the states of the analysis including initial and newly-formed cracks. Inspection, repair and replacement of structural components during the life of a structure are also included. Due to the initial assumptions and comprehensive use of the concept of Poissonian random sets, final equations are rather simple and apparent even in the case of non-homogeneous and/or non-stationary stress-states. Moreover, they are in good agreement with the acknowledged results commonly used in simpler situations.

Henceforth, we use the term 'crack' in a generalized sense covering, for example, delaminations in composite structures (BOLOTIN, 1984b). At least two types of cracks are to be distinguished: initial cracks formed during fabrication, transportation and construction stages, and cracks initiated during the service stage. Further differentiation is needed when inspection and non-destructive testing during the life of a structure are planned. We distinguish non-detected and detected cracks and among the latter, admitted and non-admitted ones. If any of the structural components has to be replaced during the service of a structure, a new set of initial cracks enters into consideration, etc.

The set of initial cracks is characterized by a certain, a priori known distribution. Undetected cracks form a set defined by means of an a posteriori distribution depending on the reliability of crack detection methods. Detected cracks in principle can be identified and followed during the life of a structure. Loading is given probabilistically at the design state, and hence the predicted sizes of detected cracks are also randomly distributed. Therefore, we should treat this set of cracks as random, too. Newly-formed cracks are in principle of stochastic nature even if loading is deterministic. In fact, initial conditions corresponding to equations of growth of newly-formed cracks are stochastic. Flow-chart showing the relationship between different sets of cracks is given in Fig. 28.

6.2. Poissonian sets of cracks

Since crack detection is included in the study, we call a crack macroscopic if its characteristic dimension l is not less than the lower detection threshold. This makes the order of the size of initial cracks $l_* \sim 1$ mm and even higher if conventional non-destructive, say, ultrasonic methods are applied. Such cracks appear to be rather rare and one may treat them by assuming a Poissonian distribution in the volume of a structural component. This approach is quite natural and, therefore, not new. However, it is developed here in the most comprehensive way with application to initial newly-formed, etc. cracks.

Let us discuss briefly the notion of a Poissonian set of objects distributed in a certain region M_0. The discussion will be carried out in terms of cracks.

Assume that all cracks in M_0 are non-interacting; numbers of cracks located in non-intersecting subregions ΔM of M_0 are independent random variables; the probability of locating a crack in a sufficiently small subregion ΔM depends only on the measure of the subregion, and with the error $\sim(\Delta M)^2$ is proportional to ΔM. Here the same notation is used both for the regions and their measures.

Manifolds of different dimensions are considered in structural reliability. For example, composite pipes or tubes may be treated as one-dimensional manifolds. The measure M_0 is the length of a segment of a tube, say, a unit length. In the case of composite pressure vessels the manifold is two-dimensional if crack distribution is referred to the middle surface, and three-dimensional if the position of cracks with respect to the vessel's thickness is of significance.

The probability that k cracks are located in the region M_0 at time t is

$$Q_k(t) = [\mu^k(t)/(k!)] \exp[-\mu(t)] \quad (k = 0, 1, \ldots) . \tag{6.1}$$

Here $\mu(t)$ is the mathematical expectation of the number of cracks in M_0 at time t. It is assumed in (6.1) that the nominal stress–strain state and crack distribution are homogeneous in M_0. Later on, when (6.1) will be generalized to non-homogeneous nominal fields, M_0 will be interpreted as a reference measure, in particular, a unit length, surface area or volume. The Poisson distribution is defined by means of a single parameter equal to the mathematical expectation of the number of objects in the reference region. This parameter becomes a function of time or coordinates when non-stationary or non-homogeneous problems are considered, respectively.

It is known that Poissonian sets are stable with respect to a rather wide class of transformations such as independent screening of some elements with respect to a certain property, or mixing of independent sets. For example, let the property A be distributed homogeneously and independently among elements of a Poissonian set with the probability $P(A)$. It is easy to show that the probability of locating k elements with the property A in the region M_0 is

$$Q_k(A; t) = [\mu^k(t)P^k(A)/(k!)] \exp[-\mu^k(t)P(A)] \quad (k = 0, 1, \ldots) . \tag{6.2}$$

In fact, (6.2) is (6.1) with $\mu(t)$ replaced by $\mu(t)P(A)$. In a more general case, when m Poissonian sets are distributed in M_0 with expected values $\mu_1(t), \ldots, \mu_m(t)$ and probabilities of the property A equal to $P_1(A), \ldots, P_m(A)$, (6.1) is valid upon putting $\mu_1(t)P_1(A) + \cdots + \mu_m(t)P_m(A)$ in place of $\mu(t)$.

Later, all sets of cracks are supposed to be non-interacting and remain Poissonian after performing all the procedures including inspections, decision making, repair, replacement and operation. For example, we assume that instruments or operators should not to be changed during inspection if the current results appear to be unsatisfactory.

6.3. Reliability functions

Let a structure or structural component consist of I manifolds $M_1, \ldots, M_I$, with J types of cracks distributed in these manifolds. Failure occurs when just one of the cracks reaches a given deterministic value. Then the reliability function is defined as

$$R(t) = \boldsymbol{P}\{\max_{M_i} l_{ij}(t) < l_{ij}^* ; \quad i = 1, \ldots, I; \quad j = 1, \ldots, J\} . \tag{6.3}$$

Here $l_{ij}(t)$ are characteristic crack sizes, and l_{ij}^* are the corresponding limit sizes. A crack may be characterized by two or more parameters of length. For instance, elliptical cracks and delaminations are given with the length of two half-axes which are, generally, mutually dependent. The account of this relation involves joint probability density functions, cross-correlation factors, etc. To simplify the following analysis we treat all parameters l_{ij} and l_{ij}^* as independent.

Suppose we know the mathematical expectations $\mu_{ij}(l_{ij}; t)$ of the number of cracks of each set located in the reference area M_i^0 whose sizes exceed l_{ij} at time t. Using (6.1) and taking into account the basic properties of Poissonian sets, (6.3) results in

$$R(t) = \exp\left(-\sum_{i=1}^{I} \sum_{j=1}^{J} \int_{M_i} \mu_{ij}(l_{ij}^* ; t) \frac{\mathrm{d}M_i}{M_i^0}\right) . \tag{6.4}$$

To obtain (6.4) each area M_i is divided into a number of small parts ΔM_i, (6.1) is applied to each part, independence of events is postulated, and the summation of M_i's is replaced by integration. Similar procedure was applied to obtain (4.18) when a Poissonian model was used to estimate the reliability function with respect to loss of integrity.

If the total expected numbers $\mu_{ij}(t)$ of macrocracks and the cumulative distribution functions $F_{ij}(l_{ij}; t)$ of their sizes are given, then

$$\mu_{ij}(l_{ij}; t) = \mu_{ij}(t)[1 - F_{ij}(l_{ij}; t)] . \tag{6.5}$$

After substituting (6.5) into (6.4), the differences $1 - F_{ij}(l_{ij}; t)$ play the same roles as the probability $P(A)$ in (6.2).

If the limit sizes l_{ij}^* depend on the load level at the considered time moment, the situation becomes more complicated. Similarly to the reliability problems with respect to single crack propagation to instability, the first excursion problem should be considered. Instead of (6.3) we should take the reliability function as

$$R(t) = \boldsymbol{P}\left\{\begin{matrix} l_{ij}(x, \tau) < l_{ij}^*(x, \tau) ; & x \in M_i ; & \tau \in [t_0, t] \\ i = 1, \ldots, I ; & j = 1, \ldots, J . & \end{matrix}\right\} . \tag{6.6}$$

With the use of Poissonian model, the reliability function can be presented in the form

$$R(t) = \exp\left[-\sum_{i=1}^{I} \sum_{j=1}^{J} \int_{t_0}^{t} \int_{M_i} \lambda_{ij}(x, \tau)\, \mathrm{d}x\, \mathrm{d}\tau \right], \tag{6.7}$$

where $\lambda_{ij}(x, \tau)$ are intensities of streams of events equal to mathematical expectations of the number of violations of the inequalities $l_{ij}^*(x, \tau) - l_{ij}(x, \tau) > 0$ per units of time and space. Eq. (6.7) may be interpreted as a far-reaching generalization of (1.14) to the case of a time–space Poissonian model. The values $\lambda_{ij}(x, \tau)$ are to be calculated with the use of (1.16). It is also obvious that (6.4) and (6.7) generalize the Weibull law of failure to multi-dimensional, nonstationary and nonhomogeneous problems.

6.4. *Newly-formed cracks*

Macrocrack initiation under long-acting quasistatic or cyclic loading is the result of dispersed damage accumulation. A probabilistic model for the initiation of a single macrocrack was developed in Subsections 2.5 and 3.8 with special application to composites. The damage measure ψ used in that model is equal to the ratio of the number of ruptured primary elements to the total number of these elements in a certain reference volume. The probability of initiation of the first, and, in that context, single macrocrack is defined by (2.12) and (2.13). On the other hand, phenomenological damage measures are used widely in engineering fatigue design. The common damage measure introduced by the Miner rule is

$$\psi_n = \sum_{k=1}^{n} \frac{1}{N(\Delta\sigma_k)} . \tag{6.8}$$

Here, k is the current number of cycles, $\Delta\sigma_k$ is the stress range in the kth cycle, and $N(\Delta\sigma_k)$ is the ultimate number of cycles in fatigue tests with the constant stress range $\Delta\sigma_k$. Summation in (6.8) covers all cycles of loading up to the nth cycle. The damage measure ψ_n is defined in the interval $[0, 1]$ and the value $\psi_n = 1$ corresponds to a failed specimen of standard size.

To correlate the damage measures ψ entering into the probabilistic model with that defined by (6.8), it is sufficient to re-normalize the former measure as ψ/ψ_*. Indeed, setting $\psi = \psi_*$ in (2.12) or (2.13) we see that the probability of macrocrack initiation in the reference volume V_0 is $1 - e^{-1} = 0.632\ldots$. This probability is of the order of unity. Consequently, the equality $\psi = \psi_*$ means that a macrocrack will almost surely initiate. Later on we will assume that the first macrocrack initiation is related to the value $\psi = 1$.

To extend the model of single crack initiation to sets of cracks, we should take the damage measure to be distributed along the interval $[0, \infty)$. Experimentalists interpret that a specimen has failed when they observe the first macrocrack. Denoting the number of initiated macrocracks in the reference volume by k, consider the step-wise process $k(t)$ and its mathematical expectation $\mu(t) = \boldsymbol{E}[k(t)]$. It has been proposed by BOLOTIN (1980) that the equality $\psi = 1$ is equivalent to the situation when $\mu = 1$ for a representative sample of specimens. Moreover, the following relationship was assumed

$$\mu(t) = h[\psi(t)] \tag{6.9}$$

with the differentiable function $h(\psi)$ satisfying the conditions $h'(\psi) > 0$, $h(0) = 0$, $h(1) = 1$. Eq. (6.9) is illustrated in Fig. 29.

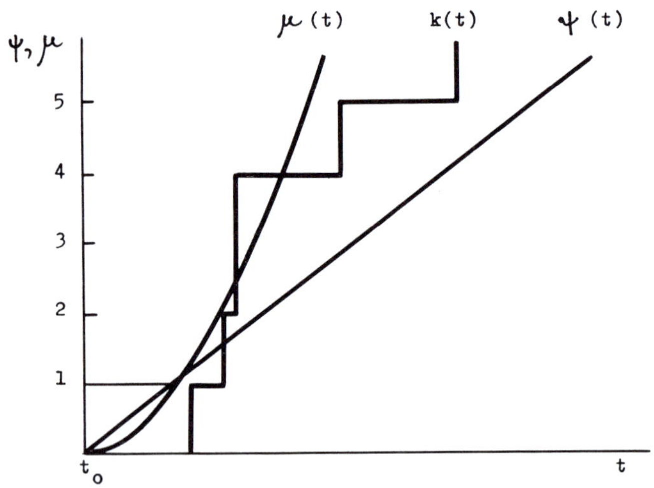

FIG. 29. Relationship between the damage measure $\psi(t)$ and the expected number of macrocracks $\mu(t)$. A typical realization of the process of macrocrack initiation is denoted by $k(t)$.

We take the equation of damage accumulation in the conventional form

$$\mathrm{d}\psi/\mathrm{d}t = f(\psi, \Delta\sigma)\,. \tag{6.10}$$

Eq. (6.10) is analogous to (2.2). The function $f(\ldots)$ is subjected to the same conditions as those of (2.2). The initial condition is $\psi(0) = \psi_0$, and in the case of virgin material $\psi_0 = 0$. In the case of deterministic loading the solution of (6.10) may be written as $\psi = \Psi(t \mid t_0, \psi_0)$ with initial data included on the right-hand side. Taking account of (6.9), the intensity of crack initiation $\lambda(t) = \mu'(t)$ is

$$\lambda(t) = \frac{\mathrm{d}}{\mathrm{d}t}\,(h[\Psi(t \mid t_0, \psi_0)])\,. \tag{6.11}$$

If loading and/or material properties are random, the process $\psi(t)$ becomes stochasitic. On the other hand, $\mu(t)$ is a deterministic function. In order to generalize (6.9) to be able to treat a general sample we should introduce into the right-hand side an averaging procedure of specimens or structural components. It may be performed in two ways: by averaging the process $\psi(t)$ or by averaging the whole of the right-hand side. Hereafter the first method is chosen by transforming (6.9) as follows

$$\lambda(t) = \frac{\mathrm{d}}{\mathrm{d}t}\,(h\{\boldsymbol{E}[\psi(t)]\})\,. \tag{6.12}$$

In the case $h(\psi) \equiv \psi$ both methods of generalization result in the same relationships.

The solution of (6.10) can be presented by means of the conditional cumulative distribution function $F_\psi(\psi, t \,|\, \psi_0, t_0)$. Its properties are similar to those of $F_l(l, t \,|\, l_0, t_0)$ discussed in connection with macrocrack propagation under stochastic loading. If the initial damage ψ_0 is a random value with a probability density function $p_0(\psi_0)$, an unconditional cumulative distribution function is determined by an equation similar to (1.3) and (4.31):

$$F_\psi(\psi; t) = \int_0^\psi F_\psi(\psi, t \,|\, \psi_0, t_0)\, p_0(\psi_0)\, \mathrm{d}\psi_0\,. \tag{6.13}$$

The mathematical expectation $\mu(t)$ of the number of newly-formed macrocracks up to time t is

$$\mu(t) = \int_0^\infty p_\psi(\psi; t)\,\psi\, \mathrm{d}\psi\,, \tag{6.14}$$

where $p_\psi(\psi; t)$ is the corresponding probability density. Eqs. (6.13) and (6.14) results in

$$\lambda(t) = \frac{\mathrm{d}}{\mathrm{d}t}\left\{h\left[\int_0^\infty [1 - F_\psi(\psi; t)]\, \mathrm{d}\psi\right\}\right. \tag{6.15}$$

with the use of $F_\psi(\psi; t)$ defined by (6.13). The main difficulty is to calculate the conditional function $F_\psi(\psi, t \,|\, \psi_0, t_0)$. Approximate and asymptotic approaches may be applied to this problem in the same way as was performed in Subsections 2.3, 3.2, etc. When the stress–strain state is inhomogeneous but slowly varying in the considered volume, λ and μ should be treated as functions of time t and the reference vector $\boldsymbol{x}$.

6.5. Growing cracks

Reliability with respect to a single growing crack was discussed in Subsection 4.7. In the presence of a set of cracks (4.25) should be replaced by (6.3), (4.26) with (6.6), etc. In the framework of the model based on the concept of Poissonian sets, the reliability functions are defined by (6.4) and (6.7). Mathematical expectations of the number of cracks whose sizes exceed certain values become the leading probabilistic parameters instead of the cumulative distribution function $F_l(l, t \mid l_0, t_0)$ entering (4.33) and (4.34).

Consider one of the sets of cracks specified by the parameter l_{ij}. For simplicity, we omit both indexes and denote the size of cracks by l. If loading is deterministic, then $l = L(t \mid l_0, t_0)$ in accordance with (4.29). Under stochastic loading, the distribution of sizes is given by the conditional cumulative distribution function $F_l(l, t \mid l_0, t_0)$. Then the unconditional function is defined analogously to (6.13) as

$$F(l, t) = \int_{l_*}^{l} F_l(l, t \mid l_0, t_0) p_0)(l_0) \, \mathrm{d}l_0 \,. \tag{6.16}$$

The lower limit l_* is the assumed size of a nucleus of macrocracks, and $p_0(l_0)$ is the probability density function of crack sizes at $t = t_0$.

The mathematical expectation of the number of cracks in M_0 with the size larger than l can be found using (6.5):

$$\mu(l; t) = \mu_0 \left[1 - \int_{l_*}^{l} F_l(l, t \mid l_0, t_0) p_0(l_0) \, \mathrm{d}l_0 \right] . \tag{6.17}$$

Here μ_0 is the mathematical expectation of the number of all the cracks from the considered set in M_0. For deterministic loading

$$F_l(l, t \mid l_0, t_0) = \eta[l - L(t \mid l_0, t_0)] \,.$$

Eq. (6.17) thus results in

$$\mu(l; t) = \mu_0 \left[1 - \int_{l_*}^{L_0(l,t,t_0)} p_0(l_0) \, \mathrm{d}l_0 \right] , \tag{6.18}$$

with the condition that $l_0 = L_0(l, t, t_0)$ is the solution of (4.29) with respect to l_0.

Eqs. (6.16)–(6.18) concern cracks with a given initial distribution. Turning to newly-formed cracks, we should introduce the intensities of crack initiation given by (6.11), (6.12) and (6.15). The product $\lambda(t) \, \mathrm{d}t$ is equal to the expected number of cracks initiating during a time segment $\mathrm{d}t$. Hence, the expected number of newly-formed cracks in M_0 whose size exceeds the value l at time t, is determined as follows

$$\mu(l; t) = \int_{t_0}^{t} [1 - F_l(l, t \mid l_*, t_*)]\lambda(t_*)\,\mathrm{d}t_* . \tag{6.19}$$

6.6. Reliability of crack detection

Suppose that the inspection procedure does not influence the properties of sets of cracks which under other conditions would be Poissonian. Then both the detected and undetected cracks form Poissonian sets, too. For example, the probability of not detecting exactly k cracks in the region M_0 with sizes larger than l is

$$H_k(l) = [\nu^k(l)/(k!)]\exp[-\nu(l)] \quad (k = 0, 1, \ldots) . \tag{6.20}$$

Here $\nu(l)$ is the mathematical expectation of the number of missed cracks, and the reference to time is omitted.

To estimate $\nu(l)$, reliability of crack detection must be defined quantitatively. One of the reliability functions concerning crack detection is the probability $P(D \mid l)$ that a crack of a given size l will be detected under the condition that the crack is located within the detection field of an instrument. The probability $P(D \mid l)$ can be estimated by statistical treatment of test results using a sample of specimens with planted cracks of known sizes. This probability characterizes the properties of an instrument as well as the efficiency of an operator. Major parameters of the probability $P(D \mid l)$ are the detection threshold l_* and the size l_s of the almost certain detection. The following forms are suitable

$$P(D \mid l) = \begin{cases} 0 , & l \leqslant l_* , \\ (l - l_*)^{\beta}/(l_s - l_*)^{\beta} , & l_* < l < l_s , \\ 1 , & l \geqslant l_s ; \end{cases} \tag{6.21}$$

$$P(D \mid l) = \begin{cases} 0 , & l \leqslant l_* , \\ 1 - (l_*/l_c)^{\beta} , & l > l_* ; \end{cases} \tag{6.22}$$

$$P(D \mid l) = \begin{cases} 0 , & l \leqslant l_* , \\ 1 - \exp[-(l - l_*)^{\beta}/l_c^{\beta}] , & l > l_* . \end{cases} \tag{6.23}$$

Eq. (6.22) corresponds to the Pareto distribution with a positive exponent β. Eq. (6.23) is the three-parameter Weibull distribution characteristic constant l_c and shape parameter β. In the two latter cases, the size l_s of almost certain detection may be associated with a sufficiently high quantile, let say, $P(D \mid l) = 0.99$.

Another reliability function of crack detection is the probability $P_D(l)$ that an arbitrarily chosen crack of size larger than l will be detected if the crack is located within the detection field of an instrument. Denote $F(l)$ the a priori cumulative distribution function of the size l. The unconditional probability of detecting a crack of size larger than l is $P(l)[1 - F(l)]$. On the other hand, this probability can be obtained by integrating of the product $P(D \mid l_l)p(l_l)$ in the interval $[l, \infty)$. Here $p(l)$ is the a priori probability density corresponding to $F(l)$. Hence,

$$P_D(l) = \left[\int_l^\infty P(D \mid l_1)p(l_1)\,\mathrm{d}l_1\right][1 - F(l)]^{-1}\,. \tag{6.24}$$

The inequality $P_D(l) \geqslant P(D \mid l)$ always holds. If the size l is almost certainly detectable, then $P_D(l) \approx P(D \mid l) \approx 1$. If a certain size $l = l_\mathrm{d}$ is deterministic, (6.24) results in $P_D(l_\mathrm{d}) = P(D \mid l_\mathrm{d})$. In general, the relationship between $P_D(l)$ and $P(D \mid l)$ depends on the a priori size distribution. This is illustrated in Fig. 30 where diagrams of $P_D(l)$ are presented for a certain function $P(D \mid l)$ and three cases of the a priori probability density. In the first case the sizes of most cracks are less than the detection threshold l_*. In the second case the maximum of the probability density $p(l)$ is located close to l_*. In the last case almost all cracks are larger than l_*. The detection system happens to be most reliable in the third case although the reliability function $P(D \mid l)$ is the same in all three cases.

Special tests are required to estimate the probability $P_D(l)$. It is necessary to test samples of specimens with cracks distributed according to the assumed

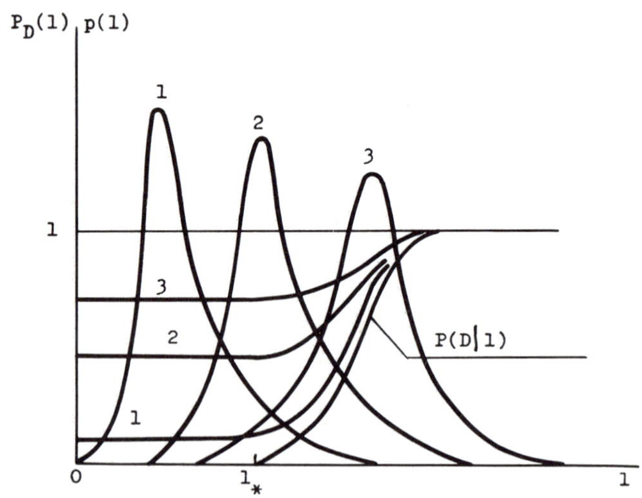

FIG. 30. Reliability functions of crack detection $P_D(l)$ for given $P(d \mid l)$ and various probability densities $p(l)$.

function $F(l)$. Most data relating to the post-mortem analysis of undetected crack-like defects in pressure vessels, pipings, etc. are related to the probability $P_D(l)$, but not to the primary reliability function $P(D \mid l)$. Such data partially reflect the distribution of crack sizes in the analysed structures. Therefore, they do not characterize by themselves the reliability of instruments of operators.

6.7. *Detected and undetected cracks*

Consider a region M_0 with a homogeneous Poissonian distribution of cracks. If the expected number of all cracks in M_0 is μ_0, the expected number of cracks of size larger than l is $\mu(l) = \mu_0[1 - F(l)]$. The expected number of missed cracks is

$$\nu(l) = \mu(l)[1 - P_D(l)] . \tag{6.25}$$

Introduce the risk function $H(l)$ equal to the probability of missing even just one crack with the size larger than l. This function is analogous to the hazard function $H(t)$ defined by (1.5). As $H(l) = 1 - H_0(l)$, where $H_0(l)$ is determined by (6.20) at $k = 0$, we obtain

$$H(l) = 1 - \exp\{-\mu(l)[1 - P_D(l)]\} . \tag{6.26}$$

If cracks are distributed nonhomogeneously in a region M, (6.26) is generalized as follows

$$H(l) = 1 - \exp\left\{-\int_M \mu(l)[1 - P_D(l)(\mathrm{d}M/M_0)]\right\} . \tag{6.27}$$

All functions entering the integral, depend generally, on the reference vector, but are supposed to vary sufficiently slowly in M. If only a part M_c of the region M is subjected to inspection, we must set $P_D(l) = 0$ in M/M_c. Assuming that other properties are homogeneous in M, (6.27) takes the form

$$H(l) = 1 - \exp\{-(M/M_0)\mu(l)[1 - (M_c/M)P_D(l)]\} . \tag{6.28}$$

The ratio M_c/M from (6.28) is the geometrical probability of missing a crack because of partial inspection.

A numerical example is presented in Fig. 31. The probability $P(D \mid l)$ is taken in the form of (6.21). It is supposed that crack sizes follow the two-parameter Weibull distribution, i.e.

$$F(l) = 1 - \exp[-(l/l_c)^\alpha] . \tag{6.29}$$

The heavy lines correspond to $P(D \mid l)$, and the light lines to $P_D(l)$. The magnitudes of the power exponents α and β in (6.21) and (6.29) are indicated near the curves. Probabilities $H_k(l)$ of missing k cracks are plotted in Fig. 32

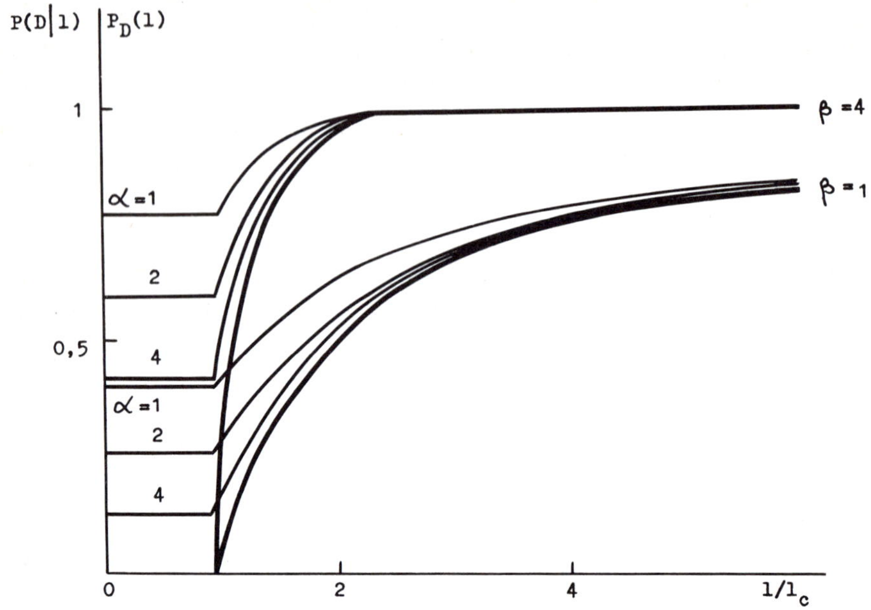

FIG. 31. Reliability functions $P(D\,|\,l)$ (heavy lines) and $P_D(l)$ (light lines) versus nondimensional crack size l/l_c for various α and β.

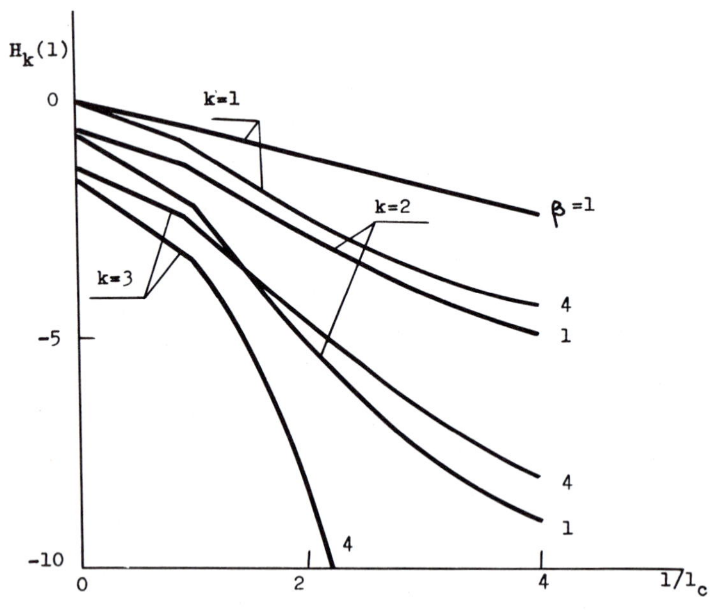

FIG. 32. Probabilities $H_k(l)$ to miss k cracks in the reference region M_0 versus nondimensional size l/l_c for $\mu = \alpha = 1$; $\beta = 1$ and $\beta = 4$.

for $\mu = \alpha = 1$, and $k = 1, 2, 3$. Since $H_1(l)$ is much larger than $H_2(l)$, $H_2(l)$ is much larger than $H_3(l)$, etc., the risk function $H(l) = H_1(l) + H_2(l) + \cdots$ is of the same order as $H_1(l)$. Indeed, the function $H(l)$ which is plotted in Fig. 32 on the logarithmic scale is very close to $H_1(l)$.

To estimate the reliability function, cumulative distribution functions $F_{\text{nd}}(l)$ for missed (undetected) cracks and $F_{\text{d}}(l)$ for detected cracks should be calculated. Let $0 < P_D(l_*) < 1$. It is evident that $F_{\text{nd}}(l) = 1 - \nu(l)/\nu(l_*)$ where $\nu(l)$ is defined by (6.25). Since by definition $\mu(l_*) = \mu_0$ it follows that

$$F_{\text{nd}}(l) = 1 - [1 - F(l)][l - P_D(l)][1 - P_D(l_*)]^{-1} . \tag{6.30}$$

Similarly, $F_{\text{d}}(l) = 1 - \mu_{\text{d}}(l)/\mu_{\text{d}}(l_*)$ where $\mu_{\text{d}}(l) = \mu(l) - \nu(l)$. This results in

$$F_{\text{d}}(l) = 1 - [1 - F(l)]P_D(l)[P_D(l_*)]^{-1} . \tag{6.31}$$

6.8. *Cracks admissibility*

Let R_* be the limit value of the reliability function for a structural component in the presence of a set of stable growing cracks. The criterion for crack admissibility at the design stage given by $R(T) \geqslant R_*$ was discussed in Subsection 4.7 in terms of a single crack. Here T is the design life-time or the time up to the first inspection. The reliability function $R(t)$ is determined with (6.4) and (6.7) taking account of the given a priori distribution functions of initial cracks and crack-like defects. All macrocracks initiated in the interval $[0, t]$ should also be included in the reliability analysis. If a structural component is designed to be inspected before being allowed into service, the function $R(t)$ is to be calculated by taking into account detected, undetected and newly-created cracks.

One of the reliability problems concern the magnitudes of admissible cracks. These magnitudes characterize the acceptance requirements in inspection. Consider a homogeneous nominal stress–strain state and a one-parameter set of cracks with characteristic size l. Denote by l_{a} the admissible size of cracks. In the case of deterministic loading we find l_{a} from the condition that $L(T \mid l_{\text{a}}, t_0) \leqslant l^*$. The left-hand side is the function from (4.29), and l^* is the limit size of cracks at time T. In the case of stochastic loading (6.17) is to be applied. Setting $p_0(l_0) = \delta(l_0 - l_{\text{a}})$ we obtain $\mu_{\text{a}}(l; t) = \mu_0[1 - F(l, t \mid l_{\text{a}}, t_0)]$. Denote by $R_{\text{a}*}$ the limit value of the reliability function with respect to the following cracks. Evidently, $R_{\text{a}*} \geqslant R_*$. The inequality $R(T) \geqslant R_{*\text{a}}$, taking account of (6.4), results in

$$\mu_{\text{a}}(l^*; T) \leqslant \ln(1/R_{\text{a}*}) . \tag{6.32}$$

Taking (6.32) with the equals sign we obtain an equation to evaluate the acceptance level l_{a}. Generalization to sets of cracks is more or less transparent since all cracks are assumed to be mutually independent.

If detection reliability is not very high there is a risk that the effect of undetected cracks on general unreliability would be significant. The expected number of cracks whose size for $t > t_0$ exceeds the level l can be found from an equation similar to (6.17):

$$\mu_{nd}(l; t) = \mu_0 \left[1 - \int_{l_*}^{l} F(l, t \mid l_0, t_0) p_{nd}(l_0) \, dl_0 \right].$$

Here $p_{nd}(l_0)$ is the probability density corresponding to the cumulative function given by (6.30). To include undetected cracks, (6.32) should be replaced with

$$\mu_a(l^*; T) \leq \ln(1/R_{a*}) - \mu_{nd}(l^*; T). \tag{6.33}$$

The ways of generalization to inhomogeneous and multi-parametric problems can be seen from (6.4) and (6.5), etc. As an example, let all the fields in M be homogeneous, with only a part M_c of the region M subjected to inspection. Instead of (6.33) we obtain

$$\mu_a(l^*; t) \leq (M_0/M) \ln(1/R_{a*}) - \mu_{nd}(l^*; T). \tag{6.34}$$

The right-hand side contains the expected number of undetected cracks in the whole region related to the reference region M_0, i.e.

$$\mu_{nd}(l; t) = \mu_0 \left[1 - (M_c/M) \int_{l_*}^{l} F(l, t \mid l_0, t_0) p_{nd}(l_0) \, dl_0 \right]$$

It is evident that both parameters l_a and M_c/M enter (6.34). Choosing them properly, a rational inspection procedure can be planned as a compromise between the acceptance level and the portion of a structural component subjected to inspection.

References

AMBARTSUMYAN, S.A. (1974), *General Theory of Anisotropic Shells* (Nauka, Moscow) (in Russian).

ARGON, A. (1974), Statistical aspects of fracture, in: *Composite Materials*, Vol. 5, *Fracture and Fatigue* (Academic Press, New York, London).

BARLOW, R.E. and F. PROSHAN (1965), *Mathematical Theory of Reliability* (John Wiley, New York, London, Sydney).

BENJAMIN, J.B. and C.A. CORNELL (1970), *Probability, Statistics and Decisions for Civil Engineers* (McGraw-Hill, New York).

BOLOTIN, V.V. (1961), *Statistical Methods in Structural Mechanics* (Stroyizdat, Moscow) (in Russian); Engl. transl., (Holden-day, San Francisco, 1969).

BOLOTIN, V.V. (1965), *Mech. Polymers* **2**, 27 (in Russian).
BOLOTIN, V.V. (1966), *Mech. Polymers* **1**, 11 (in Russian).
BOLOTIN, V.V. (1971), *Application of the Methods of the Theory of Probability and the Theory of Reliability to Analysis of Structures* (Moscow, Stroyizdat) (in Russian); Engl. transl., (NTIS, US Dept of Commerce, Springfield, 1974).
BOLOTIN, V.V. (1972), *Mech. Polymers* **3**, 529 (in Russian).
BOLOTIN, V.V. (1976), *Mech. Polymers* **2**, 247 (in Russian).
BOLOTIN, V.V. (1977), *Mashinovedenie* **5**, 86 (in Russian).
BOLOTIN, V.V. (1979), *Random Vibrations of Elastic Systems* (Nauka, Moscow) (in Russian); Engl. transl., (Martinus Nijhoff, The Hague, 1984).
BOLOTIN, V.V. (1980), *J. Appl. Math. Techn. Phys.* **5**, 149 (in Russian).
BOLOTIN, V.V. (1981a), *Mech. Composite Mater.* **3**, 405 (in Russian).
BOLOTIN, V.V. (1981b), *Probl. Strength* **7**, 3 (in Russian).
BOLOTIN, V.V. (1982) *Mech. Composite Mater.* **4**, 608 (in Russian)
BOLOTIN, V.V. (1984a) *Prediction of Life for Machines and Structures* (Mashinostroenie, Moscow) (in Russian).
BOLOTIN, V.V. (1984b), *Mech. Composite Mater.* **2**, 239 (in Russian).
BOLOTIN, V.V. and Yu.N. NOVICHKOV (1978), *Mechanics of Multilayered Structures* (Mashinostroenie, Moscow) (in Russian).
BOLOTIN, V.V. and E.V. LOBANOV (1983), Acoustic emission and damage accumulation in solids, in: *Nonlinear Deformation Waves*, IUTAM Symp., Tallin, 1982 (Springer, Berlin, Heidelberg) p. 349.
CHELIDZE, T.L. (1983), *Mech. Composite Mater.* **2**, 238 (in Russian).
CHOU T.-W. and A. KELLY (1980), *J. Mater. Sci.* **15**, 327.
FREUDENTHAL, A.M. (1968), Statistical approach to brittle fracture, in: H. LIEBOWITZ, ed., *Fracture—an Advanced Treatise* Vol. 2 (Academic Press, New York, London) p. 592.
GNEDENKO, B.V., Yu.K. BELYAEV and A.D. SOLOVYEV (1965), *Mathematical Methods in the Theory of Reliability* (Nauka, Moscow) (in Russian).
IBRAHIMOV, I.A. and Yu.V. LINNIK (1965), *Independent and Stationary Connected Values* (Nauka, Moscow) (in Russian).
JONES, B.H. (1975), Probabilistic methods and structural reliability, in: L.J. BROUTMAN and R.H. KROCK, eds., *Composite Materials* Vol.8, *Structural Design and Analysis* (Academic Press, New York, San Francisco, London).
KELLY A. (1973), *Strong Solids* (Clarendon Press, Oxford).
KOPYEV I.M. and A.S. OVTCHINSKIY (1977), *Fracture of Reinforced Metals* (Nauka, Moscow) (in Russian).
MALMEISTERS A.K., V.P. TAMUZS and G.A. TETERS (1980), *Strength of Polymer and Composite Materials*, 3rd edn. (Zinatne, Riga) (in Russian).
OBRAZTSOV I.F. and V.V. VASIL'EV, eds. (1982), *Mechanics of Composites* (Mir Publishers, Moscow) (in English).
PANASYUK V.V., A.E. ANDREIKIV and S.E. KOVCHIK (1977), *Methods of Estimation of Fracture Toughness of Structural Materials* (Naukova Dumka, Kiev) (in Russian).
PIGGOT, M.R. and B. HARRIS (1980), *J. Mater. Sci.* **15**, 2523.
PROTASOV V.D., A.F. YERMOLENKO, A.A. FILIPENKO and I.P. DIMITRENKO (1980), *Mech. Composite Mater.* **2**, 254 (in Russian).
RABOTNOV Yu.N. (1979), *Mechanics of Deformable Solids* (Nauka, Moscow) (in Russian).
ROSEN, B. (1965), Mechanics of reinforcement of composites, in: *Fiber Composite Materials* (ASM, Metals Park, Ohio) p. 37.
SMITH R.L., S.L. PHOENIX, M.R. GREENFIELD, R.B. HENSTENBURG and R.E. PITT (1983), *Proc. Roy. Soc. Lond.* **A388**, 353.
TAMUZS, V.P. (1982), Some pecularities of fracture in heterogeneous materials, in: *Fracture of Composite Materials* (Martinus Nijhoff, The Hague) p. 131.
TARNOPOL'SKI, Yu.M. and T.Ya. KINCIS (1981), *Methods of Static Testing of Reinforced Plastics* (Khimiya, Moscow) (in Russian). English edition: *Static Test Methods for Composites* (Van Nostrand Reinhold, New York, 1985).
TIKHONOV, V.I. (1970), *Excursions of Random Processes* (Nauka, Moscow) (in Russian).

PART 2

CHAPTER 7

Woven Fabric Aerospace Structures

J.A. Bailie

Structures Department
Missile Systems Division
Lockheed Missiles and Space Company
Sunnyvale, CA 94086
U.S.A.

Contents

HANDBOOK OF COMPOSITES, VOL. 2 – Structure and Design
Edited by C.T. HERAKOVICH and Y.M. TARNOPOL'SKII

1. Summary

Fabric composite structures are usually selected for aerospace use on the basis of rather pragmatic reasons such as manufacturing economics, ease of laying up complex geometries, and damage containment. Hence, there is little literature on the subject, in spite of considerable usage. The intent here is to provide a heuristic review of the materials creation, properties and utilization. After introducing the bare essentials of weaving and the parameters relevant to aerospace structural fabrics, typical data for fiberglass, graphite and aramid fabric laminates are presented. Reasons leading to the potential choice of fabric reinforced composites are discussed. To illustrate them a number of applications to aircraft, helicopters and radomes, plus a few others, are outlined to demonstrate the versatility of these materials.

2. Introduction

In their present form of very stiff, strong fibers woven into fabric and preimpregrated with resins, laminated fabric structures are considered a relatively modern development in aerospace structures. However, woven products in another form go back to the dawn of heavier than air vehicles. The Wright Brothers' machines were covered with a linen fabric, to provide the aerodynamic profiles, leaving interior wooden members as the primary structure. These fabrics coated with 'dope' remained a standard covering of many airplane components until the general acceptance of aluminum alloy structures well after the First World War. Nevertheless doped fabric remained as the aerodynamic surface components of fighter aircraft well into the Second World War. Even the Spitfire fighter had fabric covered ailerons until 1942.

The next serious utilization of fabric structures was brought about by the advent of radar and the need for a radome material that was transparent to the electromagnetic waves of the early airborne radar sets used from roughly 1941 onwards. As told by POULLOS (1970) the first operational radome was made of plexiglass, frosted to preclude visibility of the antenna and flown on a Boeing B-18A. At about the same time WOOD (1961) explains that those made in Britain consisted of dry fiberglass laid up on a male mold which was then inserted into a female mold and polyester resin poured between them and cured to create the composite. These were true load carrying structures. Since that era the use of woven fiberglass with resin matrices for electromagnetic transparencies, fairings and secondary structures have been common and are discussed in some detail in Section 7.

The transition to primary composite structures was brought about by the development of fibers having greatly increased stiffness, relative to glass. The first of these to enter service as a standard production item was boron filaments in epoxy resin, used as the skins of the horizontal tail on the Grumman F-14 and documented by LUBIN and DASTIN (1971). This significant step forward was rapidly followed by graphite and aramid Kevlar® fibers, both of which, unlike boron are flexible enough to be woven into a wide range of fabrics.

Any searching review of the literature on modern composite aerospace structures, shows it to be almost exclusively devoted to unidirectional tape. Nearly all the analysis, design data and fabrication technology address this form of the materials. For highly loaded structures having a need for markedly greater strength and/or stiffness differentials in specified directions, there is little doubt that tape or unidirectional fabric structures offer the most effective design solution. Also, because of their lower uniaxial strength to weight ratios, fabric structures tend to look less attractive than their counterparts in idealized minimum weight analyses. This conclusion is even more definite in the case of relatively flat structures amenable to fabrication by tape laying machines. However, there are numerous situations where the practical realities of fabrication complexity and cost, lead to their being chosen over tape laminates. Our intent here is to review the bare essentials of fabric characteristics and manufacture, plus illustrate the differences between fabric and tape reinforced composites. Three dimensional weaves such at those used in carbon/carbon materials are a special category deserving attention elsewhere.

3. Some elements of weaving

Before addressing the different styles of fabric, their properties and applications, it is desirable to present a small part of the nomenclature of the textile industry as outlined by MORTON (1982). This can be partially accomplished with the aid of Fig. 1 which shows the yarns (often called threads by weavers) interlaced at 90° to each other. The warp yarns are in the long direction (roll direction) and they are known as 'ends'. Filling (often abbreviated to 'fill') yarns are at right angles to these and are called 'picks'. One of the defining characteristics of any fabric construction is the number of ends per unit length times the number of picks per unit length. A weave is described as balanced if the same yarn and weight is used in both directions. All other weaves are unbalanced. Depending on the style, the strength of balanced weaves may or may not exhibit identical strength and stiffness in its two primary directions. As we shall see later, the differences in warp and fill directions can be slight or very great, depending on the intended purpose. The extreme cases of unbalanced weaves are known as 'unidirectional' because only a very small amount of fill yarn is used to keep what are effectively the unidirectional load carrying yarns together during creation and handling of the prepreg. From the structural engineering viewpoint these materials are essentially tape in all but this one

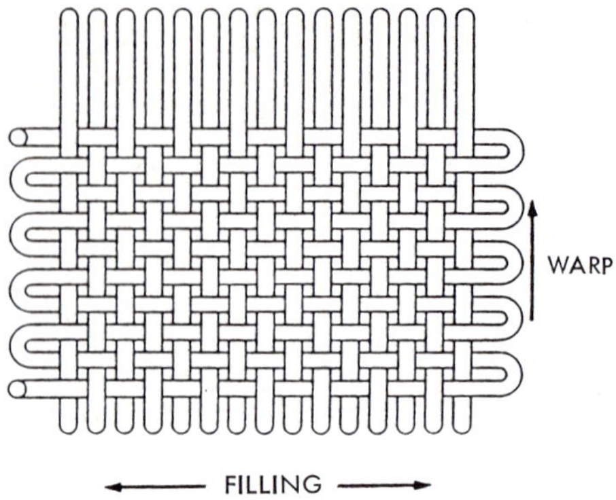

FIG. 1. Plain weave woven on a shuttle loom to illustrate fabric nomenclature.

characteristic. Also they may make a more uniform product as their widths are often an order of magnitude greater than narrow tape. Hence there are fewer seams in the completed structure.

There are basically three weave styles used in the aerospace industry. They are the plain, twill and satin. The plain weave, illustrated in more detail in Fig. 2 consists of the picks and ends going over and under each other, resulting in an equal amount of warp and fill yarns on both sides of the fabric, so it is theoretically balanced. Crimp is the term used to describe the curvature or deformation of the yarn produced during the weaving. In the more conventional weaves it is obvious that applying tension in the plane of the fabric will tend to straighten out the crimp and this manifests itself as a reduction in stiffness and strength of the fabric composite, relative to unidirectional tape of the same materials. Obviously a cured resin surrounding the fabric reduces this relative stiffness loss, but it is still present. On the other hand, in unidirectional or 'crimpless' weaves the load carrying yarns remain straight and loss of mechanical properties from this source does not arise. An obvious way to increase fabric stiffness is to reduce the amount of crimp by having as much of the warp and fill fibers be as straight as feasible. Some modern weaves are being created with this objective in mind. They have both warp and fill yarns straight and a secondary yarn binds together these two load carrying yarns.

Regardless of weave style, the mechanics of woven fabrics remains ill understood at this time. ISHIKAWA and CHOU (1982, 1983) seem to have initiated work in this area, but more research is needed.

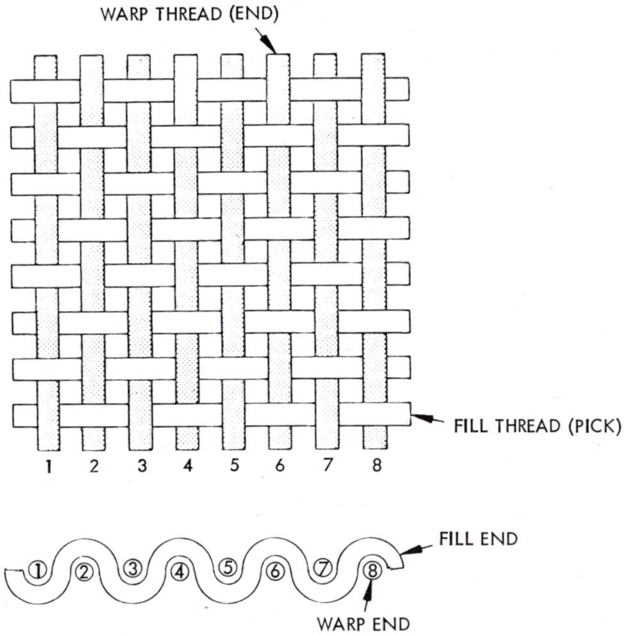

FIG. 2. Details of a plain weave.

In weaving, a float is the term used to describe the length of yarn between the crimped intersections. The 8 Harness Satin (H.S.) style is illustrated in Fig. 3. It consists of each fill pick passing over seven warp ends and then under one end. This 'seven over/one under' pattern is repeated throughout the fabric in a special fashion in which no float is repeated in the same pick in one repeat of the weave. The numerical value preceeding the 'H.S.' descriptor is always one greater than the number of warp ends over which the pick passes before crimping under a single end. These weaves obviously look different from their opposite sides and in laying them up in laminates, a complete description must include whether each ply is to be laid 'warp face up' or 'fill face up'.

Another important characteristic of fabric is its ability to conform to complex or highly curved surfaces. The tightness of the weave construction is a function of its resistance to draping over such geometries. This property can be described with the aid of the idea of the number of interlacings (binders) per unit area. In weaving, an interlacing refers to the crimping of one yarn over another and it is the interlacings which hold the fabric together. To clarify the point, consider the five satin weave styles in Fig. 4, each having 8 ends and 8 picks per unit length. The number of interlacings per unit area in the table can be counted in each figure and is a strong indication of the fabric resistance to drape and inplane distortion, assuming the same construction. Of the five styles in this figure, it is obvious that the 8 H.S. will be the easiest to drape over highly complex molds. Clearly a plain weave has the most interlacings and

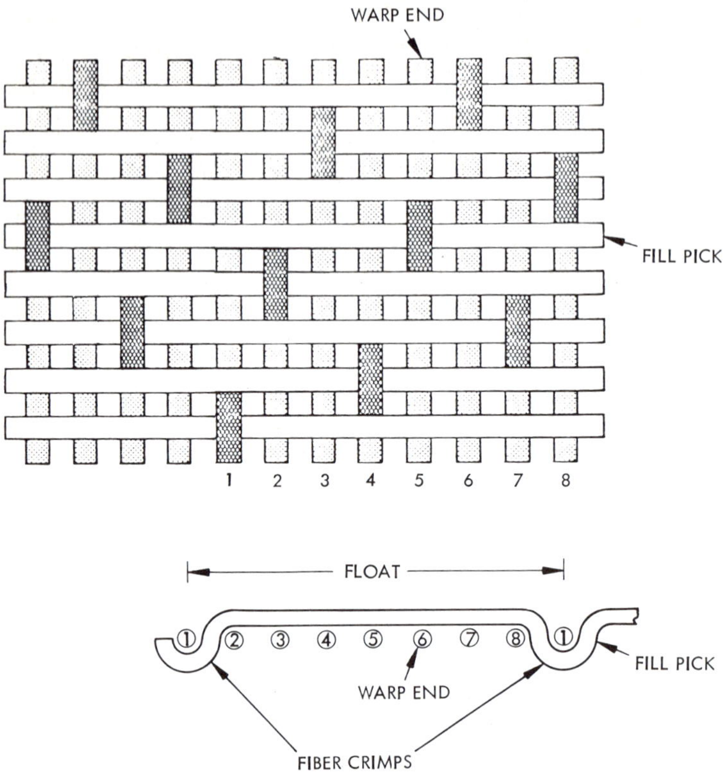

FIG. 3. Details of an 8 harness satin weave.

hence is the least drapable fabric. Now let us turn to a qualitative discussion of the characteristics of a fabric that influence its load carrying capabilities.

Strength and size of the yarn itself is important, but so also are the number of picks and ends per unit length. The larger the yarn diameter, the smaller is the maximum warp and fill count and the greater the thickness and weight per unit area. The choice of these parameters depends on the required ply thickness, strength and stiffness. While not specifically stating it so far, it might have been noticed that we have been discussing situations in which the fabrics were not very seriously unbalanced and had relatively similar properties in the two primary directions. There are other classes of severely unbalanced fabrics that are likely to see increased future use. There are basically two ways to create them. Either the same yarn can be used throughout but the ratio of picks to ends can be selected to be well removed from unity. In this case the strength and stiffness ratios in the two primary directions will tend to follow the pick to end ratio. The second option is to use different yarns in the warp and fill directions and there is a natural subdivision in this category. It is not necessary to change the basic material. For example, it may be desirable for some applications to use thin 3K filament graphite yarns in one direction while

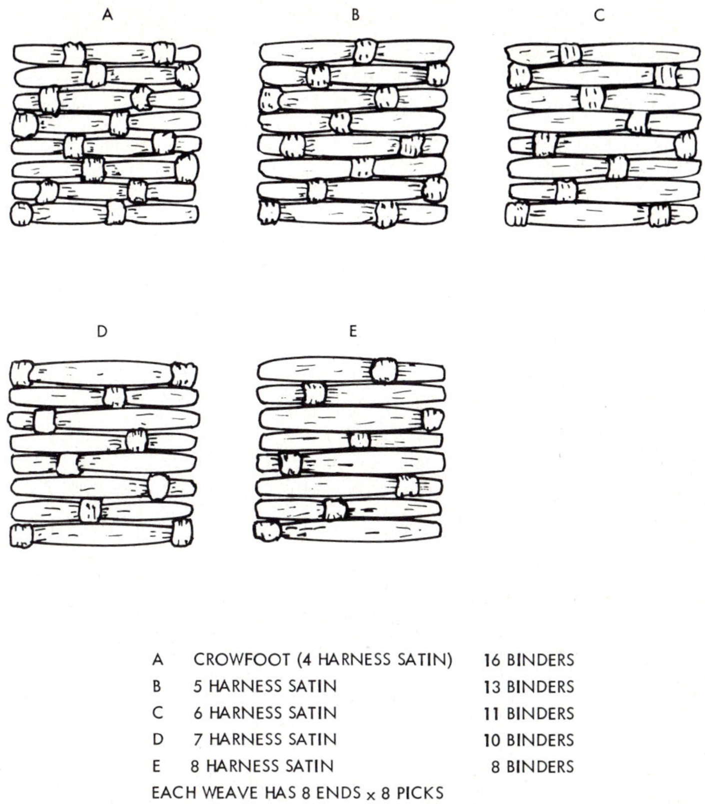

FIG. 4. Five satin weaves to illustrate the different number of binders and fabric drapability (MORTON, 1982).

using 6K filament yarns of the same material in the other. The second category is to go to hybrid weaves where the yarns are different in the two orthogonal directions. Sometimes it might be advantageous to use graphite yarns of different stiffnesses. In another application aramid or fiberglass fibers in one direction and graphite in the other may be desirable.

Having introduced some of the basic ideas and terminology, let us now address a few practical issues pertinent to the creation of high modulus fabrics, to highlight where they differ from routine textile weaving. LEVIN (1975) describes why fibers relevant to aerospace structures were not, until recently, woven on looms optimized for the garment trade. Rapier, airjet and other more modern looms are capable of handling a much wider range of yarns. Relative to the yarns used in the garment industry, all three fibers of interest exhibit much higher stiffness, lack of ductility and such low natural lubricity that special precautions must be taken to prevent serious damage during weaving. To help overcome some of these problems it is essential that before reaching the loom, the yarn have a sizing or finish on it. Its purpose is to reduce abrasion to an acceptable level and ensure the yarn remains a unitized bundle during weaving.

The treatment depends upon the yarn. For graphite it typically consists of roughly 1–2% by weight of a slightly staged epoxy resin. It is applied by passing the yarn through a solvent rich bath and allowing the volatiles to escape. Fiberglass yarns often use potato starch to hold the filaments together and other ingredients such as anti-static compounds. Kevlar® utilizes a proprietary finish.

In most phases of weaving there is a considerable amount of art involved and many details are proprietary to the weaver. Also numerous compromises need to be made and an obvious one involves the sizing. Too little results in the problems just mentioned. Too much makes the yarn excessively stiff, also increasing weaving difficulty. In addition extra cured sizing would tend to reduce the wetting of the fiber bundles by the prepregging resin and thereby reduce structural properties of the composite.

The sizings used for weaving are sometimes incompatible with the laminating resin to be applied during prepregging. In these cases they must sometimes be removed after weaving and replaced with a finish, whose purpose is to optimize the fabric–resin bond. If necessary, to clean the fabric, graphite sizing is burned off and replaced by an epoxy finish that is similar to the sizing. Sometimes the original sizing is just left on to act as the finish. With fiberglass the sizing must be burnt off as it precludes adequate fabric–resin bonding and the finish is basically a compound such as Silane or Volan. The temperatures required to burn off the sizing are too high for Kevlar®, so it is scoured in a soap and water solution. For fabrics to be used in conjunction with high temperature resins such as the polyimides, there are sound reasons for using such a resin as the sizing. This avoids any problems that might arise due to the eventual breakdown of the epoxy sizing at elevated temperatures.

Another issue deserving close attention is the degree of twist given the yarn prior to weaving. While untwisted yarns can certainly be woven, twisting them tends to minimize the loose filaments which might otherwise catch on the warp as the fill yarn traverses the loom. This basic phase of the weaving operation is accomplished in one of two main methods, depending on the type of loom. In a fly shuttle loom the shuttle contains a bobbin of tightly woven fill yarn. As it traverses back and forth between the warp yarns, they are moved up or down so the fill yarn passes between them in the manner required to produce the desired weave style. A concern about this method is the potential damage to the brittle fill yarns being wound into the small bobbin. Also the shuttle rides on the bottom yarns. The more modern rapier looms use a rapier shaped member to grasp the fill fiber from the outside of the warp fibers, draw it across them, after which it is cut off. Then the rapier grasps another fill yarn from a spool on the opposite side of the loom and draws it back across the warp fibers. Thus, in fabric woven by a shuttle loom the fill is continuous as shown in Fig. 1. That from a shuttleless (rapier) loom has each pick cut at the ends. This brings us naturally to the subject of selvages, the reinforced woven edges of a fabric which are necessary to provide strong edges which allow handling in subsequent operations. Fiberglass is usually used as the selvage reinforcement due to its lower cost. Selvages must be removed prior to layup.

4. Rationale for using fabric structures

There are essentially three situations which invite the employment of woven products in aerospace structures. They are:
- Complex geometries
- Reduced manufacturing costs
- Increased resistance to damage

In their prepreg form, unidirectional tapes have extremely low strength in the direction normal to the fibers. Thus, any attempt to stretch them in this direction to conform with doubly curved surfaces, often results in tape splitting. The ensuing gaps and overlaps lead to unacceptable variations in strength, thickness and susceptibility to damage in the cured laminate. The answer to this problem is to select a woven product with sufficient 'drape' to conform to the contoured surface. The amount of drape in fabrics has been discussed but is difficult to quantify. The plain is the tightest, has the least drape of the weaves of practical interest and hence is often used in relatively flat components. The loosest and hence most drapable common style is the 8 H.S. A second situation which calls for the serious consideration of fabric is where geometry is sufficiently complex to preclude the use of tape laying machines. Hand layup, with narrow unidirectional tape, ensuring no gaps or overlaps, becomes a very labor intensive operation. Replacement by fabric widths in excess of a meter reduces the layup time and the number of seams considerably. Also, if the fabric is close to being balanced, a single fabric ply replaces two orthogonal tape plies, thereby further reducing the amount of hand layup. Economic considerations alone strongly dictate the use of fabric in these situations as the manhours for its successful layup can be markedly less than with tape and result in much less statistical scatter in the resulting laminate strength. A few comparisons of the hours to layup tape and fabric structures are given in Section 7.1.

The lack of strength of unidirectional tape prepreg at right angles to the fibers is also a handicap in the fabrication of honeycomb sandwich structures when the facesheets are bonded to the core and themselves processed in a single cocuring operation. This method offers the considerable economic advantage of one autoclave run rather than the three required if the two facesheets are cured separately and secondarily bonded to the core. The problem with attempting to cocure tape facesheets is brought about by the cure cycle pressure tending to split it as it flexes across each cell of the core, resulting in badly dimpled thin facesheets having degraded strength, particularly compressive. The bidirectional strength in typical weaves intended for this purpose, reduces this problem to an entirely acceptable magnitude. In some flat panels there may even be an advantage to pretensioning the prepreg and maintaining this during the cure.

A subject sometimes given inadequate attention in the early tape structures was their damage propagation modes and how they were to be repaired. Again, quantitative data is not readily available, but there is little doubt that when subjected to impacts, the damage to the outer plies can propagate a long way from the damage site. In the currently used fabric, the damage is more localized to the neighborhood of the impact, as indicated by WATSON (1982). When discussing the use of graphite fabric in the AV-8B forward fuselage, he states 'a prime advantage of the eight harness satin weave cloth is its resistance to splintering. Damage remains very confined and, therefore, easy to repair.'

5. Differences in the manufacturing of tape and fabric structures

In spite of using the same fibers and resin types as tape prepregs there are a number of differences that must be borne in mind when preparing to fabricate laminates. A most obvious one is the expense of weaving which must be accounted for in any cost study. To understand the essential subtleties involved it is necessary to be aware of certain potential dissimilarities betwen tape and fabric prepregs.

In many current production prepregs the tape material is created using the hot melt process while the same fiber–resin combination in fabric utilizes the solvent impregnation method. This means the fabric prepreg contains volatiles from the solvents used that do not exist in the tape material. Hence, greater care may need to be taken in using the vacuum system to remove these volatiles during the cure of fabric laminates. The amount of resin in the two prepreg types may be different. Therefore, the optimum number of bleeder plies installed to absorb the excess resin, may not be the same in the two cases.

Before discussing the influence of prepreg bulk factor, the term must be explained. Bulk factor is the ratio of prepreg ply thickness to molded ply thickness. In tape materials, this ratio is typically 1.1 meaning the prepreg is only 10% thicker than the cured structure. In woven fabric materials, bulk factor may well be 1.7 or even higher due to the natural springiness of the crimping which results in considerably more air being trapped in fabric layups. The stiffer and more viscous the resin, the less is the bulk factor and air entrapped. In laying up plies on a convex surface, the developed length between radial lines is greater for the outer plies than it is for the inner. Hence, as the layup is compacted by air removal, the radial movement of the outer plies means they may tend to wrinkle as their freedom to slide tangentially is severely limited by prepreg viscosity. To preclude this undesirable result it may well be desirable to debulk fabric layups more often than with the lower bulk factor tape. This process consists of placing a separator ply and vacuum bag after the layup of a few plies, followed by application of vacuum pressure. The resulting removal of air and compaction ensures a much smoother layup. After each debulking, the bag is removed and further plies layed up. The frequency of debulking depends strongly on part complexity and particularly the amount of double curvature.

5.1. Typical mechanical properties of woven fabric laminates

Since there are so many variables involved in the creation of the various yarns, sizing, weaves and resins, a direct comparison of all combinations is not feasible. Each fabric–resin is fine tuned for its particular use, but what follows will illustrate certain trends.

5.1.1. Fiberglass, high silica and quartz

The great breadth of fiberglass reinforced composities for a wide range of industral uses, is outlined by MORGAN (1961) and KNOX (1982). The technology of Silica and Quartz fibers and fabrics is reviewed by SHULOCK and SAFFADI (1982). Constituents and representative strength properties are given in Table 1. Rows 1 and 2 demonstrate the typical result that higher strength is to be expected with epoxy resins. Rows 2 and 3 show that the higher tension strength *S* glass produces higher tension strength fabric laminates than E glass, but the compression strength may not follow this relationship, due to the strong influence of the resin acting as the 'elastic foundation' supporting the fibers. It cannot be stated with certainty, but the low strength of the high silica relative to quartz may be partially due to absence of sizing. Strength properties also vary as a function of the resin content, tending to produce optimum all around strength at roughtly 30% by weight. Table 8.13 and Fig. 8.4 of KNOX (1982) are typical of the data supporting the previous statement.

TABLE 1. Glass fabric properties.

Ref.	Yarn	Count	Weave	Sizing	Matrix resin	F^{tu} (MPa)	F^{cu} (MPa)	F^{f} (MPa)
a	E glass	57 × 54	8HS	Volan	Polyester	344	331	505
a	E glass	57 × 54	8HS	Volan	Epoxy	398	488	558
b	S glass	57 × 54	8HS	Volan	Epoxy	487	423	585
c	Quartz	57 × 54	8HS	Silane	Phenolic	497	—	656
d	High silica	57 × 54	8HS	None	Phenolic	159	—	228

[a]Table 8.8 from KNOX (1982).
[b]Table 8.11 from KNOX (1982).
[c]Table 9.8 from SHULOCK et al. (1982).
[d]Table 9.8 from SHULOCK et al. (1982).

The degree to which weave style influences properties in two representative glass fabrics is shown in Table 2. As expected, increasing the float, raises strength with the increment being greater for compression than tension. The obvious result of markedly unbalancing the weave is evident in Table 3 which

TABLE 2. Weave pattern influence on mech properties of glass fabric laminates (Table 8.12 of KNOX, 1982).

Designation	Weave style	Yarn		F^{tu} (MPa)	F^{cu} (MPa)	F_f (MPa)	Plies
		Type	wt (g/m^2)				
7628	44 × 32 Plain	ECE 150	142.0	317	177	371	18
7628	44 × 32 Crowsfoot (4HS)	ECE 150	142.0	408	393	584	18
16–149	60 × 54 5HS	ECDE 75	—	505	331	436	12
7781	60 × 54 8HS	ECDE 75	210.0	414	444	600	12

TABLE 3. Compression of balanced and unidirectional 8 harness satin weaves of fiberglass (Table 8.14 of KNOX, 1982).

Fabric type	Test direction	Count (%)	Yarn type	Yarn percentage	F^{tu} (MPa)	F^{cu} (MPa)	F^{f} (MPa)	E^{f} (GPa)
Bidirectional	Warp	57	ECG 150 1/2	52	400	433	581	27.8
Style 1581	Fill	54	ECG 150 1/2	48	378	437	540	25.2
Unidirectional	Warp	120	ECDE 75 1/2	90	649	445	655	34.2
Style 7743	Fill	20	ECG 150 1/2	10	83	199	161	17.0

compares a virtually balanced bidirectional weave (1581) with one having similar strength fibers in the two directions but with 90% of the yarn in the warp direction. It should be noted that the ratio of tension strengths in warp and fill directions is closely proportional to yarn percentages. However, compression strength is markedly less sensitive to yarn percentage. This arises because of the more complex cross-coupling inherent in compression response.

5.1.2. Graphite

The first paper on the use of graphite fabric laminates for flight vehicles appears to be by YATES et al. (1975). Additional early data is provided by SHIBATA et al. (1976). The strength and stiffness relationship between tape and an 8HS fabric using the same yarn and resin is presented by EKVALL and GRIFFIN (1982) and contained in Table 4. The fabric to tape tension and compression strength ratios are in the range of 0.40 to 0.45 while the stiffness ratios are somewhat higher but below 0.5. The strain ratios are in the neighborhood of 0.90. The relative degradation of fabric in tension and compression is due at least in part, to the crimping of the yarns during the weaving process. These trends are typical of other fabrics.

TABLE 4. Graphite tape and cloth comparisons (from EKVALL and GRIFFIN, 1982).

Load type	Property	Tape		Cloth		$\left(\frac{0° \text{ cloth}}{0° \text{ tape}}\right)$
		0°	90°	0°	90°	
Tension	Strength (MPa)	1386	58	574	483	0.41
	Modulus (GPa)	141	11.7	68	65	0.48
	Strain (%)	0.98	0.50	0.85	0.74	0.87
Compression	Strength (MPa)	1284	195	569	517	0.44
	Modulus (GPa)	128	11	60	58	0.47
	Strain (%)	1.12	1.84	1.01	0.94	0.90
In-plane shear	Strength (MPa)	92	—	96	—	1.04
	Modulus (GPa)	6	—	5	—	0.84
	Strain (%)	2.75	—	3.3	—	1.19

5.1.3. Aramid (Kevlar®)

Representative data for this material as tape and a fabric in epoxy resin is shown in Table 5, taken from DUPONT (1974). Neither tape nor fabric tension strength is quite as high as the graphite data of Table 4, but their ratios are very similar. Kevlar compression strengths are known to be markedly lower than graphite but the fabric to tape ratio is very similar. This difference appears to be almost wholly attributable to the inherently lower fiber compression capability. This lower compression strength may well also be the reason for the lower shear strength, since shear is equivalent to equal tension and compression at 45° to the shear loading. The original supposition of the less effective fiber/resin bond being the source of this problem has been recently

TABLE 5. Kevlar tape and fabric comparisons. (DuPont, 1974).

Load type	Property	Tape 0°	Tape 90°	Cloth 0°	Cloth 90°	(0° fabric / 0° tape)
Tension	Strength (MPa)	1380.0	30.0	517.0	—	0.38
↓	Modulus (GPa)	76.0	5.5	31.0	—	0.41
	Strain (%)	1.80	0.55	1.60	—	0.89
Compression	Strength (MPa)	276.0	138.0	172.0	—	0.63
↓	Modulus (GPa)	76.0	5.5	31.0	—	0.41
	Strain (%)	0.36	2.50	0.55	—	1.53
In-plane	Strength (MPa)	60.0	—	110.0	—	1.84
shear	Modulus (GPa)	2.0	—	2.0	—	1.00

shown to be incorrect. MORGAN et al. (1980) show that Kevlar® fibers have a rather brittle outer sheath but a softer core. This results in the fibers kinking at relatively low loads.

6. Laminate design charts for a range of fabrics

There do not appear to be any published results for laminates in which all parameters are held constant except the percentage of fibers in the warp and fill directions. However, the trends can be indicated by creating a hypothetical example. To ensure some relevance to actual laminates, the data for graphite tape and fabric provided by EKVALL and GRIFFIN (1982) will be used to influence the parameters chosen. Their fabric, like all real material is not quite balanced, and being a satin weave has different properties in the warp (0°) and fill (90°) directions. We assume three different weaves with different fractions of graphite fiber in the two directions. Number 1 has 100% in the warp direction and hence is structurally the same as unidirectional tape. Number 2 has 75% in the warp with the remainder in the fill direction, while the third one has the same strength and stiffness in the two directions. Their assumed properties are given in Table 6. Using this data in any of the laminate design programs such as that by FLAGGS (1982), design charts can be routinely produced for balanced symmetric laminates. In practice the first ply failure criterion used here can be rather conservative for many applications involving

TABLE 6. Hypothetical properties of the three different graphite weaves used to create Figs. 5 and 6.

Warp (%)	Fill (%)	F_{11}^{tu} (MPa)	F_{22}^{tu} (MPa)	F_{12}^{su} (MPa)	E_{11} (GPa)	E_{22} (GPa)	G_{12} (GPa)	ν_{12}
100	0	1380	69	83	138	1.38	6.9	0.25
75	25	966	310	83	104	41.4	6.9	0.18
50	50	552	552	83	69	69.0	6.0	0.10

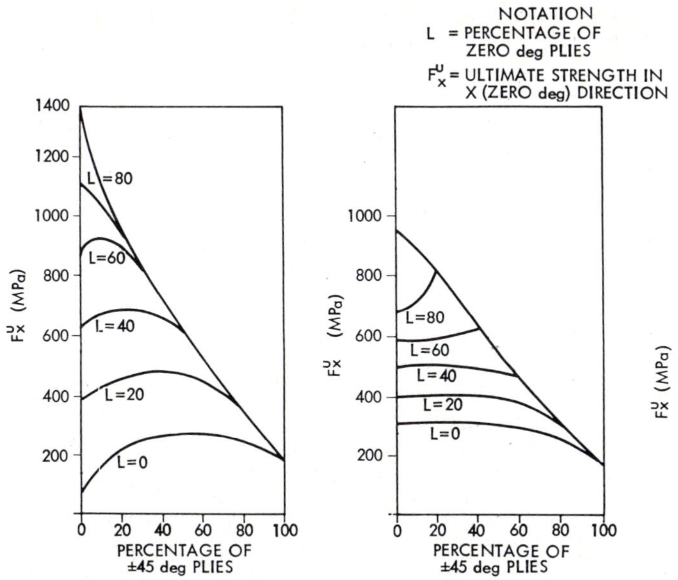

FIG. 5. Strength of laminates fabricated from plies with three different warp/fill strengths.

thick laminates as the failure mode is that of the resin. For unidirectional tape this matrix failure mode is ignored, as is frequently done in practice, including the Air Force Design Guide (ARVIN, 1973). The three design strength charts for varying fractions of 0°, 90 ± 45° plies are shown in Fig. 5. Those for the modulus E_x are shown in Fig. 6.

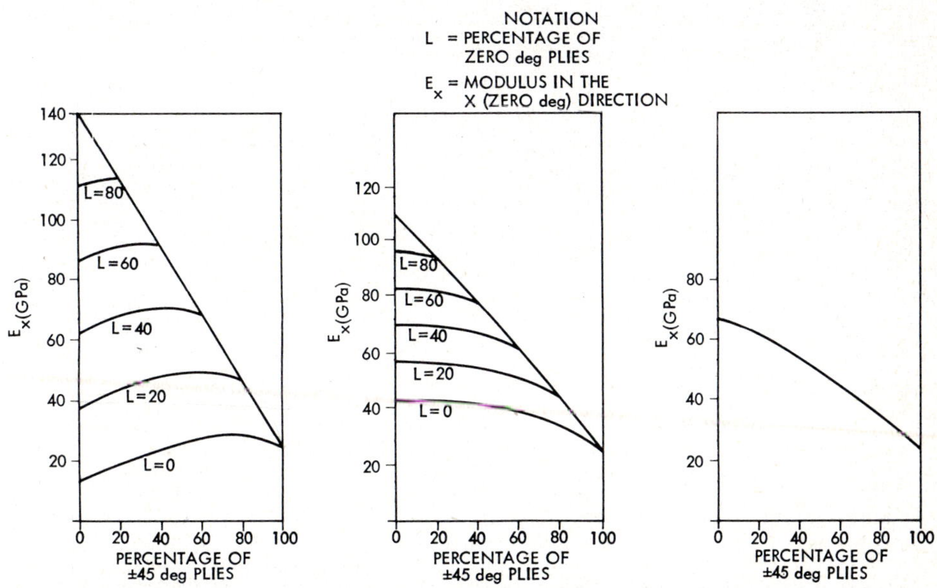

FIG. 6. Modulus of laminates fabricated from plies with three different warp/fill stiffnesses.

The most interesting feature of both the strength and modulus design charts for the three types of material is the way in which the design space closes up as the property difference in the two major directions is reduced. In the limit of the third case with the same warp and fill properties the design space for each property collapses to a single line. Note that all three materials give the same laminate property for 100 ± 45° plies. As expected there is very little difference in quasi-isotropic layups made from the three materials. It is also no surprise that the more anisotropic the fabric, the more extreme are the variations attainable for layups having a small percentage of ±45° plies. Once the basic warp and fill properties for the chosen fabric composite are measured, the relevant design charts can be easily produced and laminate design proceeds in the usual manner.

6.1. Additional data needed for the structural design of fabric structures

In addition to the lamination design charts, any designer of composite structures must have at hand many types of data to have adequate confidence in the structural integrity of the product. Among these categories are the influence of free edges, stress concentration around holes, fatigue strength and damage induced by impacts plus data on bolted and bonded joints. There is a great deal of published information on these subjects for tape laminates, but very little for their fabric counterparts. In what follows, attention is drawn to typical samples of this data and its similarities or differences relative to the copious tape data. Potential delamination at the edges of tape laminates is a thoroughly researched problem and is reviewed by HERAKOVICH (1979) and WANG and CROSSMAN (1980). It is well known that a major contributor to this failure mode is the very large difference of the two inplane Poisson's ratios with ν_{12} often being ten times ν_{21}. For those fabric materials with reasonably large fraction of their tows in the fill direction, the Poisson's ratios differ by much less, as do the moduli. This and what appears to be greater toughness in this failure mode, markedly reduces their propensity to delaminate and the literature does not appear to contain any cases of their edge delamination. This suggests that provided there are similar strengths in both primary directions, the designer can choose the fabric laminate stacking sequence without regard to this phenomenon. Those with a small fraction of the yarn in one direction should be treated as though they were unidirectional tape.

It is well known (SUAREZ, 1972; HONG and CREWS, 1979) that for tape laminates, containing a very large percentage of 0° plies, the stress concentration factor at hole edges can reach above 5 or 6. However, such layups are not likely to be chosen in practice and cannot be created using relatively balanced fabrics. The fraction of ±45° plies is likely to be made quite high, to reduce the stress concentration factor. Then there does not appear to be a marked difference between tape and fabric, as indicated by BAILIE et al. (1982) who provide data for graphite fabric laminates. In the tests reported there, the stress concentration relief manifested around small holes in composites is only slightly greater than for similar tape laminates. This small reduction in notch sensitivity of fabric laminates appears to be a general characteristic. The same situation occurs in bolted joints where the available evidence such as that by BAILIE et al. (1981) indicate the fabric laminates again exhibit strengths in the same range as similar tape reinforced materials, but are a little less notch sensitive.

In the most important areas of fatigue and impact damage, there do not appear to be any directly correlated studies between tape and fabric structures in the literature. However, the frequent use of fabric in a wide range of applications, such as those discussed in Section 7, suggests substantiating data is available to those responsible for structural intregrity of the components.

Thus, while there are clearly exceptions, it appears defensible, from a development engineer's perspective, to suggest that a great deal of the design data from tape laminates can also usefully serve as preliminary design data for similar fabric based structures. Where justified, this suggests that the structural design of fabric composites is not likely to present any obstacles to engineers familar with tape composities. Also there is a considerable data bank of experience with fabric laminates. Typical examples are included in the following sections on applications.

7. Fabric composite applications

It is desirable to discuss separately, each of a number of distinct categories.

7.1. Civil aircraft

Naturally glass fabrics were the first woven material utilized, beginning in the forties in a range of secondary, nonstructural components. In the first of these, relatively narrow strips of glass cloth, often impregnated with polyester resin, were wound over plaster molds of complex air conditioning ducts, cured in an oven and the mold broken to liberate the component. This method described by WOOD (1961) allowed geometries such as 'T' section ducts to be made very cheaply without thought of part removal from permanent tooling. In addition to their economic advantages, these ducts had longer fatigue lives than their metal counterparts. This was important as they were often attached close to compressors producing a severe vibration environment. These items were followed by geometrically complex wing and empennage tips involving severe double curvatures. Laying up glass cloth/polyester as a single piece on male tools provided fabrication advantages relative to the expensive tools needed for the deep drawing of light alloy components. Discrete reinforcing was implemented by grooves in the tool, into which the prepreg was forced until level with the tool surface. Then the surface plies were layed up, followed by the separator/breather ply, bag application of the vacuum and curing in an oven, as outlined by WOOD (1961).

By the late 1950s the early Boeing 707 aircraft appeared with roughly 22 square meters of glass cloth laminates making up lightly loaded fairings at fuselage-lifting surface attachments. As explained by THOMPSON (1981), the next step was to take advantage of cost and weight savings of roughly 25% which were attained relative to the original metal components such as rudder panels and control tabs. The early components were made by the labor intensive wet layup method of installing dry fabric, pouring on the resin, spreading it, bagging the laminate and curing. The resulting strength was satisfactory only for lightly loaded structures.

By 1962 the existence of fiberglass cloth epoxy prepregs brought about the introduction of cocured honeycomb structures without the need for film adhesives attaching the facesheets to the HRP (Heat Resistant Phenolic) honeycomb cores which also used glass fabric. The prepreg contained enough resin to ensure reliable bonds. The growth of fiberglass fabric usage on Boeing commercial aircraft is shown in Fig. 7 reaching a maximum of 1100 m^2 or 25% of the wetted area on the 747. Wing-fuselage fairings, rudder and elevator surfaces, wing leading and trailing edge panels plus the radome are responsible for 1200 lb of cured honeycomb sandwich on each aircraft. They are illustrated in Fig. 8. This fabrication and flying of over a million square meters (THOMPSON, 1981) of fiberglass fabric on Boeing commercial aircraft served as significant data base for the later development of similar graphite and aramid fabric components.

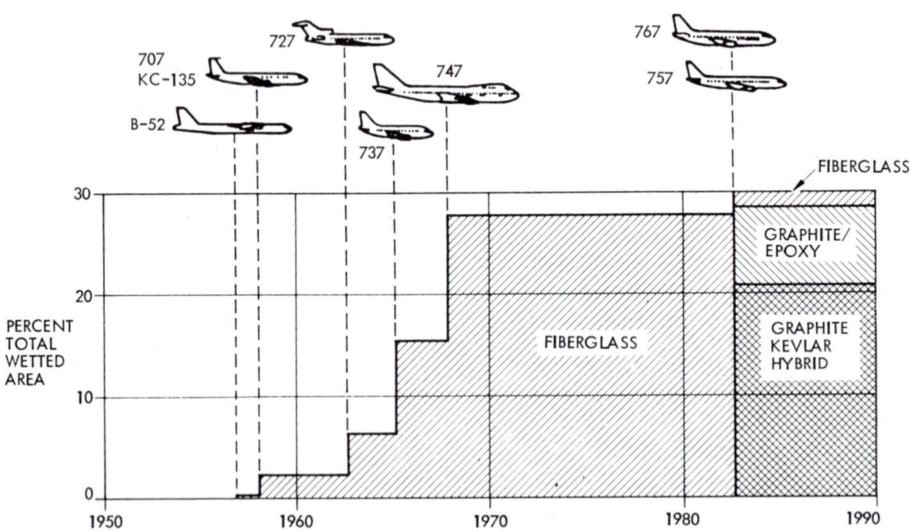

AIRPLANE	707	727	737	747
WETTED AREA (ft^2)	200	1800	3000	10,000

FIG. 7. Evolution of fabric composite surface area usage of Boeing aircraft.

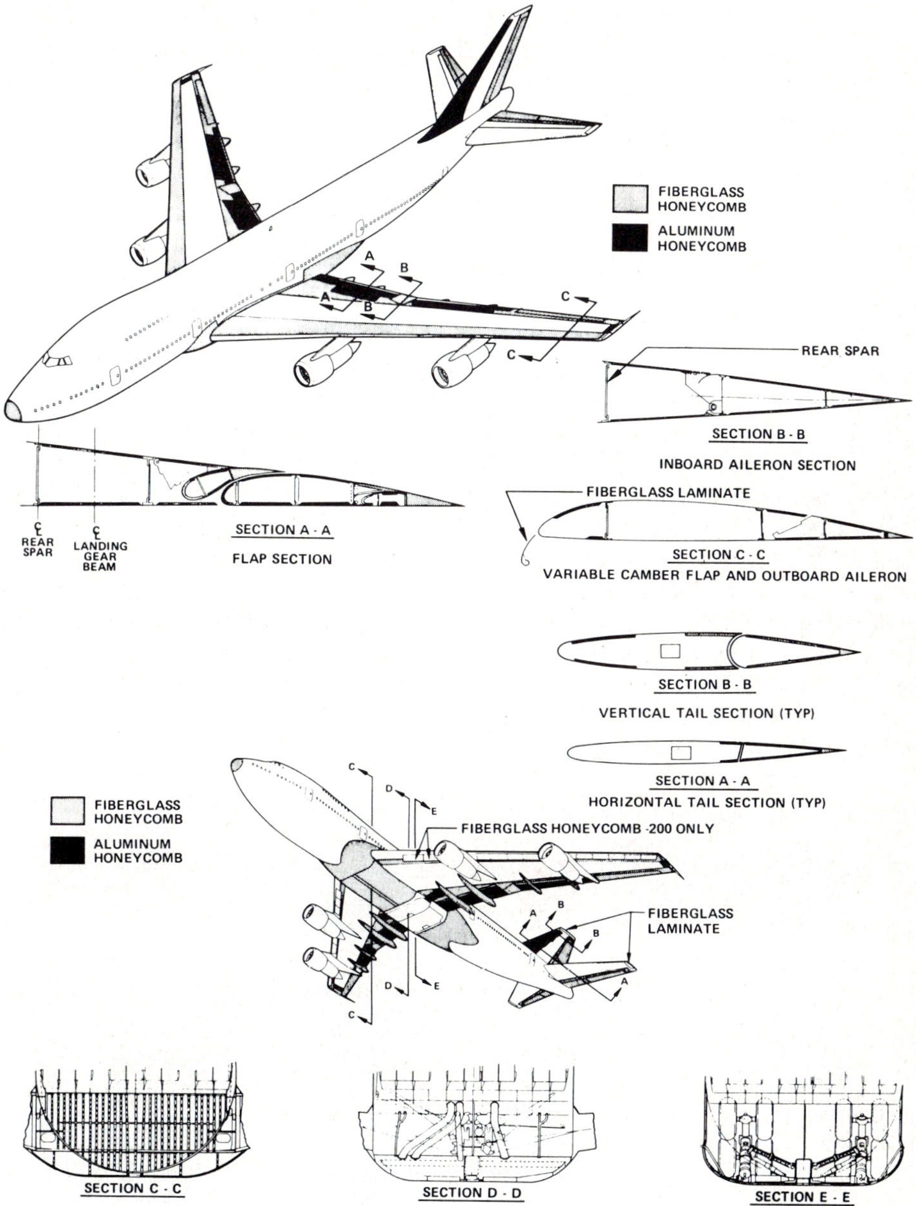

FIG. 8. Glass fabric facesheet honeycomb sandwich components on the Boeing 747.

Concurrently glass fabric was finding its way into similar secondary structural applications in a wide range of aircraft. Large wing/fuselage fairings were fabricated for the Lockheed L1011 (STONE, 1983) and the original Airbus (KOSHORST, 1983). In addition, an even greater amount of glass cloth was used in the cabin interiors of commercial aircraft. Non-structural parts such as the overhead luggage compartments, ceilings, galleys, bulkheads, and toilet compartments are routinely made of glass fabric laminates.

By the early seventies graphite and aramid components were being developed world wide for a significant array of components. Company funded 'in house' projects, government contracts and combinations of the two, lead to the rapid confirmation of the advantages attained by the judicious use of both aramid and graphite woven fabrics to supplement the existing use of glass. During this timespan many secondary and a few primary structures of existing aircraft were created using composites to replace the metal ones. Many of these contained fabrics. Using their fiberglass cloth experience, Boeing made significant use of 0.003 mm graphite fabric/epoxy in the 727 Elevator and 737 Stabilizer described by OHGI (1978) as well as fiberglass dielectric fabrics used for lightning protection. These materials were combined with graphite tape to build sandwich structures for skins, spars and ribs which were assembled into flight hardware by means of bolted joints. The weight gains of 29% relative to the metal components are summarized in Fig. 9. Replacing the existing fiberglass plies by aramid fabric on a ply by ply basis on wing-fuselage and engine fuselage fairing panels on the L1011 Tristar produced the typical 25% weight reduction. All are proving to be trouble free in service with one flight having exceeded 24 000 h service by late 1982 (STONE, 1983). The long range L1011-500 utilizes roughly 1130 kg of Kevlar® fabric for a weight savings of 360 kg on the fairings and panels shown in Fig. 10. Nomex honeycomb covered by no thicker than three ply facesheets is the predominant fabrication method, with the largest panel being 5.8 by 1.45 m in planform.

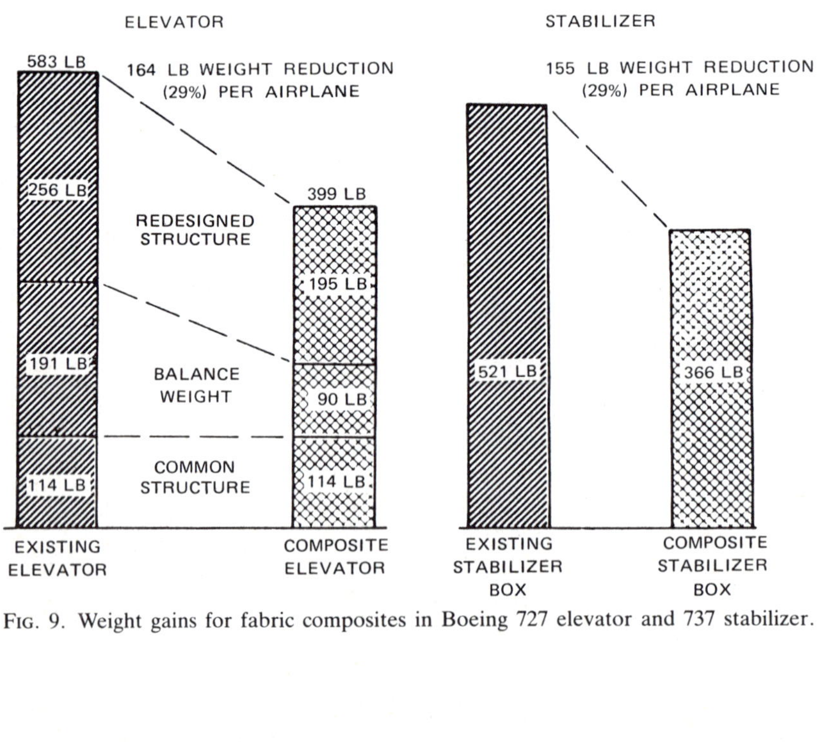

FIG. 9. Weight gains for fabric composites in Boeing 727 elevator and 737 stabilizer.

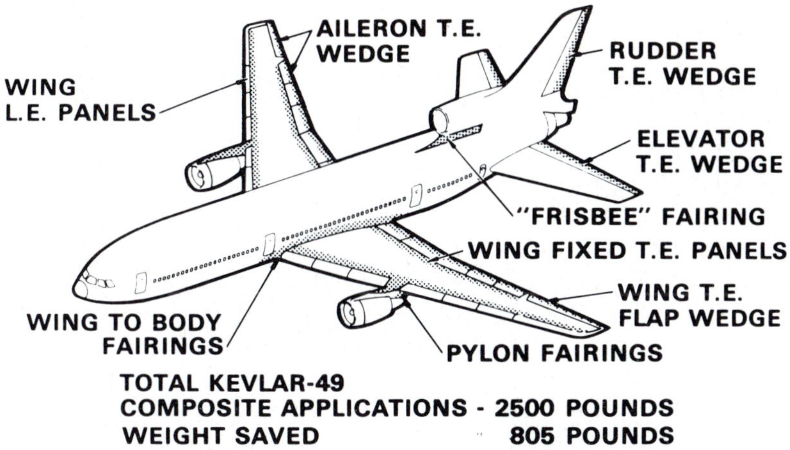

FIG. 10. Aramid fabric utilization on the Lockheed L1011-500.

Using an effective mix of glass, graphite and aramid fabrics on the fin leading and trailing edges, wing-body flap track and engine pylon fairings, the designers of the Airbus A310 have demonstrated a weight saving of 122 kg. As is frequently the case, a clear majority of these components, illustrated in Fig. 11 are sandwich honeycomb construction.

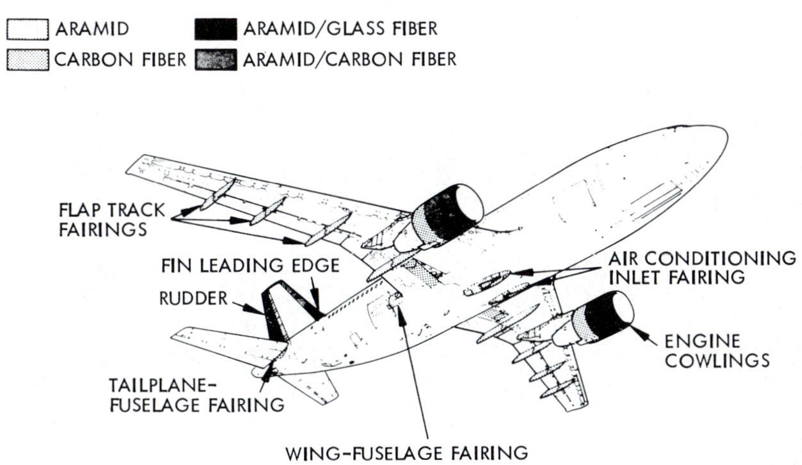

FIG. 11. Fabric facesheet sandwich components on the Airbus Industries A3.

DeHavilland's 50 passenger Dash 7 airliner uses roughly 227 kg of style 281 aramid fabric as the facesheets of honeycomb panels for cabin sidewalls, bulkheads and the flooring (JOHN, 1983). The original aluminum facesheet–balsa core sandwich intended for the floor panels weighed 4.9 kg/m while the fabric honeycomb design shown in Fig. 12 weighed only 2.74 kg/m resulting in

FIG. 12. Aramid fabric epoxy facesheets of the honeycomb core floor panels of the DeHavilland's Dash 76 transport (JOHN, 1983).

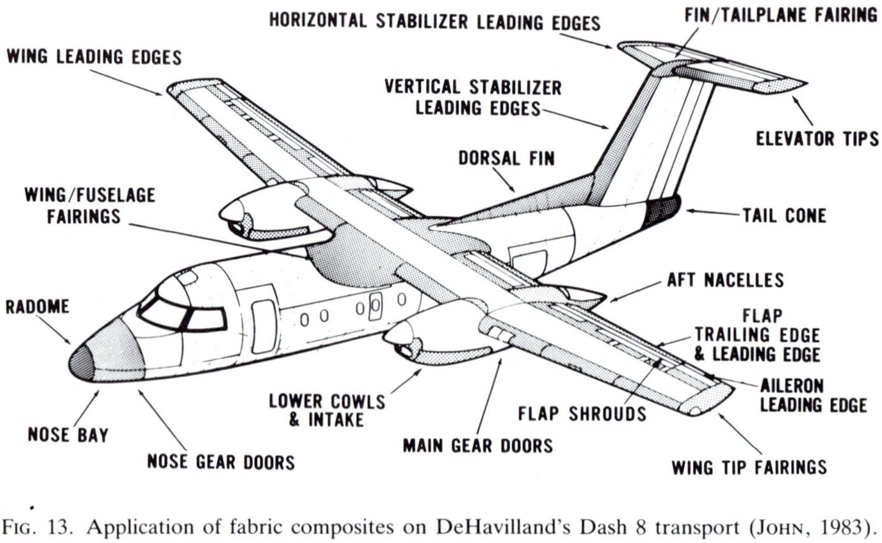

FIG. 13. Application of fabric composites on DeHavilland's Dash 8 transport (JOHN, 1983).

an airplane weight saving of 54 kg. Starting with a resin rich prepreg containing (52 + 2)% resin by weight, good surface finish and satisfactory concuring to the honeycomb core is obtained using only vacuum pressure for these very lightly loaded applications. This provides a clear economic advantage over the more conventional use of film adhesives and autoclave pressures which appear to be mandatory for flight structures, including the avionics compartment of this aircraft. It is yet another honeycomb sandwich (JOHN, 1978) with two ply style 281 aramid fabric inner and outer facesheets built to contain the 272 kg of avionics equipment. Expanded use of predominantly fabric facesheet honeycomb core sandwiches has been made to the landing gear doors, wing and stabilizer leading edges and fairings of the same company's Dash 8 aircraft, as shown in Fig. 13. Their use in the leading edges demonstrates the advances being made in composite structures whose major design condition is a severe impact. A hybrid of Kevlar® style 281 fabric and unidirectional tape is cocured with style 181 glass fabric and unidirectional glass. Naturally this sandwich is significantly unbalanced with the more impact prone forward outer facesheet being as much as five times thicker than the inner facesheet.

A detailed cost comparison of using fabrics versus unidirectional tapes in the DC-10 composite rudders is provided by COMINSKY (1979). The first ten rudders were made with unidirectional tape while the second ten used two different fabrics. Layup and mold assembly for the tenth tape rudder took

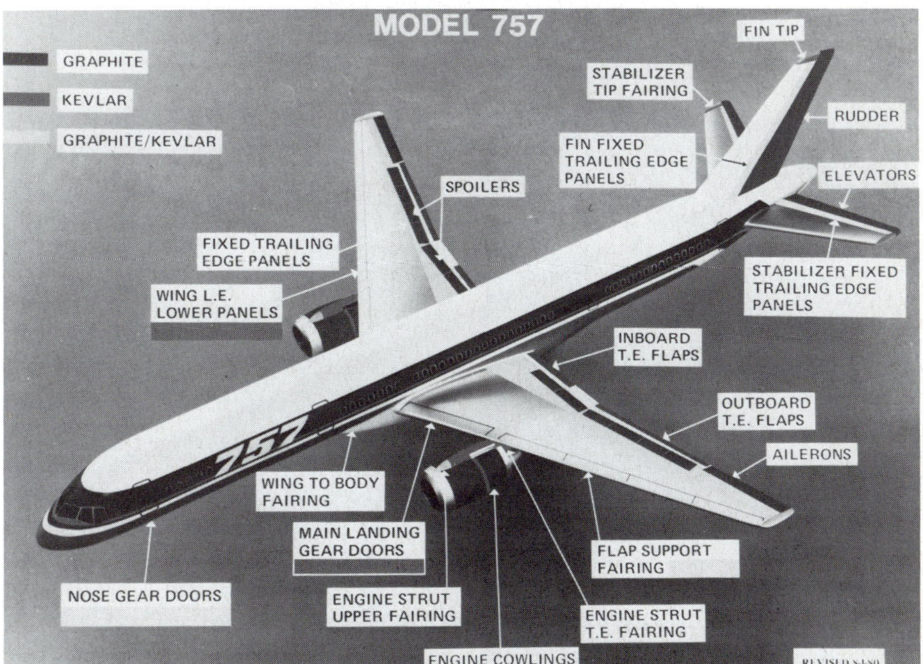

FIG. 14. Advanced composite applications on Boeing 757 aircraft.

484 h while the tenth fabric one absorbed 317 h for a 35% savings. This size of economic advantage accrued in a component whose surfaces were mostly flat. In highly curved parts even greater savings have been attained.

A very significant step in the use of composites in flight vehicles has been the commitment to all composite elevators, ailerons, rudders, spoilers and engine cowlings on the Boeing 757 and 767. The former uses 1520 kg of composites for a weight saving relative to metal of 680 kg while for the latter these weights are 1540 and 567 kg, respectively (BROWER, 1980). The major material types in all these components are glass, aramid and graphite fabrics used as facesheets of

FIG. 15. Three phases in the creation of the 767 rudder: (a) layup to the outer skin and honeycomb core; (b) cured cover panel; (c) installation of the complete rudder.

honeycomb sandwiches. To combine the most effective properties of each, many hybrids are used in layups that depend on the relative importance of factors such as impact resistance and compression strength. The locations of each type on the 757 is evident from Fig. 14 while various phases of 767 rudder construction are shown in Fig. 15.

A most prolific user of graphite fabric as a fraction of its structure was the Learfan 2100 shown in Fig. 16. Roughly 70% of its structural weight is composites. Of that 544 kg or 90% is fabric. Wing, fuselage and empennage primary structures are predominantly graphite fabric, while the control surfaces are mainly aramid fabric. Their fabrication is reviewed by NOYES (1983).

Engine nacelles, although not considered an airframe structure, are an integral part of the power plant installation. As noted by Medewar (1983) they often share load carrying capability with the engine frame and mounting system. The use of composite fabric, both graphite and aramid, is becoming more intensive, mainly for minimum weight considerations. Although unidirectional tape is used where load paths are clearly defined, composite fabrics comprise the main structural laminate shells for these components. Most of the aerodynamic surfaces that form the nacelle part of the power plant installation are of compound curvature and the cross-section throughout the structure and may be noncircular. Although strength and stiffness requirements might indicate unidirectional material would be the most desirable, the difficulty of hand laying-up these materials onto compound female molds and maintaining the unidirectional properties is neither practical nor cost effective. In most cases

FIG. 16. Learfan 2100 aircraft.

fabric material equals the resulting stiffness of the hand layed-up unidirectional structure. A typical example of the many predominantly fabric composite nacelles is that built by Rohr Industries for the Pratt and Whitney JT9D-7R4 engine which is shown in Fig. 17.

FIG. 17. Fabric composite elements of the nacelle for the Pratt and Whitney JT9D-7R4 engine.

7.2. Military aircraft

Unlike civil aircraft, much of the early production composite structures for fighters has been unidirectional tape, starting with the boron epoxy horizontal stabilizer of the Grumman F-14 and the McDonnell F-15. This was followed by a significant commitment to graphite tape in the same components of the General Dynamics F16 and wing skins, horizontal stabilizer, fins, rudders and speed brake on the McDonnell/Northrop F18. The absence of significant curvature, highly directional loading and stiffness requirements of these lifting surfaces, all designed for operation at much higher dynamic pressure than civil aircraft, made tapes the preferred choice. In the AV-8B, there are substructure components with different design criteria which resulted in the choice of bidirectional woven graphite cloth (RILEY, 1979) for the eight wing spars and 15 chordwise ribs, illustrated in Fig. 18. Tape is used for the skins. Conventional flat spar and rib webs with large numbers of separately fabricated stiffeners would have led to many parts and expensive secondary manufacturing and

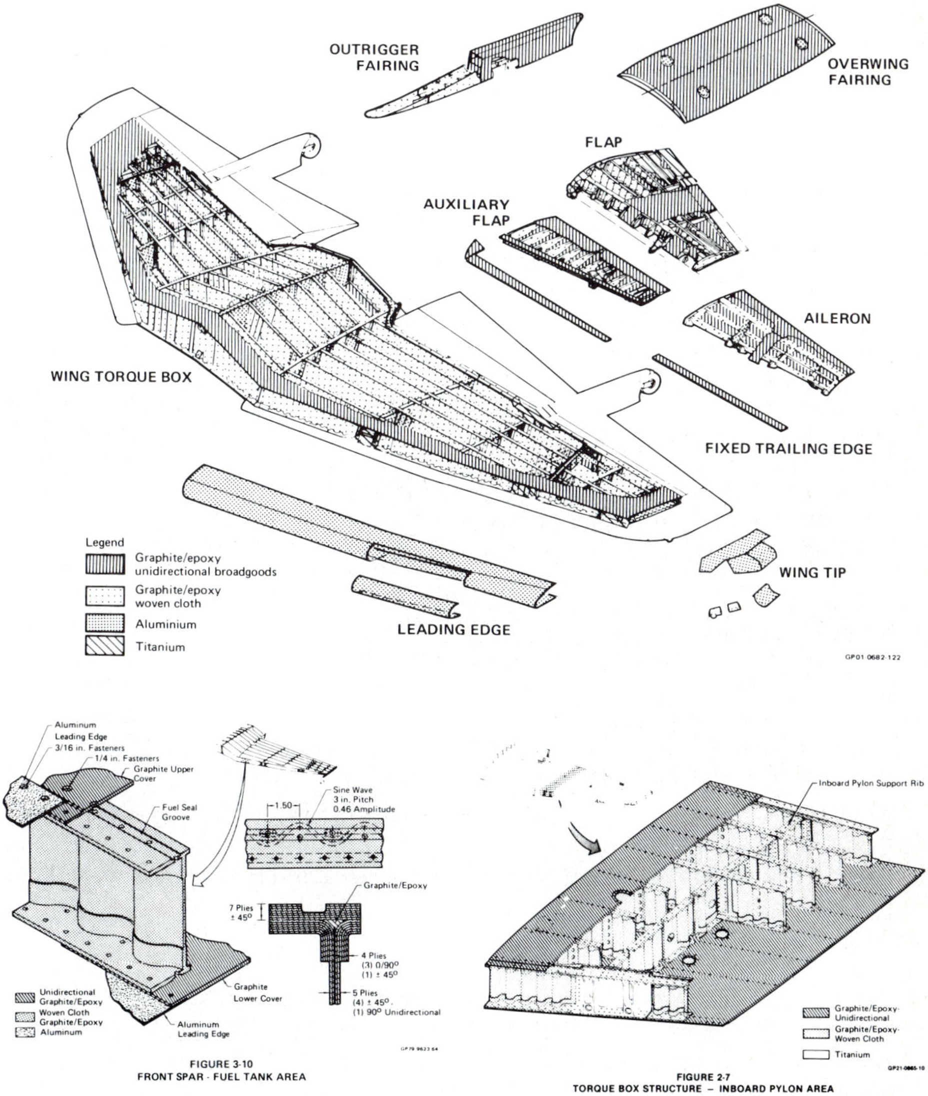

FIG. 18. Graphite fabric spars and ribs in the wing of the McDonnell AV-8B (RILEY, 1983).

assembly. A simpler and hence superior solution was the use of bidirectional fabric to create the sine wave spars shown in Fig. 18 and made in matched metal dies.

Apart from tape stiffener and longeron caps, graphite fabric was chosen for the forward fuselage for some of the same but some different reasons, explained by WATSON (1982) and illustrated in Fig. 19. Manufacturing of this type of highly curved structure is hardly feasible with existing tape laying

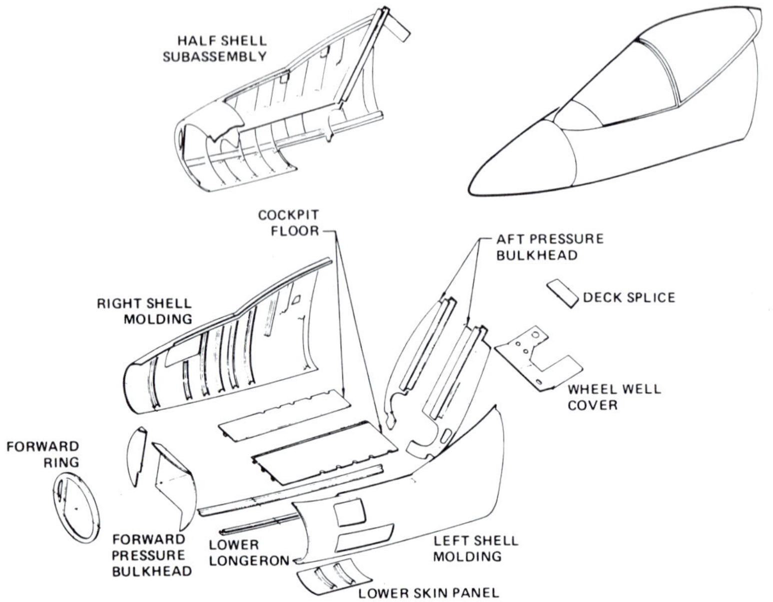

FIG. 19. Forward fuselage of the McDonnell AV-8B fuselage which utilizes graphite fabric composites extensively (WATSON, 1982).

machines. At least five 8 harness satin fabric plies are used in the skin sidewalls with cocured hat stiffeners. Similar construction is used for the floor, both bulkheads and frames. Table 7 quantifies the simplicity of this type of integral construction, as demonstrated by the almost three fold reduction of parts and fasteners relative to the metal structure it replaces at a weight saving of 25%.

TABLE 7. AV-8B Forward fuselage part and fastener count (WATSON, 1982).

	Conventional riveted aluminum	Cocured composite
Parts	237	88
Fasteners	6440	2450

7.3. *Helicopters*

In addition to the usual reasons for choosing composites, namely lightweight and reduced manufacturing costs, they are now frequently used in helicopter flooring to circumvent the serious corrosion that afflicts their metal counterparts. Honeycomb sandwiches with considerable if not total use of fabrics in the facesheets are installed in the Boeing Vertol CH46, CH47 and YUH61A which use fiberglass. They are configured as described by MARCHINSKI and

GRECO (1979). As frequently occurs in integrally stiffened composites a cost advantage was obtained by elimination of thousands of rivets and bolts attaching the discrete stiffeners in the metal design.

The multiple curvature of many fuselage components, combined with relatively low loading make them naturals for fabric composites. Both discretely reinforced and honeycomb sandwich configurations are naturally popular. Forward and aft fairings of the Sikorsky S-76 are discussed by RAY (1981) and are shown as Fig. 20. They are typical examples of discrete reinforced aramid cloth components. Another interesting use of fabric (MARCHINSKI, 1983) is the two fuel pods containing 6350 kg of fuel on the Boeing Model 234 illustrated in Fig. 21. Apart from the fore and aft bulkheads which use mostly tape and only one cloth ply, the tank structure, shown in Fig. 21(c), is 1.5 in deep honeycomb core with the facesheet layup of 4 aramid fabric plies surrounding graphite tape plies in the hoop (90°) direction.

Significant use of fabric brought about by geometric complexity is the shrouded tail rotor or 'Fenestron' of helicopters such as the 'Dauphin' depicted in Fig. 22 and discussed by BEZIAC (1983). All the sandwich components of the fin/fenestron in the exploded view of Fig. 22 use fabric in their facesheets. Economic savings brought about by the reduction in number of parts and fasteners in the fin led to a 44% reduction in production costs, relative to the original metal structure (Table 8). This was obtained while concurrently

FIG. 20. Typical complex geometry aramid fabric facesheet components on the Sikorsky S76 helicopter fuselage.

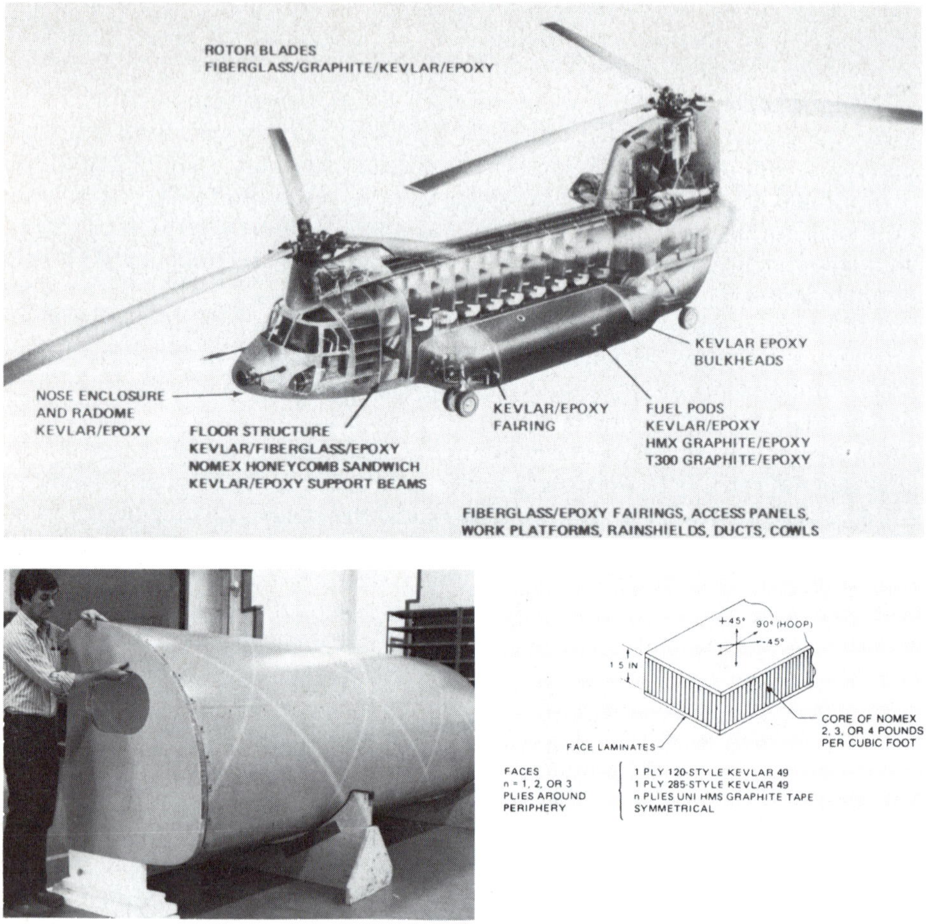

FIG. 21. Fuel pods on the Boeing Vertol Model 234: (a) the vehicle showing fuel pods on fuselage exterior; (b) completed subassemblies; (c) sandwich construction of fuel pod walls.

reducing the weight by 22%. Future generations of helicopters, as exemplified by the Bell Textron and Sikorsky entries in the US Army's Advanced Composite Airframe Program (ACAP) utilize a considerable amount of fabric reinforcing, woven from all three major fibers. The Bell configuration is reviewed by ALSMILLER et al. (1983).

TABLE 8. Part and fastener count of Dauphin helicopter fin (BEZIAC, 1983).

	Conventional riveted aluminum	Composite sandwich
Parts	231	88
Fasteners	5900	None–2 phase bonding

FIG. 22. Fenestron of the Aerospatiale Dauphin helicopter. All sandwich components have fabric facesheets.

7.4. *Radomes*

Conflicting demands on these components of the radar system and their structural performance place these electromagnetic windows in a unique class which is reviewed by WALTON et al. (1970). While frequently thought to be secondary structures, their failure can initiate devastating events in the airframe. Wall thickness is dictated by the radar wavelength and either solid (half wave) wall or sandwiches with two facesheets ('A' Sandwiches) are most often used. Many aspects of their use are described by CHASE (1970). Fabric reinforcement is generally used for sandwich radomes of all configurations and in solid laminate geometries other than some very highly streamlined geometries. Due to the double curvature 8 H.S. weaves are the most commonly chosen. 'E' glass fabric was the standard material for the first decades of airborne radar and continues to be very frequently used. The 'A' sandwich radome of the Grumman F14 fighter shown in Fig. 23, is a well known example

FIG. 23. The radome on the Grumman F-14 is a honeycomb sandwich with glass fabric facesheets.

of its use. Quartz fabric is employed where its reduced dielectric constant and loss tantent produce worthwhile gains in radar performance. Most recently Kevlar fabrics have been applied in particularly weight critical radomes. There are also various fabrication techniques. As an illustration of the matched die method, the half wave radome for the Concorde SST is made by first fitting about twenty-five dry glass fabric socks over the male mold. Then the female mold is located on it and the space between them evacuated to about 25 in Hg. Then epoxy resin pumped in at 0.028 MPa over a 4 h period prior to curing. Vacuum bag molding is a popular choice for 'A' sandwich domes. Typically the fabric outer facesheet in prepreg form or with separately applied resin is laid up on the female mold defining the aerodynamic profile. It can be cured separately or the honeycomb and inner facesheet layed up prior to cocuring. This type finds a wide range of applications of aircraft such as the LTV-A7 attack aircraft and virtually all the civil jet airliners.

The Lockheed C5 transport has used two different radome designs, both involving Kevlar® fabric epoxy. The original C5A version had a 3 m long, 5 m base width multilayer sandwich radome. This configuration was required to satisfy the needs of multiple antennas operating over a wide range of frequencies. The newer versions of the aircraft have only weather radar in the nose so the redesigned radome is just 1 m long with a 2.7 m width. It is an A sandwich with Kevlar facesheets. The structural 'plug' between the radome and the fuselage is a monolithic fabric skin with discrete stiffners. Both elements are shown in Fig. 24. In the lower photograph the radome–plug interface is more clearly visible.

Another radome of significantly different geometry houses the Airborne Warning and Control Radars mounted on the rear fuselage of the Boeing E-3A Sentry aircraft, shown in Fig. 25. The 10 m diameter, 2 m maximum depth section is a type 'C' sandwich distinguished by three facesheets of fiberglass fabric epoxy, separated by two honeycomb cores. As so often occurs the choice of fabric rather than tape was due, in large measure, to the severe double

FIG. 24. Multilayer sandwich radome for the Lockheed C-5A transport.

FIG. 25. Thirty foot diameter radome on the Boeing E3A Sentry AWACS aircraft.

curvature around the outer perimeter. The unique manufacturing problems posed by this component and their solution are summarized by ROGERS (1979).

A special type of radome, outside the mainstream of mass produced units are those subject to extreme thermal environment. Nose mounted domes on supersonic aircraft and missiles are also subjected to severe erosion in rain and have led to the development of inorganic materials including ceramics. These are reviewed by WALTON (1970, Ch. 5). There is an intermediate class of radomes housing Electronic Warfare antennas whose power output is sufficient to contribute to the heating of the radome. One such example is flown aboard the Grumman EF-111A aircraft shown in Fig. 26 and is discussed by PAEZ et

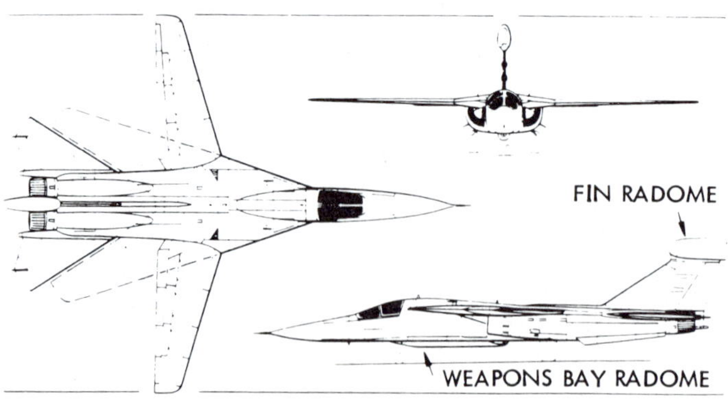

FIG. 26. Weapons bay and fin radomes on the General Dynamics EF-111A.

al. (1978). In this case the combined aerodynamic and RF induced heating leads to radome structural temperatures close to 260°C. This necessitated the development of condensation polyimide resins to replace epoxy. Their prepregging into fiberglass, the resulting laminate strength data and the construction of the complex geometry honeycomb sandwich weapons bay and fin mounted radomes are also contained in the work by PAEZ et al. (1978).

7.5. Spacecraft and missiles

Undoubtedly the best known use of fabric composites in spacecraft is the 60 ft long payload bay doors of the Space Shuttle. In Fig. 27 they are shown open during the flight of Challenger in July 1983. Each of the two doors consists of four panel segments easily distinguishable in the photograph. Expansion joints between them to allow for the body bending distortions and the larger thermal distortions of the vehicle's aluminum structure. Two thicknesses (0.0028 cm and 0.0051 cm) of graphite epoxy fabric are used in conjunction with unidirectional tape. Each panel segment illustrated in Fig. 28 consists of eight honeycomb sandwich sections, all fabricated at the same time to create the panel. Nine hat section frames are bolted to the panel to provide the required strength and stiffness. BRUNER (1976) includes the 1975 preliminary design data for the two fabrics. Since then the design allowables have been raised to exceed the specification requirements quoted.

FIG. 27. Space shuttle Challenger with cargo bay doors open.

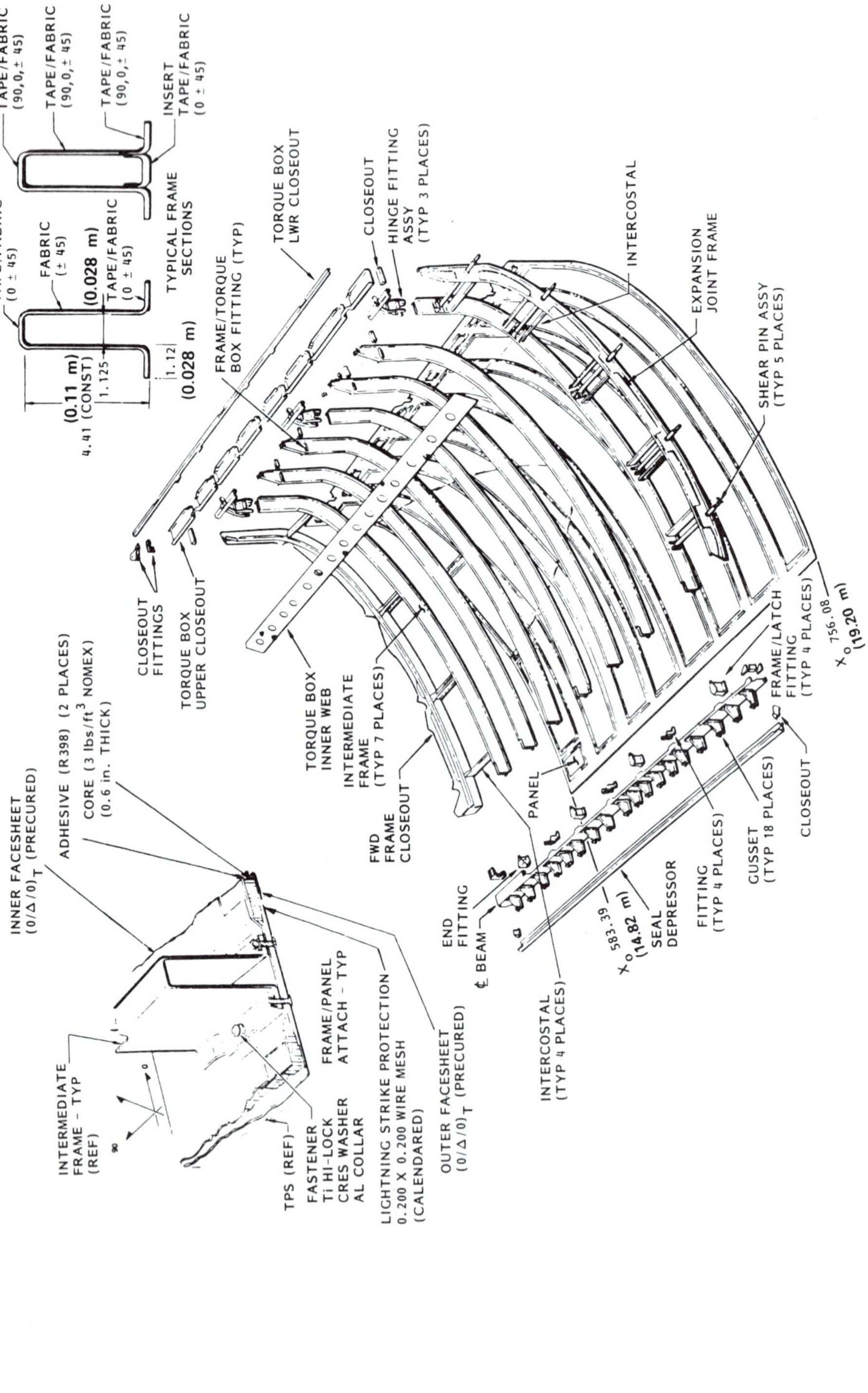

FIG. 28. Construction details of the predominantly fabric space shuttle cargo bay doors.

The multidimensional curvature of relatively small radius components in the upper stage of the TRIDENT I missile resulted, in the early seventies, in manufacturing difficulties using unidirectional graphite epoxy tape. This provided the stimulus to produce graphite fabric, whose original data is reviewed by YATES et al. (1975). Substitution of fabric throughout the components except for a few tape plies to provide extra hoop stiffness in one frame, eased the layup problems considerably. The ensuing gain in structural integrity of complex geometries led to the adoption of fabric for virtually all the upper stage primary structure, which is reviewed by WONG and VOLLERSEN (1979).

7.6. Tooling

There are three basic reasons why fabric composite tools offer distinct advantages over their metal counterparts in the manufacture of some parts. One is just basic economics. Layup molds or tools for large area, doubly curved fairings can be very expensive to machine from metals. A potentially cost effective solution is to build a plaster mold as explained by BENJAMIN (1972), spray on a gel coat, layup tooling prepreg and cure the mold. Usually the best structural support is provided by honeycomb 'egg crate' supports which allow autoclave gas to reach the rear of the mold face and thereby maximize heating efficiency of the part. In addition to the economic advantages of this type of tooling, its second advantage is its reduced thermal mass and rear face heating offer distinct improvements in cure cycle control, relative to massive metal tools. There is always concern about the sensitivity of this type of tooling to shop damage. However, experience has shown repairs to be easily accomplished. In the case of disastrous damage new tools can be fabricated in a few days, using the original mold. Special tooling prepregs using graphite fabrics are now readily available from major prepreggers. The Learfan was one of the more extensive early users of this type of material for its tooling. The third reason for considering composite tooling is in those situations where it is desirable to match the coefficient of thermal expansion of the part and tool. The large difference in this property between metal tools and graphite composites can lead to severe difficulties in complex geometries. Using graphite composite tools resolves this problem.

8. Conclusions

Fabric composities are a viable choice for numerous aerospace structural and non-structural configurations. In some components they offer distinct economic and structural integrity advantages over competing materials. They are widely used and there is no doubt they will continue to be employed in the future, in an expanding range of applications. One of their great advantages is their design flexibility. Fabrics can be configured through specific choice of weave style, yarn size and count, and other parameters under the designer's control, to fulfill the design objectives. Attention has been drawn to the large data base that already exists.

Acknowledgement

The author would like to express his sincere appreciation to J.T. Quinleven, J. Levin, R. Davis and D.W. Wood for their help and constructive comments.

References

ALSMILLER, G.R., G.O. SORREBORN and R.W. METZGER (1983), The all composite rotocraft, *Proc. 39th Annual Forum of the American Helicopter Society*, 9–11 May, 1983.

ARVIN, G.H. (Editor in Chief) (1973), *Advanced Composite Design Guide* (Air Force Materials Laboratory).

BAILIE, J.A., M.F. DUGGAN, N.C. BRADSHAW and T.G. MCKENZIE (1981), Design data for graphite cloth epoxy bolted joints at temperatures up to 450 K, in: K.T. KEDWARD, eds., *Joining of Composite Materials*, ASTM STP 749 (American Society for Testing and Materials, Philadelphia) p. 165.

BAILIE, J.A., M.F. DUGGAN, L.M. FISHER and R.C. YEE (1982), Effects of holes on graphite cloth epoxy laminates tension strength, *J. Aircraft* **19**, No. 7, 559.

BENJAMIN, W.P. (1972), *Plastic Tooling–Techniques and Applications* (McGraw Hill, New York).

BEZIAC, G. (1983), Applications of composite materials in helicopter fabrication, *Proc. Fourth 'Kevlar in Aircraft' Symposium* (DuPont, Geneva).

BROWER, G.L. (1980), Advanced graphite composites in the 757/767, Paper No. 801212 in: *Society of Automotive Engineers Inc. Aerospace Congress*, Los Angeles, October 1980.

BRUNER, R.C. (1976), Material selection for the Space Shuttle Orbiter payload bay doors, *Proc. 21st National SAMPE Symposium and Exhibition*, April 1976, p. 459.

CHASE, V.A. (1970), Organic radomes, in: J.D. WALTON, ed., *Radome Engineering Handbook* (Marcel Dekker, New York) p. 127.

COMINSKY, A. (1979), Manufacturing development of the DC-10 advanced composite rudder, *NASA CR 159060*.

DUPONT Company Inc. (1974), Kevlar Data Sheets.

EKVALL, J.C. and C.F. GRIFFIN (1982), Design allowables for T300/5208 graphite/epoxy composite materials, *J. Aircraft* **19**, No. 8, p. 661. Errata in *J. Aircraft* **20**, No. 6 (1983) p. 576.

FLAGGS, D.L. (1982), ADVLAM—an advanced composite laminate analysis code, *Lockheed Report No. LMSC D812170*.

HERAKOVICH, C.T., A. NAGARKAR and D.A. O'BRIEN (1979), Failure analysis of modern composite laminates with free edges, in: J.R. VINSON, ed., *Modern Developments in Composite Materials and Structures* (American Society of Mechanical Engineers, New York).

HONG, C.S. and J. CREWS (1979), Stress concentration factors for finite orthotropic laminates with a circular hole and uniaxial loading, *NASA TP1469*.

ISHIKAWA, T. and T.W. CHOU (1982), Elastic behavior of woven hybrid composites, *J. Comp. Mater*. **6**, 2.

ISHIKAWA, T. and T.W. CHOU (1983), One-dimensional micromechanical analysis of woven fabric composites, *AIAA J.* **21**, No. 12, 1714.

JOHN, L.K. (1983), Aramid composite applications at DeHavilland—an update, *Proc. Fourth 'Kevlar in Aircraft' Symposium* (DuPont, Geneva).

JOHN, L.K. and L.L. BRYSON (1978), The design and fabrication of the DeHavilland DHC-7 nose avionics compartment using aramid composites, *Canadian Aeronautics and Space J.*, **24**, No. 6, 354.

KNOX, C.E. (1982), Fiberglass reinforcement, in: G.E. LUBIN, ed., *Handbook of Composites* (Van Nostrand Reinhold, New York) p. 136.

KOSHORST, J. (1983), Advanced composite structures in commerical transport aircraft, *Proc. Fourth 'Kevlar in Aircraft' Symposium* (DuPont, Geneva).

LEVIN, J. (1975), Processing aerospace textiles into fabric composite reinforcements: 'The Weaver's Viewpoint' *SAMPLE Quarterly*, p. 20.

LUBIN, G. and S. DASTIN (1971), First boron composite structural production part, *Proc. 26th Annual Conf.* (SPI Reinforced Plastics/Composites Institute, New York).

MARCHINSKI, L.J. (1983), Advanced composites in production at Boeing Vertol, *Proc. Fourth 'Kevlar in Aircraft' Symposium* (DuPont, Geneva).

MARCHINSKI, L.J. and C.A. GRECO (1979), The application of composites to secondary structures, *J. Am. Hel. Soc.* p. 19.

MEDEWAR, G.E. (1983), Unpublished Rohr Industries Data.

MORGAN, P., ed. (1961), *Glass Reinforced Plastics* (Iliffe Books, London).

MORGAN, R.J. (1980), The failure modes and durability of Kevlar/epoxy composites, *Proc. 12th National SAMPE Technical Conf.*, p. 368.

MORTON, J.H. (1982), *Lecture Notes on Weaving*, unpublished DuPont Company Report.

NOYES, J.V. (1983), Composites in the construction of the Learfan 2100 Aircraft, *Composites* **14**, No. 2, 129.

OHGI, G.Y. (1978), Advanced composite 727 elevator and 737 stablizer programs, *Proc. 23rd National SAMPE Symposium*, p. 1131.

PAEZ, C.A., L.M. POVEROMO and A.R. Sarraffe (1978), Radome design/fabrication criteria for supersonic EW aircraft, *Proc. 10th National SAMPE Technical Conf.* p. 166.

POULLOS, N.E. (1970), Historical development of radomes, in: J.D. WALTON, ed., *Radome Engineering Handbook* (Marcel Dekker, New York) p. 1.

RAY, J.D. (1981), Status of advanced composites use on helicopters at Sikorsky Aircraft, *Proc. Third 'Kevlar in Aircraft' Symposium* (DuPont, Seattle, Los Angeles, Philadelphia) p. 33.

RILEY, B.L. (1979), Composite wing technology on the AV-8B advanced aircraft, *J. Am. Hel. Soc.* April, p. 29.

ROGERS, C.B. (1979), E3A sentry radome fabrication and tooling, *Proc. 11th National SAMPE Technical Conf.*

SHIBATA, N., A. NISHIMURA and T. NORITA (1976), Graphite fiber's design and composite properties, *SAMPE Quarterly*, July, p. 25.

SHULOCK, H. and R.R. SAFFADI (1982), High silica and quartz, in: G.E. LUBIN, ed., *Handbook of Composites* (Van Nostrand Reinhold, New York) p. 160.

STONE, R.H. (1983), Composite flight service experience at Lockheed–California Company, *Proc. 15th National SAMPE Technical Conference*, p. 277.

SUAREZ, J.S. (1972), Vulnerability of composite aircraft structures, *AFFDL-TR-72-8*.

THOMPSON, V.S. (1981), Composite applications on Boeing Commerical Aircraft, *Proc. 26th National SAMPE Symposium*, SAMPE, p. 181.

WALTON, J.D., ed. (1970), *Radome Engineering Handbook* (Marcel Dekker, New York) p. 229.

WANG, A.S.D. and F.W. CROSSMAN (1980), Initiation and growth of transverse cracks and edge delaminations in composite laminates, *J. Composites Suppl.* **14**, p. 71.

WATSON, J.C. (1982), AV-8B composite fuselage design, *J. Aircraft* **19**, No. 3, 235.

WONG, D.G. and C.A. VOLLERSEN (1979), Advanced material applications on the Trident I missile, in: S.W. TSAI, ed., *Composite Materials: Testing and Design* (*Fifth Conference*), ASTM STP 674 (American Society for Testing and Materials, Philadelphia) p. 14.

WOOD, R. (1961), Glass reinforced plastics in the aircraft industry, in: P. MORGAN, ed. (Iliffe, London) p. 267.

YATES, D.N., R.D. TORCZYNER and D.R. SIDWELL (1975), Development and application of woven graphite products for aerospace structures, Paper No. 75-774, *16th Structures, Structural Dynamics and Materials Conf.*, Denver. Co. 27–29 May.

CHAPTER 8

Composite Beam Structures

A.M. Skudra and F.Ya. Bulavs

Department of Structural Mechanics
Riga Polytechnic Institute
226355 Riga
U.S.S.R.

Contents

HANDBOOK OF COMPOSITES, VOL. 2 – Structure and Design
Edited by C.T. HERAKOVICH and Y.M. TARNOPOL'SKII

List of Symbols

x, y, z – arbitrary axes
$1, 2, 3$ – axes elastic symmetry of a laminate
$\|, \perp, \perp\!\!\perp$ – axes of elastic symmetry of a unidirectionally reinforced plastic
$\langle\!\langle \sigma \rangle\!\rangle, \langle\!\langle \tau \rangle\!\rangle$ – mean stresses in a laminated plastic
$\langle\!\langle \varepsilon \rangle\!\rangle, \langle\!\langle \gamma \rangle\!\rangle$ – mean strains of a laminated plastic
$\langle \sigma \rangle, \langle \tau \rangle$ – mean stresses in a unidirectionally reinforced plastic
$\langle \varepsilon \rangle, \langle \gamma \rangle$ – mean strains of a unidirectionally reinforced plastic
k – curvature
$\bar{\sigma}_{rz}, \bar{\sigma}_r, \bar{\sigma}_\theta, \bar{\sigma}_z$ – non-dimensional structural parameters of a composite characterizing stress concentration
$\nu_{\perp, \perp\!\!\perp}$ – Poisson's ratio determining lateral strain in direction $\perp\!\!\perp$ with the load acting in direction $\perp$
$E_{\mathrm{f}z}, E_{\mathrm{f}r}$ – elasticity moduli of anisotropic fibres in axial and radial directions
$G_{\mathrm{f}rz}$ – shear modulus of anisotropic fibres
$\nu_{\mathrm{f}zr}$ – Poisson's ratio of anisotropic fibres
$E_{\mathrm{m}}, \sigma_{\mathrm{m}}, \nu_{\mathrm{m}}$ – elastic properties of the matrix
Q_{ij} – component of the reduced stiffness matrix of a unidirectionally reinforced plastic
Q_{ij} – component of the reduced stiffness matrix of a unidirectionally reinforced plastic in the coordination system rotated through angle β
$\overset{+}{\sigma}_{\mathrm{m}}, \overset{\mathrm{u}}{\tau}_{\mathrm{m}}$ – matrix tensile and shear strengths
$\overset{\mathrm{u}}{\sigma}_{\mathrm{b}}, \overset{\mathrm{u}}{\tau}_{\mathrm{b}}$ – tensile and shear strengths of the fibre matrix bond
$\overset{+}{e}_{\mathrm{fu}}, \overset{-}{e}_{\mathrm{fu}}$ – ultimate strains of fibres under tension and compression
$\overset{+}{\sigma}_{\perp}, \overset{-}{\sigma}_{\perp}$ – tensile and compressive strengths of a unidirectionally reinforced plastic in the transverse direction
M, S, N – bending moment, shear and axial force
β – fibre orientation angle
$\Delta_{i\mathrm{p}}$ – displacement of the applied load
δ_{ij} – displacement of unit load
X_{i} – redundant forces in statically indeterminate systems
S_{ij} – component of the compliance matrix of a unidirectionally reinforced plastic

$\bar{S}_{ij}$ – component of the compliance matrix of a unidirectionally reinforced plastic in the coordinate system rotated through angle β

A_{ij}, B_{ij}, D_{ij} – components of membrane, membrane-bending and bending stiffness inverted matrices

A'_{ij}, B'_{ij}, D'_{ij} – components of membrane, membrane-bending and bending compliance matrices of a laminated plastic

T_{ij} – transposed matrix components of a unidirectionally reinforced plastic

$k_{ij}(t-\tau)$ – creep kernels of a unidirectionally reinforced plastic

$\tilde{A}'_{ij}(t-\tau)$, $\tilde{B}'_{ij}(t-\tau)$, $\tilde{D}'_{ij}(t-\tau)$ – creep kernels of a laminated reinforced plastic

Δ_T, X_T, M_T – displacements, redundant forces and bending moments in statically determinate and indeterminate laminated beam systems determined without considering shear forces

h – depth of the cross section

1. Introduction

The strength of reinforced plastics depends on the structure of the material. In order to take into account the effect of the microstructure it is necessary to apply structural mechanics methods as discussed by SKUDRA (1984). It ought to be kept in mind that at present, structural mechanics methods for predicting the strength of reinforced plastics in bending and axial loading are far from sufficiently developed. This is partly explained by the fact that accurate results are extremely difficult to obtain even in the simplest cases of loading. Results obtained by TARNOPOLSKII and ROZE (1969), CHRISTENSEN (1979) and MOSER (1982) for evaluating the stress state of an anisotropic laminated beam are clear evidence of this.

The classical bending theory is applicable to laminated plastics only when the structure of a laminated beam is symmetric about its midplane and the axes of orthotropy in the individual plies are parallel with the beam's edges. Hence, if the above conditions are met, the stiffness E in the classical bending theory can be replaced by the stiffness $E_1^* J$ which is obtained in the following way:

$$E_1^* J = \sum_{k=1}^{n} (E_1)_k J_k \,,$$

where

E_1^* – elasticity modulus of the beam in bending

$(E_1)_k$ – elasticity modulus of the kth layer in the axial direction of beam 1

J – moment of inertia with respect to the beam's midplane

J_k – moment of inertia of the kth ply with respect to the beam's midplane

It should be noted that the above relations have a rather limited use even in the case of symmetric structures. This is explained by the fact that failure of laminated bars occurs in a stepwise manner. The most disadvantageously oriented plies are the first to undergo failure, thus destroying the original symmetry of the structure with respect to the midplane of the laminate.

The purpose of the present paper is to develop simplified methods for stiffness, stress state, displacement and strength evaluation of laminated beams having an arbitrarily prescribed structure and taking into account the stepwise character of laminate failure.

2. Stress state of a laminated beam

2.1. Normal stresses in the individual plies

The geometric lay-up configuration of a beam is shown in Fig. 1. The stresses $\langle \tau_{yz} \rangle_k$ and $\langle \sigma_z \rangle_k$ are neglected as being insignificant.

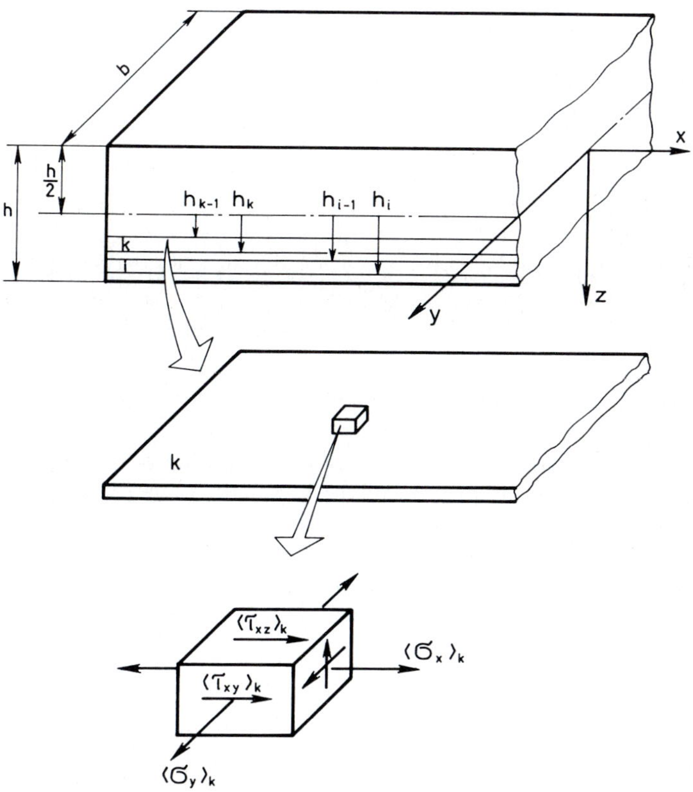

FIG. 1. Geometric lay-up configuration of a laminated beam.

A generalized laminate deformation law (JONES, 1975; SKUDRA and BULAVS, 1982) is written as:

$$\left\{\frac{N}{M}\right\}=\begin{bmatrix} A & B \\ B & D \end{bmatrix}\left\{\frac{\langle\varepsilon^0\rangle}{k}\right\}, \tag{2.1}$$

$$N=\int_{-h/2}^{+h/2}\begin{bmatrix}\langle\sigma_x\rangle\\ \langle\sigma_y\rangle\\ \langle\tau_{xy}\rangle\end{bmatrix}_k \mathrm{d}z\ , \tag{2.2}$$

$$M=\int_{-h/2}^{+h/2}\begin{bmatrix}\langle\sigma_x\rangle\\ \langle\sigma_y\rangle\\ \langle\tau_{xy}\rangle\end{bmatrix}_k z\,\mathrm{d}z\ , \tag{2.3}$$

$$A_{ij};\ B_{ij};\ D_{ij}=\sum_{k=1}^{n}(\bar{Q}_{ij})_k[(h_k-h_{k-1});\ \tfrac{1}{2}(h_k^2-h_{k-1}^2);\ \tfrac{1}{3}(h_k^3-h_{k-1}^3)]\ . \tag{2.4}$$

The deformation law (2.1) can be used in an inverted form:

$$\begin{Bmatrix}\langle\varepsilon_x^0\rangle\\ \langle\varepsilon_y^0\rangle\\ \langle\gamma_{xy}^0\rangle\\ \hline \langle k_x\rangle\\ \langle k_y\rangle\\ \langle k_{xy}\rangle\end{Bmatrix}=\left[\begin{array}{ccc|ccc} A'_{11} & A'_{12} & A'_{16} & B'_{11} & B'_{12} & B'_{16}\\ A'_{21} & A'_{22} & A'_{26} & B'_{21} & B'_{22} & B'_{26}\\ A'_{61} & A'_{62} & A'_{66} & B'_{61} & B'_{62} & B'_{66}\\ \hline B'_{11} & B'_{21} & B'_{61} & D'_{11} & D'_{12} & D'_{16}\\ B'_{12} & B'_{22} & B'_{62} & D'_{21} & D'_{22} & D'_{26}\\ B'_{16} & B'_{26} & B'_{66} & D'_{61} & D'_{62} & D'_{66}\end{array}\right]\begin{Bmatrix}N_x\\ N_y\\ N_{xy}\\ M_x\\ M_y\\ M_{xy}\end{Bmatrix}, \tag{2.5}$$

where A'_{ij}, B'_{ij} and D'_{ij} are the elements of the inverted matrices A_{ij}, B_{ij} and D_{ij}.

It follows from formula (4) that for angle-ply laminated plastics the stiffness elements D_{16} and D_{26} are not equal to zero, and hence the matrices D_{ij} and D'_{ij} will be fully populated.

When transverse bending is coupled with an axial force, $N_y=N_{xy}=M_y=M_{xy}=0$ whereas $M_x\neq 0$ and $N_x\neq 0$. From the deformation law (2.5) the following relationships are obtained for evaluating the midplane strains and curvature k:

$$\begin{aligned}\langle\varepsilon_x^0\rangle&=A'_{11}N_x+B'_{11}M_x \qquad & k_x&=B'_{11}N_x+D'_{11}M_x\ ,\\ \langle\varepsilon_y^0\rangle&=A'_{21}N_x+B'_{21}M_x \qquad & k_y&=B'_{12}N_x+D'_{21}M_x\ ,\\ \langle\gamma_{xy}^0\rangle&=A'_{61}N_x+B'_{61}M_x \qquad & k_{xy}&=B'_{16}N_x+D'_{61}M_x\ .\end{aligned} \tag{2.6}$$

According to the lamination theory, stress in the kth layer is calculated from the relationship

$$\{\langle\sigma\rangle_k\} = [\bar{Q}]_k[\langle\varepsilon^0\rangle + zk]\,. \tag{2.7}$$

In the fully developed form, the relationship (2.7) for the stress $\langle\sigma_x\rangle_k$, for instance, is written as:

$$\begin{aligned}\langle\sigma_x\rangle_k &= (\bar{Q}_{11})_k\langle\varepsilon_x^0\rangle + (\bar{Q}_{12})_k\langle\varepsilon_y^0\rangle + (\bar{Q}_{16})_k\langle\gamma_{xy}^0\rangle \\ &\quad + z[(\bar{Q}_{11})_k k_x + (\bar{Q}_{12})_k k_y + (\bar{Q}_{16})_k k_{xy}]\,.\end{aligned} \tag{2.8}$$

In order to calculate $\langle\sigma_x\rangle_k$ under combined transverse bending and axial load, the expression for $\langle\varepsilon^0\rangle$ and k from (2.6) should be inserted into relationship (2.8). Considering that $M = bM_x$ and $N = bN_x$ we obtain

$$\langle\sigma_x\rangle_k = \frac{N}{b}(q_{1k} + z_k q_{3k}) + \frac{M}{b}(q_{2k} + z_k q_{4k}) = NW_1 + MW_2\,, \tag{2.9}$$

where

$$q_{1k} = (\bar{Q}_{11})_k A'_{11} + (\bar{Q}_{12})_k A'_{21} + (\bar{Q}_{16})_k A'_{61}\,,$$

$$q_{2k} = (\bar{Q}_{11})_k B'_{11} + (\bar{Q}_{12})_k B'_{21} + (\bar{Q}_{16})_k B'_{61}\,,$$

$$q_{3k} = (\bar{Q}_{11})_k B'_{11} + (\bar{Q}_{12})_k B'_{12} + (\bar{Q}_{16})_k B'_{16}\,,$$

$$q_{4k} = (\bar{Q}_{11})_k D'_{11} + (\bar{Q}_{12})_k D'_{21} + (\bar{Q}_{16})_k D'_{61}\,,$$

$$W_1 = \frac{q_{ik} + z_k q_{3k}}{b}\,,$$

$$W_2 = \frac{q_{2k} + z_k q_{4k}}{b}\,.$$

Formula (2.9) is the normal stress distribution law for the individual plies across the thickness of an arbitrarily configured laminated beam. It is significant that in applying formula (2.9) it is not necessary to calculate the position of the neutral axis. This position may be found from (2.8) when $\langle\sigma_x\rangle_k = 0$.

It should be noted that in cases when $B'_{ij} = 0$ the maximum values of the normal stress $\langle\sigma_x\rangle_k$ will coincide with the maximum of $z[(\bar{Q}_{11})_k D'_{11} + (\bar{Q}_{12})_k D'_{21} + (\bar{Q}_{16})_k D'_{61}]$. This feature is characteristic of the stress state in laminated beams in contrast with homogeneous beams where the maximum normal stresses always take place in the fibres located farthest away from the neutral axis.

If the beam is symmetrically laminated about its midplane and consists of a sufficiently large number of plies n, then $B_{ij} = 0$, $A'_{16} = 0$ and $D'_{61} = 0$. Formula (2.9) is thus simplified:

$$\langle \sigma_x \rangle_k = NW_3 + MW_4 , \tag{2.10}$$

where

$$W_3 = b^{-1}[(\bar{Q}_{11})_k A'_{11} + (\bar{Q}_{12})_k A'_{21}] ,$$
$$W_4 = (z/b)[(\bar{Q}_{11})_k D'_{11} + (\bar{Q}_{12})_k D'_{21}] .$$

In the above formulas $N = 0$ should be used for $\langle \sigma_x \rangle_k$ in the case of simple transverse bending.

If the ratio between the span (l) and width (b) of the beam's cross-section is reasonably large, the effect of D'_{21} in bending may be neglected:

$$\langle \sigma_x \rangle_k = \frac{M}{b} z(\bar{Q}_{11})_k D'_{11} . \tag{2.11}$$

To evaluate the compliance element D'_{11} we shall make use of the expressions for N_x and M_x given in (2.2) and (2.3). Inserting $\langle \sigma_x \rangle_k$ into these relationships, where $\langle \sigma_x \rangle_k$ is calculated from the formula

$$\langle \sigma_x \rangle_k = (\bar{Q}_{11})_k (\langle \varepsilon_x^0 \rangle + zk_x) \tag{2.12}$$

we obtain after integration:

$$N_x = A_{11} \langle \varepsilon_x^0 \rangle + B_{11} k_x , \tag{2.13}$$

$$M_x = B_{11} \langle \varepsilon_x^0 \rangle + D_{11} k_x , \tag{2.14}$$

where A_{11}, B_{11} and D_{11} are calculated according to (2.4). If no axial force is applied to the beam, i.e., $N_x = 0$, then, taking into account that $M = bM_x$, (2.13) and (2.14) yield

$$M = \left[\left(D_{11} - \frac{B_{11}^2}{A_{11}} \right) b \right] k_x . \tag{2.15}$$

The expression between the brackets is the effective bending stiffness D^*_{11} of a beam:

$$D^*_{11} = \frac{b}{D'_{11}} \approx \left(D_{11} - \frac{B_{11}^2}{A_{11}} \right) b = E^* J . \tag{2.16}$$

From formula (2.16) we obtain the following result

$$D'_{11} = \frac{A_{11}}{A_{11} D_{11} - B_{11}^2} .$$

If the beam width $b = 1$, (2.16) yields another formula for determining the effective elasticity modulus E^* for a laminated beam in free bending:

$$E^* = \frac{12}{h^3}\left(D_{11} - \frac{B_{11}^2}{A_{11}}\right). \tag{2.17}$$

It should be remembered that the above relationships apply to the case when the length to width ratio of a beam is significantly large. In this case, Poisson's effect (D_{12}) and bending anisotropy (D_{16}) may be neglected.

If the beam is symmetrically laminated about its midplane, then $B_{11} = 0$ and

$$D_{11}^* = D_{11}b\ , \tag{2.18}$$

$$E^* = \frac{12}{h^3} D_{11}\ , \tag{2.19}$$

$$\langle \sigma_x \rangle_k = \frac{z(\bar{Q}_{11})_k}{bD_{11}} M\ . \tag{2.20}$$

Many types of reinforced plastics are characterized by two- or three-stage failure under tension and one-stage failure under compression.

Figs. 2 and 3 show characteristic deformation diagrams for cross-ply reinforced plastics and also fabric-reinforced plastics. According to the diagrams the elasticity moduli of the material are different in tension and compression after the first stage of failure, i.e. the material has become bimodular and the structure of the beam is no longer symmetric with respect to its midplane. For such a beam $B_{11} \neq 0$. Therefore, formula (2.18) is applicable only for the first stage of failure in the tensioned region of the beam. As the load continues to increase a more general formula (2.16) will be used in which the effect of bimodularity has been included.

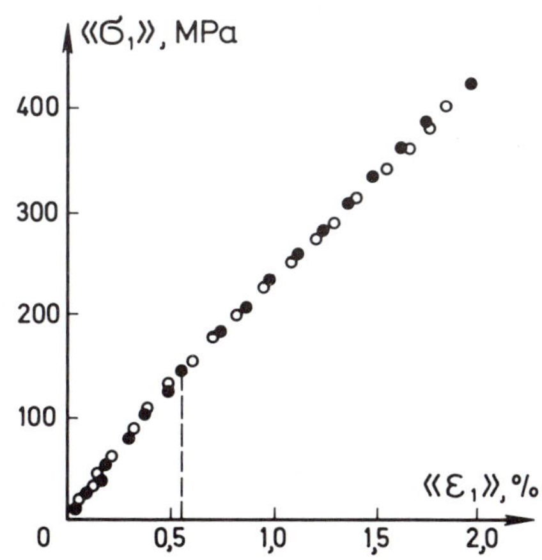

FIG. 2. Deformation diagram of an orthogonally (1:1) reinforced glass/epoxy.

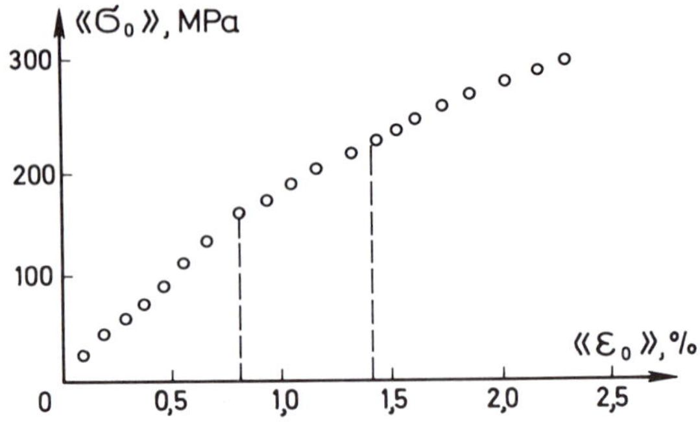

FIG. 3. Deformation diagram of a fabric reinforced glass/epoxy.

In addition to the stress $\langle \sigma_x \rangle_k$ (2.7) yields formulas for calculating the distribution of the stresses $\langle \sigma_y \rangle_k$ and $\langle \tau_{xy} \rangle_k$ across the beam's thickness:

$$\langle \sigma_y \rangle_k = \frac{N}{b}(q_{5k} + zq_{7k}) + \frac{M}{b}(q_{6k} + zq_{8k}) = NW_5 + MW_6 , \qquad (2.21)$$

where

$$W_5 = \frac{q_{5k} + zq_{7k}}{b} ,$$

$$W_6 = \frac{q_{6k} + zq_{8k}}{b} ,$$

$$q_{5k} = (\bar{Q}_{12})_k A'_{11} + (\bar{Q}_{22})_k A'_{21} + (\bar{Q}_{26})_k A'_{61} ,$$

$$q_{6k} = (\bar{Q}_{12})_k B'_{11} + (\bar{Q}_{22})_k B'_{21} + (\bar{Q}_{26})_k B'_{61} ,$$

$$q_{7k} = (\bar{Q}_{12})_k B'_{11} + (\bar{Q}_{22})_k B'_{12} + (\bar{Q}_{26})_k B'_{16} ,$$

$$q_{8k} = (\bar{Q}_{12})_k D'_{11} + (\bar{Q}_{22})_k D'_{21} + (\bar{Q}_{26})_k D'_{61} ,$$

$$\langle \tau_{xy} \rangle_k = \frac{N}{b}(q_{9k} + zq_{11k}) + \frac{M}{b}(q_{10k} + zq_{12k}) = NW_7 + MW_8 , \qquad (2.22)$$

where

$$W_7 = \frac{q_{9k} + zq_{11k}}{b} ,$$

$$W_8 = \frac{q_{10k} + zq_{12k}}{b} ,$$

$$q_{9k} = (\bar{Q}_{16})_k A'_{11} + (\bar{Q}_{26})_k A'_{21} + (\bar{Q}_{66})_k A'_{61} ,$$

$$q_{10k} = (\bar{Q}_{16})_k B'_{11} + (\bar{Q}_{26})_k B'_{21} + (\bar{Q}_{66})_k B'_{61} ,$$

$$q_{11k} = (\bar{Q}_{16})_k B'_{11} + (\bar{Q}_{26})_k B'_{12} + (\bar{Q}_{66})_k B'_{16} ,$$

$$q_{12k} = (\bar{Q}_{16})_k D'_{11} + (\bar{Q}_{26})_k D'_{21} + (\bar{Q}_{66})_k D'_{61} .$$

2.2. *Shear stresses in the individual plies*

To derive the distribution law for the shear stresses $\langle \tau_{xz} \rangle$ across the beam's thickness we shall use the equilibrium equation for an infinitely small volume of the kth ply shown in Fig. 1. Assuming that the kth ply is in a state of plane stress and that stresses do not depend on the coordinate y, the equilibrium equation is

$$\frac{\partial \langle \sigma_x \rangle_k}{\partial x} + \frac{\partial \langle \tau_{xz} \rangle_k}{\partial z} = 0 .$$

Integration of the equilibrium equation is performed to obtain an approximate expression for $\langle \tau_{xz} \rangle_k$:

$$\langle \tau_{xz} \rangle_k = - \int_{h/2}^{z} \frac{\partial \langle \sigma_x \rangle_k}{\partial x} \, \mathrm{d}z .$$

For a laminated beam of width b this relationship is written as

$$\langle \tau_{xz} \rangle_k = - \frac{1}{b} \sum_{i=1}^{k} \int_{h_{i-1}}^{h_i} \frac{\partial \langle \sigma_x \rangle_i}{\partial x} \, \mathrm{d}z . \tag{2.23}$$

The $\langle \sigma_x \rangle_i$ is obtained from a formula similar to (2.8). Utilizing the familiar relationship between the bending moment M and shear force S

$$S = \partial M / \partial x ,$$

we arrive at the following relationship for calculating derivative of the stress $\langle \sigma_x \rangle_i$ with respect to x:

$$\partial \langle \sigma_x \rangle_i / \partial x = S(q_{1i} + z q_{2i}) , \tag{2.24}$$

where

$$q_{1i} = (\bar{Q}_{11})_i B'_{11} + (\bar{Q}_{12})_i B'_{21} + (\bar{Q}_{16})_i B'_{61} ,$$
$$q_{2i} = (\bar{Q}_{11})_i D'_{11} + (\bar{Q}_{12})_i D'_{21} + (\bar{Q}_{16})_i D'_{61} .$$

Taking into account (2.24), formula (2.23) becomes

$$\langle \tau_{xz} \rangle_k = - \frac{1}{b} \sum_{i=1}^{k} \int_{h_{i-1}}^{h_i} (q_{1i} + z q_{2i}) S \, \mathrm{d}z . \tag{2.25}$$

Integration across individual plies yields

$$\langle \tau_{xz} \rangle_k = SW_9 , \tag{2.26}$$

where

$$W_9 = -\frac{1}{b} \sum_{i=1}^{k} [q_{1i}(h_i - h_{i-1}) + \tfrac{1}{2} q_{2i}(h_i^2 - h_{i-1}^2)] .$$

The shear stresses $\langle \tau_{xz} \rangle_k$ will reach their maximum at a distance h_k from the midplane of a beam which is determined from the condition that the sum

$$\sum_{i=1}^{k} [q_{1i}(h_i - h_{i-1}) + \tfrac{1}{2} q_{2i}(h_i^2 - h_{i-1}^2)]$$

attains its maximum value.

If the beam is symmetrically laminated with respect to its midplane and the number of plies is reasonably large, then

$$B'_{ij} = 0 \quad D'_{61} = 0 ,$$
$$q_{1i} = 0 \quad q_{2i} = (\bar{Q}_{11})_i D'_{11} + (\bar{Q}_{12})_i D'_{21} .$$

and formula (2.26) is simplified as follows

$$\langle \tau_{xz} \rangle_k = -\frac{S}{2b} \sum_{i=1}^{k} [(\bar{Q}_{11})_i D'_{11} + (\bar{Q}_{12})_i D'_{21}](h_i^2 - h_{i-1}^2) = -SW_{10} , \tag{2.27}$$

where

$$W_{10} = \frac{1}{2b} \sum_{i=1}^{k} [(\bar{Q}_{11})_i D'_{11} + (\bar{Q}_{12})_i D'_{21}](h_i^2 - h_{i-1}^2) .$$

Further simplification of (2.27) is obtained when $D'_{21} = 0$

$$\langle \tau_{xz} \rangle_k = -\frac{S}{2b} \frac{D_{22}}{D_{11}D_{22} - D_{12}^2} \sum_{i=1}^{k} (\bar{Q}_{11})_i (h_i^2 - h_{i-1}^2) . \tag{2.28}$$

Neglecting transverse effects, i.e., assuming that $D_{12} = 0$, we obtain

$$\langle \tau_{xz} \rangle_k = -\frac{S}{2bD_{11}} \sum_{i=1}^{k} (\bar{Q}_{11})_i (h_i^2 - h_{i-1}^2) . \tag{2.29}$$

It follows from the above that in order to determine the shear stresses it is not necessary to also determine the position of the beam's neutral axis.

We should note that for laminated structures the so-called edge effect induces a three-dimensional state of stress in the region adjacent to the free edge of the beam which is approximately equal in length to the beam thickness. In order to reduce this edge effect the width to depth ratio of the beam should be reasonably large.

3. Determination of elastic displacements

3.1. Potential energy of a laminated beam

The potential energy U of an elastic laminated beam is expressed as

$$U = \tfrac{1}{2} \int_0^l N_x \, \mathrm{d}\delta + \tfrac{1}{2} \int_0^l M_x \, \mathrm{d}\varphi + \tfrac{1}{2} \int_0^l S_x \, \mathrm{d}\gamma \,, \tag{3.1}$$

with the displacements $\mathrm{d}\delta$, $\mathrm{d}\varphi$ and $\mathrm{d}\gamma$ for a beam element of length $\mathrm{d}x$ given by

$$\begin{aligned} \mathrm{d}\delta &= \mathrm{d}\langle\!\langle \varepsilon_x^0 \rangle\!\rangle = A'_{11} N_x \, \mathrm{d}x + B'_{11} M_x \, \mathrm{d}x \,, \\ \mathrm{d}\varphi &= \mathrm{d}k_x = B'_{11} N_x \, \mathrm{d}x + D'_{11} M_x \, \mathrm{d}x \,, \\ \mathrm{d}\gamma &= \frac{\kappa}{A_{55}} S_x \, \mathrm{d}x \,, \end{aligned} \tag{3.2}$$

where $A_{55} = G_{xz} h$.

The following symbols have been used in formulas (3.2):

N_x, M_x and S_x – internal force and moment resultants acting on a beam of unit width
κ – non-dimensional coefficient of shearing strain depending on the cross-sectional shape and the lay-up configuration of the beam
G_{xz} – laminate shear modulus
A_{55} – component of laminate stiffness matrix

Inserting the expressions for the displacements $\mathrm{d}\delta$, $\mathrm{d}\varphi$ and $\mathrm{d}\gamma$ into formula (3.1) we obtain

$$\begin{aligned} U = {} & \tfrac{1}{2} \int_0^l N_x^2 A'_{11} \, \mathrm{d}x + \int_0^l N_x M_x B'_{11} \, \mathrm{d}x \\ & + \tfrac{1}{2} \int_0^l M_x^2 D'_{11} \, \mathrm{d}x + \tfrac{1}{2} \int_0^l \frac{\kappa}{A_{55}} S_x^2 \, \mathrm{d}x \,. \end{aligned} \tag{3.3}$$

Formula (3.3) is valid for the case when the width of the rectangular cross-section of a beam is unity. If the width is b, then

$$U = \frac{1}{2b}\int_0^l N_p^2 A'_{11}\,dx + \frac{1}{B}\int_0^l N_p M_p B'_{11}\,dx + \frac{1}{2b}\int_0^l M_p^2 D'_{11}\,dx + \frac{1}{2b}\int_0^l \frac{\kappa}{A_{55}} S_p^2\,dx\,. \tag{3.4}$$

If the beam is symmetrically laminated with respect to its midplane, (3.4) is written as

$$U = \frac{1}{2b}\int_0^l N_p^2 a_{11}\,dx + \frac{1}{2b}\int_0^l M_p^2 d_{11}\,dx + \frac{1}{2b}\int_0^l \frac{\kappa}{A_{55}} S_p^2\,dx\,, \tag{3.5}$$

where

$$a_{11} = \frac{1}{\Delta_A}(A_{22}A_{66} - A_{26}^2)\,,\quad d_{11} = \frac{1}{\Delta_D}(D_{22}D_{66} - D_{26}^2)\,,$$

$$\Delta_A = \begin{vmatrix} A_{11} & A_{12} & A_{16} \\ A_{21} & A_{22} & A_{26} \\ A_{61} & A_{62} & A_{66} \end{vmatrix}\,,\qquad \Delta_D = \begin{vmatrix} D_{11} & D_{12} & D_{16} \\ D_{21} & D_{22} & D_{26} \\ D_{61} & D_{62} & D_{66} \end{vmatrix}\,.$$

The methods for determination of the shearing strain coefficient κ have been discussed, for example, in the works by TEH and HUANG (1979) and DHARMARJAN and MCCUTCHEN (1973).

3.2. *Castigliano's method*

In accordance with the theorem of Castigliano the generalized displacements Δ_i are equal to the partial derivatives of the potential energy U taken with respect to the corresponding generalized forces P_i:

$$\Delta_i = \partial U/\partial P_i \tag{3.6}$$

When calculating displacements according to the theorem of Castigliano, the potential energy should be expressed as a function of generalized forces only.

It will be observed that the theorem of Castigliano allows for the determination of displacements only at those points of the system at which generalized forces are applied (e.g. a concentrated force or a concentrated moment). Determination of the displacement at an arbitrary point in an elastic system is made in the following artificial way. An imaginary additional concentrated force or moment is applied to the point 'k' where the displacement is to be determined and the potential energy is expressed as a function of the given and additionally superimposed load P_g (or M_g). The unknown displacement is obtained from the equations

$$\Delta_k = \left(\frac{\partial U}{\partial P_g}\right)_{P_g=0} \tag{3.7}$$

or

$$\theta_k = \left(\frac{\partial U}{\partial M_g}\right)_{M_g=0} .$$

Using the potential energy expression for a laminated plastic in the form of (3.4), (3.6) may be developed in the following manner:

$$\begin{aligned}\Delta_i = {} & \frac{1}{b}\int_0^l N_p A'_{11} \frac{\partial N_p}{\partial P_i}\,dx + \frac{1}{b}\int_0^l N_p B'_{11} \frac{\partial M_p}{\partial P_i}\,dx \\ & + \frac{1}{b}\int_0^l M_p B'_{11} \frac{\partial N_p}{\partial P_i}\,dx + \frac{1}{b}\int_0^l M_p D'_{11} \frac{\partial M_p}{\partial P_i}\,dx \\ & + \frac{1}{b}\int_0^l S_p \frac{\kappa}{A_{55}} \frac{\partial S_p}{\partial P_i}\,dx .\end{aligned} \tag{3.8}$$

For a symmetric configuration, formula (3.8) is simplified as follows:

$$\begin{aligned}\Delta_i = {} & \frac{1}{b}\int_0^l N_p a_{11} \frac{\partial N_p}{\partial P_i}\,dx + \frac{1}{b}\int_0^l M_p d_{11} \frac{\partial M_p}{\partial P_i}\,dx \\ & + \frac{1}{b}\int_0^l S_p \frac{\kappa}{A_{55}} \frac{\partial S_p}{\partial P_i}\,dx .\end{aligned} \tag{3.9}$$

3.3. *Mohr's method*

Mohr's method is based on the principle of virtual work. Let us assume that a unit load is applied to a laminated beam with the load acting at the location and in the direction corresponding to the displacement which is to be determined. Subsequently, a virtual displacement is applied to the beam resulting in the displacement of the unit load by an increment Δ and in the virtual work $1 \cdot \Delta$. The virtual work of the unit load will be equal to the total work of internal forces (resulting from the action of unit load) due to infinitesimally small deformations of the beam when the virtual displacement is applied:

$$1 \cdot \Delta = \int_0^L \bar{N}_x\,d\delta + \int_0^L \bar{M}_x\,d\varphi + \int_0^L \bar{S}_x\,d\gamma , \tag{3.10}$$

where $\bar{N}$, $\bar{M}$ and $\bar{S}$ are internal force and moment resultants caused by the action of the unit load; $d\delta$, $d\varphi$ and $d\gamma$ are infinitesimal displacements of the beam elements. Relationships (3.2) will be used to determine the displacements of a laminated beam of unsymmetric construction.

Inserting expressions for $d\delta$, $d\varphi$ and $d\gamma$ into expressions (3.10) and taking into account that the width of the beam's cross-section is b, we obtain the following formula for the determination of displacements in a laminated beam:

$$\Delta_{\mathrm{ip}} = \sum \frac{A'_{11}}{b} \int_0^l \bar{N}_{\mathrm{i}} N_{\mathrm{p}}\, \mathrm{d}x + \sum \frac{B'_{11}}{b_i} \int_0^l \bar{N}_{\mathrm{i}} M_{\mathrm{p}}\, \mathrm{d}x$$

$$+ \sum \frac{B'_{11}}{b} \int_0^l \bar{M}_{\mathrm{i}} N_{\mathrm{p}}\, \mathrm{d}x + \sum \frac{D'_{11}}{b} \int_0^l \bar{M}_{\mathrm{i}} M_{\mathrm{p}}\, \mathrm{d}x \qquad (3.11)$$

$$+ \sum \frac{\kappa}{A_{55} b} \int_0^l \bar{S}_{\mathrm{i}} S_{\mathrm{p}}\, \mathrm{d}x\, .$$

For transversel bending, formula (3.11) is given by

$$\Delta_{\mathrm{ip}} = \sum \frac{B'_{11}}{b} \int_0^l \bar{N}_{\mathrm{i}} M_{\mathrm{p}}\, \mathrm{d}x + \sum \frac{D'_{11}}{b} \int_0^l \bar{M}_{\mathrm{i}} M_{\mathrm{p}}\, \mathrm{d}x + \sum \frac{\kappa}{A_{55} b} \int_0^l \bar{S}_{\mathrm{i}} S_{\mathrm{p}}\, \mathrm{d}x\, . \qquad (3.12)$$

If the beam is symmetrically laminated about its midplane, formulas (3.11) and (3.12) simplify to:

$$\Delta_{\mathrm{ip}} = \sum \frac{a_{11}}{b} \int_0^l \bar{N}_{\mathrm{i}} N_{\mathrm{p}}\, \mathrm{d}x + \sum \frac{d_{11}}{b} \int_0^l \bar{M}_{\mathrm{i}} M_{\mathrm{p}}\, \mathrm{d}x + \sum \frac{\kappa}{A_{55} b} \int_0^l \bar{S}_{\mathrm{i}} S_{\mathrm{p}}\, \mathrm{d}x \qquad (3.13)$$

$$\Delta_{\mathrm{ip}} = \sum \frac{d_{11}}{b} \int_0^l \bar{M}_{\mathrm{i}} M_{\mathrm{p}}\, \mathrm{d}x + \sum \frac{\kappa}{A_{55} b} \int_0^l \bar{S}_{\mathrm{i}} S_{\mathrm{p}}\, \mathrm{d}x\, . \qquad (3.14)$$

3.4. Calculation of Mohr's integrals by means of diagram multiplication

Calculation of Mohr's integrals given in the previous section may be carried out in a much simpler way by using the method of diagram multiplication suggested by VERESCHAGIN (1925). The essence of this method is the following: if any one of the diagrams is confined by a straight line and the other has an arbitrary shape, Mohr's integral may be calculated as the product of the area of an arbitrarily shaped diagram and the ordinate corresponding to the centre of gravity of the rectilinear diagram. As an example, consider the calculation of $\int \bar{M}_{\mathrm{i}} M_{\mathrm{p}}\, \mathrm{d}x$. The diagrams M_{p} and $\bar{M}_{\mathrm{i}}$ are included in the calculation scheme shown in Fig. 4.

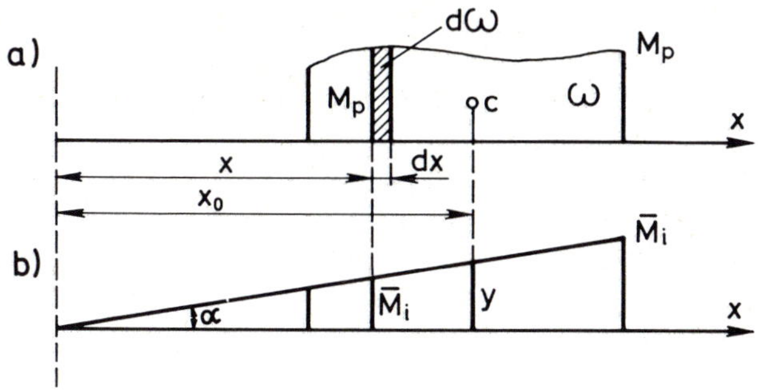

FIG. 4. Calculation scheme for arbitrary (a) and rectilinear (b) diagram shapes.

Let us consider two states of any linear beam. Let the diagram M_p (due to prescribed load) be shaped as a curve in the first state (Fig. 4(a)) and let the diagram $\bar{M}_i$ (due to unit force) be bounded by a straight line in the second case (Fig. 4(b)).

Assuming that the origin of coordinates is the point of intersection of the straight line confining the straight-line diagram $\bar{M}_i$ and the abscissa (Fig. 4(b)), the ordinate of this diagram will be expressed in terms of an arbitrary abscissa x:

$$\bar{M}_i = x \operatorname{tg} \alpha \ .$$

This gives us

$$\int_0^l M_p \bar{M}_i \, dx = \int_0^l M_p x \operatorname{tg} \alpha \, dx = \operatorname{tg} \alpha \int_0^l x M_p \, dx \ .$$

It is evident from Fig. 4(a) that $M_p \, dx$ is an element of an area $d\omega$ of the diagram M_p and the product $xM_p \, dx = x \, d\omega$ is an elementary section modulus of this area in relation to the vertical axis. Consequently

$$\int_0^l x M_p \, dx = L = \omega x_0 \ ,$$

where L is the section modulus of the area ω of the diagram M_p in relation to the vertical axis, and x_0 is the abscissa of the centre of gravity of this area.

Now

$$\int_0^l M_p \bar{M}_i \, dx = \operatorname{tg} \alpha \omega x_0 = \omega y \ . \qquad (3.15)$$

Application of formula (3.15) for calculating Mohr's integrals is specified in the following way:

(1) The product ωy should be considered to be positive if the area ω and ordinate y are located on the same side of the beam's axis, i.e., they have the same sign, and the product is negative if they are located on opposite sides of the axis.
(2) The diagram confining the area ω may have any shape whereas the diagram confining the ordinate y is to be bounded by a straight line.
(3) If both diagrams are bounded by straight lines, any one of them confines the area ω whereas the other confines the ordinate y.
(4) When one of the diagrams has a complicated shape it is to be divided into simple shapes for which the formulas for the area and the centre of gravity coordinates are known.

Taking into account relationship (3.15), formula (3.11) for determination of displacements is written as

$$\Delta_{\text{ip}} = \frac{1}{b}\Big[\sum A'_{11}\omega_N y_{\bar{N}} + \sum B'_{11}\omega_M y_{\bar{N}} + \sum B'\omega_N y_{\bar{M}} + \sum D'_{11}\omega_M y_{\bar{M}} + \sum \frac{\kappa}{A_{55}}\,\omega_S y_{\bar{S}}\Big]. \tag{3.16}$$

In the case of a symmetric structure, formula (3.16) simplifies to:

$$\Delta_{\text{ip}} = \frac{1}{b}\Big[\sum a_{11}\omega_N y_{\bar{N}} + \sum d_{11}\omega_M y_{\bar{M}} + \sum \frac{\kappa}{A_{55}}\,\omega_S y_{\bar{S}}\Big]. \tag{3.17}$$

4. Strength criteria of a ply

Assume that a laminated beam consisting of unidirectionally reinforced plies obeys general strength criteria as described in the papers of SKUDRA and BULAVS (1978) and SKUDRA (1984). The possible various orientations of an arbitrary kth ply are shown in Fig. 5. In general, the kth ply will be subjected to the stresses $\langle\sigma_x\rangle_k$, $\langle\sigma_y\rangle_k$, $\langle\tau_{xy}\rangle_k$ and $\langle\tau_{xz}\rangle_k$.

When the reinforcement of the kth ply coincides with the beam's longitudinal axis (Fig. 5(a)) there are two failure modes possible: (a) fibre failure under tension or compression and (b) matrix failure or failure of the bond between the fibres and the matrix.

In the case of fibre failure the effect of $\langle\tau_{\parallel\perp}\rangle_k$ may be disregarded, and subsequently the strength condition for the kth ply can be expressed as

$$\langle\varepsilon_x\rangle_k = e_{\text{fu}}\,, \tag{4.1}$$

where $\langle\varepsilon_x\rangle_k$ and e_{fu} are the ultimate strains of the kth ply and of the fibres, respectively.

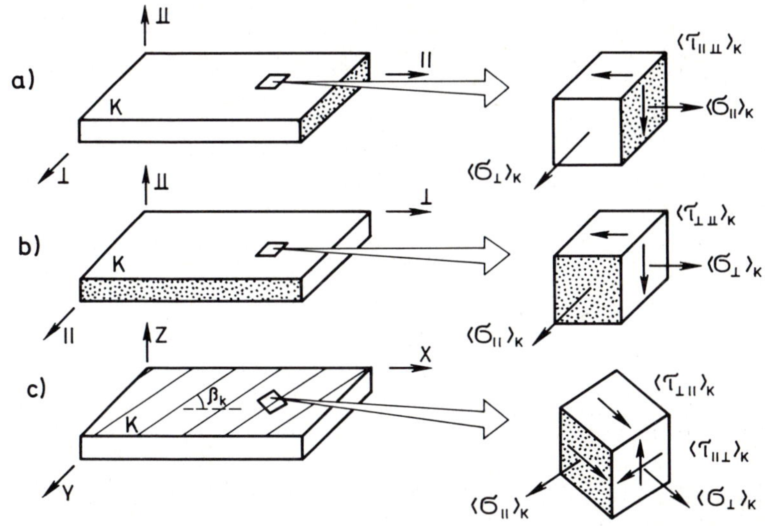

FIG. 5. States of stress in individual plies: (a) axially reinforced ply; (b) ply reinforced perpendicular to the axis; (c) ply reinforced at an angle β to the beam axis.

The total strain of the kth ply is expressed as

$$\langle \varepsilon_x \rangle_k = \langle \varepsilon_x^0 \rangle + z_k k_x \,. \tag{4.2}$$

In a generalized way

$$\langle \varepsilon_x^0 \rangle = A'_{11} N_x + B'_{11} M_x \,,$$
$$k_x = B'_{11} N_x + D'_{11} M_x \,.$$

By inserting the expressions for $\langle \varepsilon_x^0 \rangle$ and k_x into (4.2) and taking into account that $N = bN_x$ and $M = bM_x$ criterion (3.18) becomes:

$$A'_{11} N + B'_{11} M + z_k (B'_{11} N + D'_{11} M) = e_{\mathrm{fu}} b \,. \tag{4.3}$$

In the case of matrix failure in the kth ply due to the action of the stresses $\langle \sigma_\parallel \rangle_k$, $\langle \sigma_\perp \rangle_k$ and $\langle \tau_{\parallel \perp\!\!\perp} \rangle_k$ (Fig. 5(a)) it is advisable to make the following assumptions:

(a) the stress $\langle \sigma_\parallel \rangle_k$ has a negligible effect on the stress state of the matrix;
(b) the interaction between the stresses $\langle \sigma_\perp \rangle_k$ and $\langle \tau_{\parallel \perp\!\!\perp} \rangle_k$ has a negligible effect on the matrix strength.

With these assumptions, the kth ply will fail due to longitudinal shear and its strength can thus be written as

$$SW_9 \approx \tau_{\mathrm{m}}^{\mathrm{u}} \,, \tag{4.4}$$

where τ^{u}_{m} is the shear strength of the matrix. Characteristic shear failure of the matrix is illustrated in Fig. 6.

Certain types of reinforced plastics, such as carbon/epoxy, are characterized by somewhat lower bond strength between the fibres and the matrix in comparison to matrix strength. Applying the criterion of the bond strength proposed by SKUDRA et al. (1977) and SKUDRA (1984), relationship (4.4) is written as

$$SW_9 = \tau^{u}_{b}/\bar{\sigma}_{rz} \, . \tag{4.5}$$

If the kth ply is reinforced perpendicular to the beam's longitudinal axis (Fig. 5(b)), it will be subjected to the stresses $\langle \sigma_{\parallel} \rangle_k$, $\langle \sigma_{\perp} \rangle_k$ and $\langle \tau_{\perp \perp\!\perp} \rangle_k$.

Assuming that the stress $\langle \sigma_{\parallel} \rangle_k$ has practically no effect on the stress state of the matrix, the strength criterion for the kth ply is given by

$$\langle \sigma^2_{\perp} \rangle_k + 2(1 + \nu_{\perp \perp\!\perp})\langle \tau^2_{\perp \perp\!\perp} \rangle_k \pm \langle \sigma_{\perp} \rangle_k (\langle \sigma^2_{\perp} \rangle_K + 4\langle \tau^2_{\perp \perp\!\perp} \rangle_k)^{1/2} = 2(\sigma^{+}_{\perp})^2_k \, , \tag{4.6}$$

where $\nu_{\perp \perp\!\perp}$ and $(\sigma^{+}_{\perp})_k$ are the Poisson's ratio and the transverse tensile strength of a unidirectionally reinforced kth ply, respectively. In this particular case $\beta_k = 90°$, and

$$\begin{aligned} \langle \sigma_{\perp} \rangle_k &= NW_1 + MW_2 \\ \langle \tau_{\perp \perp\!\perp} \rangle_k &= SW_9 \, , \end{aligned} \tag{4.7}$$

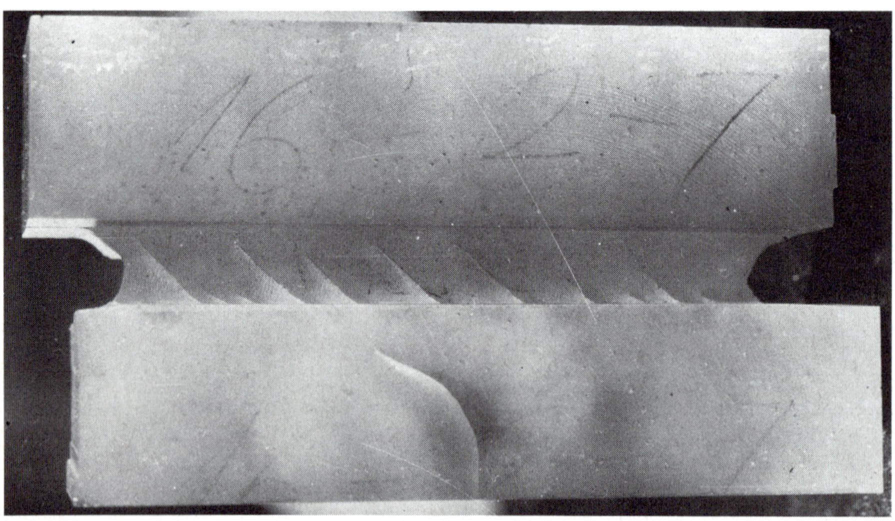

FIG. 6. Failure mode of the matrix under shear.

Substitution of the expressions for the stresses $\pm\langle\sigma_\perp\rangle_k$ and $\langle\tau_{\perp\parallel}\rangle_k$ in criterion (4.6) yields

$$(NW_1 + MW_2)^2 + 2(1 + \nu_{\perp\parallel})S^2W_9^2 \pm (NW_1 + MW_2)[(NW_1 + MW_2)^2 + 4S^2W_9^2]^{1/2} = 2(\sigma_\perp^+)_k^2\,, \tag{4.8}$$

where

$$(\sigma_\perp^+)_k = \frac{\sigma_m^+}{[\bar{\sigma}_r^2 + \bar{\sigma}_\theta^2 + \bar{\sigma}_z^2 - 2\nu_m(\bar{\sigma}_r\bar{\sigma}_\theta + \bar{\sigma}_r\sigma_z + \bar{\sigma}_\theta\bar{\sigma}_z)]^{1/2}}\,. \tag{4.9}$$

A characteristic mode of the matrix failure of a glass/epoxy under transverse tension is shown in Fig. 7.

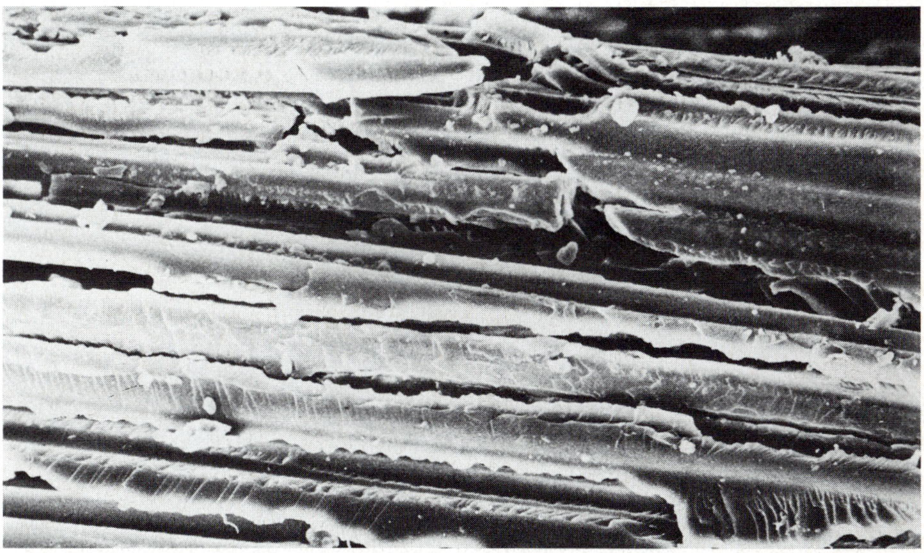

FIG. 7. Failure mode of a unidirectionally reinforced glass/epoxy under transverse tension (enlarged ×500).

If the bond strength is lower than the matrix strength, the strength in transverse tension included in criterion (4.9) is determined according to the formula

$$\sigma_\perp^+ = \sigma_b^u/\bar{\sigma}_r\,. \tag{4.10}$$

For the failure of the *k*th in transverse compression the strength criterion is given by

$$(NW_1 + MW_2)^2 + 2(1 + \nu_{\perp ⫫})S^2W_9^2 + (NW_1 + MW_2)[(NW_1 + MW_2)^2 + 4S^2W_9^2]^{1/2} = 2(\sigma_{\perp}^{-})_k^2 , \tag{4.11}$$

where the strength of the unidirectionally reinforced kth ply in transverse compression $(\sigma_{\perp}^{-})_k$ is determined according to the formula (SKUDRA, 1984)

$$(\sigma_{\perp}^{-})_k = 3.5 \frac{(\sigma_{\perp}^{+})_k}{1 + \nu_{\perp ⫫}} . \tag{4.12}$$

If the kth ply is reinforced at an arbitrary angle β_k with respect to the beam's axis, then, as seen from Fig. 5(b), the stresses $\langle \tau_{xy} \rangle_k$, $\langle \sigma_y \rangle_k$, $\langle \sigma_x \rangle_k$ and $\langle \tau_{xz} \rangle_k$ will result in the following stresses in the material principal coordinate system of the ply:

$$\langle \sigma_{\parallel} \rangle_k = \langle \sigma_x \rangle_k \cos^2 \beta_k + \langle \sigma_y \rangle_k \sin^2 \beta_k + 2\langle \tau_{xy} \rangle_k \cos \beta_k \sin \beta_k , \tag{4.13}$$

$$\langle \sigma_{\perp} \rangle_k = \langle \sigma_x \rangle_k \sin^2 \beta_k + \langle \sigma_y \rangle_k \cos^2 \beta_k - 2\langle \tau_{xy} \rangle_k \cos \beta_k \sin \beta_k , \tag{4.14}$$

$$\langle \tau_{\perp ⫫} \rangle_k = (\langle \sigma_y \rangle_k - \langle \sigma_x \rangle_k) \cos \beta_k \sin \beta_k - \langle \tau_{xy} \rangle_k (\cos^2 \beta_k - \sin^2 \beta_k) , \tag{4.15}$$

$$\langle \tau_{\parallel ⫫} \rangle_k \approx \langle \tau_{xz} \rangle_k \cos \beta_k , \tag{4.16}$$

$$\langle \tau_{\perp ⫫} \rangle_k \approx \langle \tau_{xz} \rangle_k \sin \beta_k . \tag{4.17}$$

In the relationships for $\langle \sigma_{\parallel} \rangle_k$, $\langle \sigma_{\perp} \rangle_k$ and $\langle \tau_{\parallel \perp} \rangle_k$ it is necessary to take into account the sign of the shear stress $\langle \tau_{xy} \rangle_k$ as shown schematically in Fig. 8.

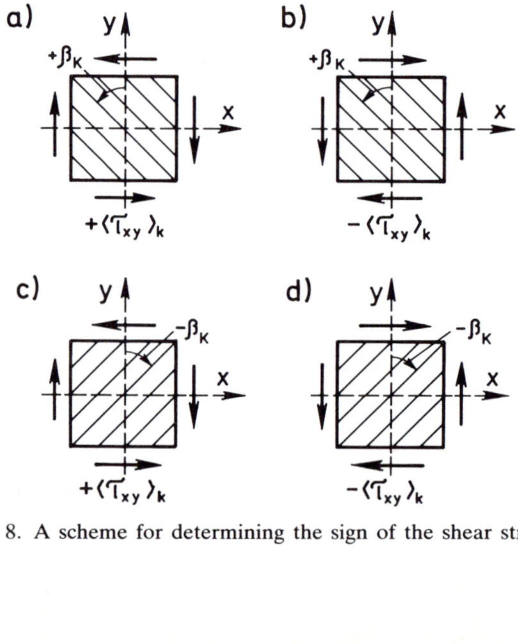

FIG. 8. A scheme for determining the sign of the shear stress.

Substituting the expressions for stresses $\langle \sigma_x \rangle_k$, $\langle \sigma_y \rangle_k$ and $\langle \tau_{xy} \rangle_k$ in formulas (4.13) through (4.17) we obtain

$$\langle \sigma_{\perp} \rangle_k = NF_1 + MF_2 \,, \tag{4.18}$$

where

$$F_1 = W_1 \sin^2 \beta_k + W_5 \cos^2 \beta_k - 2W_7 \cos \beta_k \sin \beta_k \,,$$
$$F_2 = W_2 \sin^2 \beta_k + W_6 \cos^2 \beta_k - 2W_8 \cos \beta_k \sin \beta_k \,,$$

$$\langle \tau_{\parallel\perp} \rangle_k = NF_3 + MF_4 \,, \tag{4.19}$$

where

$$F_3 = (W_5 - W_1) \cos \beta_k \sin \beta_k - W_7 \sin^2 \beta_k \,,$$
$$F_4 = (W_6 - W_2) \cos \beta_k \sin \beta_k - W_8 \sin^2 \beta_k \,,$$

$$\langle \sigma_{\parallel} \rangle_k = NF_5 + MF_6 \,, \tag{4.20}$$

where

$$F_5 = W_1 \cos^2 \beta_k + W_5 \sin^2 \beta_k + 2W_7 \cos \beta_k \sin \beta_k \,,$$
$$F_6 = W_2 \cos^2 \beta_k + W_6 \sin^2 \beta_k + 2W_8 \cos \beta_k \sin \beta_k \,,$$

$$\langle \tau_{\parallel\perp\perp} \rangle_k \approx SW_9 \cos \beta_k \,, \tag{4.21}$$

$$\langle \tau_{\perp\perp\perp} \rangle_k \approx SW_9 \sin \beta_k \,. \tag{4.22}$$

The contribution of the stress $\langle \sigma_{\parallel} \rangle_k$ to the matrix failure may be neglected in the case of polymeric matrices. Then the actual strength of the beam will be the lowest value determined from the following criteria:

(1) Failure due to the action of $\langle \tau_{\parallel\perp} \rangle_k$ and $\langle \sigma_{\perp} \rangle_k$

(a) tensile failure of the matrix

$$(1 - \nu_m^2)\left(\frac{\bar{\sigma}_r}{\sigma_m^+}\right)^2 (NF_1 + MF_2)^2 + \frac{1}{(\tau_m^u)^2}(NF_3 + MF_4)^2 = 1 \,. \tag{4.23}$$

(b) shear failure of the matrix

$$\begin{aligned} &(NF_1 + MF_2)^2 + 2(1 + \nu_m)(NF_3 + MF_4)^2 + (NF_1 + MF_2) \\ &\quad [(NF_1 + MF_2)^2 + 4(NF_3 + MF_4)^2]^{1/2} = 2(\sigma_m^+)^2 \,. \end{aligned} \tag{4.24}$$

(c) bond failure

$$(NF_1 + MF_2)(\tau_b^u)^2 \bar{\sigma}_r + (NF_3 + MF_4)^2 \sigma_b^u \bar{\sigma}_{rz}^2 = (\tau_b^u)^2 \sigma_b^u \,. \tag{4.25}$$

(2) Failure due to the action of $\langle \sigma_\perp \rangle_k$ and $\langle \tau_{\perp \perp\!\!\perp} \rangle_k$

(a) tensile failure of the matrix

$$(1-\nu_{\mathrm{m}}^2)\left(\frac{\bar{\sigma}_r}{\overset{+}{\sigma}_{\mathrm{m}}}\right)^2 (NF_1 + MF_2)^2 + \frac{1}{(\tau_{\mathrm{m}}^{\mathrm{u}})^2} S^2 W_9^2 \sin^2 \beta_k = 1 \,. \tag{4.26}$$

(b) shear failure of the matrix

$$\begin{gathered}(NF_1 + MF_2)^2 + 2(1+\nu_{\mathrm{m}}) S^2 W_9^2 \sin^2 \beta_k + (NF_1 \\ + MF_2)[(NF_1 + MF_2)^2 + 4S^2 W_q^2 \sin^2 \beta_k]^{1/2} = 2(\overset{+}{\sigma}_{\mathrm{m}})^2 \,. \end{gathered} \tag{4.27}$$

(c) bond failure

$$(NF_1 + MF_2)(\tau_{\mathrm{b}}^{\mathrm{u}})^2 \bar{\sigma}_r + SW_9^2 \sin^2 \beta_k \sigma_{\mathrm{b}}^{\mathrm{u}} \bar{\sigma}_{rz}^2 = (\tau_{\mathrm{b}}^{\mathrm{u}})^2 \sigma_{\mathrm{b}}^{\mathrm{u}} \,. \tag{4.28}$$

(3) Failure due to the action of $\langle \sigma_\| \rangle_k$

$$\begin{gathered}(NW_1 + MW_2)\cos^2 \beta_k + (NW_5 + MW_6)\sin^2 \beta_k \\ + 2(NW_7 + MW_8)\sin \beta_k \cos \beta_k = E_{\mathrm{fz}} \overset{+}{e}_{\mathrm{fu}} \,. \end{gathered} \tag{4.29}$$

The above relationships pertain to the case when the kth ply lies in the tensile region of the laminated beam. In the compressed region the stresses $\langle \sigma_x \rangle_k$ and $\langle \sigma_\perp \rangle_k$ are compressive and thus the strength criteria for the kth ply take the following form:

(1) For the failure of the kth ply under transverse compression due to the joint action of $-\langle \sigma_\perp \rangle_k$ and $\langle \tau_{\|\perp} \rangle_k$

$$\begin{gathered}(NF_1 + MF_2)^2 + 2(1+\nu_{\perp \perp\!\!\perp})(NF_3 + MF_4)^2 + (NF_1 \\ + MF_2)[(NF_1 + MF_2)^2 + 4(NF_3 + MF_4)^2]^{1/2} = 2(\overset{-}{\sigma}_\perp)^2 \,. \end{gathered} \tag{4.30}$$

(2) For the shear failure of the matrix due to the joint action of $\langle \sigma_\perp \rangle_k$ and $\langle \tau_{\|\perp} \rangle_k$

$$\begin{gathered}(NF_1 + MF_2)^2 + 2(1+\nu_{\mathrm{A}})(NF_3 + MF_4)^2 - (NF_1 \\ + MF_2)[(NF_1 + MF_2)^2 + 4(NF_3 + MF_4)^2]^{1/2} = 2(\overset{+}{\sigma}_{\mathrm{m}})^2 \,. \end{gathered} \tag{4.31}$$

(3) Criterion (4.1) for the axial compression failure of the fibres is

$$\begin{gathered}(NW_1 + MW_2)\cos^2 \beta_k + (NW_5 + MW_6)\sin^2 \beta_k \\ + 2(NW_7 + MW_8)\sin \beta_k \cos \beta_k = E_{\mathrm{fz}} \overset{-}{e}_{\mathrm{fu}} \,. \end{gathered} \tag{4.32}$$

The above strength criteria are not applicable to plastics reinforced with organic fibres. This is explained by the fact that the transverse tensile and shear strengths of organic fibres are lower than the respective matrix strengths and therefore the failure of organic-fibre reinforced plastics is always due to fibre failure. Characteristic types of failure of an unidirectionally reinforced organic plastic under tension and compression are shown in Fig. 9.

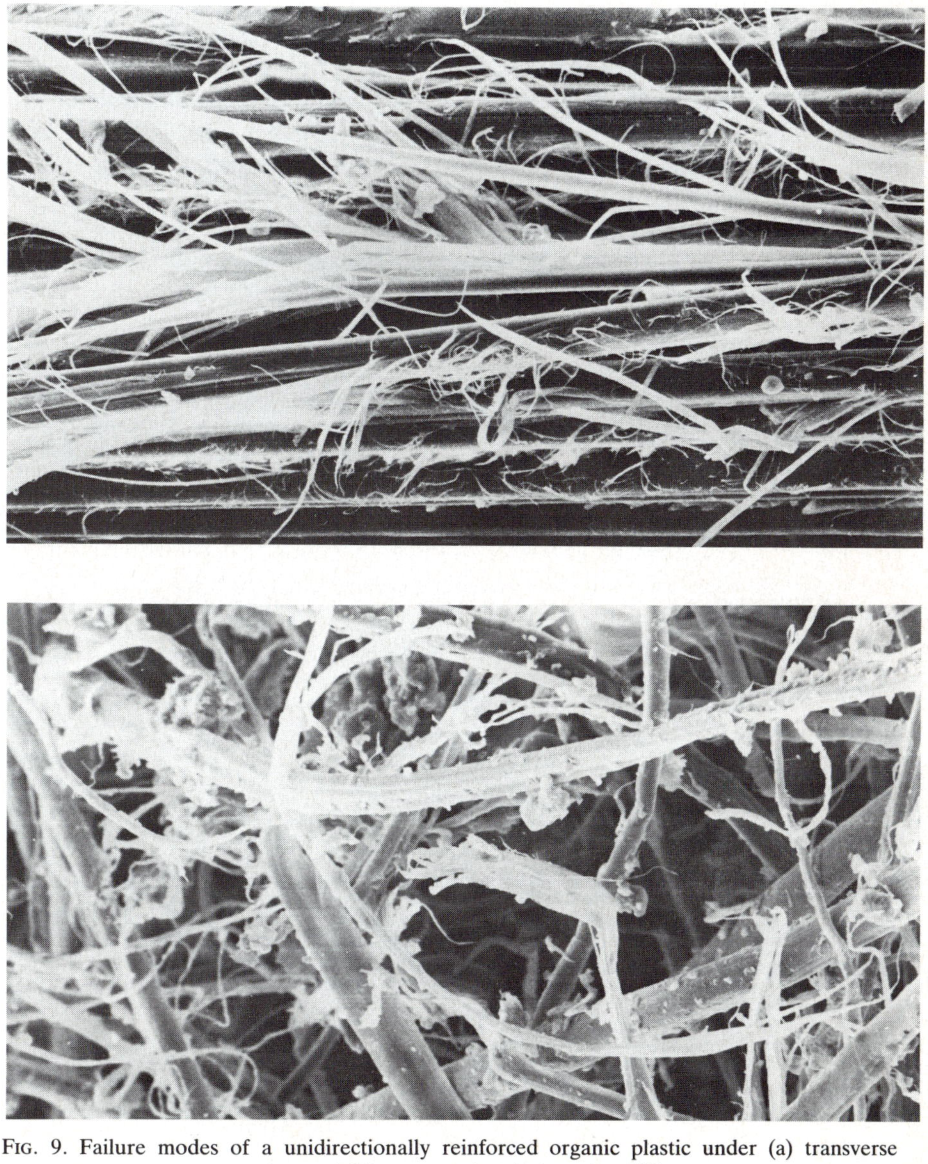

FIG. 9. Failure modes of a unidirectionally reinforced organic plastic under (a) transverse tension and (b) compression (enlarged ×500).

5. Strength of a laminated beam

5.1. Bar strength under axial load

If the stress state and strength is the same in all plies, then the assumptions made permit us to conclude that the failure of a ply will coincide with the failure of the whole beam. If the stress state and strength is not the same in different plies, the failure will occur in a stepwise manner. In the latter case the failure of the first ply will be determined by means of the relationships given in Section 4. After the first ply has failed, the laminate stiffness elements A_{ij}, B_{ij} and D_{ij} which are employed in the strength criteria of a ply will be recalculated; the axial force which causes failure in the next ply will be determined in a likewise manner. The above procedure will be repeated until the axial force causing failure of the next ply turns out to be smaller than the force causing failure of the preceding ply.

As an example we shall consider the tension of an orthogonally reinforced tubular bar. The wall structure of this cylinder is shown in Fig. 10.

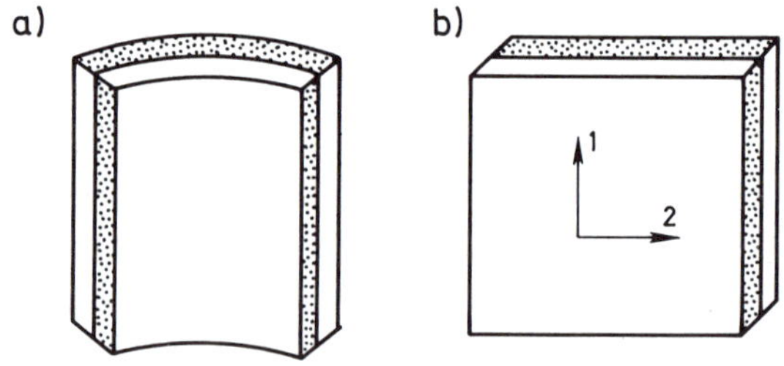

FIG. 10. Cross-section of two beams: (a) a cylindrical bar; (b) a flat beam.

The transversely reinforced ply of the cross-ply tube will be the first to fail under tension. For a tube:

$$k_1 = B'_{11}N_1 + D'_{11}M_1 = 0 \quad \text{and} \quad M_1 = -B'_{11}/D'_{11} \cdot N_1,$$

$$\langle \varepsilon_1^0 \rangle = \left(A'_{11} - \frac{(B'_{11})^2}{D'_{11}} \right) N_1 = a_1 N_1 ,$$

$$\langle \varepsilon_2^0 \rangle = \left(A'_{21} - \frac{B'_{21}B'_{11}}{D'_{11}} \right) N_1 = a_2 N_1 .$$

In this case formula (2.8) for $\langle \sigma_x \rangle_k = \langle \sigma_1 \rangle_k$ is written as

$$\langle \sigma_1 \rangle_k = \left[(Q_{22})_k \left(A'_{11} - \frac{(B'_{11})^2}{D'_{11}} \right) + (Q_{12})_k \left(A'_{21} - \frac{B'_{21}B'_{11}}{D'_{11}} \right) \right] \frac{N}{2\pi R}$$

$$= [(Q_{22})_k a_1 + (Q_{12})_k a_2] \frac{N}{2\pi R} , \qquad (5.1)$$

where

$$a_1 = A'_{11} - \frac{\langle B'_{11}\rangle^2}{D'_{11}} = \frac{A_{22}}{A_{11}A_{22} - A_{12}^2} \qquad a_2 = A'_{21} - \frac{B'_{21}B'_{11}}{D'_{11}} = -\frac{A_{12}}{A_{11}A_{22} - A_{12}^2} .$$

Neglecting the influence of the stress $\langle \sigma_y \rangle_k = \langle \sigma_2 \rangle_k$ on the stress state of the matrix, formula (4.14) yields

$$\langle \sigma_\perp \rangle_k = \langle \sigma_1 \rangle_k .$$

At the moment of matrix failure $\langle \sigma_\perp \rangle_k = (\overset{+}{\sigma}_\perp)_k$. Then, with $N^+ = 2\pi R N_1$ and $M = 0$ and $S = 0$ formula (4.8) yields

$$N^+ = \frac{(\overset{+}{\sigma}_\perp)_k 2\pi R}{(Q_{22})_k A'_{11} + (Q_{12})_k A'_{21}} , \tag{5.2}$$

where R is the midplane radius of the tube. The strength $(\overset{+}{\sigma}_\perp)_k$ is determined from formula (4.9).

For a flat beam, taking into account condition (4.8) with $M = 0$ and $S = 0$, we obtain

$$N^+ = \frac{(\overset{+}{\sigma}_\perp)_k}{W_1} = \frac{(\overset{+}{\sigma}_\perp)_k b}{q_{1k} + z q_{3k}} , \tag{5.3}$$

where

$$q_{1k} = (Q_{22})_k A'_{11} + (Q_{12})_k A'_{21} ,$$

$$q_{3k} = (Q_{22})_k B'_{11} + (Q_{12})_k B'_{12} .$$

After failure of the laterally reinforced plies it is necessary to recalculate the laminate stiffness properties A_{ij}, B_{ij}, D_{ij} assuming $E_\perp = G_{\parallel\perp} = \nu_{\parallel\perp} = \nu_{\perp\parallel} = 0$ for the plies which have undergone failure.

After the loss of material continuity, the whole load will be taken by the plies reinforced in the direction of the applied load.

In the case of fibre failure the relationship for $\langle \varepsilon_1^0 \rangle$ yields

$$N^+ = \frac{\overset{+}{e}_{\mathrm{fu}} 2\pi R}{A'_{11} - \dfrac{(B'_{11})^2}{D'_{11}}} = \frac{2\pi R}{a_1} \overset{+}{e}_{\mathrm{fu}} . \tag{5.4}$$

In the case of a flat bar, condition (4.3) permits us to write

$$N^+ = \frac{\overset{+}{e}_{\mathrm{fu}} b}{A'_{11} + z B'_{11}} . \tag{5.5}$$

If the beam is symmetrically laminated about its midplane, formula (5.5) simplifies to:

$$N^{+} = \frac{e^{+}_{fu} b}{A'_{11}} . \tag{5.6}$$

After comparing formulas (5.4) and (5.5) when $b = 2\pi R$, it becomes evident that in the case of unsymmetric lay-up the tube processes a larger tensile strength than the flat beam. The higher tensile strength of the tube is explained by the effect of the moment M caused by the delayed bending of the cylinder walls. A similar effect is observed under compression. In such a case, the ultimate compression strain of fibres e^{-}_{fu} should be inserted into formulas (5.4) and (5.5) instead of e^{+}_{fu}.

5.2. Strength under transverse bending

As an example consider the failure of a beam whose cross section and loading scheme are given in Fig. 11.

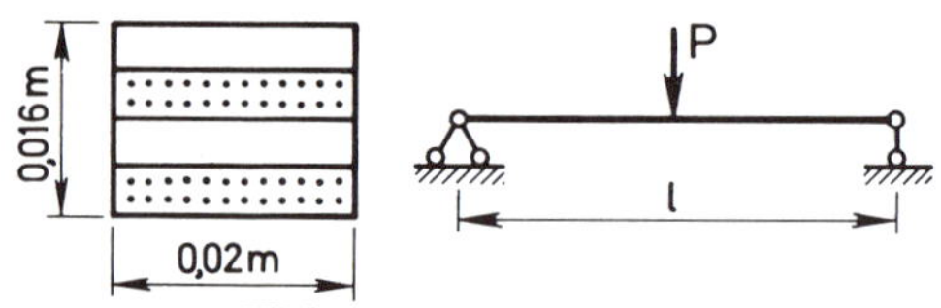

FIG. 11. Loading scheme and cross-section shape of a beam.

As seen in Fig. 11, the laminated beam consists of unidirectionally reinforced plies with some of the plies reinforced perpendicular and some parallel to the beam axis. For the sake of comparison, we shall consider two cases—beams made of carbon fibre reinforced plastic (CFP) and glass fibre reinforced plastic (GEP). The elastic and strength properties of the individual ply components are

(a) for carbon fibres

$E_{fz} = 230\,000$ MPa $\quad E_{fr} = 16\,100$ MPa ,
$G_{fzr} = 20\,000$ MPa $\quad \nu_{fzr} = 0.30$,
$e^{+}_{fu} = 0.008 \quad e^{-}_{fu} = 0.0064;\ v_f = 0.60$,

(b) for glass fibres

$E_f = 70\,000$ MPa $\quad e^{-}_{fu} = 0.021$,
$\nu_f = 0.23 \quad v_f = 0.50$,
$e^{+}_{fu} = 0.036$

(c) for the matrix

$$E_{\mathrm{m}} = 3500\ \mathrm{MPa} \qquad \nu_{\mathrm{m}} = 0.36\,,$$
$$\sigma^{+}_{\mathrm{m}} = 70\ \mathrm{MPa} \qquad \tau^{\mathrm{u}}_{\mathrm{m}} = 60\ \mathrm{MPa}\,.$$

For the above data the stress concentration coefficients (SKUDRA, 1984) and elastic properties of a ply have the following values:

(a) for a CFP

$$Q_{11} = 13.79 \times 10^4\ \mathrm{MPa} \quad Q_{66} = 0.64 \times 10^4\ \mathrm{MPa}\,,$$
$$Q_{22} = 0.81 \times 10^4\ \mathrm{MPa} \quad \nu_{\perp\perp\!\!\perp} = 0.32\,,$$
$$Q_{12} = 0.27 \times 10^4\ \mathrm{MPa} \quad \bar{\sigma}_r = 1.40,\ \bar{\sigma}_\theta = 0.62,\ \bar{\sigma}_z = 0.70\,,$$

(b) for a GFP

$$Q_{11} = 3.72 \times 10^4\ \mathrm{MPa} \quad Q_{66} = 0.27 \times 10^4\ \mathrm{MPa}\,,$$
$$Q_{22} = 0.87 \times 10^4\ \mathrm{MPa} \quad \nu_{\perp\perp\!\!\perp} = 0.13\,,$$
$$Q_{12} = 0.25 \times 10^4\ \mathrm{MPa} \quad \bar{\sigma}_r = 1.70,\ \bar{\sigma}_\theta = 0.92,\ \bar{\sigma}_z = 0.94\,.$$

For the case under consideration the deformation laws for a four-ply material in the direct and the inverted form are as follows:

(a) for a CFP

$$\begin{Bmatrix} N_x \\ N_y \\ N_{xy} \\ M_x \\ M_y \\ M_{xy} \end{Bmatrix} = \begin{bmatrix} 1168.0 & 43.2 & 0 & 2.08 & 0 & 0 \\ 43.2 & 1168.0 & 0 & 0 & -2.08 & 0 \\ 0 & 0 & 102.4 & 0 & 0 & 0 \\ 2.08 & 0 & 0 & 0.0249 & 0.0009 & 0 \\ 0 & -2.08 & 0 & 0.0009 & 0.0249 & 0 \\ 0 & 0 & 0 & 0 & 0 & 0.0022 \end{bmatrix} \begin{Bmatrix} \langle\!\langle \varepsilon^0_x \rangle\!\rangle \\ \langle\!\langle \varepsilon^0_y \rangle\!\rangle \\ \langle\!\langle \gamma^0_{xy} \rangle\!\rangle \\ k_x \\ k_y \\ k_{xy} \end{Bmatrix},$$

$$\begin{Bmatrix} \langle\!\langle \varepsilon^0_x \rangle\!\rangle \\ \langle\!\langle \varepsilon^0_y \rangle\!\rangle \\ \langle\!\langle \gamma^0_{xy} \rangle\!\rangle \\ k_x \\ k_y \\ k_{xy} \end{Bmatrix} = \begin{bmatrix} 0.00100 & -0.00004 & 0 & -0.084 & 0 & 0 \\ -0.00004 & 0.00100 & 0 & 0 & 0.084 & 0 \\ 0 & 0 & 0.00970 & 0 & 0 & 0 \\ -0.084 & 0 & 0 & 47.23 & -1.74 & 0 \\ 0 & 0.084 & 0 & -1.74 & 47.23 & 0 \\ 0 & 0 & 0 & 0 & 0 & 458.71 \end{bmatrix} \begin{Bmatrix} N_x \\ N_y \\ N_{xy} \\ M_x \\ M_y \\ M_{xy} \end{Bmatrix},$$

(b) for a GFP

$$
\begin{Bmatrix} N_x \\ N_y \\ N_{xy} \\ \hline M_x \\ M_y \\ M_{xy} \end{Bmatrix}
=
\left[\begin{array}{cccccc}
367.9 & 41.1 & 0 & 0.46 & 0 & 0 \\
41.1 & 367.9 & 0 & 0 & -0.46 & 0 \\
0 & 0 & 44.7 & 0 & 0 & 0 \\
\hline
0.46 & 0 & 0 & 0.00784 & 0.00087 & 0 \\
0 & -0.46 & 0 & 0.00087 & 0.00784 & 0 \\
0 & 0 & 0 & 0 & 0 & 0.00095
\end{array}\right]
\begin{Bmatrix} \langle\!\langle \varepsilon_x^0 \rangle\!\rangle \\ \langle\!\langle \varepsilon_y^0 \rangle\!\rangle \\ \langle\!\langle \gamma_{xy}^0 \rangle\!\rangle \\ \hline k_x \\ k_y \\ k_{xy} \end{Bmatrix},
$$

$$
\begin{Bmatrix} \langle\!\langle \varepsilon_x^0 \rangle\!\rangle \\ \langle\!\langle \varepsilon_y^0 \rangle\!\rangle \\ \langle\!\langle \gamma_{xy}^0 \rangle\!\rangle \\ \hline k_x \\ k_y \\ k_{xy} \end{Bmatrix}
=
\left[\begin{array}{ccc|ccc}
0.00297 & -0.00033 & 0 & -0.173 & 0 & 0 \\
-0.00033 & 0.00297 & 0 & 0 & 0.173 & 0 \\
0 & 0 & 0.02230 & 0 & 0 & 0 \\
\hline
-0.173 & 0 & 0 & 139.20 & -15.57 & 0 \\
0 & 0.173 & 0 & -15.57 & 139.20 & 0 \\
0 & 0 & 0 & 0 & 0 & 1049.00
\end{array}\right]
\begin{Bmatrix} N_x \\ N_y \\ N_{xy} \\ \hline M_x \\ M_y \\ M_{xy} \end{Bmatrix}.
$$

The through-the-thickness stress diagrams constructed from formulas (2.9) and (2.26) for the pre-failure state of the individual plies are shown in Fig. 12.

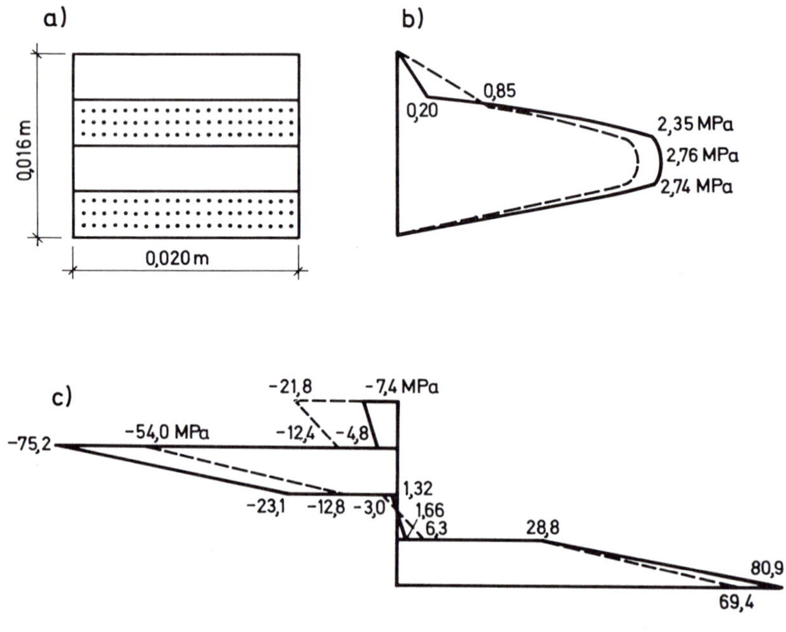

FIG. 12. Through-the-thickness (a) normal and (b) shear stress diagrams for glass/epoxy beams (dotted lines) and carbon/epoxy beams (continuous lines).

The first value to be determined is the magnitude of the force $P = P_1$ which causes fibres in the first ply to fail under tension. In the case of transverse bending, i.e. when $N = 0$, criterion (4.3) yields

$$M = \frac{\overset{+}{e}_{\text{fu}} b}{B'_{11} + z_{1k} D'_{11}} . \tag{5.7}$$

For the given case of loading with $M = P_1\, l/4$ and $z_k = h/2$ we obtain:

$$P_1 = \frac{4b\overset{+}{e}_{\text{fu}}}{l(B'_{11} + \frac{1}{2} h D'_{11})} \tag{5.8}$$

We proceed to determine $P = P_2$ which causes compressive failure of fibres in the 3rd ply with $z_k = -\frac{1}{4}h$ and $M = P_2\, l/4$

$$P_2 = \frac{4b\overset{-}{e}_{\text{fu}}}{l(B'_{11} - \frac{1}{4} h D'_{11})} . \tag{5.9}$$

For the failure of the 4th ply under transverse compression when $P = P_3$, $N = 0$, $S = P_3/2$ and $M = P_3 l/4$, criterion (4.11) yields

$$P_3 = \frac{2^{1/2}(\sigma_\perp^-)_4}{\left[\frac{l^2 W_2^2}{16} + (1 + \nu_{\perp\perp\!\!\perp}) \frac{W_9^2}{2} + \frac{l W_2}{4}\left(\frac{l_2^2 W_2^2}{16} + W_9^2\right)^{1/2}\right]^{1/2}} . \tag{5.10}$$

It should be noted that the strength $(\sigma_\perp^-)_k$ is obtained from formula (4.12). In our case, when $z_k = -\frac{1}{2}h$

$$q_{24} = Q_{22} B'_{11} + Q_{12} B'_{21} ,$$

$$q_{44} = Q_{22} D'_{11} + Q_{12} D'_{21} ,$$

$$W_2 = \frac{q_{24} - \frac{1}{2} h q_{44}}{b} ,$$

$$W_9 = 0 ,$$

and

$$P_3 = 4(\sigma_\perp^-)_4 / l W_2 . \tag{5.11}$$

It is also necessary to determine the magnitude of the force $P = P_4$ which will cause matrix failure in the 2nd ply under the combined action of transverse tension ($\langle \sigma_1 \rangle$) and shear ($\langle \tau_{13} \rangle$). Subsequently, provided that $N = 0$, criterion (4.8) becomes

$$(MW_2)^2 + 2(1 + \nu_{\perp\perp\!\!\perp})(SW_9)^2 + MW_3[(MW_2)^2 + 4(SW_9)^2]^{1/2} = 2(\sigma_\perp^+)_k^2 . \tag{5.12}$$

Inserting $M = P\,l/4$ and $S = P/2$ into criterion (5.12) we obtain the following expression for P_4:

$$P_4 = \frac{2^{1/2}(\overset{+}{\sigma}_{\perp})_2}{\left[\dfrac{l^2 W_2^2}{16} + (1+\nu_{\perp\perp})\dfrac{W_9}{2} + \dfrac{lW_4}{4}\left(\dfrac{l_2 W_2^2}{16} + W_9^2\right)^{1/2}\right]^{1/2}} . \tag{5.13}$$

It follows from the $\langle \sigma_1 \rangle$ and $\langle \tau_{13} \rangle$ stress diagrams that in the present case

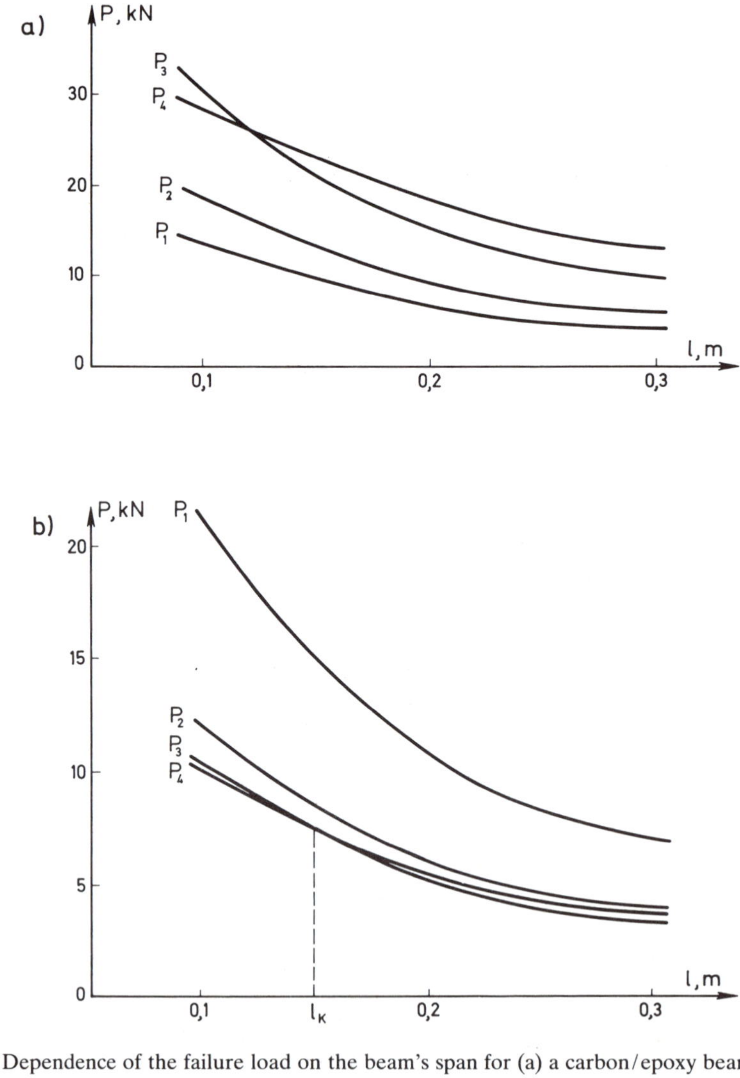

FIG. 13. Dependence of the failure load on the beam's span for (a) a carbon/epoxy beam and (b) a glass/epoxy beam.

the ply at the distance $z_2 = \frac{1}{4}h$ from the midplane will be in the most disadvantageous position.

Thus

$$W_2 = \frac{q_{22} + \frac{1}{4}hq_{42}}{b} \; .$$

For our case $q_{22} = q_{24}$ and $q_{42} = q_{44}$. The parameters q_{1i} and q_{2i} for the first ply in the formula for W_9 are found from the following relationships:

$$q_{11} = Q_{11}B'_{11} = 13.79 \times 10^4(-0.084) = -115.83 \times 10^2 \; ,$$

$$q_{21} = Q_{11}D'_{11} + Q_{12}D'_{21} = 13.79 \times 10^4 \times 47.231 + 0.27 \times 10^4(-1.74)$$
$$= 650.84 \times 10^4 \; .$$

In order to find W_9, the first ply should be divided schematically, say, into four sublayers.

Fig. 13 shows the dependence of the forces P_1, P_2, P_3 and P_4 causing failure of individual plies on the beam span l. Fig. 13(a) allows us to conclude that in the given case, as the span changes from 0.1 m to 0.3 m the strength of the CFP beam is governed by the force $P = P_1$. Formula (5.6) leads to the statement that, when $P = P_1$, carbon fibre failure takes place under tension in the first ply (Fig. 11).

It follows from Fig. 13 that for a GFP beam the mechanism of failure depends on the beam's span. When $l < l_k$, the matrix of the second ply is the first to undergo failure under the combined action of transverse tension and shear, and at the moment of beam failure $P = P_4$. When $l > l_k$, the fourth ply will be the first to fail under transverse compression; at the moment of beam failure $P = P_3$.

5.3. Strength under combined axial load and transverse bending

Bars constructed of reinforced plastic laminates are often used in systems such as flat statically determinate or indeterminate frames or arches. In these systems, such bars are generally acted upon by a combination of bending moment, shear and axial forces.

To illustrate the construction of strength envelopes for the combined action of the axial force N, bending moment M and shear S let us consider a bar for which the geometry and the loading scheme are shown in Fig. 11. The bar is made of orthogonally reinforced GFP and it possesses the following elastic properties of the individual plies:

$$Q_{11} = 3.72 \times 10^4 \text{ MPa} \quad Q_{66} = 0.27 \times 10^4 \text{ MPa} \; ,$$
$$Q_{22} = 0.87 \times 10^4 \text{ MPa} \quad \nu_{\perp \perp\!\perp} = 0.13 \; .$$
$$Q_{12} = 0.25 \times 10^4 \text{ MPa}$$

For the fibre volume content $v_f = 0.60$ the stress concentration coefficients are as follows (SKUDRA, 1984):

$$\bar{\sigma}_r = 1.70 \quad \bar{\sigma}_\theta = 0.92 \quad \bar{\sigma}_z = 0.94 \,.$$

The strength characteristics of the fibres and matrix are taken as

$$\sigma^+_{\mathrm{m}} = 70\,\mathrm{MPa} \quad e^+_{\mathrm{fu}} = 0.036\,,$$
$$\tau^{\mathrm{u}}_{\mathrm{m}} = 60\,\mathrm{MPa} \quad e^-_{\mathrm{fu}} = 0.021\,.$$

In order to construct the strength envelope in the N–M coordinate system the following relationship between S and M is used:

$$S = \mathrm{d}M/\mathrm{d}x \,.$$

It is assumed that the axial force N is tensile whereas the bending moment may be positive ($+M$) or negative ($-M$).

Fig. 14 represents strength envelopes for the bar shown in Fig. 11. Numbers 1^+, 2^+, 3^+ and 4^+ indicate here tensile failure of the plies 1, 2, 3 and 4; 1^-, 2^-, 3^- and 4^- refer to compressive failure. The straight lines 1^+, 1^-, 3^+ and 3^- are based on formula (4.3) and the straight line 4^+ and the curve 2^+ are plotted from formula (4.8). The basic peculiarity of failure of a laminated bar is the fact that every section of the strength envelope of Fig. 14 is characterized by its own mechanism of failure. Thus, section A–B represents complete failure caused by compressive failure of fibres in the 3rd ply after the 2nd ply has undergone transverse tensile failure and the 4th ply has undergone transverse compressive failure;

Section B–C corresponds to complete failure caused by tensile failure of fibres in the 1st ply after transverse tensile failure of the 2nd ply and transverse compressive failure of the 4th ply;

Section C–D shows simultaneous failure of the 1st ply caused by fibre rupture and of the 4th ply under transverse compression after the 2nd ply has failed due to transverse tension;

Section D–E shows fibre rupture in the 1st ply after pre-cracking of the 2nd ply due to transverse tension;

Section E–F shows fibre rupture in the first ply after cracking the 2nd and 4th plies due to transverse tension;

Section F–H corresponds to fibre rupture in the 3rd ply after pre-cracking of the 2nd and 4th plies due to transverse tension;

Section H–I shows simultaneous failure of the 3rd ply due to fibre rupture and the 2nd ply under transverse tension after the 4th ply has pre-cracked;

Section I–K shows fibre rupture in the 3rd ply after pre-cracking of the 4th ply;

Section K–L illustrates compressive failure of fibres in the first ply after the transverse tensile failure of the fourth ply.

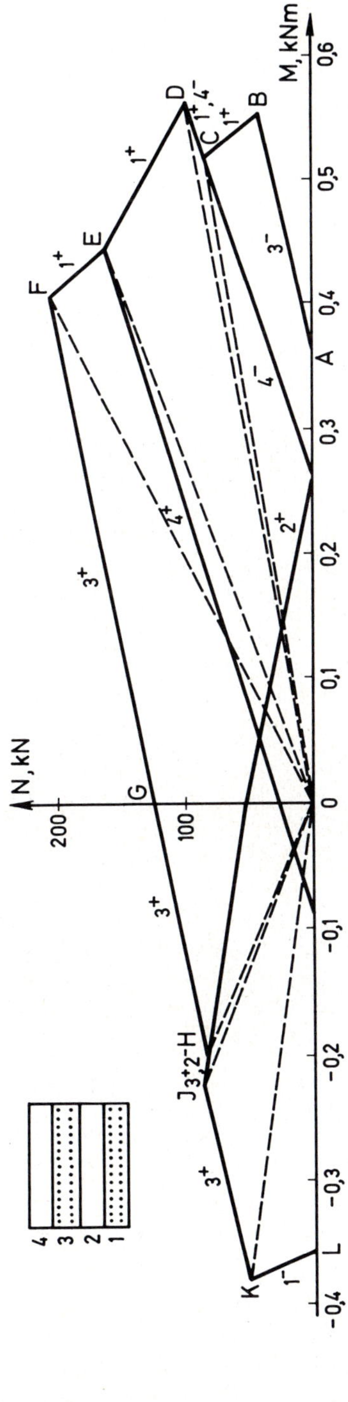

FIG. 14. Ultimate strength envelopes.

The above analysis of the strength envelope shown in Fig. 14 gives convincing evidence of the complex nature of failure mechanism in laminated bars. It should be noted that the suggested structural approach to strength prediction for a laminated member of prescribed structure may be also used as a theoretical basis in designing bars of prescribed strengths.

5.4. Strength under axial load and torsion

The geometric configuration of a thin-walled cylindrical bar of an arbitrary structure is shown in Fig. 15.

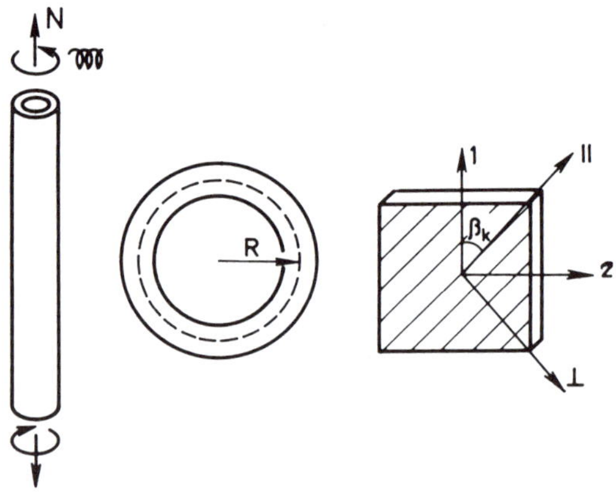

FIG. 15. Geometric configuration of a tube.

The strains in the bar are determined according to the relationships:

$$\langle\!\langle \varepsilon_1 \rangle\!\rangle = \left(A'_{11} - \frac{(B'_{11})^2}{D'_{11}} \right) \frac{N}{2\pi R} + \left(A'_{16} - \frac{B'_{11}B'_{61}}{D'_{11}} \right) \frac{\mathcal{M}}{2\pi R^2} = a_3 N + a_4 \mathcal{M} ,$$

$$\langle\!\langle \varepsilon_2 \rangle\!\rangle = \left(A'_{21} - \frac{B'_{21}B'_{11}}{D'_{11}} \right) \frac{N}{2\pi R} + \left(A'_{26} - \frac{B'_{21}B'_{61}}{D'_{11}} \right) \frac{\mathcal{M}}{2\pi R^2} = a_5 N + a_6 \mathcal{M} ,$$

$$\langle\!\langle \gamma_{12} \rangle\!\rangle = \left(A'_{61} - \frac{B'_{61}B'_{11}}{D'_{11}} \right) \frac{N}{2\pi R} + \left(A'_{66} - \frac{(B'_{61})^2}{D'_{11}} \right) \frac{\mathcal{M}}{2\pi R^2} = a_7 N + a_8 \mathcal{M} .$$

The mean stresses in an arbitrary kth ply along the directions 1 and 2 are obtained in the following way:

$$\langle \sigma_1 \rangle_k = r_1 N + s_1 \mathcal{M} , \tag{5.14}$$

where

$$r_1 = (\bar{Q}_{11})_k a_3 + (\bar{Q}_{12})_k a_5 + (\bar{Q}_{16})_k a_7 ,$$

$$s_1 = [(\bar{Q}_{11})_k a_4 + (\bar{Q}_{12})_k a_6 + (\bar{Q}_{16})_k a_8] ,$$

$$\langle \sigma_2 \rangle_k = r_2 N + s_2 \mathcal{M} , \tag{5.15}$$

where

$$r_2 = (\bar{Q}_{12})_k a_3 + (\bar{Q}_{22})_k a_5 + (\bar{Q}_{26})_k a_7 ,$$

$$s_2 = [(\bar{Q}_{12})_k a_4 + (\bar{Q}_{22})_k a_6 + (\bar{Q}_{26})_k a_8] ,$$

$$\langle \tau_{12} \rangle_k = r_3 N + s_3 \mathcal{M} , \tag{5.16}$$

where

$$r_3 = (\bar{Q}_{16})_k a_3 + (\bar{Q}_{26})_k a_5 + (\bar{Q}_{66})_k a_7 ,$$
$$s_3 = [(\bar{Q}_{16})_k a_4 + (\bar{Q}_{26})_k a_6 + (\bar{Q}_{66})_k a_8 .$$

In the material principal coordinate system of the kth (Fig. 15), the mean stresses are obtained from the formulas

$$\langle \sigma_{\parallel} \rangle_k = \langle \sigma_1 \rangle_k \cos^2 \beta_k + \langle \sigma_2 \rangle_k \sin^2 \beta_k + 2\langle \tau_{12} \rangle_k \sin \beta_k \cos \beta_k = F_7 N + F_8 \mathcal{M} , \tag{5.17}$$

where

$$F_7 = r_1 \cos^2 \beta_k + r_2 \sin^2 \beta_k + 2r_3 \sin \beta_k \cos \beta_k ,$$
$$F_8 = s_1 \cos^2 \beta_k + s_2 \sin^2 \beta_k + 2s_3 \sin \beta_k \cos \beta_k ,$$
$$\langle \sigma_{\perp} \rangle_k = \langle \sigma_1 \rangle_k \sin^2 \beta_k + \langle \sigma_2 \rangle_k \cos^2 \beta_k$$
$$- 2\langle \tau_{12} \rangle_k \sin \beta_k \cos \beta_k = F_9 N + F_{10} \mathcal{M} , \tag{5.18}$$

where

$$F_9 = r_1 \sin^2 \beta_k + r_2 \cos^2 \beta_k - 2r_3 \sin \beta_k \cos \beta_k ,$$
$$F_{10} = s_1 \sin^2 \beta_k + s_2 \cos^2 \beta_k - 2s_3 \sin \beta_k \cos \beta_k ,$$
$$\langle \tau_{\parallel\perp} \rangle_k = (\langle \sigma_2 \rangle_k - \langle \sigma_1 \rangle_k) \sin \beta_k \cos \beta_k$$
$$+ \langle \tau_{12} \rangle_k (\cos^2 \beta_k - \sin^2 \beta_k) = F_{11} N + F_{12} \mathcal{M} , \tag{5.19}$$

where

$$F_{11} = (r_2 - r_1) \sin \beta_k \cos \beta_k + r_3 (\cos^2 \beta_k - \sin^2 \beta_k) ,$$
$$F_{12} = (s_2 - s_1) \sin \beta_k \cos \beta_k + s_3 (\cos^2 \beta_k - \sin^2 \beta_k) .$$

Inserting the obtained expressions for the stresses $\langle \sigma_{\perp} \rangle_k$ and $\langle \tau_{\parallel\perp} \rangle_k$ into the general strength criteria for the kth (SKUDRA, 1984) we obtain:

(1) In the case of tensile failure of the matrix in the kth ply

$$(1-\nu_{\mathrm{m}}^{2})\left(\frac{\bar{\sigma}_r}{\sigma_{\mathrm{m}}^{+}}\right)^2(NF_9+\mathcal{M}F_{10})^2+\frac{1}{(\tau_{\mathrm{m}}^{\mathrm{u}})^2}(NF_{11}+\mathcal{M}F_{12})=1\,. \tag{5.20}$$

(2) In the case of shear failure of the matrix

$$(NF_9+\mathcal{M}F_{10})^2+2(1+\nu_{\mathrm{m}})(NF_{11}+\mathcal{M}F_{12})^2 + (NF_9+\mathcal{M}F_{10})[(NF_9+\mathcal{M}F_{10})^2+4(NF_{11}+\mathcal{M}F_{12})^2]^{1/2}=2(\sigma_{\mathrm{m}}^{+})^2\,. \tag{5.21}$$

(3) In the case of bond failure

$$(NF_9+\mathcal{M}F_{10})(\tau_{\mathrm{b}}^{\mathrm{u}})^2\bar{\sigma}_r+(NF_{11}+\mathcal{M}F_{12})^2\sigma_{\mathrm{b}}^{\mathrm{u}}\bar{\sigma}_{rz}^2=(\tau_{\mathrm{b}}^{\mathrm{u}})^2\sigma_{\mathrm{b}}^{\mathrm{u}}\,. \tag{5.22}$$

(4) In the case of transverse compressive failure of the kth ply

$$(NF_9+\mathcal{M}F_{10})^2+2(1+\nu_{\perp\perp})(NF_{11}+\mathcal{M}F_{12})^2+(NF_9+ + \mathcal{M}F_{10})[(NF_9+\mathcal{M}F_{10})^2+4(NF_{11}+\mathcal{M}F_{12})^2]^{1/2}=2(\sigma_{\perp}^{-})^2\,. \tag{5.23}$$

(5) In the case of fibre failure due to tension or compression

$$\langle\!\langle \varepsilon_1\rangle\!\rangle\cos^2\beta_k+\langle\!\langle \varepsilon_2\rangle\!\rangle\sin^2\beta_k+\langle\!\langle \gamma_{12}\rangle\!\rangle\sin\beta_k\cos\beta_k = r_4N+s_4\mathcal{M}=e_{\mathrm{fu}}^{\pm}\,, \tag{5.24}$$

where

$$r_4=a_3\cos^2\beta_k+a_5\sin^2\beta_k+a_7\sin\beta_k\cos\beta_k\,,$$
$$s_4=[a_4\cos^2\beta_k+a_6\sin^2\beta_k+a_8\sin\beta_k\cos\beta_k]\,.$$

If the beam consists of unidirectionally reinforced plies oriented at various angles β, the plies do not fail simultaneously and the stepwise character of laminate failure should be taken into account when determining A'_{ij}.

6. Analysis of statically indeterminate systems by the method of forces

Consider an arbitrary statically indeterminate system consisting of elements made up of a reinforced laminated plastic. In order to achieve a statically determinate primary system, by the 'extra' constraints are replaced by unknown redundant reactions $X_1, X_2, \ldots, X_n$ (where n is the degree of statical indeterminacy of the system by the method of forces). Consequently, the force and moment resultants can be written as

$$M = M_p + \sum_{i=1}^{n} X_i \bar{M}_i ,$$

$$N = N_p + \sum_{i=1}^{n} X_i \bar{N}_i , \tag{6.1}$$

$$S = S_p + \sum_{i=1}^{n} X_i \bar{S}_i .$$

Here $\bar{M}_i$, $\bar{N}_i$ and $\bar{S}_i$ are the corresponding internal moment and force resultants due to the action of the unit force $X_i = 1$.

The system of canonical equations for determining the redundant reactions is written as follows

$$\Delta_{ii} = 0 \quad i = 1, 2, \ldots, n , \tag{6.2}$$

where Δ_{ii} is the displacement of the primary system in the direction of the ith reaction acting at the ith point.

In the fully developed form (6.2) becomes

$$\begin{aligned} X_1\delta_{11} + X_2\delta_{12} + \cdots + X_n\delta_{1n} + \Delta_{1p} &= 0 , \\ X_1\delta_{21} + X_2\delta_{22} + \cdots + X_n\delta_{2n} + \Delta_{2p} &= 0 , \\ \cdots\cdots\cdots\cdots\cdots\cdots\cdots\cdots\cdots\cdots\cdots & \\ X_1\delta_{n1} + X_2\delta_{n2} + \cdots + X_n\delta_{nn} + \Delta_{np} &= 0 . \end{aligned} \tag{6.3}$$

As an example, consider the frame illustrated in Fig. 16(a).

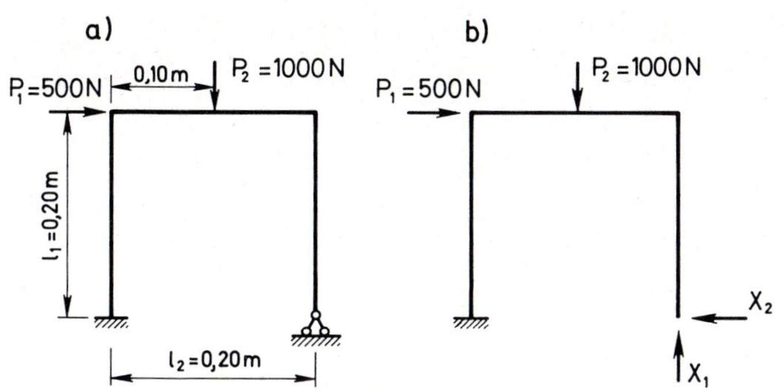

FIG. 16. Geometric configuration of (a) a frame and (b) its primary system.

For such a frame the system of canonical equations (6.3) is given by the following:

$$X_1\delta_{11} + X_2\delta_{12} + \Delta_{1p} = 0\,,$$
$$X_1\delta_{12} + X_2\delta_{22} + \Delta_{2p} = 0\,. \tag{6.4}$$

In order to find the coefficients of these equations it is necessary to construct the corresponding internal force diagrams.

The internal force diagrams of the primary system shown in Fig. 16(b) and caused by the action of $X_1 = 1$, $X_2 = 1$ and the applied load, are given in Fig. 17.

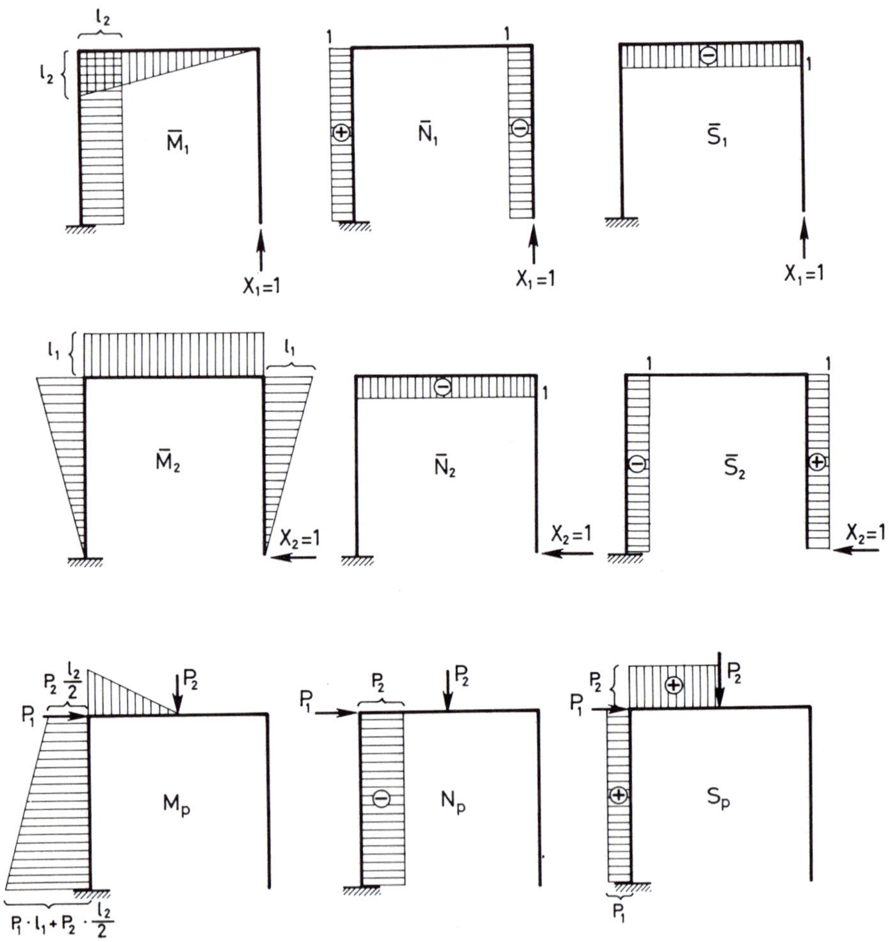

FIG. 17. Internal force diagrams of a primary system due to the action of $X_1 = 1$ and the superimposed load.

First, we shall determine the displacements due to the unit forces:

$$\delta_{11} = \frac{1}{b}\Big[A'_{11}\sum\int \bar{N}_1^2\,dx + 2B'_{11}\sum\int \bar{N}_1\bar{M}_1\,dx + D'_{11}\sum\int \bar{M}_1^2\,dx$$

$$+ \frac{\kappa}{A_{55}}\sum\int \bar{S}_1^2\,dx\Big] = \frac{1}{b}\Big[A'_{11}\sum \omega_{\bar{N}_1}y_{\bar{N}_1} + 2B'_{11}\sum \omega_{\bar{N}_1}y_{\bar{M}_1}$$

$$+ D'_{11}\sum \omega_{\bar{N}_1}y_{\bar{M}_1} + \frac{\kappa}{A_{55}}\sum \omega_{\bar{S}_1}y_{\bar{S}_1}\Big] = \frac{1}{b}\Big\{A'_{11}[1\cdot l_1\cdot 1$$

$$+(-1\cdot l_1)(-1)] + 2B'_{11}(1\cdot l_1\cdot l_2) + D'_{11}(l_1\cdot l_2\cdot l_2 + l_2\cdot l_2\tfrac{1}{2}\cdot\tfrac{2}{3}l_2)$$

$$+ (\kappa/A_{55})1\cdot l_2\cdot 1\} = \frac{1}{b}\,[A'_{11}\cdot 2l_1 + 2B'_{11}\cdot l_1\cdot l_2 + D'_{11}(l_1\cdot l_2^2 + \tfrac{1}{3}l_2^3)$$

$$+ (\kappa/A_{55})l_2]\,,$$

$$\delta_{12} = b^{-1}\Big[A'_{11}\sum\int \bar{N}_1N_2\,dx + B'_{11}\sum\int \bar{N}_1\bar{M}_2\,dx$$

$$+ B'_{11}\sum\int \bar{M}_1\bar{N}_2\,dx + D'_{11}\sum\int \bar{M}_1\bar{M}_2\,dx + \frac{\kappa}{A_{55}}\int \bar{S}_1\bar{S}_2\,dx\Big]$$

$$= b^{-1}\Big[A'_{11}\sum \omega_{\bar{N}_1}y_{\bar{N}_2} + B'_{11}\sum \omega_{\bar{N}_1}y_{\bar{M}_2} + B'_{11}\sum \omega_{\bar{M}_1}y_{\bar{N}_2}$$

$$+ D'_{11}\sum \omega_{\bar{M}_1}y_{\bar{M}_2} + \frac{\kappa}{A_{55}}\sum \omega_{\bar{S}_1}y_{\bar{S}_2}\Big]$$

$$= b^{-1}\Big[-B'_{11}\tfrac{1}{2}l_2^2 + D'_{11}(-\tfrac{1}{2}l_1^2l_2 - \tfrac{1}{2}l_2^2l_1)\big]\,;$$

$$\delta_{22} = \frac{1}{b}\Big[A'_{11}\sum\int \bar{N}_2^2\,dx + 2B'_{11}\sum\int \bar{N}_2\bar{M}_2\,dx + D'_{11}\sum\int \bar{M}_2^2\,dx$$

$$+ (\kappa/A_{55})\sum\int \bar{S}_2^2\,dx\Big] = b^{-1}\Big[A'_{11}\sum \omega_{\bar{N}_2}y_{\bar{N}_2} + 2B'_{11}\sum \omega_{\bar{N}_2}y_{\bar{M}_2}$$

$$+ D'_{11}\sum \omega_{\bar{M}_2}y_{\bar{M}_2} + (\kappa/A_{55})\sum \omega_{\bar{S}_2}y_{\bar{S}_2}\Big]$$

$$= b^{-1}[A'_{11}l_2 + 2B'_{11}l_2l_1 + D'_{11}(\tfrac{2}{3}l_1^3 + l_1^2l_2) + (\kappa/A_{55})2l_1]\,;$$

The displacements due to the applied load are

$$\Delta_{1p} = b^{-1}\Big[A'_{11}\sum\int \bar{N}_1 N_p\,dx + B'_{11}\sum\int \bar{N}_1 M_p\,dx$$

$$+ B'_{11}\sum\int \bar{M}_1 N_p\,dx + D'_{11}\sum\int \bar{M}_1 M_p\,dx + (\kappa/A_{55})\int \bar{S}_1 S_p\,dx\Big]$$

$$= b^{-1}\Big[A'_{11}\sum \omega_{\bar{N}_1} y_{N_p} + B'_{11}\sum \omega_{\bar{N}_1} y_{M_p} + B'_{11}\sum \omega_{\bar{M}_1} y_{N_p}$$

$$+ D'_{11}\sum \omega_{\bar{M}_1} y_{M_p} + \frac{\kappa}{A_{55}}\sum \omega_{\bar{S}_1} y_{S_p}\Big] = b^{-1}\Big\{A'_{11}(-1\cdot l_1\cdot P_2)$$

$$+ B'_{11}1\cdot l_1\tfrac{1}{2}\Big(\frac{P_2 l_2}{2} + \frac{P_2 l_2}{2} + P_1 l_1\Big) + B'_{11}(-l_2\cdot l_1\cdot P_2)$$

$$+ D'_{11}\Big[-l_2 l_1\tfrac{1}{2}\Big(\frac{P_2 l_2}{2} + \frac{P_2 l_2}{2} + P_1 l_1\Big) - \tfrac{1}{4}\Big(l_2 + \frac{l_2}{2}\Big)l_2\tfrac{2}{3}\,\frac{P_2 l_2}{2}\Big]$$

$$+ \frac{\kappa}{A_{55}}\Big(-1\,\frac{l_2}{2}\,P_2\Big)\Big\} = \frac{1}{b}\Big\{-A'_{11} l_1 P_2 + B'_{11}\tfrac{1}{2}(l_1^2 P_1 - l_2 l_1 P_2)$$

$$- D'_{11}\Big[P_1\frac{l_1^2 l_2}{2} + P_2\Big(\frac{l_2^3}{8} + \frac{l_1\cdot l_2^2}{2}\Big)\Big] - \frac{\kappa}{A_{55}}\,\frac{P_2 l_2}{2}\Big\}\,;$$

$$\Delta_{2p} = \frac{1}{b}\Big[A'_{11}\sum\int \bar{N}_2 N_p\,dx + B'_{11}\sum\int \bar{N}_2 M_p\,dx + B'_{11}\sum\int \bar{M}_2 N_p\,dx$$

$$+ D'_{11}\sum\int \bar{M}_2 M_p\,dx + \frac{\kappa}{A_{55}}\sum\int \bar{S}_2 S_p\,dx\Big] = \frac{1}{b}\Big[A'_{11}\sum \omega_{\bar{N}_2} y_{N_p}$$

$$+ B'_{11}\Big(\sum \omega_{\bar{N}_2}\cdot y_{M_p} + \sum \omega_{\bar{M}_2} y_{N_p}\Big) + D'_{11}\sum \omega_{\bar{M}_2} y_{M_p}$$

$$+ \frac{\kappa}{A_{55}}\sum \omega_{\bar{S}_2} y_{S_p}\Big] = \frac{1}{b}\Big\{B'_{11} P_2\tfrac{1}{2}\Big(l_1^2 + \frac{l_2^2}{4}\Big)$$

$$+ D'_{11}\Big[P_1\frac{l_1^3}{6} + P_2\Big(\frac{l_1^2 l_2}{4} + \frac{l_1 l_2^2}{8}\Big)\Big] - \frac{\kappa}{A_{55}}\,l_1 P_1\Big\}\,.$$

After all the displacements in (6.4) have been found in terms of coefficients, we may proceed to find X_1 and X_2. Thus

$$X_1 = \frac{\delta_{12}\Delta_{2p} - \Delta_{1p}\delta_{22}}{\delta_{11}\delta_{22} - \delta_{12}^2}\,,$$

$$X_2 = \frac{-X_1\delta_{12} - \Delta_{2p}}{\sigma_{22}}\,.$$

The final diagrams of the internal moment and force resultants M, S and N are constructed according to formulas (6.1), and the failure of individual plies is determined from the criteria of Section 4.

7. Stress–strain response of laminated plastics under sustained loading

The stress–strain condition of laminated plastics under sustained loading changes constantly due to the viscoelastic characteristics of the individual plies. Thus the time-dependent redistribution of stresses in the individual plies should be considered in theories describing deformations of such plastics with time.

7.1. Construction and solution of defining systems of equations

Consider an n-ply reinforced plastic subjected to the combined action of the force and moment resultants $N_i(t)$ and $M_i(t)$, where $i = x, y, xy$ (x, y are arbitrary orthogonal axes in the reinforcement plane of the plastic). The functions $N_i(t)$ and $M_i(t)$ are time-dependent in the general case and are characterized by the magnitude of the force (moment) resultants per unit length acting at the mid-plane of the material. Every ply of the material is identified by the coordinates characterizing the thickness of the particular ply (z-axis coordinates) $-h_{k-1}$, h_k for ($k = 1, \ldots, n$) and by the orientation of the fibres with respect to the x-axis indicated by angle β_k ($k = 1, \ldots, n$). Let us introduce the following matrix notation:

(a) deformations and curvatures of the laminate's mid-plane along the x and y axes:

$$[\langle\langle \varepsilon^0(t) \rangle\rangle] = \begin{bmatrix} \langle\langle \varepsilon_x^0(t) \rangle\rangle \\ \langle\langle \varepsilon_y^0(t) \rangle\rangle \\ \langle\langle \gamma_{xy}^0(t) \rangle\rangle \end{bmatrix}; \quad [k(t)] = \begin{bmatrix} k_x(t) \\ k_y(t) \\ k_{xy}(t) \end{bmatrix}; \tag{7.1}$$

(b) force and moment resultants acting per unit length of the cross-section of the laminate along the x and y axes:

$$[N(t)] = \begin{bmatrix} N_x(t) \\ N_y(t) \\ N_{xy}(t) \end{bmatrix}; \quad [M(t)] = \begin{bmatrix} M_x(t) \\ M_y(t) \\ M_{xy}(t) \end{bmatrix}; \tag{7.2}$$

(c) stresses in the kth ply along the x and y loading axes of the laminate and along the principal material directions $\parallel$ and $\perp$:

$$[\langle \sigma(t) \rangle_k] = \begin{bmatrix} \langle \sigma_x(t)_k \rangle \\ \langle \sigma_y(t)_k \rangle \\ \langle \tau_{xy}(t)_k \rangle \end{bmatrix}; \quad [\langle \sigma(t) \rangle_k] = \begin{bmatrix} \langle \sigma_\parallel(t)_k \rangle \\ \langle \sigma_\perp(t)_k \rangle \\ \langle \tau_{\parallel\perp}(t)_k \rangle \end{bmatrix} \tag{7.3}$$

In order to perform the time-dependent deformation analysis of the whole laminated packet by analogy with the short-term response we shall resort to the flat cross-section hypothesis which states that the deformations of a ply at the coordinate z are determined as follows

$$[\langle \varepsilon_x(t, z)\rangle_k] = [\langle\!\langle \varepsilon_x^0(t)\rangle\!\rangle] + z[k_x(t)] \,. \tag{7.4}$$

The equilibrium condition of a laminate allows us to write

$$[N_x(t)] = \int_{-h/2}^{h/2} [\langle \sigma_x(t)\rangle]\,\mathrm{d}z \,, \tag{7.5}$$

$$[M_x(t)] = \int_{-h/2}^{h/2} [\langle \sigma_x(t)\rangle] z\,\mathrm{d}z \,, \tag{7.6}$$

where h = total thickness of the laminate.

The deformation law for an unidirectionally reinforced visco-elastic kth ply along the principal material directions is given by

$$[\langle \varepsilon_{\|}(t)]_k = [S]_k \cdot [\langle \sigma_{\|}(t)\rangle]_k + \int_0^t [Sk(t-\tau)]_k[\langle \sigma_{\|}(\tau)\rangle]_k\,\mathrm{d}\tau \,. \tag{7.7}$$

Here $[S]_k$ is the elastic compliance matrix of the kth unidirectionally reinforced ply.

$[SK(t-\tau)]_k$ is the matrix describing the visco-elastic properties of the kth ply.

We should note that the creep kernels in the principal material coordinate system of a ply may be determined by experiment as well as by analysing the properties and geometry of the individual components of a ply (BULAVS and RADINSH, 1980). The set of equations (7.4)–(7.7) in conjunction with the known relations for determining stresses and strains in an arbitrary coordinate system enables one to completely characterise the visco-elastic stress–strain state of laminated reinforced plastics since we have $(6n+6)$ equations with $(6n+6)$ unknowns (in the matrix form – $(2n+2)$ unknowns and equations, respectively). The computation is best carried out by applying the Laplace transformation. Let $[\bar{N}(P)_x]$, $[\bar{M}(P)_x]$, $[\langle\!\langle \bar{\varepsilon}(P)_x^0\rangle\!\rangle]$, $[\bar{k}(P)_x]$ be the Laplace transforms of the corresponding functions. The solution of the time-dependent problem can thus be expressed in the transformed domain as:

$$\left[\frac{\bar{N}(P)_x}{\bar{M}(P)_x}\right] = \left[\begin{array}{c|c} \bar{A}(P) & \bar{B}(P) \\ \hline \bar{B}(P) & \bar{D}(P) \end{array}\right]\left[\frac{\langle\!\langle \bar{\varepsilon}(P)_x^0\rangle\!\rangle}{\bar{k}(P)_x}\right] \tag{7.8}$$

or

$$\left[\frac{\langle\!\langle \bar{\varepsilon}(P)_x^0\rangle\!\rangle}{\bar{k}(P)_x}\right] = \left[\begin{array}{c|c} \bar{A}(P) & \bar{B}(P) \\ \hline \bar{B}(P) & \bar{D}(P) \end{array}\right]^{-1}\left[\frac{\bar{N}(P)_x}{\bar{M}(P)_x}\right], \tag{7.9}$$

where the matrices $[\bar{A}(P)]$, $[\bar{B}(P)]$, $[\bar{D}(P)]$ are functions of the transform variable P and of the laminate configuration, and are obtained according to the principle:

$$[\bar{A}(P)] = \sum_{k=1}^{n} [S_k^*(1 + \bar{k}_k^*(P)](h_k - h_{k-1}) ,$$

$$[\bar{B}(P)] = \tfrac{1}{2} \sum_{k=1}^{n} [S_k^*(1 + \bar{k}_k^*(P))](h_k^2 - h_{k-1}^2) , \tag{7.10}$$

$$[\bar{D}(P)] = \tfrac{1}{3} \sum_{k=1}^{n} [S_k^*(1 + \bar{k}_k^*(P))](h_k^3 - h_{k-1}^3) ,$$

where

$$[S_k^*(1 + \bar{k}_k^*(P)] = [T_k] \cdot [S_k(1 + \bar{k}_k(P))] \cdot [T_k]^{-1} . \tag{7.11}$$

Here $[T_k]$ is the transposed matrix of the kth ply with respect to the x and y axes. The inverse transformations of equations (7.9) to the time domain (7.10) into account, enables us to find the dependence of deformations and curvatures at the midplane of the laminate on the mode and duration of the applied loading. This, in its turn, with the help of expressions (7.4) and (7.6), makes it possible to determine the stress–strain state at any point in the laminate.

7.2. Dependence of the stress–strain response of a ply on the laminate structure

Application of the inverse Laplace transform to expressions (7.9) is generally quite labourious, though it can be considerably simplified for a number of materials and loading modes. To this end, the deformation in the reinforcement direction of glass-, carbon- and boron-reinforced plastics may be regarded as elastic with a sufficient degree of accuracy (SKUDRA et al., 1971; SCHAPERY, 1978). Consequently, it may be assumed that $k_{11}(t-\tau) = k_{12}(t-\tau) = 0$, i.e. the analysis may be limited to visco-elastic characteristics of the plies under longitudinal shear and transverse loading ($k_{22}(t-\tau)$, $k_{66}(t-\tau) \neq 0$). As an example, consider a balanced reinforced structure which generally occurs in shells made by spiral winding (RADINSH and GURVICH, 1981). The individual plies in such a laminated plastic are laid at angles $\pm\beta$ with respect to the axis of symmetry of the material with the volumetric proportion of the plies oriented at the angle β is equal to those oriented at the angle $-\beta$. In uniaxial tension along the axis of symmetry the plies are symmetrically loaded by the force N_x since the material is orthotropically symmetric. Therefore the stress state of the plies is expressed by the relations

$$\langle \sigma_x(t) \rangle_1 = \langle \sigma_x(t) \rangle_2 = N_x/h ,$$

$$\langle \sigma_y(t) \rangle_1 = \langle \sigma_y(t) \rangle_2 = 0 , \tag{7.12}$$

$$\langle \tau_{xy}(t) \rangle_1 = -\langle \tau_{xy}(t) \rangle_2 .$$

Thus, in order to analyse the time-dependent stress–strain state of an angle-ply reinforced plastic structure subjected to axial loading, the following set of equations has to be solved:

$$\begin{cases} \langle\!\langle \bar{\varepsilon}_x(P)^0 \rangle\!\rangle = S_{11}^*[1+\bar{k}_{11}^*(P)\cdot \bar{N}_x(P)/h + S_{16}^*[1+\bar{k}_{16}^*(P)]\langle \bar{\tau}_{xy}(P)\rangle_1 \,, \\ \langle\!\langle \bar{\varepsilon}_y(P)^0 \rangle\!\rangle = S_{12}^*[1+\bar{k}_{12}^*(P)]\cdot \bar{N}_x(P)/h + S_{26}^*[1+\bar{k}_{26}^*(P)[\langle \bar{\tau}_{xy}(P)\rangle_1 \,, \\ \langle\!\langle \bar{\gamma}_{xy}(P)^0 \rangle\!\rangle = S_{16}^*[1+\bar{k}_{16}^*(P)]\cdot \bar{N}_x(P)/h + S_{66}^*[1+\bar{k}_{66}^*(P)]\langle \bar{\tau}_{xy}(P)\rangle_1 \,, \end{cases} \tag{7.13}$$

leading to the solution of:

$$\begin{aligned} \langle \bar{\tau}_{xy}(P)\rangle_1 &= -\frac{S_{16}^*[1+k_{16}^*(P)]}{S_{66}^*[1+k_{66}^*(P)]}\,\frac{\bar{N}_x(P)}{h}\,, \\ \langle\!\langle \bar{\varepsilon}_x(P)^0 \rangle\!\rangle &= \left\{ S_{11}^*[1+\bar{k}_{11}^*(P)] - \frac{[S_{16}^*[1+\bar{k}_{16}^*(P)]]^2}{S_{66}^*[1+\bar{k}_{66}^*(P)]} \right\} \frac{\bar{N}_x(P)}{h}\,, \\ \langle\!\langle \bar{\varepsilon}_y(P)^0 \rangle\!\rangle &= \left\{ S_{12}^*[1+\bar{k}_{12}^*(P)] \right. \\ &\quad \left. - \frac{S_{16}^*[1+\bar{k}_{16}^*(P)][S_{26}^*[1+\bar{k}_{26}^*(P)]}{S_{66}^*[1+\bar{k}_{66}^*(P)]} \right\} \frac{\bar{N}_x(P)}{h}\,. \end{aligned} \tag{7.14}$$

Here, the expressions $S_{ij}^*[1+\bar{k}_{ij}^*(P)]$ are calculated from (7.11). When the creep kernels of the individual plies take the following exponential forms

$$k_{22}(t) = c_{22}\,\mathrm{e}^{-\alpha_{22}}\,, \quad k_{66}(t) = c_{66}\,\mathrm{e}^{-\alpha_{66}t} \tag{7.15}$$

the functional forms of the components of the compliance matrices $S_{ij}[1+\bar{k}_{ij}(P)]$ are obtained from the relationships

$$\begin{aligned} S_{22}[1+\bar{k}_{22}(P)] &= \frac{S_{22}(P+\alpha_{22}+c_{22})}{(P+\alpha_{22})}\,, \\ S_{66}[1+\bar{k}_{66}(P)] &= \frac{S_{66}(P+\alpha_{66}+c_{66})}{(P+\alpha_{66})}\,. \end{aligned} \tag{7.16}$$

When $N_x = \text{constant}$, $\bar{N}_x(P) = N_x/P$ and after applying the inverse Laplace transform to relations (7.14) we arrive at

$$\langle \tau_{xy}(t)\rangle_1 = \frac{N_x}{hB_1}\left\{ \sum_{i=1}^{2} \frac{[A_1P_i^2 + A_2P_i + A_3]}{P_i(P_i-P_j)}\,\mathrm{e}^{P_it} + \frac{A_3}{P_1P_2} \right\}, \quad i \neq j\,, \tag{7.17}$$

where P_1 and P_2 are the roots of the quadratic equation

$$B_1P^2 + B_2P + B_3 = 0\,,$$

$$B_1 = S^*_{66}\,,$$

$$B_2 = S^*_{66}(\alpha_{22} + \alpha_{66}) + S_{66}(\sin^2\beta - \cos^2\beta)^2 c_{66} + 4S_{22}\sin^2\beta\cos^2\beta c_{22}\,,$$

$$B_3 = S^*_{66}\alpha_{22}\alpha_{66} + 4S_{22}\sin^2\beta\cos^2\beta c_{22}\alpha_{66} + S_{66}(\sin^2\beta - \cos^2\beta)^2\alpha_{22}c_{66}\,,$$

$$A_1 = S^*_{16}\,,$$

$$A_2 = S^*_{16}(\alpha_{22} + \alpha_{66}) + S_{66}(\sin^2\beta - \cos^2\beta)\sin\beta\cos\beta c_{66} - 2S_{22}\sin^3\beta\cos\beta c_{22}\,,$$

$$A_3 = S^*_{16}\alpha_{22}\alpha_{66} + S_{66}(\sin^2\beta - \cos^2\beta)\sin\beta\cos\beta c_{66}\alpha_{22} - 2S_{22}\sin^3\beta\cos\beta c_{22}\alpha_{66}\,.$$

Here S^*_{ij} is obtained from (7.11) when $\bar{k}^*_{ij}(P) = 0$.

According to relations (7.14), the creep deformations along the axes of elastic symmetry of angle-ply reinforced plastics are given by

$$\begin{aligned}\langle\langle \varepsilon^0_x(t)\rangle\rangle = \frac{N_x}{h}\Big\{ & S^*_{11} + \frac{S_{22}c_{22}}{\alpha_{22}}\sin^4\beta(1 - e^{-\alpha_{22}t}) \\ & + \frac{S_{66}c_{66}}{\alpha_{66}}\cos^2\beta\sin^2\beta(1 - e^{-\alpha_{66}t}) + \frac{A_3}{B_1P_1P_2}\Big[S^*_{16} \\ & - \frac{2S_{22}c_{22}}{\alpha_{22}}\sin^3\beta\cos\beta(1 - e^{-\alpha_{22}t}) + \frac{S_{66}c_{66}}{\alpha_{66}}(\sin^2\beta \\ & - \cos^2\beta)\sin\beta\cos\beta(1 - e^{-\alpha_{66}t})\Big] \\ & + \sum_{i=1}^{2}\frac{[A_1P_i^2 + A_2P_i + A_3]}{B_1P_i(P_i - P_j)}\Big[S^*_{16}\,e^{P_it} \\ & + \frac{S_{66}(\sin^2\beta - \cos^2\beta)\sin\beta\cos\beta c_{66}}{(\alpha_{66} + P_i)}(e^{P_it} - e^{-\alpha_{66}t})\,\cdot \\ & - \frac{2S_{22}c_{22}\sin^3\beta\cos\beta}{(\alpha_{22} + P_i)}(e^{P_it} - e^{-\alpha_{22}t})\Big]\Big\}\quad i \neq j\,,\end{aligned} \tag{7.18}$$

$$\langle\langle \varepsilon_y(t)^0 \rangle\rangle = \frac{N_x}{h} \Big\{ S^*_{12} + \frac{S_{22}c_{22}}{\alpha_{22}} \cos^2 \beta \sin^2 \beta (1 - e^{-\alpha_{22}t}) - \frac{S_{66}c_{66}}{\alpha_{66}} (1 - e^{-\alpha_{22}t}) \cos^2 \beta \sin^2 \beta$$
$$+ \frac{A_3}{B_1 P_1 P_2} \Big[S^*_{26} - \frac{2S_{22} \cos^2 \beta \sin^2 \beta c_{22}}{\alpha_{22}} (1 - e^{-\alpha_{22}t})$$
$$+ \frac{S_{66}(\cos^2 \beta - \sin^2 \beta) \sin \beta \cos \beta}{\alpha_{66}} (1 - e^{-\alpha_{66}t}) \Big] \tag{7.19}$$
$$+ \sum_{i=1}^{2} \frac{[A_1 P_i^2 + A_2 P_i + A_3]}{B_1 P_i (P_i - P_j)} \Big[S^*_{26}\, e^{P_i t}$$
$$+ \frac{S_{66}(\cos^2 \beta - \sin^2 \beta) \sin \beta \cos \beta}{(\alpha_{66} + P_i)} (e^{P_i t} - e^{-\alpha_{66}t}) - \frac{2S_{22}c_{22} \sin \beta \cos^3 \beta}{(\alpha_{22} + P_i)} (e^{P_i t} - e^{-\alpha_{22}t}) \Big] \Big\} \quad i \neq j \, .$$

Let us consider the time-dependent stress–strain state of a balanced angle-ply glass/epoxy AG–4S of the following structural characteristics: $v_f = 0.52$; $E_m = 3.5$ GPa; $E_f = 70$ GPa, $\nu_m = 0.35$; $v_f = 0.23$. The creep kernel of the polymeric matrix has been chosen to have the form $k_m(t) = c_m\, e^{-\alpha_m t}$; $c_m = 0.0872$ days^{-1}; $\alpha_m = 0.0435$ days^{-1}. The strain characteristics of a unidirectionally reinforced ply calculated according to the relationships suggested by BULAVS and RADINSH (1980) are then: $S_{11} = 0.2703 \times 10^{-1}$ GPa^{-1}; $S_{22} = 0.7936 \times 10^{-1}$ GPa^{-1}; $S_{12} = -0.8108 \times 10^{-2}$ GPa^{-1}; $S_{66} = 0.2777$ GPa^{-1}; $c_{22} = 0.02702$ days^{-1}; $c_{66} = 0.02333$ days^{-1}; $\alpha_{22} = 0.0416$ days^{-1}; $\alpha_{66} = 0.0333$ days^{-1}.

Fig. 18 shows creep curves of the above fibre-reinforced plastic with various angles of fibre orientation with respect the applied load. Plastics with their fibres oriented at 60° with respect to the loading direction show the maximum creep deformation in the x-direction. Relative creep deformations $\langle\langle \varepsilon^0_x(t) \rangle\rangle / \langle\langle \varepsilon^0_x(0) \rangle\rangle$ within the angle range from 45° to 90° remain practically the same. The reinforced plastic with fibres oriented at 45° ($\beta = 45°$) shows maximum creep deformations in the y-direction. For this case $\langle\langle \varepsilon^0_y(\infty) \rangle\rangle / \langle\langle \varepsilon^0_y(0) \rangle\rangle = 1.85$. This is explained by the fact that, according to (7.11), the component S^*_{12} of the compliance matrix is symmetric about the angle $\beta = 45°$, i.e. $S^*_{12}(\beta) = S^*_{12}(90° - \beta)$.

The stresses in the principal material directions for the considered angle-ply laminate as a function of time and the orientation angle of the individual plies (β), calculated on the basis of (7.18) and (7.19), are presented in Fig. 19. We should note that the structural parameter – angle β – has a considerable effect on the stress state of the individual plies. It follows that the fibre orientation angle is one of the principal parameters on which the design of a rational

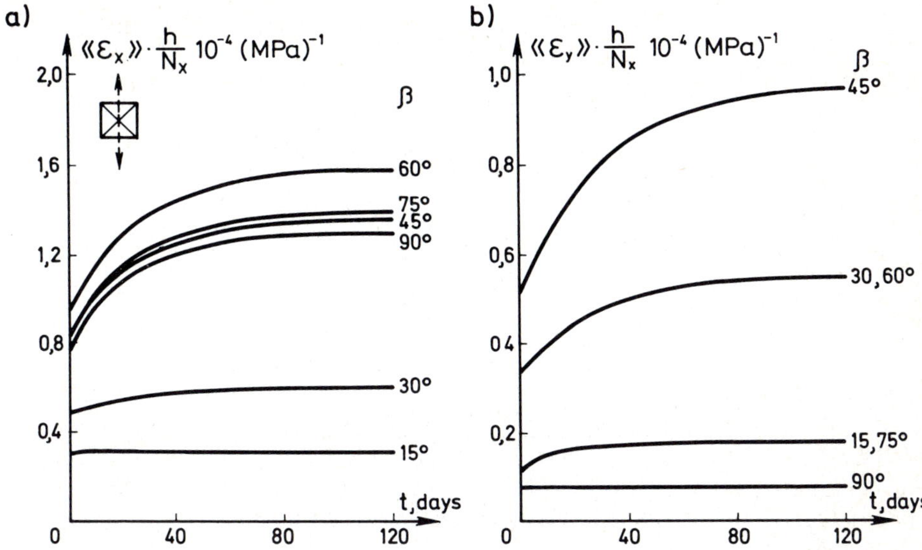

FIG. 18. Creep curves of an angle-ply glass/epoxy.

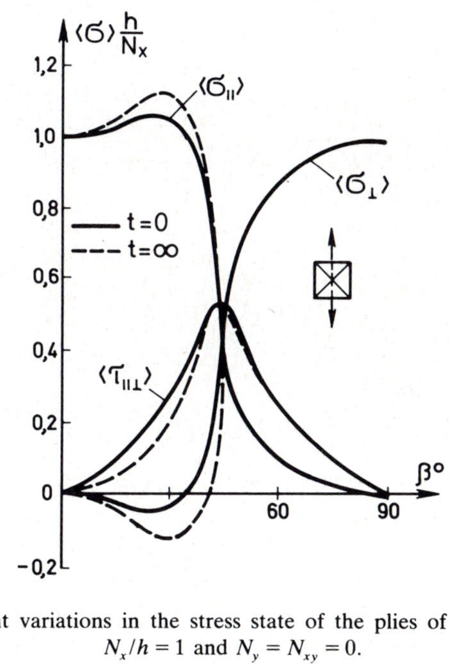

FIG. 19. Time-dependent variations in the stress state of the plies of angle-ply glass/epoxy for $N_x/h = 1$ and $N_y = N_{xy} = 0$.

configuration of an angle-ply plastic can be based. The time-dependent variations in the stress state of the individual plies are quite essential for design considerations. Thus when $\beta = 40°$, the stress component perpendicular to the fibre direction $\langle \sigma_\perp(\infty) \rangle / \langle \sigma_\perp(0) \rangle = 0.40$ whereas when $\beta = 30°$ it becomes $\langle \sigma_\perp(\infty) \rangle / \langle \sigma_\perp(0) \rangle = 2.51$. The character of the functions $\langle \sigma_i(t) \rangle_1$ ($i = \|, \perp, \|\perp$) is not necessarily monotonic, and the values $\langle \sigma_i(t) \rangle$ may increase depending on the loading time and decrease afterwards. For instance, if $\beta = 60°$ then the stress $\langle \sigma_\|(t) \rangle$ increases reaching the value of 1.04, $\langle \sigma_\|(0) \rangle$ at $t = 17$ days. Subsequently, $\langle \sigma_\|(t) \rangle$ decreases to 0.95 $\langle \sigma_\|(0) \rangle$ as $t \to \infty$.

The analysis of the effect of loading time on the stress–strain state of laminated plastic structures as such as the angle-ply glass/epoxy AG–4S system considered above leads to the conclusion that the time-dependent variations in the stress state of the individual plies are significant and cannot be ignored.

An investigation aimed at studying the stress–strain state of orthogonally reinforced balanced plastics by the method of Laplace transforms was carried out by GURVICH (1985). Similarly to the present observations, redistribution of stresses in individual monolayers was observed to be time-dependent even under constant loading.

7.3. Numerical determination of the time-dependent stress–strain response

The analytical method of determining the time-dependent stress–strain response of laminated plastics considered above makes it possible to obtain a solution in the Laplace-transformed domain for any reinforcing scheme and any kind of creep kernels of the various compliance matrix components of a given ply. Performing the inverse Laplace transformation 'by hand' is generally quite a laborious process. Another considerable disadvantage of the analytical method is its inadaptability to an electronic computer for the purpose of analysing arbitrary reinforcement schemes. This is because the compilation of a computational algorithm for performing the inverse Laplace transformation of the obtained functions is extremely difficult, probably too difficult to be implemented with the present-day computers. For this reason the analysis of multi-directionally reinforced plastics, which is general may be unbalanced, may involve the choice of a numerical procedure for the determination of the time-dependent response of individual plies. Here, the analysis will be carried out according to the methods proposed by GURVICH (1985). For reinforced plastics whose individual plies exhibit time-dependent response in accordance with (7.7), the relationship between deformations and stresses may be taken as

$$[\langle \varepsilon_\|(t) \rangle]_k = [S(t)]_k \cdot [\langle \sigma_\|(t) \rangle]_k \tag{7.20}$$

where the compliance matrix $[S(t)]_k$ is constantly changing with time, this change being governed by the following relation:

$$S_{ij}(t)_k = S_{ij}(0)_k \left(1 + \frac{1}{\sigma_i(t)} \int_0^t \sigma_i(\tau)_k k_{ij}^{(k)}(t-\tau)\, d\tau \right), \tag{7.21}$$

where $S_{ij}(0)_k$ are the instantaneous values of the kth compliance matrix. The matrix $[S_{ij}(t)]_k$ will be called pseudoelastic in contrast with the quasi-elastic matrix $[S'_{ij}(t)]_k$ for which evaluation of the individual components does not involve time-dependent changes in the stress field:

$$S'_{ij}(t)_k = S_{ij}(0)_k\left[1 + \int_0^t k_{ij}(t-\tau)\,\mathrm{d}t\right]. \tag{7.22}$$

Utilizing the pseudoelastic properties of the monolayer compliances and referring to equations (7.4)–(7.7) the relationship between the deformations of the laminate and loads takes the form

$$\left[\frac{N_x(t)}{M_x(t)}\right] = \left[\begin{array}{c|c} A(t) & B(t) \\ \hline B(t) & D(t) \end{array}\right]\left[\frac{\langle\!\langle \varepsilon_x^0(t) \rangle\!\rangle}{k_x(t)}\right], \tag{7.23}$$

or

$$\left[\frac{\langle\!\langle \varepsilon_x^0(t) \rangle\!\rangle}{k_x(t)}\right] = \left[\begin{array}{c|c} A(t) & B(t) \\ \hline B(t) & D(t) \end{array}\right]^{-1}\left[\frac{N_x(t)}{M_x(t)}\right], \tag{7.24}$$

where $[A(t)]$, $[B(t)]$, $[D(t)]$ are the corresponding matrix analogues of membrane, membrane-bending and bending stiffnesses of the laminate, respectively, their determination involving the time factor according to the scheme:

$$\begin{aligned}
[A(t)] &= \sum_{k=1}^{n} [[T]_k[S(t)]_k[T]_k^{-1}]^{-1}(h_k - h_{k-1}), \\
[B(t)] &= \tfrac{1}{2}\sum_{k=1}^{n} [[T]_k[S(t)]_k[T]_k^{-1}]^{-1}(h_h^2 - h_{k-1}^2), \\
[D(t)] &= \tfrac{1}{3}\sum_{k=1}^{n} [[T]_k[S(t)]_k[T]_k^{-1}]^{-1}(h_k^3 - h_{k-1}^3).
\end{aligned} \tag{7.25}$$

Knowing the deformations of the whole laminate from (7.24), we proceed to determine the stress state in the individual plies as follows

$$[\langle \sigma_{\|}(t) \rangle_k] = [S(t)]_k^{-1}[[\langle\!\langle \varepsilon^0(t) \rangle\!\rangle + z[k_{\|}(t)]]. \tag{7.26}$$

The numerical realization of the pseudoelastic method consists of dividing the given time range into small intervals Δt_m $(1 \leqslant m \leqslant M)$, within which $(t_{m-1} \leqslant t \leqslant t_m)$ the changes in the stress state may be neglected and the stress state regarded as constant. Then, expression (7.22) allows us to write

$$S_{ij}(t)_k = S_{ij}(0)_k\left(\frac{1 + \sum\limits_{m=1}^{M} \langle \sigma_i(t_m) \rangle_k \int\limits_{t_{m-1}}^{t_m} k_{ij}(t-\tau)_k\,\mathrm{d}\tau}{\langle \sigma_i(t_M) \rangle_k}\right). \tag{7.27}$$

It is evident that with $M \to \infty$, an accurate solution is reached and consequently, by varying the number of divisions M (or the value of Δt_m), a solution of any degree of accuracy can be obtained. It should be remarked that the value of Δt_m need not always be the same within the entire range of $1 \leq m \leq M$. As shown above, the greatest changes in the stress state of the plies take place during initial static loading with the stress state tending to its asymptotic value with increasing time. Therefore, it is possible to increase the values of Δt_m with increasing loading time and to lose practically nothing in the way of accuracy. Thus, when selecting the creep kernels $k_{ij}(t-\tau)_k$ it is advisable to assign the exponent a logarithmic function of the loading time.

Let us consider more closely the exponential creep kernels in the form of (7.15). In this case, when $t = t_m$, the expression (7.27) takes the form

$$S_{ij}(t_M)_k = S_{ij}(0)_k \left(1 + \frac{\mathrm{e}^{\alpha_{ij} t_M}}{\langle \sigma_i(t_{M-1}) \rangle_k} \Phi_{ij}(t_M)_k \right), \tag{7.28}$$

where

$$\Phi_{ij}(t_M)_k = \sum_{m=1}^{M} \langle \sigma_i(t_{m-1}) \rangle_k \int_{t_{m-1}}^{t_m} c_{ij}\, \mathrm{e}^{-\alpha_{ij}\tau}\, \mathrm{d}\tau . \tag{7.29}$$

Then, if $t = t_{m+1}$, we obtain

$$\begin{aligned} S_{ij}(t_{m+1})_k = S_{ij}(0)_k \Bigg\{ 1 + \frac{\mathrm{e}^{\alpha_{ij} t_{M+1}}}{\langle \sigma_i(t_M) \rangle_k} \Big[\Phi_{ij}(t_M)_k \\ + \langle \sigma_i(t_M) \rangle_k \int_{t_M}^{t_{M+1}} c_{ij}\, \mathrm{e}^{-\alpha_{ij}\tau}\, \mathrm{d}\tau \Big] \Bigg\} . \end{aligned} \tag{7.30}$$

Substitution of the value $\Phi_{ij}(t_M)_k$ obtained from equation (7.28) into (7.30) gives the following recurrent expression for computing $S_{ij}(t_{m+1})_k$:

$$\begin{aligned} S_{ij}(t_{M+1})_k = S_{ij}(0)_k \Bigg\{ 1 + \frac{c_{ij}}{\alpha_{ij}} [1 - \mathrm{e}^{-\alpha_{ij}(t_{M+1} - t_M)}] \\ + \mathrm{e}^{-\alpha_{ij}(t_{M+1} - t_M)} \frac{\langle \sigma_i(t_{M-1}) \rangle_k}{\langle \sigma_i(t_M) \rangle_k} \left[\frac{S_{ij}(t_M)_k}{S_{ij}(0)_k} - 1 \right] \Bigg\} . \end{aligned} \tag{7.31}$$

Utilization of the recurrent expression in the form of (7.31) will not only considerably reduce the computation time; the computer implementation will require much less storage capacity since all the partial results need not be stored and it is sufficient to store only the initial and preceding data.

The particular algorithm for the numerical calculation of time-dependent stress–strain state in laminated plastics under combined load was implemented in a real program prepared in FORTRAN-IV and compiled on an ES series computer. The program algorithm is defined by the following steps. After the input and listing of the initial data characterizing the structure of the plastic and the prescribed stress state, and construction of the operational files, the pseudoelastic compliance matrix is assigned the initial values. Subsequently, computation of the elastic response of the plastic laminated structure is carried out according to expression (7.24). Here, the instantaneous stress values in the individual plies are computed as well as the deformations of the laminate with $S_{ij}(t) = \text{constant}$. After the print-out of the final data has been obtained, the component values of the pseudoelastic compliance matrix are recomputed according to expressions (7.31) corresponding to the loading time $t_2 = t_1 + \Delta t$, assuming in the first step ($t_1 = 0$) that $\langle \sigma_i(t_0) \rangle_k = \langle \sigma_i(t_1) \rangle_k$. The computation procedure is repeated until $t = t_{\max}$ is reached.

To illustrate the use of the proposed method we shall consider time-dependent stress–strain response of angle-ply and cross-ply glass/epoxy AG–4S laminates possessing the reinforcement ratio of $v_f = 0.52$ and structural parameters as enumerated previously. The comparison between the analytical and numerical solutions, with $\Delta t = 5$ days for balanced angle-ply and cross-ply laminated plastics under uniaxial tension ($N_x = 1$) gave practically similar results – the agreement being within three significant figures, which testifies to the validity of the proposed method. The convergence of the numerical solutions is quite high in relation, to the iteration step Δt. Thus the numerical solution for deformations and curvatures at the midsurface of cross-ply and angle-ply glass/epoxy AG/4S laminates of various fibre orientations under uniaxial tension and bending when computed for $\Delta t = 10$; 5; 1 days of loading, lack of agreement of three significant figures. Taking into account that the greatest changes in the deformation of the material (95% of the creep deformation) are observed in the given loading time interval ($t_{\max} = 100$ days), the agreement of analytical and numerical solutions with $\Delta t = 0.1 t_{\max}$ illustrates a sufficiently high degree of accuracy of the numerical method.

The midsurface deformations computed by means of the proposed methods for balanced bidirectional plastics AG–4S under pure bending ($M_x = 1$) are illustrated in Fig. 20. The obtained curves are in fact creep curves for laminated plastics obtained by analytical methods which enable us to determine the visco-elastic properties of such materials according to their geometric lay-up configuration. The computation of analogous creep curves for different modes of loading (pure tension, pure shear, etc.) makes it possible to evaluate all the time-dependent functions, such as the various matrix components of $[A(t)]$, $[B(t)]$ and $[D(t)]$ and their derivatives which make up the corresponding relaxation kernels for the laminate, by analytical means. This is especially important as it permits the forecasting of visco-elastic deformations of the entire laminate on the basis of structural analysis of the individual components; this is possible not only for constant loading but also for the generalized case of time-varying combined loading. Furthermore, the developed methods enable one to obtain the solution of an inverse problem, namely, the problem of

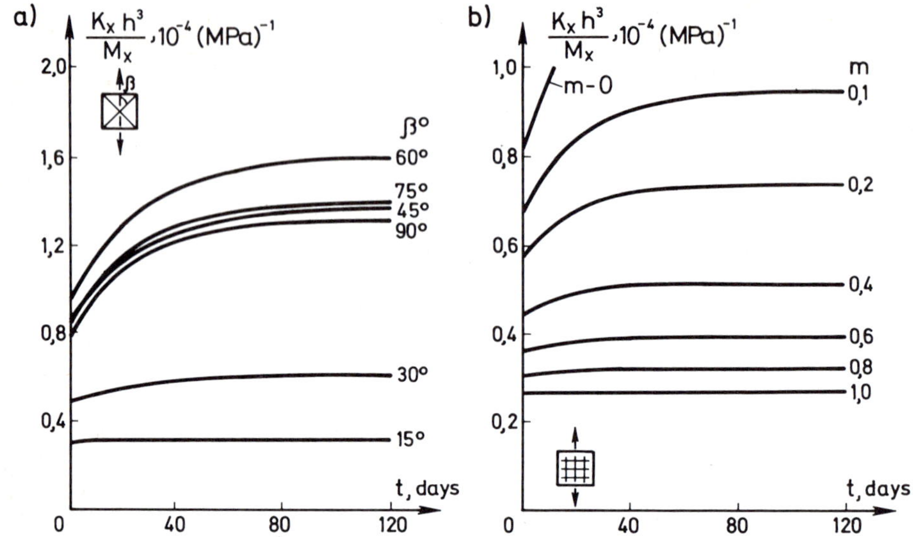

FIG. 20. Creep curves of (a) angle-ply and (b) cross-ply glass/epoxy AG–4S under pure bending ($M_x/h^2 = 1$).

selecting an optimum material structure with the aim of reducing creep deformations or changing the structural characteristics of the material.

Knowledge of stresses in the individual plies, which depend on loading time and are evaluated by numerical methods with the use of the prescribed iteration step Δt, is very useful in the analysis of time-dependent strength characteristics of laminated plastics. The illustrative example of time-dependent variation of the stresses $\langle \sigma_{\parallel} \rangle$, $\langle \sigma_{\perp} \rangle$ and $\langle \tau_{\parallel\perp} \rangle$ in angle-ply and cross-ply reinforced balanced plastics AG–4S under uniaxial tension shows that this variation may depend largely on the geometric lay-up configuration and should be taken into account in the corresponding structural criteria of the loss of material continuity. Note that the character of the time-dependent changes in the stress state of the individual plies is not always monotonic, different geometric configurations can cause either increasing or decreasing stresses.

Visco-elastic deformation analysis which takes into account the structural arrangment and response characteristics of the individual plies makes it also possible to estimate the effect of loading time on the overall structural properties of different types of laminated plastics such as unbalanced laminates. As an example consider an angle-ply reinforced plastic: for $n = 2$ the laminate is completely unbalanced ($[B] \neq 0$), whereas as $n \to \infty$ it becomes practically balanced ($[B] = 0$).

Fig. 21 presents the results of calculation of the relative difference in the deformation of unbalanced and balanced angle-ply glass – (a) and carbon/epoxy (b) (for $v_f = 0.6$), where the pure bending response ($M_x = 1$) at $t = \infty$ depends on the degree of the polymer matrix creep $E_m(0)/E_m(\infty)$. Here

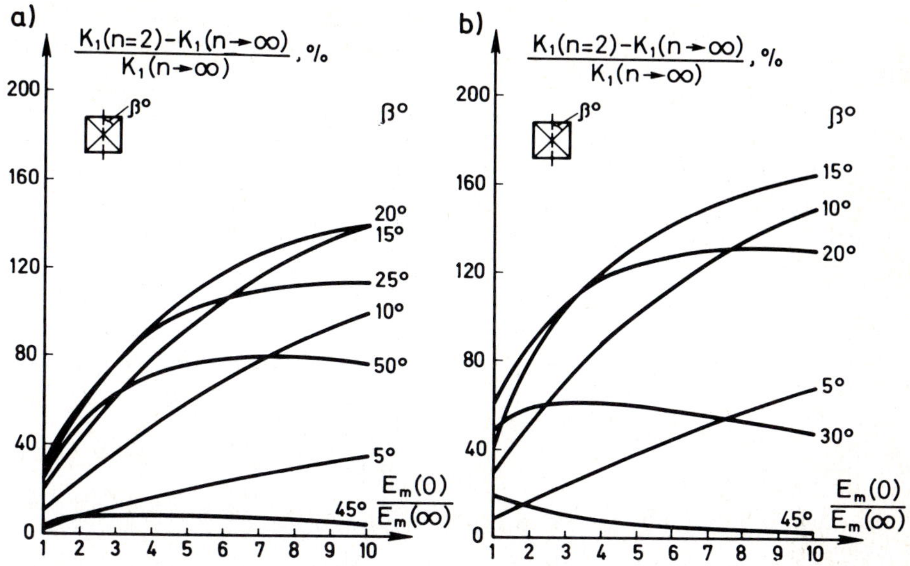

FIG. 21. Dependence of the difference in the midsurface deformations at ($t \to \infty$) on the degree of matrix creep and fibre orientation for (a) unbalanced and balanced angle-ply glass/epoxy and (b) carbon/epoxy beams subjected to pure bending.

$k_x(n=2)$ is the midsurface curvature of an angle-ply reinforced plastic made up of one pair of plies oriented at $\pm\beta$ and $k_x(n \to \infty)$ is the corresponding midsurface curvature of a similar angle-ply laminate made of an infinitely large number of angle-ply pairs oriented at angle $\pm\beta$.

The orientation angle of the individual plies is the primary factor governing the effect of the extent of the laminate's structural imbalance on its deformation. Thus, for angle-ply reinforced materials with $\beta \geqslant 45°$ the difference in midsurface curvatures of laminated plastics of balanced and unbalanced configurations does not exceed 20%, respectively. For fibre orientation angles in the range $10° \leqslant \beta \leqslant 25°$ this difference may be as large as 160%. The influence of the unbalanced structures of laminated plastics on their deformation characteristics will be more pronounced for carbon/epoxy (the material with higher anisotropy of the individual plies) than for glass/epoxy.

8. Structural method for the determination of rheological response characteristics

In investigating the stress–strain state of laminated structures subjected to continuous loading it is necessary to know the rheological properties of the materials used, which may be determined either experimentally or analytically. The response functions obtained experimentally are valid only for one particular laminate configuration and are thus not applicable to laminated plastics composed of similar unidirectionally reinforced plies if the lay-up scheme is different. This explains why it is necessary to determine the visco-elastic properties of laminated reinforced plastics by structural methods, i.e., taking into account of the deformation characteristics of the individual layers and their stacking geometry. We should note that the structural method can be applied at the microstructural level of the individual layers. In fact, the problem of determining the visco-elastic properties of unidirectionally reinforced plies on the basis of the microstructural arrangement of the individual components has been successfully solved, for instance, by BULAVS and RADINSH (1980). With this in mind, it will be further assumed for simplicity that the elastic and rheological deformation characteristics of the individual plies are known.

8.1. *Defining equations*

According to (7.24) the deformation law for a linear visco-elastic laminate can be written in a generalized way as

$$\left[\frac{\langle\!\langle \varepsilon_x^0(t) \rangle\!\rangle}{k_x(t)}\right] = \left[\begin{array}{c|c} A'_* & B'_* \\ \hline B'_{T*} & D'_* \end{array}\right]\left[\frac{N_x(t)}{M_x(t)}\right]. \tag{8.1}$$

The components of the matrices A'_*, B'_* and D'_* are linear operators determined from the following relations:

$$\begin{aligned} A'_{*ij}[F(t)] &= A'_{ij}\left\{F(t) + \int_0^t F(\xi)\cdot \tilde{A}'_{ij}(t-\xi)\,d\xi\right\}, \\ B'_{*ij}[F(t)] &= B'_{ij}\left\{F(t) + \int_0^t F(\xi)\tilde{B}'_{ij}(t-\xi)\,d\xi\right\}, \\ D'_{*ij}[F(t)] &= D'_{ij}\left\{F(t) + \int_0^t F(\xi)\tilde{D}'_{ij}(t-\xi)\,d\xi\right\}, \quad i, j = 1, 2, 6. \end{aligned} \tag{8.2}$$

The values of A'_{ij}, B'_{ij}, D'_{ij} $(i, j = 1, 2, 6)$ are obtained from the elastic problem and represent the elastic matrix elements of membrane, membrane-bending and bending compliances of laminated reinforced plastics.

For uniaxial loads we have

$$\begin{aligned} \langle\!\langle \varepsilon_i^0(t) \rangle\!\rangle_{N_j} &= A'^*_{ij}[N_j(t)], \\ \langle\!\langle \varepsilon_i^0(t) \rangle\!\rangle_{M_j} &= B'^*_{ij}[M_j(t)], \\ k_i(t)_{N_j} &= B'^*_{ji}[N_j(t)], \\ k_i(t)_{M_j} &= D'^*_{ij}[M_j(t)]. \end{aligned} \tag{8.3}$$

Here $\langle\langle \varepsilon_i^0(t) \rangle\rangle_{N_j}$; $\langle\langle \varepsilon_i^0(t) \rangle\rangle_{M_j,\ldots}$ corresponds to the ith component of the laminate deformation caused by the action of the jth loading component (N_j, M_j). Hence, it is easy to determine the rheological properties of laminated plastics subjected to continuous loading ($N_j(t) = \text{constant}$; $M_j(t) = \text{constant}$) with the aid of expressions (8.2)

$$\int_0^t \tilde{A}'_{ij}(t-\xi)\,\mathrm{d}\xi = \left(\frac{\langle\langle \varepsilon_i^0(t) \rangle\rangle_{,N_j}}{A'_{ij} N_j} - 1 \right), \tag{8.4}$$

$$\int_0^t \tilde{B}'_{ij}(t-\xi)\,\mathrm{d}\xi = \left(\frac{\langle\langle \varepsilon_i^0(t) \rangle\rangle_{,M_j}}{B'_{ij} \cdot M_j} - 1 \right), \tag{8.5}$$

$$\int_0^t \tilde{B}'_{ij}(t-\xi)\,\mathrm{d}\xi = \frac{k_i(t)_{,N_j}}{B'_{ji} \cdot N_j} - 1, \tag{8.6}$$

$$\int_0^t \tilde{D}'_{ij}(t-\xi)\,\mathrm{d}\xi = \frac{k_i(t)_{,M_j}}{D'_{ij} \cdot M_j} - 1. \tag{8.7}$$

By differentiating expressions (8.4) through (8.7) with respect to the variable t and taking into account the following:

$$\int_0^b F(b-x)\,\mathrm{d}x = \int_0^b F(x)\,\mathrm{d}x, \tag{8.8}$$

we arrive at

$$\tilde{A}'_{ij}(t) = \frac{\partial \langle\langle \varepsilon_i^0(t) \rangle\rangle_{,N_j}}{\partial t} \frac{1}{A'_{ij} \cdot N_j}, \tag{8.9}$$

$$\tilde{B}'_{ij}(t) = \frac{\partial \langle\langle \varepsilon_i^0(t) \rangle\rangle_{,M_j}}{\partial t} \frac{1}{B'_{ij} M_j}, \tag{8.10}$$

$$\tilde{B}'_{ij}(t) = \frac{\partial k_i(t)_{,N_j}}{\partial t} \frac{1}{B'_{ji} N_j}, \tag{8.11}$$

$$\tilde{D}'_{ij}(t) = \frac{\partial k_i(t)_{,M_j}}{\partial t} \frac{1}{D'_{ij} M_j}. \tag{8.12}$$

Note that equations (8.10) and (8.11) possess the same type of kernels $\tilde{B}'_{ij}(t-\xi)$ and are chosen as defining expressions depending on the availability of experimental (computational) data, convenience of computation and other specific conditions.

8.2. Analytical calculation approach to rheological response characteristics

The determination of creep kernels representing the time-dependent deformation characteristics of laminated reinforced plastics on the basis of expressions

(8.9) through (8.12) requires knowledge of the corresponding creep curves $\langle\langle \varepsilon_i(t) \rangle\rangle_{,N_j}$; $\langle\langle \varepsilon_i(t) \rangle\rangle_{M_j}$; . . . . If the time-dependent visco-elastic deformation relation for plastics are analytically prescribed, the corresponding creep kernels are obtained as a result of analytical differentiation of (8.9) through (8.12). An illustrative example will be a balanced angle-ply reinforced plastic; its time-dependent stress–strain response under constant uniaxial tension has been analysed in Section 7.2. Hence, expression (8.9) for, say, $\tilde{A}'_{11}(t-\xi)$ yields:

$$\tilde{A}'_{11}(t-\xi) = c'_{22}\,\mathrm{e}^{-\alpha_{22}(t-\xi)} + c'_{66}\,\mathrm{e}^{-\alpha_{66}(t-\xi)} + \sum_{i=1}^{2} c'_i\,\mathrm{e}^{P_i(t-\xi)}\,. \tag{8.13}$$

Here

$$\begin{aligned}
c'_{22} &= \frac{\sin^2\beta}{hA'_{11}}\Big\{S_{22}c_{22}\sin^2\beta - \frac{A_3}{B_1P_1P_2}\,S_2c_2\sin 2\beta \\
&\quad - \frac{S_{22}c_{22}\sin 2\beta\alpha_{22}}{B_1}\sum \frac{[A_1P_i^2 + A_2P_i + A_3]}{P_i(P_i-P_j)(\alpha_{22}+P_i)}\Big\}, \quad i \neq j\,, \\
c'_{66} &= \frac{1}{hA'_{11}}\Big\{S_{66}c_{66}\frac{\sin^2 2\beta}{4} + \frac{A_3}{B_1P_1P_2}\,S_{66}c_{66}\frac{\sin 4\beta}{4} \\
&\quad - \frac{S_{66}c_{66}\alpha_{66}\sin 4\beta}{4B_1}\sum_{i=1}^{2}\frac{[A_1P_i^2 + A_2P_i + A_3]}{P_i(P_i-P_j)(\alpha_{66}+P_i)}\Big\}, \quad i \neq j\,, \\
c'_i &= \frac{[A_1P_i^2 + A_2P_i + A_3]}{B_1hA'_{11}}\Big\{S^*_{16} - \frac{S_{66}c_{66}\sin 4\beta}{4(\alpha_{66}+P_i)} - \frac{2S_{22}c_{22}\sin^3\beta\cos\beta}{\alpha_{22}+P_i}\Big\}.
\end{aligned} \tag{8.14}$$

The values of the terms P_1, P_2, A_1, D_1, . . . , A_3, D_3 have been determined in Section 7.2.

The determination of the creep kernels $\tilde{A}'_{22}(t-\xi)$, $\tilde{A}'_{12}(t-\xi)$, $\tilde{A}'_{66}(t-\xi)$ is analogous to the determination of $\tilde{A}'_{11}(t-\xi)$.

The rheological properties computed on the basis of (8.13) and (8.14) hold true for any balanced cross-ply reinforced plastic regardless of the properties of the individual plies. Knowledge of these properties permits us to describe the deformation of a material in any type of plane stress state – both time-dependent and time-independent – by applying the deformation law given by (8.1).

8.3. Numerical calculation approach to rheological response characteristics

Analytical solutions to the problem of visco-elastic deformation of cross-ply and angle-ply reinforced plastics given in the previous subsection were obtained by Laplace transforms. In the general case of arbitrarily unbalanced configurations subjected to different combined loads, the application of the

Laplace transforms to the varous expressions is a laborious task. The solution becomes considerably more complicated when the creep kernels of the individual layers have been represented by a more complicated form than the exponential form, as in the representation A.R. Rzhanitsin and Yu.N. Rabotnov etc. Therefore, the analysis of three-directional schemes of reinforcement, or schemes involving an even larger number of directions will be more conveniently carried out by numerical computation creep, curves for laminated reinforced plastics with a prescribed lay-up geometry. These methods will also permit numerical evaluation of the corresponding rheological properties of the material.

By using the numerical approach for determination of the stress–strain response of laminated reinforced plastics under combined continuously applied loads where the distribution of time-dependent stresses within the individual plies is taken into account, we obtain a set of discrete values which correspond to deformations at prescribed periods of time: $\langle\!\langle \varepsilon_i^{(\prime)}(t_1)\rangle\!\rangle$, $\langle\!\langle \varepsilon_i^{(\prime)}(t_2)\rangle\!\rangle, \ldots, [\langle \varepsilon_i^{(\prime)}(t_N)\rangle\!\rangle$, where N is the number of calculated points. The above values may be regarded as being experimental, (that is, they may be viewed as the results of a computer experiment which differs from a real experiment in that the actual properties of the prescribed configuration are predicted by means of numerical simulation instead of using real material). Therefore, the mechanism of computation of the rheological properties of reinforced plastics does not differ, in principle, from that used for experimentally obtained creep kernels.

Let the unknown kernels be some functions of the material constants prescribed earlier:

$$
\begin{aligned}
\tilde{A}_{ij}(t-\xi) &= \tilde{A}_{ij}(a_k, t-\xi)\,,\\
\tilde{B}_{ij}(t-\xi) &= \tilde{B}_{ij}(b_k, t-\xi)\,,\\
\tilde{D}_{ij}(t-\xi) &= \tilde{D}_{ij}(d_k, t-\xi)\,,
\end{aligned}
\tag{8.15}
$$

where a_k, b_k, d_k $(k = 1, \ldots, M)$ are the unknown parameters of the rheological properties of the laminate. Subsequently, using expressions (8.3), it is possible to obtain the corresponding deformation values for any prescribed time period in terms of the parameters a_k, b_k, d_k:

$$
\langle\!\langle \varepsilon_i(a_k, b_k, d_k, t_1)\rangle\!\rangle,\ \langle\!\langle \varepsilon_i(a_k, b_k, d_k, t_2)\rangle\!\rangle, \ldots,\ \langle\!\langle \varepsilon_i(a_k, b_k, d_k, t_N)\rangle\!\rangle\,.
$$

By comparing the above deformations with those computed by the numerical approach we find the unknown parameters of the rheological response functions of laminated reinforced plastics. The requirement for minimizing the error vector s is used as a condition for determining a_k, b_k, d_k:

$$
s = \left[\sum_{j=1}^{N}\left(\frac{\langle\!\langle \varepsilon_i^{(\prime)}(t_j)\rangle\!\rangle - \langle\!\langle \varepsilon_i(a_k, b_k, d_k, t_j)\rangle\!\rangle}{\langle\!\langle \varepsilon_i^{(\prime)}(t_j)\rangle\!\rangle}\right)^2\right]^{1/2} \rightarrow \mathrm{MIN}\,.
\tag{8.16}
$$

Equation (8.16), which is analogous to the least squares method, permits the determination of the unknown parameters a_k, b_k, d_k for any type of creep kernels. An appropriate standard program in FORTRAN-IV and compiled on an ES-series computer has been developed using a series representation for the creep kernels in terms of exponential functions. As an algorithm for finding the minimum value of the error vector s and its respective values a_k, b_k and d_k, a grid search method has been employed which proved more effectve in saving computer time compared with the minimax methods which use derivative computation.

9. The effect of loading duration on the stress–strain response of laminated beam systems

The dependence of deformation characteristics of polymeric matrices on the duration of loading is the cause of the ensuing rheological properties of the individual plies and of the whole laminate itself. This explains why the duration and regime of loading are important factors for investigating the stress–strain response of member systems composed of laminated reinforced plastic beams, plates, etc.

9.1. Visco-elastic displacements of laminated beam systems

The determination of displacements in laminated beam systems consisting primarily of elements with rheological characteristics subjected to sustained loads is conveniently carried out by employing the Volterra principle which states that the short-term solution can be obtained by replacing the elastic constants with the corresponding operators. Let the load applied to the bar system be a function of time which results in the following force and moment resultants:

$$M_\mathrm{p}(t, x) = \sum_{i=1}^{m_2} f_i(t) M_\mathrm{p}^{(i)}(x) ,$$

$$N_\mathrm{p}(t, x) = \sum_{i=1}^{m_2} f_i(t) N_\mathrm{p}^{(i)}(x) , \tag{9.1}$$

$$S_\mathrm{p}(t, x) = \sum_{i=1}^{m_2} f_i(t) S_\mathrm{p}^{(i)}(x) ,$$

where m_2 is the number of various time-varying generalized loads acting on the system.

Consequently, taking expression (3.11) into account, the functional of the time-varying of the displacements $\Delta_{ij}(t)$ will be determined as

$$\Delta_{ij}(t) = \sum \int \Big\{ \bar{N}_i(x) \sum_{k=1}^{m_2} N_{\mathrm{p}}^{(k)}(x) A_{11}^{\prime *}[f_k(t)]$$

$$+ \bar{N}_i(x) \sum_{k=1}^{m_2} M_{\mathrm{p}}^{(k)}(x) B_{11}^{\prime *}[f_k(t)] + \bar{M}_i(x) \sum_{k=1}^{m_2} N_{\mathrm{p}}^{(k)}(x) B_{11}^{\prime *}[f_k(t)] \tag{9.2}$$

$$+ \bar{M}_i(x) \sum_{k=1}^{m_2} M_{\mathrm{p}}^{(k)}(x) D_{11}^{\prime *}[f_k(t)] + \bar{S}_i(x) \sum_{k=1}^{m_2} S_{\mathrm{p}}^{(k)}(x) F^*[f_k(t)] \Big\} \frac{\mathrm{d}x}{b} \, .$$

Here $A_{11}^{\prime *}$, $B_{11}^{\prime *}$, $D_{11}^{\prime *}$ are operators defined by (8.2) whereas the operator F^* is given by

$$F^*[f_k(t)] = F\Big[f_k(t) + \int_0^t f_k(\tau) \tilde{F}(t-\tau)\, \mathrm{d}\tau \Big] , \tag{9.3}$$

where the kernel $\tilde{F}(t-\tau)$ is found by analogy with the kernels $\tilde{A}_{11}^{\prime}(t-\tau)$, $\tilde{B}_{11}^{\prime}(t-\tau)$ and $\tilde{D}_{11}^{\prime}(t-\tau)$ from tests (experimental and analytical) for transverse shear of laminated reinforced plastics.

For time-independent (i.e. constant) loading ($f_i(t) = \text{constant} = 1$), (9.2) yields

$$\Delta_{ij}(t) = \sum \int \{ \bar{N}_i(x) N_{\mathrm{p}}(x) A_{11}^{\prime *}[1] + \bar{N}_i(x) M_{\mathrm{p}}(x) B_{11}^{\prime *}[1]$$

$$+ \bar{M}_i(x) N_{\mathrm{p}}(x) B_{11}^{\prime *}[1] + \bar{M}_i(x) M_{\mathrm{p}}(x) D_{11}^{\prime *}[1] \tag{9.4}$$

$$+ \bar{S}_i(x) S_{\mathrm{p}}(x) F^*[1] \} \frac{\mathrm{d}x}{b} \, .$$

Relationship (9.2) permits us to analyse the relative contributions of the individual continuously acting internal forces to the evolution of time-dependent displacements in beam systems constructed of reinforced plastics. Such an analysis leads to the following conclusion: the influence of the shear force increases with increasing ratio of the axial modulus of elasticity to the longitudinal shear modulus. This ratio becomes larger with time, and so does, consequently the contribution of the shear force to the deformation of the system. By way of an example consider a simply-supported beam acted upon by the centrally applied concentrated force *P*. Fig. 22 shows the results of calculation of the relative difference in the vertical displacements of the beams centre when the action of the shear forces is neglected. The results are presented for two types of unidirectional laminates with $\nu_{\mathrm{f}} = 0.60$: glass- and carbon/epoxy. $\Delta_T(\infty)$ are deflexions at $t \to \infty$ calculated without considering shear forces and $\Delta(\infty)$ are deflexions at $t \to \infty$ calculated by including the action of shear forces). It follows from Fig. 22 that the relative difference in the vertical displacement introduced by neglecting the weak shear resistance of laminated reinforced plastics actually increases in a linear way that depends on the degree of creep in the polymer matrix. This difference can reach considerable magnitudes. The definition of a 'short beam' changes with increased

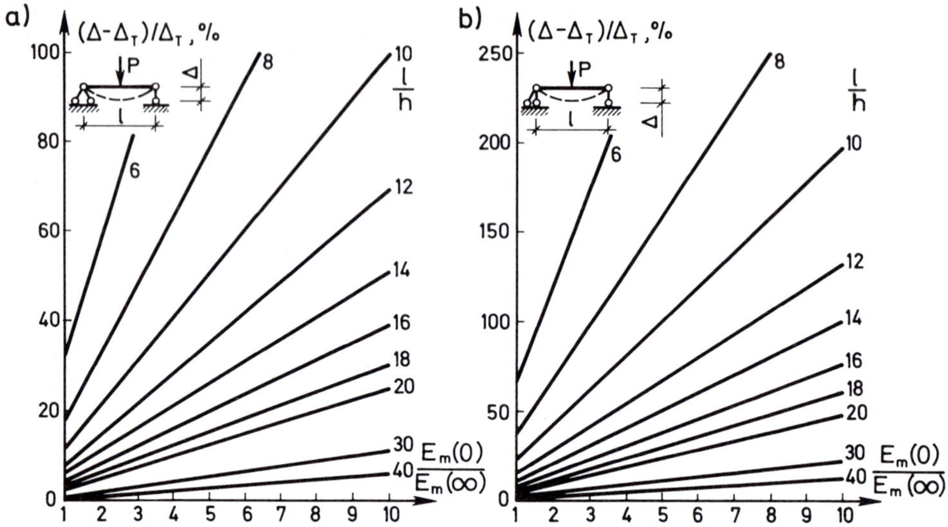

FIG. 22. Dependence of the relative difference in the ultimate ($t=\infty$) vertical displacements of the beam's centre under concentrated load on the degree of matrix creep and structural geometry of the beam in the case of unidirectionally reinforced (a) glass/epoxy and (b) carbon/epoxy with $v_f = 0.60$.

loading time depending on the rheological properties of the plastic components. Even for relatively long beams ($l/h = 20$ or more) it is necessary to consider the action of shear forces when calculating the deformed state at $t \to \infty$. Note that the response of plastics with higher degree of anisotropy is more sensitive to low shear resistance (in the given example, carbon/epoxy as compared to glass/epoxy).

9.2. The method of forces for continuously applied loads

The rheological properties of laminated reinforced plastics govern the time-dependence of the deformed state of bar systems composed of these materials; they do so even under continuously applied loads. Since in statically indeterminate systems the internal forces depend upon and are determined by the strained state of the whole system, the internal forces and consequently also the redundant forces are both functions of time. Therefore, for an nth degree statically indeterminate system the internal forces in any time interval, using the method of forces, will be written as

$$\begin{aligned} M(t,x) &= M_p(t,x) + \sum_{i=1}^{n} X_i(t)\bar{M}_i(x)\,, \\ N(t,x) &= N_p(t,x) + \sum_{i=1}^{n} X_i(t)\bar{N}_i(x)\,, \\ S(t,x) &= S_p(t,x) + \sum_{i=1}^{n} X_i(t)\bar{S}_i(x)\,, \end{aligned} \tag{9.5}$$

where M_p, N_p and S_p are internal moment and force resultants in the primary system in the time interval t caused by the action of the prescribed load; $\bar{M}_i$, $\bar{N}_i$ and $\bar{S}_i$ are internal moment and force resultants in the primary system caused by the action of the ith redundant force $X_i(t) = 1$.

In order to find the unknown redundant forces we shall resort to the condition stating that displacements in the primary system due to the prescribed load in the direction of redundant links are zero in any time interval. Thus, employing (9.2), taking expressions (9.5) into account, obtain a set of integral equations in terms of the unknown functions $X_i(t)$ (GURVICH and BULAVS, 1985):

$$[\delta][X(t)] + \int_0^t [\delta'(t-\tau)][X(\tau)]\,\mathrm{d}\tau = [\Delta_p(t)]\,. \tag{9.6}$$

Here, the matrix elements of $[\delta]$, $[\delta']$ and $[\Delta_p(t)]$ have the following form:

$$\begin{aligned}\sigma_{ij} = \sum \int \{&\bar{N}_i(x)\bar{N}_j(x)A'_{11} + [\bar{N}_i(x)\bar{M}_j(x) + \bar{N}_j(x)\bar{M}_i(x)]B'_{11} \\ &+ \bar{M}_i(x)\bar{M}_j(x)D'_{11} + \bar{S}_i(x)\bar{S}_j(x)F\}\,\mathrm{d}x/b\,,\end{aligned} \tag{9.7}$$

$$\begin{aligned}\delta'(t-\tau)_{ij} = \sum \int \{&\bar{N}_i(x)\bar{N}_j(x)A'_{11}\tilde{A}'_{11}(t-\tau) + [\bar{N}_i(x)\bar{M}_j(x) \\ &+ \bar{N}_j(x)\bar{M}_i(x)]B'_{11}\tilde{B}'_{11}(t-\tau) + \bar{M}_i(x)\bar{M}_j(x)D'_{11}\tilde{D}'_{11}(t-\tau) \\ &+ \bar{S}_i(x)\bar{S}_j(x)F\tilde{F}(t-\tau\}\,\mathrm{d}x/b\,,\end{aligned} \tag{9.8}$$

$$\begin{aligned}\Delta_p(t)_i = -\sum \int \Big\{&\bar{N}_i(x) \sum_{k=1}^{m_2} N_p^{(k)}(x)A'^{*}_{11}[f_k(t)] \\ &+ \bar{M}_i(x) \sum_{k=1}^{m_2} N_p^{(k)}(x)B'^{*}_{11}[f_k(t)] + \bar{N}_i(x) \sum_{k=1}^{m_2} M_p^{(k)}(x)B'^{*}_{11}[f_k(t)] \\ &+ \bar{M}_i(x) \sum_{k=1}^{m_2} M_p^{(k)}(x)D'^{*}_{11}[f_k(t)] + \bar{S}_i(x) \sum_{k=1}^{m_2} S_p^{(k)}(x)F^{*}[f_k(t)]\,\mathrm{d}x/b\,.\end{aligned} \tag{9.9}$$

For the case of constant load (9.9) yields:

$$\begin{aligned}\Delta_p(t)_i = -\sum \int \{&\bar{N}_i(x)N_p(x)A'^{*}_{11}[1] + [\bar{N}_i(x)M_p(x) + \bar{M}_i(x)M_p(x)]B'^{*}_{11}[1] \\ &+ \bar{M}_i(x)M_p(x)D'^{*}_{11}[1] + \bar{S}_i(x)S_p(x)F^{*}[1]\}\,\mathrm{d}x/b\,.\end{aligned} \tag{9.10}$$

The set of integral equations (9.6) can be solved either numerically by means of a computer or by analytical methods. In the latter approach the given problem will be most conveniently solved by using the Laplace transform method. Application of the transform to system (9.6) yields

$$[\delta][\bar{X}(P)] + [\bar{\delta}'(P)][\bar{X}(P)] = [\bar{\Delta}_p(P)] \,. \tag{9.11}$$

Hence

$$[\bar{X}(P)] = [[\delta] + [\bar{\delta}'(P)]]^{-1}[\bar{\Delta}_p(p)\rangle \,. \tag{9.12}$$

The solution is obtained by taking the inverse Laplace transform of (9.12). We should note that the larger the degree of static indeterminacy of the investigated bar system, the greater the effort required in taking the inverse transform. For this reason numerical methods are preferable in a number of cases.

As an illustrative example we shall consider a statically indeterminate system of the first degree where all its members are rectilinear and are constructed of the same material. For the case of constant loading, neglecting the effect of axial force (9.6) yields the following integral equation in terms of $X_i(t)$:

$$\begin{aligned}
&\sum D'_{11}\Big[\int \bar{M}^2(x)\,\mathrm{d}x\Big]\Big\{X(t) + \int_0^t X(\tau)\tilde{D}'_{11}(t-\tau)\,\mathrm{d}\tau\Big\} \\
&\quad + \sum F\Big[\int \bar{S}^2(x)\,\mathrm{d}x\Big]\Big\{X(t) + \int_0^t X(\tau)\tilde{F}(t-\tau)\,\mathrm{d}\tau\Big\} \\
&\quad = \sum D'_{11}\Big[\int \bar{M}(x)M_p(x)\,\mathrm{d}x\Big]\Big\{1 + \int_0^t \tilde{D}'_{11}(t-\tau)\,\mathrm{d}\tau\Big\} \\
&\quad + \sum F\Big[\int \bar{S}(x)S_p(x)\,\mathrm{d}x\Big]\Big\{1 + \int_0^t \tilde{F}(t-\tau)\,\mathrm{d}\tau\Big\} \,.
\end{aligned} \tag{9.13}$$

Application of the Laplace transform and subsequent separation of $\bar{X}(P)$ yields:

$$\bar{X}(P) = \frac{(S_1 + S_2) + [S_1\tilde{D}'_{11}(P) + S_2\tilde{F}(P)]}{(S_3 + S_4) + [S_3\tilde{D}'_{11}(P) + S_4\tilde{F}(P)]}\,\frac{1}{P}\,, \tag{9.14}$$

where

$$S_1 = \sum D'_{11}\Big[\int_0^l \bar{M}(x)M_p(x)\,\mathrm{d}x\Big]\,,$$

$$S_2 = \sum F\Big[\int_0^l \bar{S}(x)S_p(x)\,\mathrm{d}x\Big]\,,$$

$$S_3 = \sum D'_{11}\Big[\int_0^l \bar{M}(x)^2\,\mathrm{d}x\Big]\,,$$

$$S_4 = \sum F\Big[\int_0^l \bar{S}(x)^2\,\mathrm{d}x\Big]\,.$$

The inverse transformation of expression (9.14) is carried out upon selection of the required type of the creep kernels $\tilde{D}'_{11}(t-\tau)$ and $\tilde{F}(t-\tau)$. For the case when these kernels are specified by the following exponential functions

$$\tilde{D}'_{11}(t-\tau) = c_D\, e^{-\alpha_D(t-\tau)}\,, \qquad \tilde{F}(t-\tau) = c_F\, e^{-\alpha_F(t-\tau)}\,, \tag{9.15}$$

the corresponding transformed expressions take the following forms:

$$\tilde{D}'_{11}(P) = \frac{c_D}{P+\alpha_D}\,, \qquad \tilde{F}(P) = \frac{c_f}{P+\alpha_F}\,, \tag{9.16}$$

and thus expression (9.14) becomes:

$$X(P) = \frac{a_2P^2 + a_1P + a_0}{b_2P^2 + b_1P + b_0}\cdot\frac{1}{P}\,, \tag{9.17}$$

where

$$a_0 = (S_1+S_2)\alpha_D\alpha_F + S_1c_D\alpha_F + S_2c_F\alpha_D\,,$$
$$a_1 = S_1c_D + S_2c_F + (S_1+S_2)(\alpha_D+\alpha_F)\,,$$
$$a_2 = (S_1+S_2)\,,$$
$$b_0 = (S_3+S_4)\alpha_D\alpha_F + S_3c_D\alpha_F + S_4c_F\alpha_D\,,$$
$$b_1 = S_3c_D + S_4c_F + (S_3+S_4)(\alpha_D+\alpha_F)\,,$$
$$b_2 = (S_3+S_4)\,.$$

Application of the inverse transform to (9.17) yields

$$X(t) = A_1\,e^{l_1t} + A_2\,e^{l_2t} + A_3\,, \qquad A_i = \frac{l_i^2a_2 + l_ia_1 + a_0}{b_2l_i(l_i-l_j)} \quad (i, j = 1, 2;\ i \neq j)\,, \qquad A_3 = \frac{a_0}{b_2l_1l_2}\,. \tag{9.18}$$

Here l_1, l_2 are roots of the quadratic equation

$$b_2l^2 + b_1l + b_0 = 0\,.$$

The analysis of (9.18) makes it evident that the reactions and, consequently, the internal forces in statically indeterminate systems composed of laminated reinforced plastics subjected to a constant load vary continuously with time. This time-dependent response is caused by the rheological properties of the individual plastic components and depends both on the stress–strain properties and the lay-up geometry of the material.

9.3. *Internal forces in statically indeterminate systems*

For numerical evaluation of the time-dependent changes in the internal forces in statically indeterminate member systems composed of laminated reinforced plastic beams, we choose, for example, a fixed, simply-supported cantilever, i.e. a system indeterminate to the first degree acted upon by a uniformly distributed load.

By selecting the right-hand support reaction as an unknown reaction (see Fig. 23), we obtain

$$\begin{aligned} S_1 &= D'_{11}ql^4/8\,, \\ S_2 &= Fql^2/2\,, \\ S_3 &= D'_{11}l^3/3\,, \\ S_4 &= Fl\,, \end{aligned} \tag{9.19}$$

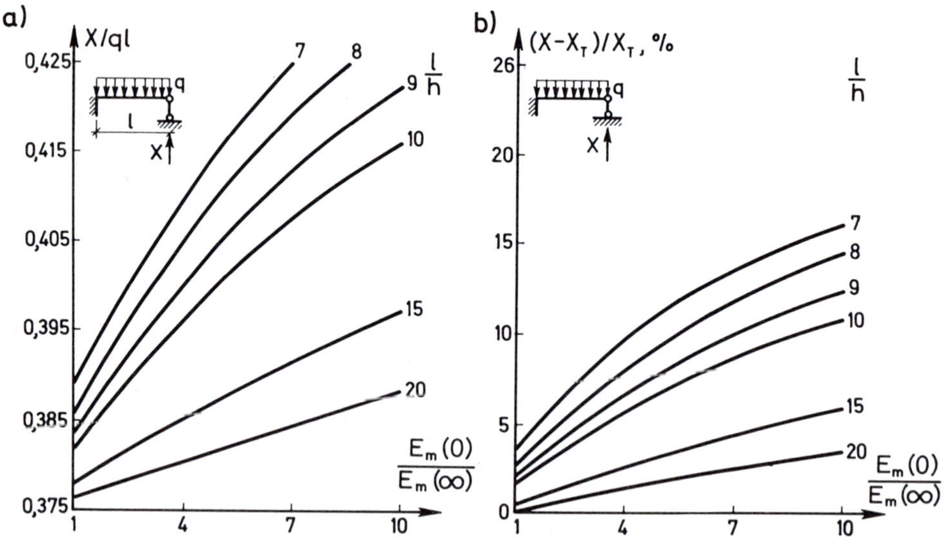

FIG. 23. Dependence of (a) the support reactions and (b) the corresponding error due to the exclusion of transverse forces at $t = \infty$ on the degree of matrix creep and structural geometry of the beam for unidirectionally reinforced carbon/epoxy material.

whereas by selecting the reaction of the restraint (Fig. 24) we have

$$\begin{aligned} S_1 &= D'_{11} q l^3/24 , \\ S_2 &= 0 , \\ S_3 &= D'_{11} l/3 , \\ S_4 &= F/l . \end{aligned} \tag{9.20}$$

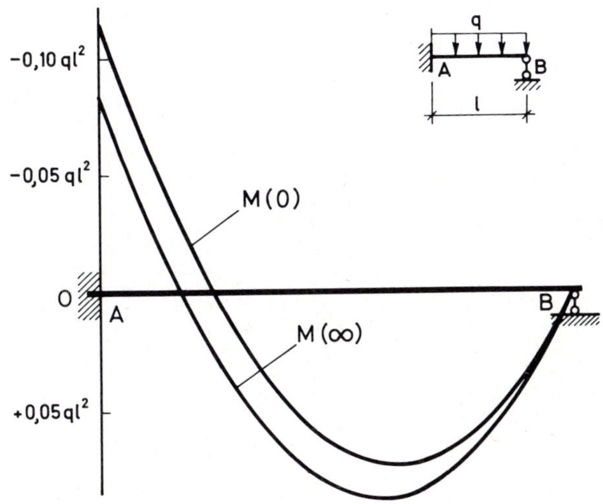

FIG. 24. Time-dependent variation of the bending moment for a carbon/epoxy beam with $l/h = 10$ and $E_m(0)/E_m(\infty) = 10$.

The ultimate values of the redundant $X(t)$, taking into account (9.18), are

when $t = 0$ $\quad X(0) = A_1 + A_2 + A_3$,
when $t \to \infty$ $\quad X(\infty) = A_3$.

Fig. 23 shows the results of computation of the dependence of the support reaction X for the given statically indeterminate beam at $t \to \infty$ on the degree of creep in the polymer matrix (the material is a unidirectionally reinforced carbon/epoxy: $E_{fr}/E_m = 2$; $E_{fz}/E_{fr} = 20$; $\nu_m = 0.35$; $\nu_{fr} = 0.6$) and on the span-to-depth ratio (l over h) of the beam. For $E_m(0)/E_m(\infty) = 1$ there are short-term values $X(\infty) = X(0)$, ($M(\infty) = M(0)$). According to the data given in Figs. 23 and 24 we may conclude that the change in the values of $X(\infty)$ as compared to $X(0)$ depends not only on the rheological properties of the polymer matrix but also on the beam geometry. Fig. 24 shows the time-variation of the bending moment diagram for a similar beam with $l/h = 10$ and $E_m(0)/E_m(\infty) = 10$.

The obtained results permit us to conclude that for the considered case the vertical support reaction X increases with time whereas the support moment M

decreases. In the actual case of a carbon/epoxy beam with $l/h = 10$ and $E_m(0)/E_m(\infty) = 10$ the support moment M decreases with time by 33%, and the reaction increases by 11%. For such variations in the support reactions the maximum bending moment in the span increases by 18.5%. Fig. 25 shows the time-dependent variation of the bending moment diagram for a statically indeterminate carbon/epoxy frame. These time-dependent variations of the forces are explained by the fact that in reinforced plastic members the energy contribution of the transverse forces will sometimes be comparable to the strain energy of the bending moments and, as was already shown above, it increases with loading time. It should be noted, however, that quite frequently in determining the response of statically indeterminate systems constructed of reinforced plastics the effect of shear forces and loading time on the forces in the redundant links may be discarded. Therefore, the effect of loading time on the variations of X_1 is to be considered separately for any specific situation involving the application of the Volterra principle.

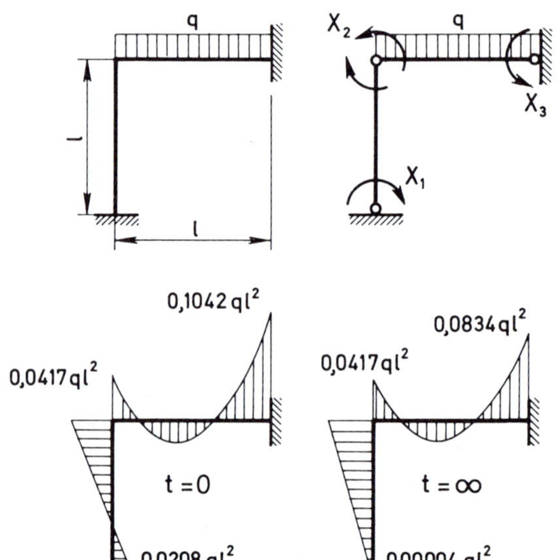

FIG. 25. Time-dependent variation of the bending moment for a statically indeterminate carbon/epoxy frame with $l/h = 10$ and $E_m(0)/F_m(\infty) - 10$.

10. Conclusions

The results discussed in the present paper lead to the conclusion that the principal peculiarities of the structural mechanics of systems consisting of reinforced plastic laminate bars are the following.

(1) Structural mechanics of reinforced plastic beam structures consists of two parts – structural mechanics of the composite material and structural mechanics of the system.

(2) For specified M, S and N the stress state of the individual plies depends largely on the structural composition of the laminate, loading time, stress–strain properties of the matrix and the fibres and the fibre volume content.
(3) The causes of failure of the most unfavourably oriented plies may be

 (a) fibre failure;
 (b) tensile or shear failure of the matrix;
 (c) ply failure in transverse compression;
 (d) bond failure between the fibres and the matrix.

(4) Strength criteria for the individual plies should include the joint effect of M, S and N, and also the subsequent stepwise character of the laminate failure leading to changes in the symmetry of its structure.
(5) In the case of nonsymmetric structures the strength of a laminated reinforced plastic beam under axial load depends on its cross-sectional shape: under axial loading the tubular cross-section provides a higher strength than a flat cross-section.
(6) In transverse bending the failure mechanism of a laminated beam depends on the ratio l/h and the type of laminate composition.
(7) Under the combined effect of M, S and N the various regions of the ultimate curve are associated with various failure mechanisms of material in a laminated beam.
(8) Under continuously applied load the energy contribution of the shear forces is comparable to the strain energy of the bending moments and its influence should be taken into account in determining the displacements in laminated reinforced plastics beams.
(9) The notion of a 'short bar' needs to be differently understood in the case of continuously loaded laminated reinforced plastic beams. For instance, it follows from Fig. 22(b) that the effect of shear on the elastic flexure of a beam (at $t = 0$) constructed of carbon/epoxy will be only 5% whereas its effect on the visco-elastic flexure ($t = \infty$) is already 50%.
(10) The effect of the structural imbalance in laminated beams varies significantly with continuously applied loading.
(11) In a continuously loaded statically indeterminate system composed of laminated beams it is necessary to estimate the extent of possible time-dependent variations in the magnitude of the redundant unknowns.

References

BULAVS F.YA. and I.G. RADINSH (1980), Time-dependent changes in stress–strain state of unidirectionally reinforced plastic components under sustained load, in: *Mechanics of Compositie Materials* (RPI, Riga) pp. 5–18 (in Russian).

CHRISTENSEN R.M. (1979), *Mechanics of Composite Materials* (John Wiley and Sons, New York) p. 348.

DHARMARJAN S. and H. MCCUTCHEN (1973), Shear coefficients for orthotropic beams, *J. Comp. Mater.* **7**, 530–535.

GURVICH M.R. (1985), Numerical method of studying viscous-elastic properties of laminated reinforced plastics under combined loading, in: *Mechanics of Composite Materials* (RPI, Riga) pp. 49–60 (in Russian).

GURVICH M.R. and F.YA. BULAVS (1985), The effect of loading duration on the stress–strain state of statically indeterminate reinforced plastic member systems, in: *Mechanics of Composite Materials* (RPI, Riga) pp. 61–71 (in Russian).

JONES R.M. (1975), *Mechanics of Composite Materials* (Scripta Book Company, Washington, D.C.) p. 355.

MOSER K. (1982), Theorien und FE-Ausätze für Plattentragwerke mit geschichtetem Aufbau, in: *Verstärkte Plaste-82. 9. wissenschaftliche Tagung mit internationalen Beteiligung* (Berlin, DDR) F4/1–F4/8.

RADINSH I.G. and M.R. GURVICH (1981), Viscous-elastic properties of laminated reinforced plastics under sustained plane state of stress, in: *Mechanics of Reinforced Plastics* (RPI, Riga) pp. 76–89 (in Russian).

SCHAPERY R.A. (1978), Viscous-elastic behaviour of composites, in: G.P. Sendeckyi, eds., *Composite Materials*, Vol. 2, *Mechancis of Composite Materials* (MIR Publishers, Moscow) pp. 102–195 (in Russian).

SKUDRA A.M., F.YA. BULAVS and K.A. ROCENS (1971), *Creep and Static Fatigue of Reinforced Plastics* (Zinatne, Riga) p. 235 (in Russian).

SKUDRA A.M., B.A. KIRULIS and A.V. ZAHAROV (1977), Fiber–matrix bond strength in reinforced plastics, in: *Mechanics of Composite Materials*, Vol. 1 (RPI, Riga) pp. 30–37 (in Russian).

SKUDRA A.M. and F.YA. BULAVS (1978), *Structural Theory of Reinforced Plastics* (Zinatne, Riga) p. 192 (in Russian).

SKUDRA A.M. and F.YA. BULAVS (1982), Strength and creep micro-mechanics of composites, in: I.F. Obraztsov and V.V. Vasil'ev. eds., *Mechanics of Composites*, (MIR Publishers, Moscow) pp. 77–109.

SKUDRA A.M. and F.YA. BULAVS (1982), *Strength of Reinforced Plastics* (Khimia, Moscow) p. 213 (in Russian).

SKUDRA A.M. (1984), Micromechanics of failure of reinforced plastics, in: *Handbook of Composites*, Vol. 3 (Elsevier, Amsterdam) pp. 1–69.

TARNOPOLSKII, YU.M. and A.V. ROZE (1969), *Calculation Peculiarities of Reinforced Plastic Elements* (Zinatne, Riga) p. 274 (in Russian).

TEH K.K. and C.C. HUANG (1979), Shear deformation coefficient for generally orthotropic beam, *Fibre Science and Technology* **12**, 73–80.

VERESCHAGIN A.N. (1925), *Stroitelnaia Promishlenost* **9**, 654–657 (in Russian).

CHAPTER 9

Composite Pressure Vessels

V.A. Bunakov and V.D. Protasov

Moscow Institute of Aviation Technology
27 Petrovka Street
103767 Moscow
U.S.S.R.

Contents

HANDBOOK OF COMPOSITES, VOL. 2 – Structure and Design
Edited by C.T. HERAKOVICH and Y.M. TARNOPOL'SKII

1. Introduction

The efficiency of reinforced composites as structural materials for pressure vessels is now generally recognized.

The development of composite pressure vessels is attributed to some principles related to certain specific properties of composite materials. There are two points which deserve to be mentioned.

First, the traditional design procedure of determining the design form and geometrical parameters of a pressure vessel for the given mechanical characteristics of the material cannot be realized strictly in the case of composites, because the properties of the material depend on its structural parameters (fibre orientations, layer thicknesses, etc.) which are not known beforehand. It follows that we cannot design a composite pressure vessel without constructing the material itself.

Second, the composite material comes into existence at the same time as the fabricating process, and the properties of the material depend largely on the parameters of the process. Furthermore, the continuous winding technique, employed for the production of pressure vessels and promising great advantages, does not allow, at present, any fibre orientations whatsoever to be implemented, including sufficiently effective ones, i.e. this shows that limitations have to be taken into account in composite pressure vessels design.

In short, optimal design, structural analysis and fabrication process form, in effect, three aspects of one problem and cannot be solved separately as is usually done in the manufacture of metal pressure vessels.

What makes the use of composite materials so effective for pressure vessels is their high strength along the fibres. It has become common to estimate the efficiency of pressure vessels by the following criterion (RIVLIN and PIPKIN, 1963)

$$p\frac{V}{G} = \frac{1}{k_\mathrm{f}}\left(\frac{\sigma}{\rho}\right),$$

where p = is the internal pressure; V = is the vessel volume; G = is the mass of the shell; σ/ρ = is the specific strength of the material; k_f = is the coefficient which describes the rational structure and is equal to $k_\mathrm{f} = \frac{3}{2}$ for metal spherical vessels and to $k_\mathrm{f} = 3$ for optimal composite vessels.

Using the data in Table 1, it is easy to see what advantages the composite pressure vessels can offer.

Now, as far as economic efficiency is concerned, we can say that the price of the initial materials makes the use of composite materials generally more expensive. However, this is overshadowed by the benefits from improved technical characteristics such as mass, corrosion resistance, etc., and more convenient maintenance. Also, as there are automatic machines for the production of pressure vessels by the winding technique, labour productivity can be higher and the cost of the finished products lower. Composite pressure

TABLE 1. Comparison of strength and rigidity characteristics of composite materials and metals.

N	Material	σ^u (MPa)	*E* (GPa)	σ^u/ρ (m × 10^3)	E/ρ (m × 10^6)
1.	Steel	1200	210	15.0	2.60
2.	Titanium	1800	110	40.0	2.50
3.	Aluminium	320	70	12.0	2.60
4.	Fibreglass reinforced plastic fabric	500	25	24.0	1.15
	unidirectional fibers	1600	60	72.4	2.71
5.	Graphite–epoxy unidirectional composite	1100	170	73.3	11.3
6.	Boron–epoxy unidirectional composite	1200	200	60.0	10.0
7.	Organic fibers–epoxy unidirectional composite	1500	80	110.0	5.93

vessels have recently become useful as storage tanks for corrosive gases and liquids, as well as for cryogenic media and deep vacuum.

Essentially, a pressure vessel is usually fabricated in the form of a cylindrical shell with specially shaped domes at each end (Figs. 1 and 2). The following major components can be identified in a pressure vessel:
- the pressure vessel shell;
- junctures or fixtures;
- nozzles for attaching pipes, fittings, manways, etc.

The manufacturing process consists of two stages:
- winding a pressure vessel shell;
- making junctures and local reinforcements.

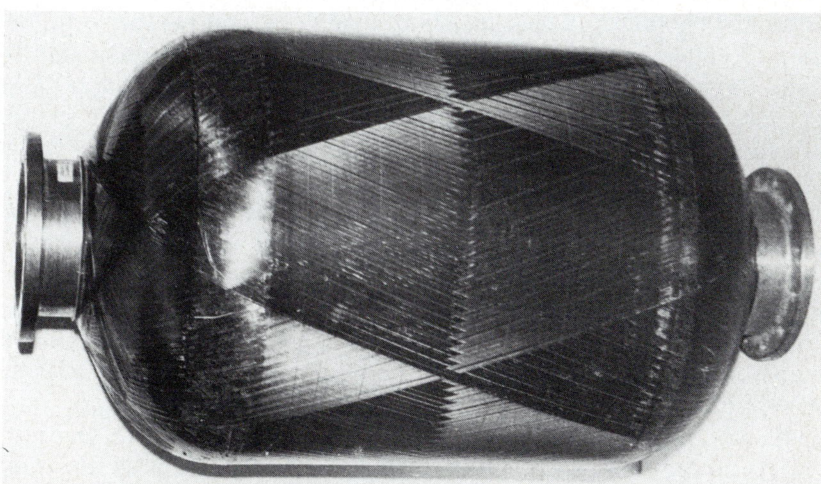

FIG. 1. An experimental pressure vessel, diameter 400 mm.

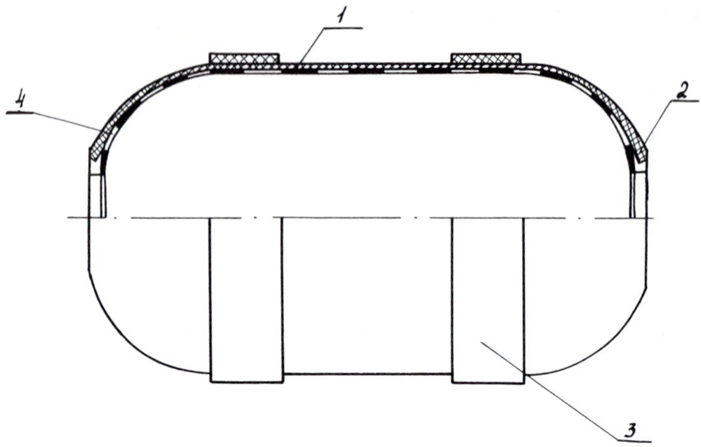

FIG. 2. A filament wound composite pressure vessel: 1, shell; 2, embedded flanges; 3, fixtures; 4, airtight shell.

Composite pressure vessels are fabricated by winding threads or tapes impregnated with epoxy resin onto a properly shaped mandrel. As the problems of the winding process have been discussed in detail elsewhere (ABIBOV, 1971; ROSATO and GROVE, 1964; TSYPLAKOV, 1968), dealing with process parameters such as relative fibre content in a composition, preliminary tape tension, etc., we shall here touch upon the specific features of the winding process in a general form.

There exist two fundamental methods of filament winding which affect the fabrication process.

In the *dry method*, winding is performed with pre-impregnated and properly treated tapes (pre-pregs), which makes it possible to control the resin volume content more effectively and have the matrix distributed over the wall thickness more uniformly during the process of winding and rigidization. This, of course, adds to the quality of the product. Furthermore, with its higher intrastructural friction coefficient, the dry method allows us to align the tapes within a much wider range of geodesic deviations.

However, the use of more complex fabricating machinery is required for the dry method. Therefore the *wet method* has found more popularity where continuous programmed winding is necessary for fabricating pressure vessels. In this method, the fibres are impregnated with the resin immediately before being applied to the mandrel. Junctures are made by additional winding of fabric tapes, and non-polar openings are cut directly and reinforced with pieces of fabric over the perimeter.

In summary, we must say that manufacturing techniques have made somewhat more progress than analytical methods of calculation in the field of composite pressure vessels.

2. Design of the pressure vessel shell

Consider a shell of revolution consisting of K composite layers, each layer being reinforced by fibres oriented at angles $+\varphi_i$ and $-\varphi_i$ to the meridian (Fig. 3). The shell is loaded by a uniform internal pressure p and axial forces Q_0, distributed uniformly over the edge of the polar opening with radius r. The latter forces may vary from $Q_0 = 0$ for vessels with open polar openings to $Q_0 = pr_0/2$, a value corresponding to a polar opening closed with a rigid pressure cover. Each layer is considered within the framework of the netting analysis but the interaction between the layers responsible for an additional tangential load for an individual layer is taken into account.

The layer equilibrium equations for the case of axially symmetric loading (BUNAKOV et al., 1982) are

$$\begin{aligned}
&\frac{\mathrm{d}}{\mathrm{d}r}(BN_{1i}^{\pm}) - N_{2i}^{\pm}\frac{\mathrm{d}B}{\mathrm{d}r} + ABp_{1i}^{\pm} = 0\,,\\
&\frac{\mathrm{d}}{\mathrm{d}r}(BN_{12i}^{\pm}) + N_{12i}^{\pm}\frac{\mathrm{d}B}{\mathrm{d}r} + ABp_{2i}^{\pm} = 0\,,\\
&\qquad\qquad \frac{N_{1i}^{\pm}}{R_1} + \frac{N_{2i}}{R_2} = p_{3i}^{\pm}\,.
\end{aligned} \tag{2.1}$$

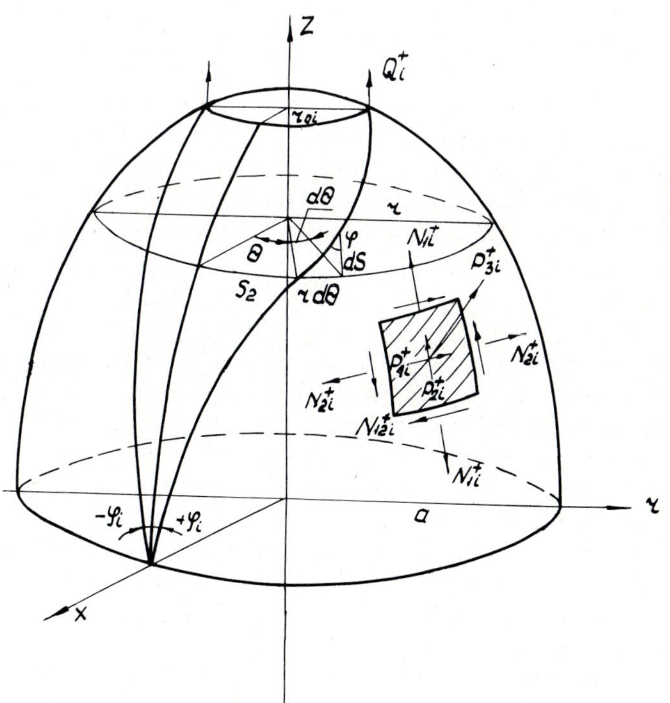

FIG. 3. Loading conditions and the element of the ith layer.

The '+' and '−' signs in (2.1) correspond to the angles $+\varphi_i$ and $-\varphi_i$ (Fig. 3 shows the forces corresponding to the layer $+\varphi_i$);

$$A=[1+(Z')^2]^{1/2}\,,\quad B=r\,,$$

are the coefficients of the first quadratic form of the surface of revolution as given by the equation, $Z=Z(r)$, R_1R_2 are the principal radii of curvature as calculated by means of the relations

$$\frac{1}{R_1}=-\frac{Z''}{[1+(Z')^2]^{3/2}}\,;\quad \frac{1}{R_2}=-\frac{Z'}{r[1+(Z')^2]^{1/2}}\,.$$

In the general case of a shell having a few pairs of symmetric layers, each of the pairs will be under the loads q_i and p_i. The following relations will hold in this case

$$p_{1i}^{\pm}=\frac{q_i}{2}\,,\quad \sum q_i=0\,;\quad p_{3i}^{\pm}=\frac{p_i}{2}\,,\quad \sum p_i=p\,.$$

Between layers, there will be a circumferential load self-balanced for each pair of layers, $\pm\tau_i$, that is $p_{2i}^{\pm}=\pm\tau_i$. With the loading conditions and the accepted notations, System (2.1) can thus be reduced, after a few manipulations, to

$$\begin{aligned}&\frac{\mathrm{d}}{\mathrm{d}r}(rN_{1i}^{\pm})-N_{2i}^{\pm}+r[1+(Z')^2]^{1/2}\frac{q_i}{2}=0\,,\\ &\frac{1}{r}\frac{\mathrm{d}}{\mathrm{d}r}(r^2N_{12i}^{\pm})\pm r[1+(Z')^2]^{1/2}\tau_i=0\,.\\ &\frac{rZ''}{1+(Z')^2}N_{1i}^{\pm}+Z'N_{2i}^{\pm}-r[1+(Z')^2]^{1/2}\frac{p_i}{2}=0\,.\end{aligned}\tag{2.2}$$

Using the third equation of system (2.2) to express $N_{2i}^{\pm}$ as follows

$$N_{2i}^{\pm}=\frac{rZ''}{Z'[1+(Z')^2]}N_{1i}^{\pm}-\frac{r[1+(Z')^2]^{1/2}}{Z'}\frac{p_i}{2}\tag{2.3}$$

and inserting the above in the first equation, we obtain, after some transformation,

$$\frac{\mathrm{d}}{\mathrm{d}r}\left(\frac{rZ'N_{1i}^{\pm}}{[1+(Z')^2]^{1/2}}\right)+\frac{r}{2}(Z'q_i+p_i)=0\,.$$

Integrating this equation and considering the axial forces $Q_i/2$, acting on a layer with radius r_{0i} along the parallel, we find

$$N_{1i}^{\pm}=\frac{[1+(Z')^2]^{1/2}}{2rZ'}\left[Q_ir_{0i}+\int_{r_{0i}}^{r}(rZ'q_i+rp_i)\,\mathrm{d}r\right].\tag{2.4}$$

Inserting (2.4) in (2.3), we find at last

$$N_{2i}^{\pm} = \frac{[1+(Z')^2]^{1/2}}{2rZ'} \left\{ rp_i - \frac{Z''}{Z'[1+(Z')^2]} \left[Q_i r_{0i} + \int_{r_{0i}}^{r} (rZ'q_i + rp_i)\,\mathrm{d}r \right] \right\}. \tag{2.5}$$

Now consider the static relationships of the mechanics of filamentary systems. The thickness of a layer placed at an angle $+\varphi_i$ or $-\varphi_i$ can be expressed in terms of the number of filaments $n_i/2$ which make up an individaul layer and the reduced cross-sectional area of a filament f (allowing for the matrix) as follows

$$\frac{h_i}{2} = \frac{n_i f}{4\pi r \cos\varphi_i} \tag{2.6}$$

Eq. (2.6) is, in effect, a technological constraint as it expresses the requirement of continuous reinforcement: the number of filaments passing through any cross section of the shell must be constant.

Now let us relate the membrane stress resultants $N_{1i}^{\pm}$, $N_{2i}^{\pm}$, $N_{12i}^{\pm}$ to the stresses in the tape. Ignoring the load-carrying capacity of the matrix and taking into account the symmetry of loading, we thus have with the aid of (2.6)

$$\begin{aligned} N_{1i}^{\pm} &= \sigma_{\|i}\frac{h_i}{2}\cos^2\varphi_i = \frac{\sigma_{\|i} n_i f}{4\pi r}\cos\varphi_i\,, \\ N_{2i}^{\pm} &= \sigma_{\|i}\frac{h_i}{2}\sin^2\varphi_i = \frac{\sigma_{\|i} n_i f}{4\pi r}\sin\varphi_i \operatorname{tg}\varphi_i\,, \\ N_{12i}^{\pm} &= \pm\sigma_{\|i}\frac{h_i}{2}\sin\varphi_i\cos\varphi_i = \pm\frac{\sigma_{\|i} n_i f}{4\pi r}\sin\varphi_i\,. \end{aligned} \tag{2.7}$$

Here $\sigma_{\|i}$ is the stress in the filaments of the ith layer (which is the same for the layers oriented at $\pm\varphi_i$ because of the symmetry of winding). Comparing the expressions (2.4) and (2.5) with the first two equations (2.7), it is easy to see that the stress resultants N_{1i} and N_{2i} cannot be taken up by one family of filaments unless, between the shape of the shell $Z(r)$, the fibre orientations and the loads Q_i, $p_i(r)$, $q_i(r)$, there exists a relationship specified by the equation

$$\frac{Z''}{Z'[1+(Z')^2]} = \frac{p_i r}{Q_i r_{0i} + \int_{r_{0i}}^{r} (rZ'q_i + rp_i)\,\mathrm{d}r} - \frac{1}{r}\operatorname{tg}^2\varphi_i\,. \tag{2.8}$$

Relation (2.8) may be referred to as the condition under which a shell of revolution formed by a system of flexible filaments can exist. If the condition fails to be met, the shell will become a mechanism, that is, it will be unable to take up external load, as long as the hypotheses of a filamentary system hold. If the relationship between the shell shape, fibre orientations and loading conditions does correspond to (2.8), then the stresses in the filaments can be found by comparing (2.4) with the first of the static relations (2.7) or (2.5) with the second relation (2.7), with (2.8) being taken into account, and will thus be given by

$$\sigma_{\|i} = \frac{2\pi[1+(Z')^2]^{1/2}}{n_i f Z' \cos\varphi_i}\left[Q_i r_{0i} + \int_{r_{0i}}^{r} r(Z'q_i + p_i)\,\mathrm{d}r\right]. \tag{2.9}$$

The interaction forces inside the pairs of layers can be calculated directly from the second equation in (2.2), with the equalities (2.7) and (2.9) taken into account. They are given by

$$\tau_i = \frac{n_i f}{4\pi r^2[1+(Z')^2]^{1/2}}\,\frac{\mathrm{d}}{\mathrm{d}r}(\sigma_{\|i} r \sin\varphi_i)\,. \tag{2.10}$$

The second technological condition is that the tape must be properly balanced on the mandrel during winding (the filament no-slip condition). It is given by the inequality

$$\max|\mathrm{tg}\,\theta_i| \leqslant \kappa\,, \tag{2.11}$$

where κ is the available friction coefficient for the filament on the surface of the mandrel; θ_i is the angle of geodesic deviation of the filament of the ith layer (the angle between the normal to the trajectory and the normal to the surface of the mandrel) whose tangent can be expressed by the formula

$$\mathrm{tg}\,\theta_i = R_{\mathrm{n}i}/R_{\mathrm{g}i}\,,$$

where

$$\frac{1}{R_{\mathrm{n}i}} = \frac{\cos^2\varphi_i}{R_1} + \frac{\sin^2\varphi_i}{R_2}\,;\quad \frac{1}{R_{\mathrm{g}i}} = \frac{\cos\varphi_i}{[1+(Z')^2]^{1/2}}\,\frac{\mathrm{d}\varphi_i}{\mathrm{d}r} + \frac{\sin\varphi_i}{r[1+(Z')^2]^{1/2}}\,,$$

are the normal and geodesic curvatures of the curve on a surface of revolution. Taking into account expressions for the principal radii of curvature R_1 and R_2, we finally have

$$\mathrm{tg}\,\theta_i = \frac{\dfrac{\mathrm{d}}{\mathrm{d}r}(r\sin\varphi_i)}{\dfrac{rZ''\cos^2\varphi_i}{1+(Z')^2} + Z'\sin^2\varphi_i}\,. \tag{2.12}$$

The expressions (2.8)–(2.12) offer some general relationships which permit us to design a wide class of pressure vessels for various reinforcement patterns. Moreover, the condition for the shell to exist as a filamentary system (2.8) is complemented by the strength conditions: for the fibres

$$\left| \frac{2\pi[1+(Z')^2]^{1/2}}{n_i f Z' \cos\varphi_i} \left[Q_i r_{0i} + \int_{r_{0i}}^{r} r(Z' q_i + p_i)\,\mathrm{d}r \right] \right| \leqslant \sigma^{\mathrm{u}}_{\|i} \tag{2.13}$$

for the matrix under interlaminar shear

$$\left| \frac{n_i f}{4\pi r^2 [1+(Z')^2]^{1/2}} \frac{\mathrm{d}}{\mathrm{d}r} (\sigma_{\|i} r \sin\varphi_i) \right| \leqslant \tau^{\mathrm{u}}_{\|\perp i};\ \frac{q_i}{2} \leqslant \tau^{\mathrm{u}}_{\|\perp i}, \tag{2.14}$$

where $\sigma^{\mathrm{u}}_{\|i}$ is the strength of the material of the ith layer in tension along the fibres; $\tau^{\mathrm{u}}_{\|\perp i}$ is the interlaminar shear strength; and by the conditions of technological feasibility: for the continuity of fibres

$$n_i f = \text{constant}, \tag{2.15}$$

and for equilibrium of the filament during winding

$$\left| \frac{\dfrac{\mathrm{d}}{\mathrm{d}r}(r \sin\varphi_i)}{\dfrac{r Z'' \cos^2\varphi_i}{1+(Z')^2} + Z' \sin^2\varphi_i} \right| \leqslant \kappa\,. \tag{2.16}$$

In the optimum design problems for pressure vessels the objective function is usually the mass of the construction, which is given by

$$G = \sum_{i=1}^{k} \rho_i n_i f S_{1i}\,,$$

where S_{1i} is the filament length in the ith family, and ρ_i is the density of the material.

It is to be noted that when shells of minimum mass are designed, the condition for the uniform strength of such constructions using a netting model corresponds to the sufficient condition for optimality because all the conditions of the Levi theorem for statically determinate systems are now satisfied. For this reason, the design criterion of optimality – uniform strength – has been used very often for composite structures.

Now we must define some general properties of uniformly stressed composite shells. Let the formula for stresses (2.9) be rewritten as follows

$$Q_i r_{0i} + \int_{r_{0i}}^{r} r(Z' q_i + p_i)\,\mathrm{d}r = \frac{\sigma_{\|i} n_i f}{2\pi} \cos\varphi_i \frac{Z'}{[1+(Z')^2]^{1/2}}\,. \tag{2.17}$$

Differentiating this with respect to r and setting $\sigma_{\|i}$ = constant, we have

$$r(Z'q_i + p_i) = -\frac{\sigma_{\|i} n_i f}{2\pi} \frac{Z'}{[1+(Z')^2]^{1/2}} \left\{ -\sin\varphi_i \frac{d\varphi_i}{dr} + \frac{Z''\cos\varphi_i}{Z'[1+(Z')^2]} \right\}. \tag{2.18}$$

Now let the denominator on the right hand side of the equation for the existence of an equilibrium shape (2.8) be eliminated by means of (2.17). We then have

$$\frac{Z''}{Z'[1+(Z')^2]} = -\frac{1}{r}\,\mathrm{tg}^2\,\varphi_i - \frac{2\pi}{\sigma_{\|i} n_i f} \frac{1}{\cos\varphi_i} \frac{[1+(Z')^2]^{1/2}}{Z'}\, r p_i\,. \tag{2.19}$$

Substituting (2.19) in the right hand side of (2.18), we finally have

$$r\cos\varphi_i \frac{d\varphi_i}{dr} + \sin\varphi_i = \frac{2\pi r^2 [1+(Z')^2]^{1/2}}{\sigma_{\|i} n_i f} \frac{\cos\varphi_i}{\sin\varphi_i}\, q_i\,. \tag{2.20}$$

Equation (2.20), in conjunction with (2.8), will define the shape of the meridian and the fibre orientation providing uniform distribution of stresses in the shell (namely, in the pairs of layers with angles $\pm\varphi_i$).

Now let us note some features of uniformly stressed system. The condition for uniform stress distribution (2.20) can be written as

$$\frac{d}{dr}(r\sin\varphi_i) = \frac{2\pi r^2 [1+(Z')^2]^{1/2}}{\sigma_{\|i} n_i f}\, q_i\,\mathrm{ctg}\,\varphi_i\,. \tag{2.21}$$

Substituting the above condition in (2.10) for tangential stresses between the layers, we have

$$\tau_i = \frac{q_i\,\mathrm{ctg}\,\varphi_i}{2} \tag{2.22}$$

if $\sigma_{\|i}$ = constant.

Finally, the condition for filament equilibrium on the mandrel surface during winding (2.16) can be simplified as follows. Substituting the formulae (2.19) and (2.21) in the expression for the tangent of the angle of geodesic deviation (2.12), we can modify it to obtain

$$\mathrm{tg}\,\theta_i = -\frac{q_i}{p_i \sin\varphi_i}\,. \tag{2.23}$$

We can show that for uniformly stressed systems the existence equation (2.8) is identically satisfied when the conditions (2.17) and (2.20) or (2.21) are fulfilled simultaneously. Thus, a uniformly stressed shell of revolution formed by filament winding at angles $\pm\varphi_i$ to the generator can be defined by the following system of equations, in which we must set $\sigma_{\|i} = \sigma^{u}_{\|i} = \text{constant}$,

$$\left.\begin{aligned}
&\frac{\sigma^{u}_{\|i} n_i f}{2\pi} \cos\varphi_i \frac{Z'}{[1+(Z')^2]^{1/2}} + Q_i r_{0i} + \int_{r_{0i}}^{r} r(Z' q_i + p_i)\,dr = 0\,, \\
&\frac{d}{dr}(r \sin\varphi_i) = \frac{2\pi r^2 [1+(Z')^2]^{1/2}}{\sigma^{u}_{\|i} n_i f} q_i \operatorname{ctg}\varphi_i\,, \\
&\left|\tau_i = \frac{q_i \operatorname{ctg}\varphi_i}{2}\right| \leqslant \tau^{u}_{\|\perp i}\,, \\
&\left|\operatorname{tg}\theta_i = -\frac{q_i}{p_i \sin\varphi_i}\right| \leqslant \kappa\,.
\end{aligned}\right\} \tag{2.24}$$

Now let us consider a pressure vessel consisting of k families of filaments. For a shell of k layers the interaction forces between layers must satisfy the following equalities

$$\sum_{i=1}^{k} p_i = p\,; \quad \sum_{i=1}^{k} Q_i = Q\,; \quad \sum_{i=1}^{k} q_i = 0\,. \tag{2.25}$$

Summing the first of the equations in (2.24) for all k layers and taking into account (2.25), we have

$$\frac{\sigma^{u}_{\|}}{2\pi} \frac{Z'}{[1+(Z')^2]^{1/2}} \sum_{i=1}^{k} n_i f \cos\varphi_i + Q r_0 + \frac{p(r^2 - r_0^2)}{2} = 0\,, \tag{2.26}$$

assuming that all the layers are of the same material and setting $p = \text{constant}$.

Thus we obtain a class of optimal structures differing from each other by the law of force distribution $q_i(r)$ between layers which must satisfy the last two conditions (2.24). When $q_i(r)$ is given, the fibre orientation in all layers and the meridian shape of a uniformly stressed pressure vessel can be determined by integrating the equations in (2.24). However, we must note that as today's materials actually have very low interlaminar shear strength, the value of $q_i(r)$ can be varied only within definite limits. Therefore only the case when $q_i = 0$ has become practical nowadays (ZIKKEL, 1962; SCHUERCH and BURGGRAF, 1964; CHEREVATSKII, 1970; ELPAT'EVSKII and VASIL'EV, 1972; OBRAZTSOV et al., 1977). This case corresponds to the winding along geodesic lines of the surface, fibre orientations being given by the well known Clairaut theorem

$$r \sin\varphi_i = \text{constant}\,. \tag{2.27}$$

In the case of geodesic winding, as can be seen from equations (2.24), the tangential forces inside a pair of layers are equal to zero and the tape on the mandrel is in equilibrium even if there is no friction between them.

Consider some cases illustrating realization of the solutions we have obtained.

First of all, consider a shell of revolution (Fig. 4) loaded with uniform internal pressure p and axial forces Q_0 ($0 \leqslant Q_0 \leqslant pr_0/2$), distributed uniformly over the edge of the polar opening with a radius r_0. Also, there is a metal flange having a radius $r = b$ round the polar opening. The shell is formed by winding a single family of filaments along the geodesic lines of the surface. Note that the filament is tangential to the contour of the polar opening ($r = r_0$, $\varphi = \pi/2$) and at the equator it forms a certain angle φ_a ($r = a$, $\varphi = \varphi_a$) to the meridian. Directly from (2.27) we find

$$\sin \varphi_a = r_0/a \,. \tag{2.28}$$

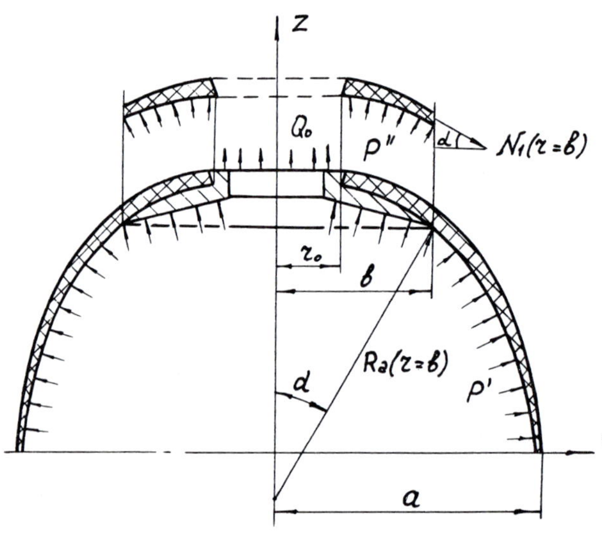

FIG. 4. Loading conditions and geometrical parameters of the shell.

The meridian of the vessel surface must consist of two sections. In the first section, outside the flange, the load applied to the shell will be

$$p' = p \,, \quad Q' = Q_0 \,, \tag{2.29}$$

whereas in the second section the loading of the part adjacent to the flange is determined by the reaction of the flange. If we assume that the axial forces Q_0 are applied to the flange and the contact pressure between the shell and the flange is distributed uniformly over the radius, then we find from the condition of flange equilibrium that

$$p'' = p + \frac{2Q_0 r_0}{b^2 - r_0^2}, \quad Q'' = 0. \tag{2.30}$$

Taking into account the reinforcement law (2.27) and the equalities (2.28) and (2.29), we can write the equation for the meridian shape (2.8) in the first section

$$\frac{Z''}{Z'[1 + (Z')^2]} = \frac{2r}{r^2 - t_0^2} - \frac{r_0^2}{r(r^2 - r_0^2)}, \tag{2.31}$$

where $t_0^2 = r_0^2 - (2Q_0/p)r_0$.

Equation (2.31) allows the variables to be separated and its first integral is

$$Z' = \frac{r(r^2 - t_0^2)[a^2 - r_0^2]^{1/2}}{[a^2(r^2 - r_0^2)(a^2 - t_0^2)^2 - r^2(a^2 - r_0^2)(r^2 - t_0^2)^2]^{1/2}}. \tag{2.32}$$

The integration constant in (2.32) has been obtained from the condition that at $r = a$ the tangential to the meridian is parallel to the axis of revolution $0z$.

Equation (2.32) can be solved in terms of elliptic integrals

$$\bar{Z}_1 = \frac{\bar{n}_2 - \bar{t}_0^2}{(1 - \bar{n}_2)^{1/2}} F(k, \eta) + (1 - \bar{n}_2)^{1/2} E(k, \eta), \tag{2.33}$$

where $F(k, \eta)$ and $E(k, \eta)$ are the first- and second-order elliptic integrals;

$$k = \left(\frac{1 - \bar{n}_1}{1 - \bar{n}_2}\right)^{1/2} \quad \text{is the modulus of elliptic integral;}$$

$$\eta = \arcsin\left(\frac{1 - \bar{r}^2}{1 - \bar{n}_1}\right)^{1/2};$$

$$\bar{n}_{1,2} = \frac{1 - 2\bar{t}_0^2}{2}\left\{-1 \pm \left[\frac{1}{1 - \bar{r}_0^2}\left(1 + r_0^2 \frac{3 - 4\bar{t}_0^2}{1 - 2\bar{t}_0^2}\right)\right]^{1/2}\right\},$$

$$\bar{r} = r/a; \quad Z_1' = Z_1/a, \quad \bar{r}_0 = r_0/a; \quad \bar{t}_0 = t_0/a.$$

Recall that to ensure continuity of the winding process the filament must be tangential to the contour of the polar opening having a radius $r_0 = a \sin \varphi_a$. However, the generator of the shell can be constituted according to (2.33) only up to the radius $r_1 = a\sqrt{\bar{n}_1}$ which gives the minimal distance from the generator to the axis of revolution and coincides with the radius r_0 only for a shell having a free opening ($Q_0 = 0$). This makes it necessary to complete the form of the meridian at the flange. In addition to this, we can suggest that the size of the flange should be chosen such as to be equal to the radius of the inflection point of the main generator. Using (2.31) to find b from $Z_1''(b)=0$, we have

$$b \geqslant r_0 \frac{\sqrt{3}}{2}\left[1+\left(1-\frac{8}{9}\frac{t_0^2}{r_0^2}\right)^{1/2}\right]^{1/2}.$$

Now consider construction of the generator in the section $r_0 \leqslant r \leqslant b$, where the shell is in contact with the flange. In accordance with (2.30), the shape of the generator is defined by (2.31), in which we must set $Z = Z_2$, $t_0 = r_0$. The first integral of this equation will thus be

$$Z_2' = \frac{r(r^2-r_0^2)^{1/2}(b^2-t_0^2)(a^2-r_0^2)^{1/2}}{[a^2(b^2-r_0^2)^2(a^2-t_0^2)^2 - r^2(r^2-r_0^2)(b^2-t_0^2)^2(a^2-r_0^2)^2]^{1/2}}\,. \tag{2.34}$$

The integration constant in this case has been determined from the condition that the sections of the generator at $r=b$ are conjugate, i.e. $Z_1'(b)=Z_2'(b)$. Eq. (2.34) can also be solved in terms of elliptic integrals. The solution takes the form

$$\bar{Z}_2 = -\frac{\bar{m}_1}{(\bar{m}_1+\bar{m}_2)^{1/2}}F(l,\lambda) + (\bar{m}_1+\bar{m}_2)^{1/2}E(l,\lambda) + C\,, \tag{2.35}$$

where the modulus of the elliptic integral is

$$l = \left(\frac{\bar{m}_2}{\bar{m}_1+\bar{m}_2}\right)^{1/2};$$

$$\lambda = \arccos\left(\frac{\bar{r}^2-\bar{r}_0^2}{\bar{m}^2}\right)^{1/2};$$

$$\bar{m}_{1,2} = \left(\frac{\bar{r}_0^4}{4} + \frac{(\bar{b}^2-\bar{r}_0^2)^2(1-\bar{t}_0^2)^2}{(\bar{b}^2-\bar{t}_0^2)^2(1-\bar{r}_0^2)^2}\right)^{1/2} \pm \frac{\bar{r}_0^2}{2};$$

$$\bar{b} = b/a\,; \quad \bar{Z}_2 = Z_2/a\,.$$

The constant C can be determined from the condition that the two parts of the shell are conjugate: at $r=b$, $Z_1=Z_2$

$$C - \frac{\bar{n}_2-\bar{t}_0^2}{(1-\bar{n}^2)^{1/2}}F(k,\eta_1) + \frac{\bar{m}_1}{(\bar{m}_1+\bar{m}_2)^{1/2}}F(l,\lambda_1) + (1-\bar{n}_2)^{1/2}E(k,\eta_1)$$
$$-(\bar{m}_1+\bar{m}_2)^{1/2}E(l,\lambda_1)\,,$$

where

$$\eta_1 = \arcsin\left(\frac{1-\bar{b}^2}{1-\bar{n}_1}\right)^{1/2}; \quad \lambda_1 = \arccos\left(\frac{\bar{b}^2-\bar{r}_0^2}{\bar{m}_2}\right)^{1/2}.$$

Thus, (2.33) and (2.35) yield a family of meridians for a filament wound shell of revolution. Eq. (2.33) will hold if $\bar{b} \leq \bar{r} \leq 1$, i.e. $0 \leq \eta \leq \eta_1$ and (2.35) – if $\bar{r}_0 \leq \bar{r} \leq \bar{b}$, i.e. $\lambda_1 \leq \lambda \leq \pi/2$.

Fig. 5 shows the curves corresponding to different values of the radius of the polar opening and the parameter Q_0/pr_0. Fig. 6 shows the maximum height of

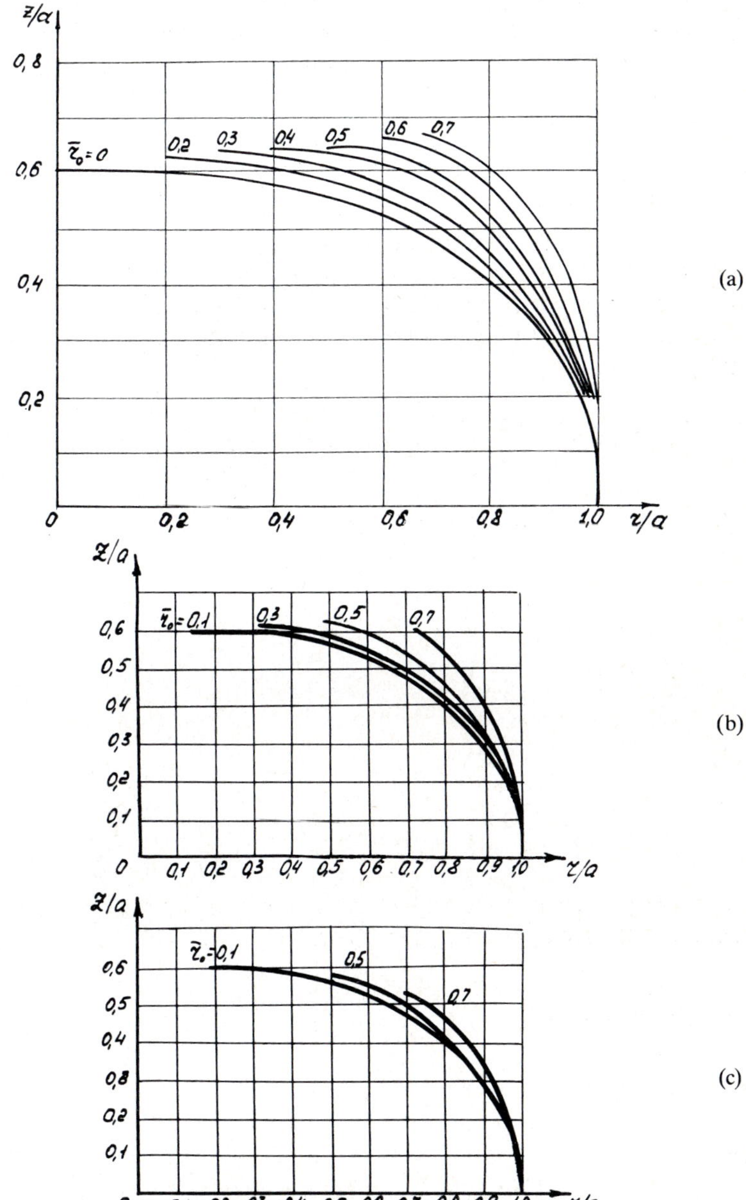

FIG. 5. (a) meridians of the shells with different radii of polar openings at $Q_0/pr_0 = 0.5$ (polar opening closed); (b) meridians at $Q_0/pr_0 = 0.4$; (c) meridians at $Q_0/pr_0 = 0.3$; (d) meridians at $Q_0/pr_0 = 0.2$; (e) meridians at $Q_0/pr_0 = 0.1$; (f) meridians at $Q_0/pr_0 = 0$ (polar opening open).

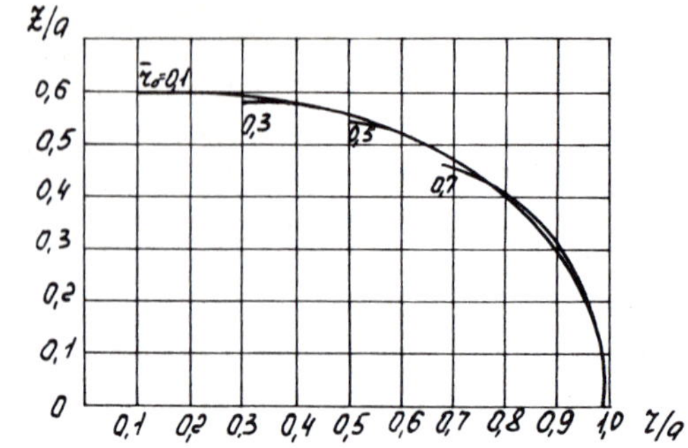

(d)

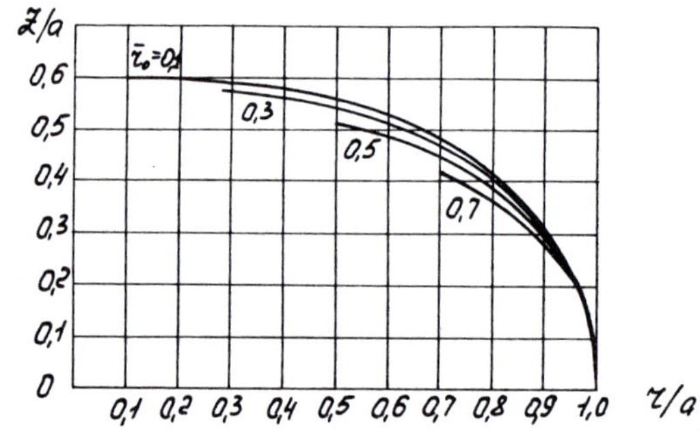

(e)

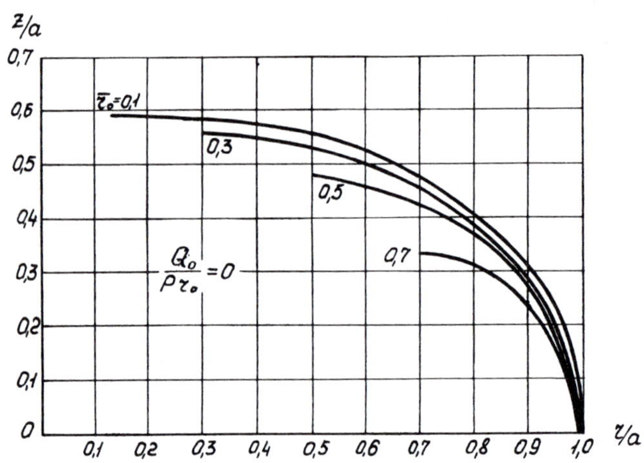

(f)

FIG. 5. *Continued.*

the shell Z_0, necessary for designing a pressure vessel, as a function of the parameters $\bar{r}_0$ and Q_0/pr_0.

Upon determination of the stress $\sigma_\parallel$ from (2.9) it is easy to see that the stresses are constant and equal, as one could have expected before,

$$\sigma_\parallel = \pi \frac{p(a^2 - r_0^2) + 2Q_0 r_0}{nf \cos \varphi_a}$$

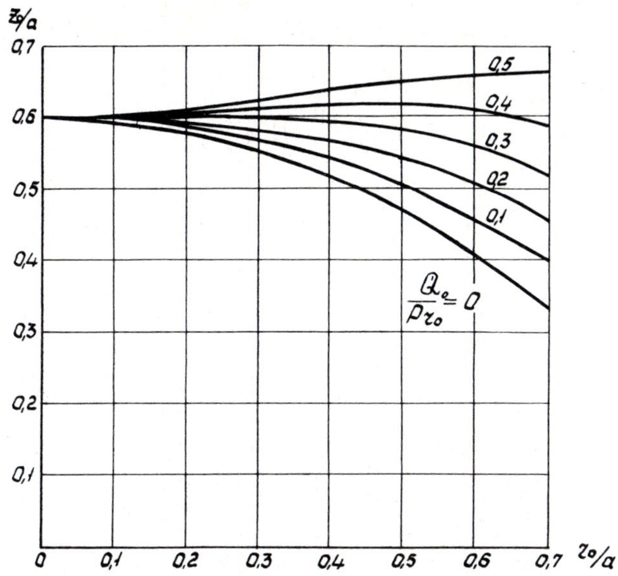

FIG. 6. Height of a shell versus radius of polar opening and load parameter Q_0/pr_0.

and the design thickness of the shell at the equator is

$$h_a = \frac{p(a^2 - r_0^2) + 2Q_0 r_0}{2a\sigma_{\|}^{u} \cos^2 \varphi_a} .$$

The inner volume of the shell is given by

$$V = \pi \int_{Z_2(r_0)}^{Z_1(a)} r^2 \, \mathrm{d}Z = \pi\left(\int_{r_0}^{b} r^2 Z_2' \, \mathrm{d}r \int_{b}^{a} r^2 Z_1' \, \mathrm{d}r\right) .$$

Substituting Z_1' and Z_2' in this equation according to (2.33) and (2.35), we finally have

$$\begin{aligned} V = \frac{\pi a^3}{3} \Big\{ & [(1 - \bar{b}^2)(\bar{b}^2 - \bar{n}_1)(\bar{b}^2 - \bar{n}_2)]^{1/2} - [(\bar{b}^2 - \bar{r}_0^2)(\bar{m}_1 - \bar{b}^2)(\bar{m}_2 + \bar{b}^2)]^{1/2} \\ & + \frac{\bar{n}_2(\bar{n}_2 + \bar{t}_0^2) - \bar{n}_1}{(1 - \bar{n}_2)^{1/2}} F(k, \eta_1) + 3\bar{t}_0^2(1 - \bar{n}_2)^{1/2} E(k, \eta_1) \\ & + \frac{\bar{m}_1(\bar{m}_2 - \bar{r}_0^2)}{(\bar{m}_1 - \bar{m}_2)^{1/2}} [\mathbb{K}(l) - F(l, \lambda_1)] + 3\bar{r}_0^2(\bar{m}_1 + \bar{m}_2)^{1/2} [\mathbb{E}(l) - E(l, \lambda_1)] \Big\} , \end{aligned} \tag{2.36}$$

where $\mathbb{K}(l)$, $\mathbb{E}(l)$ are the complete elliptic integrals of the first and second kind.

Fig. 7 shows the calculations for the volume of domes by the numerical integration method. The mass of the shell, neglecting that of the flange, can be found as follows

$$G = \rho n f S_1 , \tag{2.37}$$

where

$$nf = \frac{\pi}{\sigma_{\parallel}^{u} \cos \varphi_a} [p(a^2 - r_0^2) + 2Q_0 r_0] ,$$

$$S_1 = \int_{S(r_0)}^{S(a)} \frac{dS}{\cos \varphi} = \int_{r_0}^{b} [1 + (Z_2')^2]^{1/2} \frac{dr}{\cos \varphi} + \int_{b}^{a} [1 + (Z_1')^2]^{1/2} \frac{dr}{\cos \varphi} .$$

The results of numerical integration according to (2.37) are shown in Fig. 8.

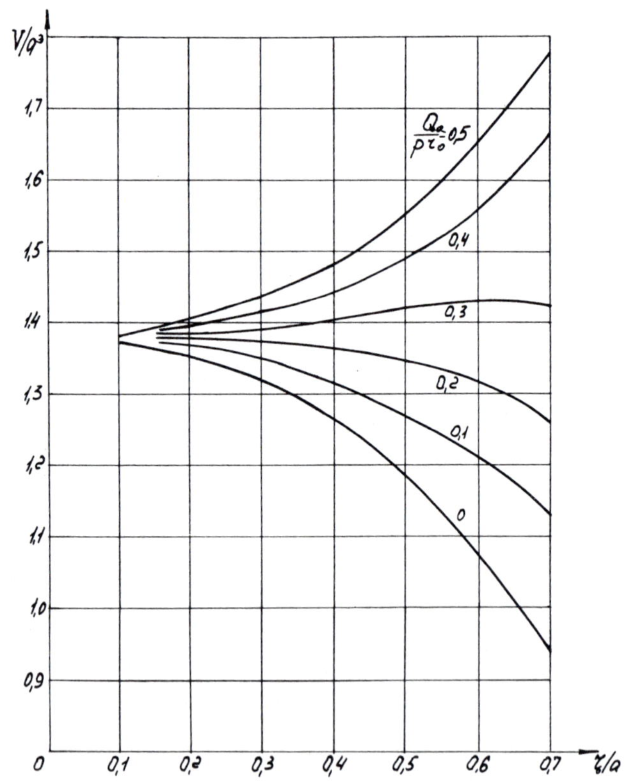

FIG. 7. Inner volume of a shell formed by geodesic winding versus radius of polar opening r_0 and load and parameter Q_0/pr_0.

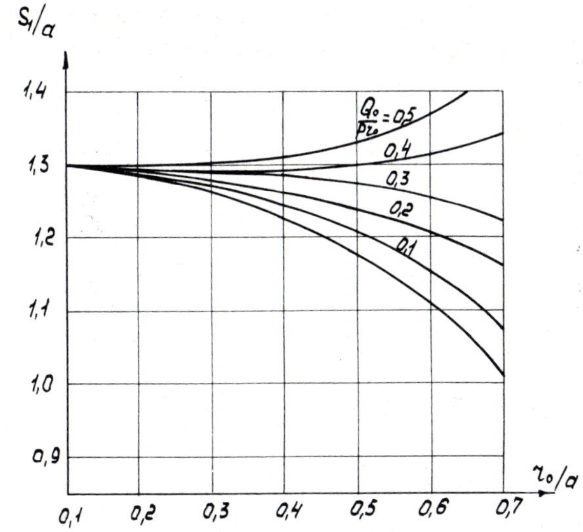

FIG. 8. Length of a filament for the shell versus radius of polar opening r_0 and load parameter Q_0/pr_0.

One can modify (2.37) to obtain

$$G = \frac{pV\rho}{\sigma_{\parallel}^{u}}(1 - \bar{r}_0^2)k_f ,$$

where

$$k_f = \frac{\pi a^3 S_1}{V(a^2 - r_0^2)^{1/2}} .$$

The dimensionless parameter (the shape factor) k_f is given in Fig. 9. As can be seen from the curve, $k_f = 3$ only when the radius of the polar opening is small.

One of the features of a construction wound with a single family of filaments is that the shell at the flange gets much thicker with the increase in internal pressure. In this case, the initial relationships of the membrane shell theory are no longer valid. Furthermore, in some cases the form of the meridian has to be varied with the objective of bringing it closer to the preassigned one. To solve the problem, one has to design a uniformly stressed shell of revolution consisting of a number of families of filaments, each family being defined by the angle of winding φ_{ai} and the thickness h_{ai} at the equator and terminating at the radius $r_{0i} = a \sin \varphi_{ai}$.

The latter equation has resulted from the winding being done along geodesic lines of the surface – a condition, which, as has been shown before, makes a uniformly stressed system feasible. With the increase of the angle φ_{ai} from one

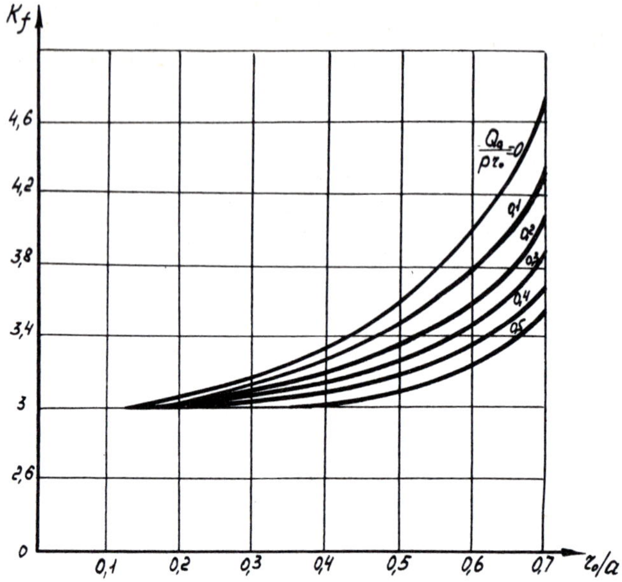

FIG. 9. Shape factor of a shell made by geodesic winding versus radius of polar opening and load parameter Q_0/pr_0.

family to another a system of layers is formed (Fig. 10). Winding of this kind does not enable us, as will be shown later, to obtain a shell of any preassigned shape, but offers much wider possibilities than the single system of fibres we have investigated before.

Now consider the problem of constructing the generator of the uniformly stressed shell shown in Fig. 10. For the sake of simplicity, let us imagine that

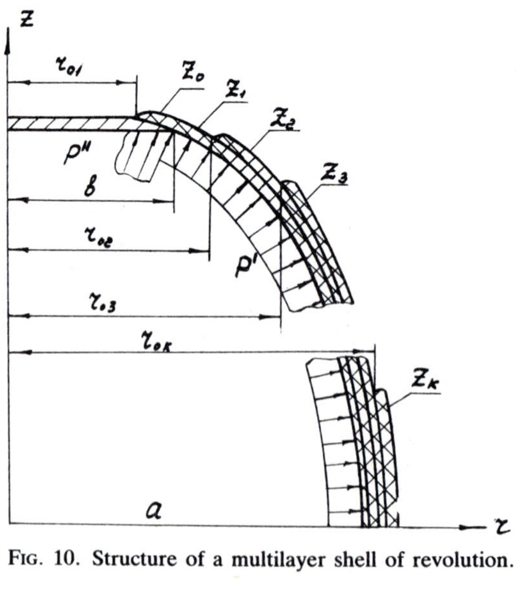

FIG. 10. Structure of a multilayer shell of revolution.

the polar opening is closed with a rigid cover. Suppose that the shell above the first layer is composed of two parts, the first contacting the flange and loaded with pressure p'' (2.30) and the second loaded with internal pressure p (2.29). In accordance with (2.26) by putting $Q_0 = 0$, $p'' = pb^2/(b^2 - r_{01}^2)$, $r_{0j} = r_{01}$. We have for the first part

$$Z_0' = -\frac{r(r^2 - r_{01}^2)^{1/2}}{[C_0^2 - r^2(r^2 - r_{01}^2)]^{1/2}},$$

where

$$C_0 = \frac{\sigma_{\|}^{u} n_1 f}{\pi p} \frac{b^2 - r_{01}^2}{b^2}; \quad r \sin \varphi_0 = r_{01}.$$

Analogously, we have for the second part

$$Z_1' = -\frac{r^3}{[C_1^2 - r^2(r^2 - r_{01}^2)]^{1/2}},$$

where

$$C_1 = \frac{\sigma_{\|}^{u} n_1 f}{\pi p}; \quad r \sin \varphi_1 = r_{01}.$$

For an arbitrary part of the shell, which consists of j layers and is located between parallels $r = r_{0j}$, $r = r_{0,j+1}$ we have

$$Z_j' = -\frac{r^3}{\left\{\left[\sum_{i=1}^{j} C_i(r^2 - r_{01}^2)^{1/2}\right]^2 - r^6\right\}^{1/2}},$$

where

$$C_i = \frac{\sigma_{\|}^{u} n_i f}{\pi p}; \quad r \sin \varphi_i = r_{0i}.$$

It can be directly verified that the conditions of conjugation for all shell parts are identically satisfied. Since the shells under consideration will be treated as domes of a pressure vessel, we can require that the tangential to the generator at the equator be parallel to the axis of the shell, i.e. $r = a$, $1/Z_k' = 0$. Equating the denominator in the equation for the meridian in the section $j = k$ to zero, we have

$$\frac{\sigma_{\|}^{u}}{\pi p} \sum_{i=1}^{k} n_i f(a^2 - r_{0i}^2)^{1/2} = a^3.$$

This latter equation is the condition for strength at the equator and can be written as follows

$$\sum_{i=1}^{k} n_i f \cos \varphi_{ai} = \frac{\pi p a^2}{\sigma_{\parallel}^{u}} . \tag{2.38}$$

The number of filaments in each layer of the shell can be selected arbitrarily within certain limits. The limits for n_j are found by selecting the number of filaments in all the preceding $j-1$ layers, provided that, on the one hand, the ultimate pressure is taken up only by the filaments of the jth layer

$$(n_j f)^{\max} = \frac{a}{(a^2 - r_{0j}^2)^{1/2}} \left[\frac{\pi p a^2}{\sigma_{\parallel}^{u}} - \sum_{i=1}^{j-1} n_i f \left(1 - \frac{r_{0i}^2}{a^2}\right)^{1/2} \right]$$

and, on the other hand, in the section of the $j-1$ layer the shell degenerates into a cylinder

$$(n_j f)^{\min} = \frac{1}{(r_{0,j+1}^2 - r_{0j}^2)^{1/2}} \left(\frac{\pi p r_{0,j+1}^3}{\sigma_{\parallel}^{u}} - \sum_{i=1}^{j-1} n_i f (r_{0,j+1}^2 - r_{0i}^2)^{1/2} \right) .$$

It is important to note that, first, the number of filaments forming each layer must be an integer and, second, the number of filaments for one layer, say, the last one, cannot be chosen arbitrarily, but must be found unambiguously from (2.38).

Fig. 11 shows how the contour of the shell changes with the number of its layers. It is easy to see that one can control the form of the shell over a fairly wide range by properly selecting the number of layers, their arrangement and the number of filaments in them. This allows one to pose the problem of constructing the shell of preassigned shape (RIVLIN and PIPKIN, 1963; CHEREVATSKII, 1970; OBRAZTSOV et al., 1977). Theoretically, the solution of such a problem calls for the assumption that there is a continuous relationship between thickness and fibre orientation angle, that is, the number of layers is infinite.

Consider the problem of optimum design for a shell of revolution of preassigned shape. For simplicity, we assume that the polar opening at the top of the vessel is closed. Putting $\sigma_{\parallel} = \text{constant} = \sigma_{\parallel}^{u}$, we can find from (2.7)

$$\begin{aligned} N_1 &= 2 \sum_{i=1}^{\infty} N_{1i}^{\pm} = \sigma_{\parallel}^{u} \int \cos^2 \varphi \, dh , \\ N_2 &= 2 \sum_{i=1}^{\infty} N_{2i}^{\pm} = \sigma_{\parallel}^{u} \int \sin^2 \varphi \, dh . \end{aligned} \tag{2.39}$$

For a closed shell loaded with internal pressure, we can find an expression for the overall stress resultants from the system (2.2). Without indicating the

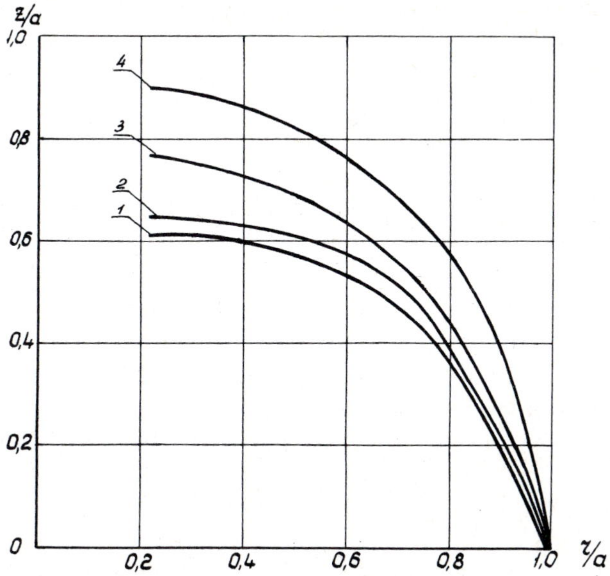

FIG. 11. Meridians of a shell with different numbers of layers k and angles of orientation: 1, $k = 1$, $\varphi_{a1} = 17°$; 2, $k = 2$, $\varphi_{a1} = 17°$, $\varphi_{a2} = 30°$; 3, $k = 3$, $\varphi_{a1} = 17°$, $\varphi_{a2} = 30°$; $\varphi_{a3} = 45°$; 4, $k = 5$, $\varphi_{a1} = 17°$; $\varphi_{a2} = 30°$, $\varphi_{a3} = 45°$; $\varphi_{a4} = 60°$, $\varphi_{a5} = 75°$.

intermediate calculations, we write the final result as

$$N_1 = \frac{pR_2}{2}\,; \quad N_2 = \frac{pR_2}{2}\left(2 - \frac{R_2}{R_1}\right), \tag{2.40}$$

where R_1 and R_2 are the principal radii of curvature of the surface.

As before, we assume that the shell is wound along geodesic lines, that is, the relationships (2.27) and (2.28) are valid

$$r \sin \varphi = a \sin \varphi_a = r_0\,, \tag{2.41}$$

where, as before, a is the radius of the equatorial cross section; $\varphi_a = \varphi\ (r = a)$; $r_0(\varphi_a)$ is the radius of the parallel to which the filaments with the initial angle are tangential.

Variation of the shell thickness can be obtained with the help of (2.6) and (2.41). If the thickness $\mathrm{d}h_a$ is taken to correspond to the angle φ_a in the equatorial cross section and the thickness $\mathrm{d}h$ – in any arbitrary cross section, then, taking into account (2.41), we find from (2.6) that

$$\mathrm{d}h = \frac{\mathrm{d}h_a \cos \varphi_a}{\left(\dfrac{r^2}{a^2} - \sin^2 \varphi_a\right)^{1/2}} \tag{2.42}$$

and the total thickness of the shell

$$h = \int \mathrm{d}h = \int_{\varphi_{a_1}}^{\arcsin r/a} \frac{\cos \varphi_a}{\left(\frac{r^2}{a^2} - \sin^2 \varphi_a\right)^{1/2}} \frac{\mathrm{d}h_a}{\mathrm{d}\varphi_a} \mathrm{d}\varphi_a , \tag{2.43}$$

where φ_{a_1} is the minimum angle of reinforcement at the equator which, according to (2.41), determines the radius of the opening at the top of the shell.

When we derived (2.43) we assumed that the thickness h_a is continuously dependent on φ_a and the points at which filaments are tangential to the parallels of radius r_0 (Fig. 2), which, according to formula (2.41), give the upper limit of the integral, are distributed uniformly along the meridian. The total thickness of the shell can be written by means of (2.39) and (2.40) as

$$h = \frac{N_1 + N_2}{\sigma_{\|}^{\mathrm{u}}} = \frac{pR_2}{2\sigma_{\|}^{\mathrm{u}}}\left(3 - \frac{R_2}{R_1}\right) \tag{2.44}$$

Thus we find from formulae (2.43) and (2.44)

$$\frac{pR_2}{2\sigma_{\|}^{\mathrm{u}}}\left(3 - \frac{R_2}{R_1}\right) = \int_{\varphi_{a_1}}^{\arcsin r/a} \frac{\cos \varphi_a}{\left(\frac{r^2}{a^2} - \sin^2 \varphi_a\right)^{1/2}} \frac{\mathrm{d}h_a}{\mathrm{d}\varphi_a} \mathrm{d}\varphi_a . \tag{2.45}$$

The integral equation (2.45) gives the law of distribution of the material with angle φ_a in a shell having the principal main radii of curvature R_1 and R_2 dependent on the form of the meridian as defined by equality $Z = Z(r)$.

Let us consider (2.45). Setting $\sin^2 \varphi_a = x$, $r/a = \bar{r}$, we get

$$F(\bar{r}^2) = \int_C^{r^2/a^2} \frac{(1 - x)^{1/2}}{(\bar{r}^2 - x)^{1/2}} \frac{\mathrm{d}h_a}{\mathrm{d}x} \mathrm{d}x ,$$

where

$$F(\bar{r}^2) = \frac{pR_2}{2\sigma_{\|}^{\mathrm{u}}}\left(3 - \frac{R_2}{R_1}\right); \quad C = \sin^2 \varphi_{a_1} .$$

By putting $\bar{r}^2 = t$, the latter equation can be reduced to the following form

$$F(t) = \int_C^t \frac{f(x)\,\mathrm{d}x}{(t - x)^{1/2}} ,$$

where

$$f(x) = (1 - x)^{1/2} \frac{\mathrm{d}h_a}{\mathrm{d}x} .$$

This integral equation is the Abel equation and its solution is

$$f(x)=\frac{1}{\pi}\frac{\mathrm{d}}{\mathrm{d}x}\int\limits_{C}^{x}F(\alpha)\,\frac{\mathrm{d}\alpha}{(x-\alpha)^{1/2}}\,.$$

Returning, finally, to the variables φ_a and $\bar{r}$, we have

$$\frac{\mathrm{d}h_a}{\mathrm{d}\varphi_a}=\frac{2p}{\pi\sigma_{\parallel}^{\mathrm{u}}\cos\varphi_a}\,\frac{\mathrm{d}}{\mathrm{d}\varphi_a}\int\limits_{\sin\varphi_{a_1}}^{\sin\varphi_a}\frac{R_2(\bar{r})\left(3-\dfrac{R_2(\bar{r})}{R_1(\bar{r})}\right)\bar{r}\,\mathrm{d}\bar{r}}{(\sin^2\varphi_a-\bar{r}^2)^{1/2}}\,. \tag{2.46}$$

Consider two examples to illustrate the solution we have arrived at.

For a spherical shell $R_1=R_2=a$ the solution (2.46) takes the form

$$\frac{\mathrm{d}h_a}{\mathrm{d}\varphi_a}=\frac{2pa}{\pi\sigma_{\parallel}^{\mathrm{u}}}\,\frac{\sin\varphi_a}{(\sin^2\varphi_a-\sin^2\varphi_{a_1})^{1/2}}\,.$$

Hence,

$$h_a=\int\limits_{\varphi_{a_1}}^{\varphi_a}\frac{\mathrm{d}h_a}{\mathrm{d}\varphi_a}\,\mathrm{d}\varphi_a=\frac{2p_a}{\pi\sigma_{\parallel}^{\mathrm{u}}}\left(\frac{\pi}{2}-\arcsin\frac{\cos\varphi_a}{\cos\varphi_{a_1}}\right). \tag{2.47}$$

The latter equation can be written to define the law of change of the number of filaments versus the angle φ_a at the equator as follows

$$nf=\frac{4pa^2}{\sigma_{\parallel}^{\mathrm{u}}}\cos\varphi_a\left(\frac{\pi}{2}-\arcsin\frac{\cos\varphi_a}{\cos\varphi_{a_1}}\right). \tag{2.48}$$

Since h_a corresponds to the total thickness of layers with angles from φ_{a_1} to φ_a and φ_a changes continuously from φ_{a_1} to $\pi/2$, then we must put $\varphi_a=\pi/2$ in (2.47) in order to find out the total thickness of the shell. Then

$$h_a^{\Sigma}=\frac{pa}{\sigma_{\parallel}^{\mathrm{u}}}\,,$$

which, as may be expected, requires that the thickness of the pressure shell of a spherical vessel should be constant.

Now consider a conical shell having a radius at the base a and the angle of the vertex 2β. The radii of curvature of the surface are $R_1\to\infty$; $R_2=r/\cos\beta$. Now write the solution (2.46) in the following fashion

$$\frac{\mathrm{d}h_a}{\mathrm{d}\varphi_a}=\frac{3pa}{2\pi\sigma_{\parallel}^{\mathrm{u}}\cos\beta\cos\varphi_a}\,\frac{\mathrm{d}}{\mathrm{d}\varphi_a}\left[\sin\varphi_{a_1}(\sin^2\varphi_a-\sin^2\varphi_{a_1})^{1/2}\right.$$
$$\left.+\sin^2\varphi_a\left(\frac{\pi}{2}-\arcsin\frac{\sin\varphi_{a_1}}{\sin\varphi_a}\right)\right]. \tag{2.49}$$

With $\varphi_{a_1} = 0$ (2.49) takes the form

$$\frac{\mathrm{d}h_a}{\mathrm{d}\varphi_a} = \frac{3pa \sin \varphi_a}{2\sigma_{\parallel}^{\mathrm{u}} \cos \beta}$$

and

$$h_a = \frac{3pa}{2\sigma_{\parallel}^{\mathrm{u}} \cos \beta} (1 - \cos \varphi_a)$$

or

$$nf = \frac{3\pi pa^2}{\sigma_{\parallel}^{\mathrm{u}} \cos \beta} \cos \varphi_a (1 - \cos \varphi_a)$$

the total thickness of the shell being given by

$$h_a^{\Sigma} = \frac{3p_a}{2\sigma_{\parallel}^{\mathrm{u}} \cos \beta} \, .$$

It should be noted that one can hardly have the relationship $h_a(\varphi_a)$ continuous when implementing the solutions in practice because the thickness of an elementary ply is finite. In order to find out the optimum structure of the material in a shell of preassigned shape it seems reasonable to preset the number of layers likely to be realized technologically, divide the range of variation of angle φ_a into the respective number of sections and regard the angle φ_a in each of the sections as equal to some average value. The average values of φ_a can be chosen by distributing the layers uniformly throughout the length of the meridian, so that the number of filaments in each layer might be given by $nf = nf(\varphi_a)$. The shell is thus formed with a finite number of layers and its equilibrium shape is defined by the above equations. Of course, it will differ from the preassigned shape, but will get closer to it as the number of layers increases. Also, it is to be noted that when the number of layers increases with the thickness of the elementary layer being finite, this causes the total thickness of the shell to increase. It will therefore be reasonable to use the obtained results in the design of high pressure vessels.

The above solutions will be sufficient in almost all cases to obtain uniformly stressed pressure vessels as minimum-mass constructions, as far as the netting model of the material goes. However, there are some cases when one has to make do with a construction which fails to meet this requirement, though it might possess some advantages as far as technological or structural considerations are concerned.

We may refer to one such technique – the planar winding (HARTUNG, 1963) remarkable for its adaptability to manufacture and simplicity in practical application. In this technique, the spiral circuit lies in one plane, forming an angle γ to the axis of revolution Z (Fig. 12). To define the fibre orientation let

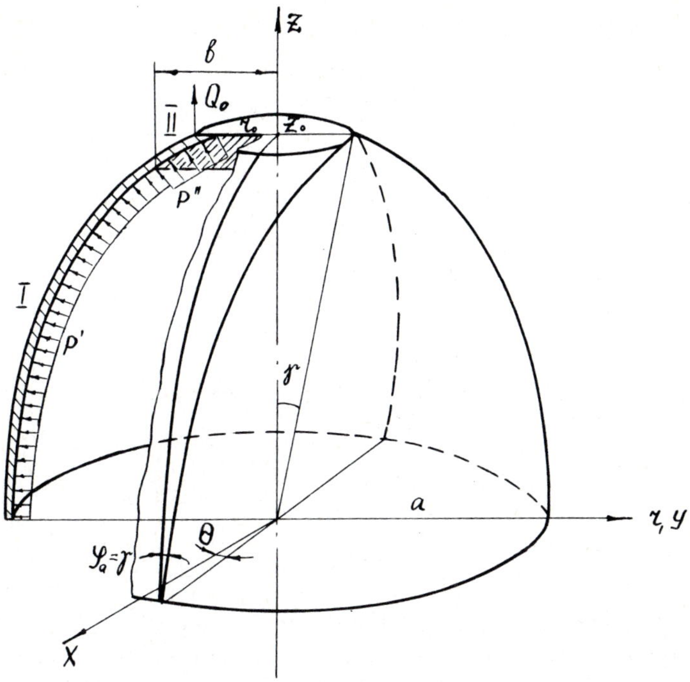

FIG. 12. A pressure vessel fabricated by a planar winding method.

us consider the obvious relationships (Fig. 12)

$$x = r\cos\theta\,, \quad y = r\sin\theta\,, \quad Z = y\,\mathrm{ctg}\,\gamma\,.$$

Supplementing these with the differential equations (Fig. 12)

$$\frac{r\,\mathrm{d}\theta}{\mathrm{d}S} = \frac{r}{[1+(Z')^2]^{1/2}}\,; \quad \frac{\mathrm{d}\theta}{\mathrm{d}r} = \mathrm{tg}\,\varphi$$

after simple transformations we shall have

$$\mathrm{tg}\,\varphi = \frac{rZ' - Z}{[1+(Z')^2]^{1/2}(r^2\,\mathrm{ctg}^2\,\gamma - Z^2)^{1/2}}\,. \tag{2.50}$$

Eq. (2.8) defining the equilibrium shape of the shell subjected to uniform internal pressure and axial force can be presented as

$$\frac{Z''}{Z'[1+(Z')^2]} = \frac{2r}{r^2 - t_0^2} - \frac{1}{r}\,\mathrm{tg}^2\,\varphi\,, \tag{2.51}$$

where the parameter t_a^2 is obtained from

$$t_0^2 = r_0^2 - \frac{2Q_0}{p} r_0$$

and, in accordance with (2.29) and (2.30), takes the values $(t_0')^2 = 0$ in the first section and $(t_0'')^2 = r_0^2$ in the second section. The equilibrium shape of the meridian is finally given by the integrated equation

$$\frac{Z''}{Z'[1 + (Z')^2]} = \frac{2r}{r^2 - t_0^2} - \frac{1}{r} \cdot \frac{(rZ' - Z)^2}{(r^2 \operatorname{ctg}^2 \gamma - Z^2)[1 + (Z')^2]} \tag{2.52}$$

The optimal generator is constructed by numerical integration of (2.52) with the initial conditions in the first section

$$r = a\,, \quad 1/Z_1' = 0\,, \quad Z_1 = 0\,, \quad \operatorname{tg} \varphi = \operatorname{tg} \gamma$$

and in the second section

$$r = b\,, \quad Z_2' = Z_1'\,, \quad Z_2 = Z_1\,.$$

Note that the integral for $Z(r)$ is improper at $r = a$, although its convergence can be proven. Also, to find the tangent of inclination angle γ for a shell with a given radius of the polar opening it is necessary to employ the successive approximation method to solve the equation

$$\operatorname{ctg} \gamma = Z_0(\gamma)/r_0\,,$$

where $Z_0 = Z(r_0)$. The technological feasibility condition is controlled by (2.16) and can be written as follows

$$\left| \operatorname{tg} \theta = \left(\frac{[1 + (Z')^2](r^2 + Z^2) - (rZ' - Z)^2}{[1 + (Z')^2](r^2 \operatorname{ctg}^2 \gamma - Z^2) + (rZ' - Z)^2} \right)^{1/2} \right| \leqslant \kappa\,.$$

The stress in the tape, according to (2.9), is given by

$$\sigma_{\|} = -\frac{\pi p r^2}{nf} \frac{[1 + (Z')^2]^{1/2}}{Z' \cos \varphi}\,.$$

The calculated results for a pressure vessel with the parameters $r_0/a = 0.25$ and $tg\,\gamma = 0.4$ are given in Fig. 13. Fig. 14 gives the values of the angle of geodesic deviation at different points on the meridian and Fig. 15 the variation of stresses in the tape. The respective parameters for a uniformly stressed shell formed by geodesic winding are also presented in these figures. Note, however, that the differences depend largely on the value of the radius of polar opening

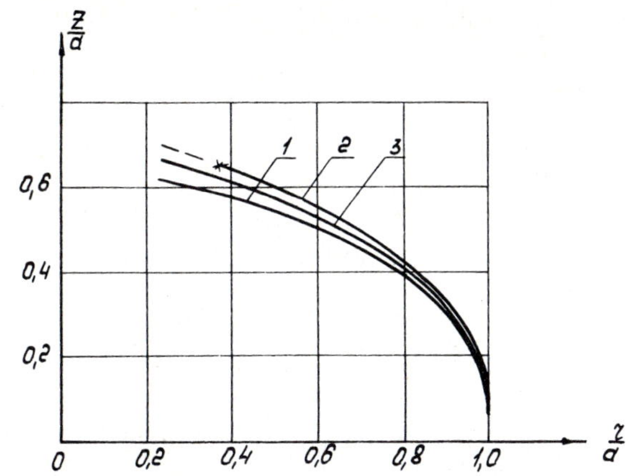

FIG. 13. Meridians of a pressure vessel: (1) planar winding tg $\gamma = 0.4$; $r_0/a = 0.25$; (2) winding along constant deviation lines: tg $\theta = 0.2$, sin $\varphi_a = 0.4$, $r_0/a = 0.25$ (A change-over point to geodesic winding), (3) geodesic winding $r_0/a = 0.25$.

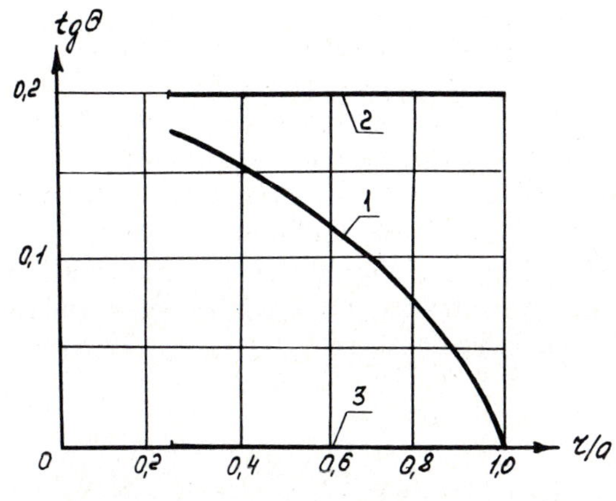

FIG. 14. Tangent of the geodesic deviation angle: (1) planar winding; (2) winding along constant deviation lines tg $\theta = 0.2$; (3) geodesic winding.

and become insignificant at $r_0/a < 0.1$. That is, at the small relative radii of the polar opening the pressure vessel formed by planar winding retains its technological advantages and is indeed as effective as a shell of uniform strength.

Another possible way of obtaining a pressure vessel is by winding along the lines of constant deviation from the geodesic lines (MITKEVICH and PROTASOV, 1975). Such a winding technique enables extreme technologically feasible fibre orientations to be achieved, i.e. at the prescribed friction coefficient it allows the maximum deviation from the pattern suitable for geodesic winding. According to (2.12), the equation for the line of constant deviation from a geodesic one is

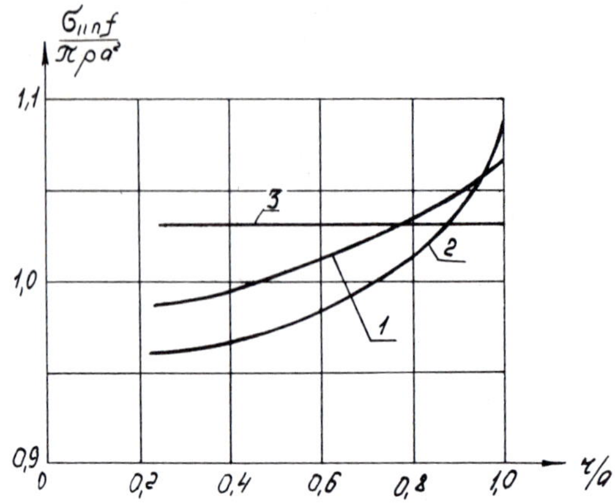

FIG. 15. Stresses in the tape: (1) planar winding $\varphi = 22°$, tg $\gamma = 0.4$, $r_0/a = 0.25$; (2) winding along constant deviation lines: $\varphi_a = 23.5$, tg $\theta = 0.2$, $r_0/a = 0.25$; (3) geodesic winding $\varphi_a = 14.5$, $r_0/a = 0.25$.

$$\operatorname{tg}\theta = \pm \frac{1}{Z'} \frac{r^2\xi'}{\dfrac{rZ''}{Z'[1+(Z')^2]}(r^2-\xi^2)+\xi^2} . \tag{2.53}$$

Here, the Clairaut function has been introduced, which at $\xi =$ constant expresses the well known equation for the geodesic line on the surface of revolution. On the other hand, the equilibrium shape of a pressure vessel must satisfy (2.51). Transforming (2.53) and taking into account (2.51), we finally procure a system of equations which makes it possible to define the meridian of a vessel and find the fibre orientation on its surface. The system can be written as follows

$$\operatorname{tg}\theta = \pm \frac{1}{Z'} \frac{\xi'(r^2-t_0^2)}{2(r^2-\xi^2)} ,$$
$$\frac{Z''}{Z'[1+(Z')^2]} = \frac{2r}{r^2-t_0^2} - \frac{1}{r}\frac{\xi^2}{r^2-\xi^2} . \tag{2.54}$$

To make the numerical realization more convenient we can introduce the arc equatorial coordinate S_2 and the meridian length S as given by the relations (Fig. 3)

$$S_2 = a\theta = a\int_a^r \frac{\xi}{r}\left(\frac{1+(Z')^2}{r^2-\xi^2}\right)^{1/2} \mathrm{d}r$$

$$S = \int_a^r [1+(Z')^2]^{1/2}\,\mathrm{d}r$$

Combining the latter expressions and adding the relationship (2.54), we finally obtain a governing system of equations for the meridian shape and coordinates of the fibres

$$\frac{d^2Z}{dr^2} = \frac{2rZ'[1+(Z')^2]}{r^2 - t_0^2} - \frac{Z'\xi^2[1+(Z')^2]}{r(r^2-\xi^2)} ,$$

$$\frac{d\xi}{dr} = \pm \frac{2Z'(r^2-\xi^2)\,\mathrm{tg}\,\theta}{r^2 - t_0^2} ,$$

$$\frac{dS_2}{dr} = \frac{a\xi}{r}\left(\frac{1+(Z')^2}{r^2-\xi^2}\right)^{1/2} , \tag{2.55}$$

$$\frac{dS}{dr} = [1+(Z')^2]^{1/2} .$$

The boundary conditions of the Cauchy problem are

$$r = a , \quad Z = 0 , \quad 1/Z' = 0 , \quad \xi = \xi_a , \quad S_2 = S = 0 .$$

As in the case of planar winding, the integral for the meridian shape is improper and the generator consists of two sections with $t_0' = 0$ in the first and $t_0'' = r_0$ in the second section. The sign of deviation from the geodesic line must be chosen depending on the value of preassigned angle of winding at the equator $\varphi = \varphi_a$. Thus, $\mathrm{tg}\,\theta > 0$ with $\varphi_a > \arcsin(r_0/a)$ and $\mathrm{tg}\,\theta < 0$ with $\varphi_a < \arcsin r_0/a$. Also, it may be necessary to pass over to geodesic winding in order to make the polar opening with a given radius $r = r_0$. The change-over point is given by

$$\xi = \xi_0 = r_0 .$$

Fig. 13 shows the form of the generator of a pressure vessel, while Fig. 15 displays the stress variation in a shell having the parameters $r_0/a = 0.25$; $\varphi_a = 23.5°$; $\xi/a = 0.4$.

The relations can be used in finding the fibre orientation and the meridian of a pressure vessel when geodesic winding is not applicable on some structural grounds.

The general problem of optimum design is to define the fibre orientation and the shape of the shell providing its minimum mass and satisfying the constraints expressed by (2.13)–(2.16). In this statement, the problem can be solved by a well developed technique – the optimum control theory. But there arise some essential difficulties such as, first, that both the minimizing functional and the imposed constraints are nonlinear and, second, the constraints are expressed in the form of inequalities. To realize the solution in practice, therefore, one can employ the method of nonlinear variable transformation.

So consider the optimum design of a pressure vessel in the case when there is no solution within the framework of uniformly stressed systems of fibres, that is $a \sin \varphi_a \neq r_0$ (MITKEVICH and PROTASOV, 1975). The analysis of stresses in the filaments of a shell wound along the lines of constant deviation allows us to assume that the stresses will be maximum on the equator and decrease monotonically toward the pole (Fig. 15). From (2.9) with $q = 0$ we find

$$\sigma_{\|} = -\frac{\pi[1+(Z')^2]^{1/2}}{nfZ' \cos \varphi} p(r^2 - t_0^2) . \tag{2.56}$$

The equation for an equilibrium meridian shape (2.8) can be written as

$$\left[\ln \frac{[1+(Z')^2]^{1/2}}{Z'}\right]' = \frac{2r}{r^2 - t_0^2} - \frac{1}{r} \operatorname{tg}^2 \varphi \tag{2.57}$$

and the no-slip condition for filament during winding (2.16), with (2.8) taken into account, as

$$\left| \operatorname{tg} \theta = \frac{1}{Z'} \frac{(r \sin \varphi)'(r^2 - t_0^2)}{2r^2 \cos^2 \varphi} \right| \leqslant \kappa . \tag{2.58}$$

The condition for monotonic stress decrease is equivalent to the condition $(\ln \sigma_{\|})' \leqslant 0$. Combining (2.56) and (2.57), we find that the assumption can be valid at

$$\frac{\sin \varphi (r \sin \varphi)'}{r \cos \varphi} \leqslant 0 .$$

For the problem under consideration the latter inequality will be satisfied if $(r \sin \varphi)' \leqslant 0$. Then, according to (2.58),

$$0 \leqslant \operatorname{tg} \theta \leqslant \kappa . \tag{2.59}$$

In this case, the condition for filament strength (2.13) is satisfied automatically. Note also that the strength restriction in interlaminar shear is insignificant, for the patterns satisfy the condition (2.59).

To state the problem in terms of optimum control, the following convention has been introduced:

$$x_1 = S_1 = \int_a^r \left(\frac{1+(Z')^2}{\cos^2 \varphi} \right)^{1/2} \mathrm{d}r$$

is the element of the filament length;

$$x_2 = r \sin \varphi = \xi$$

is the Clairaut function; $x_3 = \theta$ is the polar angle; $x_4 = r$ is an additional coordinate introduced because of the non-stationary constraints; $x_5 = Z'$, $u_1 = \theta'$, $u_2 = \xi'$. Here, x_n are coordinates of the state of the system, u_m are those of control.

The constraint (2.59) can be normally written as

$$|g(x_n, u_m)| \leq 1\,, \tag{2.60}$$

where

$$g = \frac{2\,\mathrm{tg}\,\theta}{\kappa} - 1 = \frac{1}{\kappa}\,\frac{u_2(x_4^2 - t_0^2)}{x_4^2 - x_3^2} - 1\,.$$

The inequality (2.60) can be transformed into an equality by introducing an additional control function v by means of nonlinear transformation

$$\sin v - g = 0\,.$$

Now we can write the system of equations characterizing the behavior of the system in the following normal form

$$\frac{\mathrm{d}x_n}{\mathrm{d}r} = f_n(x_n, u_m)$$

or, taking the adopted convention into account,

$$\frac{\mathrm{d}x_1}{\mathrm{d}r} = (1 + x_5^2 + x_4^2 u_1^2)^{1/2}\,,$$

$$\frac{\mathrm{d}x_2}{\mathrm{d}r} = u_2\,; \quad \frac{\mathrm{d}x_3}{\mathrm{d}r} = u_1\,, \quad \frac{\mathrm{d}x_4}{\mathrm{d}r} = 1\,,$$

$$\frac{\mathrm{d}x_5}{\mathrm{d}r} = x_5(1 + x_5^2)\left(\frac{2x_4}{x_4^2 - t_0^2} - \frac{1}{x_4}\,\frac{x_2^2}{x_4^2 - x_2^2}\right).$$

Using the Lagrangian multipliers, we can form an extended function

$$\Phi = -\sum_1^5 f_n \lambda_n + g\mu\,.$$

The variational equations, from which the additional control functions v have been eliminated, can be presented in the form

$$\frac{\partial \Phi}{\partial x_n} = \frac{\mathrm{d}\lambda_n}{\mathrm{d}r}\,, \quad \frac{\partial \Phi}{\partial u_m} = 0\,,$$

$$\mu(1 - g^2)^{1/2} = 0 \quad (n = 1, \ldots, 5;\ m = 1, 2)\,.$$

After transformation, we have

$$\left.\begin{aligned}
&\frac{d\lambda_1}{dr}=0\,,\quad \frac{d\lambda_3}{dr}=0\,,\quad \frac{d\lambda_2}{dr}=\lambda_5\,\frac{2x_2x_4x_5(1+x_5^2)}{(x_4^2-x_2^2)^2}+\frac{2\mu u_2x_2(x_4^2-t_0^2)}{\kappa(x_4^2-x_2^2)^2}\,,\\
&\frac{d\lambda_4}{dr}=-\frac{\lambda_1u_1^2x^4}{(1+x_5^2+x_4^2u_1^2)^{1/2}}+\lambda_5x_5(1+x_5^2)\left(\frac{2(x_4^2-t_0^2)}{(x_4^2-t_0^2)^2}-\frac{x_2^2(3x_4^2-x_2^2)}{x_4^2(x_4^2-x_2^2)}\right)\\
&\qquad -\frac{2\mu u_2x_4}{\kappa}\,\frac{x_2^2-t_0^2}{(x_4^2-x_2^2)^2}\,,\\
&\frac{d\lambda_5}{dr}=-\frac{\lambda_1x_5}{(1+x_5^2+x_4^2u_1^2)^{1/2}}-\lambda_5(1+3x_5^2)\left(\frac{2x_4}{x_4^2-t_0^2}-\frac{1}{x_4}\,\frac{x_2^2}{x_4^2-x_2^2}\right),\\
&\frac{-\lambda_1u_1x_4^2}{(1+x_5^2+x_4^2u_1^2)^{1/2}}-\lambda_3=0\,,\\
&-\lambda_2+\frac{\mu}{\kappa}\,\frac{x_4^2-t_0^2}{x_4^2-x_2^2}=0\,.
\end{aligned}\right\}\qquad (2.61)$$

As long as the minimizing functional is the filament length from the equator to the polar opening, we find $\lambda_1=-1$ from the transversality condition $\delta J\,(1+\lambda_1)=0$. It follows from the second equation of the system (2.61) that $\lambda_3=C$, then, taking into account the boundary condition, at $r=r_0$, $\varphi=\pi/2$ we have $\lambda_3=r_0$.

Now let us define the internal arc for the extremal by putting $\mu=0$. It follows directly from the system (2.61) that $\lambda_2=\lambda_5=0$

$$\frac{u_1x_4^2}{(1+x_5^2+x_4^2u_1^2)^{1/2}}=r_0=r\sin\varphi\,.$$

Thus, the internal arc of the extremal corresponds to the geodesic winding and it follows from the condition $\lambda_5=0$ that optimum control is independent of the shell shape. The boundary arcs correspond to $g=\pm1$. With $g=+1$ the equation for the winding pattern is the line of constant geodesic deviation, for which the fibre orientation and the meridian shape have been defined by numerical solution of the system (2.55). With $g=-1$ we have $u_2=0$ or $r\sin\varphi=$ constant, that is, the lower boundary arc coincides with the internal arc of the extremal. At the control switching point, according to the Weierstrass–Erdman condition, $\lambda_-=\lambda_+$. Consequently, passing over to the geodesic section of winding is defined by the equation $r^*\sin\varphi_*=r_0$.

Fig. 16 shows the fibre orientation in a pressure vessel having the following parameters: $r_0/a=0.3$; $\sin\varphi_a=0.5$; and the friction coefficient $\kappa=0.2$. The assumption, as before, is that the generator of the vessel consists of two sections, the corresponding values of the parameters t_0^2 being found from the conditions (2.29) and (2.30). For comparison, the geodesic line is also shown in the figure.

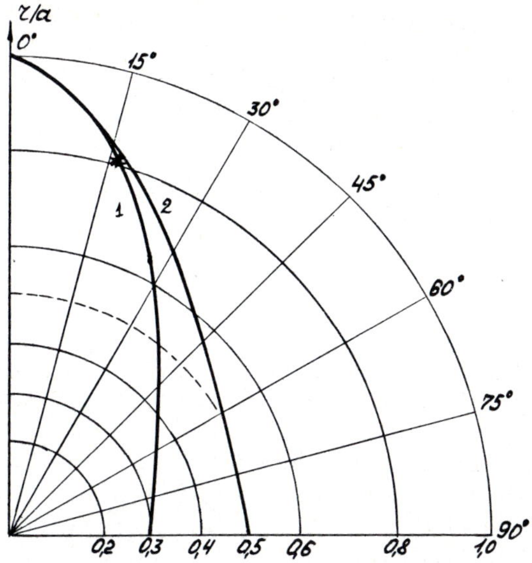

FIG. 16. Filament patterns for different winding methods: (1) combined winding $\varphi_a = 30°$, $\varphi_0 = 90°$, $\kappa = 0.2$, * switching point of optimum control function; (2) geodesic winding.

In a similar way we can solve the problem of determining the optimum material structure for a pressure vessel shaped as a conical shell with domes at each end; the construction and loading conditions are shown in Fig. 17. For conical vessels with relatively small semivertex angle it is reasonable to select a system consisting of spiral and hoop filaments, the parameters of which can be determined by a mathematical apparatus – the theory of optimum control.

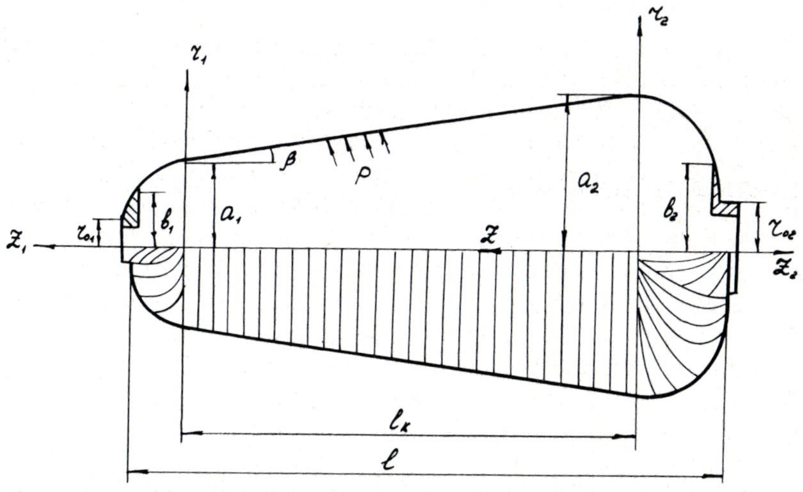

FIG. 17. A conical shell with domes.

Let us note some of the peculiarities of the construction. As the large dome is severely loaded, it seems reasonable to have it also uniformly stressed and define the shell shape in accordance with equations (2.33) and (2.36). Since the filament layup is geodesic, the angle on the equator at $r = a_2$ (Fig. 17) is determined unambiguously and is equal to $\varphi_{a_2} = \arcsin(r_{02}/a_2)$. Unambiguously, too, we can determine the number of filaments in the spiral layer as calculated from the formula (2.37)

$$nf = \frac{\pi}{\sigma_{\|}^{u} \cos \varphi_{a2}} \left[p(a_2^2 - r_{02}^2) + 2Q_{02} r_{02} \right] .$$

The requirement of winding continuity nf = constant makes the small dome essentially underloaded. The shape of this dome is found, therefore, from the conditions for technological feasibility of the winding process along the lines of ultimate geodesic deviation in accordance with the solution of (2.55) so that the point of passing over to geodesic winding is as close as possible to the bottom flange ($r^* = b$). This ensures the maximum possible angle of orientation on the equator of the small dome φ_{a1}, which defines, on the whole, the possibility of obtaining a rational conical part of the vessel.

Consider a conical shell with a semivertex angle β made by spiral and circumferential winding. Since the coefficients of the first quadratic form in this case are given by

$$A = 1/\sin \beta \quad B = r ,$$

and the main radii of curvature are

$$R_1 \to \infty \quad R_2 = r/\cos \beta ,$$

the equilibrium equations (2.1) for a spiral layer can be written as

$$\begin{aligned} &\frac{d}{dr}(rN_{1c}^{\pm}) - N_{2c}^{\pm} + \frac{rq_c}{2 \sin \beta} = 0 , \\ &\frac{d}{dr}(rN_{1,2c}^{\pm}) + N_{1,2c}^{\pm} + \frac{r\tau_c}{\sin \beta} = 0 , \\ &N_{2c}^{\pm} = rp_c/2 \cos \beta ; \end{aligned} \tag{2.62}$$

and for a circumferential layer as ($\varphi_k = \pi/2$)

$$\begin{aligned} &-N_{2k} + \frac{rq_k}{\sin \beta} = 0 , \\ &N_{2k} = \frac{rp_k}{\cos \beta} . \end{aligned} \tag{2.63}$$

The averaged stress resultants in the conical shell can be expressed in terms of stresses in the filaments of each of the layers with (2.7) taken into account

$$N_1 = \frac{pr}{2\cos\beta} = N_{1c}^{+} + N_{2c}^{-} = \sigma_{\|c}\frac{nf}{2\pi r}\cos\varphi\,,$$
$$N_2 = \frac{pr}{\cos\beta} = N_{2c}^{+} + N_{2c}^{-} + N_{2k} = \sigma_{\|c}\frac{nf}{2\pi r}\sin\varphi\,\mathrm{tg}\,\varphi + \sigma_{\|k}h_k\,, \tag{2.64}$$

where h_k is the variable thickness of the circumferential layer. By solving the system (2.62)–(2.64) we can obtain expressions for filament stresses and tangential stresses of interlayer interaction. These expressions can be written in the form of strength constraints for the spiral layer

$$\sigma_{\|c} = \frac{\pi p r^2}{nf\cos\beta}\cdot\frac{1}{\cos\varphi} \leqslant \sigma_{\|}^{u}\,;$$

and the circumferential layer

$$\sigma_{\|k} = \frac{pr}{2h_k\cos\beta}(2-\mathrm{tg}^2\,\varphi) \leqslant \sigma_{\|}^{u}\,; \tag{2.65}$$

as well as in the form of a strength constraint on the tangential interactions of spiral plies

$$\tau_c = \frac{p\,\mathrm{tg}\,\beta}{4}\left(3\,\mathrm{tg}\,\varphi + \frac{r}{\cos^2\varphi}\,\frac{\mathrm{d}\varphi}{\mathrm{d}r}\right) \leqslant \tau_{\|\perp}^{u}$$

and interactions between spiral and circumferential layers

$$\tau_k = \frac{p\,\mathrm{tg}\,\beta}{2}(2-\mathrm{tg}^2\,\varphi) \leqslant \tau_{\|\perp}^{u}\,.$$

There are some other restrictions, namely, the condition for the existence of the construction for which the thickness of the hoop layer is not negative

$$h_k = \frac{pr}{2\sigma_{\|}^{u}\cos\beta}(2-\mathrm{tg}^2\,\varphi) \geqslant 0\,, \tag{2.66}$$

as well as technological constraints, defining continuity of the winding process

$$nf = h_c 2\pi\cdot r\cos\varphi\,,$$

and the condition of no-slip filament during winding (2.67) of spiral and hoop layers

$$\left.\begin{aligned} |\mathrm{tg}\,\theta_c| &= \left|\frac{\mathrm{tg}\,\beta}{\sin\varphi}(1 + r\,\mathrm{ctg}\,\varphi)\right| \leqslant \kappa\,,\\ |\mathrm{tg}\,\theta_k| &= \mathrm{tg}\,\beta \leqslant \kappa\,. \end{aligned}\right\} \tag{2.67}$$

Taking the mass of the shell as the objective function, as before, we finally have

$$G = \frac{\rho n f}{\sin \beta} \int_{a_2}^{a_1} \frac{\mathrm{d}r}{\cos \varphi} + \frac{\rho \cdot 2\pi}{\sin \beta} \int_{a_2}^{a_1} h_k r \,\mathrm{d}r = \min . \tag{2.68}$$

It is possible to reduce the system of constraints if the circumferential layer is set in advance to be uniformly stressed and the strength limitations with interlayer interaction are neglected. By checking the optimum patterns it is easy to see that this is indeed so. The objective function (2.68) in this case can be transformed into

$$\begin{aligned} \bar{G} &= \frac{G \sin \beta \cos \beta \sigma_{\|}^{\mathrm{u}}}{\pi \rho p a_2^3} \\ &= \frac{1}{\cos \varphi_0} \int_{\bar{r}_0}^{\bar{r}_1} \frac{\mathrm{d}\bar{r}}{\cos \varphi} + \bar{r}_0^{-2} \int_{\bar{r}_0}^{\bar{r}_1} \bar{r}^2 (2 - \mathrm{tg}\, \varphi^2) \,\mathrm{d}r \end{aligned} \tag{2.69}$$

where

$$\varphi_0 = \varphi(r = a_2), \quad \bar{r}_0 = 1; \quad \bar{r}_1 = a_1/a_2, \quad r = r/a_2 .$$

Similarly, we can write the system of equations in terms of optimum control as follows

$$\frac{\mathrm{d}x_1}{\mathrm{d}t} = u, \quad \frac{\mathrm{d}x_2}{\mathrm{d}t} = 1; \quad \frac{\mathrm{d}x_3}{\mathrm{d}t} = \frac{1}{\cos \varphi_0} (1 + x_1^2)^{1/2} + \frac{x_2^2}{r_0^2} (2 - x_1^0) . \tag{2.70}$$

With the constraints written in the normal form

$$|g_l| \leqslant 1 \quad (l = 1, 2, 3), \tag{2.71}$$

where

$$g_1 = 2 \frac{\cos \varphi_0}{r_0^2} x_2^2 (1 + x_1^2)^{1/2} - 1 ,$$

$$g_2 = \frac{\mathrm{tg}\, \beta (1 + x_1^2)^{1/2}}{\kappa x_1} \left(1 + \frac{1}{x_1} \cdot \frac{x_2 u}{1 + x_1^2}\right),$$

$$g_3 = x_1/\sqrt{2} .$$

Here, as the system state coordinates we have taken $x_1 = \mathrm{tg}\, \varphi$, $x_2 = r$ is an additional coordinate introduced because of the non-stationary characteristics x_3 which is also an additional coordinate which makes it possible to reduce the

Lagrangian problem to the main variational problem ($\bar{G}_1 = x_3^1 - x_3^0$); and $t = r$. As the main control function we have taken $u = dx_1/dt$. The boundary conditions are $x_1^0 = \mathrm{tg}\ \varphi_0$, $x_1^1 = \mathrm{tg}\ \varphi_1$, $x_2^0 = r_0$ and $x_2^1 = r_1$.

By omitting the intermediate transformations, we obtain, analogous to (2.61), a set of extremals for the solution of the problem: for uniformly stressed fibres

$$\frac{\cos \varphi}{r^2} = \text{constant} \quad \text{or} \quad \cos \varphi = \frac{r^2}{r_0^2} \cos \varphi_0 ,$$

for winding along the lines of ultimate deviation from the geodesic

$$\sin \varphi = \frac{r_0 \sin \varphi_0}{r \pm (r_0 - r) \dfrac{\kappa}{\mathrm{tg}\ \beta} \sin \varphi_0} ,$$

for winding in the absence of circumferential reinforcement

$$\varphi = 54°44'(\mathrm{tg}^2\ \varphi = 2) .$$

Note that one has to solve a cumbersome system of equations in order to find the switching points according to the Weierstrass–Erdman conditions. However, this problem can be simplified by performing numerically the following simple algorithm. Since the sufficient condition for optimality depends on the trajectory being uniformly stressed, one must take the maximum advantage of the present winding condition. Therefore, all the constraints are checked at each step of the coordinate r, and if all of them are fulfilled, the winding is carried out according to the equal-strength condition. Should at least one of the restrictions happen to be unrealizable, the reinforcement scheme is defined by the condition of the respective boundary extremal. The process goes on until the second boundary condition is reached or the extremal, defining the condition of technological feasibility on change-over to the small dome, is crossed

$$\sin \varphi = \frac{r_1 \sin \varphi_1}{r \pm (r_1 - r) \dfrac{\kappa}{\mathrm{tg}\ \beta} \sin \varphi_1} .$$

Fig. 18 shows an example of the optimum reinforcement of a conical shell having the following parameters: $\beta = 5°$, $\kappa = 0.2$, $\bar{l} = l/a_2 = 3.9$, $\bar{r}_0 = 1$, $\bar{r}_1 = 0.686$, $\varphi_0 = 37°$, $\varphi_1 = 43°$. The law of reinforcement along the line of constant deviation from the geodesic one is also shown by a dotted line.

Of special interest are combined-construction pressure vessels, which consist of a metal shell reinforced by means of a wound unidirectional composite tape. One reason why vessels of this kind emerged is that fully composite construc-

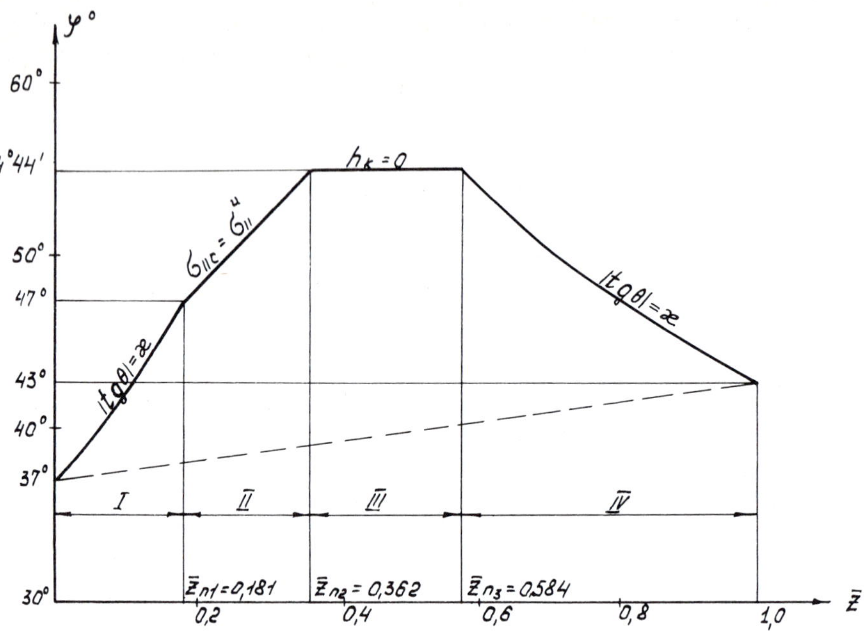

FIG. 18. Variation of optimal angle of orientation along the length of the shell (h_k min = 0).

tions, when subjected to prolonged storage, usually fail as airtight vessels. On the other hand, the use of high pressures calls for a great thickness of the load-carrying shells, so that the composite material has to work under three-dimensional stress and, as practice shows, is much less effective than a fully metal vessel. However, there is a range of pressures in which the mass of a vessel can be reduced by reinforcing the metal shell with a layer of composite material and the thickness of this layer is such that the composite material can work under plane stress. Furthermore, the problem of airtightness for constructions of this kind can be solved in a simple way, while the airtight isotropic shell itself acts as a mandrel in the winding process to make the production cycle shorter.

Having obtained some experience in fabricating composite pressure vessels, we are now able to put forward two approaches for extremizing the load carrying capacity, depending on the properties of the material used. In the first approach, suggested for metals with low reinforcement content and a well pronounced yield area, we may suppose that even at the moment of failure the stresses remain equal to the yield value, that is, we can use the rigidly plastic model here (BUNAKOV et al., 1976). In the second approach, the real strain pattern of the material of the airtight layer is taken into account (KOLEROV, 1978).

As far as the adopted hypotheses go, we will assume that the stresses in all the filaments are constant and equal to the ultimate ones, i.e. $\sigma_{\|}(r) =$

constant $= \sigma_{\|}^{u}$, and the stress in the metal shell of thickness h_a is equal to the yield strength, i.e. $\sigma_1^0 = \sigma_2^0 = \sigma^y$, the stresses in the matrix being neglected, putting $\sigma_\perp = \tau_{\|\perp} = 0$. Then the equilibrium equation for a vessel having a polar opening closed with a rigid pressure cover can be written as follows

$$N_1 = \frac{pR_2}{2} = \sigma^y h_0 + \sigma_{\|}^{u} h_1 \cos^2 \varphi ,$$
$$N_2 = \frac{pR_2}{2}\left(2 - \frac{R_2}{R_1}\right) = \sigma_y h_0 + \sigma_{\|}^{u} h_1 \sin \varphi . \tag{2.72}$$

Taking into consideration that for a system of uniform strength the fibres must be placed along geodesic lines of the surface that satisfy the Clairaut theorem $r \sin \varphi = \text{constant} = r_0$ and accepting the law of thickness variation of the composite layer which follows from the condition of continuity $h_1 = h_a \cos \varphi_a / \cos \varphi$, where h_a and φ_a are the thickness and the angle of the fibres at the vessel equator, respectively, we proceed to transform (2.72).

Next, taking into account that the principal radii of curvature are expressed through the equation for the shell generator $Z(r)$, we finally have

$$N_1 = \frac{p}{2}\frac{r}{Z'}[1 + (Z')^2]^{1/2} = \sigma^y h_0 + \sigma_{\|}^{u} h_a \frac{(a^2 - r_0^2)^{1/2}}{a}\frac{(r^2 - r_0^2)^{1/2}}{r} ,$$
$$N_2 = \frac{p}{2}\frac{r}{Z'}[1 + (Z')^2]^{1/2}\left\{1 - \frac{rZ''}{Z'[1 + (Z')^2]}\right\} \tag{2.73}$$
$$= = \sigma^y h_0 + \sigma_{\|}^{u} h_a \frac{(a^2 - r_0^2)^{1/2}}{r_0^2}\frac{r^2}{(r^2 - r_0^2)^{1/2}} .$$

Dividing the second of equations (2.73) by the first and introducing the symbol

$$\lambda = \frac{\sigma^y h_0 a}{\sigma_{\|}^{u} h_a (a^2 - r_0^2)^{1/2}} .$$

we obtain, after a few modifications, an equation for the meridian shape of the optimum shell

$$\frac{Z''}{Z'[1 + (Z')^2]} = \frac{1}{r} + \frac{1}{r}\frac{(r^2 - r_0^2) - r_0^2}{\lambda \frac{r^2}{a}(r^2 - r_0^2)^{1/2} + (r^2 - r_0^2)} . \tag{2.74}$$

Integrating twice (2.74) and taking into account the boundary conditions $r = 0$, $Z = 0$; $1/Z' = 0$, we finally have for the meridian

$$\bar{Z} = -\int_1^{\bar{r}} \frac{\bar{r}^3 \, d\bar{r}}{\left(\left[\frac{\lambda \bar{r}^2 + (\bar{r}^2 - \bar{r}_0^2)^{1/2}}{\lambda + (1 - r_0^2)^{1/2}}\right]^2 - \bar{r}^6\right)^{1/2}} \tag{2.75}$$

where $\bar{Z} = Z/a$, $\bar{r} = r/a$ are dimensionless coordinates referred to the radius of the vessel equator.

The integral (2.75) is improper with $\bar{r} \to 1$, but its convergence is easy to prove. Similarly, for the equator region with $\bar{r} = 1 - \varepsilon$ we have

$$\bar{Z}(1 - \varepsilon) = (2\varepsilon)^{1/2} \left(\frac{(1 - \bar{r}_0^2)^{1/2} [\lambda + (1 - \bar{r}_0^2)^{1/2}]}{\lambda (1 - \bar{r}_0^2)^{1/2} + 2 - 3\bar{r}_0^2} \right)^{1/2} .$$

Let us note some particular forms in which the solution can be written. With $\lambda = 0$, which corresponds to a shell without a metal layer, the form of the generator is given by

$$\bar{Z} = -\int_1^{\bar{r}} \frac{\bar{r}^3 \, d\bar{r}}{\left(\frac{\bar{r}^2 - \bar{r}_0^2}{1 - \bar{r}_0^2} - \bar{r}^6\right)^{1/2}} .$$

A similar result has been obtained in the form of (2.32). With $\lambda \to \infty$, which corresponds to a fully metal vessel, the integral (2.75) is

$$\bar{Z} = -\int_1^{\bar{r}} \frac{\bar{r} \, d\bar{r}}{(1 - \bar{r}^2)^{1/2}}$$

and defines a spherical shell. Fig. 19 shows a series of contours of optimum pressure vessels for different parameters λ with $\varphi = 5°$.

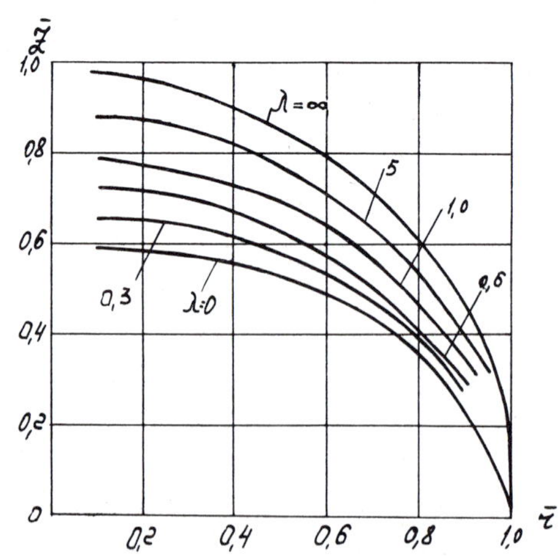

FIG. 19. Meridians of optimum vessels for different parameters λ at $\varphi_a = 5°$.

Note that the shell is of uniform strength if the equatorial strength condition is satisfied

$$\frac{p \cdot a}{2} = \sigma^{y} h_0 + \sigma_{\|}^{u} h_a \cos^2 \varphi_a \,. \tag{2.76}$$

In order to be able to develop the second approach, let us write the constitutive equations for an isotropic layer using the formation of the deformation theory of plasticity

$$\begin{aligned} \varepsilon_1 &= \frac{A_2 + A_1}{2} \sigma_1 + \frac{A_2 - A_1}{2} \sigma_2 \,, \\ \varepsilon_2 &= \frac{A_2 - A_1}{2} \sigma_1 + \frac{A_2 + A_1}{2} \sigma_2 \,, \\ A_2 &= \frac{1}{2E_c(\sigma_i)} + \frac{1 - 2\nu}{2E} \,; \quad A_1 = \frac{3}{2E_c(\sigma_i)} - \frac{1 - 2\nu}{2E} \,, \end{aligned} \tag{2.77}$$

where E and ν are, respectively, the modulus of elasticity and Poisson's ratio of the material used, σ_i is the intensity of stresses as given by

$$\sigma_i^2 = \sigma_1^2 - \sigma_1 \sigma_2 + \sigma_2^2 \,.$$

$E_c(\sigma_i) = \sigma_i / \varepsilon_i$ is the secant modulus as defined by the effective stress–strain curve of the material of the metal layer. For the composite layer, we will assume, in accordance with the netting analysis, that

$$\sigma_{\|} = E_{\|} \varepsilon_{\|} \quad \sigma_{\|} \leqslant \sigma_{\|}^{u} \,. \tag{2.78}$$

As has been shown, for a shell to be optimal, the following relation must hold

$$\varepsilon_1 = \varepsilon_2 = \varepsilon_{\|} = \varepsilon = \text{constant} \,. \tag{2.79}$$

Let us require, in addition, that at a certain pressure $p = p^*$ the condition for uniform stressing of the metal layer be satisfied, that is,

$$\sigma_1^* = \sigma_2^* = \sigma_i^* = \text{constant} \,. \tag{2.80}$$

Substituting the conditions (2.80) and (2.79) in the equilibrium equation, with (2.77) taken into account, we have

$$\begin{aligned} N_1 &= \frac{pR_2}{2} = \sigma_{\|} \left(\frac{1}{E_{\|} A_2^*} h_0 + h_1 \cos^2 \varphi \right) , \\ N_2 &= \frac{pR_2}{2} \left(2 - \frac{R_2}{R_1} \right) = \sigma_{\|} \left(\frac{1}{E_{\|} A_2^*} h_0 + h_1 \sin^2 \varphi \right) . \end{aligned} \tag{2.81}$$

where $A_2^* = A_2(\sigma_i^*)$.

Using the expression for the principal radii of curvature R_1, R_2 and regarding the winding as geodesic, we obtain an equation for the meridian shape of the optimum shell, provided the winding continuity condition is met,

$$\frac{Z''}{Z''[1+(Z')^2]} = \frac{1}{r} + \frac{1}{r}\,\frac{(r^2-r_0^2)-r_0^2}{\lambda^* \frac{r^2}{a}(r^2-r_0^2)^{1/2}+(r^2-r_0^2)}\,, \qquad (2.82)$$

where

$$\lambda^* = \frac{h_0 a}{E_\parallel h_a A_2^* (a^2-r_0^2)^{1/2}}\,.$$

It is easy to see that (2.82) is similar to (2.74), however, in (2.74) the coefficient λ included the strength characteristics of structural materials, whereas in (2.82) the coefficient λ^* now includes the rigidity characteristics and corresponds to a certain degree of strengthening of the material of the metal layer with $p = p^*$. Thus the integral (2.75) will provide the solution of (2.82) and all the arguments that follow remain valid. The relations (2.77), (2.78) and (2.80) allow us to find stresses in the layers of the two-layer shell, for which the expressions, after a few modifications, are

$$\sigma_1^* = \sigma_2^* = \sigma_i^* = \frac{p^* a}{2[h_0 + h_a E_\parallel A_2^* \cos^2 \varphi_a]}\,,$$

$$\sigma_\parallel^* = \frac{p^* \cdot a E_\parallel A_2^*}{2[h_0 + h_a E_\parallel A_2^* \cos^2 \varphi_a]}\,.$$

As could be expected, the stresses do not depend on the coordinates of the surface, that is, the shell turns out to be uniformly stressed.

To illustrate the second design approach, let us consider a two-layer vessel with given geometrical characteristics a, r_0, given material properties and with a preassigned (say, for some technological reasons) thickness of the metal layer h_0. The condition of uniform strength of the construction can be realized at a certain pressure p^* or at a certain degree of strengthening of the material of the metal layer as given by the value A_2^*. We may, for example, require that a uniform stress state should exist at the ultimate pressure ($p^* = p_{\max}$; $A_2^* = A_{2\,\max}$) or at the working pressure. We may pose the problem of choosing the optimum values for p^* and A_2^* in such a way that the construction, while remaining strong enough up to the ultimate pressure, should have the composite layer of a minimum thickness. For this reason, we may construct a series of generators as the integrals of (2.82), for given values of A_2^*. For each of the generators, we can find the maximum stress values and also the required value for the thickness of the composite layer h_a by using the von-Mises criterion for the metal layer $\sigma_i \leqslant \sigma^{\mathrm{u}}$ and the strength criterion for the fibres $\sigma_\parallel < \sigma_\parallel^{\mathrm{u}}$.

Fig. 20 shows the dependence $h_a = f(A_2^*)$ as plotted for a vessel with the parameters $a = 0.2 \times 10^{-3}$ m, $r_0/a = 0.1125$, $h_a = 0.5 \times 10^{-3}$ m and designed to

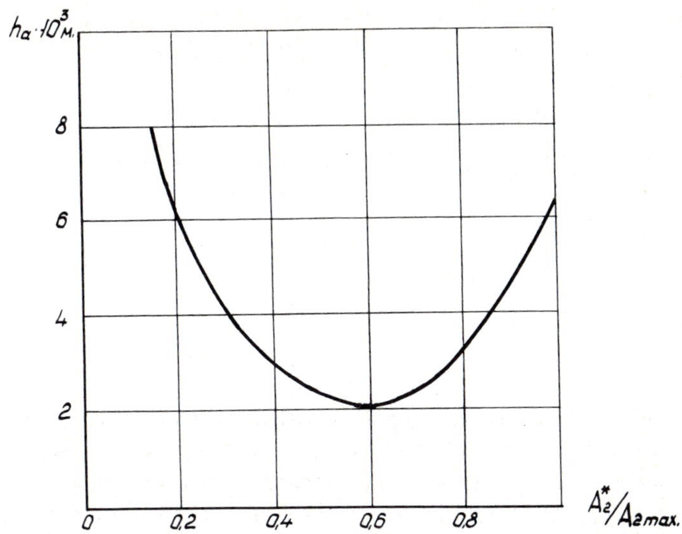

FIG. 20. Thickness of a composite layer versus parameters A_2^*/A_2 max.

withstand the pressure $p_{max} = 300 \times 10^5$ Pa. The metal layer is made of a material with $E = 201 \times 10^3$ MPa, $\nu = 0.33$, $\sigma^y = 670$ MPa, $\sigma^u = 1100$ MPa; its stress–strain diagram being shown in Fig. 21. The composite material has the following characteristics: $E_{\|} = 6.03 \times 10^3$ MPa, $\sigma_{\|}^u = 1400$ MPa. It follows from Fig. 20 that the minimum thickness of the composite layer will be obtained for a shell designed according to some intermediate state as defined by the values $A_{2\,opt}^* = 0.667 A_{2\,max}^*$; $p_{opt}^* = 0.89\, p_{max}$ and not by the ultimate pressure. Fig. 22

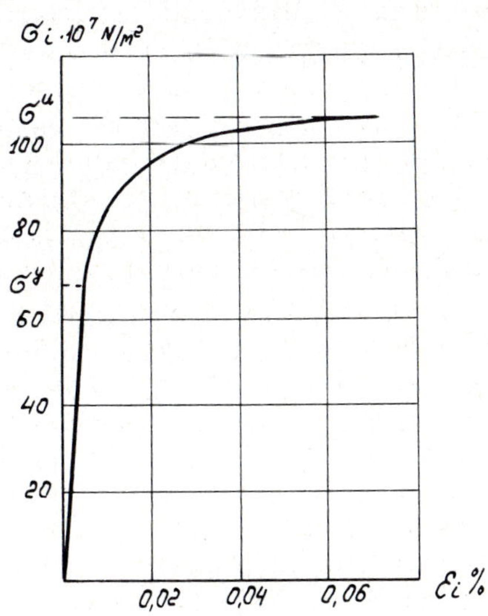

FIG. 21. Stress–strain curve for the metal layer.

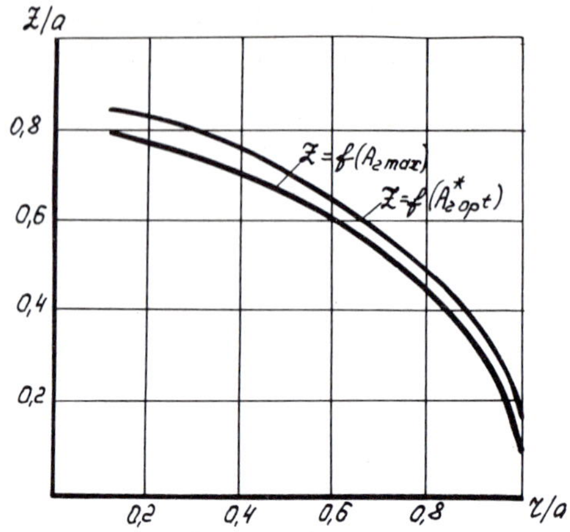

FIG. 22. Meridians of optimum combined composite–metal shells.

shows the shape of an optimum vessel and that of a vessel constructed according to (2.75). For materials with a high degree of strengthening, then, it is reasonable to design metal–composite vessels taking the strengthening of the metal into account.

All the above results have been based on the netting model for a composite material, the food carrying capacity of the matrix being neglected. That we are correct in using the model has been confirmed by the fact, typical of most of the existing materials, that the resin tends to fail at the initial stage of loading. The pressure which initiates cracking depends largely on the rigidity ratio of a unidirectional material along and across the fibres ($E_{\|}/E_{\perp}$). The use of high-modulus carbon, boron and other fibres has shown that in some cases the matrix can fail simultaneously with the fibres. A similar situation may take place in the case of a more plastic, say, metal matrix. These facts enable us to conclude that as far as some of the existing materials and some future materials are concerned, it seems important, from a practical point of view, to take into account the load-carrying capacity of the matrix when solving the problems of optimum pressure vessel design (JONG, 1972; BUNAKOV and RADOVINSKII, 1975). There is yet another result which seems essential if one wants to design a minimum mass construction. Since the shell, in which the matrix is taken into account, is redundant with respect to the stresses in the composite layers, the condition of uniform strength is no longer sufficient for an optimum structure. There has to be an additional requirement, namely, that the fibres should be placed along the directions of principal stresses in each layer, which is equivalent to

$$\tau^{i}_{\|\perp} = 0 , \tag{2.83}$$

where $\tau^{i}_{\|\perp}$ are shear stresses in the tape of the ith layer.

Consider the design of a pressure vessel wound with a high-modulus material. Assuming, for the sake of simplicity, that the polar openings are closed with rigid pressure covers, we can write a few equations which will relate stress resultants to the strains of the shell. For axisymmetric loading we have

$$N_1 = \frac{pR_2}{2} = A_{11}\varepsilon_1 + A_{12}\varepsilon_2 ,$$
$$N_2 = \frac{pR_2}{2}\left(2 - \frac{R_2}{R_1}\right) = A_{12}\varepsilon_1 + A_{22}\varepsilon_2 . \tag{2.84}$$

Here, ε_1 and ε_2 are the meridian and circumferential strains; A_{ij} are the generalized stiffnesses as given by

$$A_{11} = h(\bar{E}_{\parallel} \cos^4 \varphi + 2\bar{E}_{\parallel}\nu_{\parallel\perp} \sin^2 \varphi \cos^2 \varphi + \bar{E}_{\perp} \sin^4 \varphi + G_{\parallel\perp} \sin^2 2\varphi) ,$$
$$A_{12} = h[(\bar{E}_{\parallel} + \bar{E}_{\perp}) \sin^2 \varphi \cos^2 \varphi + E_{\parallel}\nu_{\parallel\perp}(\sin^4\varphi + \cos^4 \varphi) - G_{\parallel\perp} \sin^2 2\varphi] ,$$
$$A_{22} = h(\bar{E}_{\parallel} \sin^4 \varphi + 2E_{\parallel}\nu_{\parallel\perp} \sin^2 \varphi \cos^2 \varphi + \bar{E}_{\perp} \cos^4 \varphi + G_{\parallel\perp} \sin^2 2\varphi) , \tag{2.85}$$
$$\bar{E}_{\parallel,\perp} = \frac{E_{\parallel,\perp}}{1 - \nu_{\parallel\perp}\nu_{\perp\parallel}} .$$

R_1 and R_2 are the principal radii of curvature of the shell.

The equations relating the shell strains ε_1 and ε_2, to the strains along ($\varepsilon_{\parallel}$) and across (ε_1) the fibres, and to the in-plane shear strain ($\varepsilon_{\parallel\perp}$) are

$$\varepsilon_{\parallel} = \varepsilon_1 \cos^2 \varphi + \varepsilon_2 \sin^2 \varphi ,$$
$$\varepsilon_{\perp} = \varepsilon_1 \sin^2 \varphi + \varepsilon_2 \cos^2 \varphi , \tag{2.86}$$
$$\varepsilon_{\parallel\perp} = (\varepsilon_2 - \varepsilon_1) \sin 2\varphi .$$

Subsequently, from the optimality condition (2.83) we can find

$$\varepsilon_{\parallel\perp} = 0 , \quad \varepsilon_1 = \varepsilon_2 = \varepsilon_{\parallel} = \varepsilon_{\perp} = \varepsilon \tag{2.87}$$

and, using the constitutive equations for an orthotropic material, we obtain, taking (2.87) into account

$$\sigma_{\parallel} = \bar{E}_{\parallel}(1 + \nu_{\parallel\perp})\varepsilon , \quad \sigma_{\perp} = \bar{E}_{\perp}(1 + \nu_{\perp\parallel})\varepsilon . \tag{2.88}$$

Finally, solving the system (2.84) with (2.87) and substituting the result in (2.88), we find the stresses in the tape of a shell under consideration

$$\sigma_{\parallel} = \frac{\bar{E}_{\parallel}(1+\nu_{\parallel\perp})(N_1+N_2)}{h[\bar{E}_{\parallel}(1+\nu_{\parallel\perp})+\bar{E}_{\perp}(1+\nu_{\perp\parallel})]}\,,$$

$$\sigma_{\perp} = \frac{\bar{E}_{\perp}(1+\nu_{\perp\parallel})(N_1+N_2)}{h[\bar{E}_{\parallel}(1+\nu_{\parallel\perp})+\bar{E}_{\perp}(1+\nu_{\perp\parallel})]}\,. \qquad (2.89)$$

Thereby, $\tau_{\parallel\perp}=0$ according to (2.83).

Let us note some important results following from the above solution.

Directly from (2.89) we can deduce that the tape stress ratio

$$\frac{\sigma_{\perp}}{\sigma_{\parallel}} = \frac{E_{\perp}(1+\nu_{\perp\parallel})}{E_{\parallel}(1+\nu_{\parallel\perp})} = k$$

is constant and depends only on the properties of the material. It follows from the linearity of stress–strain response of composite materials that this relation will hold until one of the stresses $\sigma_{\parallel}$ or $\sigma_{\perp}$ has attained its maximum value, that is, the ultimate strength. Here, it is assumed that the condition for in-plane tape strength $(\sigma_{\parallel}, \sigma_{\perp})$ is defined by the sides of a rectangle $\sigma_{\parallel}=\sigma_{\parallel}^{u}$, $\sigma_{\perp}=\sigma_{\perp}^{u}$.

Now let us introduce another material characteristic

$$\sigma_{\perp}^{u}/\sigma_{\parallel}^{u} = k_1\,.$$

It follows from comparison of the coefficients k and k_1 that for composites with $k_1 < k$ (Fig. 23) the matrix fails before fibre breakage. For rigid materials with $k_1 \geqslant k$ the shell fails due to fibre breakage. For this reason, the netting model will be sufficient for some materials such as fibreglass plastics, while for high-modulus fibres the design model in the form of a netting system is not fully acceptable.

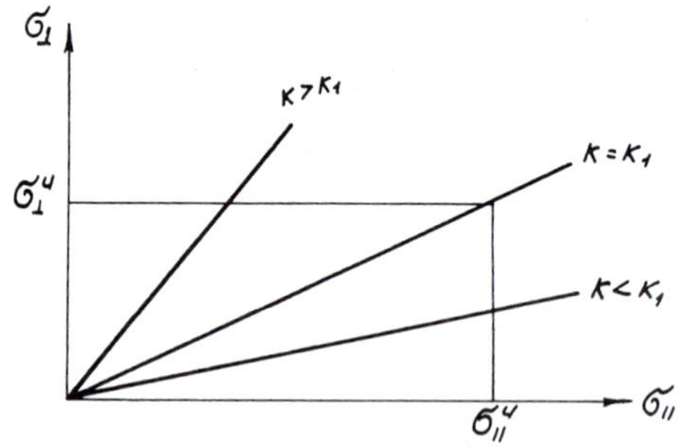

FIG. 23. Limit interaction curve and types of loading for composite shells.

For an optimum vessel design, as far as the choice of an optimal fibre orientation of a shell shape is concerned, let us consider the ratio (2.84), which, with (2.87) and the expressions for the radii of curvature taken into account, assumes the form

$$\frac{rZ''}{Z'[1+(Z')^2]} = 2 - \frac{1-(1-k)\cos^2\varphi}{k+(1-k)\cos^2\varphi}\,, \tag{2.90}$$

while, on the other hand, after transformations similar to (2.17)–(2.20), we obtain the equation

$$r\frac{\mathrm{d}\varphi}{\mathrm{d}r}\frac{\sin\varphi}{\cos\varphi}\frac{k-(1-k)\cos^2\varphi}{1-(1-k)\cos^2\varphi} = 1$$

the integral of which

$$r\cos^k\varphi[1-(1-k)\cos^2\varphi]^{(1-k)/2} = \text{constant}\,, \tag{2.91}$$

defines the fibre orientation for a shell of uniform strength. The constant in (2.91) can be expressed through the angle of reinforcement φ_a at $r=a$. Note that with $k\to 0$ Eq. (2.91) specifies the geodesic pattern.

Dropping the intermediate operations, we write the final result. The optimum shape of the generator is obtained by numerical integration of the equation

$$Z(r) = -\int_a^r \frac{\mathrm{d}r}{\left(\left[\dfrac{k+(1-k)\cos^2\varphi}{k+(1-k)\cos^2\varphi_a} - \dfrac{a^2\cos\varphi_a}{r^2\cos\varphi}\right]^2 - 1\right)^{1/2}}\,.$$

in conjunction with the fibre orientation law

$$r = a\frac{\cos^k\varphi_a[1-(1-k)\cos^2\varphi_a]^{(1-k)/2}}{\cos^k\varphi[1-(1-k)\cos^2\varphi]^{(1-k)/2}}\,.$$

Note that since the filaments are not laid along the geodesic lines of the surface, the technological constraint that the filaments should not slip off during winding becomes essential. For this reason, there has to be change-over to a geodesic path of winding in the vicinity of the polar opening.

Fig. 24 shows a pressure vessel generator for the relative radius of a polar opening $\bar{r}_0 = r_0/a = 0.2$, the friction coefficient between the tape and the mandrel $\kappa = 0.5$ and the relative rigidity parameter of the material $k = 0.1$ (curve 1). The figure also shows, for comparison, the meridian shape as constructed by integration of (2.90) and corresponding to geodesic winding (curve 2) and, finally, a vessel generator as constructed by the netting model (curve 3). If we introduce the shape factor (similar to that shown in Fig. 9)

$$k_\mathrm{f} = \frac{G/V}{\rho/\sigma_\parallel^\mathrm{u} p}$$

then for the obtained solutions we shall have $k_{\mathrm{f}1} = 2.53$; $k_{\mathrm{f}2} = 2.65$ and $k_{\mathrm{f}3} = 3.0$. That is, the saving in mass will be greater than 15%.

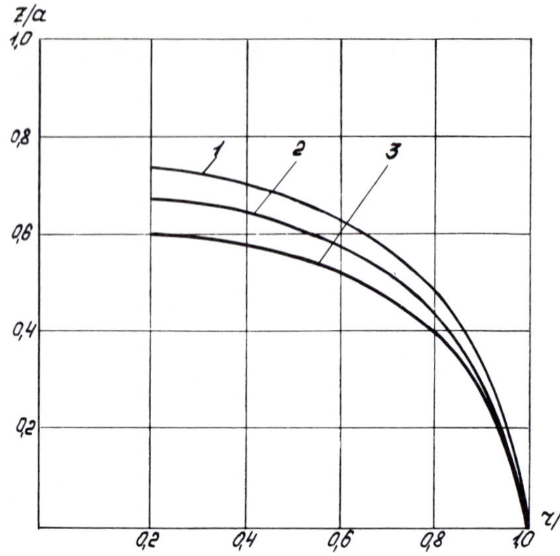

FIG. 24. Meridians of pressure vessels for different design versions: (1) a uniformly stressed shell ($k = 0.1$); (2) geodesic winding ($k = 0.1$); (3) geodesic winding ($k = 0$).

3. Strength and rigidity of pressure vessels

Filament wound pressure vessels are composed of variable thickness layers of anisotropic and heterogeneous material – a feature rendering the analysis of deformation and fracture of the construction rather complex.

In assessing the strength and deformability of such vessels, the problems of choosing a proper design model and an adequate model of the material appear predominant. The membrane shell of revolution which can successfully be taken as a model for constructions made by the continuous winding technique is usually used for calculations. It is to be noted that in shells of revolution under internal pressure the induced moments are localized and occur in the vicinity of fixed edges and lines where the geometrical parameters or rigidity of the construction become discontinuous. As mentioned above, it is also necessary to choose an appropriate model for the material response because fracture of laminated composites has a complicated character – the matrix fails first in the initial stage, loss of integrity takes place and in the final stage the fibres fail. The loss of integrity of the material occurs because of brittle failure of layers in tension across the fibres as well as in shear and results in considerable reduction in the material rigidity, so that proper allowance for the fracture process has to be made in the constitutive relations. For this reason, it seems reasonable to estimate the load-carrying capacity of a pressure vessel by employing the model (VASIL'EV and ELPAT'EVSKII, 1967) in which the failure of the matrix of an individual layer is taken into account. According to this model, the tape splits up into a number of strips and upon further loading the number of cracks increases, although neither the tensile stress across the fibres nor the shear stress can exceed its ultimate values. In the process of deformation the failing layer is maintained at the ultimate state due to its interaction with the neighbouring layers. According to this model, the stress–strain relation can be defined by the law similar to that of an ideally plastic body. The failure of the matrix of an individual layer results in considerable reduction in rigidity but affects very little the average ultimate strength of the material. The reduction in rigidity can be found, if for the layer in which the matrix fails, the value of elastic modulus in tension across the fibres or in shear is taken to be zero. In this case, breaks (knees) appear in the strain diagram for a whole packet which are well confirmed by experimental data.

For practical analysis of pressure vessels as axisymmetric shells of revolution we can suggest the following scheme for determining the stress–strain for a composite material.

Our assumption is that at any point of the vessel the stress resultants N_1, N_2 are known. Now, we can write how these stress resultants are related to the strains of the shell by means of equations similar to (2.84) and (2.85)

$$\begin{aligned} N_1 &= A_{11}\varepsilon_1 + A_{12}\varepsilon_2 , \\ N_2 &= A_{12}\varepsilon_1 + A_{22}\varepsilon_2 , \end{aligned} \tag{3.1}$$

where the generalized stiffnesses in the initial stage of loading are given as

$$\begin{aligned} A_{11} = \sum_{i=1}^{n} h_i(\bar{E}^i_{\parallel} \cos^4 \varphi_i + 2\bar{E}^i_{\parallel}\nu^i_{\parallel\perp} \sin^2 \varphi_i \cos^2 \varphi_i + \bar{E}^i_{\perp} \sin^4 \varphi_i \\ + G^i_{\parallel\perp} \sin^2 2\varphi_i) , \end{aligned}$$

$$\begin{aligned} A_{12} = A_{21} = \sum_{i=1}^{n} h_i[(\bar{E}^i_{\parallel} + \bar{E}^i_{\perp}) \sin^2 \varphi \cos^2 \varphi_i + \bar{E}^i_{\parallel}\nu^i_{\parallel\perp}(\sin^4 \varphi_i + \cos^4 \varphi_i) \\ - G^i_{\parallel\perp} \sin^2 2\varphi_i] \end{aligned}$$

$$\begin{aligned} A_{22} = \sum_{i=1}^{n} h_i(\bar{E}^i_{\parallel} \sin^4 \varphi_i + 2\bar{E}^i_{\parallel}\nu^i_{\parallel\perp} \sin^2 \varphi_i \cos^2 \varphi_i + E^i_{\perp} \cos^4 \varphi_i \\ + G^i_{\parallel\perp} \sin^2 2\varphi_i) . \end{aligned}$$

Solving the system (3.1) with respect to the strains and taking into account the geometrical relations given by

$$\begin{aligned} \varepsilon^i_{\parallel} &= \varepsilon_1 \cos^2 \varphi_i + \varepsilon_2 \sin^2 \varphi_i , \\ \varepsilon^i_{\perp} &= \varepsilon_1 \sin^2 \varphi_i + \varepsilon_2 \cos^2 \varphi_i , \\ \varepsilon^i_{\parallel\perp} &= (\varepsilon_2 - \varepsilon_1) \sin 2\varphi_i , \end{aligned} \tag{3.2}$$

we find the strains in the tape, and using these in conjunction with Hooke's law,

$$\sigma^i_{\parallel} = \bar{E}^i_{\parallel}(\varepsilon^i_{\parallel} + \nu^i_{\parallel\perp}\varepsilon^i_{\perp}), \quad \sigma^i_{\perp} = \bar{E}^i_{\perp}(\varepsilon^i_{\perp} + \nu^i_{\perp\parallel}\varepsilon^i_{\parallel}), \quad \tau^i_{\parallel\perp} = G^i_{\parallel\perp}\varepsilon^i_{\parallel\perp},$$

we can estimate stresses in the tape. In order to find out the value of the internal pressure p_1 at which the loss of integrity of the most highly stressed layer (for instance, kth layer) takes place, we can use one of the well known phenomenological failure criteria for a unidirectional layer. With the assumption that the tensile and compressive strengths are the same, the fracture surface in the most general form can be given by the equation

$$F_{11}\sigma^2_{\parallel} + F_{12}\sigma_{\parallel}\sigma_{\perp} + F_{22}\sigma^2_{\perp} + F_{66}\tau^2_{\parallel\perp} = 1,$$

where F_{ij} is the strength tensor of the fourth rank whose components are defined by the choice of the respective strength criteria.

As the load increases, the change in the pressure level is determined starting from the level p_1 and the change in the strains from the values corresponding to $p = p_1$. To find out the rigidity of the laminate when estimating the generalized stiffnesses A_{ij}, one must put $E^k_{\perp} = G^k_{\parallel\perp} = \nu^k_{\parallel\perp} = \nu^k_{\perp\parallel} = 0$ with $\sigma^k_{\perp} > 0$ and $G^k_{\parallel\perp} = \nu^k_{\parallel\perp} = \nu^k_{\perp\parallel} = 0$ with $\sigma^k_{\perp} < 0$ in terms corresponding to the number k. Next, one must find the stresses in the tape with $p > p_1$ and, using these, set the pressure level p_2 at which the matrix will fail in the next layer numbered m, also putting for that layer $E^m_{\perp} = G^m_{\parallel\perp} = \nu^m_{\parallel\perp} = \nu^m_{\perp\parallel} = 0$ with $\sigma^m_{\perp} > 0$ and $G^m_{\parallel\perp} = \nu^m_{\parallel\perp} = \nu^m_{\perp\parallel} = 0$ with $\sigma^m_{\perp} < 0$.

Thus, the deformation–pressure relationship is a piecewise linear curve with the breaks (knees) corresponding to the instances of failure of the matrix in individual layers. The process goes on until the matrix has failed in all the layers. In this case the generalized stiffnesses are

$$A_{11} = \sum_{i=1}^{n} h_i E^i_{\parallel} \cos^4 \varphi_i\,; \quad A_{12} = A_{21} = \sum_{i=1}^{n} h_i E^i_{\parallel} \sin^2 \varphi_i \cos^2 \varphi_i\,;$$
$$A_{22} = \sum_{i=1}^{n} h_i E^i_{\parallel} \sin^4 \varphi_i\,. \tag{3.3}$$

It is to be noted that it is possible to obtain the same expressions for the generalized stiffnesses if the structure of the shell material is presented as a system of flexible filaments. Also, at the final stage of calculation the following two cases are most likely to occur: either the strains are finite (in this case the ultimate strength is determined by fibre breakage) or the strains are infinite, which means that the filament system becomes a mechanism (that is, the shell for which the matrix has been neglected does not take up the load). In the latter case one has to apply nonlinear relationships of the mechanics of fibrous shells in order to find out their load-carrying capacity.

One feature of the deformation process in pressure vessels is the possibility of creep of the material under prolonged loading, especially when organic fibres are used. In order to describe the process attributed to the viscoelastic behavior of composite materials, the linear theory is most often used resulting in the following integral equation of the Boltzmann–Volterra type

$$\varepsilon_{ij}(t) = \lambda_{ijkl}\sigma_{kl}(t) + \int_0^t k_{ijkl}(t-\tau)\sigma_{kl}(\tau)\,d\tau\,. \qquad (3.4)$$

In order to determine the kernels (creep compliances) creep tests are usually performed under loading in the reinforcement direction, across the fibres and in shear. In this case the relation (3.4) takes the form

$$\varepsilon(t) = \frac{1}{E}[\sigma_0 + F(\sigma_0, t)] \qquad (3.5)$$

where $F(\sigma_0, t)$ is usually referred to as the creep function. Figs. 25–27 shows creep curves obtained from loading along the fibres, across them and in shear

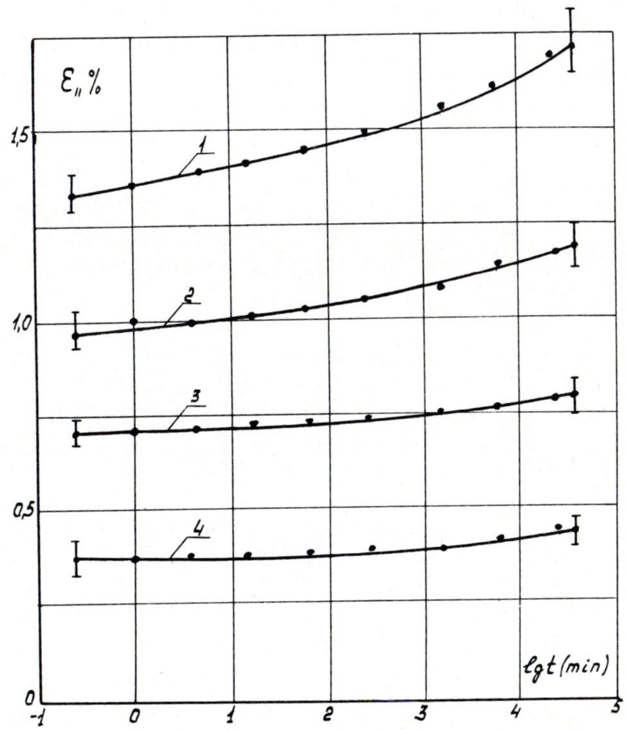

FIG. 25. Creep of a unidirectional composite reinforced with organic fibres ($E_{\parallel} = 64.7$ GPa): dots – experimental, full line – theoretical approximation: (1) $\sigma_{\parallel} = 841$ MPa; (2) $\sigma_{\parallel} = 630$ MPa; (3) $\sigma_{\parallel} = 450$ MPa; (4) $\sigma_{\parallel} = 238$ MPa.

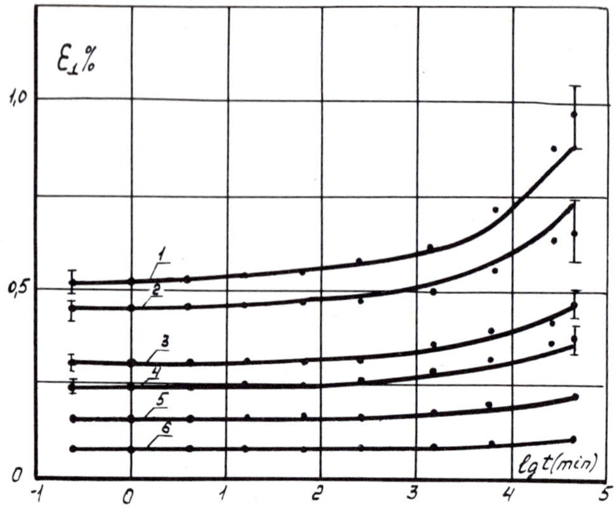

FIG. 26. Transversal creep of a unidirectional composite reinforced with organic fibres ($E_\perp$ = 3.3 GPa): dots – experimental; full line – theoretical approximation: (1) $\sigma_\perp = 16.9$ MPa; (2) $\sigma_\perp$ = 14.7 MPa; (3) $\sigma_\perp = 10$ MPa; (4) $\sigma_\perp = 8.1$ MPa; (5) $\sigma_\perp = 5.3$ MPa; (6) $\sigma_\perp = 2.7$ MPa.

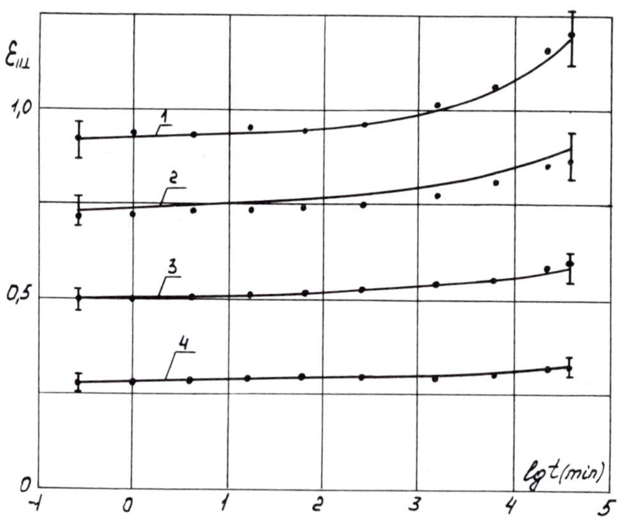

FIG. 27. Shear creep curves of a unidirectional composite reinforced with organic fibres ($G_{\parallel\perp}$ = 2.08 GPa): dots – experimental; full line – theoretical approximation: (1) $\tau_{\parallel\perp} = 18.7$ MPa; (2) $\tau_{\parallel\perp} = 15.4$ MPa; (3) $\tau_{\parallel\perp} = 10.5$ MPa; (4) $\tau_{\parallel\perp} = 5.5$ MPa.

for 30 days at different load levels. Five specimens were used in each type of tests. The same figures show the theoretical approximation of the curves as found from relations of the kind

$$F(\sigma_0, t) = f(\sigma_0) + \sum_{m=1}^{M} q_m(\sigma_0) \int_0^t k_m(\tau)\, d\tau , \tag{3.6}$$

where exponential functions have been used for the creep compliances:

$$k_m(t) = \gamma_m \exp(-\gamma_m t)$$

The choice for the functions $f(\sigma_0)$, $q_m(\sigma_0)$ was governed by the condition that the corresponding creep curves be similar.

The results for a unidirectional material have been used to find out the creep strains in cylindrical shells wound with unidirectional tape. When deriving the constitutive relations we employed the deformation model for an orthotropic materials based on the assumption that creep strains in a given direction are only governed by stresses acting in this direction. Then the constitutive relations are as follows

$$\begin{aligned} \varepsilon^i_{\parallel} &= \frac{1}{E^i_{\parallel}} [\sigma^i_{\parallel} + F_1(\sigma^i_{\parallel})] - \frac{\nu^i_{\perp\parallel}}{E^i_{\perp}} \sigma^i_{\perp} , \\ \varepsilon^i_{\perp} &= \frac{1}{E^i_{\perp}} [\sigma^i_{\perp} + F_2(\sigma^i_{\perp})] - \frac{\nu^i_{\parallel\perp}}{E^i_{\parallel}} \sigma^i_{\parallel} , \\ \varepsilon^i_{\parallel\perp} &= \frac{1}{G^i_{\parallel\perp}} [\tau^i_{\parallel\perp} + F_3(\tau^i_{\parallel\perp})] . \end{aligned} \tag{3.7}$$

Using the obtained formulae, we can derive relationships which will relate the stress resultants to the strains in a shell with an arbitrary laminate structure. Omitting the intermediate operations, we can write the final result as follows

$$\begin{aligned} N_1 &= A_{11}\varepsilon_1 + A_{12}\varepsilon_2 - B_1 , \\ N_2 &= A_{12}\varepsilon_1 + A_{22}\varepsilon_2 - B_2 , \end{aligned} \tag{3.8}$$

where

$$B_1 = \sum_{i=1}^{n} h_i(D^i_1 \cos^2 \varphi_i + D^i_2 \sin^2 \varphi_i - D^i_3 \sin 2\varphi_i) ,$$

$$B_2 = \sum_{i=1}^{n} h_i(D^i_1 \sin^2 \varphi_i + D^i_2 \cos^2 \varphi_i + D^i_3 \sin 2\varphi_i) ,$$

$$D^i_1 = \frac{F_1(\sigma^i_{\parallel}) + \nu^i_{\parallel\perp} F_2(\sigma^i_{\perp})}{1 - \nu^i_{\parallel\perp}\nu^i_{\perp\parallel}} , \quad D^i_2 = \frac{F_2(\sigma^i_{\perp}) + \nu^i_{\perp\parallel} F_1(\sigma^i_{\parallel})}{1 - \nu^i_{\parallel\perp}\nu^i_{\perp\parallel}} ,$$

$$D^i_3 = F_3(\tau^i_{\parallel\perp}) .$$

The results of calculation for a shell composed of a system of spiral and hoop layers are shown in Fig. 28. For comparison, the same figure shows creep curves based on the hypotheses of the netting model. The calculation was carried out by the successive approximation method.

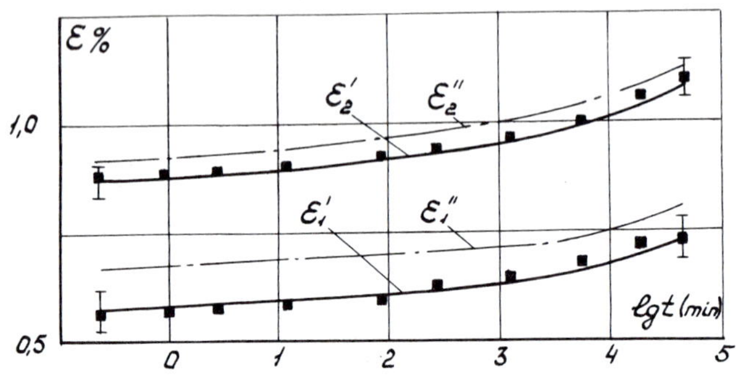

FIG. 28. Circumferential (ε_2) and meridian (ε_1) creep deformations of shells under constant internal pressure $p = 0.35p_{\text{resolv}}$: full squares – experimental: ε_2', ε_1' estimated by model (3.7); ε_2'', ε_1'' estimated by netting model.

When determining the strength and deformations of pressure vessels with a load-carrying isotropic layer, one must take into account, besides the loss of integrity and creep deformations, also the possible emergence of plastic strains in the metal layer. Within the framework of the membrane shell theory one can write the relationships between the meridian and circumferential stress resultants N_1 and N_2 and the stresses in the layers as follows

$$N_1 = h_0\sigma_1^0 + \sum_{i=1}^{n} h_{1i}(\sigma_{\|}^i \cos^2 \varphi_i + \sigma_{\perp}^i \sin^2 \varphi_i - \tau_{\|\perp}^i \sin 2\varphi_i)\,,$$

$$N_2 = h_0\sigma_2^0 + \sum_{i=1}^{n} h_{1i}(\sigma_{\|}^i \sin^2 \varphi_i + \sigma_{\perp}^i \cos^2 \varphi_i + \tau_{\|\perp}^i \sin 2\varphi_i)\,, \tag{3.9}$$

where σ_1^0 and σ_2^0 are meridian and circumferential stresses in the isotropic layer; h_0 is the thickness of the isotropic layer; $\sigma_{\|}^i$, $\sigma_{\perp}^i$, $\tau_{\|\perp}^i$ are the stresses in the composite tape oriented at angle φ_i to the generator; h_{1i} is the thickness of the ith layer. It is easy to see that although the stress resultants N_1 and N_2 do not depend on the material characteristics, the shell under consideration is redundant because the two stresses in the isotropic layer σ_1^0, σ_2^0 and stresses in the composite layers $\sigma_{\|}^i$, $\sigma_{\perp}^i$, $\tau_{\|\perp}^i$ $(i = 1, 2, \ldots, n)$ are related by two equations (3.9).

Now let us consider one of the possible solutions of the problem – a displacement formulation, with the meridian and circumferential strains ε_1 and ε_2, common for all layers, taken as the basic unknowns. Supposing that the strains in the shell are elastic in the initial stage, we can write their relation to the stresses in the isotropic layer in the form of Hooke's law

$$\sigma_1^0 = \frac{E}{1-\nu^2}(\varepsilon_1 + \nu\varepsilon_2); \quad \sigma_2^0 = \frac{E}{1-\nu^2}(\varepsilon_2 + \nu\varepsilon_1) \tag{3.10}$$

where E and ν are the elastic modulus and Poisson's ratio of the material.

The stresses in the tape of the ith layer are related to the respective strains as follows

$$\sigma_{\parallel}^i = \bar{E}_{\parallel}^i(\varepsilon_{\parallel}^i + \nu_{\parallel\perp}^i\varepsilon_{\perp}^i), \quad \sigma_{\perp}^i = \bar{E}_{\perp}^i(\varepsilon_{\perp}^i + \nu_{\perp\parallel}^i\varepsilon_{\parallel}^i), \quad \tau_{\parallel\perp}^i = G_{\parallel\perp}^i\varepsilon_{\parallel\perp}^i . \tag{3.11}$$

The strains in the tape $\varepsilon_{\parallel}^i$, $\varepsilon_{\perp}^i$, $\varepsilon_{\parallel\perp}^i$ can be expressed through the meridian and circumferential strains by means of the relations

$$\begin{aligned} &\varepsilon_{\parallel}^i = \varepsilon_1\cos^2\varphi_i + \varepsilon_2\sin^2\varphi_i, \quad \varepsilon_{\perp}^i = \varepsilon_1\sin^2\varphi_i + \varepsilon_2\cos^2\varphi_i, \\ &\varepsilon_{\parallel\perp}^i = (\varepsilon_2 - \varepsilon_1)\sin 2\varphi_i . \end{aligned} \tag{3.12}$$

Substituting the strains (3.12) in (3.11) and the stresses (3.11) and (3.10) in the relations (3.9), we obtain two equations with the two unknowns ε_2 and ε_1

$$\begin{aligned} N_1 &= \left(\frac{Eh_0}{1-\nu^2} + A_{11}\right)\varepsilon_1 + \left(\frac{E\nu h_0}{1-\nu^2} + A_{12}\right)\varepsilon_2, \\ N_2 &= \left(\frac{E\nu h_0}{1-\nu^2} + A_{21}\right)\varepsilon_1 + \left(\frac{Eh_0}{1-\nu^2} + A_{22}\right)\varepsilon_2 . \end{aligned} \tag{3.13}$$

The generalized stiffnesses A_{ij} have already been written and are given by the formulae (2.84).

Now let us turn to the analysis of deformation peculiarities of the materials used in combined composite–metal shells. As the pressure in the metal layer increases, plastic defomation arises, and the load corresponding to the elastic limit of the metal is indicated by the stress intensity σ_i attaining the yield stress of the material. For a plain state of stress the stress intensity is given by

$$\sigma_i^2 = (\sigma_1^0)^2 - \sigma_1^0\sigma_2^0 + (\sigma_2^0)^2 . \tag{3.14}$$

In the general case, the loading of the metal layer is not proportional (i.e. the stresses σ_1^0, σ_2^0 do not vary with pressure in a directly proportional way) and in order to solve the elasto-plastic problem one has to employ the plastic flow theory. Then the physical relations for the metal layer take the following form

$$\begin{aligned} d\varepsilon_1 &= \frac{1}{E}(d\sigma_1^0 - \nu\, d\sigma_2^0) + \frac{d\sigma_i}{\sigma_i}\left(\frac{1}{E_k} - \frac{1}{E}\right)(\sigma_1^0 - \tfrac{1}{2}\sigma_2^0), \\ d\varepsilon_2 &= \frac{1}{E}(d\sigma_2^0 - \nu\, d\sigma_1^0) + \frac{d\sigma_i}{\sigma_i}\left(\frac{1}{E_k} - \frac{1}{E}\right)(\sigma_2^0 - \tfrac{1}{2}\sigma_1^0), \end{aligned} \tag{3.15}$$

where $E_k(\sigma_i) = \mathrm{d}\sigma_i/\mathrm{d}\varepsilon_i$ is the tangential modulus as defined by the effective curve $\sigma_i - \varepsilon_i$ (ε_i is the deformation intensity). The effective curve, in the first approximation, can be taken to coincide with the stress–strain curve in uniaxial tension–compression.

To procure a governing system of equations, let us substitute in place of N_1 and N_2 in relations (3.9) their values for a vessel having a closed polar opening

$$N_1 = \frac{pR_2}{2}, \quad N_2 = \frac{pR_2}{2}\left(2 - \frac{R_2}{R_1}\right),$$

and express the stresses $\sigma^i_{\parallel}$, $\sigma^i_{\perp}$, $\tau^i_{\parallel\perp}$ in terms of the strains ε_1 and ε_2 using (3.11) and (3.12). Differentiating the expressions and excluding $\mathrm{d}\varepsilon_1$ and $\mathrm{d}\varepsilon_2$ by means of (3.15), we write the final system

$$\begin{aligned}\frac{R_2}{2}\,\mathrm{d}p &= \left(h_0 + \frac{A_{11} - \nu A_{12}}{E}\right)\mathrm{d}\sigma_1^0 + \frac{A_{12} - \nu A_{11}}{E}\,\mathrm{d}\sigma_2^0 \\ &\quad + \frac{\mathrm{d}\sigma_i}{\sigma_i}\left(\frac{1}{E_k} - \frac{1}{E}\right)[(A_{11} - \tfrac{1}{2}A_{12})\sigma_1^0 + (A_{12} - \tfrac{1}{2}A_{11})\sigma_2^0]\,, \\ \frac{R_2}{2}\left(2 - \frac{R_2}{R_1}\right)\mathrm{d}p &= \frac{A_{21} - \nu A_{22}}{E}\,\mathrm{d}\sigma_1^0 + \left(h_0 + \frac{A_{22} - \nu A_{21}}{E}\right)\mathrm{d}\sigma_2 \\ &\quad + \frac{\mathrm{d}\sigma_i}{\sigma_i}\left(\frac{1}{E_k} - \frac{1}{E}\right)[(A_{21} - \tfrac{1}{2}A_{22})\sigma_1^0 + (A_{22} - \tfrac{1}{2}A_{21})\sigma_2^0]\,.\end{aligned} \tag{3.16}$$

The stresses σ_1^0, σ_2^0 are determined by numerical integration of (3.16) with the effective curve $\sigma_i - \varepsilon_i$ and the initial conditions $\sigma_1^0\ (p = 0) = \sigma_2^0(p = 0) = 0$ taken into account (the initial stresses generated by pretensioning of the tape during winding are neglected here). By integrating (3.15) and by means of relations (3.12) and (3.11) we find the stresses in the layers of the composite material.

Since only the internal pressure p acts on the shell, we can regard the loading of the metal layer to be close to a proportional one. As is known, sufficiently accurate results in this case can be obained by employing the deformation theory of plasticity. The physical relationships for the metal layer are then given by

$$\begin{aligned}\varepsilon_1 &= \frac{\sigma_1^0 - \nu\sigma_2^0}{E} + \left(\frac{1}{E_c} - \frac{1}{E}\right)(\sigma_1^0 - \tfrac{1}{2}\sigma_2^0)\,, \\ \varepsilon_2 &= \frac{\sigma_2^0 - \nu\sigma_1^0}{E} + \left(\frac{1}{E_c} + \frac{1}{E}\right)(\sigma_2^0 - \tfrac{1}{2}\sigma_1^0)\,,\end{aligned} \tag{3.17}$$

where the modulus $E_c(\sigma_i) = \sigma_i/\varepsilon_i$ is defined by the effective curve $\sigma_i = \varepsilon_i$. Rearranging the system (3.17) with respect to stresses and substituting the latter in the equilibrium condition (3.9), modified by means of (3.12) and

(3.11) for a vessel with a closed polar opening, we have

$$\begin{aligned} \frac{pR_2}{2} &= A_{11}^* \varepsilon_1 + A_{12}^* \varepsilon_2 , \\ \frac{pR_2}{2}\left(2 - \frac{R_2}{R_1}\right) &= A_{21}^* \varepsilon_1 + A_{22}^* \varepsilon_2 . \end{aligned} \tag{3.18}$$

Here, the generalized stiffnesses are given by

$$\begin{aligned} A_{11}^* &= \frac{4E^2 E_c h_0}{3E^2 + (1-2\nu)E_c[2E - (1-2\nu)E_c]} + A_{11} , \\ A_{12}^* = A_{21}^* &= \frac{2EE_c[E - (1-2\nu)E]h_0}{3E^2 + (1-2\nu)E_c[2E - (1-2\nu)E_c]} + A_{12} , \\ A_{22}^* &= \frac{4E^2 E_c h_0}{3E^2 + (1-2\nu)E_c[2E - (1-2\nu)E_c]} + A_{22} . \end{aligned} \tag{3.19}$$

Note that with the elastic strains of the metal layer $E_c = E$ and (3.19) coincide with (3.13).

To solve (3.18) one can use the method of successive loading. In this case, the range of pressure variation is divided into sections, so that in the initial section the secant modulus E_c is taken to be equal to E and in each subsequent section E_c is defined by the stress intensity found in the preceding stage of loading.

It is to be noted that the parameters A_{11}, A_{12}, A_{21}, A_{22}, characterizing rigidity of a composite layer, are not generally constant during loading of a shell. As has been pointed out earlier, deformation of a composite material is accompanied by matrix failure leading to the reduction in rigidity. This can be taken into account by using the above model, according to which one must put $E_\perp = G_{\parallel\perp} = \nu_{\parallel\perp} \equiv \nu_{\perp\parallel} = 0$ when defining the generalized stiffnesses A_{ij} for layers in which the matrix has failed. Thus, with numerical integration of (3.16) or solution of (3.18) by the successive loading method, it is necessary, at each step, to determine stresses in the tape $\sigma_\parallel^i$, $\sigma_\perp^i$, $\tau_{\parallel\perp}^i$, in order to ascertain the instant of failure of the matrix in the layer by means of corresponding strength criterion and find the generalized stiffnesses of the composite material by taking into account the loss of integrity of individual layers. The ultimate pressure of the combined shell can correspond to failure of the metal or composite layer, depending on the elongation ratio of the metal and composite material. In the first case, the ultimate pressure is evaluated by means of the respective strength criterion for isotropic materials and in the second case – after the maximum stresses in fibres have attained their ultimate value.

As an example, we may show how the stress–strain state can be evaluated in a pressure vessel composed of fibreglass–epoxy spiral layers and a metal layer of thickness $h_0 = 10^{-3}$ m. The shape of the shell meridian is given by (2.75) as

$\lambda = 0.2$ and the relative radius of the polar opening $\bar{r}_0 = 0.147$. The properties of the material of the composite layer are as follows: $E_{\parallel} = 50.4$ GPa; $E_{\perp} = 10$ GPa; $G_{\parallel\perp} = 3.65$ GPa; $\nu_{\parallel\perp} = 0.0397$; $\nu_{\perp\parallel} = 0.2$; $\sigma_{\parallel}^{u} = 1200$ MPa; $\sigma_{\perp}^{u} = 27.5$ MPa; $\tau_{|\perp}^{u} = 25$ MPa. The material of the metal layer has the following mechanical characteristics: $E = 71$ GPa; $\nu = 0.3$; $\sigma^{y} = 280$; $\sigma_{0}^{u} = 450$ MPa.

Fig. 29 shows the results of calculation plotted as a curve of the meridian and circumferential strains versus internal pressure for a cross section $r/a = 0.7$. Note that the curves have no piecewise linear segments peculiar to vessels without a metal layer. The experimental results are shown by dots. Strains have been measured by resistance strain gauges which were glued to the metal mandrel before winding of the glass tape.

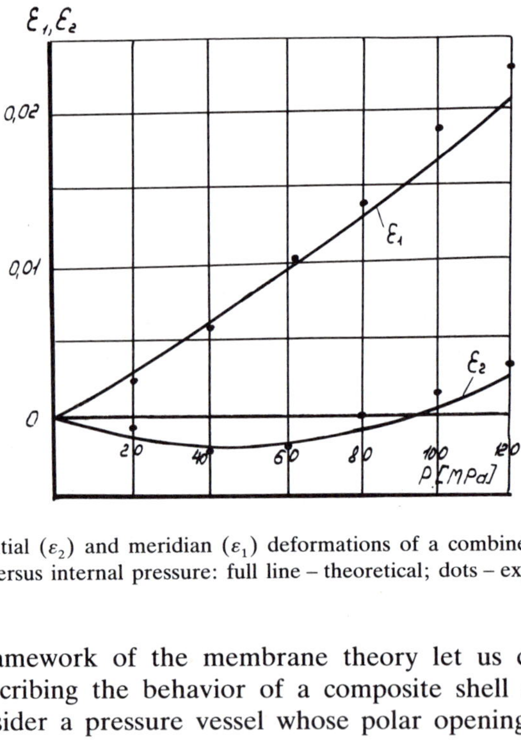

FIG. 29. Circumferential (ε_2) and meridian (ε_1) deformations of a combined vessel at the point $r/a = 0.7$ versus internal pressure: full line – theoretical; dots – experimental.

Within the framework of the membrane theory let us consider now the relationships describing the behavior of a composite shell (VASIL'EV, 1969). First of all, consider a pressure vessel whose polar opening is closed with a rigid cover. Let us relate the shell surface to the coordinates r, Z, θ before deformation and to – r_1, Z_1, θ after deformation and put

$$r_1 = r + u_1 \quad Z_1 = Z + u_2 , \tag{3.20}$$

where u_1 and u_2 are the radial and axial displacements which are taken to be the same for all layers. The equilibrium equations for the state after deformation are

$$N_1 = \frac{pR_2}{2} \quad N_2 = \frac{pR_2}{2}\left(2 - \frac{R_2}{R_1}\right), \tag{3.21}$$

where

$$R_1 = -\left(\frac{\mathrm{d}^2 r_1}{\mathrm{d}Z_1^2}\right)^{-1}\left[1 + \left(\frac{\mathrm{d}r_1}{\mathrm{d}Z_1}\right)^2\right]^{3/2}, \quad R_2 = r_1\left[1 + \left(\frac{\mathrm{d}r_1}{\mathrm{d}Z_1}\right)^2\right]^{1/2} \tag{3.22}$$

are the principal radii of curvature of the surface.

Now returning to variables r and Z in (3.22), we obtain by means of (3.20)

$$R_1 = -\frac{[(v_1')^2 + v_2^2]^{3/2}}{v_1'' v_2 - v_1' \cdot v_2'}, \quad R_2 = \frac{v_1}{v_2}[(v_1')^2 + v_2^2]^{1/2}, \tag{3.23}$$

where

$$v_1 = r + u_1, \quad v_2 = 1 + u_2', \tag{3.24}$$

and $(\ )'$ denotes the derivative with respect to Z. The stress resultants N_1 and N_2 are related to the stresses in the layers along the fibres $\sigma_{\|i}$, as well as to the layer thicknesses h_{1i} and angles of orientation φ_{1i} (Fig. 30) after deformation as follows

$$N_1 = \sum_{i=1}^{k} \sigma_{\|i} h_{1i} \cos^2 \varphi_{1i}, \quad N_2 = \sum_{i=1}^{k} \sigma_{\|i} h_{1i} \sin^2 \varphi_{1i}. \tag{3.25}$$

Here, k is the number of layers. From Fig. 30, before and after deformation we have

$$\mathrm{d}s_1 = \mathrm{d}Z[1 + (r')^2]^{1/2}; \quad \mathrm{d}s_2 = r\,\mathrm{d}\theta,$$
$$\mathrm{d}S_1 = \mathrm{d}Z_1[1 + (\mathrm{d}r'/\mathrm{d}Z_1)^2]^{1/2}; \quad \mathrm{d}S_2 = r_1\,\mathrm{d}\theta.$$

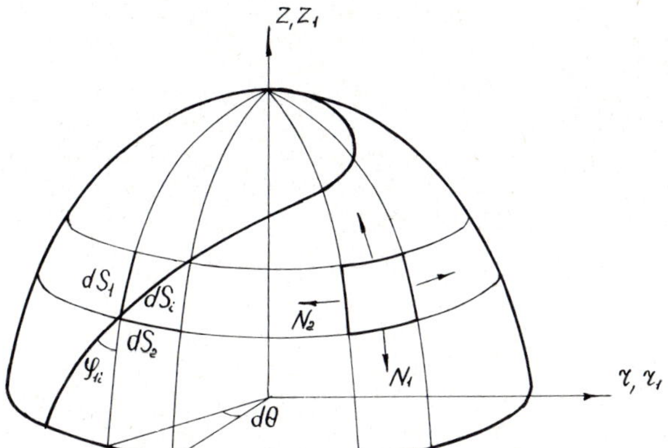

FIG. 30. A composite shell of revolution.

Taking into account the relationships (3.20), we can obtain the following expressions for the meridian and circumferential strains

$$\varepsilon_1 = \frac{dS_1}{ds_1} - 1 = \left[\frac{(v_1')^2 + v_2^2}{1 + (r')^2}\right]^{1/2} - 1\,,$$
$$\varepsilon_2 = \frac{dS_2}{ds_2} - 1 = \frac{v_1}{r} - 1\,. \tag{3.26}$$

The angle of orientation before and after deformation is given by

$$\sin\varphi_i = \frac{ds_2}{ds_i}\,; \qquad \cos\varphi_i = \frac{ds_1}{ds_i}\,,$$
$$\sin\varphi_{1i} = \frac{dS_2}{ds_{1i}} \qquad \cos\varphi_{1i} = \frac{dS_1}{ds_{1i}}\,.$$

Taking into account the relationships (3.26) and also the condition $ds_{1i} = (1 - \varepsilon_i)\,ds_i$, which relate the length of the fibre element of the ith layer before and after deformation to its relative elongation ε_i, we have

$$\sin\varphi_{1i} = \sin\varphi_i\,\frac{1+\varepsilon_2}{1+\varepsilon_i}\,, \qquad \cos\varphi_{1i} = \cos\varphi_i\,\frac{1+\varepsilon_1}{1+\varepsilon_i}\,. \tag{3.27}$$

From (3.27) we obtain

$$(1+\varepsilon_i)^2 = (1+\varepsilon_1)^2\cos^2\varphi_i + (1+\varepsilon_2)^2\sin^2\varphi_i\,. \tag{3.28}$$

All the fibres used in modern composites are linearly elastic right up to failure, so that,

$$\sigma_{\parallel i} = E^i_\parallel\,\varepsilon_i\,, \tag{3.29}$$

where $E^i_\parallel$ is the modulus of elasticity of the ith layer in tension along the fibres. Now let us find the law of variation of the layer thickness h_i during deformation. Let n_i fibres with the cross-sectional area f_i and the angle of orientation φ_i pass through the cross section of the shell. Then, before and after deformation

$$\frac{n_i f_i}{\cos\varphi_i} = 2\pi r h_i\,, \qquad \frac{n_i f_{1i}}{\cos\varphi_{1i}} = 2\pi r_1 h_{1i}\,.$$

Hence, taking into account (3.20), (3.24) and (3.27), we have

$$h_{1i} = h_i\lambda_i\,\frac{r(1+\varepsilon_i)}{v_1(1+\varepsilon_1)}\,. \tag{3.30}$$

Here, $\lambda_i = f_{1i}/f_i$; f_{1i} is the cross-sectional area of the fibre after deformation. For rigid fibres, $\lambda_i = 1$, and for polymer fibres, which can be regarded as incompressible, $\lambda_i = 1/(1+\varepsilon_i)$. Finally, by substituting (3.25) into equilibrium equations in (3.21) and taking into account the relationships (3.23), (3.26)–(3.30), we have

$$(1+\varepsilon_1)\sum_{i=1}^{k}\frac{E_{\parallel}^{i}\varepsilon_i h_i\lambda_i}{1+\varepsilon_i}\cos^2\varphi_i = \frac{pv_1^2}{2v_2 r}[(v_1')^2+v_2^2]^{1/2},$$

$$\frac{(1+\varepsilon_2)^2}{1+\varepsilon_1}\sum_{i=1}^{k}\frac{E_{\parallel}^{i}\varepsilon_i h_i\lambda_i}{1+\varepsilon_i}\sin^2\varphi_i = \frac{pv_1^2}{2v_2^2}[(v_1')^2+v_2^2]^{1/2}\left\{2+\frac{v_1(v_1''v_2 - v_1'v_2')}{v_2[(v_1')^2+v_2^2]}\right\}, \tag{3.31}$$

where

$$(1+\varepsilon_1)^2 = \frac{(v_1')^2+v_2^2}{1+(r')^2}, \quad (1+\varepsilon_2)^2 = \frac{v_1^2}{r^2},$$

$$(1+\varepsilon_i)^2 = \frac{(v_1')^2-v_2^2}{1+(r')^2}\cos^2\varphi_i + \frac{v_1^2}{r^2}\sin^2\varphi_i$$

The solution of (3.31) allows us to define the functions v_1 and v_2, i.e. the displacements u_1 and u_2, as given by (3.24).

To illustrate the above mentioned peculiarities of composite pressure vessels, we may consider an infinitely long cylindrical shell with domes at each end, reinforced at angles $\pm\varphi$ to the generator. For such a shell, v_1, v_2, ε_1, ε_2, ε_i, φ_i are constants. Putting $r = R$ in the system (3.31) and retaining one term in the sums, we obain after a few transformations

$$\frac{(1+\varepsilon_1)\varepsilon\cos^2\varphi}{(1+\varepsilon_2)^2(1+\varepsilon)} = \frac{\bar{p}}{2}, \quad \frac{\varepsilon\sin^2\varphi}{(1+\varepsilon_1)(1+\varepsilon)} = \bar{p}, \tag{3.32}$$

where

$$\bar{p} = \frac{pa}{E_{\parallel}h\lambda}, \quad (1+\varepsilon)^2 = (1+\varepsilon_1)^2\cos^2\varphi + (1+\varepsilon_2)^2\sin^2\varphi. \tag{3.33}$$

For a shell reinforced by organic fibres with $p = 100$ MPa, $E = 100$ GPa, $a = 0.1$ m, $h = 1.10^{-3}$ m, $\lambda = 1$, the dependence of ε_1 and ε_2 on the initial angle of orientation φ is shown in Fig. 31. Consider the initial angle $\varphi = \varphi_0$ which corresponds to the point of intersection of the curves $\varepsilon_1(\varphi)$ and $\varepsilon_2(\varphi)$. In this case, $\varepsilon_1 = \varepsilon_2 = l$ and, as follows from the second equation in (3.33), $\varepsilon = l$. Then we have from (3.32)

$$\varphi_0 = \text{arc tg}\,\sqrt{2}, \quad \frac{l}{(1+l)^2} = \frac{3\bar{p}}{2}.$$

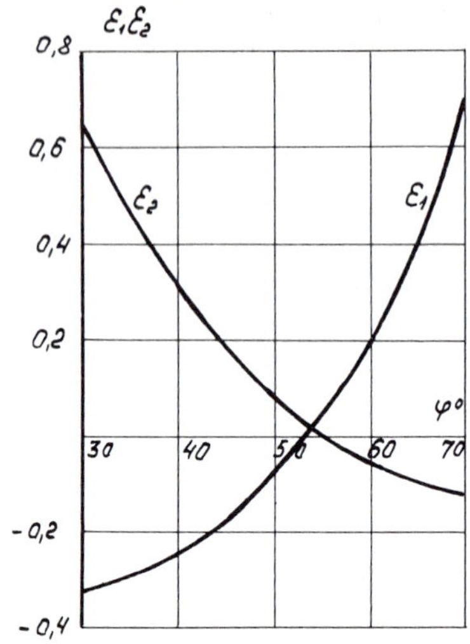

FIG. 31. Axial circumferential strains versus fibre orientation angle for a cylindrical shell.

For $\varphi = \varphi_0$, $u_1/a = u_2^1 = 0.015$, i.e. the displacements are small. Therefore the condition for existence of the shell as a netting system is fulfilled. From (3.27) with $\varepsilon_1 = \varepsilon_2 = \varepsilon$ it follows that $\varphi_1 = \varphi_0$, i.e. the angle φ_0 does not change during deformation. The angle φ_0 is the equilibrium angle for the shell under consideration. Indeed, from (3.21) and (3.25) which in this case can be written as $\sigma_{\|} h \cos^2 \varphi = 0.5pa$, $\sigma_{\|} h \sin^2 \varphi = pa$ it follows that equilibrium is possible only if $\varphi = \varphi_0$. The reason why there are large strains with $\varphi \neq \varphi_0$ is that the system is geometrically changeable. The large strains are due to rotation of the fibres. However, it is possible to eliminate the geometrical variability by introduction of one more layer, say a hoop layer, into the system. Under such conditions, deformation, rotation angles and displacements will be small.

Consider composite shells of revolution which are invariable in the above mentioned sense. Put $u_1 = u_0$ and $u_2' = w_0$ and assume that the functions u_0/r, w_0 and their derivatives are negligible as compared to unity. Then the system (3.31) takes the form

$$\frac{A_{11}}{1+(r')^2}(r'u_0' + w_0) + A_{12}\frac{u_0}{r} = \frac{pr}{2}[1+(r')^2]^{1/2},$$
$$\frac{A_{12}}{1+(r')^2}(r'u_0' + w_0) + A_{22}\frac{U_0}{r} = \frac{pr}{2}[1+(r')^2]^{1/2}\left[2 + \frac{rr''}{1+(r')^2}\right], \tag{3.34}$$

where

$$A_{11} = \sum_{i=1}^{k} E_{\|}^i h_i \cos^4 \varphi_i\,; \quad A_{12} = \sum_{i=1}^{k} E_{\|}^i h_i \sin^2 \varphi_i \cos^2 \varphi_i\,,$$

$$A_{22} = \sum_{i=1}^{k} E_{\|}^i h_i \sin^4 \varphi_i\,.$$

Equation (3.34) corresponds to the linear membrane theory of shells of revolution.

For a cylindrical shell, we have

$$u_0 = \frac{0.5A_{22} - A_{12}}{A_{11}A_{22} - A_{12}^2}\, pa^2\,; \quad w_0 = \frac{A_{11} - 0.5A_{12}}{A_{11}A_{22} - A_{12}^2}\, pa\,. \tag{3.35}$$

As an example, let us consider a fibreglass–epoxy cylindrical pressure vessel reinforced with a ring. The shell with a radius of $a = 0.1$ m consists of symmetrical spiral layers with $\varphi = 36°$, $h = 1.2 \times 10^{-3}$ m, $E_{\parallel} = 41.4$ GPa and a hoop layer with $\varphi = 90°$, $h = 0.93 \times 10^{-3}$ m and $E_{\parallel} = 53.75$ GPa. The dimensions and modulus of elasticity of the ring (Fig. 32) are: $b = 34 \times 10^{-3}$ m, $H = 2.75 \times 10^{-3}$ and $E = 17.95$ GPa, respectively. The intensity of internal pressure $p = 50$ MPa. In Fig. 32, the circumferential deformation $\varepsilon_2 = u_0/R$ is shown by the dotted line.

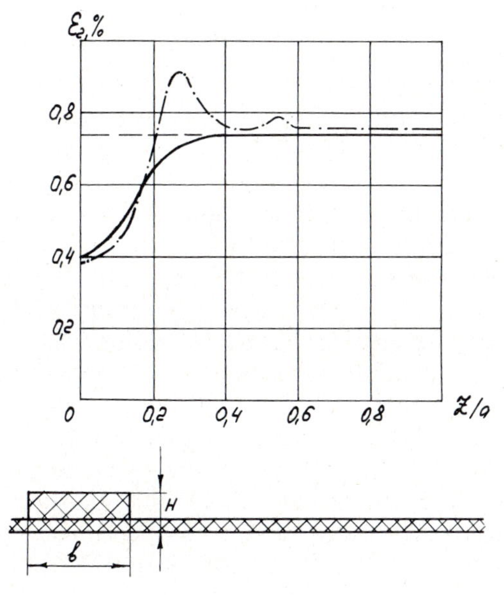

FIG. 32. Distribution of circumferential strain throughout the length of the shell.

In order to eliminate the discontinuity in displacements obtained with the membrane shell theory (in particular, in the above example – the discontinuous displacements for the shell and the ring) we shall present the functions v_1 and v_2 in the form

$$v_1 = r + u_0 + u \quad v_2 = 1 + w_0 + w \tag{3.36}$$

and adopt the following hypotheses:

(1) the functions u_0, w_0 are solutions of the system (3.34);
(2) the functions u_0/r, u/r, w_0, w are negligible as compared to unity, i.e. $v_1 \approx r$ and $v_2 \approx 1$ (except for the expression for deformation where the unity term disappears);
(3) derivatives of the functions u_0 and w_0 are small as compared with those of u and w, i.e. $v_1' \approx r' + u'$, $v_2 \approx w'$. Then, retaining only the linear terms in (3.31) and subtracting the respective equations in (3.34), we obtain

$$\frac{A_{11}}{1+(r')^2}(r'u'+w) + A_{12}\frac{u}{r} = \frac{p \cdot r}{2}\,\frac{r'u'}{[1+(r')^2]^{1/2}}\,,$$

$$\frac{A_{12}}{1+(r')^2}(r'u'+w) + A_{22}\frac{u}{r} = \frac{p \cdot r}{2}[1+(r')^2]^{1/2}\left\{\frac{rr''}{1+(r')^2}\left[\frac{1}{r''}(u''-r'w') - \frac{2r'u'}{1+(r')^2}\right] + \frac{2r'u'}{1+(r')^2}\right\}, \quad (3.37)$$

For a cylindrical shell with $r = a$, $r' = r'' = 0$

$$A_{11}w + A_{12}\frac{u}{a} = 0\,, \quad A_{12}w + A_{22}\frac{u'}{a} = \frac{pa^2}{2}u''$$

hence

$$u'' - k^2u = 0 \quad (3.38)$$

where

$$k^2 = \frac{2(A_{11}A_{22} - A_{12}^2)}{pA_{11}a^3}\,.$$

With k = constant solution of (3.38) is

$$u = C_1\,e^{-kz} + C_2\,e^{kz}\,. \quad (3.39)$$

For a long shell we must put $C_2 = 0$ and the total displacements are given by

$$u_1 = u_0 + C_1\,e^{-kz}\,, \quad u_2' = w_0 - \frac{A_{12}}{A_{11a}}C_1\,e^{-kz} \quad (3.40)$$

where u_0 and w_0 have the form (3.35). The stresses in the layers are

$$\sigma_{\|i} = E_{\|}^i\left(u_2'\cos^2\varphi_i + \frac{u_1}{a}\sin^2\varphi_i\right).$$

For the above vessel with a ring, the constant C_1 is determined from the condition that the radial displacements of the shell and the ring are the same (Fig. 32)

$$u_1\left(Z=\frac{b}{2}\right)=\frac{R^2}{B}(pb+2q), \quad q=\frac{pa}{2}u_2'\left(Z=\frac{b}{2}\right),$$

where B is the rigidity of the ring and q is the radial load transmitted from the shell to the ring. The solution given by (3.40), shown by the solid line, is in good agreement with the results of deflection measurements. For comparison the figure also shows the solution for the same problem obtained from the classical moment shell theory (the dash-dotted line).

It follows from Fig. 32 that the additional displacements fade away fairly rapidly with the distance from the edge. In this connection, in general (3.37), as in the equations of the classical edge effect, the coefficients can be taken as constant and equal to their values at the edge, i.e. $A_{ij}=A_{ij}^k$, $r=r_k$, $r'=r_k'$, $r''=r_k''$. In this case, the equations in (3.37) take the form

$$w=\left[\frac{pr_k}{2A_{11}^k}[1+(r_k')^2]^{1/2}-1\right]r_k'u'-\frac{A_{12}^k[1+(r_k')^2]}{A_{11}^k r_k}u,$$

$$\left(1+(r_k')^2-\frac{pr_k(r_k')^2}{2A_{11}^k}[1+(r_k')^2]^{1/2}\right)u''+\frac{2r_k'}{r_k}\left[1+\frac{A_{12}^k(r_k')^2}{2A_{11}^k}\right.$$

$$\left.-\frac{r_kr_k''}{1+(r_k')^2}\right]u'-\frac{2[1+(r_k')^2]^{1/2}}{pA_{11}^k r_k^3}[A_{11}^kA_{22}^k-(A_{12}^k)^2]u=0. \tag{3.41}$$

The stresses in the fibres are given by

$$\sigma_{\parallel i}=E_{\parallel}^i\left[\frac{r'u_1'-u_2'}{1+(r')^2}\cos^2\varphi_i+\frac{u_1}{r}\sin^2\varphi_i\right],$$

where $u_1=u_0+u$, $u_2=w_0+w$, u_0, w_0 being the solution of the system (3.34) and u, w the solution of thc system (3.41).

References

ABIBOV, A.L., B.V. BOYTSOV, G.A. MOLODTSOV and I.YU. SHEIDEMAN (1971), Fabrication of large-scaled counterparts made from oriented fibreglass-reinforced plastics, in: *Application of Plastics in Aircraft Production* (Mashinostroyeniye, Moscow) Ch. 2, p. 63 (in Russian).

BUNAKOV, V.A. and A.L. RADOVINSKII (1975), *Polymer Mechanics* **882** (in Russian).

BUNAKOV, V.A., V.V. VASIL'EV and V.S. PETUSHKOV (1976), Optimal design of composite pressure vessels, in: *Design of Structures*, Issue 17 (Stroyizdat, Moscow) p. 142 (in Russian).

BUNAKOV, V.A., V.D. PROTASOV and S.B. CHEREVATSKII (1982), Optimum design of membrane composite shells of revolution, in: *Mechanics of Composites* (MIR Publishers, Moscow) p. 252.

CHEREVATSKII, S.B. (1970), On the basic equilibrium equations for fibrous shells, in: *Investigations on Plate and Shell Theories*, VI–VII (Kazanskii Universitet, Kazan) p. 738 (in Russian).

ELPAT'EVSKII, A.N. and V.V. VASIL'EV (1972), *Strength of Fibre Reinforced Shells* (Mashinostroyeniye, Moscow) Ch. 2, p. 58 (in Russian).

HARTUNG, R. (1963), *AIAA Journal*, **1**, 159.
JONG, TH. (1972), in: *Contributions to Theory of Aircraft Structures* (Delft) p. 207.
KOLEROV, N.N. (1978), in: *Design, Analysis and Tests of Composite Structures*, Issue VI (Tsentral-'nij aerodinamicheskii institut, Moscow) p. 153 (in Russian).
MITKEVICH, A.B. and V.D. PROTASOV (1975), *Polymer Mechanics*, **983** (in Russian).
OBRAZTSOV, I.F., V.V. VASIL'EV and V.A. BUNAKOV (1977), *Optimum Design of Composite Shells of Revolution* (Mashinostroyeniye, Moscow) p. 111 (in Russian).
RIVLIN, R.S. and A.C. PIPKIN (1963), *Trans. ASME*, Ser. **B1**, 123.
ROSATO, D.V. and C.S. GROVE (1964), *Filament Winding* (Interscience Publishers, New York).
SHUERCH, H. and O.R. BURGGRAF (1964), *AIAA Journal*, 809.
TSYPLAKOV, O.G. (1968), Technological principles of formation of shells from fibreglass-reinforced plastics, in: *Principles of Formation of Shells from Fibreglass-Reinforced Plastics* (Mashinostroyeniye, Leningrad) Ch. 3, p. 94 (in Russian).
VASIL'EV, V.V. and A.N. ELPAT'EVSKII (1967), *Polymer Mechanics*, 915 (in Russian).
VASIL'EV, V.V. (1969), *Aviatsionnaya Tekhnika*, 8 (Izvestiya VUZov., Moscow) (in Russian).
ZIKKEL, J. (1962), *ARS Journal*, 120.

CHAPTER 10

Composite Flywheels

G.G. Portnov

Institute of Polymer Mechanics
Academy of Sciences of the Latvian SSR
23 Aizkraukles Street
226006 Riga
U.S.S.R.

Contents

HANDBOOK OF COMPOSITES, VOL. 2 – Structure and Design
Edited by C.T. HERAKOVICH and Y.M. TARNOPOL'SKII

List of Symbols

r, θ, z – cylindrical coordinates
a – inner radius of a rim
b – outer radius of a rim
h – thickness of a shell (or rim)
V – volume of a structure
J_{p} – polar moment of inertia
$\rho = r/b$ – dimensionless radius
l – length of a spoke
S – cross-sectional area of a spoke
S_{R} – area of the outer cylindrical surface of the rim
S_{S} – overall cross-sectional area of all spokes
ρ_V – mass density
$\rho_{V\mathrm{R}}$ – mass density of a rim
$\rho_{V\mathrm{S}}$ – mass density of a spoke
M – mass of a structure
ω – angular velocity (radians/sec)
$\dot{\omega}$ – angular acceleration (radians/s^2)
v_{u} – ultimate rotational speed
ω_{u} – ultimate angular velocity
$\boldsymbol{K}$ – mass forces
$\boldsymbol{P}$ – surface forces
σ_{ij} – stress tensor
$\sigma_r, \sigma_\theta, \tau_{r\theta}$ – stress tensor components
ε_{ij} – strain tensor
$\varepsilon_r, \varepsilon_\theta, \gamma_{r\theta}$ – strain tensor components
$I_1(T_\sigma)$ – the first invariant of stress tensor
E_r, E_θ – Young's modulus in the radial and circumferential directions
u_r – radial displacement
$\nu_{r\theta}, \nu_{\theta r}$ – Poisson's ratios
u, v – longitudinal and transverse displacements of a spoke
σ_0^+ – tensile strength along fibers
$\sigma_r^+, \sigma_\theta^+$ – tensile strength of a rim in the radial and circumferential directions, respectively
σ_{s}^+ – tensile strength of a spoke
τ_{su} – shear strength of a spoke
$\tau_{r\theta}^{\mathrm{u}}$ – shear strength of a rim
W – energy storage capacity
W^M – specific mass energy storage capacity
W^V – specific volumetric energy storage capacity

$\dot{W}$ – power of a flywheel
$\dot{W}^M$ – specific mass power
$\dot{W}^V$ – specific volumetric power

1. Introduction

High-strength fiber composites are the most promising materials for flywheel rotor the design of which is the main task in the development of inertial energy accumulators. The interest in inertial energy accumulators such as flywheels has increased with the growing technological needs that require to have highly efficient energy accumulators capable of producing high-power impulses. The use of composite materials for flywheel rotors has two advantages over more conventional materials: high specific strength along the fiber direction and, in a properly designed structure, relatively safe (infragmentary) mode of failure. The problems associated with the development of composite flywheels have been investigated by GULYA (1976), CHEREVATSKII et al. (1983a) and CHRISTENSEN (1979). A variety of problems arising in the design, testing and use of mechanical energy accumulators have been treated in the Proceedings of three symposia on flywheel technology (1975, 1977, 1980) and GENTA (1985).

Tests on models and experimental structures (VOROPAEV et al., 1982; KNIGHT et al., 1977) have shown that the energy storage capacity of flywheels attained in practice is still far from the potential capabilities of fiber composites. This has led to the necessity of developing design methods for the power and energy storage capacity of various rotating fiber composite structures and the evaluation of economical expediency with regard to their use as energy storing elements in flywheels. It should be emphasized that designing for energy storage capacity is not a traditional topic in the mechanics of deformable solids.

In this chapter, which is based mainly on the studies performed under the guidance of the author, various types of rotational fiber composite structures have been analyzed and methods of increasing their energy storage capacity studied. At first, the relationship between the state of stress in rotating structures and their energy storage capacity is evaluated. Subsequently, the conventionally optimal fiber composite structures which only undergo tension along the fibers are analyzed. The analysis of thick-walled rims and discs follows for composites which are also loaded across the reinforcement. The survey is completed by treatment of one of the most promising flywheel designs – a circular system consisting of a rim and hub connected by spokes–chords of a unidirectional fiber composite (the so-called chord flywheel).

2. Relationship between energy storage capacity of rotating bodies and global characteristics of the state of stress

Optimal design of rotating structures based on energy storage capacity is aimed at attaining the maximum magnitudes of specific energy stored per unit mass

$W^M = W/M$ (where W is the maximum kinetic energy stored in a rotating structure, M is mass of the structure) and/or specific energy stored per unit volume $W^V = W/V_0$ (where V_0 is the volume occupied by the rotating structure). These characteristics are complex and they depend on the ultimate rotational speed of the structure, its geometrical shape and material density. The design for maximum energy storage capacity cannot be pursued within the framework of the conventional design methods employed for machine elements. Therefore, it is useful to search for approaches which would allow one to work out general principles of the design of such structures. The approach based on the analysis of the relationship between energy storage capacity of rotating bodies and the state of stress is one of them.

Let us consider the global characteristics of the state of stress in rotating bodies. In order to do so, we shall employ the relationships derived by LANDAU and LIFSHITZ (1965) and LURIE (1970) through transformation of equations of equilibrium for a continuum over the volume V, which is subjected to the action of body forces $\boldsymbol{K}$ and surface load $\boldsymbol{P}$ acting on the surface S (or a part of it) encompassing the volume:

$$\iiint_V \sigma_{mn}\,\mathrm{d}V = \tfrac{1}{2}\left[\iiint_V (x_m K_n + x_n K_m)\,\mathrm{d}V + \iint_S (x_m P_n + x_n P_m)\,\mathrm{d}S\right]$$

$$m, n = 1, 2, 3\,. \tag{2.1}$$

where σ_{mn} are components of the stress tensor; K_n and P_n are projections of the vectors of body forces and load in the coordinate axes 1, 2, 3 and x_n the coordinates of the points of their application.

As can be seen in (2.1), the volume integral of the magnitudes of stress tensor components can be directly evaluated from the forces acting on the body without a preliminary solution of the problem.

By summing the volume integrals of the normal stresses σ_{nn} we obtain:

$$\iiint_V I_1(T_\sigma)\,\mathrm{d}V = \iiint_V \boldsymbol{Kr}\,\mathrm{d}V + \iint_S \boldsymbol{Pr}\,\mathrm{d}S\,, \tag{2.2}$$

where $\boldsymbol{r}$ is the radius-vector; $I_1(T_\sigma) = \sigma_{ij}\delta_{ij}$, where δ_{ij} is Kronecker's delta; the expression $\int\int_S \boldsymbol{Pr}\,\mathrm{d}S$ has been called the load potential and is conceived as the work performed by surface forces upon transfer of the points of their application to the origin of coordinates.

Relationship (1.2) was obtained by ALFUTOV and ZINOV'EV (1973) by a somewhat different procedure, namely, by application of the principle of virtual displacements to equations of equilibrium used for the treatment of problems of optimal design of shells loaded by a system of surface forces. The approach generalizes the results of energy storage capacity analysis carried out by JONSON and GORMAN (1980).

For a body rotating with an angular velocity ω the volume integral on the right-hand side of (2.2) is equal to twice the kinetic energy

$$\iiint_V \boldsymbol{K}\boldsymbol{r}\,\mathrm{d}V = \iiint_V \rho_V \omega^2 \boldsymbol{r}_0 \boldsymbol{r}\,\mathrm{d}V = \iiint_V \rho_V \omega^2 r_0^2\,\mathrm{d}V = 2W\,, \tag{2.3}$$

where r_0 is the distance of the point to the rotational axis; ρ_V is the mass density. In this way, we obtain

$$\iiint_V I_1(I_\sigma)\,\mathrm{d}V = 2W + \iint_S \boldsymbol{P}\boldsymbol{r}\,\mathrm{d}S\,. \tag{2.4}$$

For a rotating body free of surface loads we shall obtain the following magnitude of energy stored per unit mass

$$W^M = \frac{W}{\rho_V V} = \frac{\iiint_V I_1(T_\sigma)\,\mathrm{d}V}{2\rho_V V} = \frac{I_{1\mathrm{m}}(T_\sigma)}{2\rho_V}\,, \tag{2.5}$$

where $I_{1\mathrm{m}}$ is the mean value of the first invariant of the stress tensor over the volume V.

It follows from relationships (2.4) and (2.5) that

(a) To the specified magnitude of kinetic energy and load potential of a rotating body there corresponds one and the same magnitude of the integral of the first stress tensor invariant per unit volume of the body, which is independent of its shape, volume and material properties.

(b) The difference between the integral of the first stress invariant and load potential, divided by the square of the angular velocity, is equal to the polar moment of inertia of the rotating body. Since as $W = \frac{1}{2}J_\mathrm{p}\omega^2$, where J_p is the polar moment of inertia of a rotating body, we have

$$J_\mathrm{p} = \frac{1}{\omega^2}\left[\iiint_V I_1(T_\sigma)\,\mathrm{d}V - \iint_S \boldsymbol{P}\boldsymbol{r}\,\mathrm{d}S\right]. \tag{2.6}$$

(c) The higher the ratio of the mean magnitude of the first stress tensor invariant (per unit volume of the rotating body) to the mass density at the same load potential, the higher the specific mass energy storage capacity of the body.

(d) The maximum possible mass energy storage capacity of a structure made of homogeneous materials is attained when the maximum value of $I_1(T_\sigma)$ is ensured at each material point of the structure.

The last result is in need of refinement with respect to the value max $I_1(T_\sigma)$ in a manner analogous to that performed by ZINOV'EV (1972) in the treatment of stationary structures of minimum weight. Naturally, max $I_1(T_\sigma)$ should satisfy the strength criterion $\phi(\sigma_{ii})=0$, however, the requirement of attaining $\max I_1(T_\sigma)$ at each point is narrower than the requirement of equal strength for the structure (simultaneous failure of the structure over the whole volume). The state of stress at each point should not be merely maximum but should also correspond to a fairly definite combination of stresses on the failure surface $\phi(\sigma_{ii})=0$. This combination is defined by the point of contact of the failure surface with the plane of the first invariant ($\sigma_{11}+\sigma_{22}+\sigma_{33}=\text{constant}$) which is farthest from the coordinate origin. Under the conditions of plane state of stress the above is illustrated in Fig. 1(a). The maximum possible mass energy storage capacity will be attained in structures with stresses σ_1^*, σ_2^* at each material point. The equally stressed rotating disc of variable thickness of an isotropic material, in which $\sigma_1=\sigma_2=\text{constant}$ belongs to such structures. Such a disc will have the maximum possible mass energy storage capacity. The shape of the ultimate curve $\phi(\sigma_{ii})=0$ for an isotropic material is inessential, since for any convex strength criterion $\max I_1(T_\sigma)$ will be attained due to symmetry in the direction $\sigma_1=\sigma_2$ (see Fig. 1(b)). For an anisotropic material the profile should be chosen on the condition of generating σ_1^*, σ_2^* at each point (see Fig. 1(a)). Likewise, the efficiency of the optimally designed structure with regard to its energy storage capacity can be evaluated from the following value of the functional W:

$$W=\tfrac{1}{2}\left[\iiint_V I_1(T_\sigma)\,\mathrm{d}V-\iint_S \boldsymbol{Pr}\,\mathrm{d}S\right]. \tag{2.7}$$

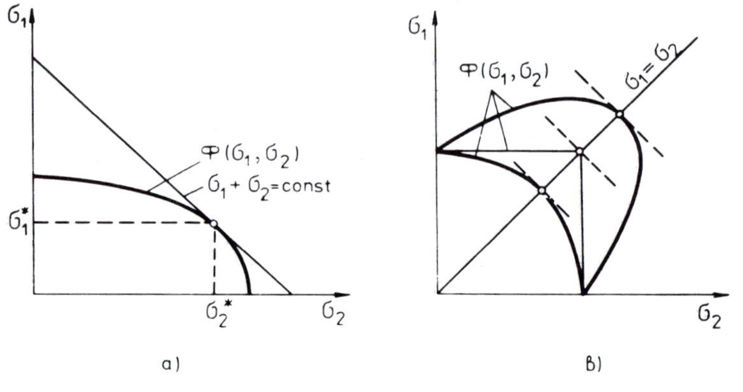

FIG. 1. Determination of components of a uniform state of stress ensuring maximum mass energy storage capacity of a rotating body (optimal combinations of σ_1 and σ_2): (a) anisotropic material; (b) isotropic material.

In designing according to W^M (2.7) should be taken over to the mass of the structure (M), in designing according to W^V it should be taken over the volume (V_0) occupied by the structure.

A study of several properties of a structure in a uniform state of stress ($I_1(T_\sigma) = \max I_1(T_\sigma) = \text{constant}$) is of interest. It follows from the analysis of (2.7) that for equally stressed structures the following conditions hold:

(e) In the case of an equally stressed structure loaded with a surface load that has a positive potential ($\int\int_S \boldsymbol{Pr}\, \mathrm{d}S > 0$), the energy stored per unit mass (W_p^M) is lower than that of an equally stressed structure of the same material (W_0^M) free of external load (self-balanced):

$$W_\mathrm{p}^M = \frac{\max I_1(T_\sigma)}{2\rho_V} - \frac{\int\int_S \boldsymbol{Pr}\, \mathrm{d}S}{2\rho_V V} < W_0 = \frac{\max I_1(T_\sigma)}{2\rho_V} \,. \tag{2.8}$$

(f) In the case when the surface loads acting on equally stressed structures are due to extra masses (a ballast) located at a radial distance r_0 the total energy storage capacity of the structure, with allowance for energy storage capacity of the ballast, is equal to the energy capacity (W_0) of an equally stressed structure of the same volume and without ballast:

$$\begin{gathered} \boldsymbol{P} = m\omega^2 \boldsymbol{r}_0 \,, \quad I_1(T_\sigma) = \text{constant} = \max I_1(T_\sigma) \,, \\ 2W_0 = \max I_1(T_\sigma)V = 2W_\mathrm{p} + \int\int_S m\omega^2 \boldsymbol{r}_0 \boldsymbol{r}\, \mathrm{d}S = 2(W_\mathrm{p} + W_m) \end{gathered} \tag{2.9}$$

where m is the mass distributed over the surface S; W_p is the energy accumulated by the structure itself; W_m is the energy accumulated by the extra mass. At $V = \text{constant}$ the sum $W_\mathrm{p} + W_m$ must also be constant (while both W_p and W_m can vary).

Thus, utilization of ballast in equally stressed structures free of other surface loads does not result in an increase of the maximum value of the energy storage capacity (provided the volume of the structure has been preserved) and, consequently, decreases its mass energy storage capacity. The use of ballast in equally stressed structures turns out to be useful only for reducing the ultimate rotational speed.

(g) If a rotating structure consists of a set of i elements in a uniform state of stress, its mass energy storage capacity is less than or equal to the sum of the mass energy storage capacities of the individual elements multiplied by their relative masses. Since $I_{1i}(T_\sigma)$ is constant throughout the volume of each of the elements, the overall energy storage capacity of the structure is equal to

$$W = \tfrac{1}{2} \sum_i V_i I_{1i} \leq \tfrac{1}{2} \sum_i V_i \max I_{1i} \,, \tag{2.10}$$

where $\max I_{1i}$ are the ultimate magnitudes of the first stress invariant ensuring the maximum mass energy storage capacity of the ith element. Hence, we obtain

$$\frac{W}{M} = \frac{W}{\sum_i V_i \rho_{V_i}} \leqslant \frac{\sum_i V_i \rho_{V_i} \left(\dfrac{V_i \max I_{1i}(T_\sigma)}{2\rho_{V_i} V_i} \right)}{\sum_i V_i \rho_{V_i}} = \frac{\sum_i M_i W_i^M}{M} , \quad (2.11)$$

where M_i and W_i^M are the masses and maximum mass energy storage capacities of the structural elements.

The obtained results allow one, in a number of cases, to obtain estimates of the energy storage capacity of structures without resorting to awkward calculations, in order to compare the efficiency of various methods of increasing the energy storage capacity and to check (with the help of the result given in (b)) the accuracy of computation of the state of stress in rotating structures. The use of (2.7) in structural optimization problems may turn out to be more convenient than the direct calculation of the maximum energy storage capacity of a given structure.

3. Structural design based on netting analysis methods

One of the basic methods of optimal design of laminated structures of fiber composites is the approach based on netting analysis (see OBRAZTSOV and VASIL'EV, 1986; BUNAKOV and PROTASOV, 1986). According to this approach the reinforcement directions in each layer coincide with the principal stress trajectories. It is further assumed that the fiber composite possesses zero stiffness in the direction transverse to the reinforcement and thus carries the load only by tension along the fibers. In this case, calculation of the volume integral of the first invariant in relationship (2.7) becomes considerably simplified, since $\sigma_2 = \sigma_3 = 0$. Evidently, the maximum energy storage capacity is attained in equally stressed structures with $\sigma_1 = \text{constant} = \overset{+}{\sigma}_0$, where $\overset{+}{\sigma}_0$ is the tensile strength of a unidirectional composite along the fibers.

The mass energy storage capacity of rotating, equally stressed, net fiber composite structures is independent of their shape and volume as well as of the type of trajectories and reinforcement patterns. If the reinforcing fibers in a self-balanced (free or surface forces) structures are equally stressed, their maximum mass energy storage capacity is equal to

$$W^M = \frac{\iiint_V \sigma_1 \, \mathrm{d}V}{2\rho_V V} = \frac{\overset{+}{\sigma}_0}{2\rho_V} . \quad (3.1)$$

Thin rings and equally stressed bars of variable cross section (PORTNOV and POLYAKOV, 1982), equally stressed filamentary discs (KYSER, 1965; BOKOV et al.,

1981) and equally stressed discs of variable cross section with radial-circumferential reinforcement (CHRISTENSEN and WU, 1977) can be regarded as such structures. The common results obtained from the analysis of their mass energy storage capacities serve as particular illustrations of the above-described property of equally stressed, net rotating structures. It should be noted that, in some cases, arriving at the above result requires rather awkward calculations.

If a structure of mass M is composed of i net elements in a uniform stage of stress, its mass energy storage capacity, as follows from (g), does not exceed the following value:

$$W^M_{\Sigma_i} \leqslant \tfrac{1}{2} \sum_i \frac{M_i}{M} \frac{\overset{+}{\sigma}_{0i}}{\rho_{Vi}} . \tag{3.2}$$

Thus, for instance, it is not useful to design a freely rotating thin rim as a multilayered structure; it is more expedient that the thin rim and the equally stressed spokes connecting the rim to the shaft be made of a material with the highest specific strength ($\max \overset{+}{\sigma}_{0i}/\rho_{Vi}$).

The attachment of mass (ballast) to equally stressed bars and thin rings, as follows from (f), does not cause an increase in the energy storage capacity. The use of ballast results merely in the reduction of the maximum rotational speed of the structure:

$$\frac{\omega_{\mathrm{m}}}{\omega_0} = \left(\frac{J_{\mathrm{p}0}}{J_{\mathrm{p}0} + J_{\mathrm{pm}}} \right)^{1/2} , \tag{3.3}$$

where ω_{m} and ω_0 are the maximum angular velocity of the ring with and without ballast; $J_{\mathrm{p}0}$ is the moment of inertia of the ring; and J_{pm} is the moment of inertia of the ballast.

Shells are the most widely used structures designed by netting analysis. Let us consider, with the help of (2.7), the energy storage capacity of a rotating filamentary shell symmetrical about the equatorial plane (Fig. 2). It follows from the equilibrium condition that the projection on the z-axis of resultant meridianal forces in all cross-sections should be constant (the index 1 designates the magnitudes of parameters at the equator of the shell):

$$2\pi rh \cos^2 \varphi \sin \alpha \cdot \sigma = \text{constant} = 2\pi a h_1 \cos^2 \varphi_1 \cdot \sigma_1 = Q . \tag{3.4}$$

By employing the relationship $\sin \alpha = h/h_r$, where h_r is the thickness of a shell in the section perpendicular to the rotational axis, and the relationship $rh \cos \varphi = ah_1 \cos \varphi_1$ which follows from the condition of continuity of the reinforcement (see OBRAZTSOV and VASIL'EV, 1986), we obtain:

$$rh_r/\sigma = ah_1/\sigma_1 . \tag{3.5}$$

With the help of (3.5), the material volume for a shell can be calculated in the

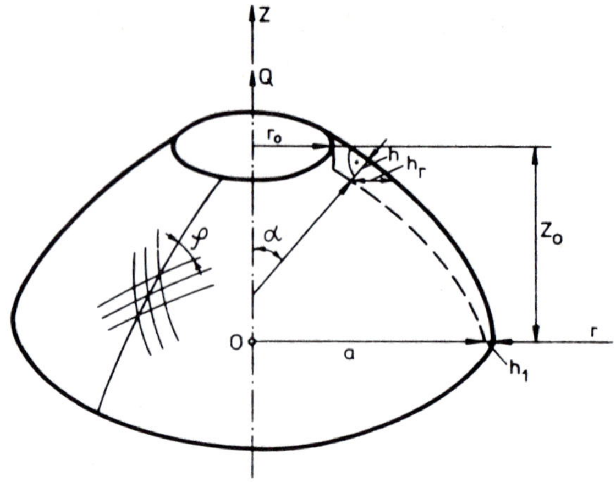

FIG. 2. Parameters of a thin-walled shell, symmetrical about the equatorial plane.

following fashion:

$$V = 2\int_0^{z_0} 2\pi r h_r \,\mathrm{d}z = 4\pi a h_1 \int_0^{z_0} \frac{\sigma}{\sigma_1}\,\mathrm{d}z\,. \tag{3.6}$$

It follows from (3.6) that the mass of an equally stressed shell (M_c) is

$$M_c = 4\pi a h_1 z_0 \rho_V\,. \tag{3.7}$$

Employing (3.4) and (3.5), let us calculate the values which enter (2.7) (the index 0 designates the values of parameters at the polar opening of the shell):

$$\iiint_V \sigma\,\mathrm{d}V = 4\pi \int_0^{z_0} r\sigma h_r\,\mathrm{d}z = 4\pi a h_1 \int_0^{z_0} \frac{\sigma^2}{\sigma_1}\,\mathrm{d}z\,, \tag{3.8}$$

$$\iint_S \boldsymbol{Pr}\,\mathrm{d}S = A_r + A_z\,,$$

$$A_z = 2Qz_0 = 4\pi a h_1 \cos^2\varphi_1 \sigma_1 z_0\,, \tag{3.9}$$

$$A_r = -4\pi r_0 P_r r_0 = -4\pi r_0^2 h_0 \cos\alpha_0 \sigma_0 \cos^2\varphi_0$$

$$= -4\pi r_0 h_1 a \cos^2\varphi_1 \sigma_1 \operatorname{ctg}\alpha_0\,, \tag{3.10}$$

where A_z is the potential of the axial component of external forces acting at the polar opening and A_r is the potential of the radial component of external forces acting at the polar opening.

Employing (3.8)–(3.10), we express the kinetic energy of a filamentary shell as follows:

$$W = 2\pi a h_1 \left[\int_0^{z_0} \frac{\sigma^2}{\sigma_1}\, dz - \cos^2 \varphi_1 \sigma_1 \left(z_0 - \frac{r_0}{\operatorname{tg} \alpha_0} \right) \right]. \tag{3.11}$$

The mass energy storage capacity of a filamentary shell is equal to

$$W^M = \tfrac{1}{2} \frac{\int_0^{z_0} \frac{\sigma^2}{\sigma_1}\, dz - \sigma_1 \cos^2 \varphi_1 (z_0 - r_0 \operatorname{ctg} \alpha_0)}{\rho_V \int_0^{z_0} \sigma/\sigma_1\, dz}. \tag{3.12}$$

For an equally stressed shell with $\sigma = \overset{+}{\sigma}_0$ we obtain

$$W_c^M = \tfrac{1}{2} \frac{\overset{+}{\sigma}_0}{\rho_V} \left(\sin^2 \varphi_1 + \frac{r_0 \cos^2 \varphi_1}{z_0 \operatorname{tg} \alpha_0} \right). \tag{3.13}$$

It follows from (3.13) that the mass energy storage capacity of equally stressed, closed ($r_0 = 0$) shells, or shells with a smooth coupling at the polar opening ($\alpha = \pi/2$) in $\sin^2 \varphi_1$ is less than the mass energy storage capacity of a thin ring.

At $\varphi_1 = \pi/2$, it follows from (3.4) that the longitudinal force in the shell section should be zero. This is possible only at $\varphi \equiv \pi/2$ or at $z \equiv 0$. In the first case, the shell should be cylindrical, in the second, the shell degenerates into an equally stressed, flat disc. For a cylinder and a self-balanced disc with fibers in contact with the polar opening ($\varphi_0 = \pi/2$) the mass energy storage capacities are equal and are given by (3.1).

In the case when $\varphi_0 = 0$ (the fibers in the polar opening are directed along the meridian) it follows from (3.4), and the relationship $rh \cos \varphi = ah_1 \cos \varphi_1$ for an equally stressed shell, that $\sin \alpha_0 = \cos \varphi_1$ and the expression for the mass energy storage capacity (3.13) takes the form

$$W_c^M(\varphi_0 = 0) = \tfrac{1}{2} \frac{\overset{+}{\sigma}_0}{\rho_V} \left(\sin^2 \varphi_1 + \frac{r_0}{2z_0} \sin 2\varphi_1 \right). \tag{3.14}$$

It follows from (3.14) that for shells with a shaft or hub in the polar opening, loaded not only in the axial but also radial direction, some gain in the mass energy storage capacity, in comparison with a ring, can be attained at

$$\varphi_1 \geqslant \operatorname{arctg} \frac{z_0}{r_0}. \tag{3.15}$$

It should be noted that the analysis of the energy storage capacity of such shells without consideration of the load-carrying and energy storage capacities of highly stressed hubs or shafts is only of theoretical interest. Shells which can be fabricated by the continuous winding method, i.e. shells with $\alpha_0 = \pi/2$ seem to be promising.

The meridian shapes and reinforcement trajectories of equally stressed shells have been studied by BOKOV et al. (1981). Their methods of calculation and analysis have been outlined by OBRAZTSOV and VASIL'EV (1986). However, it is possible to fabricate only shallow, equally stressed shells with high aspect ratios of r_0/a and z_0/a with the winding process. In the majority of cases, reinforcement trajectories for equally stressed shells differ greatly from the geodesic lines which makes their winding impossible. In view of this CHEREVATSKII et al. (1983a,b) and POLYAKOV et al. (1983) investigated the technologically important rotating shells wound along geodesic lines as well as along lines of ultimate deviation from the geodesic ones. The specific mass energy storage capacity of such shells turned out to be low. Various ways of increasing the absolute energy storage capacity of rotating shells, such as the use of ballast which entirely fills up the volume of a shell (VASIL'EV et al., 1982) and attachment of shells to a rim (POLYAKOV et al., 1983; CHEREVATSKII et al., 1983) have also been studied.

As a practical consequence of the use of shells as energy storing elements of flywheels we can emphasize the following:

(1) The mass energy storage capacity of even equally stressed shells is less than that of rings and equally stressed discs; unequally stressed shells are essentially still more inferior.

(2) Considerable forces are generated at the polar opening of rotating shells which might require massive hubs and, which consequently, will result in a further reduction of the mass energy storage capacity.

(3) Shells occupy appreciable volume.

The obtained results confirm the fact that the use of shells as energy storing elements in flywheels can be beneficial only in those cases when the requirements for specific energy storage capacities are not decisive. However, they can be used as rotating, protective housings for flywheels such as metallic ones as well as in other technological applications, involving centrifuges for instance.

4. Discs

4.1. *Estimates of energy storage capacity*

In the analysis of the energy storage capacity of composite discs let us assume material of the disc to be an anisotropic continuum. In the calculation of the state of stress in rotating, polar-orthotropic discs of finite axial thickness we can employ the formulas of the plane state of stress without incurring significant errors (KULAKOV and PORTNOV, 1981). This enables us to express (2.7) in terms of radial (σ_r) and circumferential (σ_θ) stresses which are constant throughout the thickness of the disc

$$W = \pi \int_a^b rh(\sigma_r + \sigma_\theta)\,\mathrm{d}r - \pi[p(a)h(a)a^2 + p(b)h(b)b^2]\,, \tag{4.1}$$

where h is the axial thickness of the disc; b and a are the outer and inner radii of the disc; $p(a)$ and $p(b)$ is the surface load acting at the opening and over the periphery of the disc.

At a specified surface load, the highest energy storage capacity is attained in an equally stressed disc with $\sigma_r = \sigma_r^* = \text{constant}$, $\sigma_\theta = \sigma_\theta^* = \text{constant}$, where σ_r^* and σ_θ^* are coordinates of the point of intersection of the failure envelope, characterizing the material strength of the disc, with the line $\sigma_r + \sigma_\theta = \text{constant}$ (see condition (d) of Section 2 and Fig. 1(a)). It follows from $\sigma_r = \text{constant} = \sigma_r{}^*$, $\sigma_\theta = \sigma_\theta^* = \text{constant}$ that $\varepsilon_r = \text{constant}$, $\varepsilon_\theta = \text{constant}$ and from the condition of strain compatibility one obtains the requirement that $\varepsilon_r = \varepsilon_\theta$. For a linearly-elastic material it is possible to satisfy this condition only in the case when

$$\frac{\sigma_\theta^*}{\sigma_r^*} = \frac{E_\theta(1 + \nu_{\theta r})}{E_r(1 + \nu_{r\theta})} = \alpha\,. \tag{4.2}$$

The disc with $\sigma_\theta = \sigma_\theta^* = \text{constant}$ and $\sigma_r = \sigma_r^* = \text{constant}$ fulfilling (4.2) is theoretically feasible provided that its thickness varies in accordance with the law (CHRISTENSEN, 1979):

$$h = kr^{-(1-\alpha)}\,\mathrm{e}^{\lambda r^2}\,, \tag{4.3}$$

where k and λ are constants (k is determined from the boundary condition). As can be seen from (4.3), the cases $\alpha \gtrless 1$ are not realizable. Isotropic materials as well as polar orthotropic and quasi-isotropic composites with $E_r = E_\theta$, i.e. materials with $\sigma_r^* = \sigma_\theta^* = \sigma$ correspond to the case $\alpha = 1$ which is of practical interest. The mass energy storage capacity of a disc whose profile is given by (4.3) and which is free of surface loads is equal to

$$W^M = \frac{\sigma_r^* + \sigma_\theta^*}{2\rho_V} = \frac{\sigma^*}{\rho_V}\,. \tag{4.4}$$

Thus, it is possible to realize some gain in the mass energy storage capacity upon transition from net structures to profiled fiber composite solid discs with symmetrical radial-hoop reinforcement or quasi-isotropic layer layup only in those cases, when the strength of these materials, characterized by the value σ^* (evaluated in accordance with the condition (d) of Section 2) satisfies the requirement

$$\sigma^* > \tfrac{1}{2}\sigma_0^+\,, \tag{4.5}$$

where $\overset{+}{\sigma}_0$ is the tensile strength of a unidirectional layer of the composite along the fibers.

It should be noted that fabrication and use of fiber composite flywheels in the form of profiled solid laminated discs is associated with a variety of technical problems. These problems comprise such aspects as the impossibility of exact realization of the theoretical profile ($b \to \infty$), technological complexities in the production of discs, losses in the load carrying capacity associated with attachment of discs, and the possibility of preliminary delamination along the radial planes due to stress concentration at the edges among others. Research results presented at the Flywheel Technology Symposia in 1975 and 1977 also confirm the fact that shaping rotating composite discs is of low efficiency with regard to their energy storage capacity. Therefore, in the ensuing discussion, only discs of constant thickness will be considered.

Let us employ the maximum stress criterion for the evaluation of the energy storage capacity of discs of constant thickness (in a nonuniform state of stress)

$$\sigma_r \leqslant \overset{+}{\sigma}_r \,, \quad \sigma_\theta \leqslant \overset{+}{\sigma}_\theta \,. \tag{4.6}$$

The applicability of criterion (4.6) for biaxial tension along the axes of orthotropy has been illustrated, for instance, by NIKOLAEV (1978) and JONES (1969). Substituting the expressions $\sigma_\theta = \overset{+}{\sigma}_\theta - (\overset{+}{\sigma}_\theta - \sigma_\theta)$, $\sigma_r = \overset{+}{\sigma}_r - (\overset{+}{\sigma}_r - \sigma_r)$, $\rho = r/b$, $m = a/b$, into (4.1) we obtain, for a disc with uniform properties, the following result:

$$W^M = \frac{\overset{+}{\sigma}_r + \overset{+}{\sigma}_\theta}{2\rho_V} - \frac{1}{\rho_V(1-m^2)} \left[\int_m^1 (\overset{+}{\sigma}_\theta - \sigma_\theta)\rho \, \mathrm{d}\rho + \int_m^1 (\overset{+}{\sigma}_r - \sigma_r)\rho \, \mathrm{d}\rho + p(m)m^2 + p(1) \right]. \tag{4.7}$$

For the volumetric energy storage capacity (energy storage capacity related to the volume of a solid disc) on the other hand we obtain:

$$W^V = \tfrac{1}{2}(\overset{+}{\sigma}_\theta + \overset{+}{\sigma}_r)(1-m^2) - \left[\int_m^1 (\overset{+}{\sigma}_\theta - \sigma_\theta)\rho \, \mathrm{d}\rho + \int_m^1 (\overset{+}{\sigma}_r - \sigma_r)\rho \, \mathrm{d}\rho + p(m)m^2 + p(1) \right]. \tag{4.8}$$

Thus, the upper bound of the energy stored per unit mass of a disc can be considered to be $W_\mathrm{u}^M = (\overset{+}{\sigma}_\theta + \overset{+}{\sigma}_r)/2\rho_V$, whereas the upper bound of the energy stored per unit volume – $W_\mathrm{u}^V = \tfrac{1}{2}(\overset{+}{\sigma}_r + \overset{+}{\sigma}_\theta)$. The determination of the optimal disc dimensions ensuring max W^M and max W^V can be reduced to the minimization of the bracketed expressions in (4.7) and (4.8).

4.2. *Specific energy storage capacities*

In a general case, $\max W^M$ and $\max W^V$ are attained in discs of different relative dimensions. At the same time, the design of a flywheel in which these two characteristics, though not a maximum, still have acceptable magnitudes, is of great practical interest. Therefore, a separate analysis of the specific mass energy storage capacity in discs of anisotropic materials and subsequent comparison with the volumetric energy storage capacity is needed. With this in mind, the state of stress and energy storage capacity of discs within the range of parameters $1 \leqslant \beta \leqslant 10$ and $0.1 \leqslant m \leqslant 0.9$ $[\beta = (E_\theta/E_r)^{1/2}]$ have been analyzed. Estimates of the maximum energy storage capacities of discs have been performed in accordance with the strength criterion given in (4.6). Thus it was necessary to evaluate and compare the two energy storage capacities corresponding to failure due to radial and circumferential stresses W_r, W_θ. The procedure for determining $W_{r,\theta}$ is as follows. By substituting into the expressions[1])

$$\bar{\sigma}^{\max}_{r,\theta} = \frac{\sigma^{\max}_{r,\theta}}{\rho_V \omega^2 b^2 (3 + \nu_{r\theta})}, \tag{4.9}$$

the respective ultimate strengths $\overset{+}{\sigma}_{r,\theta}$ instead of $\sigma^{\max}_{r,\theta}$ the squares of the possible maximum rotational speeds were determined from the calculated values of maximum stresses

$$(\omega^2 b^2)^{r,\theta}_{\max} = \frac{\overset{+}{\sigma}_{r,\theta}}{\rho_V \bar{\sigma}^{\max}_{r,\theta} (3 + \nu_{r\theta})}. \tag{4.10}$$

The expression for the kinetic energy of a rotating disc of unit axial thickness takes the form

$$W = \tfrac{1}{4}\pi(1 - m^4)\rho_V \omega^2 b^4 . \tag{4.11}$$

Substituting (4.10) into (4.11) we determine the magnitudes of W_r and W_θ and the respective specific mass and volumetric energy storage capacities $\bar{W}^M_{r,\theta}$ and $\bar{W}^V_{r,\theta}$

$$\bar{W}^M_{r,\theta} = \frac{\rho_V(3 + \nu_{r\theta})}{\overset{+}{\sigma}_{r,\theta}} W^M_{r,\theta} = \frac{1 + m^2}{4\bar{\sigma}^{\max}_{r,\theta}}, \tag{4.12}$$

$$W^V_{r,\theta} = \rho_V(1 - m^2) W^M_{r,\theta}\,; \quad \bar{W}^V_{r,\theta} = \frac{3 + \nu_{r\theta}}{\overset{+}{\sigma}_{r,\theta}} W^V = \frac{1 - m^4}{4\bar{\sigma}^{\max}_{r,\theta}}. \tag{4.13}$$

[1] Introduction of the coefficient $(3 + \nu_{r\theta})$ reduces the dependence of the obtained results on Poisson's ratio, which has been assumed to be constant and equal to 0.3. Expressions for $\bar{\sigma}_{r,\theta}$ correspond to the plane state of stress.

In the calculation of energy storage capacities two cases of boundary conditions, representing extreme ways of fastening the flywheel to the shaft, have been studied:
(1) free disc $\sigma_r|_{\rho=m} = \sigma_r|_{\rho=1} = 0$,
(2) a disc tightly fastened to the shaft

$$u|_{\rho=m} = \sigma_r|_{\rho=1} = 0\,, \tag{4.14}$$

where u is the radial displacement.

The dependence of $\bar{W}^M_{r,\theta}$ on m, for various measures of material anisotropy β and the boundary conditions given in (4.14) are shown in Fig. 3. Extremums in relationships $\bar{W}^M_\theta(m)$ and $\bar{W}^M_r(m)$ correspond to such m for which the mean through the thickness values of $\bar{\sigma}_\theta$ and $\bar{\sigma}_r$ are maximum.

In order to determine the maximum mass energy storage capacity, subject to condition (4.6), of a disc of optimal dimensions and with preset parameters of β, $\sigma^+_\theta/\sigma^+_r$, it is necessary to compare the relationships $\bar{W}_\theta(m)$ and $\bar{W}^M_r(m)$ reduced to the same scale along the ordinate axis. To accomplish this it is necessary to transform $\bar{W}^M_\theta(m)$ into $\bar{\bar{W}}^M_\theta = \bar{W}^M_\theta \cdot \sigma^+_\theta/\sigma^+_r$ and determine the maximum ordinate $\bar{W}^M_{\max}$ in the region bounded by curves $\bar{\bar{W}}^M_\theta(m)$ and $\bar{W}^M_r(m)$, and the respective magnitudes of m_{opt}. The expression for $\bar{W}^M_{\max}$ has the same structure as $\bar{W}^M_r$ i.e.

$$\bar{W}^M_{\max} = W^M_{\max}(3+\nu_{r\theta})\rho_V/\sigma^+_r\,.$$

The dependence of $\bar{W}^V_{r,\theta}$ on m for various values of β necessary for the investigation of the volumetric energy storage capacity is presented in Fig. 4. To determine the maximum volumetric energy storage capacity, subject to condition (4.6), of a disc of optimal dimensions with preset parameters of $\beta, \sigma^+_\theta/\sigma^+_r$, it is necessary to compare the relationships $\bar{W}^V_r$ and $\bar{\bar{W}}^V_\theta = \bar{W}^V_\theta\sigma^+_\theta/\sigma^+_r$ and evaluate the maximum ordinate $\bar{W}^V_{\max}$ in the region bounded by the curves $\bar{W}^V_r(m)$ and $\bar{\bar{W}}^V_\theta(m)$, and the respective m_{opt}.

A detailed study of the dependence of $\bar{W}^M_{\max}$ and $\bar{W}^V_{\max}$ on the parameters of anisotropy has been performed by PORTNOV and KULAKOV (1978a, 1980). The results of the study have shown that the freely rotating discs, in the majority of cases, are more effective than tightly fastened discs. Data obtained for freely rotating discs presented in Fig. 5 can contribute greatly to the design of flywheels exhibiting maximum, or sufficiently high values of the two specific characteristic of energy storage capacities.

Region I in Fig. 5 corresponds to the values of β and $\sigma^+_\theta/\sigma^+_r$ for which maximum values of both specific energy storage capacities can be attained at the occurrence of simultaneous radial and circumferential failure, i.e. they are combined in one and the same disc with $m = m_{\text{opt}}$. The upper limit of region I depends on the constraints imposed on the maximum values of m. In the case of increasing m region I becomes narrower (the dashed line in Fig. 5 corresponds to $m = 0.95$). In region II the rim of minimum thickness ($m = 0.9$) has the maximum mass energy storage capacity. Discs which exhibit simulta-

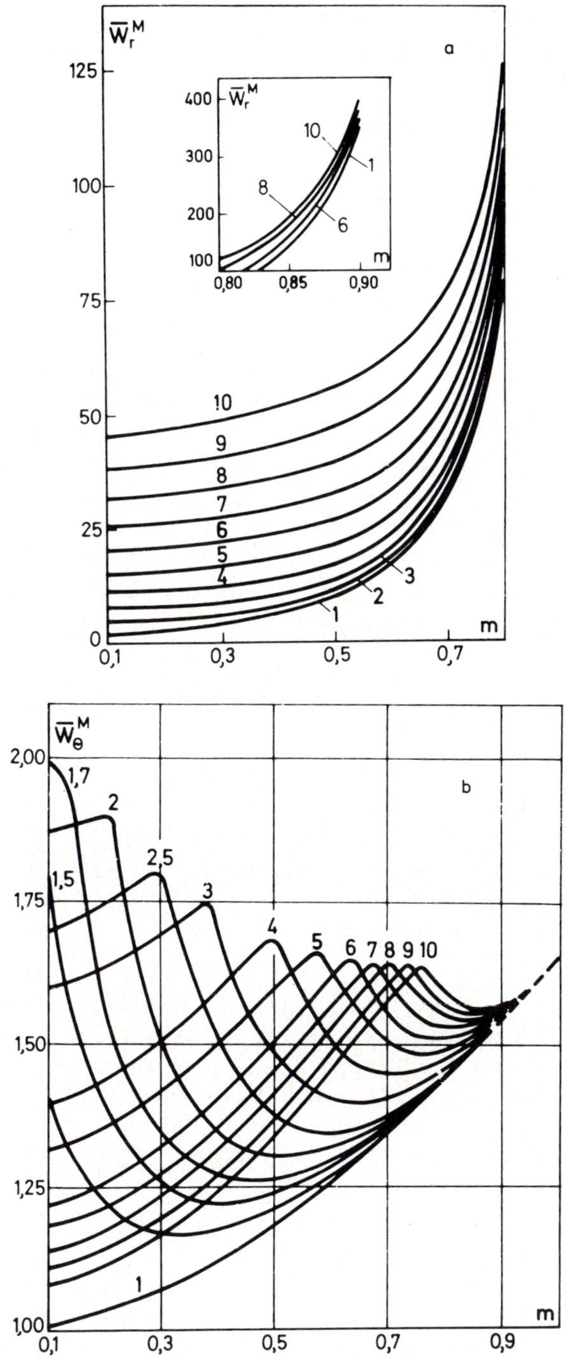

FIG. 3. Dependence of the reduced specific mass energy storage capacities of discs $\bar{W}^M_{r,\theta}$ on their relative dimensions and degree of material anisotropy, m and $\beta = (E_\theta/E_r)^{1/2}$, respectively; a, b, free disc; c, d, tightly fastened disc; figures on the curves designate the magnitudes of β.

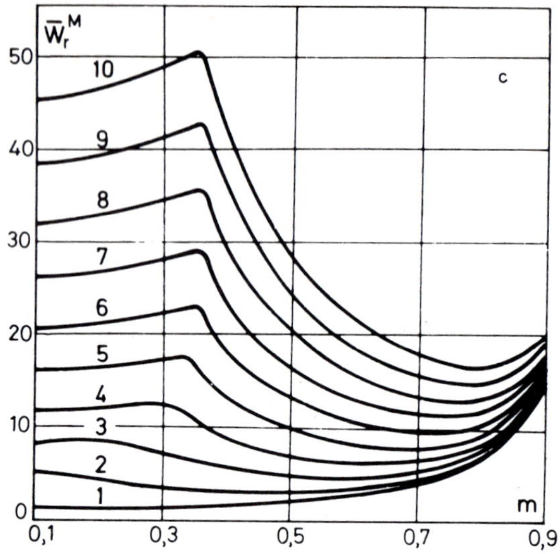

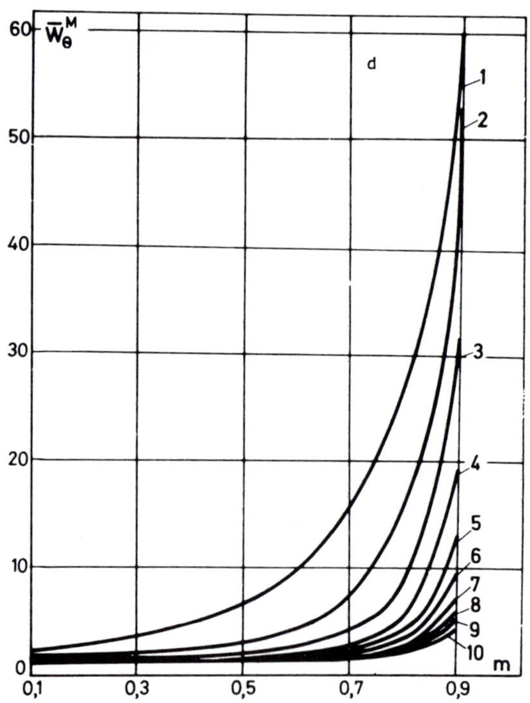

FIG. 3. *Continued.*

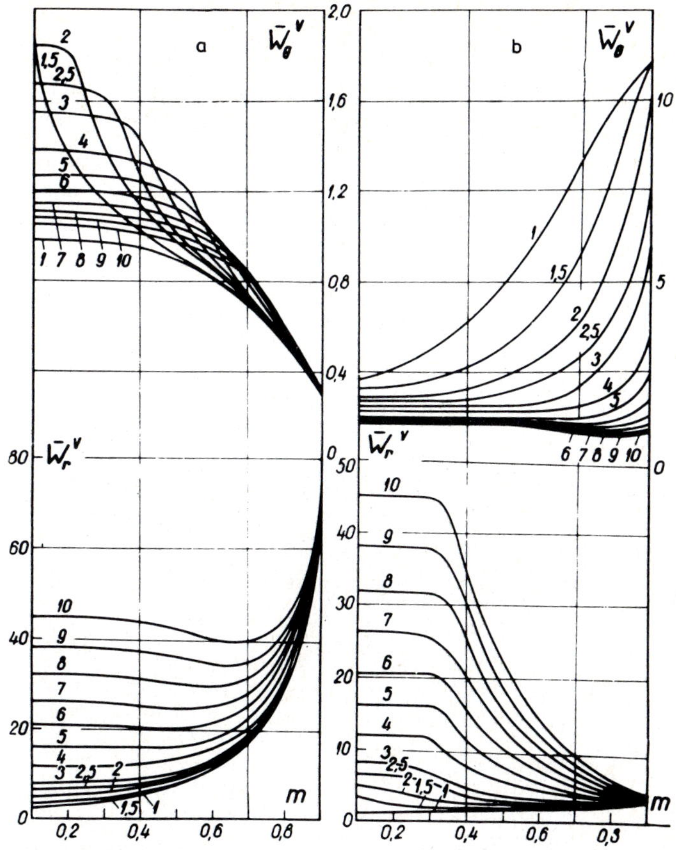

FIG. 4. Dependence of the reduced specific volumetric energy storage capacities of discs $\bar{W}^V_{r,\theta}$ on their relative dimensions and degree of material anisotropy, m and $\beta = (E_\theta/E_r)^{1/2}$, respectively; a, a free disc; b, a tightly fastened disc; figures on the curves designate the magnitudes of β.

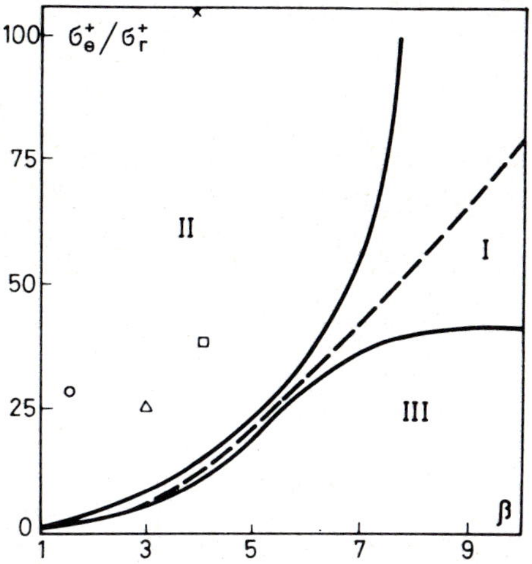

FIG. 5. Ranges of parameters β and $\sigma^+_\theta/\sigma^+_r$ corresponding to different methods of evaluating optimal relative dimensions of freely rotating discs with maximum mass and volumetric energy storage capacities; ○, fiberglass composite; △, boron fiber composite; □, carbon fiber composite; ×, organic fiber composite.

neous radial and circumferential failure are 'nearly optimal' since they ensure maximum volumetric energy storage capacity with negligible reduction in their mass energy storage capacity compared to rims with $m = 0.9$. It should be noted that the extension of the range of relative thicknesses to $m = 0.95$ increases the mass energy storage capacity by only 3% in comparison with the thickness $m = 0.9$. In region III, discs with $m = 0.9$ have the maximum energy storage capacity and exhibit circumferential failure. The optimal dimensions of discs with maximum volumetric and mass energy storage capacities can differ significantly within this range of parameters.

The values of maximum specific energy storage capacities and the respective optimal relative dimensions of discs made of typical unidirectional fiber composites (whose properties are presented in Table 1) are shown in Table 2. As is illustrated, free, comparatively thin discs–rims that exhibit simultaneous

TABLE 1. Characteristics of unidirectional fiber composites (after CHAMIS and KIRALY, 1975).

Characteristics	Dimensions	Fiberglass composite	Carbon fiber composite	Boron fiber composite	Organic fiber composite
v_f [a]	–	0.72	0.70	0.50	0.54
ρ_V	10^3 kg/m^3	2.13	1.61	2.02	1.36
E_θ	10^4 MPa	6.07	18.14	20.13	8.34
E_θ/E_r	–	2.44	17.53	9.27	17.43
$E_\theta/G_{r\theta}$	–	5.07	26.43	37.43	29.66
σ_θ^+	10^2 MPa	12.90	14.94	13.73	11.86
$\sigma_\theta^+/\sigma_r^+$	–	28.0	37.3	24.6	108.2
$\sigma_\theta^+/\tau_{r\theta}^u$	–	28.0	22.1	21.8	43.0
$\nu_{r\theta}$	–	0.23	0.28	0.17	0.32
α_θ	10^{-6} deg^{-1}	3.78	0.02	6.12	−2.88
α_θ/α_r	–	0.02	0.001	0.02	−0.05

[a] v_f is the fiber volume content.

TABLE 2. Specific energy storage capacities of fiber composite discs. (The magnitudes of the optimal relative dimensions of the disc are in the denominator.)

Disc material	Free disc		A tightly fastened disc	
	W^V_{max} $\times 10^{-6}$ J/m^3	W^M_{max} $\times 10^{-6}$ J/kg	W^V_{max} $\times 10^{-6}$ J/m^3	W^M_{max} $\times 10^{-6}$ J/kg
Fiberglass composite	266.3 / 0.717	0.284 / 0.9	50.7 / 0.1	0.097 / 0.9
Carbon fiber composite	296.6 / 0.742	0.436 / 0.9	153.4 / 0.1	0.121 / 0.9
Boron fiber composite	319.3 / 0.692	0.321 / 0.9	146.2 / 0.1	0.130 / 0.9
Organic fiber composite	149.2 / 0.852	0.409 / 0.9	41.2 / 0.1	0.039 / 0.9

radial and circumferential failure possess maximum specific volumetric energy storage capacity. The optimal relative dimensions are within the range $m_{opt} =$ 0.70–0.8; i.e. only a small portion of the structural volume is being effectively utilized. Both the mass and volumetric energy storage capacities of freely rotating discs are higher than those of discs tightly fastened to the shaft. Therefore, in the ensuing discussion, only discs with free rotation will be treated.

4.3. The effect of initial thermal stresses

During the winding of discs under low tension the initial stresses are determined mainly by the extent of heat treatment. The magnitudes of these stresses are very well estimated by solving the problem of linear thermoelasticity. The procedure of calculating the specific energy storage capacities with allowance for initial thermal stresses is described in the papers by PORTNOV and KULAKOV (1978b). Estimates of the effect of thermal stresses for four types of fiber composites (Table 1) are given in Table 3. For each case, the maximum specific energy storage capacity and the respective optimal relative dimension have been calculated. The calculated values of $W^M_{T\,\max}$ and $W^V_{T\,\max}$ with allowance for initial thermal stresses have been divided by the respective magnitudes of the energy storage capacities of discs without initial stresses. The polymerization temperature of epoxy resin employed in the calculations was varied from $T_n = 50°C$ to $T_n = 170°$ (thermophysical and elastic properties of the materials were considered to be constant). The obtained results allow one to draw a conclusion about the essential effect of initial thermal stresses on the characteristics of flywheels made of high-modulus composite materials. It is necessary to note that the consideration of initial thermal stresses can affect the

TABLE 3. The effect of initial thermal stresses on specific energy storage capacities of fiber composite discs.

ΔT	Fiberglass composite		Carbon fiber composite		Boron fiber composite		Organic fiber composite	
	$\tilde{W}^{V}$ [a]	$\tilde{W}^{M}$	$\tilde{W}^{V}$	$\tilde{W}^{M}$	$\tilde{W}^{V}$	$\tilde{W}^{M}$	$\tilde{W}^{V}$	$\tilde{W}^{M}$
−30°C	$\frac{0.99}{0.717}$	$\frac{1.00}{0.9}$	$\frac{0.86}{0.711}$	$\frac{0.92}{0.9}$	$\frac{0.85}{0.660}$	$\frac{0.95}{0.9}$	$\frac{0.87}{0.844}$	$\frac{0.90}{0.9}$
−50°C	$\frac{0.98}{0.717}$	$\frac{0.99}{0.9}$	$\frac{0.71}{0.664}$	$\frac{0.87}{0.9}$	$\frac{0.71}{0.622}$	$\frac{0.91}{0.9}$	$\frac{0.77}{0.832}$	$\frac{0.83}{0.9}$
−70°C	$\frac{0.98}{0.716}$	$\frac{0.99}{0.9}$	$\frac{0.53}{0.770}$	$\frac{0.82}{0.9}$	$\frac{0.56}{0.717}$	$\frac{0.87}{0.9}$	$\frac{0.65}{0.823}$	$\frac{0.76}{0.9}$
−150°C	$\frac{0.96}{0.716}$	$\frac{0.99}{0.9}$	$\frac{0.28}{0.9}$	$\frac{0.62}{0.9}$	$\frac{0.31}{0.830}$	$\frac{0.73}{0.9}$	$\frac{0.34}{0.9}$	$\frac{0.49}{0.9}$

[a] $\tilde{W}^{V,M} = W^{V,M}_T / W^{V,M}_0$ where $W^{V,M}_T$ are the specific energy capacities of discs with thermal stresses, but $W^{V,M}_0$ is without thermal stresses. (The optimal relative dimensions of discs with thermal stresses are in the denominator.)

initial selection of composite materials for flywheel design based merely on the analysis of stresses due to centrifugal forces.

4.4. *Methods of increasing the specific volumetric energy storage capacity*

As was shown in Section 4.2, the rotating composite discs made by the winding process exhibit maximum specific volumetric energy storage capacity in the event of simultaneous failure due to circumferential and radial stresses. The significant strength anisotropy results in the respective optimal ratios of disc radii within the range of 0.7–0.8, i.e. only a small portion of the structural volume is utilized. Further increase in the specific volumetric energy storage capacity associated with the increase in the radial thickness requires additional measures to be taken in order to increase the resistance to tensile radial stresses of the composite. Subsequently, some structural and technological ways of increasing max W^V of wound discs (where properties are presented in Table 1) will be treated.

Disc with ballast. Two ways of creating a ballast load on the bearing disc will be considered. The first one involves 'pure' ballast in the form of a ring with $E_\theta = 0$ which is not connected to the shaft. Attachment of a bearing disc to the shaft is realized along the disc periphery (for instance, by chord winding). The second involves a disc of low-modulus material, connected with both the rim and the shaft, which acts simultaneously as ballast and attachment of the bearing rim to the shaft. The procedure of solving this problem was described in the paper by KULAKOV and PORTNOV (1980). The data characterizing the effectiveness of ballast for fiber composite discs are presented in Table 4 with allowance for the energy storage capacity of the ballast itself. In the case of utilization of 'pure' ballast the energy storage capacity of the optimal structure practically does not depend on the density of the ballast. If the ballast is attached to the shaft, the maximum energy storage capacity is attained when the mass densities of the shaft and ballast are similar.

The gain in the volumetric energy storage capacity is considerable only for discs made of highly anisotropic composites such as organic fiber composites in which the state of stress is essentially nonuniform. The closer the state of stress is to being uniform, the weaker the effect of the ballast (in the uniform state of

TABLE 4. A ratio of volumetric and mass energy capacities of fiber composite discs with a ballast to energy storage capacity of the disc of optimal dimensions without a ballast ($0.1 \le m \le 0.9$), ($n_\rho = \rho_{VB}/\rho_{VR}$, where ρ_{VR}, ρ_{VB} are the material densities in a disc and ballast.)

	Fiberglass composite	Carbon fiber composite	Boron fiber composite	Organic fiber composite
A 'pure' ballast ($0.25 \le n_\rho \le 10$)	1.14	1.14	1.06	1.49
	0.61	0.71	0.71	0.55
A ballast connected with a shaft ($n_\rho = 1$)	1.14	1.14	1.07	1.49
	0.56	0.51	0.56	0.42

stress the ballast, in accordance with the conclusion (f) in Section 2, does not increase the value of the energy storage capacity). The use of ballast is accompanied, as could be expected, by a significant reduction in the mass energy storage capacity.

Multilayered discs. This method of increasing max W^V is based on constraints imposed on the radial stresses. It entails utilization of the entire volume between the periphery of a flywheel and a shaft by a system of concentric rings, the thickness of which is sufficiently small to prevent preliminary delamination. The rings are separated by compliant interlayers, which practically do not constrain the radial displacements of ring surfaces. The problem consists in the selection of an appropriate sequence of thicknesses of concentric rings, which form the multilayered disc of specified relative dimensions, such that the number of rings is a minimum and the allowable circumferential speed of the disk, a maximum. To ensure the maximum angular velocity it is necessary to select a rim of optimal relative dimensions exhibiting simultaneous radial and circumferential failure at the outer ring, and to ensure the minimum number of layers it is necessary to choose the succeeding inner layers which fail radially. Results of calculating the sequence of ring thicknesses, forming multilayered discs made of advanced composites with $m = 0.1$, and the relative values of their energy storage capacities are presented in Table 5. As can be seen, the minimum number of composite layers necessary to fill the entire structural volume is small. The calculation has shown that the level of allowable radial stresses in the last ring permits a rigid fit onto the shaft. Efficient utilization of the entire structural volume of a disc-flywheel leads to an appreciable increase in the specific volumetric energy storage capacity.

Discs with radial reinforcement. It is of interest to evaluate the effectiveness of the transition from a circumferential to a radial–circumferential reinforcement pattern, when the increase in radial strength obtained through redistribu-

TABLE 5. Optimal dimensions of multilayered discs.

Characteristics	Fiberglass composite	Carbon fiber composite	Boron fiber composite	Organic fiber composite
m_1	0.717	0.742	0.692	0.852
m_2	0.660	0.617	0.502	0.824
m_3	0.280	0.218	0.288	0.782
m_4	0.830	–	–	0.709
m_5	–	–	–	0.503
m_6	–	–	–	0.511
$\tilde{W}^{V}$ [a]	1.36	1.43	1.30	2.11
$\tilde{W}^{M}$	0.59	0.61	0.64	0.58

[a] $\tilde{W}^V = W^V/W^V_{opt}$, where W^V and W^V_{opt} are the specific volumetric energy storage capacities of a multilayered disc and one of optimal dimensions respectively. $\tilde{W}^M = W^M/W^M_{opt}$ where W^M, W^M_{opt} are the specific mass energy storage capacities of a multilayered disc with $m = 0.1$, and one of optimal dimensions, respectively.

tion of the reinforcement can be attained only at the expense of reduction in circumferential strength. Let us characterize the structure of such a composite by the fiber volume content in the circumferential direction (n_θ). Upon redistribution of fibers from the circumferential into the radial direction, i.e. at $n_\theta < 1$, both the strength and deformation characteristics of the composite change simultaneously. As an example, the case when the composite characteristics are determined by the so-called 'law of mixtures' given below will be considered:

$$\begin{aligned} E_\theta &= E_0 h_\theta + (1 - n_\theta) E_{90} , \\ E_r &= E_0(1 - n_\theta) + n_\theta E_{90} , \\ \overset{+}{\sigma}_\theta &= \overset{+}{\sigma}_0 n_\theta + (1 - n_\theta) \overset{+}{\sigma}_{90} , \\ \overset{+}{\sigma}_r &= \overset{+}{\sigma}_0 (1 - n_\theta) + n_\theta \overset{+}{\sigma}_{90} . \end{aligned} \tag{4.15}$$

Here, E_0 and $\overset{+}{\sigma}_0$ are the modulus and strength in the reinforcement direction of a unidirectional composite; E_{90} and $\overset{+}{\sigma}_{90}$ are the modulus and strength across the reinforcement.

The analysis involving the characteristics of unidirectional composites (Table 1) illustrates that the effect of redistributing the reinforcement can be appreciable. The optimal magnitudes of the parameter n_θ for the considered materials turn out to be rather close. To attain the maximum level of W^V it is sufficient to transfer 15–20% of the reinforcing fibers into the radial direction (Fig. 6(a)). Rejection of a purely circumferential reinforcement is accompanied by a decrease in the mass energy storage capacity (Fig. 6(b)).

Prestressed discs. In wound composite discs it is possible to generate a favourable profile of initial stresses with the help of a programmed, layer-by-layer winding technique. To evaluate the effectiveness of this technique the problem of generating a system of stresses $\overset{0}{\sigma}_r$ and $\overset{0}{\sigma}_\theta$ has been considered to compensate for radial tensile stresses in the disc ($\overset{c}{\sigma}_r$) at some angular velocity $\bar{\omega}$. At $\bar{\omega}$ the stresses in the disc will be

$$\begin{aligned} \sigma_r &= \overset{0}{\sigma}_r + \overset{c}{\sigma}_r = 0 \\ \sigma_\theta &= \overset{0}{\sigma}_\theta + \overset{c}{\sigma}_\theta = \rho_V \bar{\omega}^2 r^2 . \end{aligned} \tag{4.16}$$

The prescribed profile of initial stresses, though it certainly restricts the problem, allows one to perform the analysis and evaluate the effectiveness of the prestressing technique (see BEIL' et al., 1981). The level of initial stresses is defined by the parameter $æ = \rho_V \bar{\omega}^2 b^2 (3 + \nu_{r\theta}) / \overset{+}{\sigma}_r$. The higher this parameter the higher the angular velocity at which radial stresses are zero. As could be expected (Fig. 7), the maximum mass energy storage capacity is attained, as before, in a ring of minimum allowable thickness (in the range under consideration this takes place at $m = 0.9$). However, at optimal æ the energy storage

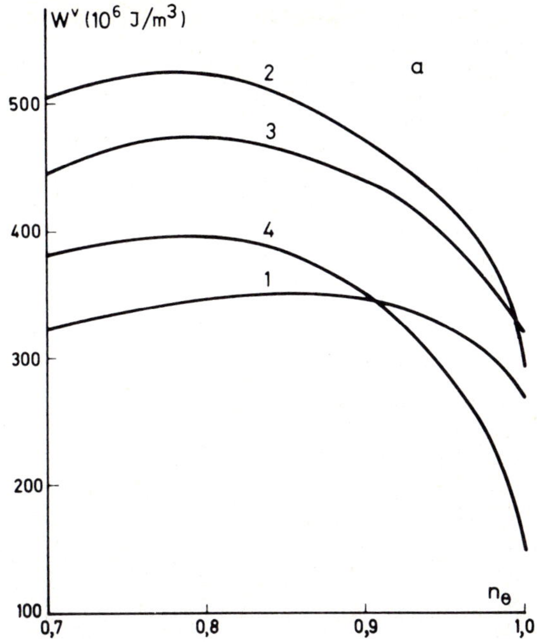

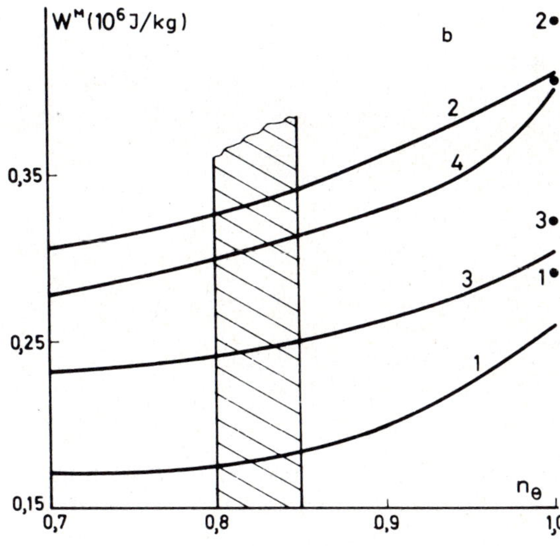

FIG. 6. The effect of fiber redistribution on (a) the maximum specific volumetric and (b) mass energy capacities of discs; 1, fiberglass composite; 2, carbon fiber composite; 3, boron fiber composite; 4, organic fiber composite. The region of magnitudes n_θ where the highest increase in specific volumetric energy storage capacity is attained is cross-hatched; the points correspond to the energy storage capacity of rims with $m = 0.9$.

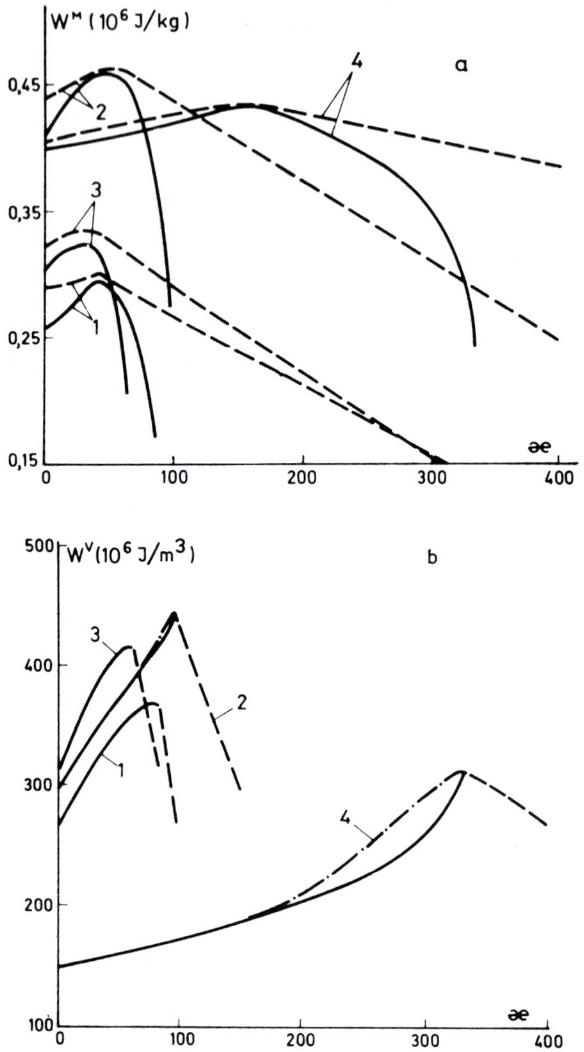

FIG. 7. Dependence of maximum specific mass (a) and volumetric (b) energy storage capacities on the parameter æ characterizing the level of initial stresses for the same composites as in Fig. 6.

capacity of thicker rings in which failure occurs simultaneously in both directions approaches the above. The maximum magnitudes of mass (Fig. 7(a)) and volumetric (Fig. 7(b)) energy storage capacities are attained at different levels of initial stresses. It is not useful to increase the initial stresses beyond this level; as can be seen, for sufficiently large æ the energy capacity of prestressed flywheels is lower than that of flywheels initially stress-free.

Comparison of the effectiveness of the considered methods. The ratios of the maximum values of W^M and W^V attainable at $m = m_{\mathrm{opt}}$, and the theoretical ultimate values of W_u^M and W_u^V (see, Section 4.1) respectively, for the problems treated are shown in Figs. 8(a, b). As is illustrated (Fig. 8(a)), the highest increase in the volumetric energy storage capacity as a result of using the described methods is attained for highly anisotropic fiber composites (carbon and organic fiber composites), while the use of radial–circumferential reinforcement yielded results closest to the maximum magnitudes.

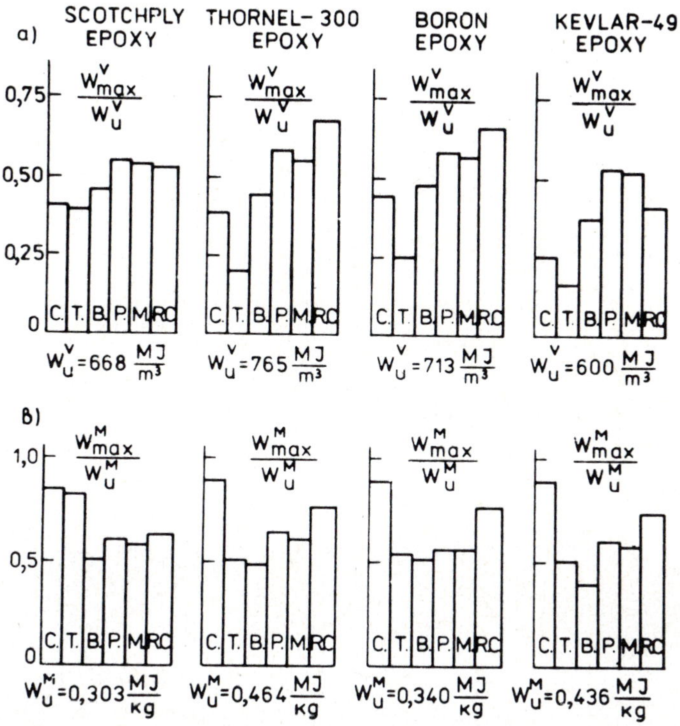

FIG. 8. Relative magnitudes of (a) volumetric and (b) mass energy storage capacities of discs of optimal dimensions, failing simultaneously due to σ_r and σ_θ(C – with circumferential reinforcement; T – with allowance for thermal stresses; B – with ballast; P – prestressed, M – a multilayered disc; R.C. – with radial-circumferential reinforcement).

The above described methods allow one to considerably increase the volumetric energy storage capacity of composite discs. The increase in the volumetric energy storage capacity is accompanied by considerable reductions in the mass energy storage capacity (Fig. 8(b)) which constitutes about 0.5 of the maximum capacity. Thus, the difficulty of incorporating high magnitudes of both characteristics into one and the same structure is one of the basic problems for fiber composite flywheel design.

5. Chord flywheels

Past investigations have shown that the most effective energy storing element in flywheels made of advanced composites is the freely rotating thick-walled rim fabricated by hoop winding. It combines the simplicity of fabrication with high magnitudes of mass energy storage capacity and sufficiently high volumetric energy storage capacity. The difficulty of fastening it to the shaft is successfully overcome in the so-called chord flywheel – a circular system consisting of a rim and a hub, which are joined together by spokes–chords made of a unidirectional composite surrounding the rim over its periphery (Fig. 9). The

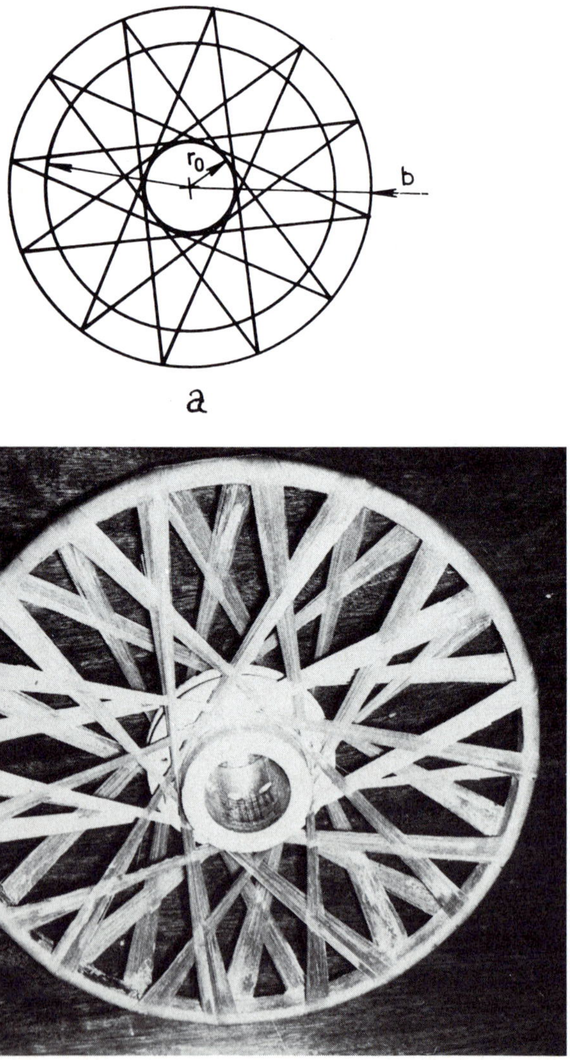

FIG. 9. A schematic (a) and photo of a chord flywheel.

concept of such a flywheel has been outlined in the Patent of Great Britain (1964). Several successful tests of circular composite systems (chord flywheels) (VOROPAEV et al., 1982) confirmed the great potential of this type of structures. A variety of easily controllable parameters such as the pattern of starlike polygons–spokes and their relative thickness, relative rim thickness, and the possibility of using various material combinations open up wide avenues for optimization of these structures. Chords can be concentrated in discrete spokes which form starlike polygons, as shown in Fig. 9, or continuous discs on each side of the rim. For structures of the first type, flywheel balancing and control of its response during service are greatly facilitated.

Chord flywheels can be fabricated entirely made by the most promising technique of composite processing, the winding process. Kinematic problems and the equilibrium of a filament during chord winding have been treated by MOORLAT et al. (1982) who have also studied the laws governing the movement of a filament winder, and performed chord winding experiments to investigate the various geometrical details. The following treatment will be devoted to a calculation of the state of stress in flywheels with chords–spokes during uniform rotation and braking as well as an analysis of their energy storage capacity and maximum power.

5.1. *The state of stress in a chord flywheel*

Let us consider a system consisting of n chords–spokes on each side of the flywheel. Let us assume that at crossover points the chords are connected rigidly and that they do not rotate towards each other. The system processes central symmetry, that is, the crossover points deform only in the radial direction during uniform rotation and in the circumferential direction during acceleration. Under the above conditions, the problem of the response of a system of chords can be reduced to the problem of the response of one continuous beam (or chord) with a specified displacement direction at the supports (i.e. at the crossover points). The continuous beam is loaded by inertia forces due to rotation with an angular velocity ω and acceleration $\dot{\omega}$, and by the force at the outer end which can be determined from the condition of displacement compatibility for the beam and rim-disc. The beam is subjected to the combined action of longitudinal–transverse bending and tension. It has been shown by MOORLAT and PORTNOV (1983) that, at small r_0 (Fig. 9(a)) at least, it is possible to disregard the longitudinal bending and to resolve the problem of the state of stress and strain in chord flywheels within the linear framework. The system of resulting equations for a continuous beam with multiple supports has the form:

$$T' + \rho_{VS}\omega^2 Sx + \rho_{VS}\dot{\omega}Sr_0 = 0\,,$$

$$P' + \rho_{VS}\omega^2 Sr_0 - \rho_{VS}\dot{\omega}Sx = 0\,, \tag{5.1}$$

$$P = -EJv''' \,, \quad T = ESu' \,,$$

where T and P are the longitudinal and transverse forces in the beam; u and v are the longitudinal and transverse displacements; x is the coordinate along the beam axis taken from the midpoint of the beam; S and J are the cross-sectional area and moment of inertia; and ρ_{VS} and E are the mass density and elastic modulus of the material. Since we are dealing with a linear system, it is possible to consider separately the problems of uniform rotation and smooth acceleration.

Uniform rotation. During uniform rotation the deformed spokes exhibit reflection symmetry. Therefore, instead of analyzing the entire beam we can consider only half of it, that is, a semichord (Fig. 10). In such a case, the crossover points form supports at which the beam can only be displaced in the radial direction without rotation.

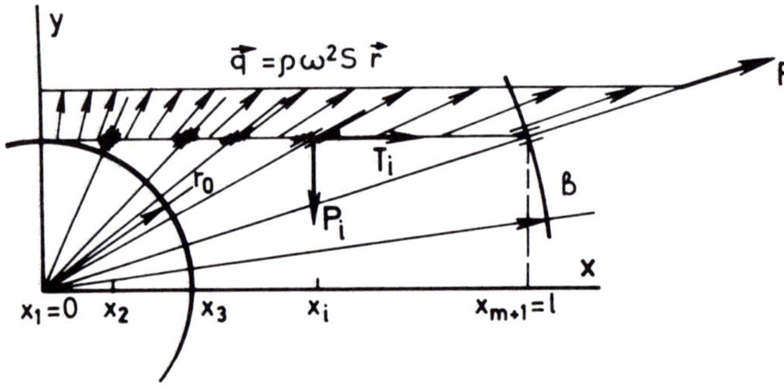

FIG. 10. A calculation scheme for stress analysis of a chord flywheel spoke during uniform rotation.

The system of equilibrium equations (5.1) for the case of uniform rotation takes the form

$$
\begin{gathered}
T' + \rho_{VS}\omega^2 Sx = 0\,, \\
P' + \rho_{VS}\omega^2 Sr_0 = 0\,, \\
P = -EJv''' \,, \quad T = ESu'\,.
\end{gathered}
\tag{5.2}
$$

Let us introduce the dimensionless variables

$$
\begin{gathered}
x = b\bar{x}\,, \quad r_0 = b\bar{r}_0\,, \\
T = \rho_{VS}\omega^2 b^2 S\bar{T}\,, \quad P = \rho_{VS}\omega^2 b^2 S\bar{P}\,, \\
u = \frac{\rho_{VS}\omega^2 b^3}{E}\bar{u}\,, \quad v = \frac{\rho_{VS}\omega^2 b^3}{E}\bar{v}\,.
\end{gathered}
\tag{5.3}
$$

The system of equilibrium equations (5.2) thus can be rewritten in terms of the dimensionless variables as follows

$$\begin{aligned} \bar{T}' + \bar{x} &= 0 , \\ \bar{P}' + r_0 &= 0 , \\ \bar{P} &= -\lambda_1 \bar{v}''' , \\ \bar{T}' &= \bar{u}' , \end{aligned} \tag{5.4}$$

where

$$\lambda_1 = J/Sb^2 .$$

Since in the ensuing discussion we shall employ only the dimensionless variables, the notation '–' over the variables will be omitted.

By integrating (5.4), we find

$$T = \tfrac{1}{2}(l^2 - x^2) + \sum_{i=k}^{m} T_i , \quad (x \in [x_{k-1}, x_k]) , \tag{5.5}$$

$$P = r_0(l - x) + \sum_{i=k}^{m} P_i , \quad (x \in [x_{k-1}, x_k]) . \tag{5.6}$$

Consequently,

$$u = \tfrac{1}{2} l^2 x - \frac{x^3}{6} + \sum_{i=k}^{m} T_i x + A_{k-1} , \tag{5.7}$$

$$\lambda_1 v''' = r_0(l - x) - \sum_{i=k}^{m} P_i . \tag{5.8}$$

By subsequent integration of (5.8) we obtain the following relations for the transverse displacement V:

$$\lambda_1 v'' = r_0 l x - \tfrac{1}{2} r_0 x^2 - \sum_{i=k}^{m} P_i x + B_{k-1} , \tag{5.9}$$

$$\lambda_1 v' = \tfrac{1}{2} r_0 l x^2 - \tfrac{1}{6} r_0 x^3 - \sum_{i=k}^{m} P_i \frac{x^2}{2} + B_{k-1} x + C_{k-1} , \tag{5.10}$$

$$\lambda_1 v = \tfrac{1}{6} r_0 l x^3 - \tfrac{1}{24} r_0 x^4 - \sum_{i=k}^{m} P_i \frac{x^3}{6} + B_{k-1} \frac{x^2}{2} + C_{k-1} x + D_{k-1} . \tag{5.11}$$

In this way, the solution of the problem is reduced to the determination of σm integration constants (T_0 and P_0 are calculated automatically after all the others have been obtained).

To reduce the number of unknown integration constants it is useful to employ the condition of continuity of the functions u, v, v'.

Let us introduce the following notations:

$$g(x) = \tfrac{1}{2}l^2 x - \tfrac{1}{6}x^3 , \tag{5.12}$$

$$f(x) = \frac{1}{2\lambda_1}\left(\tfrac{1}{3}r_0 l x^3 - r_0 x^4\right) . \tag{5.13}$$

Thus the functions $u(x) - g(x)$ and $\lambda_1[v(x) - g(x)]$ are polynomials of the first and third degree, respectively, in each interval $[x_{k-1}, x_k]$, and consequently they must coincide with their Lagrangian interpolation polynomials (for v with two-fold nodes):

$$u(x) - g(x) = (u_{k-1} - g_{k-1})\frac{x_k - x}{x_k - x_{k-1}} + (u_k - g_k)\frac{x - x_{k-1}}{x_k - x_{k-1}} , \tag{5.14}$$

$$\begin{aligned} \lambda_1[v(x) - f(x)] &= (v_{k-1} - f_{k-1})\frac{(x_k - x)^3}{(x_k - x_{k-1})^3} + (v_k - f_k)\frac{(x - x_{k-1})^3}{(x_k - x_{k-1})^3} \\ &+ \left[\frac{v'_{k-1} - f'_{k-1}}{(x_k - x_{k-1})^2} + 3\frac{v_{k-1} f_{k-1}}{(x_k - x_{k-1})^2}\right](x - x_k)^2 (x - x_{k-1}) \\ &+ \left[\frac{v'_k - f'_k}{(x_k - x_{k-1})^2} - 3\frac{v_k - f_k}{(x_k - x_{k-1})^2}\right](x - x_k)(x - x_{k-1})^2 , \end{aligned} \tag{5.15}$$

where

$$v_k = v(x_k) , \quad v'_k = v'(x_k) , \quad u_k = u(x_k) ,$$

$$g_k = g(x_k) , \quad f_k = f(x_k) , \quad f'_k = f'(x_k) .$$

The displacements and forces at the nodes are related as follows

$$\begin{aligned} T_k &= u'(x_k - 0) - u'(x_k + 0) , \\ M_k &= \lambda_1[v''(x_k - 0) - v''(x_k + 0)] , \\ P_k &= \lambda_1[v'''(x_k - 0) - v'''(x_k + 0)] . \end{aligned} \tag{5.16}$$

Upon differentiating expressions (5.14) and (5.15) and substituting into (5.16) we obtain the expressions for nodal reactions

$$\begin{aligned} T_k &= (u_k - u_{k-1} - g_k + g_{k-1})/(x_k - x_{k-1}) - (u_{k+1} - u_k \\ &- g_{k+1} + g_k)/(x_{k+1} - x_k) ; \end{aligned} \tag{5.17}$$

$$M_k = \lambda_1\{[2(v'_{k-1} - f'_{k-1}) + 4(v'_k - f'_k)]/(x_k - x_{k-1}) - 6[(v_k - v_{k-1} - f_k + f_{k+1})/(x_k - x_{k-1})^2 + (v_{k+1} - v_k - f_{k+1} + f_k)/(x_{k+1} - x_k)^2] + 4[(v'_k - f'_k) + 2(v'_{k+1} - f'_{k+1})]/(x_{k+1} - x_k)\}\ ; \tag{5.18}$$

$$P_k = \lambda_1\{12[(v_{k+1} - v_k - f_{k+1} + f_k)/(x_{k+1} - x_k)^3 - (v_k - v_{k+1} - f_k + f_{k+1})/(x_k - x_{k-1})^3] - 6[(v'_{k+1} + v'_k - f'_{k+1} - f'_k)/(x_{k+1} - x_k)^2 - (v'_k + v'_{k-1} - f'_k - f'_{k-1})/(x_k - x_{k-1})^2]\}\ . \tag{5.19}$$

Thus T_k, P_k and M_k are linear functions of displacements at the nodes. To determine nodal displacements and rotations we shall use the following conditions:

(i) Rigid attachment to the hub

$$u_0 = v_0 = v'_0 = 0\ ; \tag{5.20}$$

(ii) displacement of nodal points is possible only in the radial direction

$$u_i r_0 - v_i x_i = 0 \quad (i = 1, 2, \ldots, m)\ ; \tag{5.21}$$

(iii) resulting force in the support is perpendicular to the radius (the displacement direction).

$$T_i x_i + P_i r_0 = 0 \quad (i = 1, 2, \ldots, m-1)\ . \tag{5.22}$$

(iv) At the outer end of the beam the radial force is expressed as:

$$T_m x_m + P_m r_0 = F_r\ . \tag{5.23}$$

(v) The conditions of attachment of the beams at the crossover points are

(a) rigid $v'_i = 0 \qquad (i = 1,2, \ldots, m)$; (5.24)

(b) hinged $M_i = 0 \quad (i = 1, 2, \ldots, m)$. (5.25)

In this way, we have the required $3m + 3$ conditions necessary to determine the $3m + 3$ unknown constants. Conditions (5.20) and (5.21) allow us to eliminate $m + 3$ unknowns from the system and, in the case of rigid attachment at the crossover points, the order of the system is further reduced by m. Consequently, in the case of rigid attachment of the chords at the crossover points it is necessary to resolve an algebraic system of linear equations with $2m$ unknowns. For wound structures made of fiber composite materials rigid attachment at the nodes seems to be more realizable.

Smooth acceleration. Consider the case when the acceleration at all points of a flywheel is the same. This problem differs slightly from the problem of uniform rotation. Instead of a semichord it is necessary to consider the entire chord for which the symmetrically located, with respect to the center, crossover points have the same circumferential and radial components of displacements and the same rotation.

In the case of constant acceleration the equations of equilibrium take the form

$$\begin{gathered} T' + \rho_{VS}\dot{\omega} r_0 S = 0\,, \\ P' - \rho_{VS}\dot{\omega} x S = 0\,, \\ P = -EJv''' \,, \quad T = ESu'\,. \end{gathered} \tag{5.26}$$

The dimensionless variables can be determined from the formulas

$$\begin{gathered} x = b\bar{x}\,, \quad r_0 = b\bar{r}_0\,, \\ T = \rho_{VS}\dot{\omega} b^2 S\bar{T}\,, \\ P = \rho_{VS}\dot{\omega} b^2 S\bar{P}\,. \end{gathered} \tag{5.27}$$

i.e. formulas (5.27) differ from (5.3) only in that instead of the multiplier $\omega^2 b^2$ we employ $\dot{\omega} b^2$. Now let us express functions $f(x)$ and $g(x)$ in terms of the following relations:

$$\begin{gathered} g(x) = r_0 l x - \tfrac{1}{2} r_0 x^2\,, \\ f(x) = \frac{1}{\lambda_1}\left(\tfrac{1}{12} l^2 x^3 - \tfrac{1}{120} x^5 - \tfrac{1}{6} l^3 x^2\right). \end{gathered} \tag{5.28}$$

Formulas (5.14)–(5.19) remain valid for the above choice of functions $f(x)$, $g(x)$ whereas $\bar{u}(x)$, $\bar{v}(x)$ are solutions of the corresponding dimensionless problem. Retaining the designation of the crossover points for the right semichord, let us designate the symmetrically placed points by negative indices: $x_{-i} = -x_i$. The conditions of compatibility for the displacements and forces at these points are:

$$\begin{gathered} (u_i + u_{-i})x_i + (v_i - v_{-i})r_0 = 0\,, \\ (u_i - u_{-i})r_0 - (v_i + v_{-i})x_i = 0\,, \\ v_i' = v_{-i}'\,, \quad M_i + M_{-i} = 0\,, \\ (T_i - T_{-i})x_i - (P_i + P_{-i})r_0 = 0\,, \\ (T_i + T_{-i})r_0 - (P_i - P_{-i})x_i = 0\,. \end{gathered} \tag{5.29}$$

In the center of the chord the boundary conditions are specified by:

$$v(0) = v'(0) = u(0) = 0 ,$$

and at the outer end two kinds of boundary conditions are possible:
(a) rigid attachment of spokes

$$u'(x_m) = -u_m r_0 + v_m x_m = v_\theta , \tag{5.30}$$

where v_θ is the circumferential displacement of the rim;
(b) hinged attachment:

$$M_m = 0 , \tag{5.31}$$

At the outer end the circumferential force F_θ, given below, is due to interaction with the rim

$$F_\theta = (P_m - P_{-m})x_m + (T_m + T_{-m})r_0 . \tag{5.32}$$

Interaction between spokes and the rim. Let us assume that the number of spokes is sufficiently large so that we can consider the loads due to interaction of spokes with the rim as uniformly distributed over the outer radius of the rim-disc.

Let us consider the state of stress and strain in the rim during uniform rotation. The equation of equilibrium and Hooke's law take the form:

$$\frac{\mathrm{d}\sigma_r}{\mathrm{d}r} + \frac{\sigma_r - \sigma_\theta}{r} + \rho_{VR}\omega^2 r = 0 , \tag{5.33}$$

$$\sigma_r = A_{11}\frac{\mathrm{d}u_r}{\mathrm{d}r} + A_{12}\frac{u_r}{r} ,$$

$$\sigma_\theta = A_{21}\frac{\mathrm{d}u_r}{\mathrm{d}r} + A_{22}\frac{u_r}{r} , \tag{5.34}$$

where ρ_{VR} is the mass density of the rim.

By analogy with (5.3) let us introduce the following dimensionless variables

$$\sigma_r = \rho_{VS}\omega^2 b^2 \bar{\sigma}_r ,$$

$$\sigma_\theta = \rho_{VS}\omega^2 b^2 \bar{\sigma}_\theta ,$$

$$u_r = \rho_{VS}\omega^2 b^3 \bar{u}_r / E , \quad r = b\bar{r} ,$$

where ρ_{VS} and E are the mass density and elastic modulus of the spoke material, respectively.

In terms of the new variables the system of equations given in (5.33)–(5.34) takes the form

$$\frac{d\bar{\sigma}_r}{d\bar{r}} + \frac{\bar{\sigma}_r - \bar{\sigma}_\theta}{\bar{r}} + \gamma\bar{r} = 0\,, \tag{5.35}$$

$$\begin{aligned} \bar{\sigma}_r &= a_{11}\frac{d\bar{u}_r}{d\bar{r}} + a_{12}\frac{\bar{u}_r}{\bar{r}}\,, \\ \bar{\sigma}_\theta &= a_{12}\frac{d\bar{u}_r}{d\bar{r}} + a_{22}\frac{\bar{u}_r}{\bar{r}}\,, \end{aligned} \tag{5.36}$$

where

$$\gamma = \rho_{VR}/\rho_{VS}\,, \quad a_{ij} = A_{ij}/E\,.$$

The boundary conditions for the system (5.35)–(5.36) are

$$\begin{aligned} &\bar{\sigma}_r(1) + \alpha F_r = 0\,, \\ &\bar{u}_r(1) = u_m x_m + v_m r_0\,, \quad \bar{\sigma}_r(m) = 0\,, \end{aligned} \tag{5.37}$$

where α is the ratio of the overall cross sectional area of the spokes to the area of the outer cylindrical surface of the rim; and m is the inner radius of the rim. The force F_r is determined from (5.23).

By denoting $\beta^2 = a_{22}/a_{11}$ we can express the solution of the system (5.35)–(5.36) as follows (the upper indices are omitted):

$$\begin{aligned} u_r &= C_1 r^\beta + C_2 r^{-\beta} + \gamma r^3/(a_{22} - 9a_{11})\,, \\ \sigma_r &= C_1(a_{11}\beta + a_{12})r^{\beta-1} + C_2(a_{12} - \beta a_{11})r^{-\beta-1} \\ &\quad - \gamma(3a_{11} + a_{12})r^2/(9a_{11} - a_{22})\,, \\ \sigma_\theta &= C_1(a_{12}\beta + a_{22})r^{\beta-1} + C_2(a_{22} - \beta a_{12})r^{-\beta-1} + \frac{\gamma(3a_{11} + a_{22})r^2}{(9a_{11} - a_{22})}\,. \end{aligned} \tag{5.38}$$

Thus, boundary conditions (5.37) take the form

$$\begin{aligned} &C_1 + C_2 - u_m x_m - v_m r_0 = \gamma/(9a_{11} - a_{22})\,, \\ &C_1(a_{11}\beta + a_{12}) + C_2(a_{12} - \beta a_{11}) + \alpha F_r = \gamma(3a_{11} + a_{22})/(9a_{11} - a_{22})\,, \\ &C_1(a_{11}\beta + a_{12})m^{\beta-1} + C_2(a_{12} - \beta a_{11})m^{-\beta-1} = \gamma(3a_{11} + a_{22})m^2/(9a_{11} - a_{22})\,. \end{aligned} \tag{5.39}$$

Equations (5.39) together with equations (5.20)–(5.23) define a closed system of equations for evaluation of the integration constants.

In the case of smooth braking the tangential dimensionless stresses in the rim are determined from the equation

$$\frac{\mathrm{d}}{\mathrm{d}r}(r^2\tau) = \gamma r^3 . \tag{5.40}$$

and, consequently, for the disc with a free inner radius, as follows

$$\tau = \gamma \frac{(r^4 - m^4)}{4r^2} . \tag{5.41}$$

The tangential stresses at the outer radius of the rim are

$$\tau(1) = \frac{(1 - m^4)}{4} \gamma . \tag{5.42}$$

Now, the boundary conditions for the spokes are expressed as follows:

$$\tau(1) + \alpha F_\theta = 0 . \tag{5.43}$$

The above expression is subsequently employed in (5.32).

Thus, both problems are reduced to the solution of linear systems of algebraic equations that can be obtained with the help of an electronic computer. The suggested design methods have been experimentally verified by MOORLAT and PORTNOV (1983) and MOORLAT et al. (1985), and the results have confirmed the applicability of the outlined methods.

5.2. Analysis of energy storage capacity

Chord flywheels with spokes and rims made of fiberglass, carbon, boron and organic fiber reinforced composites will be considered. The material properties are listed in Table 1. Calculation of maximum energy storage capacities and ultimate rotational speeds of flywheels with spokes and rims of selected materials and specified geometrical parameters is based on evaluation of the load-carrying capacity of flywheels during uniform rotation. For such an evaluation it is necessary that the state of stress in the rim and spokes be calculated according to the relationships presented in Section 5.1. On the basis of the ultimate rotational speed v_u the energy storage capacity is determined and subsequently employed for calculation of the specific characteristics W^M and W^V (MOORLAT and PORTNOV, 1985).

In the course of calculating the energy storage capacity, the kinetic energy of the rim W_R is determined from the formula

$$W_R = \frac{h}{2} \int_0^{2\pi} \int_a^b \rho_{VR} \omega^2 r^3 \, \mathrm{d}r \, \mathrm{d}\varphi = \tfrac{1}{8} \rho_{VR} \omega^2 b^3 S_R (1 - m^4) , \tag{5.44}$$

where a and b are the inner and outer radii of the rim of rectangular cross section; ρ_{VR} is the mass density of the rim; h is the axial dimension of the rim; $S_R = 2\pi bh$ is the area of the outer cylindrical surface of the rim; and $m = a/b$.

The kinetic energy of a semichord-spoke W_{S1} is

$$W_{S1} = \tfrac{1}{2}\int_0^b \rho_{VS}S(x^2 + r_0^2)\omega^2\,dx = \tfrac{1}{2}\rho_{VS}S\omega^2 b^3\bar{l}\,\frac{(1+2m_0^2)}{3}\,, \tag{5.45}$$

where l is the length of a semichord; r_0 is the inner radius of the system of chords (a fit radius of the flywheel); ρ_{VS} and S are the mass density and cross sectional area of a spoke; $\bar{l} = l/b$; $m_0 = r_0/b$. The energy storage capacity of the entire system of chords–spokes is evaluated as follows

$$W_S = \rho_{VS}S_S\omega^2 b^3\bar{l}\,\frac{(1+2m_0^2)}{6}\,, \tag{5.46}$$

where S_S is the overall cross sectional area of all the spokes.

The mass of the rim (M_R) and of the spokes (M_S) is evaluated from the formulas

$$M_R = \tfrac{1}{2}\rho_{VR}S_R b(1-m^2)\,,$$
$$M_S = \rho_{VS}S_S b\bar{l}\,.$$

The increase in the overall volume V_0 of the flywheel due to the increase in the cross-sections of the spokes is taken into account in the conventional manner as follows:

$$V_0 = \tfrac{1}{2}b(S_R + S_S)\,. \tag{5.47}$$

Passing over from (5.47) to the axial dimension in accordance with the employed technology of chord winding does not cause difficulties in a specific case.

The ultimate speed v_u of the flywheel periphery is defined by the minimum value of the following expression:

$$v_u^2 = \omega_u^2 b^2 = \min\left[\frac{\sigma_r^+}{\rho_{VR}\bar{\sigma}_r^{max}}, \frac{\sigma_\theta^+}{\rho_{VR}\bar{\sigma}_\theta^{max}}, \frac{\sigma_S^+}{\rho_{VS}\bar{\sigma}_S^{max}}, \frac{\tau_{Su}}{\rho_{VS}\bar{\tau}_S^{max}}\right], \tag{5.48}$$

where σ_r^+ and σ_θ^+ are the material tensile strengths of the rim in the radial and circumferental directions; $\bar{\sigma}_{r,\theta}^{max}$ are the maximum reduced tensile radial and circumferential stresses in the rim-disc, σ_S^+ and τ_{Su} are the material tensile and shear strength of the spokes, $\bar{\sigma}_S^{max}$ are the maximum reduced tensile stresses in the spoke (with allowance for bending stresses), and $\bar{\tau}_S^{max}$ are the maximum reduced tangential stresses in the spoke due to shearing forces. The magnitudes

$\bar{\sigma}_r^{\max}$, $\bar{\sigma}_\theta^{\max}$ and $\bar{\sigma}_S^{\max}$, $\bar{\tau}_S^{\max}$ defining v_u^2 and, consequently, W_R and W_S are evaluated by the methods described in Section 5.1.

The specific mass energy storage capacity of a chord flywheel is determined from the relationship

$$W^M = \frac{W}{M} = \frac{W_R + W_S}{M_R + M_S} = \tfrac{1}{2} J^* v_u^2 , \tag{5.49}$$

where

$$J^* = \frac{4 \dfrac{S_S}{S_R} \bar{l}(1 + 2m_0^2) + 3 \dfrac{\rho_{VR}}{\rho_{VS}} (1 - m^4)}{6\left[2 \dfrac{S_S}{S_R} \bar{l} + \dfrac{\rho_{VR}}{\rho_{VS}} (1 - m^4)\right]} .$$

The specific volumetric energy storage capacity of the flywheel is determined as follows:

$$W^V = \tfrac{1}{2} J^* \rho_V^* v_u^2 , \tag{5.50}$$

where

$$\rho_V^* = \rho_{VS} \frac{2 \dfrac{S_S}{S_R} \bar{l} + \dfrac{\rho_{VR}}{\rho_{VS}} (1 - m^2)}{1 + \dfrac{S_S}{S_R}} .$$

The effect of the ratio of a chord flywheel's rim and spoke thickness on W^M, W^V, v_u has been investigated for a flywheel with the following relative inner radius of the system of chords $m_0 = 0.092$. In the considered case there were thirty four spokes – semichords on each side of the rim with eight crossover points (nodes) on the semichord. Four different fiber composites employed as materials for the rim and spokes have been analyzed: their properties are listed in Table 1. Altogether, sixteen types of fiber composite flywheels have been considered, with different material combinations for the rim and spokes. The numerical analysis consisted of the following: evaluation of v_u, W^M, W^V for various combinations of materials for the rim and spokes and for the range of geometrical parameters $0.5 \leq m \leq 0.98$ indicative of the relative dimension of the rim; ratios of the cross sectional area of the spokes to the area of the outer cylindrical surface of the rim in the range $0.05 \leq \alpha = S_S/S_R \leq 1.0$. The ranges of m and α cover the magnitudes of parameters used in practice.

The limits of variation of W^M, W^V and v_u for the range of geometrical parameters for flywheels of the above mentioned type are shown in Table 6. It can be seen that by employing combinations of various materials we cannot exceed the mass energy storage capacities attained for an optimal flywheel

TABLE 6. Variation limits of W^M (kJ/kg), W^V (MJ/m^3), v_u (m/s) for chord flywheels within the range of parameters $0.5 \leq m \leq 0.98$; $0.05 \leq \alpha \leq 1.00$.

Rim		Spokes			
		Fiberglass composite	Carbon fiber composite	Boron fiber composite	Organic fiber composite
Fiberglass composite	W^M	80–280	130–360	130–250	90–350
	W^V	60–620	80–800	60–650	70–660
	v_u	500–1100	640–1370	650–1140	540–1330
Carbon fiber composite	W^M	50–300	140–430	130–310	100–390
	W^V	40–320	70–660	60–640	60–440
	v_u	460–1380	710–1380	710–1140	670–1290
Boron fiber composite	W^M	40–240	130–400	130–300	90–290
	W^V	40–280	80–630	70–610	50–370
	v_u	440–850	660–1360	650–1150	620–1140
Organic fiber composite	W^M	30–330	60–380	50–270	50–530
	W^V	50–480	70–670	60–600	60–530
	v_u	350–1100	440–1380	440–1150	400–1330

made of one and the same fiber composite. It is possible to attain the highest W^M for flywheels made entirely of carbon and organic fiber composites and for flywheels whose spokes are also made of these fiber composites.

Since the maximum possible values of W^M and W^V correspond to flywheels with different m and α, it is useful to represent the results of numerical analyses of W^M, W^V as well as of v_u by contour lines. Such a representation allows one to quickly determine the ranges of geometrical parameters corresponding to the maximum or sufficiently high magnitudes of specific energy storage capacities. It also facilitates the search for trade-offs, ensuring acceptable combinations of W^M, W^V and v_u. The contour lines have been constructed for all the sixteen possible combinations of the fiber composites investigated; their analysis allows one to draw some general conclusions.

For flywheels with spokes and rims made of one and the same material, the pattern of contour lines does not depend on the type of fiber composite. As an example, Fig. 11 shows the patterns of contour lines for a carbon fiber composite flywheel. The breaks in the curves correspond to changes in the mode of flywheel failure. In the case of rims of small relative thickness the spokes essentially unload the rim of tensile stresses and consequently these fail first – region 1 (Fig. 11). With the increase in the relative radial thickness of the rim, provided that the value $\alpha = S_S/S_R$ is maintained constant, the mass energy storage capacity of a flywheel increases in region I, and its ultimate speed changes relatively little (Figs. 11(a), (c)) – up to the break in the contour lines. This break corresponds to transition from failure of the spokes to failure of the rim due to circumferential stresses – region 2. In this region, the increase in the rim thickness is already associated with some reduction in the mass energy storage capacity and appreciable reduction in the ultimate speed – up to the second break in the contour lines where the rim failure due to circumferential stresses is replaced by failure due to radial stresses – region 3. In this region, the increase in thickness drastically reduces the mass energy storage capacity and the ultimate speed.

The volumetric energy storage capacity increases very significantly with increasing rim thickness in region 1 (Fig. 11(b)). In region 2, W^V, up to about $\alpha = 0.3$–0.4, also increases with increasing rim thickness, however less appreciably than in region 1; at $\alpha > 0.3$–0.4, W^V in region 2 has a poorly defined maximum. In region 3, W^V decreases sharply.

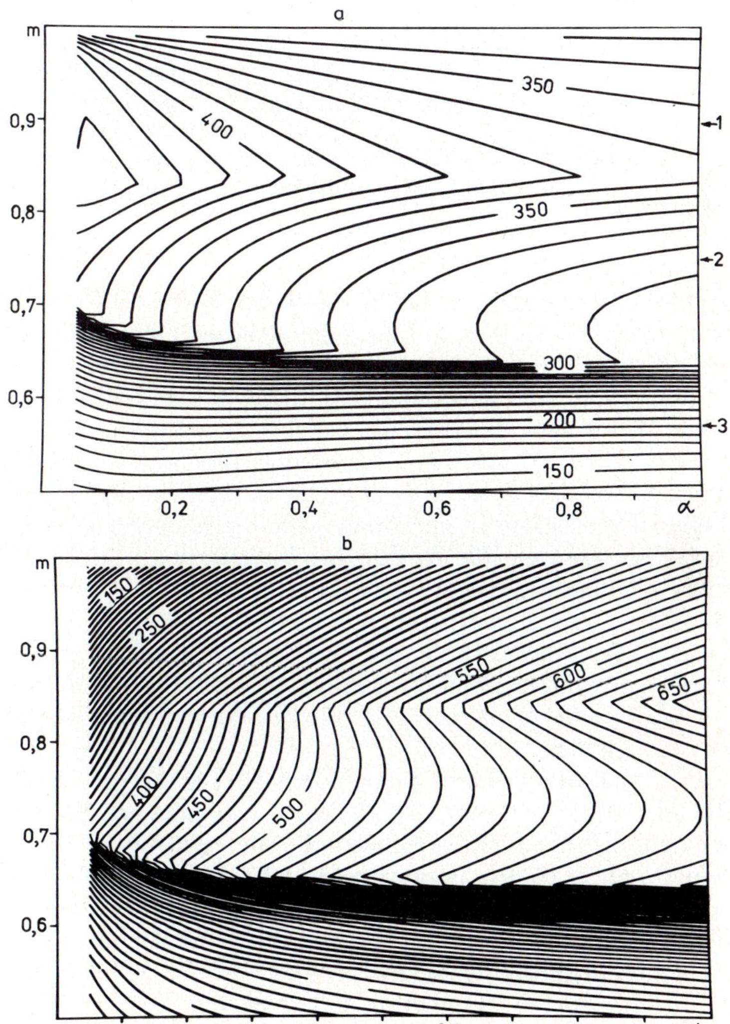

FIG. 11. Contour lines for a carbon fiber composite flywheel: a, W^M (kJ/kg); b, W^V (MJ/m^3); c, v_u (m/s).

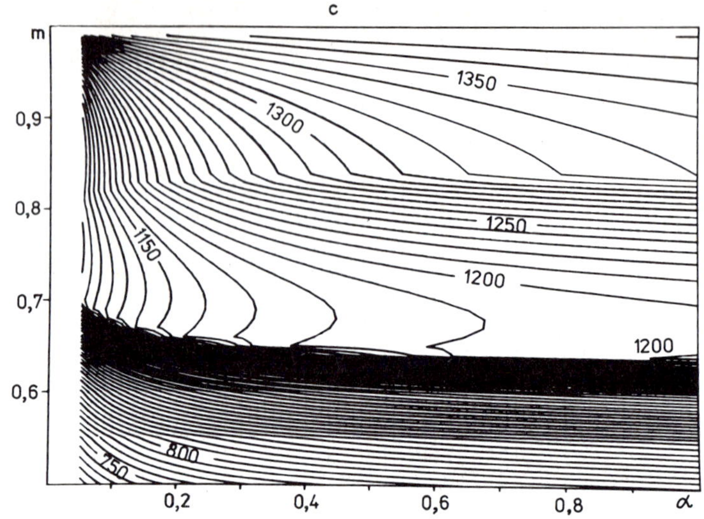

FIG. 11. *Continued.*

Thus, region 3 corresponds to small magnitudes of both W^M and W^V and in the design of flywheels this region should be avoided. Obviously, in designing homogeneous flywheels it is useful to stick to the 'ridge' which is the boundary between regions 1 and 2. Moving along the 'ridge' in the direction of increasing α we can gain significantly in W^V at the expense of a slight reduction W^M (see Figs. 11(a), (b)). It is of interest to note that the relative rim dimensions corresponding to the 'ridge' are very close to the optimal dimension m_{opt} of a freely rotating rim ensuring its maximum volumetric energy storage capacity (Table 2). Thus, it is possible to design a satisfactory, with respect to the energetic characteristics, chord flywheel if by means of the results of Section 5.2 we evaluate m_{opt} for a freely rotating disc and assume it to be the required rim dimension. We then specify the thickness of spokes from structural considerations or requirements for a specific volumetric energy storage capacity, since in the vicinity of $m = m_{\text{opt}}$ W^M is not very sensitive to the relative thickness of the spokes.

In the case of employing different fiber composites in the spokes and rim of a chord flywheel, within the range of indicated parameters, the contour line patterns become displaced and sometimes the above-mentioned characteristic regions even disappear. Thus, employing more rigid or lighter fiber composites for the spokes than the rim results in expansion of region 1, where the spokes fail first (Fig. 12). In such flywheels, it is possible to attain high magnitudes of mass and volumetric energy storage capacities and to use relatively thick-in-the-radial-direction rims in conjunction with spokes of large cross section.

By using rigid rims in conjunction with more compliant spokes we can attain significant expansion of region 2, where the rim fails first due to circumferential stresses (Figs. 13a, b). In such flywheels, the spokes are underloaded and the increase in their load-carying capacity during braking can be expected.

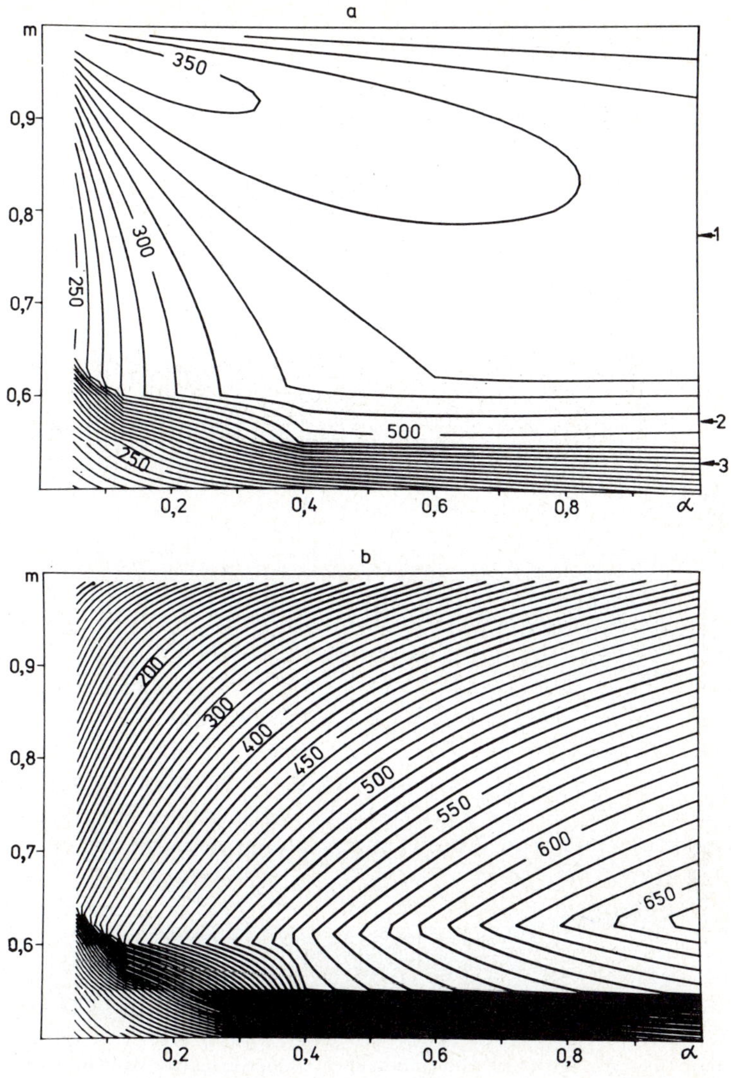

FIG. 12. Contour lines for a flywheel with organic fiber composite spokes and fiberglass composite rim; a, W^M (kJ/kg); b, W^V (MJ/m^3).

It should be emphasized as a general conclusion for all flywheels that a significant decrease in the ultimate speed cannot be attained without appreciable losses in the mass energy storage capacity. It is possible to attain the maximum decrease in the ultimate speed by varying the geometrical parameters of flywheels whose rims are made of fiberglass and spokes of carbon or organic fiber composite.

An attempt was made to evaluate the effect of the number of spokes (at given mass of spokes) on W^M and W^V. Such an analysis is complicated by the

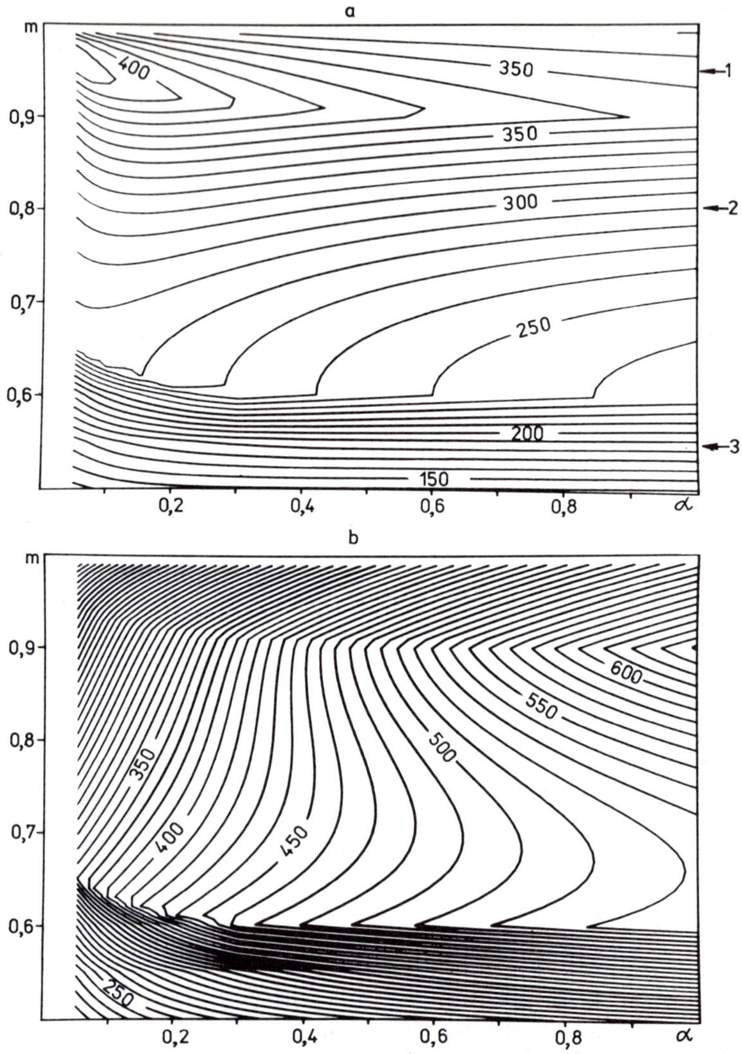

FIG. 13. Contour lines for a flywheel with carbon fiber composite spokes and boron fiber composite rim; a, W^M (kJ/kg); b, W^V (MJ/m^3).

fact that it is impossible to vary the inner radius continuously during winding of straight chords. Therefore, different inner radii correspond to different numbers of chords and this complicates the comparison of the results. An increase in the number of spokes n is accompanied by an increase in the number of nodes k (crossover points) which defines to a large extent the state of stress in the spokes. This can lead to a variation in the energy storage capacities of a flywheel. As an example, the patterns of contour lines for flywheels of approximately equal inner diameters and equal overall cross sectional area of spokes, but different number of spokes and nodes on each spoke are shown in Figs. 14a, b. As is illustrated, the increase in the number of spokes leads to the

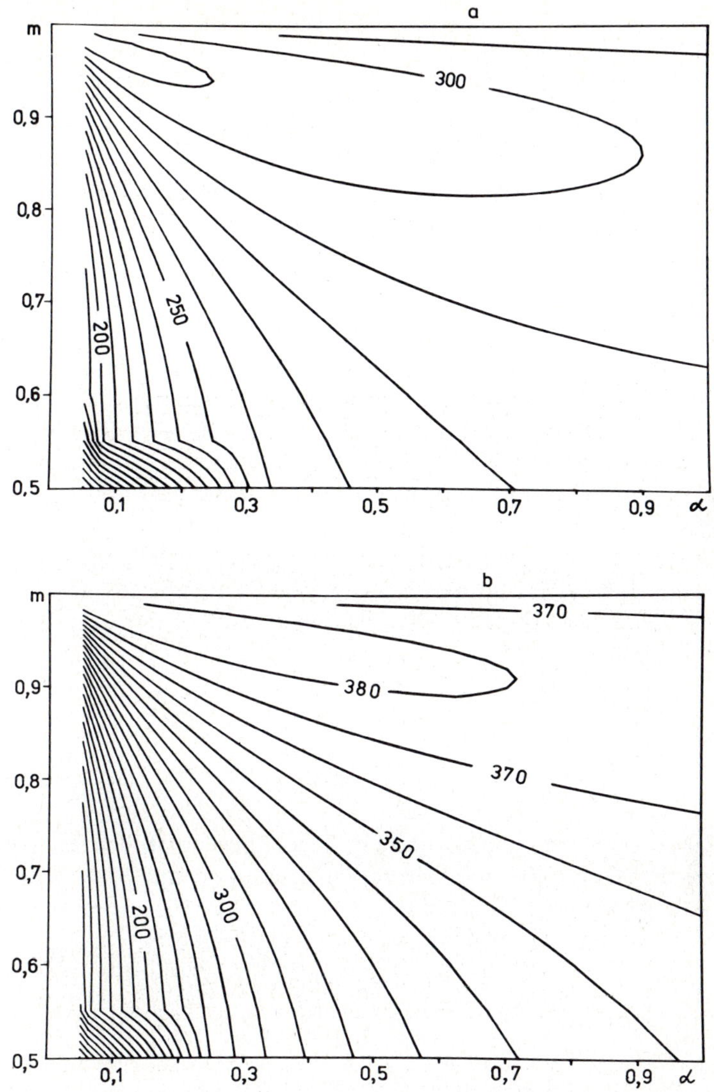

FIG. 14. Contour lines W^M (kJ/kg) for a flywheel with carbon fiber composite spokes and fiberglass composite rim; a, $m_0 = 0.309$, $n = 5$, $m = 2$; b, $m_0 = 0.294$, $n = 21$, $m = 9$.

increase in the maximum magnitudes of W^M; max W^V also increases in the same proportion. The increase is associated with a reduction of the contribution of bending stresses to the state of stress in the spoke and with some decrease in tensile forces in the spoke with increasing number of nodes. Thus, within the range of allowable variation of the inner diameter, it is expedient that those magnitudes of the inner diameter should be selected which allow one to perform the winding operation with a large number of chords.

It should be noted that the increase in the number of chords of a specified diameter can be achieved by foregoing winding with straight chords in favour of bending them around the inside opening (this is associated with more complicated kinematics of the winding process).

5.3. Analysis of maximum power with regard to energy supply and pick-up during speed-up and braking action of chord flywheels

Let us consider the procedure which allows one to evaluate the possibility of realization of a required power impulse $\dot{W}_i$ for a particular flywheel design, specified in the form $\dot{W}_i = \dot{W}_i(t)$ and transformed through integration into the form $\dot{W}_i(W_i)$. That is, we need to determine the ultimate possible combinations of the power $\dot{W}$ and the flywheel energy W. The solution of this problem is based on the calculation of the ultimate states of stress in chords–spokes and rim of a flywheel loaded with inertia forces due to the instantaneous angular velocity ω and acceleration $\dot{\omega}$ which, during speed-up and braking, is assumed to be constant over the entire flywheel volume. Calculation of the respective states of stress is given in 5.1. After summation of the states of stress due to ω and $\dot{\omega}$, the strength of the spokes is estimated by comparing the tensile (with allowance for bending σ_x^{S} and shear stresses due to bending τ_{xy}^{S} with the respective strengths:

$$\max_x \sigma_x^{\mathrm{S}} \leqslant \sigma_{\mathrm{S}}^{+} , \tag{5.51}$$

$$\max_x \tau_{xy}^{\mathrm{S}} \leqslant \tau_{\mathrm{Su}} . \tag{5.52}$$

The strength of a rim is estimated in accordance with the criterion taking into account the interaction between radial stresses σ_r (due to ω) and interlaminar shear stresses $\tau_{r\theta}$ (due to $\dot{\omega}$); the experimental confirmation is given in the work by NIKOLAEV et al. (1983) and RABOTNOV (1979):

$$\frac{\tau_{r\theta}}{\tau_{r\theta}^{\mathrm{u}}} + \frac{\sigma_r}{\sigma_r^{+}} = 1 , \tag{5.53}$$

where $\tau_{r\theta}^{\mathrm{u}}$ is the interlaminar shear strength of the rim material made by hoop winding; σ_r^{+} is the transverse tensile strength.

The evaluation of the ultimate regimes of braking action based on criteria (5.51)–(5.53) is reduced to the investigation of three equalities of the following kind

$$\rho_V \omega^2 b^2 \bar{\sigma}(\omega) + \alpha \rho_V |\dot{\omega}| b^2 \bar{\sigma}(\dot{\omega}) = \sigma_{\mathrm{u}} . \tag{5.54}$$

In (5.54)

$$\bar{\sigma}(\omega) = \frac{\sigma(\omega)}{\rho_V \omega^2 b^2} , \quad \bar{\sigma}(\dot{\omega}) = \frac{\sigma(\dot{\omega})}{\rho_V |\dot{\omega}| b^2} ,$$

are the dimensionless reduced stresses or their components defining the state of stress due to inertia forces caused by uniform rotation and acceleration, which are independent of the overall dimensions of the given flywheel design, b is the outer radius of the flywheel; in the case of using criteria (5.51) and (5.52), $\rho_V = \rho_{VS}$ is the mass density of spokes, σ_u is the longitudinal tensile strength σ_S^+ of the spokes or the shear strength τ_{Su}, respectively, $\alpha = 1$; making use of criterion (5.53), $\rho_V = \rho_{VR}$ is the material density in the rim, $\sigma_u = \sigma_r^+$, $\alpha = \sigma_r^+/\tau_{r\theta}^u$.

It follows from (5.54) that the ultimate values of $v = \omega b$ and $\dot{\omega} b^2$, defined by the strength and stress values, are invariant to shape-preserving variations in flywheel dimensions. The above makes it possible to employ invariant characteristics such as W^M and $b\dot{W}^M$ given below in the course of determining the exteme relationship between W and $\dot{W}$:

$$W^M = \frac{W}{M} = \frac{W}{\rho_V V} = \frac{\bar{J} v^2}{2\bar{V}}, \tag{5.55}$$

$$b\dot{W}^M = \frac{b\dot{W}}{\rho_V V} = \frac{\bar{J}}{\bar{V}} (\dot{\omega} b^2) v, \tag{5.56}$$

where W^M is the specific mass energy storage capacity; $b\dot{W}^M$ is the reduced specific mass power; $\bar{J} = J/(\rho_V b^5)$ is the reduced dimensionless inertia moment of the flywheel; ρ_V is the mass density of the flywheel material; $\bar{V} = V/b^3$ is the reduced flywheel volume.

Thus, $b\dot{W}^M$ and W^M characterize the specified flywheel design regardless of its dimensions. It follows from the invariance of $b\dot{W}^M$ and W^M that for a k-fold increase in the flywheel dimensions the maximum energy increases by k^3, and the power by k^2. Hence, a particular conclusion can be drawn that in order to attain higher power it is more useful to increase the number of flywheels on the shaft and not their overall dimensions, while for the energy storage capacity it makes no difference whether one concentrates all the material mass into one flywheel or distributes it between several similar flywheels of smaller size acting together.

The outlined problem is solved by means of the relationships introduced above and reduced to the following. It is possible to evaluate W^M from (5.55) by specifying the magnitudes of v within the range of $0 \leqslant v \leqslant v_u$ (v_u is the ultimate speed of uniform rotation of the specified flywheel design). Then, by using (5.54), the ultimate magnitude of the reduced acceleration $(\dot{\omega} b^2)_u$ corresponding to the specified v is determined and from (5.56) the ultimate magnitude of the reduced power $(b\dot{W}^M)_u$ in agreement with the specified magnitude of W^M is calculated. It is convenient to express the relationship between the reduced ultimate mass power and the mass energy storage capacity by using (5.54)–(5.56) as follows:

$$(b\dot{W}^M)_{\text{u}} = \frac{2W^M}{v} \min \phi_i \,, \quad \text{where}$$

$$\phi_1 = \min_x \left\{ \frac{1}{\bar{\sigma}_{\text{S}}(\dot{\omega})} \left[\frac{\overset{+}{\sigma}_{\text{S}}}{\rho_{V\text{S}}} - v^2 \bar{\sigma}_{\text{S}}(\omega) \right] \right\},$$

$$\phi_2 = \min_x \left\{ \frac{1}{\bar{\tau}_{\text{S}}(\dot{\omega})} \left[\frac{\tau_{\text{Su}}}{\rho_{V\text{S}}} - v^2 \bar{\tau}_{\text{S}}(\omega) \right] \right\}, \tag{5.57}$$

$$\phi_3 = \min_r \left\{ \left(\frac{\overset{+}{\sigma}_r}{\rho_{VR}} - v^2 \bar{\sigma}_r \right) \frac{\overset{+}{\sigma}_r}{\tau^{\text{u}}_{r\theta}} \bar{\tau}_{r\theta} \right\}.$$

The values ϕ_1 and ϕ_2 correspond to minimum magnitudes of the respective expressions along the spoke length at the specified v; ϕ_3 corresponds to minimum magnitude along the rim radius. The smallest of these minimum magnitudes is substituted into expression (5.57).

In this way, the constructed ultimate curve forms the basis for estimating the possibility of realization of the required impulse, specified in the form $\dot{W}_i = \dot{W}_i(W_i)$. To carry this out it is expedient to transform the dependence of $(b\dot{W}^M)_{\text{u}}$ on W^M into dependence of $\dot{W}_{\text{u}}$ on W in accordance with (5.55) and (5.56) by specifying the flywheel radius b. The required energy impulse expressed as $\dot{W}_i(W_i)$ (during braking as $\dot{W}_i(W_{\text{u}} - W_i)$, where W_{u} is the maximum energy capacity of the flywheel corresponding to v_{u}) is compared with the function $\dot{W}_{\text{u}}(W)$ and if it falls entirely inside the ultimate curve it can be realized. If it is outside this curve, either the flywheel dimensions should be increased or its design changed.

As an illustration the ultimate curves $(b\dot{W}^M)_{\text{u}} - W^M$ for two chord flywheel designs have been constructed in accordance with the above-described procedure. These differ principally by the character of variation of specific mass energy storage capacity with the varying thickness of the rim and spokes. The dependence of mass energy storage capacity on geometrical parameters is illustrated by the contour lines of W^M for flywheels with carbon fiber composite spokes and fiberglass reinforced composite rim in Fig. 15(a) and for flywheels with organic fiber composite spokes and organic fiber composite rim in Fig. 15(b). The change of the ultimate regimes of braking action, the relationships between $(b\dot{W}^M)_{\text{u}}$ and W^M, due to the variation of geometrical parameters along straight lines shown on the contour line patterns was analyzed. By varying the parameters along straight lines (increasing α) we can appreciably increase the specific volumetric energy storage capacity of flywheels without significant losses in the mass energy storage capacity. The straight line in Fig. 15(a), passes into the region in which the flywheel spokes fail first; in Fig. 15(b), the straight line passes into the region where the rim fails due to circumferential stresses and near the ridge where the mode of failure changes to that of the rim due to radial stresses. Thus, at the ultimate rotational speed the load-carrying capacity of spokes is either completely exhausted (the first case) or the spokes are virtually underloaded (the second case).

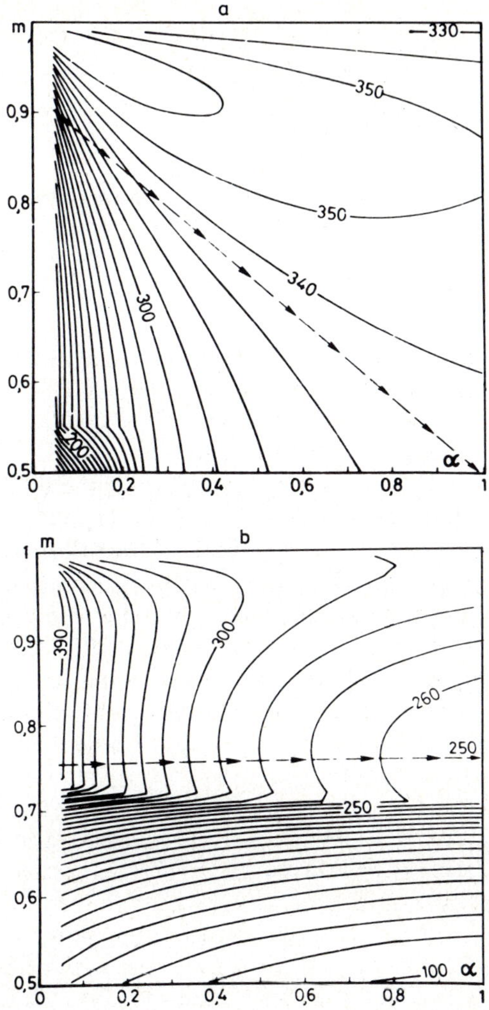

FIG. 15. Contour lines W^M (kJ/kg) for a flywheel with carbon fiber composite spokes and fiberglass composite rim (a), with organic fiber composite spokes and fiberglass composite rim (b); combinations of m and α are designated by lines, investigated in the evaluation of ultimate power of flywheels.

The ultimate curves $(b\dot{W}^M)_u - W^M$ and $(b\dot{W})_u^V - W^V$ pertaining to these cases (where $\dot{W}^V, W^V$ are power and energy per unit volume for a flywheel whose volume is determined from (5.47) are shown in Figs. 16 and 17.

For all the considered flywheel designs, the load-carrying capacity during speed-up and braking operations was defined by the strength of the spokes. Therefore, regardless of quantitative differences, the shape of the ultimate curves is the same. The maximum magnitudes of ultimate power differed

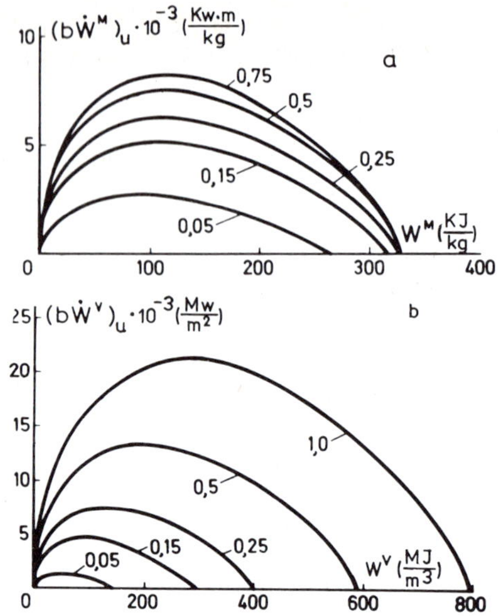

FIG. 16. Dependence of (a) specific reduced mass and (b) volumetric power on mass and volumetric energy storage capacities for a flywheel with characteristics listed in Fig. 15(a).

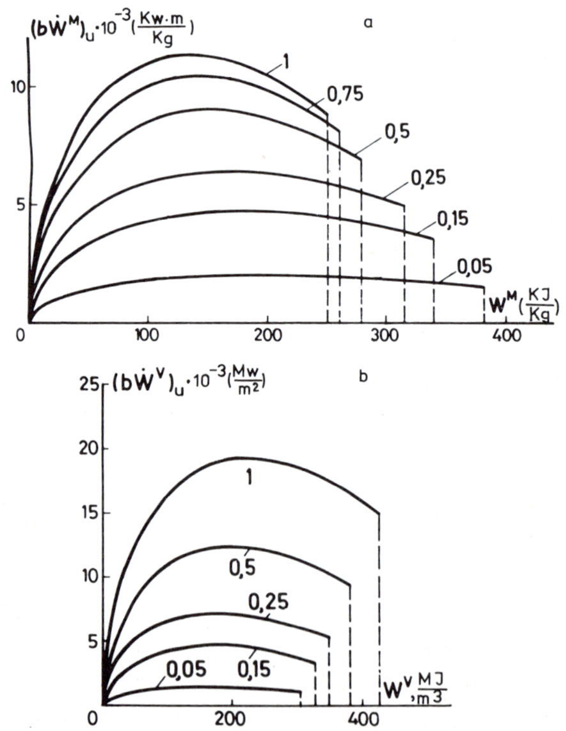

FIG. 17. Dependence of (a) specific reduced mass and (b) volumetric power on mass and volumetric energy storage capacities for flywheels with characteristics listed in Fig. 15(b).

negligibly in the range of low α (small area of spokes). In the range of high α this difference increased. The results have been obtained using condition (5.30) at the outer end of the spoke; condition (5.31) results in negligible increase (no more than 10%) in the ultimate power.

It follows from the obtained results that the power developed in chord flywheels evaluated on the basis of the spoke and rim strengths can be fairly high.

The theoretical investigations and experimental verification performed on models have convincingly demonstrated the high effectiveness of fiber composite flywheels. At the present time, the most promising design is a chord flywheel, especially with respect to impulse energy requirements. It can be assumed that the high complicance of chords will allow one to balance flywheel structures relatively easily. However, the problem of balancing and, in general, dynamics of energy storing rotating composite elements requires further investigation.

References

ALFUTOV, N.A. and P.A. ZINOV'EV (1973), *Mechanics of Rigid Body* **1**, 181–183.

BEIL', A.I., V.L. KULAKOV and G.G. PORTNOV (1981), *Mekhanika kompozitnykh Materialov* **17**(6), 1055 (*Mechanics of Composite Materials* **17**(6), 701).

BOKOV, Yu.V., V.V. VASIL'EV and G.G. PORTNOV (1981), *Mekhanika kompozitnykh Materialov* **17**(5) 846 (*Mechanics of Composite Materials* **17**(5), 572).

BUNAKOV, V.A. and V.D. PROTASOV (1986), Structures and Design, in: C.T. HERAKOVICH and Yu.M. TARNOPOL'SKII, eds., *Handbook of Composites*, Vol. 2 (North-Holland, Amsterdam).

CHAMIS, C.C. and L.I. KIRALY (1975), *Flywheel Technology Symposium Proceedings*, Berkeley, California, 1975 (U.S. Government Printing Office), p. 110.

CHEREVATSKII, S.B., Yu.P. ROMASHOV, E.M. TSENTOVSKII and S.G. SIDORIN (1983a), *Mekhanika kompozitnych Materialov* **19**(3), 510.

CHEREVATSKII, S.B., Yu.P. ROMASHOV and S.G. SIDORIN (1983b), *Mekhanika kompozitnych Materialov* **19**(6), 1116.

CHRISTENSEN, R.M. and E.M. WU (1977), *Flywheel Technology Symposium Proceedings*, San Francisco, California, 1977 (U.S. Government Printing Office), p. 363.

CHRISTENSEN, R.M. (1979), *Mechanics of Composite Materials* (John Wiley & Sons, New York).

GENTA, G. (1985), *Kinetic Energy Storage* (Butterworths, London).

GULYA, N.V. (1976), *Flywheel Power Plants* (Mashinostroyeniye, Moscow).

JONES, E.R. (1969), *SPE Journal* **25**, 50.

JONSON, D.E. and I.I. GORMAN (1980), *Flywheel Technology Symposium Proc.*, Scottsdale, Arizona, 1980 (U.S. Government Printing Office) p. 93.

KYSER, A.C. (1965), AIAA Journal **3**, 127.

KNIGHT, C.E., Jr, I.I. KELLY, R.F. HUDDLESTON and D.E. POLLARD (1977), *Flywheel Technology Symposium Proceedings*, San Francisco, California, 1977 (U.S. Government Printing Office) p. 137.

KULAKOV, V.L. and G.G. PORTNOV (1980), *Mekhanika kompozitnykh Materialov* **16**(2), 291 (*Mechanics of Composite Materials* **16**(2), 218).

KULAKOV, V.L. and G.G. PORTNOV (1981), *Mekhanika kompozitnykh Materialov* **17**(2), 267 (*Mechanics of Composite Materials* **17**(2), 184).

LANDAU, L.D. and E.M. LIFSHITZ (1965), *Theory of Elasticity* (Nauka Publishing House, Moscow).

LURIE, A.I. (1970), *Theory of Elasticity* (Nauka Publishing House, Moscow).

MOORLAT, P.A., G.G. PORTNOV and L.N. SELEZNEV (1982), *Mechanika kompozitnykh Materialov* **18**(5), 859 (*Mechanics of Composite Materials* **19**(5), 579).
MOORLAT, P.A. and G.G. PORTNOV (1983), *Mekhanika kompozitnykh Materialov* **19** (5), 853 (*Mechanics of Composite Materials* **19** (5), 631).
MOORLAT, P.A., G.G. PORTNOV, V.I. ROMASHKO and A.P. RYAZANTSEV (1985), *Mekhanika Kompozitnykh Materialov* **21** (4), 665.
MOORLAT, P.A. and G.G. PORTNOV (1985), *Mekhanika kompozitnykh Materialov* **21** (5), 861.
NIKOLAEV, V.P. (1978), *Strength Problems* **3**, 86.
NIKOLAEV, V.P., V.D. POPOV and A.K. SBOROVSKII (1983), *Strength and Reliability of Wound GFRP* (Mashinostroyeniye, Leningrad).
OBRAZTSOV, I.F. and V.V. VASIL'EV (1986), Structures and Design, in: C.T. HERAKOVICH and Yu.M. TARNOPOL'SKII, eds., *Handbook of Composites*, Vol. 2 (North-Holland, Amsterdam).
Patent of Great Britain, No. 29450 (1964).
POLYAKOV, V.A., G.G. PORTNOV and P.A. MOORLAT (1983), *Mekhanika kompozitnykh Materialov* **19**(6), 1044.
PORTNOV, G.G. and V.L. KULAKOV (1978a), *Mekhanika Polimerov* **14**(1), 73 (*Polymer Mechanics* **14**(1), 61).
PORTNOV, G.G. and V.L. KULAKOV (1978b), *Mekhanika Polimerov* **14**(4), 615 (*Polymer Mechanics* **14**(4), 503).
PORTNOV, G.G. and V.L. KULAKOV (1980), *Mekhanika kompozitnykh Materialov* **16**(5), 887 (*Mechanics of Composite Materials* **16**(5), 605).
PORTNOV, G.G. and V.A. POLYAKOV (1982), *Mekhanika kompozitnykh Materialov* **18**(2), 301 (*Mechanics of Composite Materials* **18**(2), 212).
Proceedings of the 1975 Flywheel Technology Symposium (Berkeley, California, November 1975).
Proceedings of the 1977 Flywheel Technology Symposium (San Francisco, California, October 1977).
Proceedings of the 1980 Flywheel Technology Symposium (Scottsdale, Arizona, October 1980).
RABOTNOV, Yu.N. (1979), *Mechanics of Deformable Rigid Body* (Nauka Publishing House, Moscow).
VASIL'EV, V.V., V.A. POLYAKOV, G.G. PORTNOV and Yu.V. BOKOV (1982), *Mekhanika kompozitnykh Materialov* **18**(1), 85 (*Mechanics of Composite Materials* **18**(1), 70).
VOROPAEV, V.I., G.G. PORTNOV and L.N. SELEZNEV (1982), *Mekhanika kompozitnykh Materialov* **18**(1), 159.
ZINOV'EV, P.A. (1972), *News of Higher Education Institutions* (Mashinostroyeniye) **12**, 32.

CHAPTER 11

Stability Critical Compression Members

E.E. Spier

Rohr Industries, Inc.
Chula Vista, CA 92012
U.S.A.

R.R. Arnold

Anamet Laboratories, Inc.
Hayward, CA 94545-3811
U.S.A.

K.T. Kedward

Alcoa Defense Systems, Inc.
San Diego, CA 92127
U.S.A.

Contents

HANDBOOK OF COMPOSITES, VOL. 2 – Structure and Design
Edited by C.T. HERAKOVICH and Y.M. TARNOPOL'SKII

List of Symbols

A – total area
b – width of plate
b_c – width of crown (hat-section)
b_f – width of flange
b_w – width of web
B_{ij} – bending–stretching elastic coupling constants
C – end-fixity coefficient of column
D_{ij} – flexural–twisting stiffness elastic constants
E^c – compressive modulus of elasticity
E_L^t, E_T^t – longitudinal and transverse elastic tensile moduli of elasticity, respectively, of lamina
E_L^c, E_T^c – longitudinal and transverse elastic compressive moduli of elasticity, respectively, of lamina
E_{11}^c – longitudinal elastic modulus of elasticity of laminate
F_{cl}^{cr} – classical, specially orthotropic, elastic plate-buckling stress
$F_{w,cl}^{cr}$ – classical buckling stress of web
F^{cr} – buckling stress as determined from test
F^{cc} – postbuckling failure (crippling) stress
F_w^{cc} – crippling stress of web
F^{co} – short or Euler column buckling stress
F_L^{tu}, F_T^{tu} – longitudinal and transverse tensile strength, respectively, of lamina
F_L^{cu}, F_T^{cu} – longitudinal and transverse compressive strength, respectively, of lamina
F_{11}^{cu} – longitudinal compressive strength of laminate
F^f – maximum fatigue stress
G_{LT} – shearing modulus of rigidity for lamina
L – overall length of test specimen
L' – effective column length of test specimen
M – edge moment
P_{cl}^{cr} – classical, specially orthotropic, elastic plate-buckling load
P^{cr} – buckling load as determined from test
P^{cc} – postbuckling failure (crippling) load
P_w^{cc} – crippling load of web
P^f – maximum fatigue load unless noted as minimum (min)
R – ratio of minimum to maximum fatigue loads
t – thickness of plate
t_f – thickness of flange
t_w – thickness of web

t_c – thickness of crown (hat-section)
w – displacement component of plate in normal direction
v/o – volume per cent of fiber
ε_L^{tu}, ε_L^{cu} – longitudinal ultimate tensile and compressive strain, respectively
ν_{LT} – major Poisson's rato of lamina
ρ – minimum radius of gyration

General introduction

Wide exploitation of advanced composites in stability critical structural designs will depend to a large degree on the ability of such heterogeneous anisotropic composites to support loads well beyond the initial buckling level. Unquestionably, the high stiffness-to-weight ratio of composites renders them potentially attractive up to initial buckling. However, since postbuckling design has been established over several decades for conventional metallic alloy construction we should anticipate that composites should demonstrate a similar tolerance to be competitive. Hence, this chapter and Chapter 12 address this vitally important issue as it pertains to the design of structural compression members and stiffened panels.

The treatment herein covers a range of stability phenomena that should be preceded by the following basic and general definitions:

(1) *Initial buckling*. The onset of instability as predicted by classical theory, refined theories, numerical methods, or by experiment. In this regard comparisons between the methods of prediction and experimental data are presented in both chapters. Occasionally, a design philosophy that precludes use of the structure beyond initial buckling is adopted. Sometimes initial buckling is precluded up to limit load.

(2) *Postbuckling*. The ability of a compression member or stiffened panel to carry loads well in excess of the initial buckling load is widely accepted. In both chapters we consider the 'postbuckling range' to exist between the initial buckling load and some higher load representing failure, e.g., delamination at the free edge of a compression member or the disbonding of a stiffener from the panel in a stiffened panel. For basic elements, comprising a simple structural constituent panel, the upper limit of the postbuckling range is sometimes termed 'local crippling' or simply 'crippling'.

(3) *Failure*. The load at which irreversible damage such as delamination, transverse cracking or stiffener disbonding occurs and is usually accompanied by an abrupt loss in load carrying capability. This event, a subset of which has been termed 'crippling', may or may not be predicted.

In this chapter a largely empirical discussion of compression members and one form of stiffened panel is presented; only classical expressions for predicting initial loads of flat plate elements are included. This section is extensively illustrated and several useful forms of design data presentation are contributed. Following the well established methodology adopted for metallic compression members a given structural configuration is assumed to consist of a series of flat plat elements exhibiting two basic support conditions for the unloaded edges, i.e., no-edge-free and one-edge-free. To avoid the complications contributed by transverse shear effects discussion is confined to reasonably large values of breadth-to-thickness (b/t) ratios, greater than 15. The effects of transverse shear and material nonlinearity is treated in the subsequent chapter.

In addition, the next chapter deals with both flat and curved composite panels with stiffeners (stringers and frames) that are representative of numerous aerospace structural configurations, typically, fuselage and empennage components.

Both chapters therefore contribute to the presently sparse database for composite stiffened panels and serves to identify the special attributes afforded by advanced composites. Presented in the next chapter are theoretical and experimental data for initial buckling, postbuckling and failure characteristics. Where appropriate, simple expressions are provided with graphical presentation for phenomena requiring a more complex mathematical treatment. In the latter case the key reference material is included with the discussion. Discussion includes the effects of transverse shear, edge stiffening, curvature, anisotropy, stiffness unbonding, imperfection sensitivity, boundary conditions and combined loads.

1. Introduction – compression members

The military services are presently very interested in aircraft designs that use the postbuckled strength of structures made from advanced composite materials, particularly in fuselage structures where weight savings potential is high, MINECCI and HESS (1980). The most significant weight savings would occur in composite semi-tension field panels with additional weight savings occurring in the stiffeners. This chapter deals with the stiffeners and stiffened panels in compression only. A stiffened panel under combined compression and shear is discussed in a subsequent chapter.

This chapter contains compression test results, from SPIER (1981), pertaining to structures that are typical of those found in aircraft. Subsequent evaluation of a suggested methodology for the static buckling, postbuckling, and crippling behavior is presented. Consideration is also extended to 100 000 cycle postbuckling fatigue behavior of short graphite/epoxy thin-walled plates and common stiffener configurations as well as stiffened panels. In addition, the effects of postbuckling fatigue testing on the static buckling and crippling strengths are discussed. Some static test results from SPIER (1978) and SPIER and KLOUMAN (1977) are also included herein.

The data presented in this chapter have significant impact on the static ultimate and postbuckling fatigue analyses of aircraft structures, particularly where some buckling is permitted at or below limit load. The configurations of the test specimens were chosen to have sufficient variation of width-to-thickness ratios in the plates and the plate elements of the stiffeners to be representative of a wide class of aircraft structures. Most of the test results are presented in nondimensional form, thus permitting the analysis of virtually all aircraft structures with diverse compressive load intensity. With these data, most aircraft fuselage, wing, and related structures can be conveniently analyzed based on the methodology presented herein. However, it will be observed that the scope of the present chapter is not sufficient to encompass the many potential opportunities available for minimum weight design by appropriate changes in the ply orientations of the constituent elements of either the stiffeners or the plate-elements of the stiffened structure.

Only experimental data for those stiffeners that had significant postbuckling ranges are presented; thus, the class of plates and stiffeners that fails in crippling near the initial buckling load has not been included herein. For this class, there is no requirement for the use of postbuckling fatigue methodology. However, the allowable fatigue loads for buckling critical structures may be somewhat below the initial buckling loads. The plates and stiffeners, for the most part, are designed to have significant postbuckling strength based on earlier experimental work, SPIER (1978), SPIER and KLOUMAN (1977). Significant problems relating to analytical predictions for initial buckling of plates and stiffeners have been observed, ANDERSON and MAYERS (1979), ARNOLD and MAYERS (1984), and all tests reported herein include an evaluation of such problems because of their importance in the design of compression structures. Included in the testing of stiffeners were channels, zees, hat-sections, and I-sections. However, the test results of the hat- and I-sectioned stiffeners are not reported herein because of the limited variations in configuration, but the results are included in SPIER (1981). Hat-stiffened panel test results are presented in Section 3.4 of this chapter.

All of the test specimens were fabricated from either A-S/3501-6 or T300/5208 graphite/epoxy with the stacking sequence of $[\pm 45/0_3/90]_s$ in most instances. See Table 1 for typical lamina material properties. A few test specimens had other stacking sequences for various reasons, which will become

TABLE 1. Typical lamina properties of two high strength graphite/epoxy systems.

Property	System			
	A-S/3501-5 (v/o = 62%)		T-300/5208 (v/o = 64%)	
	English units	SI units	English units	SI units
E_{11}^{t}	20.0 Msi	138 GPa	22.8 Msi	157 GPa
E_{22}^{t}	1.4 Msi	9.6 GPa	1.4 Msi	9.6 GPa
E_{11}^{c}	18.0 Msi	124 GPa	21.8 Msi	150 GPa
G_{12}	0.74 Msi	5.1 GPa	0.74 Msi	5.1 GPa
ν_{12}	0.3	0.3	0.3	0.3
F_{11}^{tu}	214 ksi	1475 MPa	214 ksi	1475 MPa
F_{22}^{tu}	5.5 ksi	38 MPa	7 ksi	48 MPa

clear in the presentation of these test results. A load-shortening curve was obtained for each test in a Tinius–Olsen test machine, which provided the best record for buckling, postbuckling, and crippling behavior. Curves from the static tests were used to estimate a reasonable and appropriate maximum load for the fatigue tests. Load-shortening curves to failure were obtained for those specimens that were successfully subjected to 100 000 cycles without failure at a rate of two cycles per second. The fatigue test specimens were C-scanned before and after the fatigue test and only specimens from successful fatigue tests were C-scanned after testing. Initial buckling in the fatigue machine could be estimated only by eye because load-shortening curves could not be plotted satisfactorily with the MTS test machine.

Some short-column buckling tests, which included stiffened panels, were performed to study potential limitations in the stability predictions of short-columns. Short-column stiffeners included two channel configurations, one hat configuration, and a hat-stiffened panel configuration. All of these configurations corresponded to other crippling tests and were only tested statically. The data from the hat-stiffener tests was of little value because of the simultaneous torsional buckling and secondary bending due to movement of the neutral axis after initial buckling.

2. Classical buckling theory

The well-known classical, specially orthotropic, elastic plate-buckling equations for narrow plates subjected to compression loading along the short sides and with all sides simply supported are used for buckling analysis of the no-edge-free plates. The specially orthotropic equations are for symmetrical laminates where the B_{ij} bending–stretching terms vanish due to symmetry and, in addition, the D_{16} and D_{26} twist–bending coupling terms are assumed to vanish. In symmetrical and balanced laminates, the B_{ij} terms vanish, but the D_{16} and D_{26} terms exist. The D_{16} and D_{26} terms were found to have little effect on buckling for specific cases of the no-edge-free plates as judged from the use of an anistropic plate-buckling computer code, STAGS, ALMROTH et al. (1975), which demonstrated that the specially orthotropic equations gave results that were less than 2% higher than those of the anisotropic solutions. Furthermore, the use of STAGS confirmed that solutions of the test configurations for simply supported and clamped boundary conditions on the loaded edges were identical. Thus, the specially orthotropic equations are valid for thin, no-edge-free plate buckling specimens with the loaded ends potted in metal plates. Accordingly, the appropriate specially orthotropic equation for no-edge-free plates where $4 < a/b < \infty$ is in the form obtained from the ADVANCED COMPOSITE DESIGN GUIDE (1973):

$$F_{\mathrm{cl}}^{\mathrm{cr}} = (2\pi^2/b^2t)[(D_{11}D_{22})^{1/2} + D_{12} + 2D_{66}] . \tag{2.1}$$

Eq. (2.1) is not valid for thick plates, where transverse shear effects become important. For b/t ratios much less than 20, the plate falls into the thick plate regime, DURLOFSKY and MAYERS (1971), NOOR (1975), ARNOLD (1983). Consequently, test results for no-edge-free specimens with b/t ratios less than 15 are not reported herein.

For the analysis of one-edge-free plates with a simply supported unloaded edge, the appropriate specially orthotropic, elastic plate-buckling equation is in the form given in the ADVANCED COMPOSITE DESIGN GUIDE (1973), see also HOLSTON (1970):

$$F_{\rm cl}^{\rm cr} = \frac{12D_{66}}{b^2t} + \frac{\pi^2 D_{11}}{(L')^2 t}, \tag{2.2}$$

where $b/t > 20$, and $L' = L/\sqrt{C}$.

Using the STAGS computer code, ALMROTH (1975), it was found that Eq. (2.2) was approximately valid for the one-edge-free tests reported herein with b/t ratios greater than 20; for assumed clamped end-conditions ($C = 4$), Eq. (2.2) yielded results that were less than 4% higher than the anisotropic solutions. A value for C was assumed equal to 3.6 rather than 4.0 for determining the initial buckling loads of the one-edge-free tests reported herein; this value for C is more representative of fixed-end test specimens loaded by platens of a testing machine.

For the development of postbuckling methodology, it is usually of importance to have the initial buckling stress included in certain parmeters, especially for the generation of design graphs. It has been found that elastic initial buckling theory for narrow no-edge-free plates with the most efficient stacking sequence does not yield results consistent with experiment, SPIER (1980). Both higher initial buckling and crippling will result when the ± 45 plies are on the outside surfaces of $[0, \pm 45]$ plates and the 0° plies are inside. It can be shown by use of approaches presented by PIPES and PAGANO (1970, 1971, 1973) and by WHITNEY and KIM (1977) that the laminates with 0° plies on the outside experience interlaminar tensile stresses along the boundary layer of the unloaded free edges. This indicates that a lower crippling strength is expected but it does not explain the lower initial buckling strength. It was shown by SPIER and KLOUMAN (1976) that the compressive stress–strain curve for $[\pm 45]_s$ laminates is highly nonlinear and that the tangent modulus becomes significantly less than the initial (elastic) modulus. It has also been shown that, in addition to the nonlinearity of the in-plane stress–strain curves, the nonlinear transverse shear stress–strain curve can significantly affect the postbuckling load–displacement behavior, ARNOLD (1983). To properly account for the effects of all nonlinear stress–strain curves and boundary layer effects to obtain true initial buckling is beyond the scope of the present chapter. Accordingly, conventional elastic buckling methods are used in this section.

The critical buckling stress of an element of a stiffener can be determined by the use of Eqs. (2.1) or (2.2), whichever is appropriate. For a constant thickness stiffener with constant stacking sequence, the critical element for buckling is readily determined. The classical critical buckling load (P^{cr}_{cl}) for the stiffener is then found by the equation:

$$P^{cr}_{cl} = F^{cr}_{cl} A \,, \tag{2.3}$$

where F^{cr}_{cl} is the lowest buckling stress of all the elements in the stiffener.

3. Experimental results

Representative load-shortening curves for the wide range of plates and stiffeners tested are presented to exemplify the effects of geometry and stacking sequence on the buckling behavior. Specifically, the experimental buckling loads (as determined by use of the load-shortening curves) are compared to predictions based on classical, specially orthotropic, elastic plate-buckling theory.

3.1. Plates

The plate test program presented herein was concerned with narrow no-edge-free and one-edge-free plates in compression. The testing of no-edge-free plates in compression was accomplished by the use of a fixture that consisted of steel plates with V-grooves to provide, as nearly as possible, simple-support to the unloaded edges up to the buckling load. The loaded ends of the specimens were potted in aluminum alloy blocks with adhesive. The V-blocks were adjusted by hand to the unloaded edges, where minimum clearances were provided so that the specimen could deflect only a small amount when loaded by normally applied hand pressure. (Thus, these tests were subject to some human error during test setup.) A typical no-edge-free test is shown in progress in Fig. 1, where the plate has buckled. A typical one-edge-free test is shown in progress in Fig. 2 where the plate has buckled in one longitudinal wave. The V-blocks closely approximate simple-support up to the onset of buckling, but there is some question as to the manner of support provided in the postbuckling range.

Prior to buckling of no-edge-free plates, the simple-support assumption requires the edge to be held straight with no other constraints. This results in zero edge moments and zero out-of-plane displacements ($M = w = 0$) along the edges. These conditions may be virtually satisfied up to the load at which buckling occurs but may be violated in the postbuckling range. Simple-support in the postbuckling range implies that the edge is held straight and, in addition, out-of-plane displacements may be resisted by lateral in-plane forces. The V-blocks neither provide such forces nor do they maintain straight edges and, therefore, the crippling tests were conservative because lower-bound crippling strengths were obtained. Due to clearances, the unloaded edges were not

FIG. 1. Typical postbuckling pattern of a no-edge-free plate test.

FIG. 2. Typical postbuckling pattern of a one-edge-free plate test.

completely restrained out-of-plane prior to buckling, and were free to wave in certain areas of the postbuckling range. If the edges had been held straight in the postbuckling range, the number of longitudinal waves may have increased. This condition would have reduced the amplitude of the waves and the resulting bending and transverse shear stresses, which may have resulted in significantly higher crippling strengths than were obtained. Based upon the analysis performed by ARNOLD (1983), it has been concluded that the precise boundary conditions are not a factor in the test evaluations when the end-shortening to critical end-shortening ratio is less than about four.

Load-shortening curves which best exemplify the loading history through initial buckling, postbuckling, and crippling were obtained for each static test to failure on a Tinius–Olsen test machine. Satisfactory load-shortening curves could not be achieved on the MTS fatigue testing machine.

3.1.1. No-edge-free plates

The configurations and test results of the no-edge-free plate tests are presented in Tables 2 and 3. Typical load-shortening curves are presented in Fig. 3, which were obtained from static tests. The abrupt change in slope coincides with initial buckling, and the buckling load is shown as P^{cr}; the crippling load is shown as P^{cc}. It was found in the testing of the plate fatigue specimens that continuous rebound against the steel V-grooves caused premature failure; both fatigue specimens 3-F1 and 3-F2 were lost. Fatigue testing of no-edge-free plates was discontinued, and the remaining plates were transferred to the one-edge-free testing. Static test results of plates and channels from SPIER and KLOUMAN (1977), are presented in Tables 4 and 5, respectively; they are used later in this section.

TABLE 2. No-edge-free plate test configurations (SPIER, 1981).

ID	t (in.)	L (in.)	b (in.)	b/t	A (in^2)
3-S1	0.0590	8.50	2.0030	33.9	0.1182
3-S2	0.0595	8.48	2.0032	33.7	0.1192
3-F1	0.0605	8.49	2.002	33.1	0.1211
3-F2	0.0590	8.47	2.002	33.9	0.1181
4-S1	0.0600	8.46	2.491	41.5	0.1495
4-S2	0.0612	8.47	2.488	40.65	0.1523
7-S1	0.0608	8.47	2.0030	32.9	0.1218
7-S2	0.0630	8.49	2.0025	31.8	0.1262

There was good agreement in the initial buckling loads for the two tests shown in Fig. 3, but specimen S-31 failed prematurely for some unknown reason. Typical scatter of the crippling loads is shown in Figs. 4 and 5. Note that the laminate tested in Fig. 5 has a slightly different stacking sequence from the rest of the plates; the 0_3 and 90° plies were interchanged. The b/t ratios for the plate tests in Figs. 3 and 5 are similar, and the initial buckling loads are

TABLE 3. No-edge-free plate test results (SPIER, 1981).

ID	P^{cr}_{cl} (kips)	F^{cr}_{cl} (ksi)	P^{cr} (kips)	F^{cr} (ksi)	P^{f} (kips)	F^{f} (ksi)	Fatigue cycles (10^3)	P^{cc} (kips)	F^{cc} (ksi)	P^{cr}/P^{cr}_{cl} or F^{cr}/F^{cr}_{cl}	P^{cc}/P^{cr}_{cl} or F^{cc}/F^{cr}_{cl}	P^{cc}/P^{cr} or F^{cc}/F^{cr}	F^{cu}/F^{cr}_{cl}	F^{cu}/F^{cr}
3-S1	3.01	25.43	2.95	24.96				4.20	35.53	0.98	1.40	1.42	5.47	5.57
3-S2	3.075	25.79	2.95	24.75				4.99	41.86	0.96	1.62	1.69	5.39	5.61
3-F1					3.60		48.81							
3-F2					3.30		54.44							
4-S1	2.536	16.97	2.45	16.39				5.27	35.25	0.97	2.08	2.15	8.19	8.48
4-S2	2.698	17.72	2.55	16.74				5.23	34.34	0.945	1.94	2.05	7.84	8.30
7-S1	3.34	27.42	2.82	23.15				4.715	38.71	0.844	1.41	1.67	5.07	6.00
7-S2	3.73	29.54	3.00	23.77				5.070	40.17	0.804	1.36	1.69	4.705	5.85

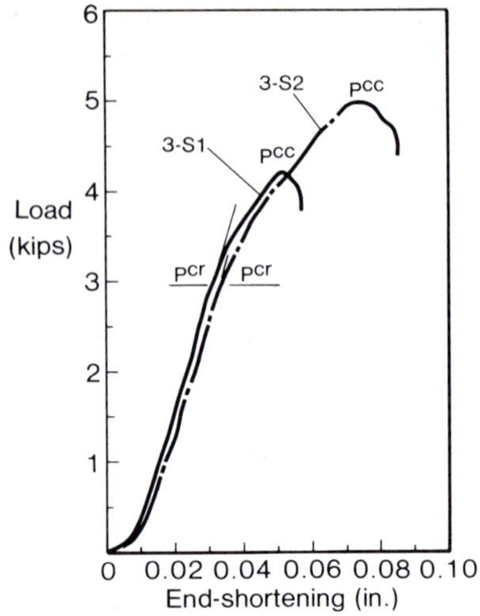

FIG. 3. No-edge-free plate 3 crippling tests – A-S/3501-6 $[\pm45/0_3/90]_s$ – $b/t = 34$.

about the same. Thus, the change in stacking sequence made little difference in initial buckling. Also, the crippling load of plate 3-S2 (Fig. 3) is within the usual scatter of the crippling loads shown in Fig. 5.

TABLE 4. No-edge-free compression test results of $[\pm45_2/0_B/\mp45_2]_T$ laminated plates (SPIER and KLOUMAN, 1977).

Geometry							Theoretical buckling	Test results			
Specimen no.	v/o (%)	B	L (in.)	b (in.)	t (in.)	$\frac{b}{t}$	P^{cr} (lb)	P_i^{cr} (lb)	P_i^{cr}/P^{cr}	P^{cc} (lb)	P^{cc}/P^{cr}
2A	60	4	8	2.002	0.0630	31.75	3990	3000	0.75	3570	0.89
2B	60	4	8	2.001	0.0642	31.17	4220	3250	0.77	4260	1.01
2C	60	4	8	2.002	0.0617	32.45	3745	3250	0.87	4115	1.10
4A	65	8	8	2.001	0.0838	23.88	9010	6300	0.70	7680	0.85
4B	65	8	8	2.001	0.0821	24.38	8470	6400	0.76	7500	0.88
4C	65	8	8	2.007	0.0849	23.64	9340	6350	0.68	7175	0.77
6A	65	12	8	2.001	0.1030	19.43	15 940	10 200	0.64	11 360	0.71
6B	65	12	8	2.001	0.1052	19.02	16 985	10 300	0.61	12 300	0.72
6C	65	12	8	2.004	0.1031	19.44	15 960	12 400	0.78	13 125	0.82
1A	60	4	12	3.000	0.0644	46.58	2840	2300	0.81	3900	1.37
1B	60	4	12	3.010	0.0644	46.74	2830	1900	0.67	4070	1.44
1C	60	4	12	3.000	0.0642	46.73	2815	2250	0.80	4375	1.55
3A	65	8	12	3.000	0.0836	35.89	5965	5100	0.85	7800	1.31
3B	65	8	12	3.000	0.0867	34.60	6655	5200	0.78	7400	1.11
3C	65	8	12	3.000	0.0860	34.88	6495	5700	0.88	7100	1.09
5A	65	12	12	3.000	0.1057	28.38	11 495	9250	0.80	11 550	1.00
5B	65	12	12	3.000	0.1066	28.14	11 790	10 400	0.85	12 250	1.04
5C	60	12	12	3.000	0.0980	30.60	9170	8000	0.87	11 825	1.29

TABLE 5. Compression test results of $[\pm 45_A/0_B/\mp 45_A]_T$ laminated channels (SPIER and KLOUMAN (1977).

Geometry							Theoretical buckling	Test results			
Specimen no.	v/o (%)	B	L (in.)	b_f (in.)	b_w (in.)	t (in.)	P_{cl}^{cr} (lb)	P^{cr} (lb)	P^{cr}/P_{cl}^{cr}	P^{cc} (lb)	P^{cc}/P_{cl}^{cr}
14-1	62	4	5.25	0.749	2.645	0.0727	7430	7500	1.01	10 050	1.35
14-2	62	4	5.25	0.750	2.649	0.0714	6820			9250	1.36
14-3	62	4	5.25	0.747	2.652	0.0756	8280	8100	0.98	10 225	1.23
18-1	62	8	5.25	0.741	2.652	0.0819	9990	9300	0.93	12 050	1.21
18-2	62	8	5.25	0.744	2.658	0.0839	10 880	11 250	1.03	13 175	1.21
18-3	62	8	5.25	0.746	2.656	0.0838	10 810	10 500	0.97	13 625	1.26
22-1	62	12	5.25	0.752	2.661	0.0995	17 590	15 000	0.85	20 750	1.18
22-2	62	12	5.25	0.747	2.654	0.1009	17 950	17 500	0.97	21 875	1.22
22-3	62	12	5.25	0.742	2.663	0.1026	18 740	17 500	0.93	22 450	1.20
16-1	58	4	10.50	1.500	5.249	0.0632	2510	2600	1.04	9625	3.83
16-2	58	4	10.50	1.502	5.254	0.0630	2590	2600	1.00	9475	3.66
16-3	58	4	10.50	1.501	5.256	0.0644	2800	2550	0.91	9325	3.33
20-1	59	8	10.50	1.497	5.265	0.0864	6500	5000	0.77	17 425	2.68
20-2	59	8	10.50	1.495	5.266	0.0880	6800	6250	0.92	17 075	2.51
20-3	59	8	10.50	1.495	5.273	0.0853	5900	6500	1.10	18 025	3.06
24-1	58	12	10.50	1.495	5.291	0.1082	12 070	11 100	0.92	27 600	2.29
24-2	58	12	10.50	1.493	5.289	0.1082	11 860	11 500	0.97	27 125	2.29
24-3	58	12	10.50	1.497	5.287	0.1089	12 270	12 500	1.02	25 950	2.11

The A-S/3501-6 channel and zee tests are reported in a later section of this chapter, and the A-S/3501-5 plate and channel test data were taken from Tables 4 and 5, respectively, SPIER (1980).

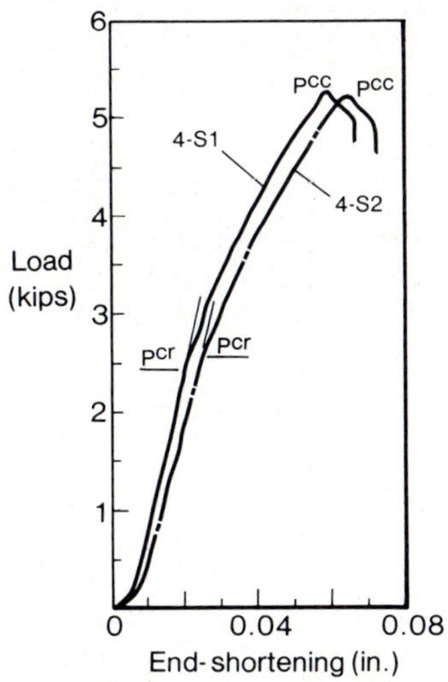

FIG. 4. No-edge-free plate 4 crippling tests – A-S/3501-6 $[\pm 45/0_3/90]_s$ – $b/t = 34$.

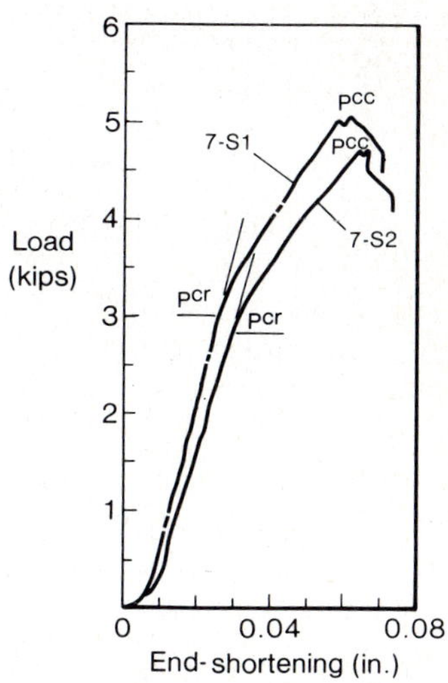

FIG. 5. No-edge-free plate 7 crippling tests – A-S/3501-6 $[\pm 45/90/0_3]_s$ – $b/t = 32$.

T-300/5208 plate test data from SPIER (1978) are included in Fig. 6. There is wide scatter in a few instances, which may be attributable to imperfect or faulty test setup as discussed earlier. For a particular stacking sequence, the parameter F^{cr}/F^{cr}_{cl} tends to be reduced at the lower b/t ratios. In all low b/t cases where ±45 plies were on the surfaces, classical buckling theory significantly overpredicted the buckling load. At larger b/t ratios, the agreement between actual and classical buckling improved in most instances.

It is of interest to exemplify the postbuckling strength (crippling) relative to the buckling strength, although the crippling strength happens to be lower-bound due to the type of testing performed. Accordingly, a nondimensional plot that provides such information is presented in Fig. 6, where the parameter F^{cc}/F^{cr} is plotted versus the parameter F^{cu}/F^{cr}. This type of plot has proven useful for presenting empirical crippling curves, SPIER (1978), on the log–log scale, where F^{cr} represented the classical buckling stress instead of test values. There is considerable postbuckling strength relative to buckling for much of the test data, which follows a general trend as illustrated in Fig. 6. Therefore, approximate lower-bound crippling strengths can be predicted for no-edge-free plates with any stacking sequence, provided F^{cu} and F^{cr} are known or can be reasonably estimated.

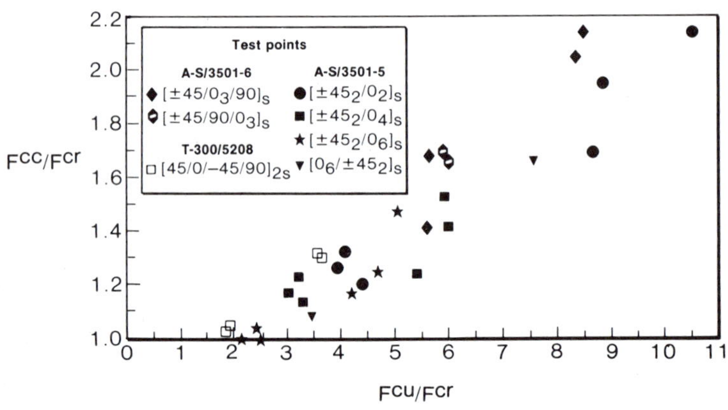

FIG. 6. Nondimensional crippling plot of no-edge-free plates.

It is of interest to plot log–log crippling test data, similar to that found in SPIER (1978), as shown in Fig. 7. The parameters are similar to those of Fig. 6 except the buckling stress used is the calculated classical (see Eq. (2.1)). An approximate mean curve is shown through the test data, and the scatter is quite small. As stated earlier, this curve represents approximately the lower-bound crippling strength of simply supported plates with the stacking sequences indicated.

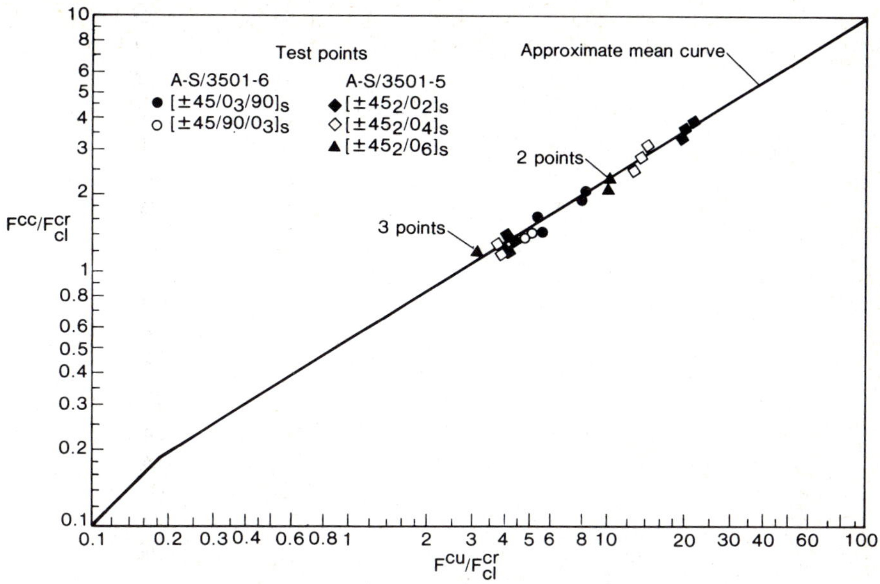

FIG. 7. Nondimensional crippling plot of no-edge-free plates.

3.1.2. One-edge-free plates

The configurations and test results of the one-edge-free plate tests are presented in Tables 6 and 7, respectively. Typical load-shortening curves are presented in Fig. 8. The fatigue tests for specimens 4-S3, 4-F1, 7-F1 and 7-F2 were originally planned as no-edge-free fatigue tests, which encountered difficulties with the crippling fixture as discussed earlier. The sudden drop in load for specimen 5-S2 in Fig. 8 was due to an increase in the number of half-waves

TABLE 6. One-edge-free plate test configurations (SPIER, 1981).

ID	t (in.)	L (in.)	L' (in.)	b (in.)	b/t	A (in^2)
5-S1	0.0605	5.47	2.88	1.2038	19.9	0.0728
5-S2	0.0598	5.47	2.88	1.2070	20.2	0.0722
5-F1	0.0663	5.47	2.88	1.203	18.1	0.0798
5-F2	0.0615	5.47	2.88	1.203	19.6	0.0740
6-S1	0.0608	8.47	4.46	2.0034	33.0	0.1218
6-S2	0.0581	8.47	4.46	2.0033	34.5	0.1164
6-F1	0.0588	8.47	4.46	2.004	34.1	0.1178
6-F2	0.0610	8.47	4.46	2.003	32.84	0.1222
4-S3	0.0610	8.47	4.46	2.502	41.0	0.1526
4-F1	0.0615	8.47	4.46	2.502	40.7	0.1539
7-F1	0.0610	8.47	4.46	2.002	32.8	0.1221
7-F2	0.0615	8.47	4.46	2.004	32.6	0.1232

Table 7. One-edge-free plate test results (Spier, 1981).

ID	P_{cl}^{cr} (kips)	F_{cl}^{cr} (ksi)	P^{cr} (kips)	F^{cr} (ksi)	P^{f} (kips)	F^{f} (ksi)	Fatigue cycles (10^3)	P^{cc} (kips)	F^{cc} (ksi)	P^{cr}/P_{cl}^{cr} or F^{cr}/F_{cl}^{cr}	P^{cc}/P_{cl}^{cr} or F^{cc}/F_{cl}^{cr}	P^{cc}/P^{cr} or F^{cc}/F^{cr}	F^{cu}/F_{cl}^{cr}	F^{cu}/F^{cr}	P^{f}/P_{cl}^{cr} or F^{f}/F_{cl}^{cr}	P^{f}/P^{cr} or F^{f}/F^{cr}	P^{f}/P^{cc} or F^{f}/F^{cc}
5-S1	0.93	12.80	1.00	13.73				2.780	38.16	1.07	2.99	2.78	10.86	10.12			
5-S2	0.90	12.44	1.00	13.85				2.610	36.15	1.11	2.90	2.61	11.17	10.04			
5-F1			≈1.0[a]		2.00	25.06	8.41										
5-F2	0.978	13.21	0.85	11.49	1.50	20.27	100.00	3.120	42.16	0.87	3.19	3.67	10.52	12.10	1.53	1.76	0.48
6-S1	0.59	4.87	0.60	4.93				2.645	21.72	1.02	4.48	4.41	28.54	28.19			
6-S2	0.515	4.43	0.75	6.44				3.125	26.85	1.46	6.06	4.17	31.30	21.57			
6-F1	0.534	4.53	0.65	5.52	1.80	15.27	100.00	2.965	25.17	1.22	5.56	4.56	30.68	25.18	3.37	2.77	0.61
6-F2	0.598	4.89	0.60	4.91	1.80	14.73	100.00	3.210	26.27	1.00	5.37	5.35	28.43	28.31	3.01	3.00	0.56
4-S3	0.561	3.675	1.06	6.95				2.980	19.53	1.89	5.31	2.81	37.82	20.00			
4-F1	0.575	3.737	1.30	8.45	1.50	9.75	100.00	2.940	19.10	2.26	5.11	2.26	37.20	16.45	2.61	1.15	0.51
7-F1	0.545	4.467	0.60	4.91	1.80	14.74	100.00	3.185	26.085	1.10	5.84	5.31	31.11	28.31	3.30	3.00	0.57
7-F2	0.577	4.686	0.60	4.87	1.80	14.61	100.00	2.815	22.85	1.04	4.87	4.69	29.66	28.54	3.12	3.00	0.64

[a] Visual observation of buckling.

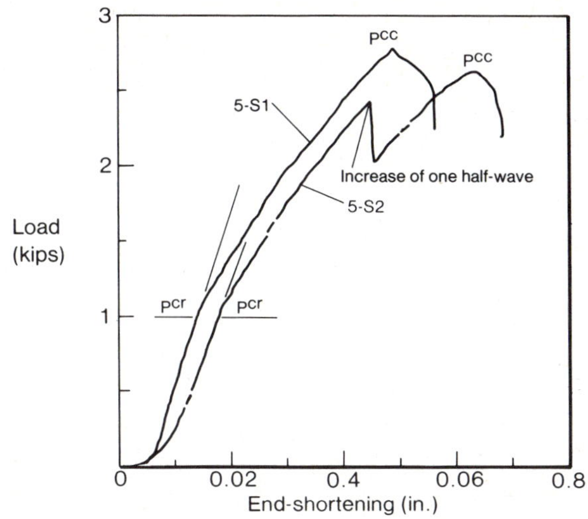

FIG. 8. One-edge-free plate 5 crippling tests – A-S/3501-6 $[\pm 45/0_3/90]_S$ – $b/t = 20$.

from two to three, which resulted in a lower energy mode with immediate reduction in buckle amplitudes. It is not known why specimen 5-S1 did not react similarly, particularly since these similar specimens buckled at the identical load. C-scanning of specimen 5-F2 after fatigue showed no changes in the laminate. Specimen 5-F1 failed prematurely in fatigue at a maximum load of 2.0 kips (see Table 7).

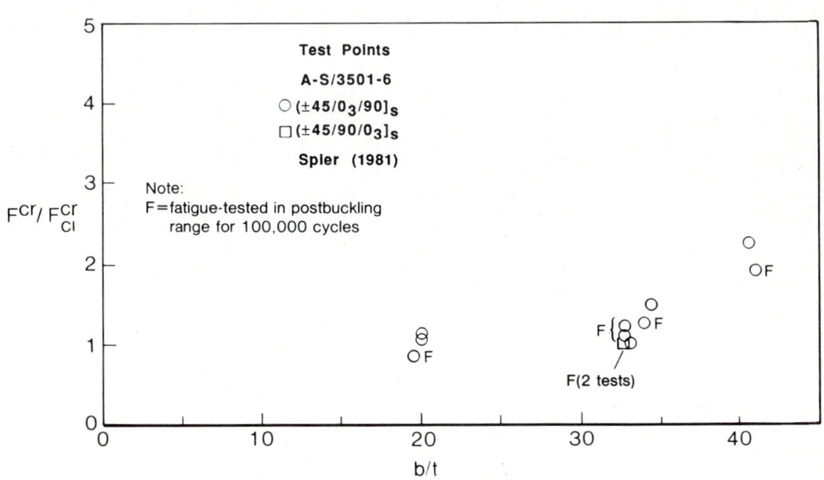

FIG. 9. Nondimensional buckling plot of one-edge-free plates.

The static test specimens 6-S1 and 6-S2 exhibited no wave change in the postbuckling range, but the similarly fatigued specimen 6-F2 did. There was considerable postbuckling strength for all test specimens, and the highest crippling strength was for the fatigued specimen 6-F2. No changes in the laminate of specimens 6-F1 and 6-F2 were detected by C-scan after the fatigue testing. There was little scatter in the initial buckling loads for all specimens, but there was some scatter in the crippling strengths. The maximum fatigue load was about three times that of initial buckling.

Note in Table 7 that the static test specimen (4-S3) and the fatigued specimen (4-F1) had virtually identical crippling strengths. There was some scatter in initial buckling loads. Higher fatigue loads than those indicated were tried, but dynamic buffeting occurred; it appeared that premature failure would occur similarly to that of specimen 5-F1. Plates 7-F1 and 7-F2, had a different stacking sequence $[\pm 45/90/0_3]_s$ and were originally to be tested no-edge-free. Both specimens were tested with one-edge-free successfully in fatigue prior to static testing. The load-shortening curves for the two specimens were quite similar except for the crippling strengths. Note that the maximum fatigue loads were approximately three times that of initial buckling, which is similar to that found for plates 6. The b/t ratios and crippling strengths of plates 6 and 7 were similar, but the 0 and 90° plies were reversed. Thus, the results for these one-edge-free plates showed that the two stacking sequences were approximately equivalent. C-scanning of the fatigued specimens showed no changes in the laminates.

The comparison between actual and classical initial buckling is shown for the test data in Fig. 9, where the parameter F^{cr}/F^{cr}_{cl} is plotted versus the b/t ratio. The actual and classical buckling stresses are nearly equal up to the b/t ratio of about 33, where the actual buckling load starts to significantly exceed the classical. It is presently unclear why this is so. The test data shown includes specimens that were successfully fatigued in the postbuckling range. The degree of postbuckling strength of the one-edge-free plates is exemplified by the plot of Fig. 10, where the ordinate is the ratio of crippling to initial buckling; this parameter is plotted versus the usual parameter F^{cc}/F^{cr}. The ratio of 4–5 times is not too uncommon, which quite graphically shows the penalty of nonbuckled designs. An empirical crippling curve plotted in the usual way is presented in Fig. 11. The test data are all A-S/3501-6 plates supported by one steel V-block, and some of the specimens were tested in fatigue for 100 000 cycles in the postbuckling range. The boundary condition for these tests is very close to simple-support, since the supported edge is held virtually straight throughout the loading cycle up to crippling. However, transverse in-plane membrane stresses cannot develop along the unloaded edge in the postbuckling range.

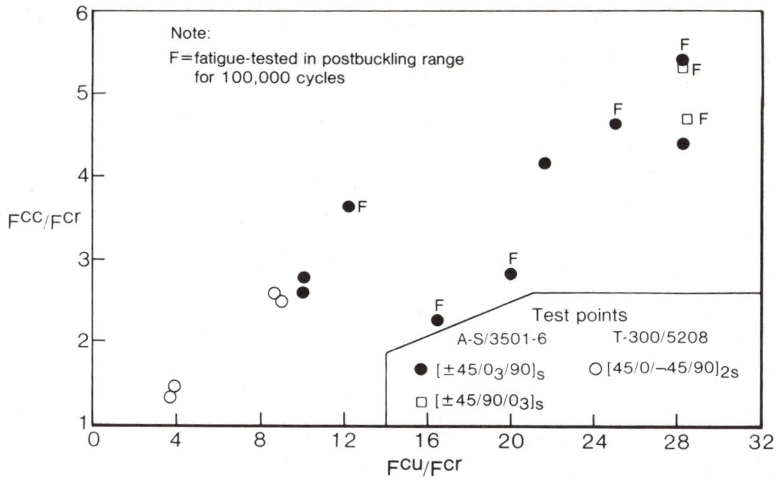

FIG. 10. Nondimensional crippling plot of one-edge-free plates.

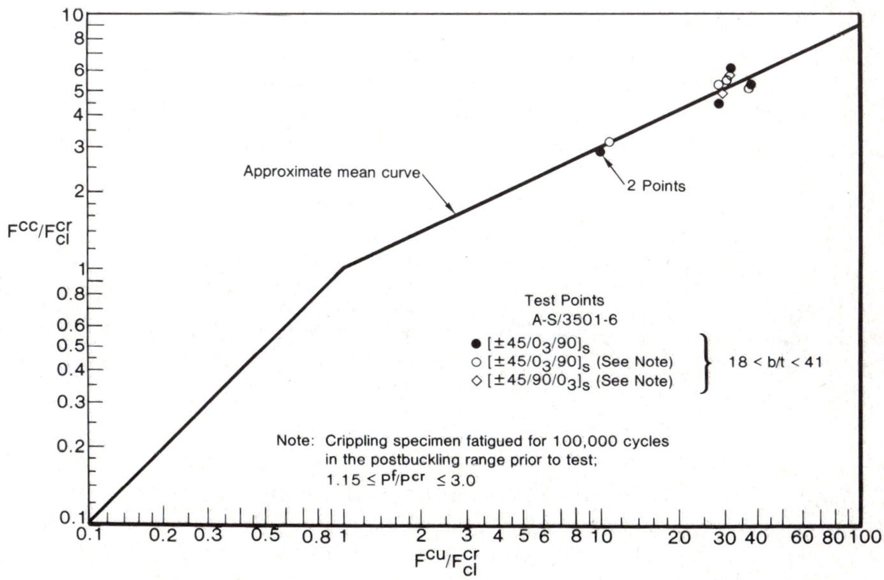

FIG. 11. Nondimensional crippling plot of one-edge-free plates.

3.2. Channel stiffeners

The testing of channels (and also zees, presented next) is limited to configurations with equal flange widths. Unequal flange widths would add another dimension of complexity that is beyond the scope of the present program. As in the design of the plate specimens, the channels (and zees) were designed to have significant postbuckling strengths. Consequently, the designs were not directed toward any particular structural application.

The channel test specimen cross-sections from SPIER (1981) are shown in Fig. 12. The section for Test 12 is critical for web buckling as were the channels tested by SPIER and KLOUMAN (1977) which were added to the database of the

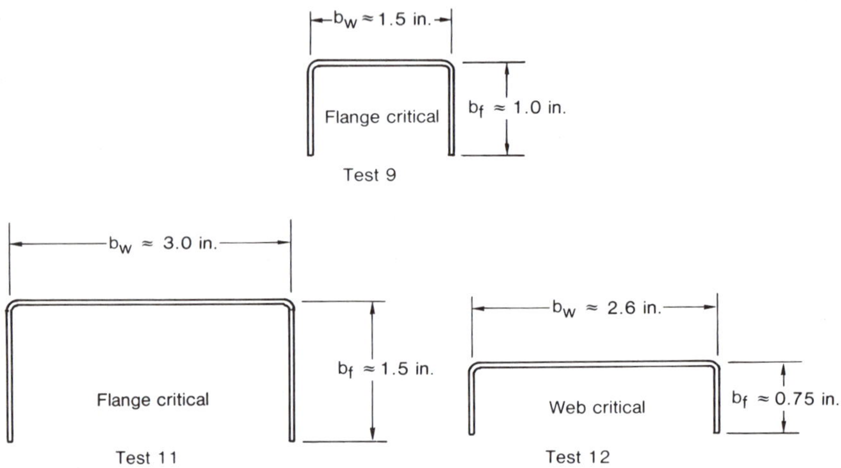

FIG. 12. Channel test specimen configurations.

present program (Table 5). Consequently, the other two configurations (Tests 9 and 11) were designed with the flanges critical in buckling. The configurations and test results of the channels from SPIER (1981) are presented in Tables 8 and 9, respectively.

Typical load-shortening curves for channels 11 are presented in Fig. 13. Note the short duration drop of load in the postbuckling range for channel 11-S1.

TABLE 8. Channel test configurations (SPIER, 1981).

ID	t_w (in.)	t_f (avg) (in.)	L (in.)	L' (in.)	ρ_{min} (in.)	L'/ρ_{min}	b_w (in.)	b_f (avg) (in.)	b_w/t_w	b_f/t_f	A (in^2)	b_f/b_w
9-S1	0.0611	0.0615	10.3	5.42	0.33	16.4	1.565	1.008	25.6	16.4	0.2195	0.64
9-S2	0.0611	0.0617	10.3	5.42	0.33	16.4	1.562	1.000	25.6	16.7	0.2189	0.64
9-F1	0.0605	0.0593	10.3	5.42	0.33	16.4	1.562	0.992	25.8	16.7	0.2120	0.64
9-F2	0.0608	0.0592	10.3	5.42	0.33	16.4	1.564	0.992	25.7	16.7	0.2126	0.64
11-S1	0.0582	0.0581	11.49	6.05	0.48	12.6	3.069	1.516	52.73	26.12	0.3546	0.49
11-S2	0.0577	0.0578	11.48	6.04	0.48	12.6	3.072	1.509	53.24	26.12	0.3515	0.49
11-F1	0.0608	0.0582	11.49	6.05	0.48	12.6	3.064	1.513	50.4	26.01	0.3622	0.49
11-F2	0.0612	0.0575	11.49	6.05	0.48	12.6	3.059	1.510	50.0	26.25	0.3608	0.49
12-S1	0.0648	0.0630	5.58	2.94	0.22	13.36	2.662	0.739	41.1	11.7	0.2657	0.28
12-S2	0.0640	0.0646	5.59	2.94	0.22	13.36	2.661	0.749	41.6	11.6	0.2671	0.28
12-F1	0.0582	0.0623	5.59	2.94	0.22	13.36	2.668	0.746	45.8	12.0	0.2482	0.28
12-F2	0.0610	0.0605	5.59	2.94	0.22	13.36	2.675	0.743	43.9	12.3	0.2530	0.28

TABLE 9. Channel test results (SPIER, 1981).

ID	P_{cl}^{cr} (kips)	F_{cl}^{cr} (ksi)	P^{cr} (kips)	F^{cr} (ksi)	P^{f} (kips)	F^{f} (ksi)	Fatigue cycles (10^3)	P^{cc} (kips)	F^{cc} (ksi)	P^{cr}/P_{cl}^{cr} or P^{cr}/F_{cl}^{cr}	P^{cc}/P_{cl}^{cr} or F^{cc}/F_{cl}^{cr}	P^{cc}/P^{cr} or F^{cc}/F^{cr}	F^{cu}/F_{cl}^{cr}	F^{cu}/F^{cr}	P^{f}/P_{cl}^{cr} or F^{f}/F_{cl}^{cr}	P^{f}/P^{cr} or F^{f}/F^{cr}	P^{f}/P^{cc} or F^{f}/F^{cc}
9-S1	3.230[c)]	14.72[c)]	4.295	19.57				6.30[c)]	28.70	1.33	1.95	1.47	9.44	7.10			
9-S2	3.253[c)]	14.86[c)]	3.525	16.10				8.00	36.55	1.08	2.46	2.27	9.35	8.63			
9-F1	2.963[c)]	13.98[c)]	3.30	15.57	5.40	25.47	100.00	7.45	35.14	1.11	2.51	2.25	9.94	8.93	1.82	1.64	0.72
9-F2			≈4.0[a)]		6.00	28.22	56.36										
11-S1	2.145[c)]	6.05[c)]	2.57	7.25				9.975	28.13	1.20	4.66	3.88	22.97	19.17			
11-S2	2.122[c)]	6.04[c)]	3.00	8.53				9.275	26.39	1.41	4.37	3.09	23.00	16.30			
11-F1	2.25[c)]	6.21[c)]	2.25	6.21	8.0	22.09	100.00	9.750	26.92	1.00	4.33	4.33	22.38	22.38	3.56	3.56	0.82
11-F2			≈2.5[a)]		7.80		9.45										
12-S1	4.61[d)]	17.32[d)]	5.03	18.91				10.00	37.61	1.09	2.17	1.99	8.02	7.35			
12-S2	4.50[b)]	16.86[b)]	4.98	18.64				10.10	37.81	1.10	2.24	2.03	8.24	7.46			
12-F1			≈5.0[a)]		8.20	3.04	2.32										
12-F2	3.86[b)]	15.24[b)]	4.70	18.58	7.00	27.67	100.00	10.5	41.50	1.22	2.72	2.23	9.12	7.48	1.81	1.49	0.67

[a)] Visual observance of buckling.
[b)] Poor test; cause uncertain.
[c)] Flange critical in buckling.
[d)] Web critical in buckling.

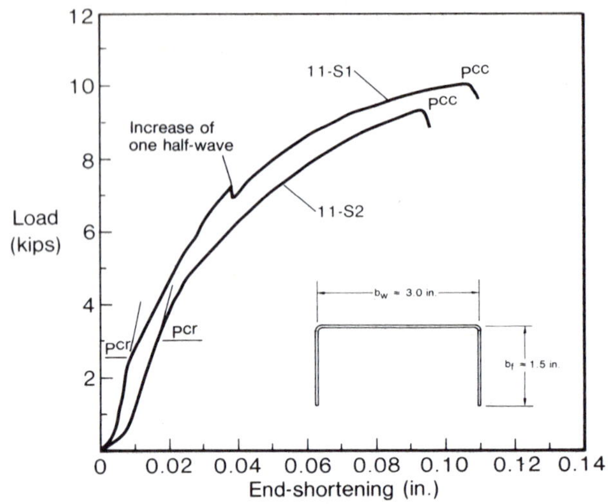

FIG. 13. Channel 11 crippling tests – A-S/3501-6 $[\pm 45/0_3/90]_S$.

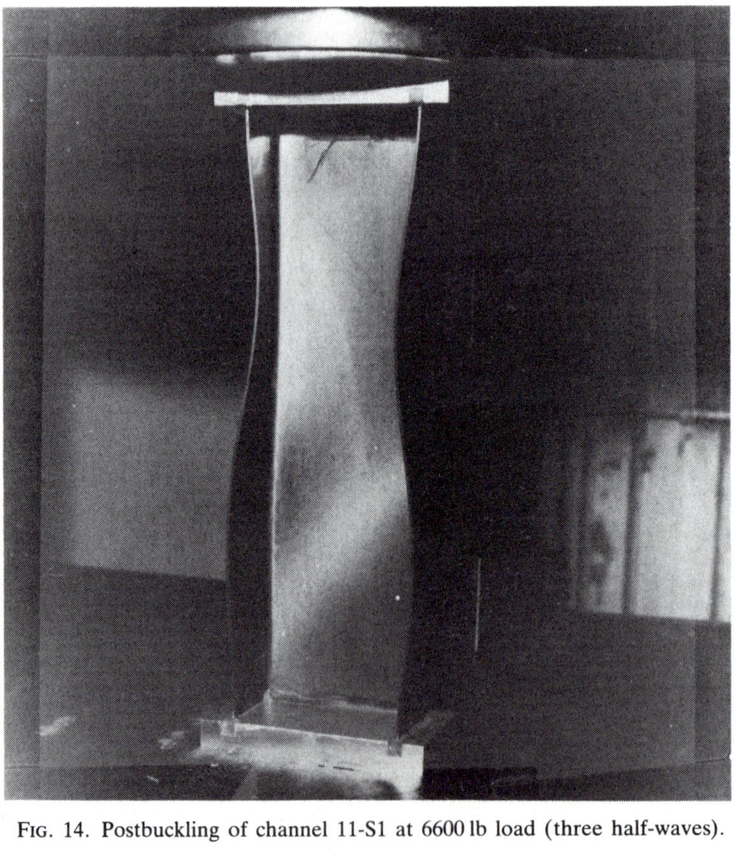

FIG. 14. Postbuckling of channel 11-S1 at 6600 lb load (three half-waves).

This occurs when the number of half-waves of the buckled pattern is increased by one, which occurs quickly and dynamically with a loud noise. It is not known why no such change took place for channel 11-S2. It was found in the fatigue testing that the maximum load had to be below this point of wavelength change; otherwise, the test specimen would never achieve 100 000 cycles. Channel 11-S1 is shown in the test machine in Fig. 14 with the postbuckling state showing three half-waves. Fig. 15 shows the specimen after the number of half-waves changed from three to four. Channel 11-F1 was tested in fatigue for 100 000 cycles in the MTS testing machine with a maximum compressive load of 8.0 kips which was 256% above initial buckling and just below the point at which the number of half-waves would increase by one.

Typical load-shortening curves for channels 12 are presented in Fig. 16 where the buckling and crippling loads show good correlation between tests.

FIG. 15. Postbuckling of channel 11-S1 at 7250 lb load (four half-waves).

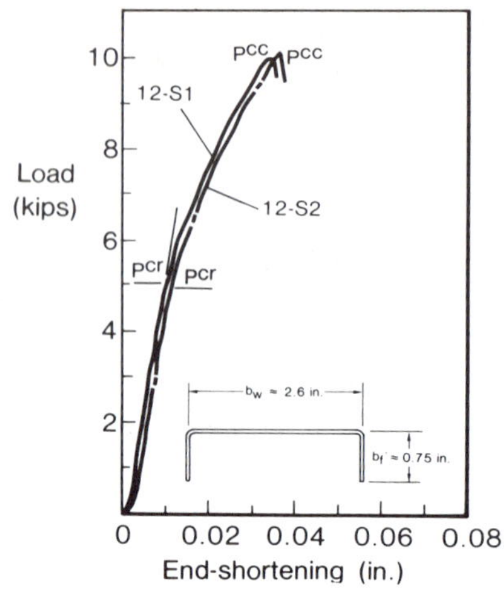

FIG. 16. Channel 12 crippling tests – A-S/3501-6 $[\pm 45/0_3/90]_s$.

FIG. 17. Postbuckling of channel 11-S1 at 6025 lb load.

Channel 12-F2 was tested in fatigue for 100 000 cycles at a maximum load of 7.0 kips, and C-scan (also by eye examination) revealed a small region of delamination at a free edge; however, the static test result showed no effect on buckling or crippling when compared to specimens 12-S1 and 12-S2. The maximum fatigue load was 49% above initial buckling. Channel 12-F1 failed at 2320 cycles with a maximum fatigue load of 8.2 kips (see Table 9). Specimen 12-S1, shown in the test machine in Fig. 17 is loaded into the postbuckling range. Crippling occurred at a load of 10.0 kips, and the failure is shown in Fig. 18.

A summary plot of the channel test results from SPIER (1981) with load versus the b_f/b_w ratio is shown in Fig. 19 where the initial buckling and crippling loads are indicated with approximate mean curves through the data points. Also shown are other approximate cross-sectional properties.

FIG. 18. Crippling of channel 12-S1 at 10 000 lb load.

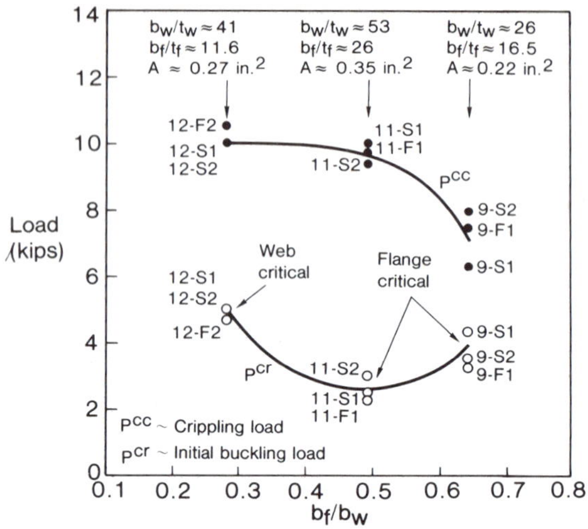

FIG. 19. Channel test results – A-S/3501-6 $[\pm 45/0_3/90]_s$.

There are three channels shown (9-F1, 11-F1, and 12-F2) in Fig. 19 that were tested in fatigue for 100 000 cycles prior to the static tests. Notice that no initial buckling or crippling loads for these specimens fell outside a reasonable scatter band for all of the test results. Also, the fatigued specimens did not produce the lowest crippling loads for their respective channel sets. However, the initial buckling loads were just slightly less than the non-fatigued specimens. Accordingly, it can be concluded that for the channels tested in fatigue without failure, no significant damage had occurred as a result of the cyclic loading. This statement may possibly be true for other channels of constant thickness and stacking sequence. However, no testing has yet been performed on channels with unequal flange widths and/or unequal stacking sequences of the elements of the channels.

3.3. Zee stiffeners

The zee test specimen cross-sections are shown in Fig. 20. The section for Test 14 is critical for web buckling, while Test 15 is flange critical. The configurations and test results are presented in Tables 10 and 11. The load-shortening curves for specimens 14-S1 and 14-S2 are shown in Fig. 21, and there is little difference between the two curves. As shown in Table 11, specimen 14-F1 failed in the fatigue test at 18 790 cycles where the maximum fatigue load was 7.5 kips. However, specimen 14-F2 survived for 100 000 cycles under a maximum fatigue load of 6.0 kips. Specimen 14-S2 is shown in the test machine in

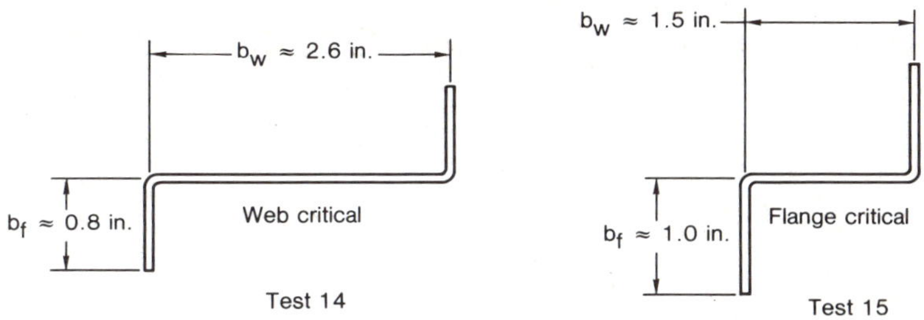

FIG. 20. Zee test specimen configurations.

TABLE 10. Zee test configurations (SPIER, 1981).

ID	t_w (in.)	t_f (avg) (in.)	L (in.)	L' [a)] (in.)	ρ_{min} (in.)	L'/ρ_{min}	b_w (in.)	b_f (avg) (in.)	b_w/t_w	b_f/t (avg)	A (in^2)	b_f/b_w
14-S1	0.0590	0.0578	5.7	2.47	0.14	21.0	2.602	0.8025	44.10	13.90	0.2462	0.308
14-S2	0.0612	0.0578	5.7	2.47	0.14	21.0	2.601	0.7972	42.50	13.85	0.2513	0.307
14-F1	0.0585	0.0588	5.7	2.47	0.14	21.0	2.601	0.8108	44.50	13.80	0.2474	0.311
14-F2	0.0595	0.0578	5.7	2.47	0.14	21.0	2.614	0.7952	43.90	13.77	0.2474	0.304
15-S1	0.0582	0.0595	6.87	3.1	0.25	14.4	1.506	1.0510	25.88	17.69	0.2127	0.698
15-S2	0.0593	0.0610	6.87	3.1	0.25	14.4	1.492	1.0250	25.16	16.80	0.2135	0.687
15-F1	0.0600	0.0588	6.87	3.1	0.25	14.4	1.612	1.0450	26.90	17.79	0.2194	0.648
15-F2	0.0595	0.0592	6.87	3.1	0.25	14.4	1.620	1.0375	27.20	17.54	0.2191	0.640

[a)] L assumed to be reduced by 1-inch because of the extra thick end blocks; then L' was calculated.

Fig. 22 in the postbuckling state. The crippling failure is shown in Fig. 23, which is located near one end.

Load-shortening curves for specimens 15-S1 and 15-S2 are shown in Fig. 24, and both tests result in similar buckling and crippling loads. The maximum fatigue load for specimen 15-F1 was 6.0 kips, but specimen 15-F2 failed at 1430 cycles at a maximum fatigue load of 7.0 kips. Specimen 15-S2 is shown in the testing machine in Fig. 25 in the postbuckling state. Thc crippling failure is shown in Fig. 26, which is located near mid-span.

A summary plot of the zee test results are shown in Fig. 27, where load is plotted versus the b_f/b_w ratio. There is a profound difference between the data of Figs. 19 and 27 where laminates and dimensions are similar. Notice that for the b_f/b_w ratio of 0.3, the zee and channel test results are comparable. However, at the b_f/b_w ratio of about 0.65, the initial buckling loads are comparable, but the channel crippling strengths are much lower than that of the zees. This difference is most likely due to the shift in neutral axis of these channels and resulting bending moments in the postbuckling range as discussed earlier. Thus, the channels would be expected to cripple at lower loads than that of the zees, since the zees have virtually no shift of the neutral axis in the postbuckling range. Obviously, as a compression strut, the zee is superior to a channel. This is not necessarily true for stiffened panels where webs are

TABLE 11. Zee test results (SPIER, 1981).

ID	P^{cr}_{cl} (kips)	F^{cr}_{cl} (ksi)	P^{cr} (kips)	P^{cr} (ksi)	P^{f} (kips)	F^{f} (ksi)	fatigue cycles (10^3)	P^{cc} (kips)	F^{cc} (ksi)	P^{cr}/P^{cr}_{cl} or F^{cr}/F^{cr}_{cl}	P^{cc}/P^{cr}_{cl} or F^{cc}/F^{cr}_{cl}	P^{cc}/p^{cr} or F^{cc}/F^{cr}	F^{cu}/F^{cr}_{cl}	F^{cu}/F^{cr}	P^{f}/P^{cr}_{cl} or F^{f}/F^{cr}_{cl}	P^{f}/P^{cr} or F^{f}/F^{cr}	P^{f}/P^{cc} or F^{f}/F^{cc}
14-S1	3.71	15.07[c]	5.00	20.31				9.99	40.58	1.34	2.69	2.00	9.22	6.84			
14-S2	4.02	16.01[c]	5.00	19.90				10.16	40.43	1.24	2.53	2.03	8.68	6.98			
14-F1			≈6.0[a]		7.5	30.31	18.79										
14-F2	3.86	15.60[c]	5.30	21.42	6.0	24.25	100.00	8.33	33.67	1.37	2.16	1.57	8.91	6.49	1.55	1.13	0.72
15-S1	3.10	14.64[b]	4.00	18.81				9.41	44.24	1.28	3.02	2.35	9.49	7.39			
15-S2	3.43	16.05[b]	4.40	20.61				9.80	45.90	1.28	2.86	2.23	8.66	6.74			
15-F1	3.17	14.46[b]	4.20	19.14	6.0	27.35	100.00	9.58	43.66	1.32	3.02	2.28	9.61	7.26	1.89	1.43	0.63
15-F2			≈5.0[a]		7.0	31.94	1.43										

a) Visual observance of buckling.
b) Flange critical in buckling.
c) Web critical in buckling.

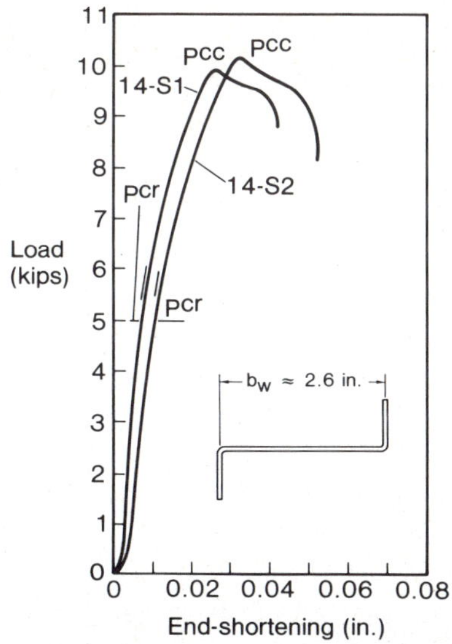

FIG. 21. Zee 14 crippling tests – A-S/3501-6 $[\pm 45/0_3/90]_s$.

FIG. 22. Postbuckling of Z-section 14-S2 at 8000 lb load.

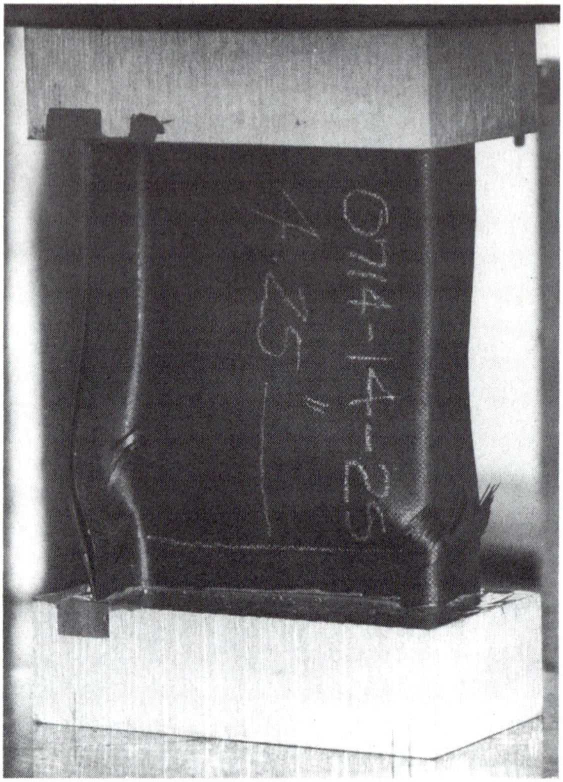

FIG. 23. Crippling of Z-section 14-S2 at 10 160 lb load.

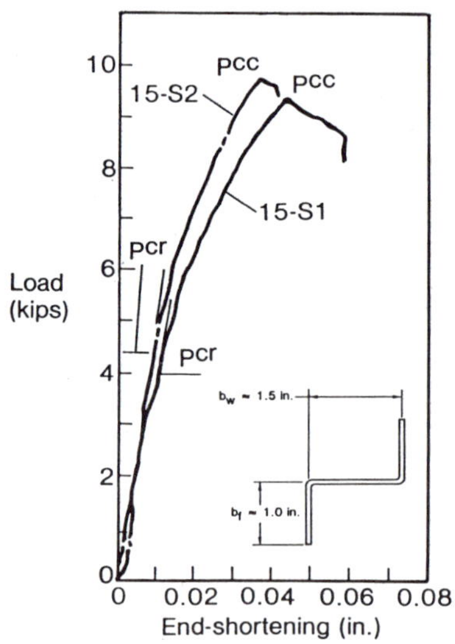

FIG. 24. Zee 15 crippling tests – A-S/3501-6 $[\pm 45/0_3/90]_s$.

FIG. 25. Postbuckling of Z-section 15-S2 at 8000 lb load.

perpendicular to the panel skin; here the allowable crippling strengths for zees would probably be appropriate for channels.

The channel and zee test results are plotted in the usual nondimensional form for crippling strengths as shown in Fig. 28. The A-S/3501-5 channels had b_f/b_w ratios of approximately 0.28, and the curve shown represents the mean of the data points. There is some scatter of the A-S/3501-6 channels with various b_f/b_w ratios relative to the curve. However, all zee data points but one are significantly higher than the curve. This graph exemplifies the superiority of zees over channels because of the bending effects in the channels as discussed earlier.

It is of primary interest to ascertain the validity of the method whereby a stiffener such as a channel or zee is divided into no-edge-free and one-edge-free plates, and the crippling strength of the stiffener is equivalent to the summation of the crippling strengths of the individual plate elements. Accordingly, it is assumed that the one-edge-free crippling curve of Fig. 11 is valid for determining the crippling loads of the flanges of the channels and zees reported herein; then the crippling load of the web (P_w^{cc}) is equivalent to the crippling

FIG. 26. Crippling of Z-section 15-S2 at 9800 lb load.

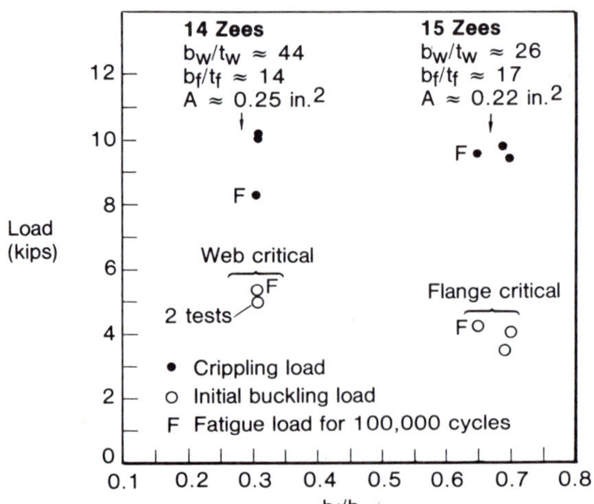

FIG. 27. Zee test results – A-S/3501-6 $[\pm45/0_3/90]_s$.

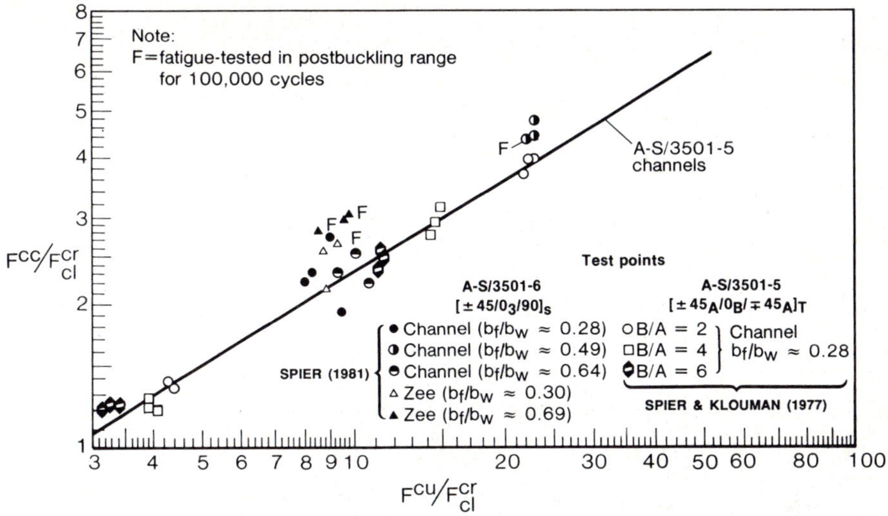

FIG. 28. Nondimensional channel – zee crippling plot.

load of the channel or zee less the calculated crippling loads of the flanges. Thus, the channel and zee test data are analyzed in this manner from Tables 5, 8, and 9 and are plotted as shown in Fig. 29. The no-edge-free plate crippling curve is shown for comparison purposes; as stated earlier, this curve is a lower bound for simply supported plates. Most of the data show this curve to be

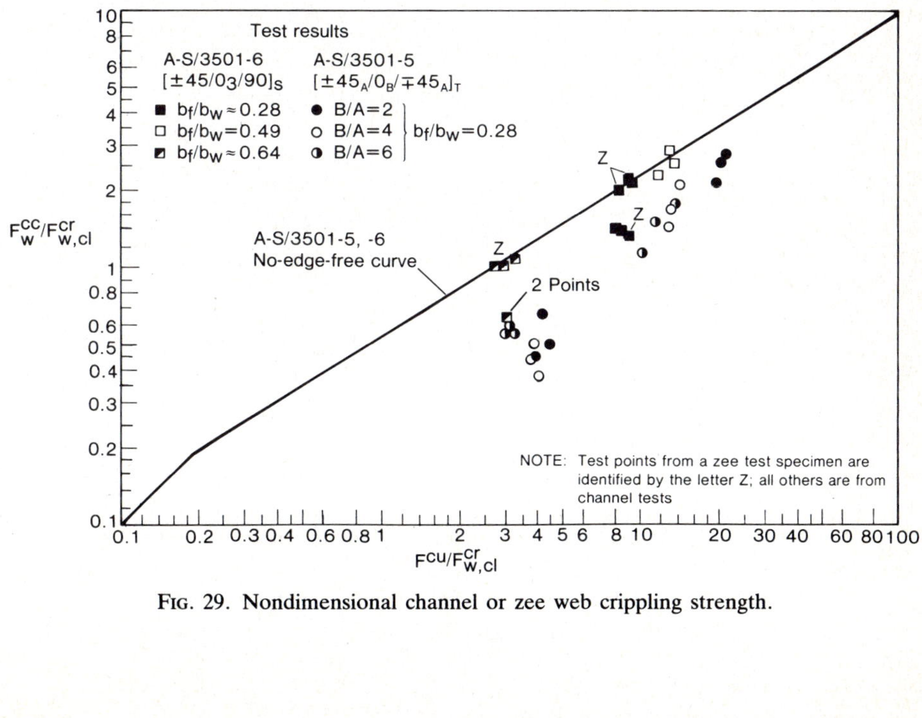

FIG. 29. Nondimensional channel or zee web crippling strength.

extremely unconservative for the webs of most channels and show it to be an upper bound for the webs of zees. Thus, this method of analysis for crippling strength of a constant thickness stiffener may be appropriate provided secondary stiffener bending prior to crippling is precluded.

3.4. Hat-stiffened panels

The cross-section of all hat-stiffened panels is shown in Fig. 30, where the hats are identical to hat 16 (Fig. 31). Crippling tests of hat 16 were performed but not reported because severe neutral axis shifting resulted in large secondary bending stresses and premature failure. The flanges of the hats were secondari-

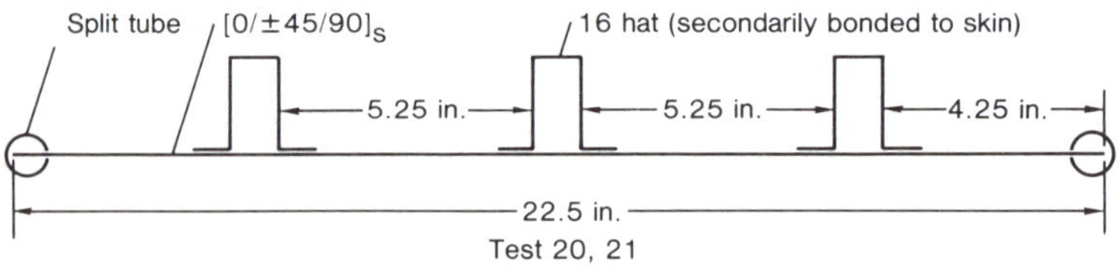

FIG. 30. Cross-section of hat-stiffened panels.

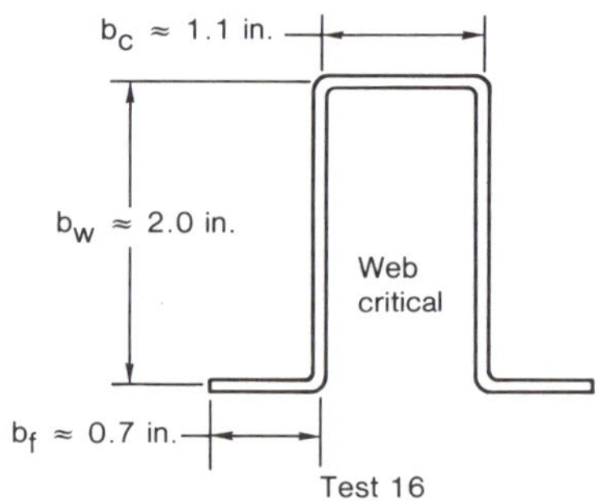

FIG. 31. Hat test specimen – A-S/3501-6 $[\pm45/0_3/90]_s$.

ly bonded with EA 934 adhesive at room temperature. The stacking sequence of the skin was $[0/\pm45/90]_s$. The stiffened panels had approximately the same total width, and the outside unloaded edges were supported by a split ring to simulate simple-support. Panels 20 were crippling tests, while panels 21 were corresponding short-column tests. The test results and the lengths of the specimens are shown in Table 12. Specimen 20-S1 failed at the lowest load of 50.6 kips, which was due to failure of the adhesive in the secondary bond without crippling failure of any of the hats. Typical load-shortening curves for the crippling tests are shown in Fig. 32; skin buckling occurred below 10 kips but no indication can be seen in the curves. The maximum crippling load was obtained for 20-F1, which had been fatigue tested for 100 000 cycles at a

TABLE 12. Hat-stiffened panel test results (SPIER, 1981).

ID	L (in.)	P^{cc} (kips)	P^{co} (kips)	Comment
20-S1	17.0	50.6		Debonded only in adhesive
20-S2	17.0	56.5		Crippled and debonded in adhesive and laminate
20-F1	17.0	59.0		Crippled and debonded in adhesive and laminate
20-F2	17.0	55.8		Crippled and bebonded in adhesive and laminate
21-S1	55.3		52.9	Crippled and debonded in adhesive and laminate
21/S2	55.3		49.5	Crippled and debonded in adhesive and laminate

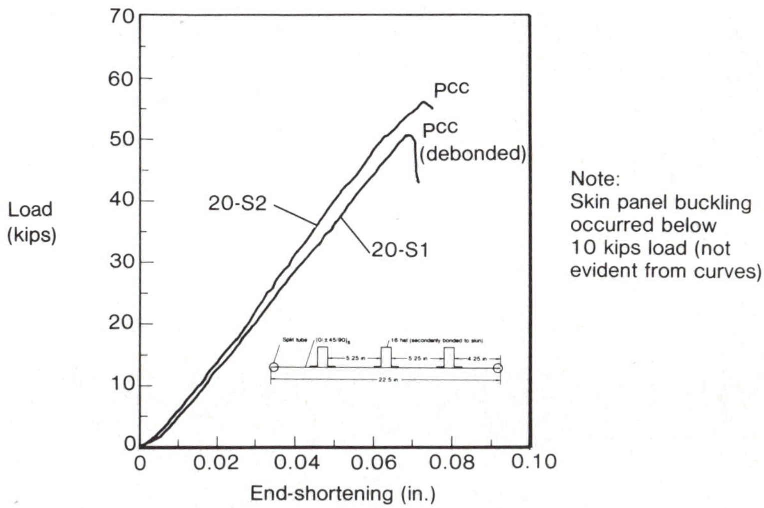

FIG. 32. Stiffened panel 20 crippling tests.

maximum load of 30 kips. C-scan of the fatigue specimens showed no variation of the laminates or bondlines before or after fatigue testing.

Panel 20-S2 is shown in the test machine in Fig. 33. Skin buckling is evident when the compressive load reached 51 kips. The specimen is shown after crippling in Fig. 34, where all three hats failed and the flange bondlines consisted of a combination of adhesive failure and composite delamination. Another view of the crippling failure is shown in Fig. 35, which shows the debonding failure more clearly.

The short-column failure for panel 21-S1 is shown in Fig. 36; the hats crippled locally, while delamination and adhesive failure of the flange bondlines occurred along the entire length of every flange. The bondline failure is clearly shown in Fig. 37, where adhesive failure was predominant over composite delamination. The backside of the panel at failure is shown in Fig.

FIG. 33. Postbuckling of hat-stiffened panel 20-S2 at 51 000 lb load.

FIG. 34. Crippling of hat-stiffened panel 20-S2 at 56 500 lb load.

FIG. 35. Crippling of hat-stiffened panel 20-S2 at 56 500 lb load.

FIG. 36. Stiffened panel 21-S1 short-column buckling load = 52 900 lb.

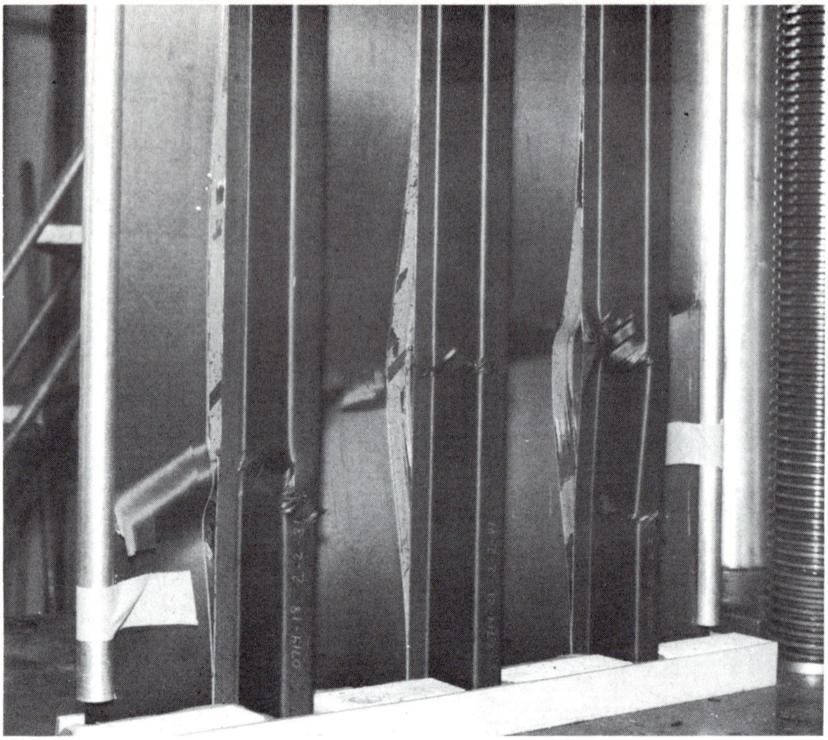

FIG. 37. Stiffened panel 21-S1 short-column buckling load = 52 900 lb.

FIG. 38. Stiffened panel 21-S1 short-column buckling load = 52 900 lb.

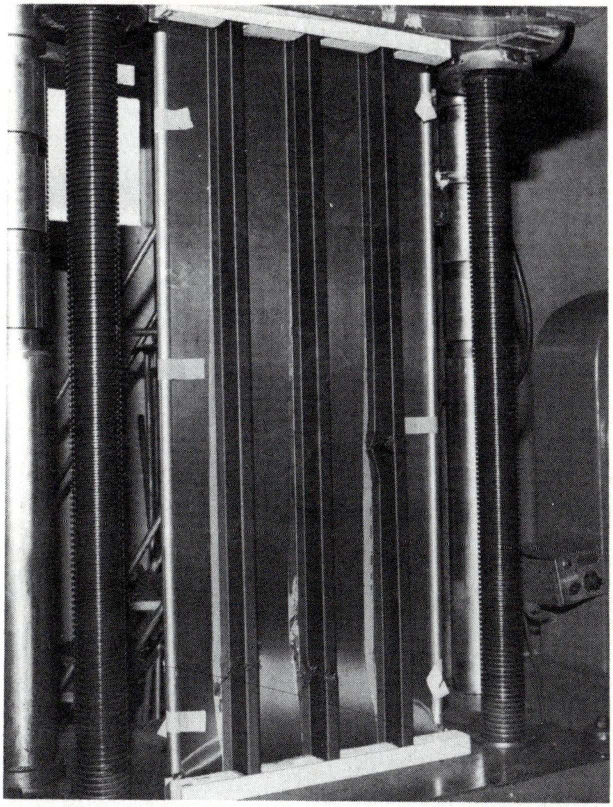

FIG. 39. Stiffened panel 21-S2 short-column buckling load = 49 500 lb.

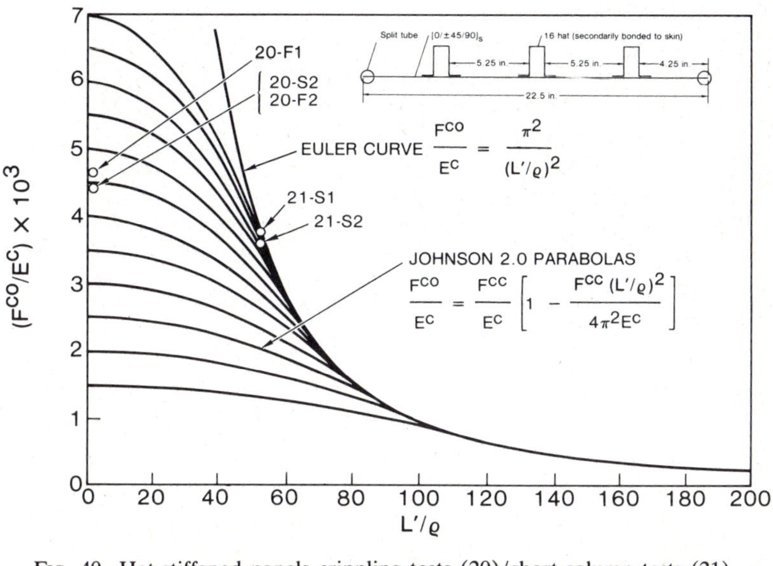

FIG. 40. Hat-stiffened panels crippling tests (20)/short-column tests (21).

38, where shear crimping of the skin is clearly shown; this failure is also evident in Figs. 36 and 37. The crippling failure of panel 21-S2 is shown in Fig. 39, where crippling of one hat was quite remote from that of the other two; in panel 21-S1, all three hats crippled near the same section. Bondline failure over considerable lengths of the flanges is clearly evident. A summary of the results is shown in Fig. 40 based on an unbuckled cross-section. Corrections for postbuckled skins would move the test points somewhat to the left (lower L'/ρ values).

References

ADVANCED COMPOSITE DESIGN GUIDE (1973), Vol. II – *Analysis* (Air Force Materials Laboratory, Advanced Development Division, Dayton, Ohio) Table 2.2.2-I.

ALMROTH, B.O., et al. (1975), Structural analysis of general shells, user instructions for STAGSC, Vol. III, Report No. LMSC D502277, Lockheed Structural Mechanics Laboratory, Lockheed Palo Alto Research Laboratory, Palo Alto, California.

ANDERSON, R.E. and J. MAYERS (1979), Effects of non-linear material behavior on postbuckling stiffness of laminated composite plates, AIAA Paper 79-1806, presented at *AIAA Aircraft Systems and Technology Meeting*, New York, N.Y., August 20–22, 1979.

ARNOLD, R.R. (1983), Buckling, postbuckling, and crippling of materially nonlinear laminated composite plates, Ph.D. Dissertation, Stanford University.

ARNOLD, R.R. and J. MAYERS (1984), Buckling, postbuckling, and crippling of materially nonlinear laminated composite plates, *J. Solids and Structures*, **20**, (9/10), 863–880.

DURLOFSKY, H. and J. MAYERS (1971), Effects of interlaminar shear on the bending and buckling of fiber-reinforced composite flat and curved plates, U.S. Army Air Mobility Research and Development Laboratory, Fort Eustis, Virginia, USAAVLABS Technical Report 71–70.

HOLSTON, Jr, A. (1970), Buckling of orthotropic plates with one free edge, *AIAA J.* **8**, (7), 1352–54.

MINECCI, J.J. and T.E. HESS (1980), Composite fuselage development for naval aircraft, *Proc. 25th National SAMPE Symposium and Exhibition*, May 6–8, 1980, pp. 188–201.

NOOR, A.K. (1975), Stability of multilayered composite plates, *Fibre Sci. Technol.* **8**, 81–89.

PAGANO, N.J. and R.B. PIPES (1971), The influence of stacking sequence on laminate strength, *J. Comp. Mater.* **5**, 50–57.

PAGANO, N.J. and R.B. PIPES (1973), Some observations on the interlaminar strength of composite laminates, *Int. J. Mechan. Sci.* **15**, 679–688.

PIPES, R.B. and N.J. PAGANO (1970), Interlaminar stresses in composite laminates under uniform axial extension, *J. Comp. Mater.* **4**.

SPIER, E.E. (1978), Stability of graphite/epoxy structures with arbitrary symmetrical laminates, *Exper. Mechan.* **18**, (11), 255–271.

SPIER, E.E. (1980), On experimental versus theoretical incipient buckling of narrow graphite/epoxy plates in compression, AIAA-80-0686-Paper, published in: *Proc. AIAA/ASME/ASCE/AHS 21st Structures, Structural Dynamics & Materials Conference*, May 12–14, 1980, pp. 187–193.

SPIER, E.E. (1981), Local buckling, postbuckling, and crippling behavior of graphite–epoxy short thin walled compression members, Naval Air Systems Command Report NASC-N00019-80-C-0174, July 1981.

SPIER, E.E. and F.L. KLOUMAN (1976), Ultimate compressive strength and non-linear stress–strain curves of graphite/epoxy laminates, *Proc. 8th National SAMPE Conf. Bicentennial of Materials Progress – Part II*, Seattle, Washington, October 1976, pp. 213–223.

SPIER, E.E. and F.L. KLOUMAN (1977), Empirical crippling analysis of graphite/epoxy laminated plates, *Composite Materials: Testing & Design (4th Conference)*. ASTM STP 617, pp. 255–271.

WHITNEY, J.M. and R.Y. KIM (1977), Effect of stacking sequence on the notched strength of laminated composites, *Composite Materials: Testing and Design (4th Conference)*, ASTM STP 617, 1977, pp. 229–242.

CHAPTER 12

Stability Critical Stiffened Panels

R.R. Arnold

Anamet Laboratories, Inc.
Hayward, CA 94545-3811
U.S.A.

K.T. Kedward

Alcoa Defense Systems, Inc.
San Diego, CA 92127
U.S.A.

Contents

HANDBOOK OF COMPOSITES, VOL. 2 – Structure and Design
Edited by C.T. HERAKOVICH and Y.M. TARNOPOL'SKII

Stiffened panels

Flat and curved panel structures stiffened by either longitudinal stiffeners (stringers) or lateral stiffeners (frames) are a very common form of construction in aircraft and missile structures. Usually, stiffened-panel elements are an integral part of a complete shell structure such as the fuselage and empennage. In other situations, the stiffened-panel is a basic element of a slightly curved structure such as the wing and horizontal and vertical stabilizers. The general geometry of the stiffened-panel structure is depicted in Fig. 1 which shows a multi-bay configuration composed of stringers (frames are normally present but not shown in the figure) and a curved panel. Any combination of biaxial loading, shear, and internal and external pressure can be found in a typical aircraft structure; although, longitudinal compression, shear, and pressure usually are the major concern from the standpoint of stability.

A rather substantial body of theory and experimental data is available for the design and analysis of stiffened-panels made from metallic materials such as aluminum and steel. To a large extent, this technology has evolved over a period of some 50 years and is adequately described in several government-sponsored design guides (GERARD and BECKER 1957a,b; ASTRONAUTIC STRUCTURES MANUAL, 1970) and the complete and authoritative guide of BRUHN (1973). With respect to composites, such a technology database scarcely exists.

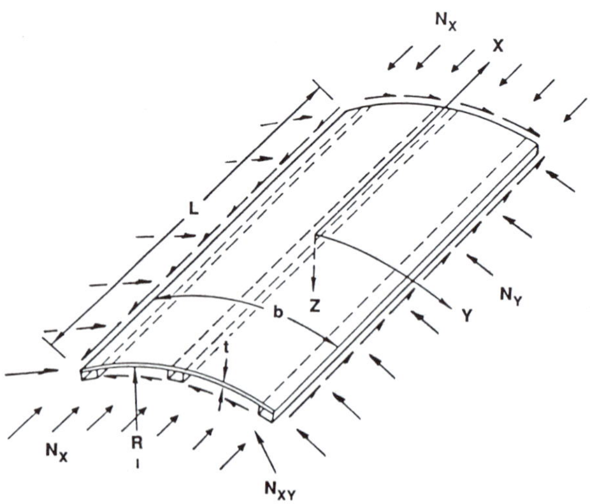

FIG. 1. General stiffened panel configuration.

Thus, this section will bring out, to the extent possible, the important attributes of composite stiffened-panel construction that have been verified by actual experimental testing and relatively recent theoretical advances.

Theoretical and experimental data for both metal and composite stiffened-panel constructions will be presented. Comparison of the two different types of materials aids in understanding the advantageous attributes of composites while simultaneously providing a point of reference. Additionally, it is noted that the next generation of aircraft structures will continue to use significant amounts of metal construction – the present efforts to develop composite primary structures notwithstanding.

To adequately design a stiffened-panel structure, it is necessary to predict initial buckling, postbuckling behavior, and final failure; consequently, this section will examine each of these behaviors separately. It should be mentioned that there are two distinct design philosophies that are applicable to stiffened-panel structures. The first, and most simple, requires that the structure not buckle at ultimate loads. Critical aerodynamic surfaces, especially for supersonic vehicles, usually employ this design requirement. Also, the non-buckling design philosophy is used when doubt exists about the behavior of the structure after initial buckling has occurred. The second philosophy allows buckling to occur at some fraction of the ultimate load and relies upon the postbuckling strength of the panel to increase the load-carrying capacity of the structure as a whole. Early aircraft of the 1930s and 1940s used the postbuckling concept to reduce the weight of the metallic structure; however, when composite panels were introduced, it was found from experimental testing that the postbuckling strength was difficult to predict analytically, and premature failures by stiffener disbonding limited the performance gains over aluminum by substantial amounts. Thus, the design and analysis of postbuckled composite panels has been largely based on experimental and semi-empirical methods.

It is convenient to discuss the stability of stiffened-panel structures in terms of their three distinct characteristics: initial buckling, postbuckling, and final failure. For non-buckling designs, only the first characteristic, initial buckling, is of concern. For postbuckling designs, all three characteristics must be fully quantified. Where possible, simple formulae for predictive purposes will be shown. In other instances, because of the complexity of the mathematical analysis, it will be necessary to show the stability characteristics in graphical form with references provided for access to the analytical methods.

1. Initial buckling

Prior to considering the complications of combined loads, boundary conditions, curvature, edge-stiffening, and other important factors, it will be useful to first examine the simplest of available theoretical solutions; namely, the simply-supported flat plate under uniform uniaxial in-plane compression. The solution to this problem can be found in several different books (LEKHNITSKII, 1968; AMBARTSUMYAN, 1970; ASHTON and WHITNEY, 1970) and is usually presented in the form

$$N_{\rm cl}^{\rm cr} = \frac{\pi^2 D_{11}}{b^2}\left[\left(\frac{mb}{L}\right)^2 + \frac{2(D_{12}+2D_{66})}{D_{11}}\,n^2 + \frac{D_{22}}{D_{11}}\,n^4\left(\frac{L}{mb}\right)^2\right]. \tag{1.1}$$

The critical buckling load, $N_{\rm cl}^{\rm cr}$, is dependent on the bending stiffnesses, length and width of flat panel, and the wavenumbers m and n which are the number of buckle half wavelengths in the x- and y-directions, respectively. This equation is sometimes referred to as the classical form for the simple reason that it was derived on the basis of the exclusion of transverse shear. The variation in critical buckling load as a function of the plate aspect ratio, L/b is shown in Fig. 2. As the aspect ratio becomes larger, that is, as the plate becomes longer, the critical buckling load curves become flatter. Also, notice that for certain values of the aspect ratio, the buckling mode changes. The parameter $\beta = mb/L$ is the ratio of the width of the buckled shape to the length of one buckled wave – for flat unstiffened isotropic plates, initial buckling occurs for a value of β equal to one when the y-direction wavenumber, n, is also equal to one. As will subsequently be shown, more complex panels have values of β which are greater than one, and in many instances the effects of curvature and edge-stiffening on initial buckling half-wavenumbers can be estimated reasonably accurately with just a knowledge of panel radius, width, and length.

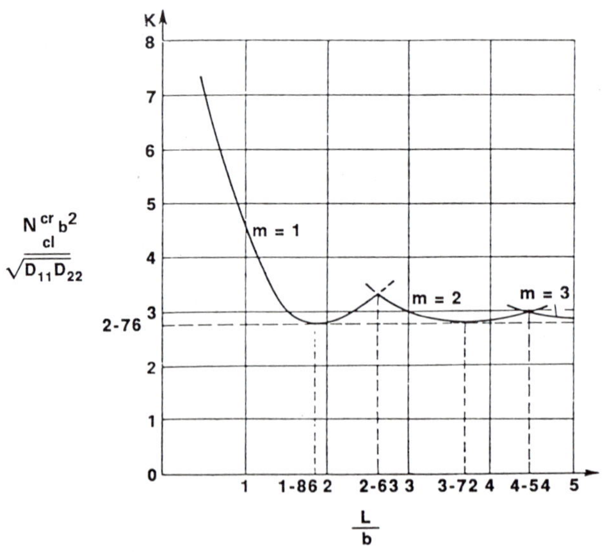

FIG. 2. Buckling coefficients for flat rectangular plates (LEKNITSKII, 1968).

There are many concise formulae available for calculating the initial buckling load of classical plates under different loading conditions, and the References (LEKHNITSKII, 1968; AMBARTSUMYAN, 1970; ASHTON and WHITNEY 1970; BAKER et al., 1972) are very handy for this purpose. Some of these cited references also provide the theoretical derivations which may be useful in understanding the underlying characteristics of the more complex stiffened-panel structures to be discussed in the remaining parts of this section. Where possible, the quantitative results presented herein will be non-dimensionalized by the buckling load for the equivalent flat unstiffened plate, Eq. (1.1).

The following subsections will discuss in some detail the characteristics of real-life stiffened-panel structures that have a significant effect on the prediction of initial buckling. The order of presentation has no particular significance inasmuch as all of the topics have a direct bearing on the stability of stiffened-panel structures. Furthermore, almost all practical applications of composite structures will be influenced by one or more of the topics discussed.

1.1. Transverse shear

For metallic materials, the effects of transverse shear are generally negligible for all but the thickest of sections. However, for composite materials, the presence of a relatively low modulus and low strength matrix material can have a significant effect on transverse shear. A comparison of the buckling loads for a flat plate with (N^{cr}) and without (N^{cr}_{cl}) transverse shear is shown as a function of the plate width to thickness (b/t) ratio in Fig. 3 obtained from ARNOLD

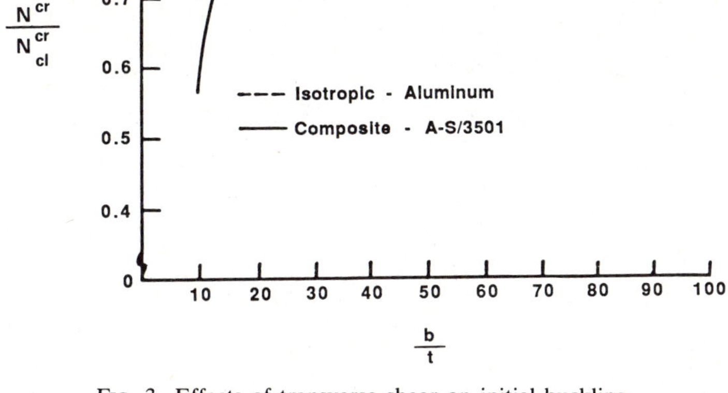

FIG. 3. Effects of transverse shear on initial buckling.

(1983). If somewhat arbitrarily it is assumed that a thick plate is defined as one which has a buckling load of less than 95% of classical, then a metal plate is thick when b/t is less than 20; whereas the corresponding composite plate is thick when the b/t is less than 50. Similar results have been obtained by WHITNEY (1969) for the bending and buckling of flat square plates of various lamination sequences although restricted to symmetric laminates with large length-to-thickness ratios.

The theoretical model used to derive the results in Fig. 3 is based on the work of DURLOFSKY (1970) which was a generalization of the work by HOFF (1950) for sandwich plates. Fig. 4, taken from DURLOFSKY (1970), shows how the buckling load varies with both the radius to thickness (r/h) ratio and the matrix shear modulus to axial extensional modulus of the composite plate. The buckling load based on inclusion of transverse shear is N^{cr}, and for classical theory the buckling load is N^{cr}_{cl}. Thus, for composite constructions which exhibit either a low matrix shear modulus or a relatively thick cross-section, the effects of transverse shear are potentially important to the prediction of initial buckling.

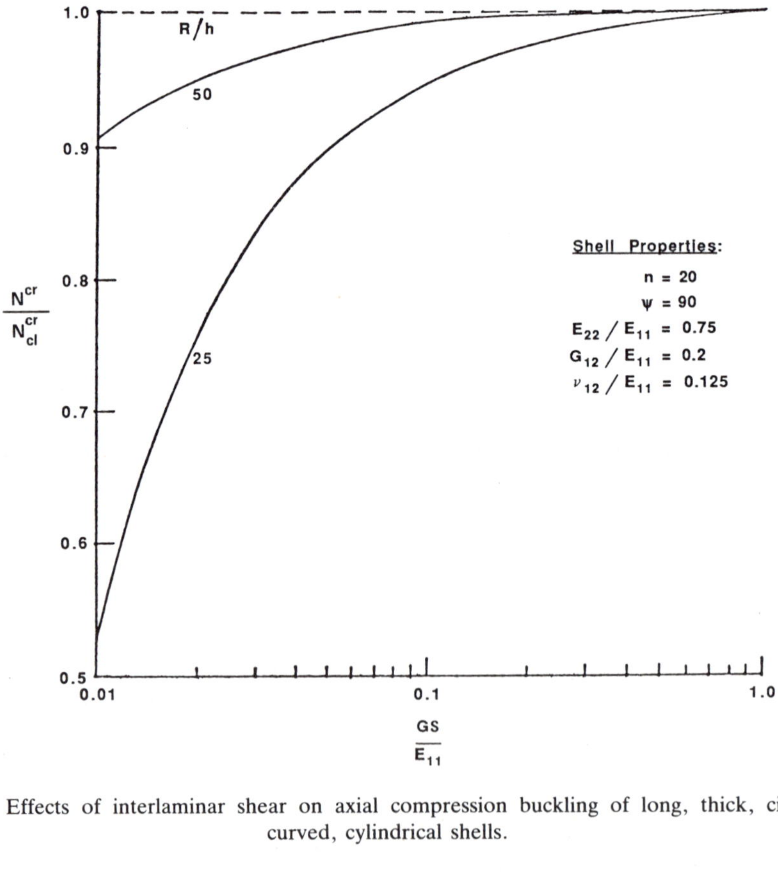

FIG. 4. Effects of interlaminar shear on axial compression buckling of long, thick, circularly curved, cylindrical shells.

An excellent and relatively recent review of the literature pertaining to the effects of transverse shear on the stability of flat unstiffened composite plates with various boundary conditions is contained in the book by VINSON and CHOU (1975), in part based on the work of SMITH (1973). Of special interest is the description of the Reissner variational theorem (REISSNER, 1950) and its application to plate stability problems with different boundary conditions under axial compression.

With respect to experimental testing, it is of interest to note that SPIER (1976) has stated '. . . in certain tests, the ultimate loads were less than the corresponding theoretical elastic buckling loads, exposing the futility of using classical buckling theory for laminated plates.' The magnitude of transverse shear effects established in Figs. 3 and 4 justifies Spier's observation. To demonstrate the importance of transverse shear, Figs. 5 and 6 are obtained

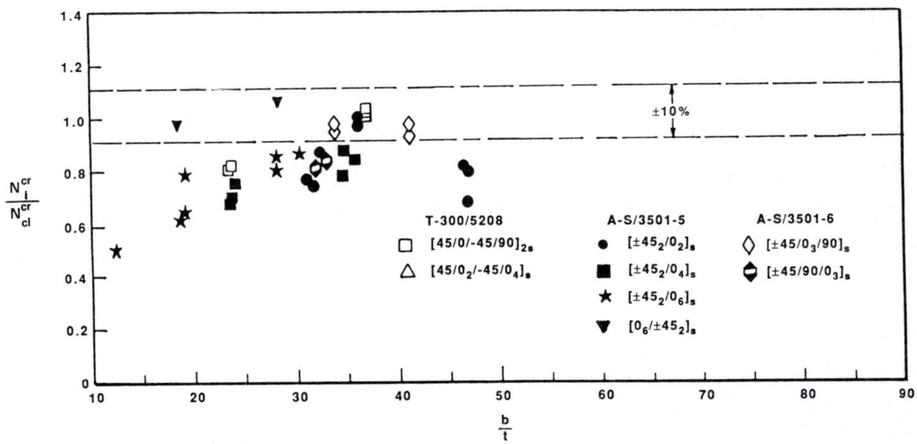

FIG. 5. Predicted classical buckling loads compared to experimental data.

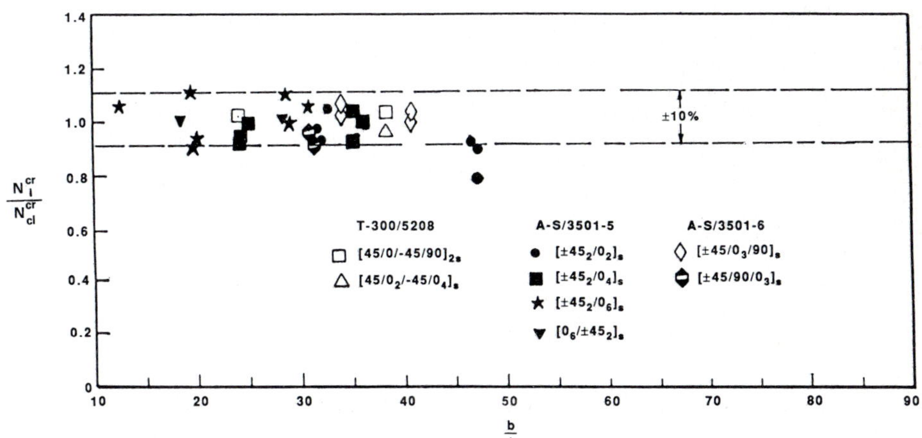

FIG. 6. Predicted buckling loads of the current theory compared to experimental data.

from ARNOLD (1983). Fig. 5 shows the experimental buckling load normalized by the classical buckling load as a function of the b/t ratio for a variety of lamination sequences and composite materials. The correlation does not speak too well for the use of classical buckling equations for composites with b/t of less than about 35. In Fig. 6 the same experimental data has been normalized by the buckling load prediction which includes the effects of transverse shear. Obviously, the correlation establishes quite convincingly the important role that transverse shear plays in the prediction of initial buckling of laminated composite plates.

The effect of transverse shear on the initial buckling behavior of stiffened-panel structures is presumed to be consistent with the data of Figs. 3 through 6 for either flat or curved, stiffened or unstiffened panels and under various loading conditions.

1.2. Edge-stiffening

Classical solutions for edge-stiffened panels generally do not exist in a closed form sense. Yoo (1977) has developed an efficient theoretical/analytical approach for prediction of buckling, postbuckling, and maximum strength of curved stiffened isotropic metal panels which is general enough to illustrate the effects of edge-stiffening on the initial buckling load. Fig. 7, obtained from Yoo (1977), shows the predicted buckling load normalized by the corresponding flat

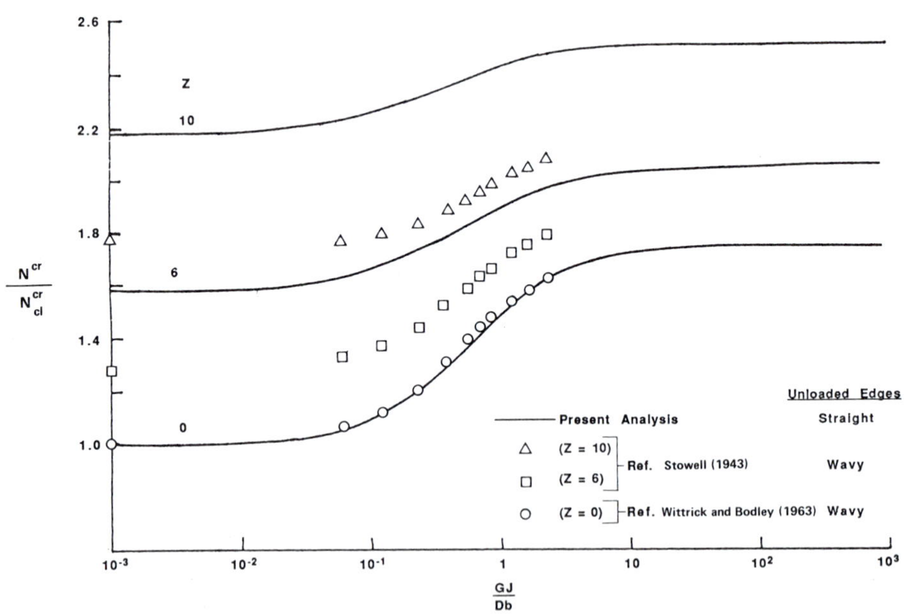

FIG. 7. Effects of rotationally restrained unloaded edges on compressive buckling load of shallow-curved plates.

plate buckling load as a function of the stiffening parameter GJ/Db. The comparison is good, especially when one considers that the experimental data had wavy edges, whereas the analytical prediction used straight edges. From Fig. 7, it is concluded that for a stiffening parameter of greater than about 10, the panel has unloaded edges which are essentially clamped.

Based on the work of Yoo for curved stiffened isotropic panels, and ANDERSON and MAYERS (1979) for flat unstiffened composite panels, Arnold and Mayers (ARNOLD and MAYERS, 1984) developed a theoretical/analytical procedure for prediction of buckling, postbuckling, and crippling of shallow curved stiffened composite panels which essentially reproduces the work of Yoo for metal panels. A typical result which demonstrates the effects of edge-stiffening in a composite panel is shown in Fig. 8. The predicted buckling load is normalized by the experimental buckling load and plotted as a function of the stiffener parameter, GJ/Db, corresponding to experimental data obtained by NASA (KNIGHT and STARNES, 1985). Similar predictions by the PASCO (ANDERSON and JEFFERSON, 1979) computer code, generated by NASA (KNIGHT and STARNES, 1985), are also shown in Fig. 8. The PASCO-generated buckling predictions correspond approximately to unloaded edges which are essentially clamped.

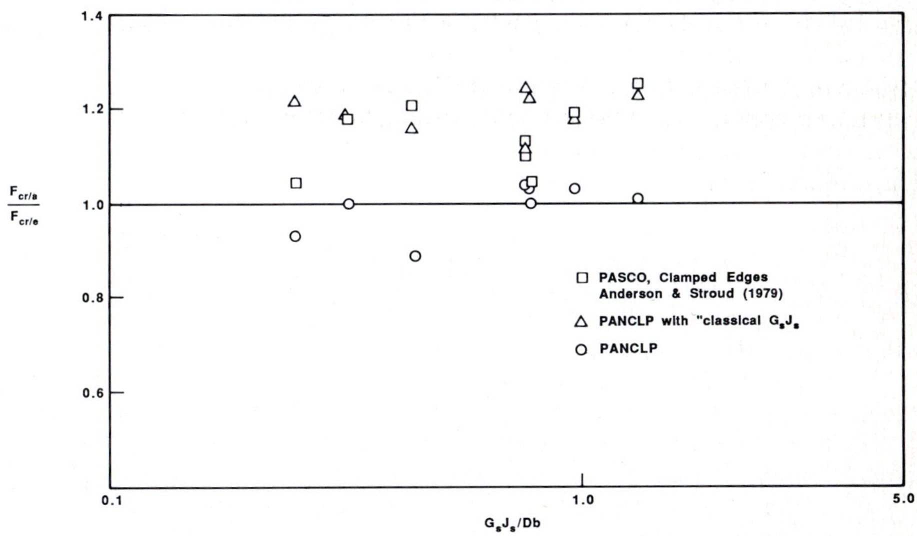

FIG. 8. Comparison of analytical prediction and experimental results for initial buckling – flat stiffened panel in compression (ARNOLD, 1986).

Fig. 9 illustrates the importance of edge restraint on initial buckling for the experimental data point most in error from the predictions shown in Fig. 8. It can be observed that even small changes in the torsional stiffness of the edge-stiffeners can have a large effect on initial buckling.

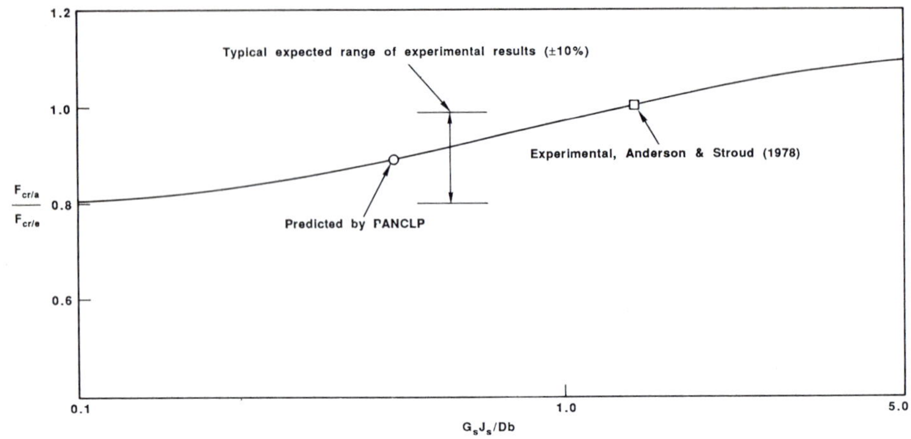

FIG. 9. Variation of predicted buckling load with stiffener torsional parameter (ARNOLD, 1986).

1.3. Curvature

As the radius of the panel decreases from an infinite to a finite value, the buckling load increases from that of the corresponding flat plate. Results for isotropic unstiffened curved panels with either simply supported or clamped boundary conditions, axial, shear or pressure loading, can be found in the References (GERARD and BECKER, 1957b) and (BRUHN, 1973). Fig. 10, obtained from ARNOLD et al. (1985), shows the buckling load of a curved panel under axial compression normalized by the flat plate buckling load as a function of the curvature parameter

$$Z = \frac{b^2}{Rh}(1-\nu^2)^{1/2}$$

and the edge-stiffening parameter GJ/Db. For a clamped panel, GJ/Db is very large; that is, GJ/Db is greater than about 10; whereas for an unrestrained panel, GJ/Db is approximately zero. Corresponding to Fig. 9, the buckle aspect ratio, β, is plotted as a function of the curvature parameter in Fig. 11. Similar results can be expected for composite panels in compression, although the exact numerical values will be dependent on the various ratios of bending stiffnesses. Thus, a rough estimate of initial buckling load and wavelength can be obtained through a knowledge of the buckling load for a flat unstiffened plate, Eq. (1.1), and the curves of Figs. 10 and 11.

In shear, results similar to those produced for axial compression are shown in Fig. 12 based upon the work of ARNOLD and PAREKH (1987). The theoretical/analytical capabilities of ARNOLD and PAREKH (1987) are numerically programmed in the computer code PANCLP (ARNOLD, 1985b); thus, curves for any particular lamination sequence, material, and panel size could be generated, and they would all be generally the same as shown in Fig. 12.

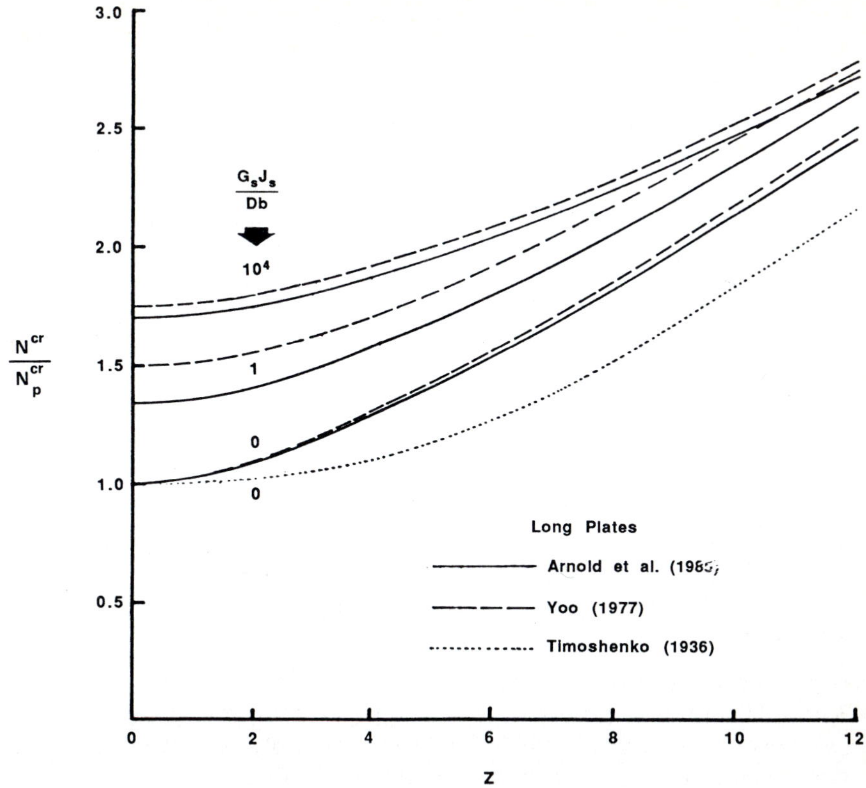

FIG. 10. Effects of curvature and unloaded edges conditions on compressive buckling load of long plates (ARNOLD et al., 1985).

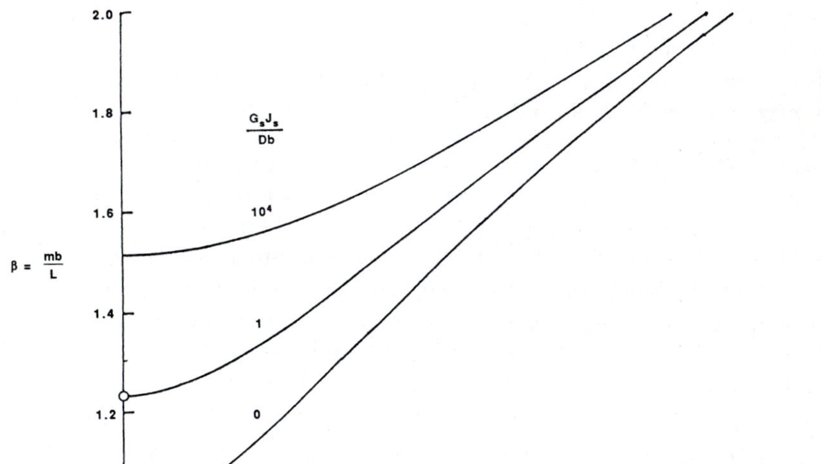

FIG. 11. Effects of curvature and rotational restraints along the unloaded edges on buckle aspect ratio of long plates (YOO, 1977).

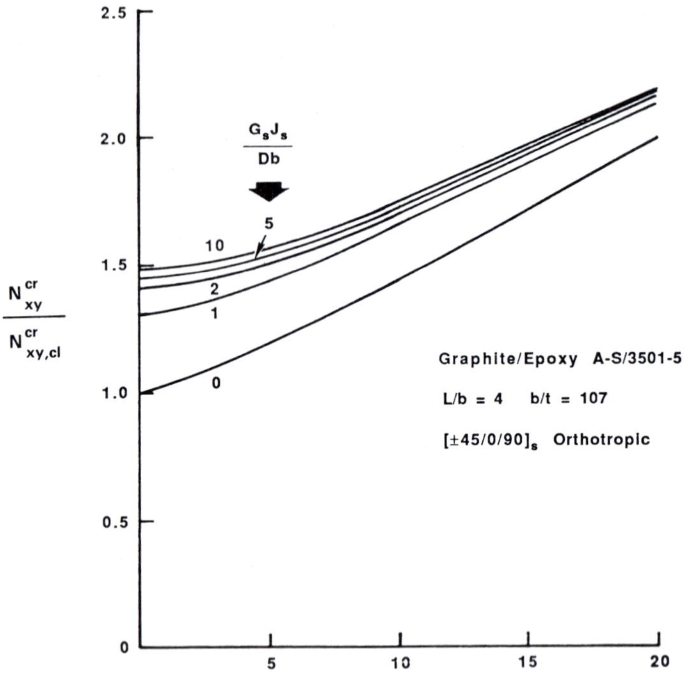

FIG. 12. Effects of curvature and unloaded edges conditions on shear buckling load of long plates (ARNOLD and PAREKH, 1986).

1.4. Combined loads

Most real structures experience a combination of loads which are neither purely axial nor purely shear. Additionally, many commercial and military aircraft are pressurized for high altitude flight. As a result of these conditions, it is frequently necessary to predict initial buckling in the presence of combined loading. Classical solutions for the simplest of boundary conditions can be found in the References (LEKHNITSKII, 1968; ASHTON and WHITNEY, 1970). Usually, the relationship between any two loads, such as axial compression and shear, can be presented by an interaction formula with the form (LEKHNITSKII, 1968):

$$\left(\frac{N_{xy}}{N_{xy}^{\mathrm{cr}}}\right)^{p} + \left(\frac{N_{x}}{N_{x}^{\mathrm{cr}}}\right)^{q} = 1\,, \tag{1.2}$$

where p and q are determined from experimental data or theoretical analysis.

For metallic materials, BRUHN (1973) provides a rough guide to the combined effects of biaxial, shear and pressure loading. Trends for composite materials are approximately the same from a qualitative standpoint; however, quantitative data on buckling under combined loading usually requires a more elaborate analysis procedure. In many instances, the use of finite element techniques is necessary to more accurately determine the combined effects.

When edge-stiffening and curvature are present in the stiffened-panel structure, no simple formula or graphical solution is available. As a representative example of the combined effects of curvature, edge-stiffening, lamination sequence, and axial compression and shear loading, several curves have been generated using the numerical program PANCLP (ARNOLD, 1985b). Fig. 13 shows an interaction diagram for Aluminum 2024-T3 with various values of curvature and edge-stiffening. Predicted buckling loads have been normalized by their respective flat plate classical equivalent loads wherein Eq. (1.1) is used for axial compression and for shear, the following is used (BOUADI, 1982).

$$N_{xy}^{cr} = \frac{\pi^2}{2b^2\alpha}\left(D_{11}\beta^2 + 2(D_{12} + 2D_{66})(1 + \alpha^2\beta^2) + D_{22}\left(\frac{1}{\beta^2} + 6\alpha^2 + \alpha^4\beta^2\right)\right.$$
$$\left. - 4\alpha[D_{16}\beta^2 + D_{26}(3 + \alpha^2\beta^2)]\right), \qquad (1.3)$$

where α is the nodal slope.

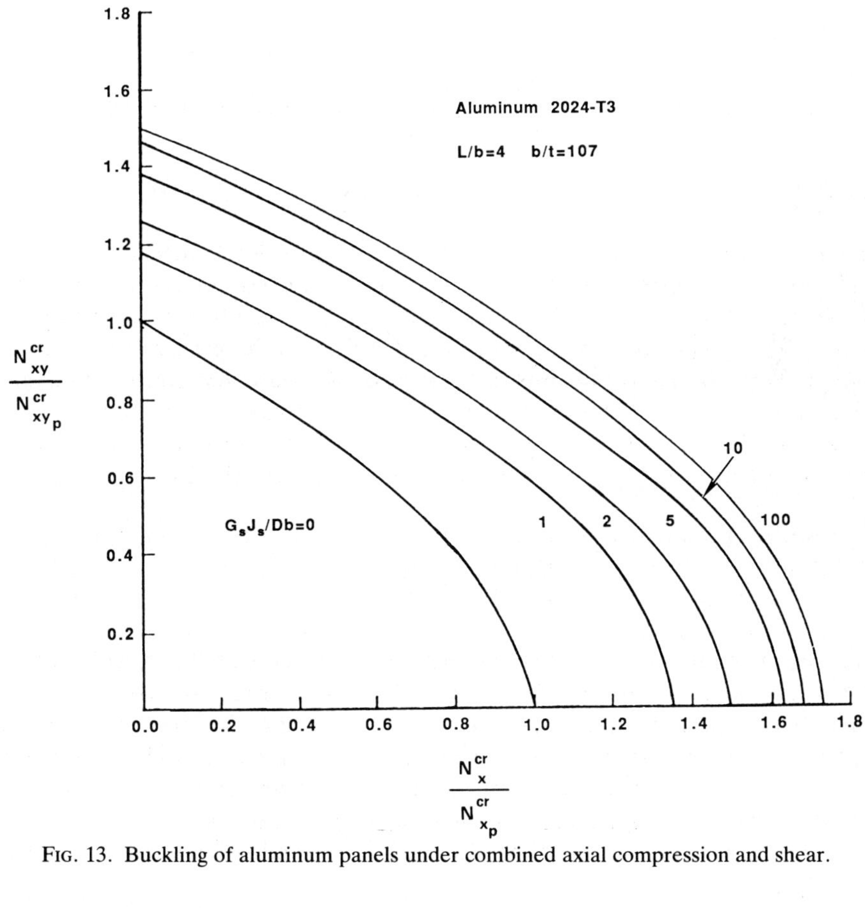

FIG. 13. Buckling of aluminum panels under combined axial compression and shear.

Fig. 14 shows a similar characteristic for a typical graphite/epoxy laminate. For either an isotropic metal or a laminated composite material, shear buckling is more sensitive to the effects of edge-stiffening and curvature.

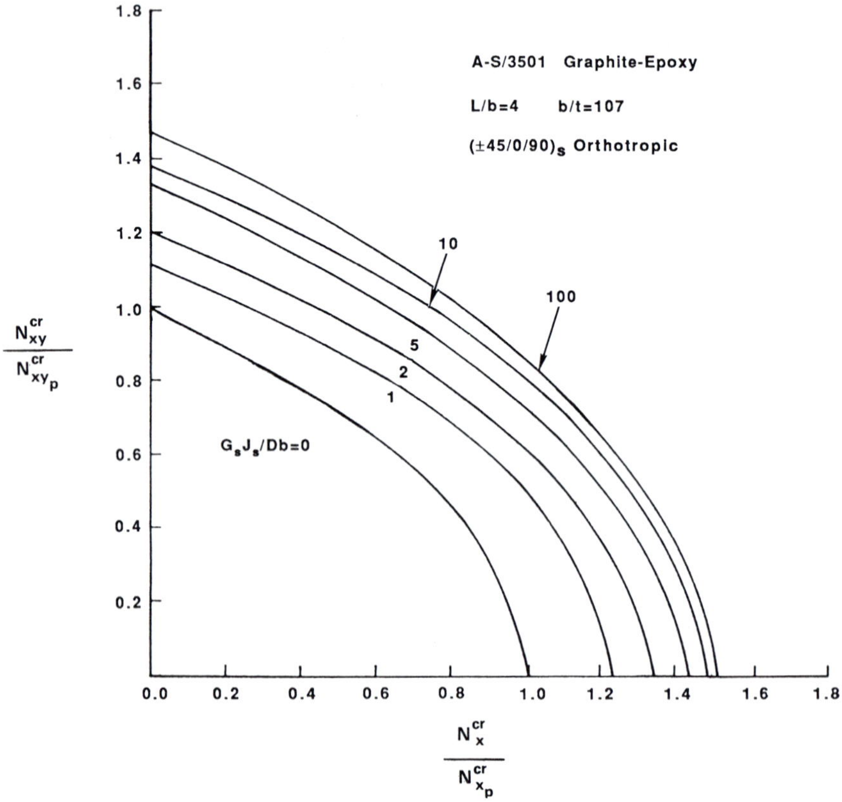

FIG. 14. Buckling of composite panels under combined axial compression and shear.

1.5. Anisotropic laminates

Anisotropic laminates are formed whenever direct and shear stresses are coupled by the presence of laminate stiffness terms A_{16}, A_{26}, B_{16}, B_{26}, D_{16} and D_{26}. Experimental evaluations of angle-plied boron/epoxy plates loaded in diagonal tension (KAMINSKI and ASHTON, 1971) have shown that the presence of bending anisotropy (D_{16} and D_{26} only) can have a pronounced effect on initial buckling. ZHANG and MATHEWS (1983) have presented solutions for the buckling of laminated composite curved plates under combined loads (compression and shear). Also WHITNEY (1984) has developed an analytical solution for cylindrical anisotropic plates. These studies clearly demonstrate the importance of anisotropy for prediction of initial buckling.

Practically speaking, most aircraft structures are made so that the resulting laminate section is symmetric; thus, A_{16}, A_{26}, B_{16} and B_{26} terms are zero. The most common laminate form, composed of various layers of alternating angle-plied materials, possesses bending anisotropy only. When the number of layers is eight or greater, the amount of bending anisotropy is usually too small to be of importance.

Fig. 15 shows the effect of anisotropy for a typical laminate construction when combined compression and shear loads are present. For uniaxial loads, the anisotropy of the laminate has a relatively small effect; however, in shear, the anisotropy has a dramatic impact on initial buckling load. The orthotropic solution is shown as a reference point. The particular laminate used to create Fig. 15 is composed of eight layers – laminates with fewer layers will exhibit an even greater effect due to the anisotropy.

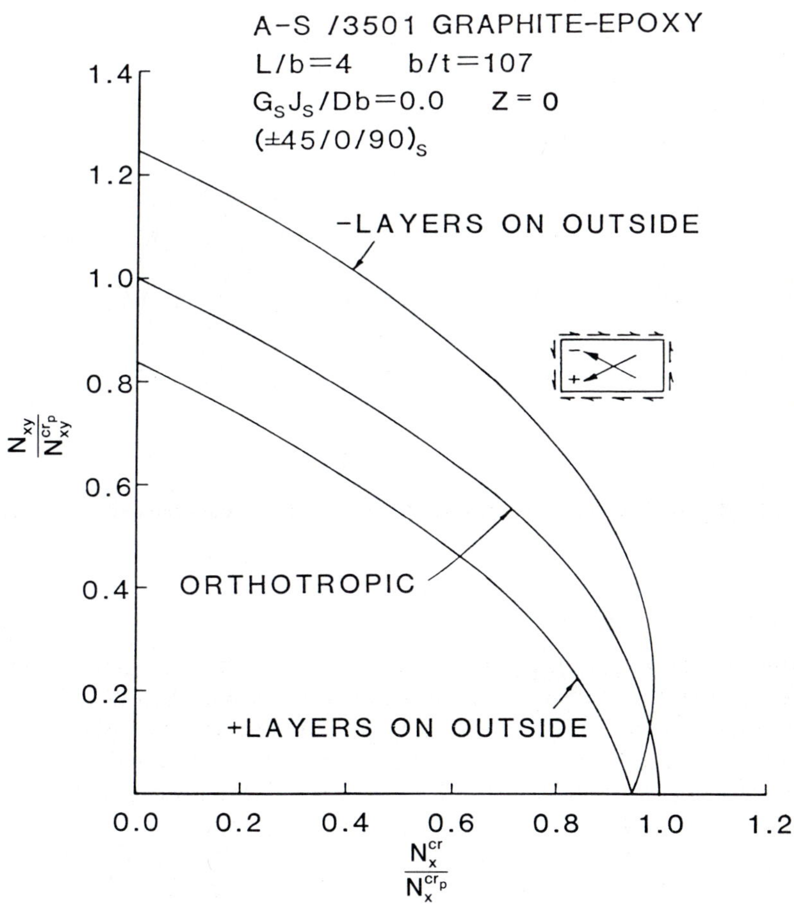

FIG. 15. Buckling of orthotropic/anisotropic flat plates under combined axial compression and shear.

1.6. Imperfection sensitivity

Inasmuch as no structure can be made perfect, the presence of imperfections, either in geometry, material, or load, can cause the panel to buckle at a load lower than what would be expected from purely theoretical considerations. When stiffeners and panel have individual buckling loads in close proximity to each other, it is possible to have a serious decrease in the buckling load capability. The most interesting theoretical works in the area of imperfection sensitivity and mode interaction have been documented by KOITER (1976), KOITER and PIGNATARO (1976), and TVERGAARD (1973) – only several papers out of many by these authors are mentioned here. Although the analyses presented in the papers cited above are necessarily constrained by the complexity of the mathematical formulation, the results for linear elastic isotropic materials are at least partially valid for laminated composite panels.

One of the important conclusions illustrated by KOITER (1976) is that '. . . the addition of small load imperfections to predominant overall imperfections has comparatively minor adverse consequences; whereas small overall imperfections in addition to predominant local imperfections may cause a further significant drop of the critical load factor . . .'. Typical metal stiffened-flat-panel geometries have been examined by KOITER, and Fig. 16, obtained from KOITER (1976), shows the effects of initial imperfections on the initial buckling load. The quantity λ^* is the ratio of the buckling load with imperfection to the classical buckling load and f_0 is an imperfections parameter. The number of stiffeners is given by N_S. As the number of stiffeners is decreased, the buckling load of the panel changes with respect to that of the overall structure, and thus,

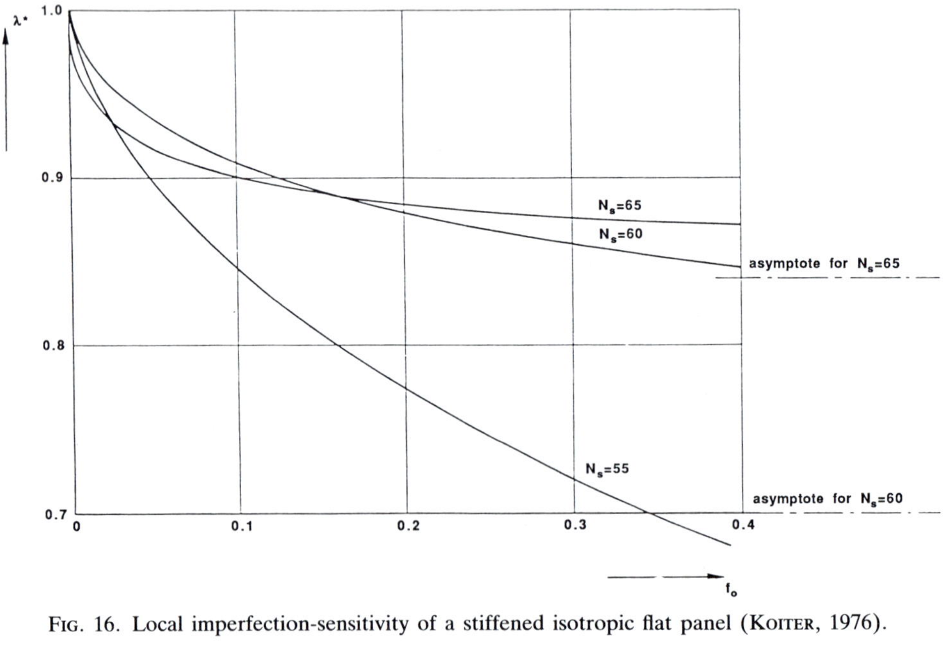

FIG. 16. Local imperfection-sensitivity of a stiffened isotropic flat panel (KOITER, 1976).

mode interaction and imperfection sensitivity is also changed. The results of Fig. 16 demonstrate how improperly designed stiffened-panel structures can buckle at loads significantly lower than predicted.

It should be kept in mind that the imperfection sensitivity of panels designed for local panel and euler buckling modes close together represents a worst case situation from a practical standpoint. Initial buckling predictions and experiments shown in Section 1.1.1 would indicate that typical aircraft stiffened panels are designed so that mode interaction is precluded; consequently, initial buckling can be predicted accurately for thin composite panels ($b/t > 50$) without shear, and for thick panels with transverse shear considered.

For curved panels, imperfections have been shown experimentally to be of some importance. KNIGHT and STARNES (1985) have obtained experimental data for the buckling and postbuckling of flat and curved stiffened panels under axial compression loads. A typical example of the data generated is provided in Fig. 17, which shows the out-of-plane displacement as a function of the applied axial compression load. The predicted initial buckling point is also indicated. The presence of imperfections can be seen from the fact that the curve 'bends over' below the predicted buckling load (transverse shear has been included in the buckling prediction). Notice, however, that the effect of imperfections is even more pronounced by the fifth cycle of loading. In these tests, the panel

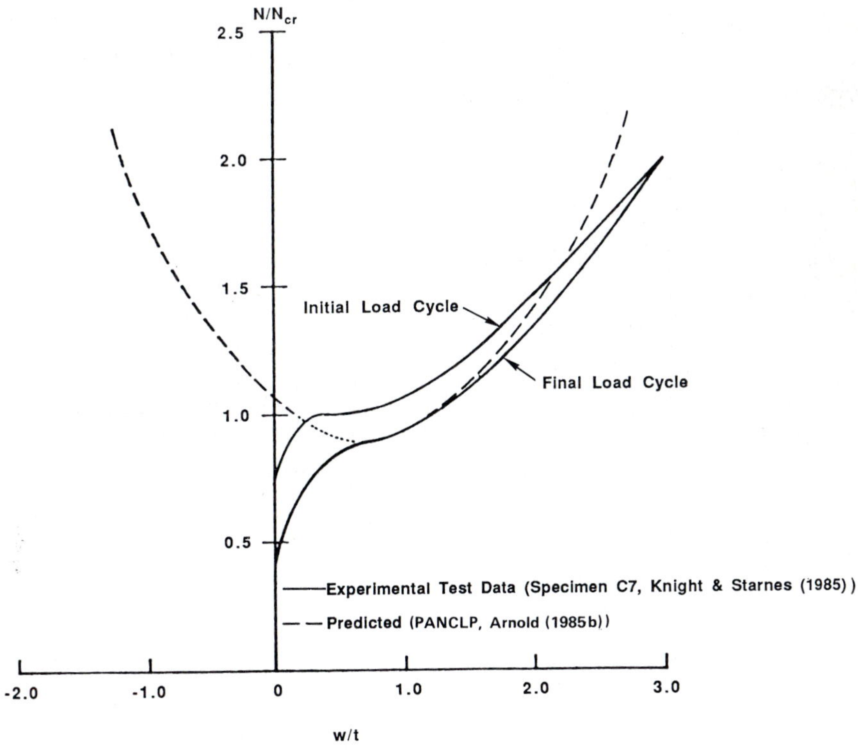

FIG. 17. Comparison of predicted and experimental out-of-plane displacements.

was loaded until damage was indicated by visual observation or sound, unloaded, and loaded again. So, part of the decreased initial buckling load is at least due to some amount of induced damage.

HU et al. (1946) has analyzed the behavior of perfect and imperfect plates subjected to a uniform edge shortening, and some of his results are presented in Fig. 18. It is interesting to note that the initial buckling load is lowered as the imperfection increases in magnitude; however, the postbuckling response is changed only slightly in the extended postbuckling range.

No general guidelines can be offered for predicting the effects of imperfections on initial buckling in flat and curved laminated composite stiffened-panels. Recent experimental efforts RENIERI and GARRETT (1982) – AGARWAL et al. (1984), correlated by the analysis procedure of ARNOLD (1985b) and documented in ARNOLD (1984b, 1985a), result in the conclusion that 'knock-down factors' of close to 1.0 are valid for design purposes for flat plates. For curved stiffened-panels with Z less than about 40, a knockdown factor of 0.95

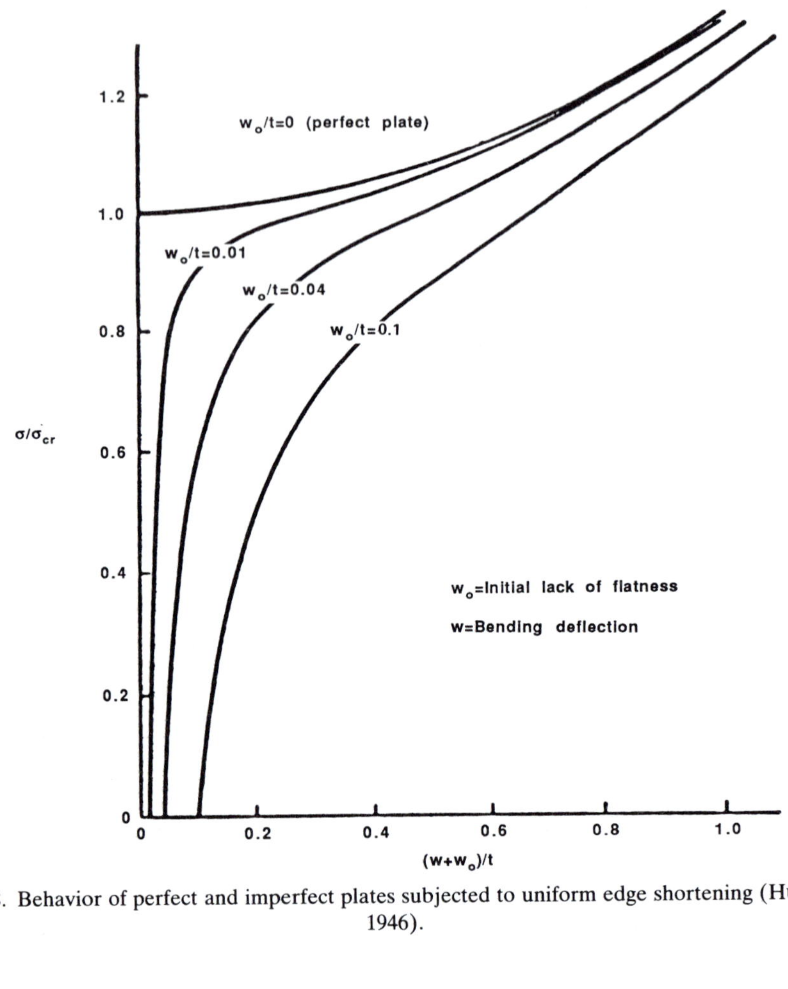

FIG. 18. Behavior of perfect and imperfect plates subjected to uniform edge shortening (HU et al., 1946).

is recommended. For panels with Z greater than 40, experimental verification would be required to assure a high confidence design. Finally, it is remarked that, in general, stiffened-shell structures exhibit a much reduced sensitivity to imperfections when compared to the analogous monocoque shell as discussed by MAYERS and MELLER (1972).

1.7. Boundary conditions

For initial buckling, elastic materials, the data by GERARD and BECKER (1957a,b) provides a comprehensive collection of curves which show the effects of various boundary conditions (simply supported, free or clamped on different edges) on the buckling of flat, square and rectangular plates. Fig. 19, obtained

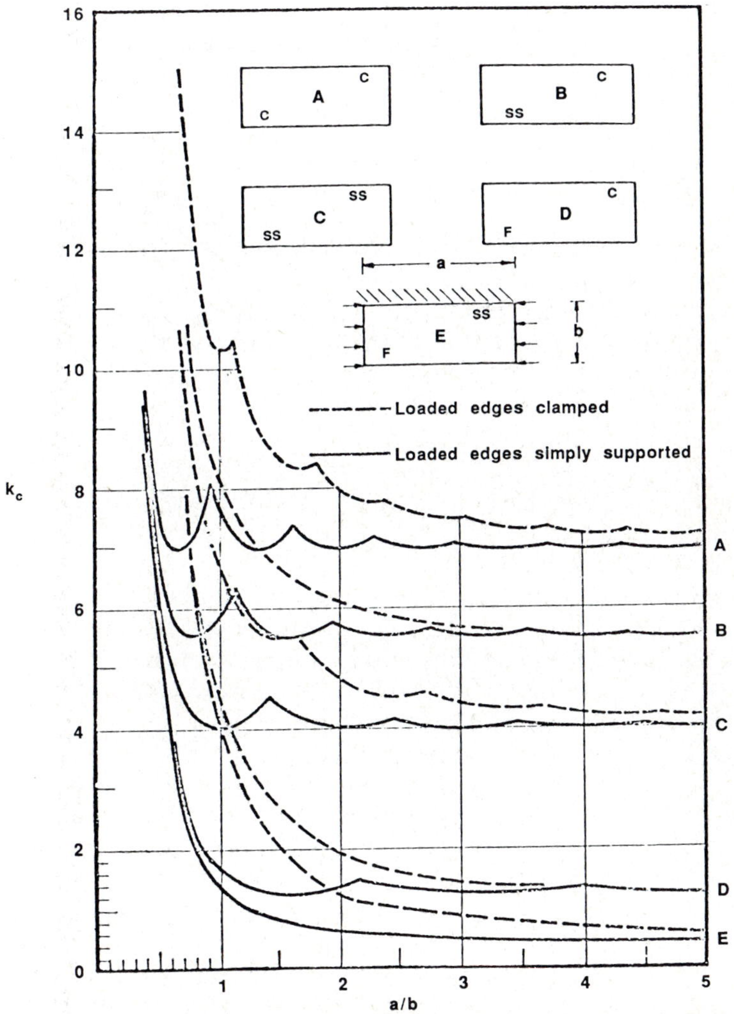

FIG. 19. Compressive-buckling coefficients for flat rectangular plates (GERARD and BECKER, 1957).

from GERARD and BECKER (1957), shows some typical results for compressive buckling with the buckling coefficient, k_c, being used in the formula

$$N_{cl}^{cr} = \frac{k_c \pi^2 E t}{12(1-\nu^2)} \left(\frac{t}{b}\right)^2 . \tag{1.4}$$

SMITH (1973) has presented solutions for the buckling of flat plates including the effects of transverse shear and, based on the Reissner variational principle for combinations of free, simply supported and clamped edges. These results are also presented by VINSON and CHOU (1975).

LEKHNITSKII (1968) presents many solutions for compression, shear, combined loads, and simply supported and clamped boundary conditions. For the most part, these solutions are in terms of buckling equations of moderate length and complexity; thus, it is recommended that this reference be consulted. Lekhnitskii's work is valid for elastic materials and does not consider the influence of edge-stiffening or transverse shear.

2. Postbuckling

The idea of permitting a panel structure to buckle and carry loads in excess of the buckling load was developed in Germany and the United States in the 1930s. For the relatively low flight speeds and operating stress levels in evidence at that time, neither aerodynamic nor structural integrity was compromised by permitting postbuckling to occur. Early theoretical efforts by von KARMAN et al. (1932) were directed towards the goal of defining the amount of buckled-plate material that was effective in supporting additional load. The excellent survey by YOO (1977) discusses many of the earlier efforts directed at the definition of effective-width from 1932 to 1955. All of these works were linear-elastic.

In 1955, the first plate-postbuckling study using actual nonlinear material characteristics was presented by MAYERS and BUDIANSKY (1955). Subsequently, the design formulae of GERARD (1957) were developed to unify the theoretical results of Mayers and Budiansky and the experimental data of ANDERSON and ANDERSON (1956), BOTMAN (1954), and BESSELING (1953) for rectangular plates with different materials and unloaded-edge boundary conditions. In 1966, MAYERS and NELSON (1968) extended the theoretical analysis of Mayers and Budiansky to rectangular plates and obtained excellent correlation with the experimental data presented by STEIN (1959) for 2024-T3 aluminum.

In spite of the theoretical effort described above, the predominant design criteria for postbuckled metal structures form the 1930s to the present day has been the use of linear-elastic effective-width criteria such as found in GERARD (1946). Even recent efforts devoted to obtaining the postbuckled strength of composite panels relies upon the use of experimental data and semiempirical formulae similar to those derived over 50 years ago.

With respect to composites, the definition of a suitable effective-width formula is not available for the simple reason that the ability to predict the postbuckling behavior of a laminated composite structure is restricted by the lack of a readily usable theoretical approach. Research on a theoretical level for the postbuckling behavior of composite plates is relatively recent with notable examples being Turvey and Wittrick (1973), Harris (1975), Banks (1976), Anderson and Mayers (1979), and Arnold et al. (1985). These cited references approach the postbuckling problem from the standpoint of predicting postbuckled stiffness, displacements, strains, and stresses with a decreased reliance on the effective-width concept. This is due, in part, to the difficulty of succinctly quantifying the effect of the 21 membrane and bending stiffness terms on the postbuckled average plate load.

It seems evident at this time that future designs of postbuckled composite panel structures will rely upon the use of modern numerical computerized analysis methods for prediction of responses with a decreased use of 'simple' formulae.

2.1. Postbuckling stiffness

As mentioned earlier, for flat plates and shallow curved plates, there is significant strength remaining in the panel after initial buckling has occured. In some parts of an aircraft structure, this basic fact can be used to reduce structural weight. In this section, several examples of postbuckling behavior will be presented.

Figs. 20 and 21 for no edge-stiffening and effectively clamped unloaded edges, respectively, show typical load-shortening trends for metallic materials (Aluminum 2024-T3 in this instance) loaded in axial compression with the free edges constrained to be straight, obtained from Yoo (1977). Several important points need to be made here. First, curvature, which has a strong influence on initial buckling, is seen to have a diminishing effect in the postbuckling region. In fact, after an end-shortening ratio of approximately 3.0, a shallow curved plate and a flat plate have the same postbuckling stiffness. Second, the load-shortening curves for assumed inelastic behavior are quite different than the elastic counterpart, particularly so when the end-shortening ratio is greater than about 3.0. Note, that if elastic material behavior is assumed, then a maximum strength in the plate cannot be predicted. In other words, prediction of maximum strength in the plate can only be predicted with an inelastic material analysis. Finally, in terms of an overall conclusion, it is noticed that increasing the rise of a curved plate and the torsional rigidity of the edge-stiffeners simply to raise the initial or bifurcation buckling load cannot be expected to produce comparable increases in crippling load. This can be an important design point for postbuckled structures.

Similarly, in compression for a laminated composite material, Figs. 22 and 23 demonstrate the significant postbuckled strength of properly designed stiffened

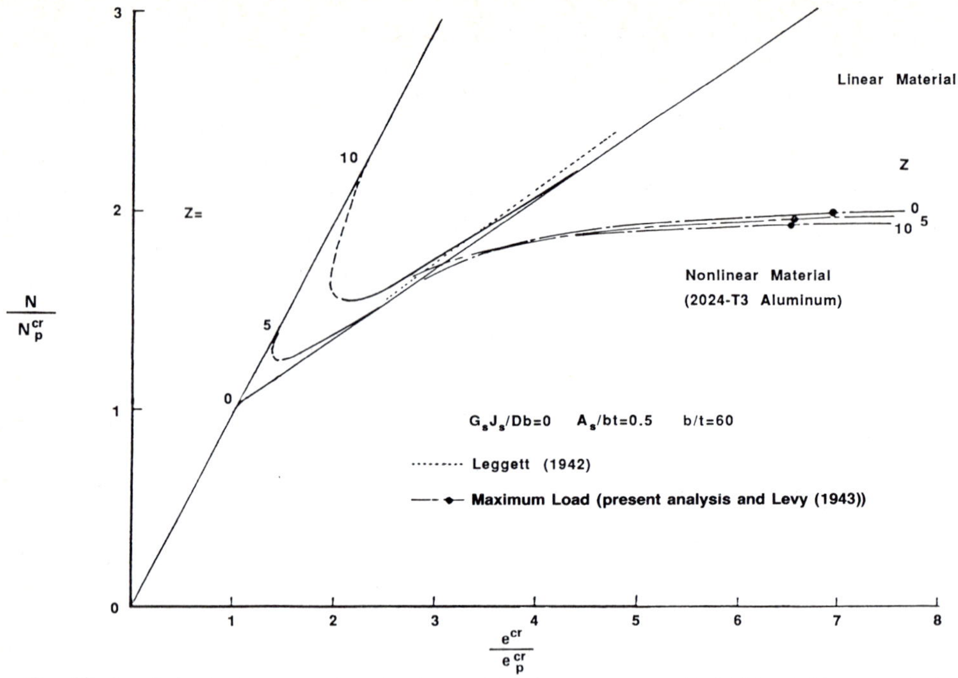

FIG. 20. Load-shortening curves for aluminum plates in compression $G_sJ_s/Db = 0$ (ARNOLD et al., 1985).

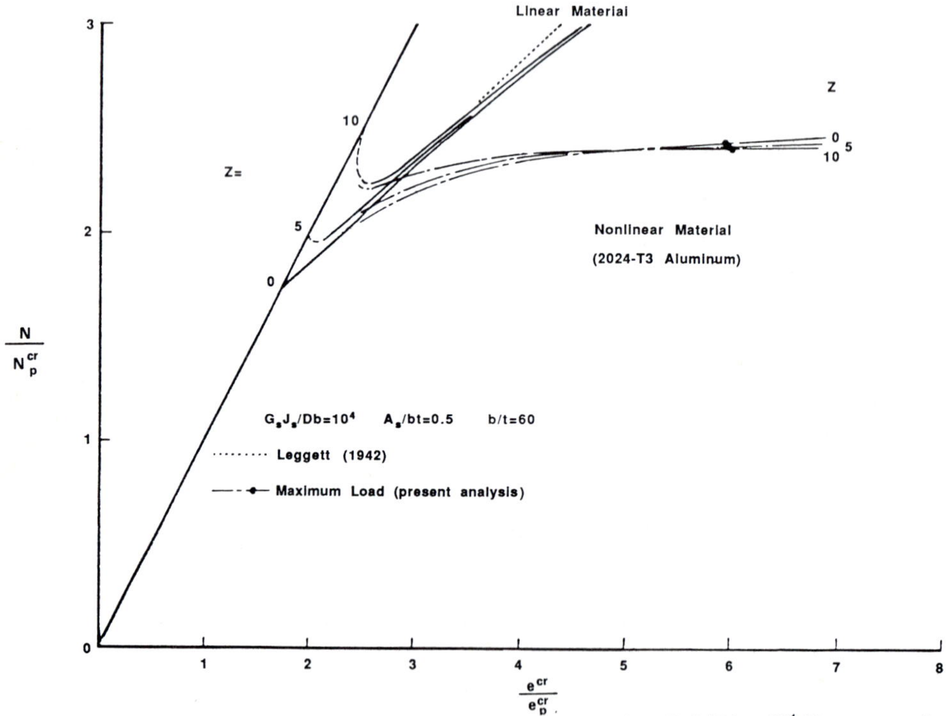

FIG. 21. Load-shortening curves for aluminum plates in compression $G_sJ_s/Db = 10^4$ (ARNOLD et al., 1985).

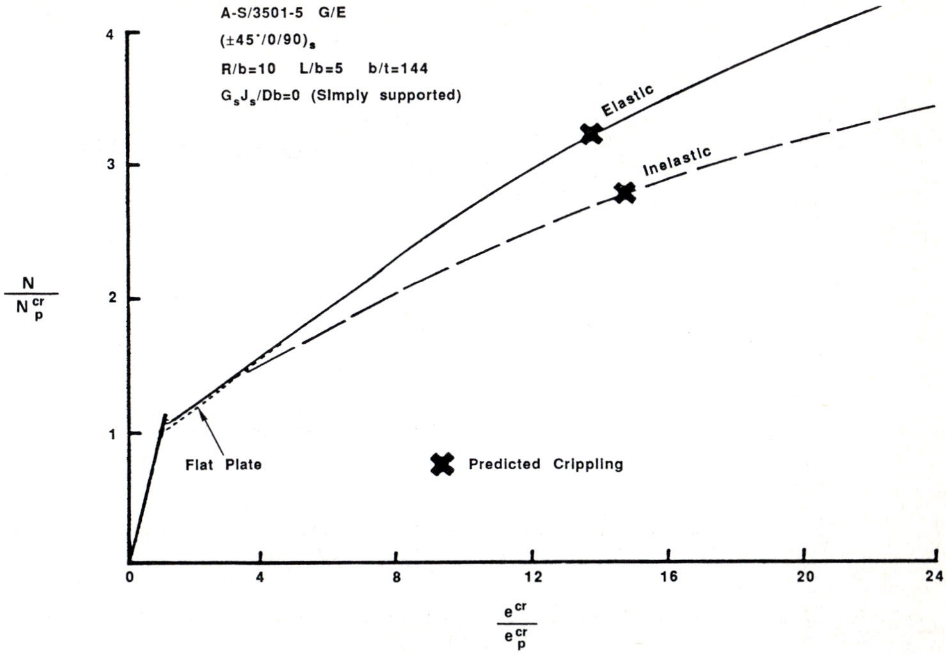

FIG. 22. Load-shortening curves for a laminated composite plate in compression, $G_sJ_s/Db = 0$ (ARNOLD et al., 1985).

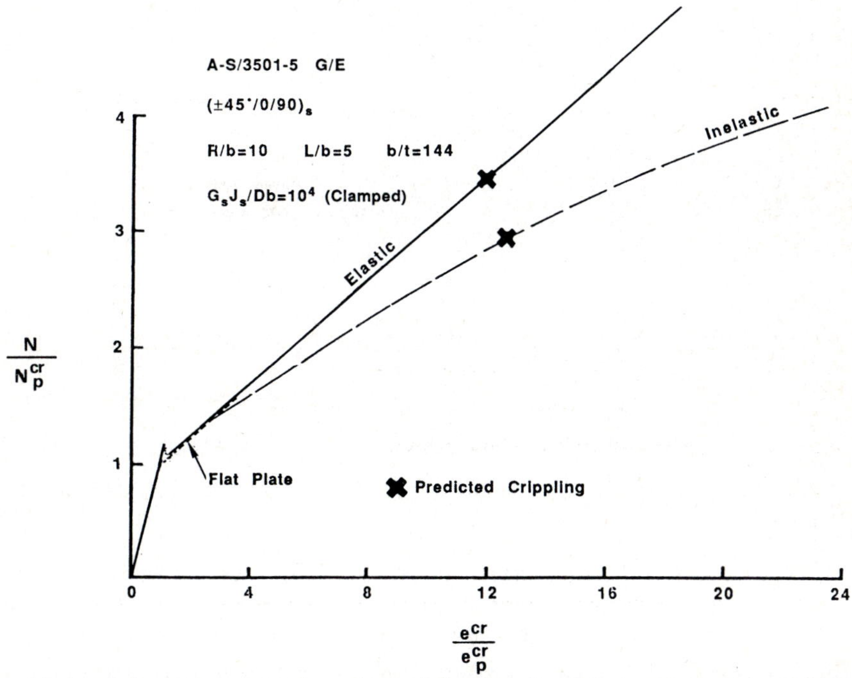

FIG. 23. Load-shortening curves for a laminated composite plate in compression, $G_sJ_s/Db = 10^4$ (ARNOLD et al., 1985).

panels. As in the case of conventional metal plates, the marked difference in linear elastic versus inelastic material behavior is evident.

In shear, Fig. 24 is presented for a typical aerospace stiffened panel construction. The influence of simultaneous axial compression (or the lack of compression load) is shown for both elastic and inelastic material assumptions. Thus, for composites, as for metals, the effect of material inelasticity causes a reduced postbuckled stiffness in the structure.

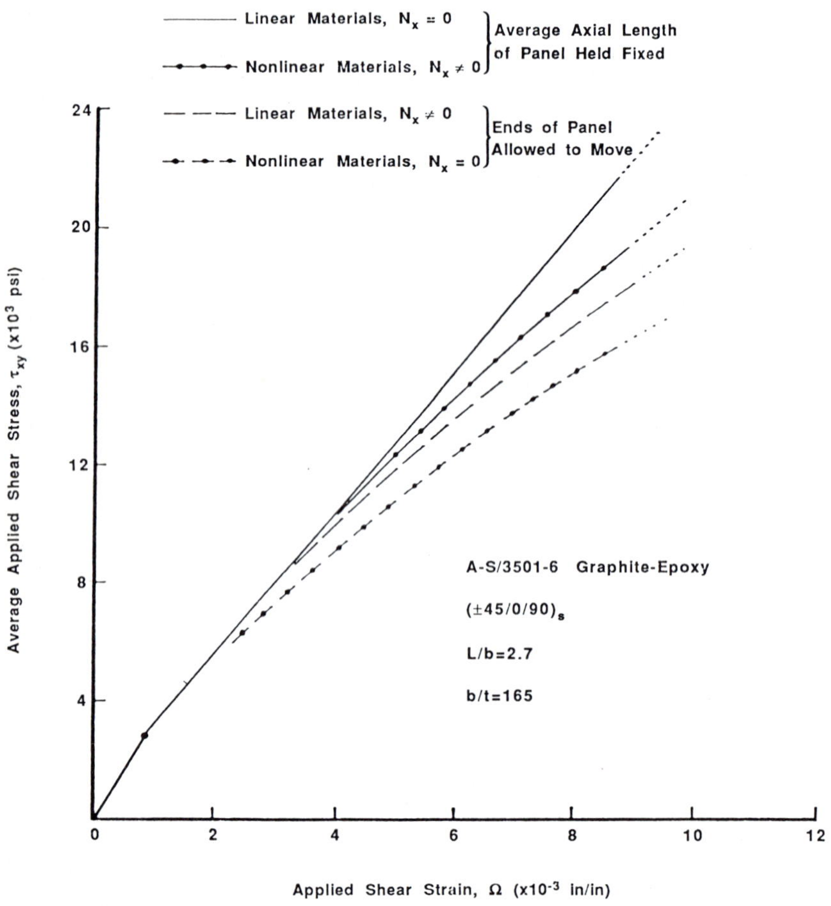

FIG. 24. Predicted load–shear strain curves (ARNOLD, 1985a).

2.2. *Effective-width*

An excellent review and discussion of effective-width for elastic and elasto-plastic isotropic edge-stiffened panels has been presented by Yoo (1977). Some of Yoo's comments are also applicable to composite structures. Prior to buckling, the stress distribution in a loaded plate is uniform; however, in the postbuckled state, the redistribution of loading causes complicated biaxial stresses to occur. It is interesting to note that the average stress carried by each transverse portion of the plate is invariant, and this average stress can be used as a measure of the load-carrying capability of the postbuckled structure. Thus, the effective-width of an elastic plate is defined as the ratio of the average stress in the postbuckled state to that of the plate were it to remain unbuckled throughout the loading history. For inelastic materials, the effective-width is defined as the ratio of average stress to edge stress.

A good example of effective-width is shown in Fig. 25, obtained from Yoo (1977). Although the figure has been derived for a metal (Aluminum 2024-T3) it, nevertheless, is a good basis for understanding the role of material inelasticity in the derivation of effective-width. In Fig. 23, the average stress in the postbuckled range has been normalized by the elastic stress corresponding to the prescribed end-shortening. For elastic materials (solid curve), the trend is higher than that for inelastic material with the inelastic material effective-width curves being a strong function of the b/t ratio of the plate. The circle symbol on these curves represents the effective-width at maximum load. All curves of Fig. 25 are for an assumed simply supported boundary condition with no edge-stiffening.

For composite materials, a representative effective-width curve is presented in Fig. 26 for boron/epoxy based on Anderson and Mayers. These curves can be used as a general guide for angle-plied laminates that are symmetric.

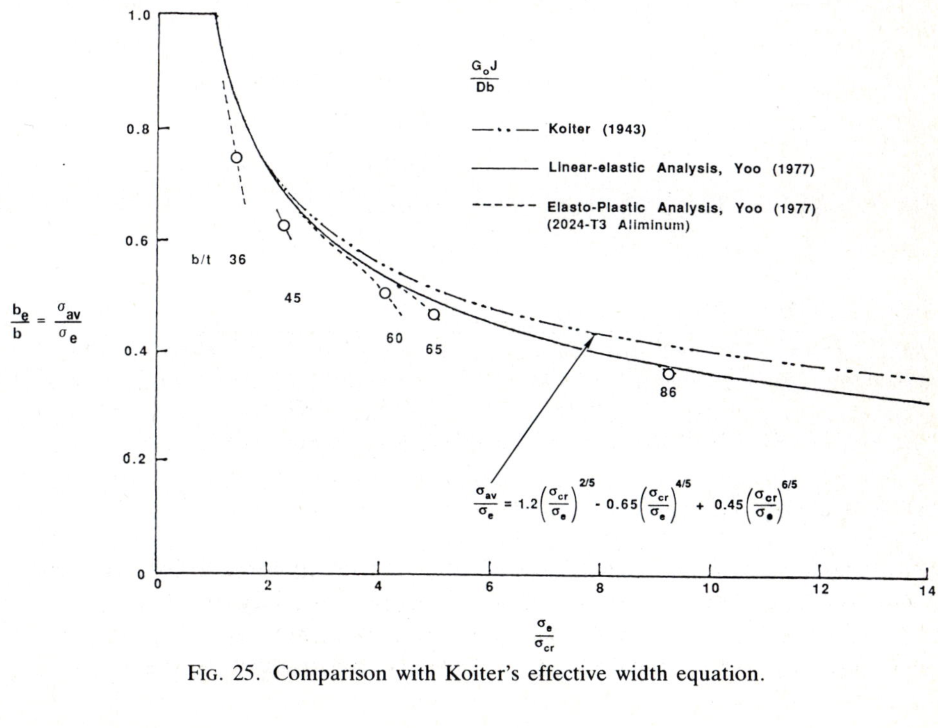

FIG. 25. Comparison with Koiter's effective width equation.

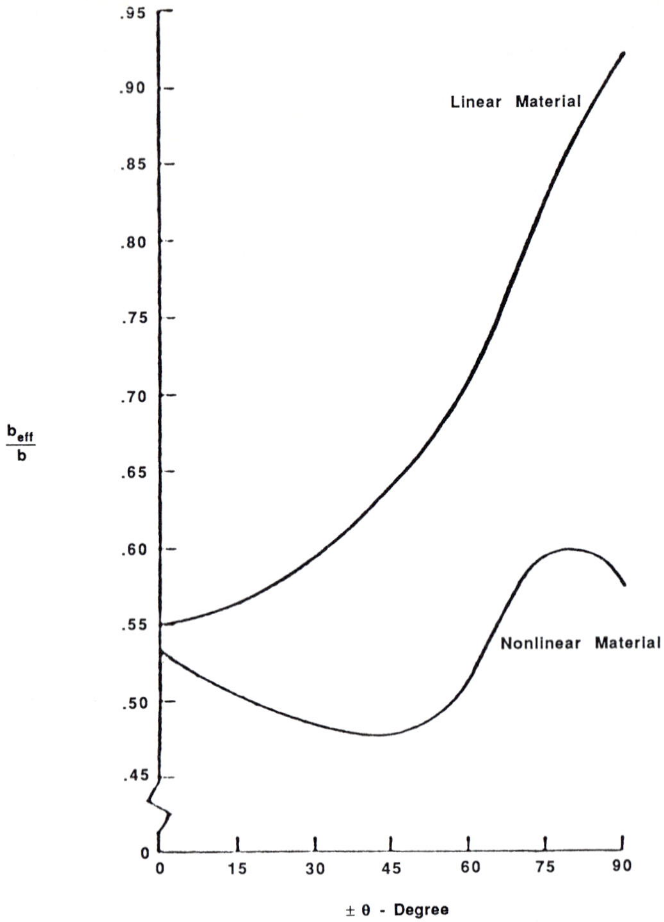

FIG. 26. Effective width of ±0° boron–epoxy laminates ($\delta/\delta_{cr} = 4.0$) (ANDERSON and MAYERS, 1979).

2.3. *Wavelength changes*

Experimental evidence for the change in wavelength in postbuckled metal structures can be found in STEIN (1959) for metal panels under uniaxial compression loading. Generally speaking, as the structure is loaded further into the postbuckled range, the wavenumber increases, or in other words, the wavelength decreases. In all cases, the wavelength decreases from the value at initial buckling. Wavelength changes occur for compression and shear loading; however, in the case of shear loading, the nodal lines (lines of zero deflection) are skewed relative to the orthogonal sides of the panel. Typical results for the laminated composite plate in shear of Fig. 24 are shown in Fig. 27 for half-wavenumber and nodal slope angle, respectively. The diagonal tension angle (α) and the nodal slope angle (α_n) are related by the following approximate formula.

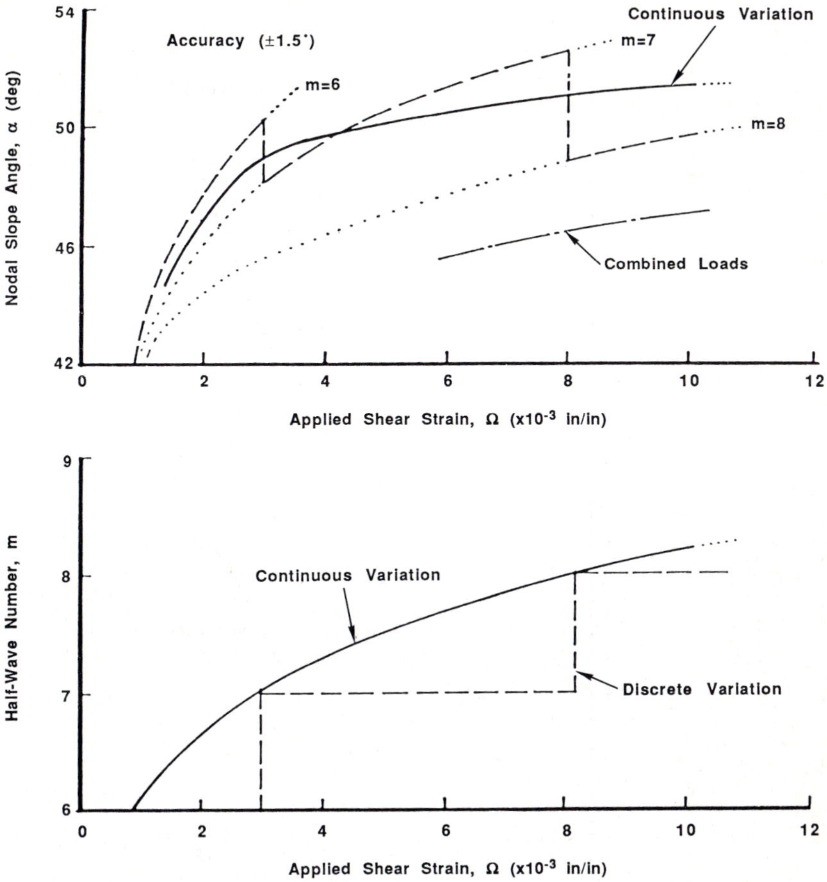

FIG. 27. Predicted half-wave number and nodal slope angle (ARNOLD, 1984a).

$$\tan \alpha \simeq -\tan \alpha_n + \frac{1}{\beta}\,.$$

Both wavelength and nodal slope angle change progressively as the panel is loaded; however, in moving from one stable configuration to another, sudden 'jerks' are common and are usually observed experimentally. Generally speaking, these sudden movements are less violent as the wavenumber increases.

2.4. *Displacements/strains*

Prior to buckling, out-of-plane displacements are either zero (perfect panel assumption) or small (imperfections present) for most typical stiffened panel constructions made from composite materials. After initial buckling, displacements increase quadratically with applied load. Some examples for metal flat and curved plates are provided in Fig. 28. For curved long panels, the inward

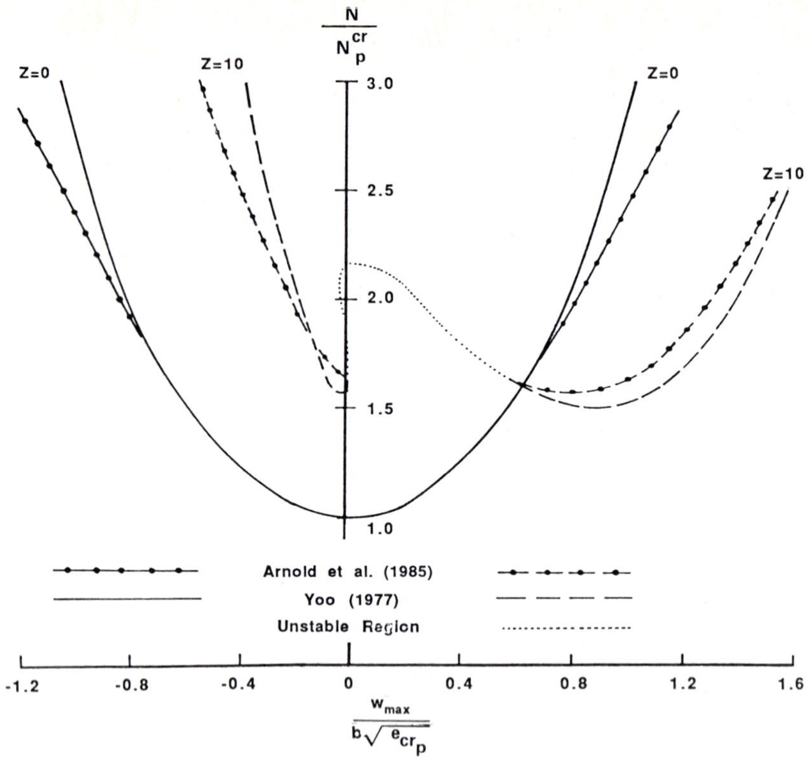

FIG. 28. Out-of-plane displacements for flat and curved isotropic panels.

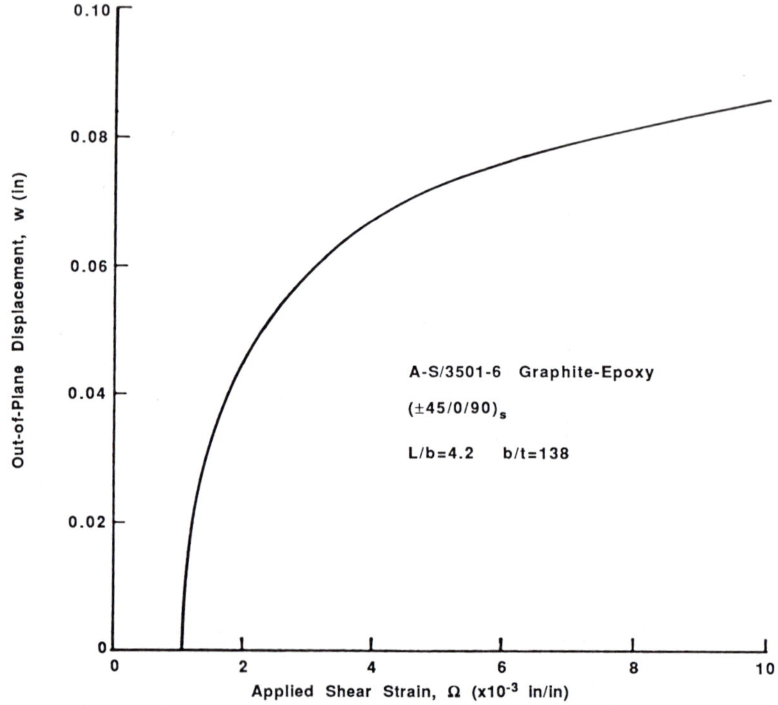

FIG. 29. Out-of-plane displacements as a function of the applied shear strain (ARNOLD, 1984a).

deflection is greater than the outward deflection. Curved panels, as mentioned earlier, exhibit a higher buckling load than the corresponding flat panel; however, in the postbuckling range, the load-shortening curves are very similar. Out-of-plane displacements follow a similar trend except the displacements for a curved panel contain a net inward displacement superimposed on the approximately sinusoidal distribution found for flat plates. Thus, curved plates attempt to 'flatten-out' in the postbuckled region.

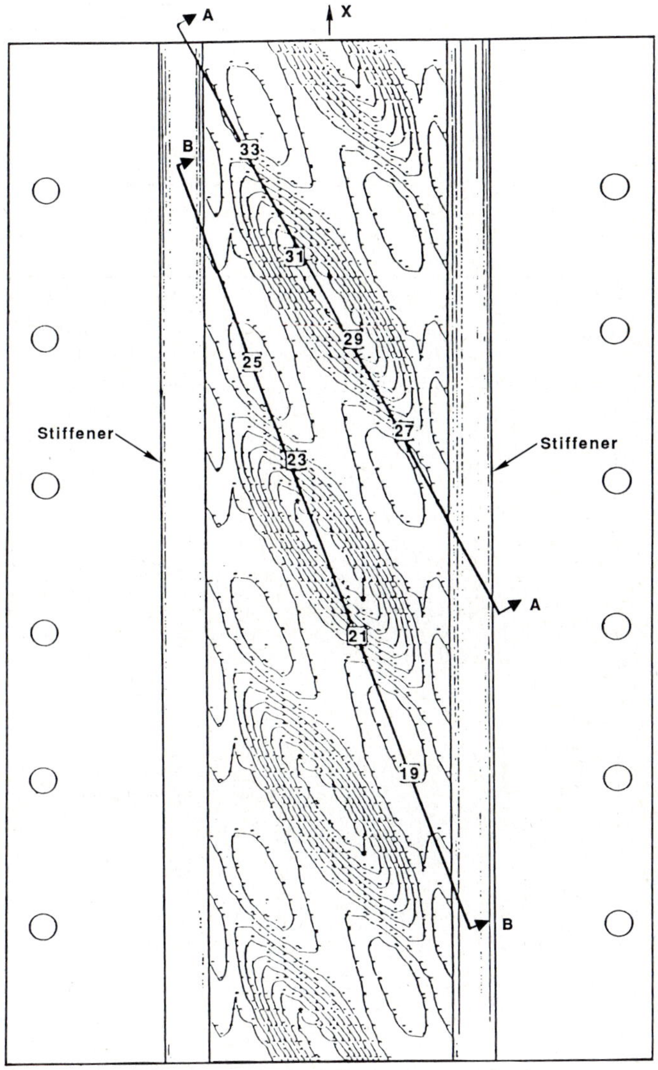

FIG. 30. Predicted axial strain distribution with location of rosette strain guages noted (ARNOLD, 1984a).

Some typical results for laminated composite panels are shown in Fig. 17 for axial compression and Fig. 29 for shear. For both kinds of load, the out-of-plane displacement varies quadratically with the end-shortening, reaching a maximum value in the far postbuckled range.

Typical plots of strain distributions across the surface of a laminated composite panel in shear are shown in Fig. 30. Comparisons with experimental data are shown in Fig. 31. Plots of experimentally determined and analytically predicted strains on the surface from back-to-back strain gages are shown in Fig. 32 for a point located near the maximum strain location. As can be observed, strains also vary quadratically with the applied load and approach a maximum in the far postbuckled range. Fig. 32 also shows some small shifts in strain as the postbuckled panel passes from one wavenumber to a higher (wavelength change).

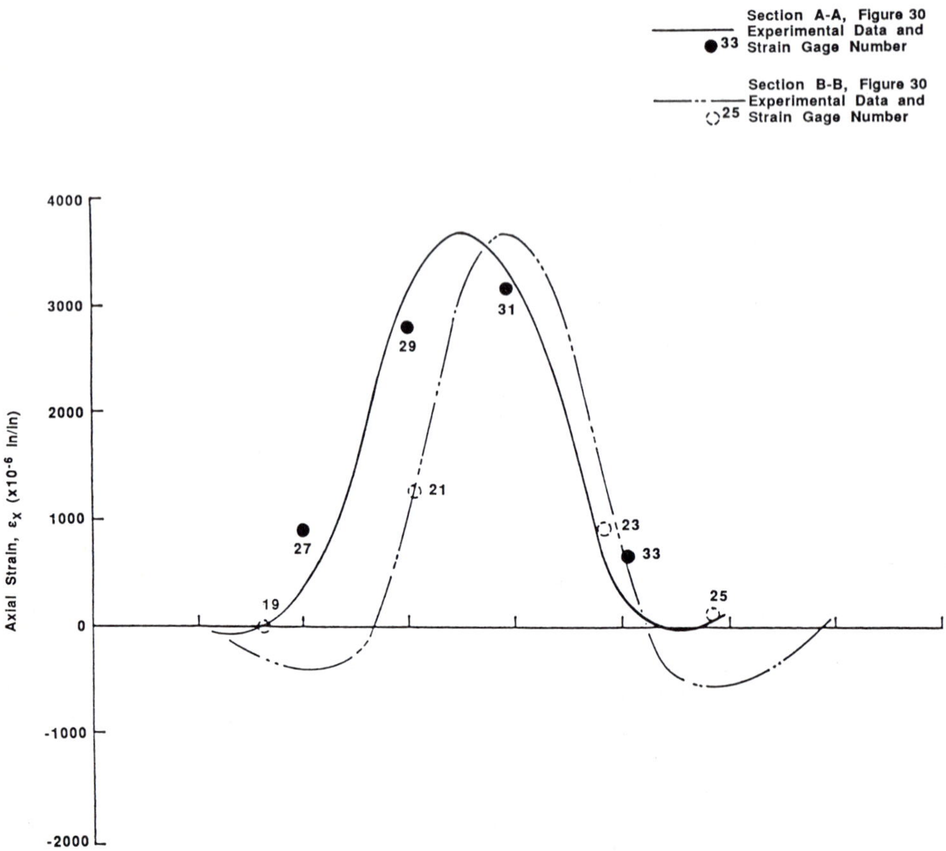

FIG. 31. Correlation of predicted axial strains with experimental strain guage data (ARNOLD, 1984a).

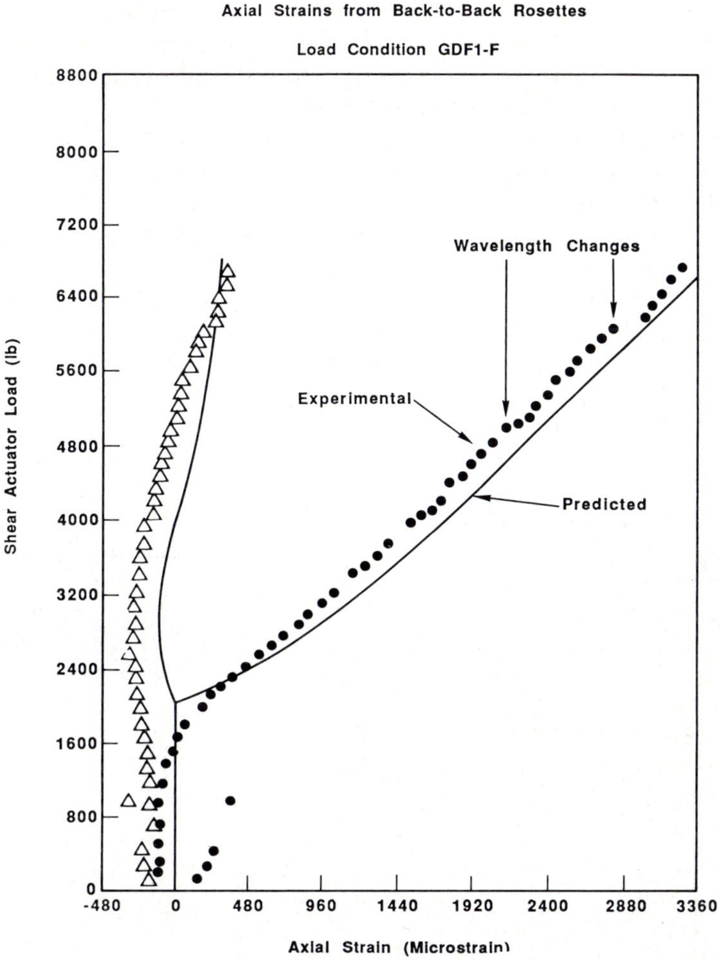

FIG. 32. Comparison of predicted and experimentally determined axial strains (guages 31A and 32A) (ARNOLD, 1984a).

3. Failure

The previous sections have described the initial buckling and postbuckling behaviour of typical metal and composite stiffened panels. However, the proper design of postbuckled panels can only be accomplished by knowing the ultimate strength of the stiffened-panel assembly. For thin panels in the near postbuckled field, the influences of transverse shear and material nonlinearity on displacements and strains are relatively small; consequently, simplified analyses are possible if sufficient confirming experimental data is available. For thicker laminates ($b/t < 50$) and in the far postbuckled range ($e/e_{cr} > 4$), the roles of transverse shear and material nonlinearity become much more important. Indeed, for prediction of failure in the most general sense, the theoretical/analytical approach must consider all of the various factors discussed previously in the sections on buckling and postbuckling.

During the last decade, the design of postbuckled structures has been hampered by a lack of experimental and analytical work that could be used to guide the designer to more optimum structures. The experimental studies of RENIERI and GARRETT (1982); OSTROM (1981); and AGARWAL et al. (1984) have been directed towards defining a basic laminated panel configuration that would be capable of withstanding specified typical aircraft loadings and environments. Unfortunately, these experimental studies have not provided a concurrent design methodology that has general application. Thus, postbuckled panel designs, including prediction of failure, are based on empirical formulae (usually proprietary to the particular airframe manufacturer).

The government has sponsored many postbuckling research projects over the last ten years – the latest 'Design Methodology and Life Analysis of Postbuckled Metal and Composite Panels' (ANON, 1984) is presently on-going; however, the basis for this latest program is the use of semi-empirical methods based on new experimental data to provide a design methodology for composite panels. It will be many years before this technology can be used and evaluated. On a theoretical/analytical level, the efforts of Arnold and Mayers, based on the work of Mayers and many of his research students, has evolved into a computational procedure, PANCLP, ARNOLD (1985b) which has been used to derive many of the figures used to illustrate the buckling, postbuckling, and failure characteristics of stiffened panel structures.

In the following paragraphs, the various potential failure modes of laminated composite structures will be reviewed. It should be kept in mind that a comprehensive database from which meaningful designs can be generated is extremely limited.

3.1. *Plate crippling (maximum load)*

Previously, load-shortening curves for metal stiffened panels in compression were presented (Figs. 20 and 21). These curves also included the point of maximum load (crippling). For metallic panels, the maximum load is achieved at the point where the tangent modulus is zero. However, for composite materials, the point of maximum load corresponds to failure of the fibers, either in tension or compression, as exemplified in Figs. 22 and 23 (ARNOLD, 1986).

Various failure criteria have been described in other parts of this book, and thus, no elaboration is required here. For laminated composite stiffened panels, it is informative to examine these various failure criteria in order to determine which may be the more appropriate. ARNOLD (1983) has compared the maximum strain criterion with the experimental and analytical data of TENNYSON et al. (1978) and those results are shown in Fig. 33. The experimental specimen is an internally pressurized filament-wound pressure vessel com-

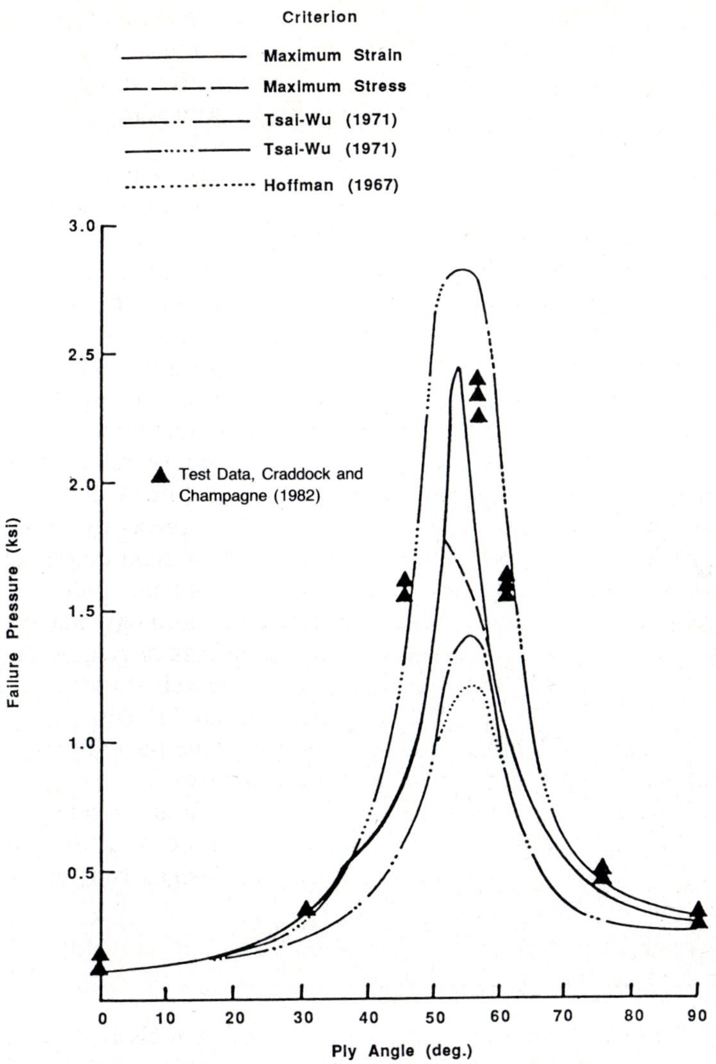

FIG. 33. Comparison of various failure criteria (ARNOLD, 1983).

prised of $\pm\theta$ laminae. The analytical comparisons of CRADDOCK and CHAMPAGNE (1982) are also included in the Fig. 33. CRADDOCK and CHAMPAGNE consider only stress-based theories; however, as can be seen from the results of Fig. 33, the maximum strain criterion fits the experimental data very well.

Using the maximum strain criterion as a failure criterion and the experimental data of SPIER and KLOUMAN (1976a,b)–SPIER (1978), Arnold has made predictions of buckling and crippling which are tabulated in Table 1 for seven different laminated composite specimens. The predicted fiber strains corre-

TABLE 1. Predicted fiber strains at experimental value of crippling load.

Specimen ID	b/t	Layup	$(\delta/\delta_{cr})_e$	N_i^{cr}/N^{cr}	N_i^{cc}/N^{cc}	$\varepsilon_f(\%)$
3-S2	32	$[\pm 45/90/0_3]_s$	2.8	0.97	1.06	−1.2
4-S2	41	$[\pm 45/0_3/90]_s$	3.4	1.01	0.95	−1.4
7-S1	34	$[\pm 45/0_3/90]_s$	2.6	1.01	0.95	−1.1
5A-N	37	$[45/0/-45/90]_{2s}$	2.1	1.04	1.07	−0.8
F7	15	$[45/0_2/-45/0_4]_{2s}$	3.0	0.96	0.98	−0.8
5B	13	$[\pm 45/0_6]_s$	1.8	1.09	1.03	−1.0
3A	36	$[\pm 45/0_4]_s$	2.4	1.00	1.02	−1.1

spond to the experimentally determined value of end-shortening at failure. The predictions are all in the range of 0.8% to 1.4% for maximum fiber strain at failure. It is important to note that the typical B-basis allowable strain (95% confidence, 90% probability) for unidirectional A-S/3501 graphite/epoxy tape presented in DoD/NASA Advanced Composites Design Guide (ANON 1983) is 0.83% in compression and 1.15% in tension. Thus, the predicted fiber strains are in reasonable agreement with the nominal material allowables. Ideally, the experimental data of Spier would have included strain gage data; unfortunately, this data is not available, and thus, comparisons with the maximum strain criterion must be evaluated cautiously. Nevertheless, the maximum strain criterion seems to perform well for laminated composites in which both fiber and matrix are relatively brittle. As pointed out by GRIMES and WHITNEY (1974), the maximum strain criterion should be accurate as long as the nonlinear shear strain components are small and do not interact with the normal strains.

It should also be noted that the maximum strain criterion is operationally simple to use, requires only limited test data, and can be used to predict the actual mode of failure; that is, either fiber failure or matrix failure in tension or shear. Other failure theories (for instance, TSAI and WU, 1971) produce a failure surface for which no physically plausible failure mode can be associated.

Typical present-day stiffened panel constructions made of composite stiffeners bonded to composite panels usually disbond prior to crippling of the panel; thus, no experimental data is available that can be used to further verify the use of the maximum strain failure criterion for postbuckled composite panels.

3.2. Stiffener disbond

The most common failure mode of composite stiffened panels is stiffener disbonding. This failure mode has been so distressing that the evolution of primary composite stiffened structures employing the postbuckling concept has been severely curtailed in recent years. This is due to the fact that postbuckled stiffened panels usually disbond at approximately 40–60% of the potential load based on the maximum strain criterion and panel crippling failure mode. This relatively large loss in potential ultimate load capability of the composite stiffened panel has made all-metal construction look favorable, especially in light of new alloys of lithium and aluminum which are just on the horizon.

Present postbuckled laminated composite panel designs are based on extensive experimental efforts whose objective was to define the maximum static and fatigue loads that could be handled by a pre-defined laminate and material. Typical examples of these costly point designs can be found in: RENIERI and GARRETT (1982); AGARWAL (1981a,b)–SCHNEIDER (1981). Simplified analysis methods are, however, being developed.

AGARWAL (1982) has developed an approach to predict stiffener disbonding in a composite shear panel tested to failure. A simple two-dimensional mathematical model is used to determine the stresses in the stiffener/web interface. Using a Rayleigh–Ritz approach, the potential energy of the model is minimized, thus providing the magnitudes of the interfacial stresses. Then, a failure criterion, based on experimental coupon testing, is used to define failure as a function of the applied membrane strain, as shown in Fig. 34 for a composite shear panel. A comparison of the predicted and experimental behavior for the composite shear panel is shown in Fig. 35. The results show that the test load was approximately 10% higher than the predicted value.

Tsai has developed a method for prediction of stiffener disbonding based on diagonal tension field theory and solution of a large-displacement beam model. The method requires an idealization of a flat panel in shear to a unit strip beam model. The predictions of ultimate load capability compare to the experimental

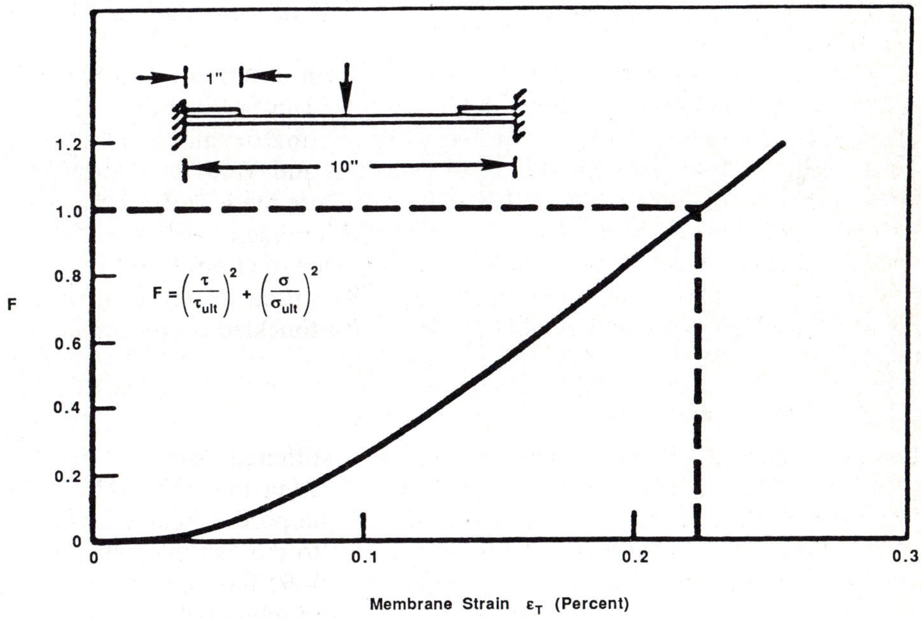

FIG. 34. Plot of membrane strain and cumulative damage index.

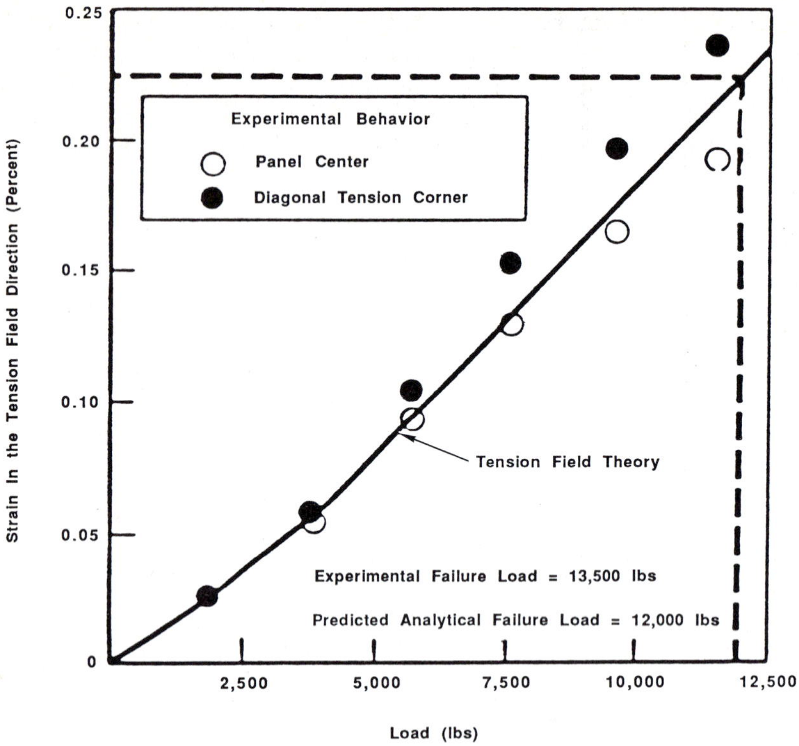

FIG. 35. Comparison of predicted and experimental behaviour for shear panels (AGARWAL, 1982).

data to within approximately ±8%, although the method requires the use of some engineering judgement for allowables and buckling loads.

Both of the methods just described address the need to define the stresses which exist between the stiffener and panel in the postbuckled range. Also, these methods, while demonstrating some capabilities for predicting disbonding based on limited experimental testing, must be evaluated and proven over a much wider range of materials and design configurations.

With respect to the determination of the stresses which exist between the stiffener and panel, recent correlative analysis by ARNOLD (1985a)–ARNOLD (1984a,b) has pointed out the necessity of accurately defining the elastic and inelastic stiffness of the panel and stiffener. For instance, in the postbuckled region, the out-of-plane displacements introduce normal (and shear) loads on the stiffener-panel interface. These interface loads are a function of the stiffener torsional rigidity and the panel geometry.

Examples of the average torsional stress in typical composite stiffeners is provided in Fig. 36, based on the previously cited references by Arnold. The determination of these stiffener stresses is based on the numerical code of ARNOLD (1985) which includes the effects of transverse shear, material non-

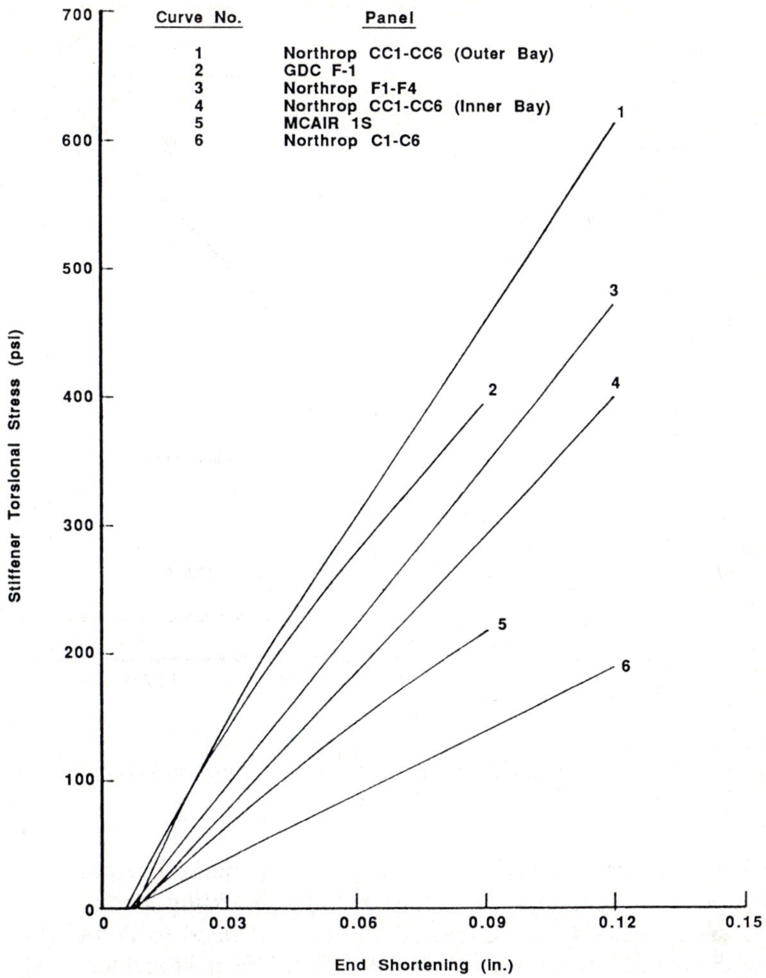

FIG. 36. Analytically predicted torsional stresses in composite compression panels.

linearity, anisotropy, curvature, and edge-stiffening. All of these mentioned factors have an important influence on the prediction of stiffener torsional stress.

In Fig. 37, the average torsional stress is used to compute the maximum bondline normal stress as a function of the applied load for the NASA-generated data of KNIGHT and STARNES (1985). The laminated composite panels had varying amounts of edge stiffening and curvature and all failed by stiffener disbond. The indicated points of failure correspond to the experimentally determined value of applied compression load and seem to indicate, when taken as a group, that the allowable normal stress acting on the bondline is approximately 25 psi. In tests performed by Lockheed to assess the pull-off

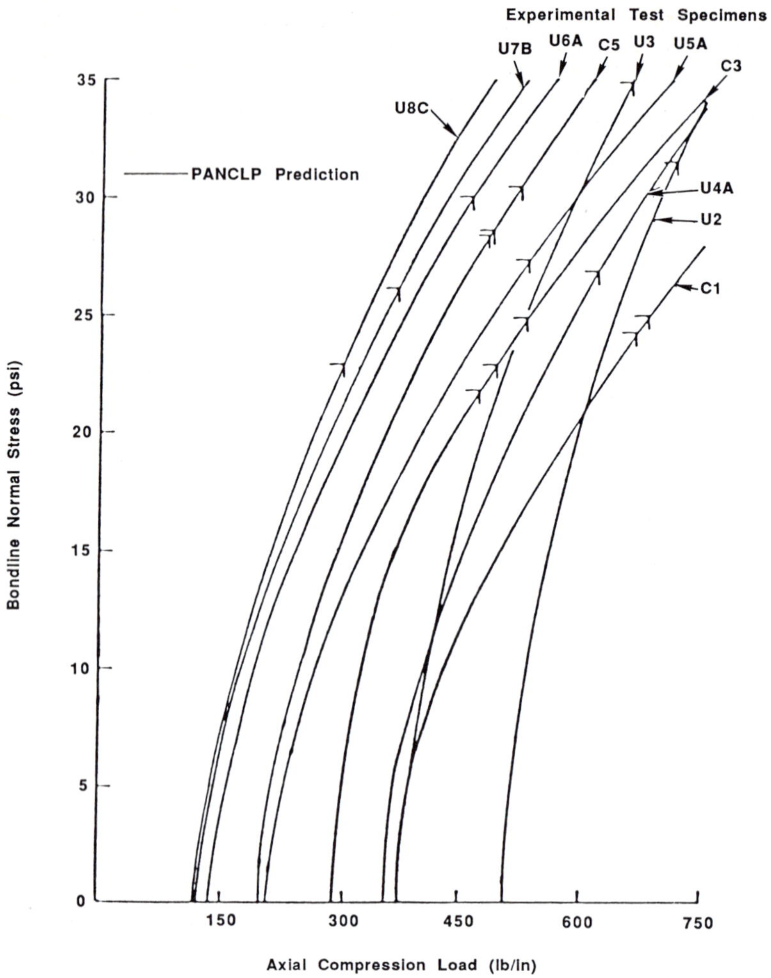

FIG. 37. Predicted bondline normal stress at experimental value of failure load.

strength of composite stiffeners, it was found that the average normal stress was 33 psi OSTROM (1981). Thus, the predicted stress has at least some quantitative merit. Note also that the stiffener disbond stress varies quadratically with the applied load, as is also the case for displacements and strains. Thus, using the theory as a guide, it is observed that improvements in the basic material that bonds the stiffener to the panel can offer substantial improvements in postbuckling strength only if the bond strength can be significantly improved–such improvements are not apparent at this time. Thus, alternative methods of attaching stiffeners to the panel offer some near-term potential. An example would be the use of 'stitching' to increase the attachment strength of the panel and stiffener RENIERI and GARRETT (1982). Alternative techniques from stitching, such as 'tailoring' or the stiffener to panel attachment plies, have also been investigated experimentally with at least partial success RENIERI and GARRETT (1982).

No simple formula or criterion exists, in the general sense, for prediction of stiffener disbonding, although most airframe manufacturers have their own predictive techniques which are a combination of semi-empirical formulae and experimental data acquired over a period of time. Thus, present-day designs are confined to minor variations of experimentally proven postbuckled designs.

3.3. Fracture and crack propagation

The subject of fracture and crack propagation in composite structures is presently an area of active research and experimentation with most of the effort directed at characterizing the most simple failure mechanisms. Research devoted to defining the damage tolerance of postbuckled composite panels is limited to experimental testing in order to define relatively rough guidelines for design purposes. Thus, theoretical efforts are either too simplified or underdeveloped for meaningful design of stiffened composite panels. A good starting point for understanding the basic failure mechanisms in composite materials is provided in 'Fracture Mechanics of Composites; (ANON, 1974b).

3.4. Fatigue

A reasonable body of experimental data exists for evaluation of fatigue of composite materials. 'Fatigue of Composite Materials' (ANON, 1974a) provides an introduction to this area of mechanics. Also, other chapters in this book address the question of fatigue in composites at a more fundamental level. With respect to postbuckled structures, the effort to date has been confined to experimental studies and verification tests of fatigue life for several restricted classes of symmetric, quasi-isotropic laminates.

AGARWAL (1981a) has performed experimental testing of laminated composite graphite panels in compression under constant amplitude fully-reversed fatigue loading. As might be expected, stiffener disbonding was the major mode of failure. An important conclusion of his work was that there was significant postbuckling strength remaining in the panel provided that the stiffeners and web could be held together. On the average, the panels failed at a load that was approximately five times the initial buckling load. Fatigue testing demonstrated that the fatigue load could be close to 80% of the static load strength and still satisfy the design requirements.

In shear, there have been many relatively recent experimental efforts devoted to defining the static and fatigue strength of postbuckling stiffened-panel structures RENIERI and GARRETT (1982)–OSTROM (1981). The experimental results showed that the panel could satisfy the design requirement for fatigue provided the fatigue loading was less than approximately 60% of the static strength. Furthermore, in some of the shear tests OSTROM (1981), the ultimate failure load was approximately eight times the initial buckling load, further proving the significant strength of stiffened-panels in the postbuckled range.

4. Closure

The objective of this chapter has been three-fold. First, and most importantly, the various geometric and material-related factors that have an influence on buckling, postbuckling, and failure of stiffened composite panels have been presented and quantified where possible. Second, exemplary references have been provided to allow both the designer and the student the opportunity to investigate and understand in more depth, the behavior of stiffened panel structures. Finally, this section has attempted to survey the current state-of-the-art for the design of buckled and postbuckled structures.

References

AGARWAL, B.L. (1981a), Flat stiffened graphite/epoxy tension field panels under constant-amplitude fully-reversed fatigue loading, Naval Air Development Center Report No. NADC-81169-60.

AGARWAL, B.L. (1981b), Postbuckling behavior of hat stiffened flat and curved composite compression panels, Naval Air Systems Command Report No. NOR 81-187.

AGARWAL, B.L. (1982), A model to simulate failure due to stiffener/web separation of composite tension field panels, *AIAA 23rd Structures, Structural Dynamics, and Materials Conf.*, AIAA Paper No. 82-0746.

AGARWAL, B.L., N.J. KUDVA and R.B. DEO (1984), Design methodology and life analysis of postbuckled metal and composite panels, Interim Technical Report Number 8, Northrop Corporation, Contract No. F33615-81-C-3208.

AMBARTSUMYAN, S.A. (1970), in: J.E. ASHTON, ed., *Theory of Anisotropic Plates—Strength, Stability, Vibration*, translated from the Russian Edition by T. CHERON (Technomic Publishing Company, Stamford, Connecticut).

ANDERSON, R.A. and M.S. ANDERSON (1956), Correlation of crippling strength of plate structures with material properties, National Advisory Committee for Aeronautics, NACA Technical Note 3600.

ANDERSON, M.S. and S.W. JEFFERSON (1979), A general panel sizing computer code and its application to composite structural panels, *AIAA J.* **17**, No. 8, 892–897.

ANDERSON, R.E. and J. MAYERS (1979), Effects of non-linear material behavior on postbuckling stiffness of laminated composite plates, *AIAA Aircraft Systems and Technology Meeting*, Paper No. 79-1806, New York, NY.

ANON (1974a), Fatigue of composite materials, *ASTM Special Technical Publication 569, ASTM Symposium*, Bal Harbour, Florida.

ANON (1974b), Fracture mechanics of composites, *ASTM Special Technical Publication 593, Symposium of High Modulus Fibers and Their Composites*, Gaithersburg, Maryland.

ANON (1983), *DoD/NASA Advanced Composites Design Guide* (Air Force Flight Dynamics Laboratory, Wright-Patterson Air Force Base).

ANON (1984), Design methodology and life analysis of postbuckled metal and composite panels, Air Force Systems Command Contract to Northrop Corporation.

ANDERSON, M.S. and W.J. STROUD (1979), A general panel sizing computer code and its application to composite structural panels, *AIAA J.* **17**, No. 8, 892–897.

ARNOLD, R.R. (1983), Buckling, postbuckling, and crippling of materially nonlinear laminated composite plates, PhD. Dissertation, Stanford University.

ARNOLD, R.R. (1984a), Correlative analysis of general dynamics convair stiffened shear panel F-1, Anamet Laboratories, Inc., Report No. 84.055.

ARNOLD, R.R. (1984b), Correlative analysis of Northrop metal compression panels MC1-4, IC1, Anamet Laboratories, Inc., Report No. 84.033.

ARNOLD, R.R. (1985a), Correlative analysis of McDonnell Aircraft Company flat stiffened composite shear panels, Anamet Laboratories, Inc., Report No. 85.001.

ARNOLD, R.R. (1985b), PANCLP Users Guide—Version 31.0, April 31, 1985, Anamet Report No. 85.033.

ARNOLD, R.R. (1986), Failure mechanisms of postbuckled metal and composite panels, *AIAA/AHS/ASEE Aircraft Systems, Design and Technology Meeting*, Paper No. 86-2685, Dayton, Ohio.

ARNOLD, R.R. (1987a), Disbond Criterion for Postbuckled Composite Panels, *AIAA/ASME/ASCE/AHS 28th Structures, Structural Dynamics and Materials Conference*, Paper No. 87-0732, Monterey, California.

ARNOLD, R.R. (1987b), Equivalent elastic torsional properties of composite stiffeners, to be published.

ARNOLD, R.R. and J. MAYERS (1984), Buckling, postbuckling, and crippling of materially nonlinear laminated composite plates, *International Journal of Solids and Structures*, **20**, No. 9/10, 863–880.

ARNOLD, R.R. and J. PAREKH (1987), Buckling, postbuckling, and failure of stiffened panels under compression and shear, *AIAA/ASME/ASCE/AHS 27th Structures, Structural Dynamics and Materials Conf.*, Paper No. 86-1027 San Antonia, Texas. *Journal of Aircraft* **24**, No. 11, 803–811.

ARNOLD, R.R., S.Y. YOO and J. MAYERS (1985), Buckling, postbuckling, and crippling of shallow curved composite plates with edge stiffeners, *AIAA/ASME/ASCE/AHS 26th Structures, Structural Dynamics and Materials Conf.*, Paper No. 85-0769-CP, Orlando, Florida. Also, *Journal of Aircraft* **23**, No. 7, 589–598.

ASHTON, J.E. and J.M. WHITNEY (1970), *Theory of Laminated Plates* (Technomic Publishing Company, Stamford, Connecticut).

ASTRONAUTIC STRUCTURES MANUAL (1970), Volumes I–III, Marshall Space Flight Center, NASA.

BAKER, E.H., L. KOVALEVSKY and F.L. RISH (1972), *Structural Analysis of Shells* (McGraw-Hill, New York).

BANKS, W.M. (1976), The post buckling behavior of composite panels, *Proc. 1975 Int. Conf. on Composite Materials*, ICCM Vol. 2, pp. 272–293.

BESSELING, J.F. (1953), De experimentele bepaling van de meevagende breedte van vlakke platen in het elastiche en het plastische gebied, (The experimental determination of the effective width of flat plates in the elastic and plastic range.) NLL Report S. 414, National Luchtvaarlaboratorium, Amsterdam.

BOTMAN, M. (1954), De experimentele bepaling van de meedvagende breedte van vlakke platen in het elastische en het plastische gebied, (deel II). (The experimental determination of the effective width of flat plates in the elastic and plastic range, Part II). NLL Report S. 438, Nationaal Luchtvaartlaboratorium, Amsterdam.

BOUADI, H. (1982), Buckling and postbuckling of composite plates—fiber orientation and stacking sequence effects, Engineer Thesis, Stanford University.

BRUHN, E.F. (1973), *Analysis and Design of Flight Vehicle Structures*, (S.R. Jacobs and Associates, Indianapolis, Indiana).

CRADDOCK, J.N. and D.J. CHAMPAGNE (1982), A comparison of failure criteria for laminated composite materials, *AIAA 23rd Structures, Structural Dynamics and Materials Conf.*, Paper No. 82-0739, p. 268.

DURLOFSKY, H. (1970), The effects of interlaminar shear on the bending and buckling of fiber-reinforced, composite, flat and curved plates, PhD Dissertation, Stanford University.

GERARD, G. (1946), Effective width of elastically supported flat plates, *J. Aeronautical Sci.* **13**, No. 10, 518–524.

GERARD, G. (1957), Handbook of structural stability, part V—compressive strength of flat stiffened panels, National Advisory Committee for Aeronautics, NACA Technical Note 3785.

GERARD, G. and H. BECKER (1957a), Handbook of structural stability, part I, buckling of flat plates, NACA TN 3781.

GERARD, G. and H. BECKER (1957b), Handbook of structural stability, part III, buckling of curved plates and shells, NACA TN 3783.

GRIMES, G.C. and J.M. WHITNEY (1974), Degradation of graphite/epoxy composite materials because of load-induced micromechanical damage, *SAMPE Quarterly*.

HARRIS, G.Z. (1975), Buckling and postbuckling of orthotropic laminated plates, AIAA Paper No. 75-813.

HINKLE, T.V. and R.A. GARRETT (1982), Examination of postbuckled compression behavior of curved panels, Naval Air Systems Command Report MDC A7264.

HOFF, N.J. (1950), Bending and buckling of rectangular sandwich plates, NACA Technical Note 2225.

HOFFMAN, O. (1967), The brittle strength of orthotropic materials, *J. Comp. Mater.* **1**.

HU, P.C., E.E. LUNDQUIST and S.B. BATDORF (1946), Effect of small deviations from flatness on effective width and buckling of plates in compression, NACA Technical Note 1124.

KAMINSKI, B.E. and J.E. ASHTON (1971), Diagonal tension behavior of boron/epoxy shear panels, *J. Comp. Mater.* **5**, 553–558.

KNIGHT, N.F. and J.H. STARNES (1985), Postbuckling behavior of selected curved stiffened graphite–epoxy panels loaded in axial compression, *AIAA/ASME/ASCE/AHS 26th Structures, Structural Dynamics and Materials Conf.*, Paper No. 85-0768-CP, Orlando, Florida.

KOITER, W.T. (1943), De meedragende breedte bij groote overschrijding der knikspanning voor verschillende inklemming der plaatrandem, (The effective width at loads far in excess of the critical load for various boundary conditions), NLL Report S. 287, National Luchtvaart-laboratorium, Amsterdam.

KOITER, W.T. (1976), General theory of mode interaction in stiffened plate and shell structures, Department of Mechanical Engineering, Delft University of Technology, The Netherlands, Report nr. 590.

KOITER, W.T. and M. PIGNATARO (1976), A general theory for the interaction between local and overall buckling of stiffened panels, Department of Mechanical Engineering, Delft University of Technology, The Netherlands, Report WTHD 83.

LEGGETT, D.M.A. (1942), The buckling of a long curved panel under axial compression, Aeronautical Research Council, Reports and Memoranda, No. 1899.

LEKHNITSKII, S.G. (1968), *Anisotropic Plates*, Translated from the second Russian edition by S.W. Tsai and T. Cheron (Gordon and Breach, New York).

LOVE, A.E.H. (1888), The small free vibations and deformations of a thin elastic shell, *Phil. Trans. Royal Soc. London*, Series A, **179**, 491–546.

MAYERS, J. and B. BUDIANSKY (1955), Analysis of behavior of simply supported flat plates compressed beyond the buckling load into the plastic range, National Advisory Committee for Aeronautics, NACA Technical Note 3368.

MAYERS, J. and E. MELLER (1972), Material nonlinearity effects in optimization considerations of stiffened cylinders and interpretation of test data scatter for compressive buckling, U.S. Army Air Mobility Research and Development Laboratory Technical Report USAAMRDL 71-70.

MAYERS, J. and E. NELSON (1968), Elastic and maximum strength analyses of postbuckled rectangular plates based upon modified versions of Reissner's variational principle, Presented at AIAA Sixth Aerospace Sciences Meeting, AIAA paper No. 68-171, New York.

OSTROM, R.B. (1981), Post-buckling fatigue behavior of flat stiffened graphite/epoxy panels under shear loading, Naval Air Development Center Report NADC-78137-60.

REISSNER, E. (1950), On a variational theorem in elasticity, *J. Math. and Phys.* **XXIV**, No. 2, 90–95.

RENIERI, M.P. and R.A. GARRETT (1982), Postbuckling fatigue behavior of flat, stiffened graphite/epoxy panels under shear loading, Naval Air Development Center Report NADC-81168-60.

RENIERI, M.P. and R.A. GARRETT (1982), Stiffener/skin interface design improvements for post-buckled composite shear panels, Naval Air Development Center Report No. NADC-80134-60.

SCHNEIDER, G.J. (1981), Initial buckling and diagonal tension behavior of kevlar/graphite epoxy flat stiffened shear panels, Sikorsky Engineering Report No. SER-510047 for the Naval Air Systems Command.

SMITH, A.P. (1973), The effect of transverse shear deformation on the elastic stability of orthotropic plates due to inplane loads, MMAE Thesis, University of Delaware.

SOOVERE, J. (1982), Effect of acoustic, thermal and shear loading on flat integrally stiffened graphite/epoxy fuselage panels, Naval Air Development Center Report No. NADC-78169-60.

SPIER, E.E. (1976), On experimental versus theoretical incipient buckling of narrow graphite/epoxy laminated plates, *Proc. Army Symposium in Solid Mechanics, 1976 – Composite Materials: The Influence of Mechanics of Failure on Design, Army Materials and Mechanics Research Centre*, Watertown, Massachusetts, AMRAC MS76-2.

SPIER, E.E. (1978), Stability of graphite/epoxy structures with arbitrary symmetrical laminates, *Experimental Mechanics* **18**, No. 11.

SPIER, E.E. and F.L. KLOUMAN (1976a), Postbuckling behavior of graphite/epoxy laminated plates and channels, *Army Symposium on Solid Mechanics*, Cape Cod, Massachusetts.

SPIER, E.E. and F.L. KLOUMAN (1976b), Ultimate compressive strength and nonlinear stress–strain curves of graphite/epoxy laminates, presented at the *Eighth National SAMPE Conf., Bicentennial of Materials Progress–Part II*, Seattle, Washington.

SPIER, E.E. and F.L. KLOUMAN (1977), Empirical crippling analysis of graphite/epoxy laminated plates, *Composite Materials: Testing and Design (4th Conf.)*, ASTM STP-617.

STARNES, J.H., N.F. KNIGHT and M. ROUSE (1982), Postbuckling behavior of selected flat stiffened graphite–epoxy panels loaded in compression, *23rd Structures, Structural Dynamics, and Materials Conf.*, AIAA Paper No. 82–0777.

STEIN, M. (1959), Loads and deformations of buckled rectangular plates, National Aeronautics and Space Administration, NASA Technical Report R-40.

STOWELL, E.Z. (1943), Critical compressive stress for curved sheet supported along edges and elastically restrained against rotation along the unloaded edges, National Advisory Committee for Aeronautics, NACA Restricted Bulletin 3107.

TENNYSON, R.C., D. MACDONALD and A.P. NANYARA (1978), Evaluation of the tensor polynomial failure criterion for composite materials, *J. Comp. Mater.* **12**, 63–68.

TSAI, H.C. (1986), Solution method for stiffener–skin separation in composite tension field panel, Naval Air Development Center, unpublished report.

TSAI, S. and E. WU (1971), A general theory of strength for anisotropic materials, *J. Comp. Mater.* **5**.

TURVEY, G.J. and W.H. WITTRICK (1973), The large deflection and postbuckling behavior of some laminated plates, *Aeronautical Quarterly* **24**, 77–84.

TVERGAARD, V. (1973), Imperfection-sensitivity of a wide integrally stiffened panel under compression, *Int. J. Solids and Structures* **9**, 177–192.

VINSON, J.R. and T.W. CHOU (1975), *Composite Materials and Their Use in Structures* (John Wiley and Sons, New York).

VON KARMAN, T., E.E. SECHLER and L.H. DONNELL (1932), The strength of thin plates in compression, *Trans. ASME* **54**, No. 2, 53–57.

WHITNEY, J.M. (1984), Buckling of anisotropic laminated cylindrical plates, *AIAA J.* **22**, No. 11, 1641–1645.

WHITNEY, J.M. (1969), The effect of transverse shear deformation on the bending of laminated plates, *J. Comp. Mater.* **3**, 534.

WITTRICK, W.H. and W.E. BODLEY (1963), Compressive buckling of infinite strips with elastically restrained edges, *The Aeronautical Quarterly* **XIV**, 158–162.

YOO, S.Y. (1977), On maximum strength of plates under axial compression effects of curvature and edge-stiffener extensional and torsional rigidity, PhD Dissertation, Stanford University.

ZHANG, Y. and F.L. MATHEWS (1983), Initial buckling of curved panels of generally layered composite materials, *Composite Structures* **1**, 969–975.

Subject Index